# Water-Soil-Plant-Animal Nexus in the Era of Climate Change

Ahmed Karmaoui
*Moulay Ismail University, Meknès, Morocco & Moroccan Center for Culture and Science, Morocco*

IGI Global
PUBLISHER of TIMELY KNOWLEDGE

A volume in the Practice, Progress, and Proficiency in Sustainability (PPPS) Book Series

Published in the United States of America by
IGI Global
Engineering Science Reference (an imprint of IGI Global)
701 E. Chocolate Avenue
Hershey PA, USA 17033
Tel: 717-533-8845
Fax: 717-533-8661
E-mail: cust@igi-global.com
Web site: http://www.igi-global.com



Library of Congress Cataloging-in-Publication Data

Names: Karmaoui, Ahmed, 1982- editor.
Title: Water-soil-plant-animal nexus in the era of climate change / edited by: Ahmed Karmaoui.
Description: Hershey, PA : Engineering Science Reference, [2024] | Includes bibliographical references and index. | Summary: "The book brings together affordable methods and techniques to optimize the use of the nexus in the context of climate change. It will present an inventory of techniques and practices in the field. This book will be an opportunity to discuss the strengths and weaknesses of these techniques. The proposed collection may be used as a basis to support adaptation to climate change"-- Provided by publisher.
Identifiers: LCCN 2023029603 (print) | LCCN 2023029604 (ebook) | ISBN 9781668498385 (hardcover) | ISBN 9781668498392 (paperback) | ISBN 9781668498408 (ebook)
Subjects: LCSH: Ecology. | Climatic changes.
Classification: LCC QH541 .W2775 2024 (print) | LCC QH541 (ebook) | DDC 577--dc23/eng/20230726
LC record available at https://lccn.loc.gov/2023029603
LC ebook record available at https://lccn.loc.gov/2023029604

This book is published in the IGI Global book series Practice, Progress, and Proficiency in Sustainability (PPPS) (ISSN: 2330-3271; eISSN: 2330-328X)

British Cataloguing in Publication Data
A Cataloguing in Publication record for this book is available from the British Library.

# Practice, Progress, and Proficiency in Sustainability (PPPS) Book Series

Ayman Batisha
International Sustainability Institute, Egypt

ISSN:2330-3271
EISSN:2330-328X

## Mission

In a world where traditional business practices are reconsidered and economic activity is performed in a global context, new areas of economic developments are recognized as the key enablers of wealth and income production. This knowledge of information technologies provides infrastructures, systems, and services towards sustainable development.

The **Practices, Progress, and Proficiency in Sustainability (PPPS) Book Series** focuses on the local and global challenges, business opportunities, and societal needs surrounding international collaboration and sustainable development of technology. This series brings together academics, researchers, entrepreneurs, policy makers and government officers aiming to contribute to the progress and proficiency in sustainability.

## Coverage

- Outsourcing
- Environmental informatics
- Knowledge clusters
- Technological learning
- Innovation Networks
- Global Content and Knowledge Repositories
- Strategic Management of IT
- Sustainable Development
- Green Technology
- Intellectual Capital

IGI Global is currently accepting manuscripts for publication within this series. To submit a proposal for a volume in this series, please contact our Acquisition Editors at Acquisitions@igi-global.com or visit: http://www.igi-global.com/publish/.

The Practice, Progress, and Proficiency in Sustainability (PPPS) Book Series (ISSN 2330-3271) is published by IGI Global, 701 E. Chocolate Avenue, Hershey, PA 17033-1240, USA, www.igi-global.com. This series is composed of titles available for purchase individually; each title is edited to be contextually exclusive from any other title within the series. For pricing and ordering information please visit http://www.igi-global.com/book-series/practice-progress-proficiency-sustainability/73810. Postmaster: Send all address changes to above address.

## Titles in this Series

*For a list of additional titles in this series, please visit: www.igi-global.com/book-series/practice-progress-proficiency-sustainability/73810*

*Opportunities and Challenges for Women Leaders in Environmental Management*
Perfecto Gatbonton Aquino, Jr. (The University of Cambodia, Cambodia) Revenio Cabanilla Jalagat, Jr. (Al-Zahra College for Women, Oman) and Mercia Selva Malar Justin (Xavier Institute of Management and Entrepreneurship, India)
Engineering Science Reference • copyright 2024 • 259pp • H/C (ISBN: 9781668459867) • US $250.00 (our price)

*The Role of Women in Cultivating Sustainable Societies Through Millets*
Raghvendra Kumar (GIET University, India) and Ishaani Priyadarshini (University of California, Berkeley, USA)
Engineering Science Reference • copyright 2024 • 350pp • H/C (ISBN: 9781668498194) • US $265.00 (our price)

*Quality of Life and Climate Change Impacts, Sustainable Adaptation, and Social-Ecological Resilience*
Kasturi Shukla (Symbiosis International University) Yogesh B. Patil (Symbiosis International University, India) Ronald C. Estoque (Forestry and Forest Products Research Institute (FFPRI), Japan) and Pedro Antonio López de Haro (Universidad Autónoma Indígena de México, Mexico)
Engineering Science Reference • copyright 2024 • 300pp • H/C (ISBN: 9781668498637) • US $250.00 (our price)

*Sustainable Development in AI, Blockchain, and E-Governance Applications*
Rajeev Kumar (Moradabad Institute of Technology, India) Abu Bakar Abdul Hamid (Infrastructure University, Kuala Lumpur, Malaysia) Noor Inayah Binti Ya'akub (Infrastructure University, Kuala Lumpur, Malaysia) Hari Om Sharan (Amity University, Kolkata, India) and Sandeep Kumar (Rightzone Technologies Pvt. Ltd., India)
Engineering Science Reference • copyright 2024 • 350pp • H/C (ISBN: 9798369317228) • US $285.00 (our price)

*Global Perspectives on Climate Change, Social Resilience, and Social Inclusion*
Aly Abd ElRazek Galaby (Alexandria University, Egypt) Mohammed Abo ElEnein (American University, UAE) and Hassan Mohamed (Alexandria University, Egypt)
Engineering Science Reference • copyright 2024 • 380pp • H/C (ISBN: 9781668489635) • US $275.00 (our price)

*Achieving the Sustainable Development Goals Through Infrastructure Development*
Cristina Raluca Gh. Popescu (University of Bucharest, Romania & The Bucharest University of Economic Studies, Romania) Poshan Yu (Xi'an Jiaotong-Liverpool University, China & European Business University of Luxembourg, Luxembourg) and Yue Wei (Institution of Public Private Partnerships, Hong Kong)
Engineering Science Reference • copyright 2023 • 303pp • H/C (ISBN: 9798369307946) • US $230.00 (our price)

IGI Global
PUBLISHER of TIMELY KNOWLEDGE

701 East Chocolate Avenue, Hershey, PA 17033, USA
Tel: 717-533-8845 x100 • Fax: 717-533-8661
E-Mail: cust@igi-global.com • www.igi-global.com

# Table of Contents

**Section 2**
**Water and Soil in the Context of Climate Change**

# Detailed Table of Contents

approaches to accomplish multiple sustainability objectives in agriculture while simultaneously promoting carbon sequestration. The relationship between carbon sequestration and sustainable agriculture is investigated, and trade-offs between carbon sequestration and other sustainability objectives are outlined. This study emphasizes the policies and governance frameworks that support carbon sequestration, as well as the opportunities and risks associated with their implementation. Overall, carbon sequestration strategies have the potential to provide substantial environmental benefits; however, their implementation requires careful consideration of socioeconomic dimensions.


*Saira Shafiq, Department of Botany, University of Agriculture, Faisalabad, Pakistan*
*Muhammad Zia Ul Haq, Department of Agronomy, University of Agriculture, Faisalabad, Pakistan*
*Amina Shahbaz, Department of Agronomy, University of Agriculture, Faisalabad, Pakistan*
*Sadaf Shafique, Department of Chemistry, Government Municipal Graduate College, Faisalabad, Pakistan*
*Maryam Riaz, Department of Botany, University of Agriculture, Faisalabad, Pakistan*
*Muhammad Tayyab Hanif, Department of Agronomy, University of Agriculture, Faisalabad, Pakistan*
*Gulam Jilani, Department of Agronomy, University of Agriculture, Faisalabad, Pakistan*
*Hina Ali, Department of Botany, University of Agriculture, Faisalabad, Pakistan*
*Wardha Sarfaraz, Department of Botany, University of Agriculture, Faisalabad, Pakistan*
*Syed Abbas Raza Naqvi, Department of Agronomy, University of Agriculture, Faisalabad, Pakistan*
*Gul Zaib Hassan, Department of Agronomy, University of Agriculture, Faisalabad, Pakistan*

This chapter offers a comprehensive overview of agrochemicals in modern agriculture. Agrochemicals, including fertilizers, herbicides, insecticides, and fungicides, are vital for boosting crop yields, managing pests, and enhancing farm productivity. While they have contributed to addressing global food security challenges since the Green Revolution, their improper and excessive use raises concerns of environmental contamination, soil degradation, and harm to nontarget organisms. Climate change adds further complexities to agrochemical use. To ensure long-term sustainability, integrated pest management, precision agriculture, and eco-friendly alternatives should be embraced. Responsible agrochemical use must be prioritized by farmers and policymakers, considering potential environmental and health impacts. By balancing the benefits and risks, we can work towards a more sustainable future in agriculture, ensuring food availability while safeguarding the environment.


*L. Gowri, SASTRA University, India*
*S. Pradeepa, SASTRA University, India*
*N. Sasikaladevi, SASTRA University, India*

Weather forecasting is an important and indispensable procedure in peoples' day-to-day lives, it evaluates the alteration happening in the current condition of the atmosphere. The major goal of this project is to create a weather-based crop prediction system to predict the crops based on weather forecasting using big data analytics. In this project, the authors gather and analyze data based on temperature, rainfall,

array of traits and characteristics encoded within their genomes, enabling them to respond to diverse environmental challenges. By subjecting crops to stress conditions, breeders can identify individuals with desirable traits that confer enhanced resilience. Agroecological approaches and sustainable farming practices also contribute significantly to crop resilience. Crop rotation, intercropping, and agroforestry systems promote biodiversity and enhance the resilience of agroecosystems. In the face of climate change, developing crop varieties with improved resilience to extreme weather events is paramount. By harnessing agroecological principles, genetic diversity, and adaptive capacity, we can bolster crop resilience, ensuring the productivity and adaptability of our agricultural systems in a changing environment.


*Jalal Kassout, National Institute of Agricultural Research, Morocco*
*Khalil Kadaoui, Faculty of Sciences of Tétouan, University of Abdelmalek Essaâdi, Morocco*
*Younes Hmimsa, Polydisciplinary Faculty of Larache, University of Abdelmalek Essaadi, Morocco*
*Salama El Fatehi, Polydisciplinary Faculty of Larache, University of Abdelmalek Essaadi, Morocco*
*Soufian Chakkour, Faculty of Sciences of Tétouan, University of Abdelmalek Essaâdi, Morocco*
*Mhammad Houssni, Faculty of Sciences of Tétouan, University of Abdelmalek Essaâdi, Morocco*
*Sara Oulbi, National Institute of Agricultural Research, Morocco*
*Rachid Azenzem, National Institute of Agricultural Research, Morocco*
*Abdelouahab Sahli, Faculty of Sciences of Tétouan, University of Abdelmalek Essaâdi, Morocco*
*Mohammed Ater, Faculty of Sciences of Tétouan, University of Abdelmalek Essaâdi, Morocco*

Plants contend with a myriad of environmental stressors that exert constraints on their performances, posing considerable concerns regarding the enduring persistence and resilience of plant populations over the long term. This apprehension is especially pertinent for Mediterranean species, exemplified by the wild olive and the carob tree. Nevertheless, a multitude of studies have illustrated that plant species exhibit the capability to endure adverse environmental conditions through the deployment of various morpho-physiological and biochemical traits that support physiological processes. Given the pivotal role of wild fruit trees as invaluable genetic resources, a comprehensive understanding of their potential responses to evolving environmental conditions becomes imperative for the development of apt strategies for resource preservation and breeding programs. In the context of this chapter, our hypothesis posits that intraspecific trait variability emerges as a central player in augmenting the resilience of these emblematic wild fruit trees in the face of climate change.

**Section 2**
**Water and Soil in the Context of Climate Change**



The chapter applied a willingness to pay ability (WPA) approach to estimate the economic value of drinking water (EVM) using primary data. The sample of 350 households are randomly taken from four wards of Kanpur Nagar, Uttar Pradesh (India). Simple OLS, probit, logit, and mixed-effect logit regression models are applied to examine the effect of social- economic variables on WPA of households to receive better water supply from suppliers. It found that the municipality is charging a very low price to provide water supply of 30 Kilolitre/household per month. However, implicit cost (IC) of water/household is found higher as compared to WPA. IC/household appeared high due to consumption of unsafe drinking water, and incidence of various waterborne health issues and diseases. Family size, occupation, and annual income showed a positive impact on WPA of households.



Agriculture in the Mediterranean region is substantially challenged by climate change, and olive growers in particular are facing difficulties because of its effects. In this regard, the olive sector can adopt deficit irrigation strategies, to achieve sustainable management practices that improve water productivity and save irrigation water. This chapter aims to investigate the potential impacts of deficit irrigation strategies on Menara olive tree. Agronomical parameters were assessed in 2021 and 2022, focusing on trees subjected to regulated (RDI) and sustained (SDI) deficit irrigation. The results showed that the performances of Menara olive trees under controlled water stress are better under RDI strategy compared to SDI one. Moreover, the RDI strategy, particularly under specific treatment, allowed to reduction of supplied water by 20% and 40% during sensitive and normal periods, respectively. These results highlight the importance of water management strategies and the need to consider their implications on flowering traits, crop yield, and fruit quality over consecutive growing seasons.

Major advances in the scientific understanding of eutrophication and the significant effects of eutrophication have been made over the decades. The effects of eutrophication include increased biomass of freshwater phytoplankton and periphyton; shifts in phytoplankton species composition to taxa that may be toxic or inedible, e. g. bloom-forming cyanobacteria; reduced biodiversity of waterbodies; changes in vascular plant production, biomass, and species composition; reduced water clarity; and decreases in the perceived aesthetic value of the water. Relative abundances of major bacterial and archaeal groups indicate high similarities at the phylum- and class-level groups across eutrophicated lakes. The impact of climate change and greenhouse gases emissions regards to affect and effect of eutrophication leads to various studies. The effect of reducing nitrogen inputs alone for controlling eutrophication by controlling inputs of phosphorus.



As the worldwide waste management problem has risen, composting has gained popularity. It turns organic trash into nutrient-rich soil for agriculture, landscaping, and environmental rehabilitation. This chapter on composting, "Composting for a Sustainable Future: Turning Waste into Nutrient-Rich Soil," is comprehensive. It covers decomposition science, composting methods, materials, and procedures. Composting's agricultural, landscaping, and environmental remediation benefits are also covered. The chapter also examines composting's role in climate change mitigation, its obstacles, and remedies. Decomposition can help people, businesses, and communities live more sustainably. It urges decomposition and trash reduction and provides information about tools to start composting, a simple yet efficient solution to worldwide waste management.

# Preface

Climate change (CC) has emerged as one of the greatest issues facing our planet today, with profound consequences on water, soil, plants, and animals, the primary components of ecosystems. These components play a crucial role in bolstering biodiversity and furnishing ecosystem services (ES). They can be reflected in the water cycle, soil quality, plant growth, and animal species and habitats. They constitute a nexus of paramount importance in the era of CC.

The first component is water availability, which is vital for sustaining all forms of life on Earth. However, changing temperatures and precipitation patterns are inducing droughts in some regions while causing damaging floods in others. The imbalance generated may affect seriously both terrestrial and aquatic ecosystems, the associated ES, and biodiversity including plants and animals. The second element is the soil considered a crucial support of life on Earth is also being jeopardized by the effect of CC. Any change in climatic parameters may accelerate some environmental issues such as soil erosion, desertification, and fertility reduction. This results in a decline in agricultural productivity and poses a threat to global food security.

The third component, plants both are exposed to the effects of CC and play a vital role in mitigating CC through carbon sequestration. The change in temperatures and precipitation patterns affect substantially their growth cycles and distribution in time and space. This not only influences the density and diversity but also disrupts entire food chains. The fourth component is represented by animals that are equally affected by the effect of CC. This impact is complex, with migration being one notable consequence. The long-term changes in temperature, precipitation, and other atmospheric parameters may disrupt ecosystems and affect the distribution, behavior, and abundance of a large number of animal species. These species are forced to migrate or adapt to new habitats, which may lead to habitat loss. The destruction of habitats can be caused by deforestation or temperature changes.

Water, soil, plants, and animals are the main pillars that support global food security. Plants grow using nutrients from water and soil resources and then used by animals which affects them consequently. Water is the essential condition of life for all living beings and soil is its support and a crucial reservoir. Water, soil, plants, and animals interact in a complex way, which requires further effort to study them and to explore the interconnections. Over the past three decades, the adverse impacts of climate change and human activities have exerted a profound influence on the components of water, soil, plants, and animals. On a practical level, urgent action must be taken at local and global scales to mitigate the effects of CC through sustainable practices. Reforestation programs, conservation efforts in forests and wetlands, land management that promotes soil health restoration, and responsible water use. To mitigate the effects of CC, several actions can be taken including the reduction of greenhouse gas emissions through sustainable practices such as renewable energy sources and reducing waste production.

The interactions between the Water-Soil-Plant-Animal nexus and CC are of increasing concern to scholars, decision-makers, and researchers. The effects of CC on these resources are various including water and soil quality degradation, infectious disease spread, water shortage, desertification, and erosion… These repercussions were accelerated due to human pressure through overuse and pollution. To highlight the importance of these interactions and impacts, a selection of studies encompassing relevant theoretical and empirical approaches, and bibliometric and bibliographic analyses were conducted. This collection of chapters may be used as a reference and a basis to support adaptation to CC. This edited book comprises 18 chapters addressing the nexus of water, soil, plants, and animals in the context of CC. Organized into four sections, the first focuses on plants in the era of CC with nine chapters. Section 2 explores water and soil in the context of CC, featuring four chapters. Section 3 consists of a single chapter, while Section 4, titled "Sustainable Alternatives in the Environment and Agriculture," encompasses four chapters.

In the section one, Chapter 1 reviewed in detail the *Aloe Vera* (AV) plant, characterizing the components and activities, and exploring some stress factors associated with CC. AV is a multi-benefit plant well known for diverse applications due to its unique gel. This gel contains high-quality oil used to produce a high range of phytochemical products where the quantity and quality depend on stress factors including climatic variables. This study presents a review of AV components and the impact of stress factors including CC through a bibliometric analysis. For the trend, research on this plant, its products, and CC tends towards the chemistry aspect and the use of scanning electron microscopy, phytotherapy, metabolism, and CC. Thus, CC was highlighted as a trending topic but not well explored. The topic of CC was well mentioned but the majority of research efforts explored it partially or generally. The results showed that temperature and precipitation are the most important climatic variables that affect the aloe species. However, water and saline stress increase the quality of the gel and reduce the growth and leaf yield.

Chapter 2 explored the carbon sequestration strategies in agricultural lands for sustainable agriculture. In fact, carbon sequestration strategies can reduce greenhouse gas emissions, enhance soil fertility, improve water dynamics, and increase soil biodiversity. This chapter investigates such nexuses through the analysis of impacts and benefits mainly from three major strategies: conservation agriculture, agroforestry and organic farming including soil amendments. Moreover, the current study emphasizes the need for integrated approaches to accomplish multiple sustainability objectives in agriculture while simultaneously promoting carbon sequestration. The relationship between carbon sequestration and sustainable agriculture is investigated, and trade-offs between carbon sequestration and other sustainability objectives are outlined. This study emphasizes the policies and governance frameworks that support carbon sequestration, as well as the opportunities and risks associated with their implementation. Overall, carbon sequestration strategies have the potential to provide substantial environmental benefits, however, their implementation requires careful consideration of socioeconomic dimensions.

Chapter 3 investigated the interconnection between agrochemicals and CC. This chapter offers a comprehensive overview of agrochemicals in modern agriculture. Agrochemicals, including fertilizers, herbicides, insecticides, and fungicides, are vital for boosting crop yields, managing pests, and enhancing farm productivity. While they have contributed to addressing global food security challenges since the Green Revolution, their improper and excessive use raises concerns of environmental contamination, soil degradation, and harm to nontarget organisms. CC adds further complexities to agrochemical use. To ensure long-term sustainability, integrated pest management, precision agriculture, and eco-friendly alternatives should be embraced. Responsible agrochemical use must be prioritized by farmers and policymakers, considering potential environmental and health impacts. By balancing the benefits and risks, we can work towards a more sustainable future in agriculture, ensuring food availability while safeguarding the environment.

Chapter 4 is entitled "Weather Based Crop Prediction Using Big Data Analytics". The authors showed that weather forecasting is an important and indispensable procedure in people's day-to-day lives, it evaluates the alteration happening in the current condition of the atmosphere. The major goal of this work is to create a weather-based crop prediction system to predict the crops based on weather forecasting using big data analytics. In this project, the authors gather and analyze data based on temperature, rainfall, soil, seed, crop production, humidity, and wind speed to assist farmers in improving agricultural yields. The data is first preprocessed in a Python environment, and then the MapReduce framework is used to further analyze and process the massive amount of data. Second, based on MapReduce findings, k-means clustering algorithm is used to produce a mean result on the data in terms of accuracy. After that, the authors examine the link between crops, rainfall, temperature, soil, and seed type using bar graphs and scatter plots.

Chapter 5 shifted focus on crop improvement strategies and principles of selective breeding. Introducing new crop varieties, crop improvement, and selective breeding enable contemporary agriculture to alleviate food poverty, increase crop production, and enhance product quality. This chapter covers selective breeding, genetically modified crops, genetic resource conservation, and upcoming technologies. Selective breeding discusses heritability, gene frequency, dominance, and epistasis to predict genetic progress. Agricultural innovation involves genetic variety and variation to develop robust and high-performing crop types. Mass selection, pedigree selection, and recurrent selection for trait enhancement are investigated. Marker-assisted and genomic selection are tested for breeding speed and accuracy. The chapter covers genetically modified crops, agricultural improvement, and genetic resource conservation for sustainable agriculture. Gene editing, synthetic biology, and machine learning may boost agricultural yields. According to the abstract, crop development and selective breeding are essential for food security, sustainability, CC, and agricultural demands.

Chapter 6 is entitled "Enhancing Crop Resilience in the Face of a Changing Climate: Strategies for Sustainable Agricultural Production". In this chapter, the authors showed that crops can adapt to shifting environmental conditions due to the diversity of traits and characteristics that genetic diversity gives. The capacity to adapt is a fundamental component of crop resilience. Breeding and natural selection are necessary for developing hardy crop types. Breeders can create more resilient crops by subjecting them to various stresses and identifying individuals with favorable features. Agroecological approaches and sustainable farming can improve crop resilience. Crop rotation, intercropping, and agroforestry increase agroecosystem resilience and promote biodiversity. Developing crop varieties with improved resilience to extreme weather events, such as droughts, floods, and heat waves, is essential to addressing the difficulties posed by CC. Crop resilience can be improved by utilizing agroecological methods, genetic variety, and adaptive ability. The productivity and adaptability of our agricultural systems in the face of shifting environmental conditions can be increased by creating resilient crop types.

Chapter 7 delved into the impacts and mitigation strategies of abiotic and biotic stresses in olive orchards in the era of CC. The authors highlighted that CC is the most critical issue threatening human survival. Droughts, heat waves, and floods are examples of extreme events that pose a threat to global ecosystems and future food security. Such changes have a significant impact on the biotic and abiotic components of agroecosystems. The olive tree (*Olea europaea L.*) is of considerable economic importance in many countries worldwide. Nowadays, its cultivation is facing new challenges related to CC, such as environmental stresses, pest outbreaks, and increased incidence of pathogens. Therefore, various abiotic stresses, such as heat, drought, flooding, and salt are hostile to the growth and development of olive trees, resulting in significant yield losses. Global climate variations can make trees vulnerable, promote the spread of pests and pathogens, accelerate their development, and weaken or eliminate their competitors and natural enemies. To address this situation, adaptable management techniques and changes in agricultural practices are needed to mitigate the effects of biotic and abiotic stresses.

Chapter 8 explored how to bridge the gap between CC and plant biology by enhancing resilience through genetic diversity and agroecological practices. The inherent genetic diversity of crops equips them with a remarkable ability to adapt to shifting environmental conditions, a crucial aspect of crop resilience. This adaptability stems from the vast array of traits and characteristics encoded within their genomes, enabling them to respond to diverse environmental challenges. By subjecting crops to stress conditions, breeders can identify individuals with desirable traits that confer enhanced resilience. Agro-ecological approaches and sustainable farming practices also contribute significantly to crop resilience. Crop rotation, intercropping, and agroforestry systems promote biodiversity and enhance the resilience of agroecosystems. In the face of CC, developing crop varieties with improved resilience to extreme weather events is paramount. By harnessing agroecological principles, genetic diversity, and adaptive capacity, it can bolster crop resilience, ensuring the productivity and adaptability of our agricultural systems in a changing environment.

Chapter 9 highlighted the missing component investigating the intraspecific traits variation enhancing wild fruit trees tolerance to CC. Plants contend with a myriad of environmental stressors that exert constraints on their performances, posing considerable concerns regarding the enduring persistence and resilience of plant populations over the long-term. This apprehension is especially pertinent for Mediterranean species, exemplified by the wild olive and the carob tree. Nevertheless, a multitude of studies have illustrated that plant species exhibit the capability to endure adverse environmental conditions through the deployment of various morpho-physiological and biochemical traits that support physiological processes. Given the pivotal role of wild fruit trees as invaluable genetic resources, a comprehensive understanding of their potential responses to evolving environmental conditions becomes imperative for the development of apt strategies for resource preservation and breeding programs. In the context of this chapter, the hypothesis posits that intraspecific trait variability emerges as a central player in augmenting the resilience of these emblematic wild fruit trees in the face of CC.

Within section 2, Chapter 10 reported a Case Study from Kanpur Nagar, Uttar Pradesh Assessing the Economic Value of Water and its Determinants. The chapter applied a willingness to pay ability (WPA) approach to estimate the economic value of drinking water (EVM) using primary data. The sample of 350 households is randomly taken from four wards of Kanpur Nagar, Uttar Pradesh (India). Simple OLS, probit, logit, and mixed-effect logit regression models are applied to examine the effect of social-economic variables on WPA of households to receive better water supply from suppliers. It found that the municipality is charging a very low price to provide water supply of 30 Kilolitre/household per month. However, the implicit cost (IC) of water/household is found higher as compared to WPA. IC/household appeared high due to the consumption of unsafe drinking water, and the incidence of various waterborne health issues and diseases. Family size, occupation, and annual income showed a positive impact on the WPA of households.

Chapter 11 elaborated on the strengthening capacity for CC adaptation of Moroccan agriculture exploring the deficit irrigation strategies for the resilient olive sector. Agriculture in the Mediterranean region is substantially challenged by CC and olive growers face particular difficulties because of its effects. In this regard, the olive sector can adopt deficit irrigation strategies, to achieve sustainable management practices that improve water productivity and save irrigation water. This chapter aims to investigate the potential impacts of deficit irrigation strategies on Menara olive tree. Agronomical parameters were assessed in 2021 and 2022, focusing on trees subjected to regulated (RDI) and sustained (SDI) deficit irrigation. The results showed that the performances of Menara olive trees under controlled water stress are better under RDI strategy compared to SDI one. Moreover, the RDI strategy, particularly under specific treatment, allowed to reduction of supplied water by 20% and 40% during sensitive and normal periods, respectively. These results highlight the importance of water management strategies and the need to consider their implications on flowering traits, crop yield, and fruit quality over consecutive growing seasons.

Chapter 12 detailed the Eutrophication under a bionomical vision. In this chapter, the authors highlighted that major advances in the scientific understanding of eutrophication and the significant effects of eutrophication have been made for decades. The effects of eutrophication include increased biomass of freshwater phytoplankton and periphyton, shifts in phytoplankton species composition to taxa that may be toxic or inedible e. g. bloom-forming cyanobacteria, reduced biodiversity of waterbodies, changes in vascular plant production, biomass, and species composition, reduced water clarity, decreases in the perceived aesthetic value of the water. Relative abundances of major bacterial and archaeal groups indicate high similarities at the phylum- and class-level groups across eutrophicated lakes. The impact of CC and greenhouse gases emissions regards to affect and effect of eutrophication leads to various studies. The effect of reducing nitrogen inputs alone for controlling eutrophication by controlling inputs of Phosphorus.

Chapter 13 presented composting as a technique for a sustainable future turning waste into nutrient-rich soil. As the worldwide waste management problem has risen, composting has gained popularity. It turns organic trash into nutrient-rich soil for agriculture, landscaping, and environmental rehabilitation. This chapter on composting, "Composting for a Sustainable Future: Turning Waste into Nutrient-Rich Soil," is comprehensive. It covers decomposition science, composting methods, materials, and procedures. Composting's agricultural, landscaping, and environmental remediation benefits are also covered. The chapter also examines composting's role in CC mitigation, its obstacles, and remedies. Decomposition can help people, businesses, and communities live more sustainably. It urges decomposition and trash reduction and provides information about tools to start composting, a simple yet efficient solution to worldwide waste management.

Section 3 is represented by one single chapter (chapter 14) entitled "Protozoan, Animal, Socio-ecological Factors, and Vector-Borne Diseases, the Leishmaniasis in the Province of Errachidia, Morocco". Leishmaniasis is a parasitosis caused by a flagellate protozoan of the genus Leishmania, transmitted by a vector (Phlebotomus) from a reservoir, rodent. It is endemic in several countries in the Mediterranean region including Morocco. This is a retrospective and descriptive study of the epidemiological profile of leishmaniasis in the province of Errachidia over the period 2012-2022. The total number of cases recorded during this period is 6726 cases. This number is very high, and despite control programs, the province of Errachidia continues to record cases, reaching 105 cases in 2022. The seasonal distribution of leishmaniasis cases was more concentrated in the autumn-winter seasons of each year. The majority of cases were recorded in rural areas and Goulmima was the most affected district in the whole province. The authorities concerned have set up a surveillance and control program aimed at reducing the number of cases and limiting the geographical spread of leishmaniasis.

In the section 4, chapter 15 reviewed the impact of agrochemicals on the environment and ecological alternatives. To promote plant development and enhance agricultural yield, farmers utilize a variety of agrochemicals. Insecticides, herbicides, fungicides, weedicides, and other pesticides are among the agrochemicals. The agrochemicals and their classification are mentioned in this paper. The pros and cons of agrochemicals are also discussed. Focus is also placed on how agrochemicals affect the ecology and how they work. The risks and protective measures related to the usage of agrochemicals are also included. These crucial agrochemical issues will be covered in full in the study. Future research into the molecular basis of agrochemical mechanisms must be done.

Chapter 16 explored the sustainable integrated farming in agriculture. Advancement of human civilization resulting in depletion of resources and the deterioration of living conditions in certain regions of the world, human activity on Earth has taken a hostile turn that threatens the existence of species and mankind. The circular agriculture focuses primarily on using minimal external input, closing nutrient cycles, regenerating soils, and minimizing environmental impact. Agriculture chains must be efficient in order to maximize the use of resources and return nutrients to the soil for a sustainable future. AI technology will lead to substantial changes in agricultural practices. The current discussion was explicitly related to United Nations Sustainable Development Goal 2: Zero Hunger, as well as other UNSDGs. To generate synergies and effectively manage trade-offs between agriculture, water, energy, land use, and CC, the international community is urged to increase the number of integrated national and regional decision-making processes.

Chapter 17 investigated the intersection of ecological thought and economics, examining the evolution of ecological economics and its potential implications for economic theory and policy. Ecological economics offers a more comprehensive framework for understanding and addressing the relationship between the natural environment and economic systems. Key concepts, such as externalities, natural capital, and ES are discussed, as well as challenges to incorporating ecological considerations into economic decision-making. By promoting more sustainable and just economic systems that prioritize environmental sustainability and social equity, ecological economics can inform and transform economic policy and governance.

Chapter 18 reported that a good assurance of food security requires a good understanding of the plant-soil-water-living being and CC interaction. In this chapter, the authors highlighted that the impacts of CC on agriculture can have repercussions on global food security. Losses in agricultural yield can lead to food shortages and increases in food prices, particularly affecting the most vulnerable populations. CC can result in food insecurity due to water resource scarcity, while soil degradation may worsen production problems. A quality harvest depends on optimal climatic conditions. However, CC, characterized by infrequent or irregular precipitation and high temperatures, leads to a decrease in yield on one hand. On the other hand, this CC has an additional negative impact by deteriorating soils and reducing water resources through evaporation, reaching yield limits. Within this chapter, the authors explore the different interactions between plants, soil, water, and CC. The objective is to design strategies specifically tailored to the changing climatic context, thus enabling the development of an agricultural sector that is both resilient and sustainable.

The outcome of these 18 chapters is a product of collaborative efforts involving professionals, academics, and scholars worldwide who have generously contributed their ideas, expertise, and insights. Ultimately, the book *Water-Soil-Plant-Animal Nexus in the Era of CC* represents a singular contribution among many essential endeavors aimed at constructing the knowledge foundation necessary to bolster the sustainable development goals in the present era of human history.

*Ahmed Karmaoui*
*Moulay Ismail University, Meknès, Morocco & Moroccan Center for Culture and Science, Morocco*

# Acknowledgment

The interactions between the water, soil, plant, animal, and CC are of increasing concern to scholars, decision-makers, and researchers. The effects of CC on these resources are various and accelerated due to human pressure. To highlight the importance of these interactions and impacts, several chapters including relevant theoretical and empirical approaches, and bibliometric analyses were selected. The success of this project owes much to the contributions of several international authors. I would like to thank all the authors for their hard work in writing and improving their manuscripts and the reviewers who assisted me in the peer-review providing constructive feedback to ensure the high quality of the entire collective book. My sincere thanks also go to the editorial team at IGI Global for their amazing editorial work. The outputs of this edited book are in harmony with the mission of the Moroccan Alliance for Climate and Sustainable Development (AMCDD), an organization of which I am proud to be a part.

## Editorial Advisory Board

# Section 1

# Plants in the Context of Climate Change

# Chapter 1

# Aloe Vera Plant:
## Characterizing the Components and Activities and Exploring Some Stress Factors Associated With Climate Change

**Ahmed Karmaoui**
https://orcid.org/0000-0003-3881-4029
*Université Moulay Ismail, Meknès, Morocco & Moroccan Center for Culture and Science, Morocco*

**Zineb Moussafir**
*Biochemistry of Natural Products Team, Faculty of Sciences and Techniques, Errachidia, Morocco & Moulay Ismail University, Meknes, Morocco*

## ABSTRACT

*Aloe vera (AV) is a multi-benefit plant well known for diverse applications due to its unique gel. This gel contains high-quality oil used to produce a high range of phytochemical products where the quantity and quality depend to stress factors including climatic variables. This study presents a review of AV components and the impact of stress factors including climate change (CC) through a bibliometric analysis. For the trend, research on this plant, its products, and CC tends towards the chemistry aspect and the use of scanning electron microscopy, phytotherapy, metabolism, and CC. Thus, CC was highlighted as a trending topic but not well explored. The topic of CC was well mentioned but the majority of research efforts explored it partially or generally. The results showed that temperature and precipitation are the most important climatic variables that affect the aloe species. However, water and saline stress increases the quality of the gel and reduces the growth and leaf yield.*

## INTRODUCTION

Aloe vera is a succulent plant with many benefits that are found in many zones throughout the world. It grows in subtropical and arid climates and is used for its medicinal properties to treat minor cuts, gastrointestinal problems, burn wounds, and genital herpes (Lanka, 2018). Regarding health applications,

DOI: 10.4018/978-1-6684-9838-5.ch001

many minerals and biological substances such as vitamins, amino acids, polysaccharides, and enzymes are concentrated in the plant leaves and the associated gel. These substances have excellent biological activities such as hypoglycemic or antidiabetic effects and immunomodulatory and gastroprotective properties (Hamman, 2008), skin elasticity (Tanaka *et al.,* 2016), antiviral (Kahlon *et al.,* 1991), anti-bacterial, antifungal, antioxidant (Hęś *et al.,* 2019), and anti-inflammatory (Talmadge *et al.,* 2004) activities, etc.

Aloe vera is well known for the diverse applications of its gel. This gel contains a high-quality oil used to produce a high range of cosmetics, food supplements, and hygiene products. In addition, it plays a socio-economic role due to its importance in job creation in agronomic and industrial processes. It is a promising succulent plant that may support the economic development of the subtropical and arid zones. In fact, aloe species have significant commercial value for use in poverty reduction and rural development (Grace, 2011).

Aloe vera is a succulent plant belonging to the Asphodelaceae family and genus Aloe. It is native to North and East Africa for over five millenniums and can be exploited according to several disciplines: biochemical, chemical, pharmacology, economic, and horticultural (Chinchilla et al., 2013). Due to its strong adaptability, it can be found or cultivated throughout the world. Historically, Greek scientists considered it the universal panacea while Egyptians give it the name, the "plant of immortality" (Surjushe et al., 2008).

It is an evergreen perennial rooted and xerophytic plant with triangular green leaves and yellow tubular flowers that grows mainly in subtropical and arid zones. Due to its ecologic, economic importance, and medical propriety, Aloe vera has been cultivated in several countries, including India, Haiti, and the United States. Concerning aloe vera uses, the most successful companies of aloe vera are located in the United States. Aloe vera gel (extracted fresh juice) became widely known for its uses in cosmetics (moisturizing), hygiene (as a shampoo and soap), and supplement products (vitamins and minerals). Skin lotion, food supplements, soaps, and toothpaste…are some of the most well-known products derived from this juice. It is considered a miraculous plant by Chinchilla et al., (2013) but it may face threats as a result of changes in climatic parameters such as temperature and precipitation that will impact the spatial distribution and the growth. The aloe vera growth, productivity, and geographical distribution are affected by several climatic and environmental factors such as temperature Kumar et al., (2017a), rainfall and groundwater level (Manvith & Bidya, 2014), water stress (Hazrati et al., 2017), and saline stress (Mota-Ituarte et al., 2023; Murillo-Amador et al., 2015). At the molecular level, it was demonstrated that the diversity of the plant's phytochemicals and its capacity as an antioxidant are influenced by various agro-climatic circumstances (Kumar et al., 2017b). In a study conducted by Mkala et al., (2022) it was reported that the current distribution of three species namely *Aloe ballyi* Reynolds, *A. classenii* Reynolds, and *A. penduliflora* Baker has been significantly impacted by precipitation, temperature, and some environmental factors, such as land cover, soil sedimentation, and solar radiation. For plants, generally, the variation of temperature and wind (linked to CC) has an effect on precipitation and thereby plant architecture, flowering, fruiting, and phytochemical components (Kumar et al., 2017a).

Ecologically, aloe species are commonly found in arid conditions in Africa, India, and other arid regions (Salehi et al., 2018). Thus, basically, like other Succulent plants, Aloe vera is heat-tolerant and cold-sensitive plant. In contrast, they are found in subtropical climates with rainy conditions. Samples from the Highland and Semi-arid zones had the highest antioxidant potential, whereas samples from the Tropical zone had the lowest (Kumar et al., 2017b). CC combined with human pressure affects negatively many components that support plant development such as soil and water. This can induce mainly hydric and saline stresses primary factors influencing the growth of plants.

The current study presents a review of the plant components, activities, and the impact of some stress factors including climate change (CC) using a specialized software, VOSViewer. This tool was used for showing maps of research structure in many fields including agriculture (Karmaoui, et al., 2023a; 2023b and Karmaoui, 2023) and environmental fields such as maize or corn (Neelam & Sood, 2020), rice (Sun & Yuan, 2020), plant leaf diseases (Patil & Kumar, 2020), alpine vegetation (Verrall & Pickering, 2020), chemicals compounds such as antioxidants (Tabatabaei-Malazy et al., 2015), natural products in diabetes research (Yeung et al., 2020), Rooibos Tea (*Aspalathus linearis*) (Elisha & Viljoen, 2021), and in fields like ethnopharmacology (Yeung et al., 2018). The discovery of the benefits of the aloe vera plant has been the subject of much research. However, to our knowledge, no study has dealt with bibliometric analysis on this plant. This method helps understand both the structure and diffusion of science, explore its importance in the field, and its stress factors.

## Material and Method

### Plant Presentation

Aloe vera is a succulent plant belonging to the Asphodelaceae family and genus Aloe. It is native to North and East Africa. It is an evergreen perennial rooted and xerophytic plant with triangular green leaves and yellow tubular flowers that grows mainly in subtropical and arid zones. Xerophytes have the capacity to survive in zones with low precipitation (Chinchilla et al., 2013). The cuticle of the thick and fleshy aloes leaves is protected by a layer of wax allowing the accumulation of gel tissue (Chinchilla et al., 2013). The extracted fresh juice constitutes a gel that contains high-quality oil. Due to its strong adaptability, it can be found or cultivated throughout the world. In regard to the chemicals elements, the pulp contains several types of polysaccharides, namely mannan, acemannan (or acetylated mannan, the principal polysaccharide of the gel) (Hamman, 2008), galactan, and arabinan (Ni et al., 2004), and pectic substance (Mandal & Das, 1980). Other carbohydrates, maloyl glucanss were identified by Esua & Rauwald (2006), chromones isolated by Okamura et al., (1998), anthraquinone (Kang et al., 2017), tetrahydroanthracenones (Chinchilla et al., 2013), anthrones (Salehi et al., 2018), phenolic compounds (Debnath et al., 2018), sterols (Tanaka et al., 2016). Inorganic elements were also identified by Rajasekaran et al., (2005). Calcium, potassium, zinc, manganese, magnesium, phosphorus, and sodium… Some trace elements can be found in this plant such as copper, selenium, arsenic, nickel, ferrous, and chromium… Anthraquinone glycosides of this plant have the propriety of cicatrization for external application and laxative effect for internal use (Rosca-Casian et al., 2007).

Aloe vera is one of the most important crops and is of high economic and ecological value (Solomou et al., 2020). In fact, it is able to be grown in poor soil conditions and promotes rural development (Liontakis & Tzouramani, 2016). Another important ecologic role of aloe vera is the ability to clean up contaminated soil and to phytoextract various metals such as chrome, nickel, copper, lead, and cadmium (Shokri et al., 2016).

### Data Acquisition and Method

This research is based on the evolution of the knowledge diffusion approach (Liu et al., 2017). In order to reflect the research structure on the aloe vera key term, the general analysis was performed on 5,111

*Figure 1. Bibliometric structure of the study, general and specific analyses*

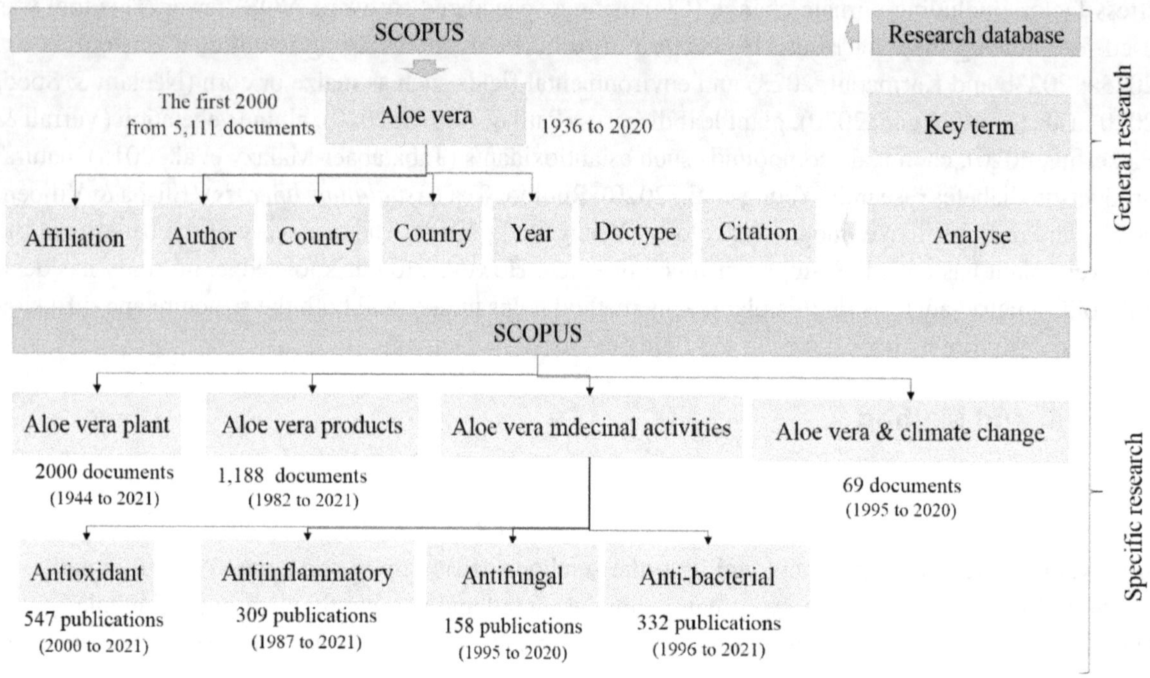


papers (Figure 1). The first 2,000 papers of the period 1936 to 2020 were retrieved from the database "Scopus".

Affiliation, Author, Country, Year, Doctype (type of publication), and Citation were used to analyze search results. For further information, a specific analysis (Figure 1) was carried out using three topics: aloe vera plant (2000 from 2,793 papers, 1944 to 2021), aloe vera products 1,188 papers, 1982 to 2021), and aloe vera medicinal activities. This last category includes the terms anti-bacterial (332 papers, 1996 to 2021), antifungal (158 papers, 1995 to 2020), antioxidant (547 papers, 2000 to 2021), and antiinflammatory (309 papers, 1987 to 2021). The specific research was conducted also using the keywords (TITLE-ABS-KEY (aloe AND vera)) AND (climate AND change), which yielded 69 document results in the period 1995-2020 (Figure 1).

The above-mentioned key terms were considered as criteria for the selection of publication groups. The gathered groups of papers that meet the criteria were analyzed separately by VOSviewer software.

## RESULTS AND DISCUSSION

1. Research structure and network analysis

Based on the keywords containing Aloe vera, three groups of studies were retrieved: aloe vera plant (plant agronomy and preparation), aloe vera products (types of products), and aloe vera activities (medicinal applications). The first group includes papers based on the keywords "aloe vera plant." The second group contains documents on products using the key term "aloe vera products" and, the third group comprises four health medicinal characteristics namely anti-bacterial, antifungal, antioxidant, anti-inflammatory:

- 332 document results: Aloe vera anti-bacterial activity in the period 1996-2021 until January 31st;
- 158 document results: Aloe vera antifungal activity in the period 1995-2020 until June;
- 547 document results: Aloe vera antioxidant in the period 2000-2021 until January 31st;
- 309 document results: Aloe vera anti-inflammatory activity in the period 1987-2021 until June;

The findings show a positive trend of publication over the period of this bibliometric analysis and that 2020 is the most influential year in the selected articles (Figure 2a). In a general search, the first study identified was published in 1936 and explores the impact of aloe vera in the treatment of roentgen ulcers and telangiectasis (Wright, 1936). Between 1936 and 2021, in Scopus database, 5,179 publications were found which were divided into three periods: 1/ from 1936 to 1982 with 33 publications (0.63%); 2/ from 1983 to 2001 (lasting 20 years) with 346 publications (6.68%); 3/ started from 2002 to 2020 with 4,736 papers (91.44%). The number of publications increased exponentially from 1994 (with 10 publications) to 2020 (the most productive year), which recorded 515 publications. India, the United States, Iran, China, and the United Kingdom produced the most research on aloe vera at the global scale for this study (Figure 2b). Tehran University of Medical Sciences, Islamic Azad University, Saveetha Dental College And Hospitals, Chulalongkorn University, and Tumkur University are the top five affiliations at the global scale (Figure 2c). The main subjects of this study are 1/ Medicine; 2/ Pharmacology, and Toxicology and Pharmaceutics; 3/ Agricultural and Biological Sciences; 4/ Biochemistry, and Genetics and Molecular Biology and ; 5/ Chemistry (Figure 2d). The most influential authors are Nagabhushana, H., Sharma, S.C., Tanaka, M., Valero, D., and Basavaraj, R.B., (Figure 2e) while article, review, and conference paper are the top three doctypes (Figure 2f).

Data acquisition was carried out on the basis of the Scopus database using VOSviewer software. To start the analysis, it is essential to define some parameters such as the minimum number of occurrences of a keyword, the most applicable keywords and the number of keywords to be selected. To identify key terms, a minimum occurrence threshold of 30 was applied to the 21,410 keywords, resulting in 308 keywords meeting this criterion. Subsequently, the total strength of co-occurrence links with other keywords was computed for each of these 308 selected terms.

According to research based on the term "aloe vera plant", the most influential activity was the antioxidant activity (Occurrence of 206, 4459 links) which is followed by the anti inflammatory activity (with a frequency of 160, 3682 links). Then comes the antibacterial activity (138 and 2795 links), the antidiabetic activity (84 and 2122 links), the antineoplastic activity (85 occurrence and 1923 links), antiviral activity (39 occurrence and 948 links), the antifungal activity (53 occurrence and 1042 links), and the biological activity (38, 664). By changing the term of activity to agent, the following results are obtained: anti infective agent (80, 1774), antidiabetic agent (59, 1587 links), anti-inflammatory agent (62, 1582 links), antineoplastic agent (63, 1497 links), antibacterial agent (39, 865 links), antibiotic agent (35, 737 links). In regard to the chemicals (chemistry: 291, 5817), several elements were found associated with the topic of the aloe vera plant, namely: flavonoid (107, 2468 links), glucose (64, 1367 links), polysacharide (69, 1046 links), insuline (39, 956 links), reactive oxygen metabolite (45, 950 links), anthraquinone derivative (44, 941 links), superoxide dismutase (42, 940 links), aloe emodin (49, 914 links), alcool (45, 892 links), alpha tocopherol (42, 844 links), resveratrol (42, 844 links), tannin derivative (32, 760 links), catalase (30, 739 links), polyphenol (31, 716 links), triacylglycerol (32, 692

*Figure 2. Bibliometric structure of the study based of the key term "Aloe vera" using Scopus database: (a) Number of papers by year; (b) Number of papers by country; (c) Number of papers by Affiliation; (d) Number of papers by subject ; (e) Number of papers by Author; (f) Number of papers by type of publication.; g, co-occurrence of words based on research using the key term "Aloe vera" analyzed by VOSviewer*

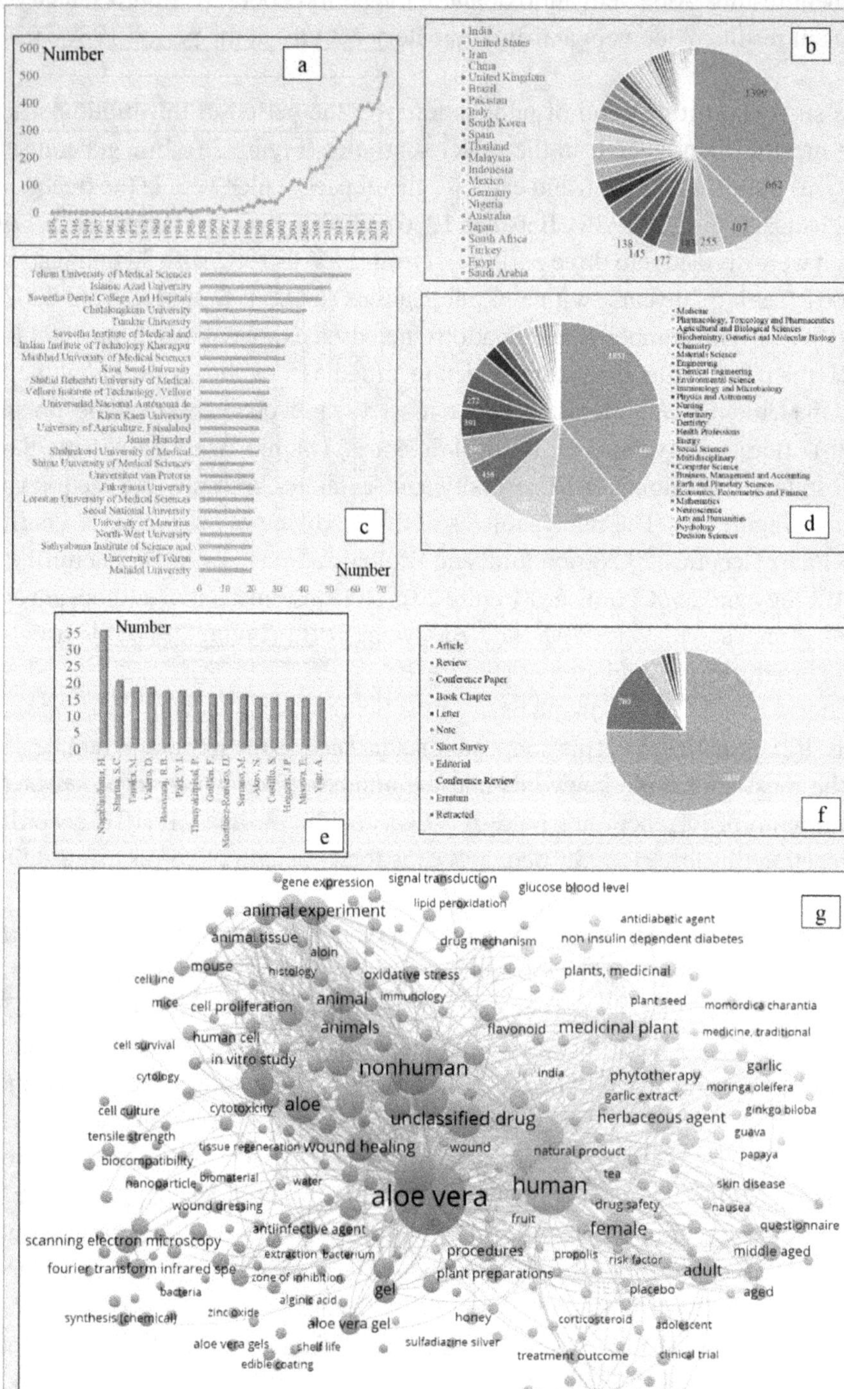

*Table 1. The publications the most cited using VOSviewer*

| Document | Citations | Document | Citations |
|---|---|---|---|
| Greenlee H. (2017) | 195 | Nanda S. (2016) | 104 |
| Izzo A. A. (2016) | 191 | Sogvar O. B. (2016) | 101 |
| Hassan B. (2018) | 183 | Hazrati S. (2016) | 86 |
| Pereira R. F. (2016) | 129 | A. N. B. (2017) | 85 |
| Ali K. (2016) | 113 | Charbgoo F. (2017) | 82 |
| Vieira J. M. (2016) | 108 | Ashraf J. M. (2016) | 78 |

links), vitamin (34, 692 links), carbohydrate (36, 686 links), tannin (30, 644 links), nanoparticles (48, 510 links), polysaccharides (39, 510 links), acemannan (40, 499 links), zinc (33, 487 links), and chitosan (31, 376 links). Kahramanoğlu et al., 2019 enumerated 13 flavonoids and their glycoside derivatives, 32 anthraquinones and their glycoside derivatives, and 29 chromone derivatives were identified from the aloe vera plant.

Regarding the diseases and health problems, most studied in the current research, around thirty terms have been identified, including digestive, skin, metabolic problems, and disease symptoms (Appendix 1). Diabetes mellitus (151, 4130), diarrhea (81, 2456), skin disease (71, 2178), inflammation (87, 2112), oxidative stress (78, 1806), and constipation (59, 1762) are the top five most cited disease symptoms in this general research (Aloe vera research).

Bibliometric analysis shows four main clusters (groups of keywords) (Figure 2g). The first group includes 103 items represented by red circles such as aloe vera, chemistry, gel, antimicrobial activity, wound healing, antibacterial activity, anti-infective activity, plant leaf, in vitro study, scanning electron microscopy, Fourier transform infrared spe, and ascorbic acid (Figure 2g). The second cluster comprises 87 items represented by green circles such as human, male/female, adult, drug efficacy, plant medicinal product, aged/middle-aged, questionnaire, plant preparation, and procedure (Figure 2g). However, the third group contains 76 items represented by blue circles. This group includes key terms such as non-human, animals, animal tissue, animal experiment, animal model, unclassified drugs, drug effect, plant extract, aloe, metabolism, and mouse. The last group (67 yellow circles) includes herbal remedies, herbaceous agents, phototherapy, garlic, ginger, flavonoids, and skin disease (Figure 2g).

Concerning the quality of publications on aloe vera research, an analysis of the citations was carried out using VOSviewer software (Table 1). In this study, a document was deemed relevant only if it had a minimum of 50 citations. Out of the 2,000 documents, 28 met this criterion. The greatest number of documents and citations was recorded in 2016 (A total number of 1,429), which represents a very great interest in this plant while 2017 recorded 10 documents, with a total number of citations of 786), and 2018 (two documents, with 256 citations).

## 2. Aloe vera plant

In terms of specific research, the query "aloe vera plant" was carried out, and the on-site tool Scopus yielded a total of 2,793 articles in the period 1944 to 2021. Only the first 2,000 papers were considered and imported to the VOSviewer software. Trend analysis of this research was conducted, resulting in two-network maps as shown in Figure 3. The first trend network represents the results of the search from

*Figure 3. Overlay visualization (research trend) using key words "Aloe vera plant" and VOSviewer software*

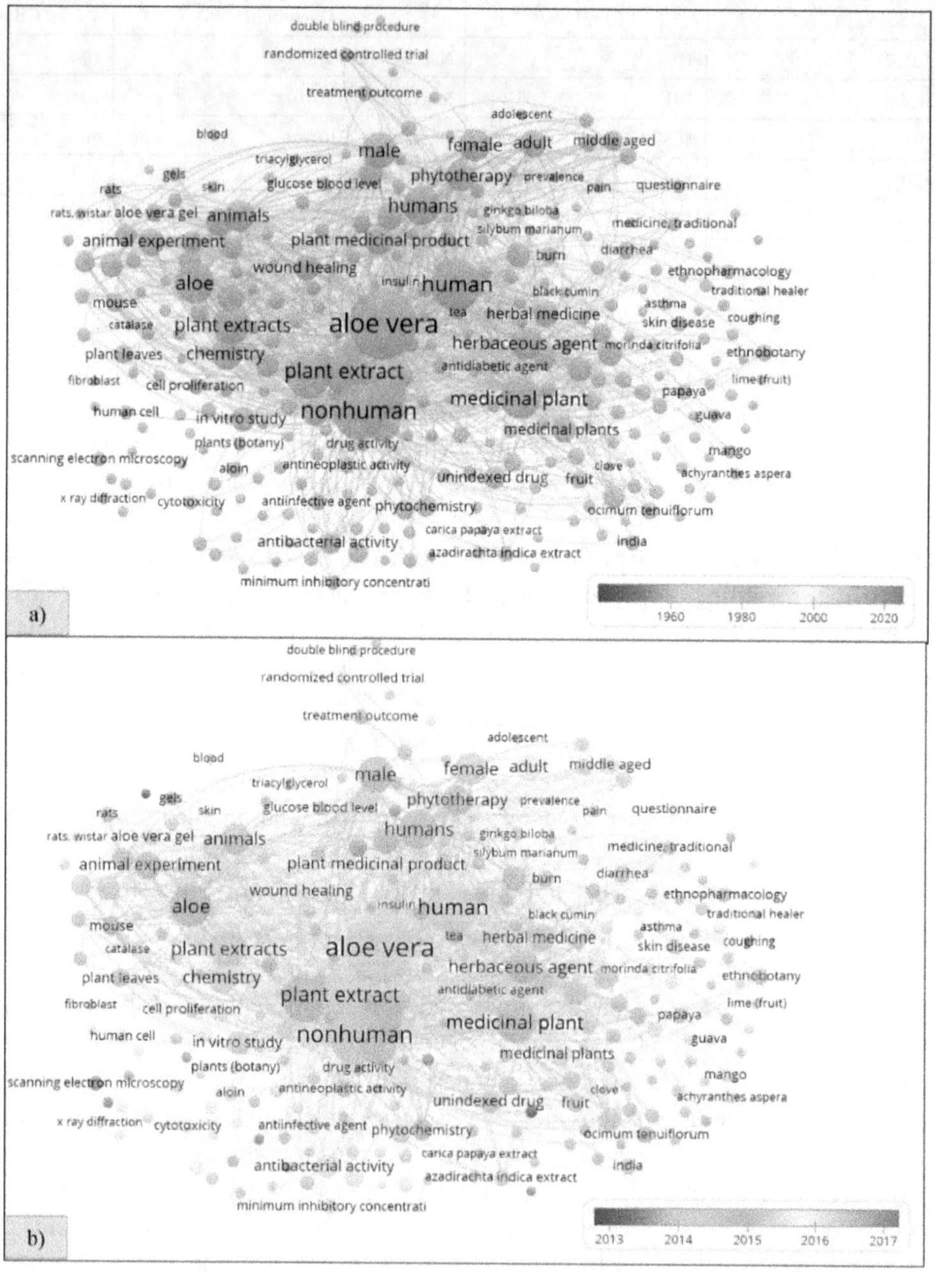


1944 to 2021 (Figure 3a). This map illustrates that the majority of the publications were produced from 2010 onwards, which means that research on this plant is current. For more details, another map has been produced based on further research on the period of publication concentration (2012- 2018) (Figure 3b). The findings on this map show a trend toward research in the field of chemistry and the use of scanning electron microscopy (SEM). An SEM is a microscope that uses a focused beam of low-energy electrons

*Figure 4. Overlay visualization (research trend) using key words "Aloe vera products" and VOSviewer software*

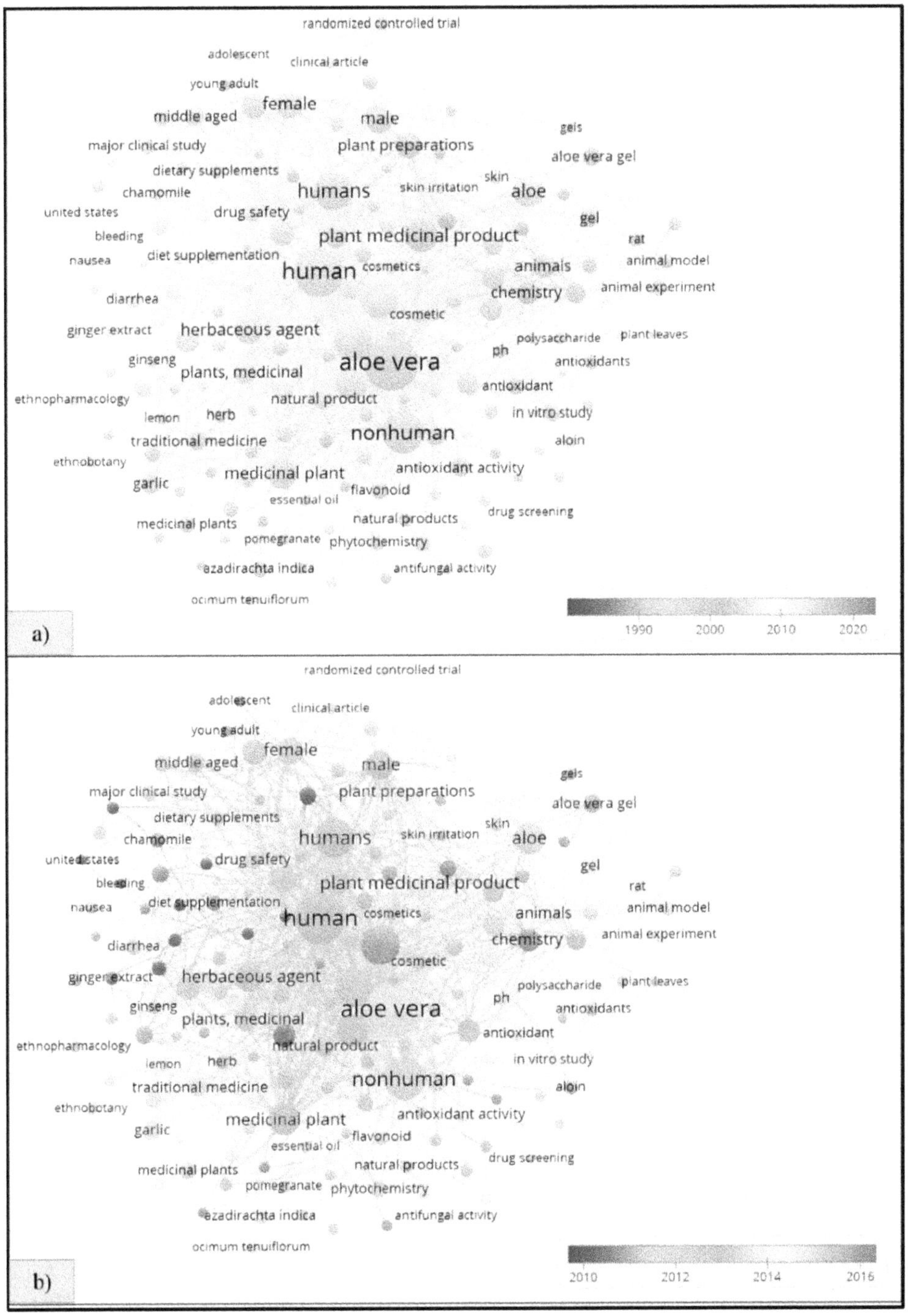


to image the surface morphology and microstructure of a sample (Omidi et al., 2017). This method was used to observe the internal microstructures of aloe vera by Choi et al. (2016).

*Figure 5. Network visualization using key words "Aloe vera antioxidant activity" and VOSviewer as a tool*

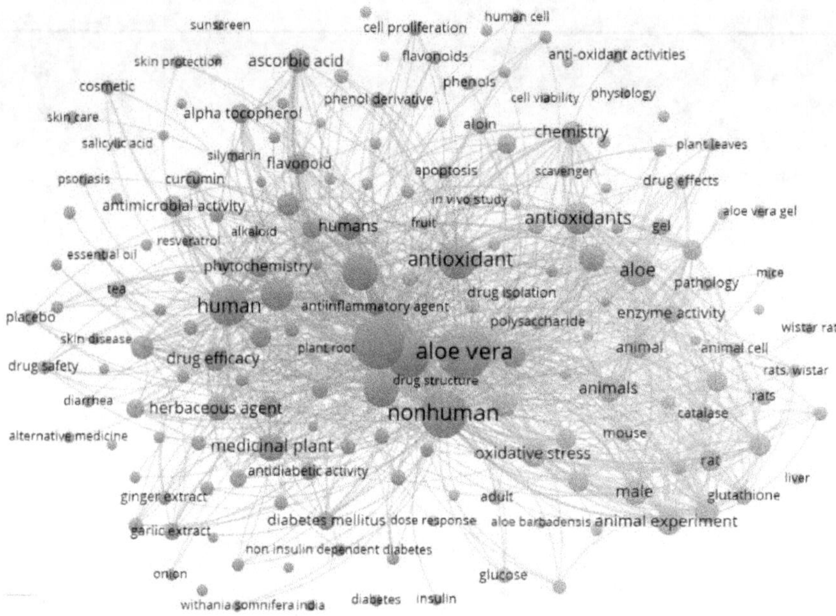


3. Aloe vera products

At the level of the query "aloe vera products", the search conducted on the Scopus database covered a total number of 1,188 papers for the period from 1982 to 2021. All these papers were taken in the account and imported to the VOSviewer tool. Trend analysis yielded the results shown in Figure 4, where the top map (Figure 4a) shows the search results of the entire period (1982 to 2021) and the bottom map (Figure 4b) depict the period 2009 to 2017 as the most concentrated period. The results of this map also show a trend towards research in the field of chemistry. This confirms the importance of the identification, purification, efficacy, and interactivity of the chemical products.

In order to highlight the medicinal importance of the aloe vera, four medicinal characteristics namely anti-bacterial, antifungal, antioxidant, and anti-inflammatory were explored in light of this bibliometric analysis.

4. Aloe vera activities (medicinal applications)

a. Aloe vera antioxidant activity

Aloe vera gel contains a large number of active substances including antioxidants. An antioxidant is a compound that protects another element against oxidation (Amorati et al., 2013). The antioxidant properties have been demonstrated in several studies (Hęś et al., 2019; Tabatabaei-Malazy et al., 2015). Embuscado (2015) has defined three classes of antioxidants; the first group is represented by free radical scavengers such as tocopherols and propyl gallate which have the function of blockcing free radicals by giving a hydrogen atom; the second class is the oxygen scavenger, ascorbic acid is the famous compound which has the ability to react with oxygen; and the third class consists of chelating agents such as citric acid which can sequester metal ions capable of catalyzing oxidation. In this study, 547 documents on aloe vera antioxidant in the period 2000-2021 until January 31st were retrieved (Figure 5). The most relevant key terms have been considered and the minimum number of occurrences of a keyword has been set at 15. Of a total number of 7,856 keywords, 190 meet the threshold.

According to the results (Figure 5 and Appendix 2), the occurrence and links of the substances cited in research on the antioxidant activity of aloe vera show a high score of the keyword, "unclassified drug" (245, 5212). This means that the research into antioxidants reveals an incomplete understanding of the composition of all molecules involved. The most important substances are flavonoids (89, 1909), followed by ascorbic acid (75, 1221), phenols and derivatives (58, 1078), and antioxidant enzymes. Superoxide dismutase (55, 1263) and catalase (55, 1012) are the most significant antioxidant enzymes that protect cells against oxidative stress. Other very important antioxidants are alpha tocopherol (53, 1047), glutathione (38, 877), curcumin (38, 848), and quercetin (30, 710). Ascorbic acid or vitamin C is one of the main natural antioxidant vitamins found in the aloe vera plant. It has the ability to protect body cells, fight bacterial infections, maintain healthy skin and teeth (Kahramanoğlu et al., 2019). Quercetin is a flavonol that has the ability to control the atherosclerotic inflammatory process, protect against mitochondrial dysfunction, and inhibit cell apoptosis (Yamagata, 2019). Glutathione has been purified and characterized from the aloe vera plant (Sabeh et al., 1993). It is another important antioxidant that plays a role in the management of redox potential at the cellular level. Indeed, it neutralizes singlet oxygen, free radicals, hydroxyl radicals, and superoxide radicals, intervenes in the regeneration of vitamins C and E, and in the transport of mercury out of cells and the brain (Pizzorno, 2014).

Curcumin is naturally occurring phenol of the curcuminoids group that is used worldwide in foods primarly as a colorant. Due to its structure and porosity, aloe vera has been used as a template for adsorbing and storing the curcumin called "self-assembled mesoporous biopolymer template" (Kitture et al., 2015). Curcumin has also been shown to control oxidative conditions, arthritis, anxiety, metabolic syndrome, and muscle pain (Hewlings & Kalman, 2017). Keyhanian & Stahl-Biskup, (2007) identified quercetin and anthranoids in dried flowers (*Aloe barbadensis Mill.*) and linked the hydrophilic antioxidant capacity to polyphenol and flavonoid. All of these compounds are known to prevent cells from toxic metabolites of oxygen.

b. Aloe vera anti-inflammatory activity

Aloe vera plant is a rich source of phytochemicals, including substances with anti-inflammatory proprieties. Research on this activity identified 309 documents in the period 1987-2021 up to January 31st. According to this research, the minimum number of occurrences of a keyword was set at 15. Of a total number of 6,068 keywords, 128 met the threshold. The number of keywords to select is 128. Anti-inflammatory activity has been studied in several studies (Talmadge et al., 2004; Yamagata, 2019). Inflammation is the body's reaction to an external or internal threat. Pahwa et al., (2020), identified two types of inflammation, acute inflammation caused by microbial invasion, trauma, or noxious compounds lasting a few days and chronic inflammation that affects a part of the body for a prolonged period up to years. According to Pahwa et al., (2020), chronic inflammatory diseases are the leading cause of death worldwide.

The network visualization generated using the Scopus database and VOSviewer separated the associated keywords into three main clusters represented by circles using colors, red, green, and blue. The

*Figure 6. Network visualization using keywords "Aloe vera antiinflammatory activity" and VOSviewer software*

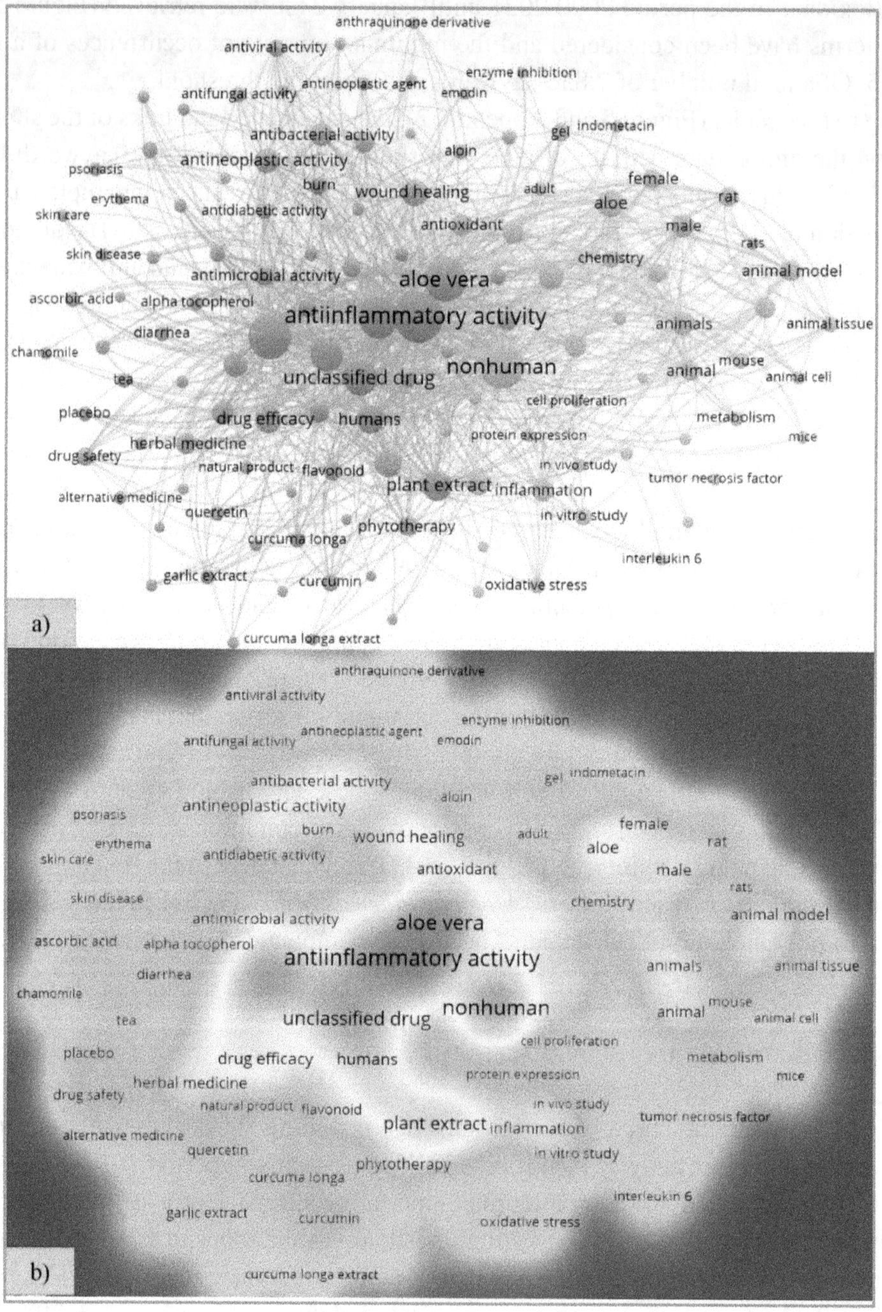


search of the words designating substances or chemical compounds related to the anti-inflammatory activity of aloe vera generated a large variability classified into the three clusters as shown in Figure 6a. All of these terms can be visualized as a heat map that shows the importance of the keywords using a color interval (Figure 6b). The results depict several substances known to have anti-inflammatory potential. In addition to the unrancked drug and high occurrence and linkage antioxidants, flavonoids scored high frequency and linkages in this search. This agent has a considerable anti-inflammatory effect on many chronic inflammatory conditions such as cardiovascular disorders, autoimmune diseases, neurodegenerative diseases, and cancer (Ginwala et al., 2019). Flavonoids are followed by curcumin (33, 674), quercetin (26, 528), ascorbic acid (27, 512), and alpha tocopherol (26, 499). These substances have also been found to be associated with antioxidant activity (see Appendix 3). It also means that oxidative stress and inflammation are correlated. These results are in agreement with the work of Biswas (2016) that found that inflammation and oxidation have reversible actions.

c. Aloe vera anti-bacterial and antifungal activities

Among the most interesting medicinal compounds of the aloe vera plant are anti-bacterial agents. This activity has been demonstrated in various studies (Hęś et al., 2019; Jain et al., 2016). In this specific search, 332 documents were retrieved and analyzed through Scopus database during the period 1996-2021, up to January 31st (Figure 5). To select only the most relevant keyword terms in this search, the minimum number of occurrences for a keyword was 15. Of a total number of 5,354 keywords, 89 met the threshold. The number of keywords that were selected is 89. The terms directly associated with the anti-bacterial activity of aloe vera can be classified into four categories: substances, microbes, techniques, and activities as follows (Appendix 4).

This information suggests the presence of substances in aloe vera that have an anti-infective effect. Jain *et al.*, Jain *et al.* (2016) explored the antibacterial activity working on the inhibitory and anti-microbial activities of Aloe Vera Gel on oral pathogenic bacteria, which suggest its use for oral health care. In the same study, gel concentrations of 50% and 100% were found to have similar Inhibition Zone with Ciprofloxacin (30mcg) and Ofloxacin (5mcg).

The antifungal activity has been reported by several studies as the medicinal property of the aloe vera plant (Hęś et al., 2019). The antifungal efficacy of a selected aloe vera leaf extract was investigated against pathogenic otomycosis fungal species, *Aspergillus niger* and *Candida albicans* (Saniasiaya et al., 2017). Documents on the antifungal activity of the aloe vera have been recovered. In this specific search, 158 publications were analyzed during the period 1995-2020 as selected by Scopus on-site tool (Appendix 5). The most relevant keywords were considered, the minimum number of occurrences of a keyword was 5. Of a total number of 3,114 keywords, 173 met the threshold. The number of keywords to select is 173. The key terms related to antifungal activity of the aloe vera have been presented Appendix 5. Aloe vera extracts have shown antifungal activity against several fungal species. De Rodrıguez et al. (2005) explored its antifungal activity against three phytopathogenic fungi, *Fusarium oxysporum*, *Colletotrichum coccodes*, and *Rhizoctonia solani*. This antifungal activity was also found by the total hydroalcoholic extract against *fusarium oxysporum* f.sp., *penicillium gladioli, botrytis gladiolorum*, and *heterosporium pruneti* (Rosca-Casian et al., 2007).

5. Climate change and Aloe vera plant and components

A total number of 69 document results was retrieved and analyzed using the key terms (TITLE-ABS-KEY (aloe AND vera)) AND (climate AND change). The analysis using Scopus online tool allowed to show the structure of research in terms of evolution of produced documents by year, by author, by country/territory, by subject area, and affliation (Figure 7).The results dipect a positive evolution of

*Figure 7. Structure of research based on the key terms (aloe AND vera)) AND (climate AND change): (a) Documents by year; (b) Documents byauthor; (c) Documents bycountry/territory;(d) Documents by subject area; (e) Documents by affliation*

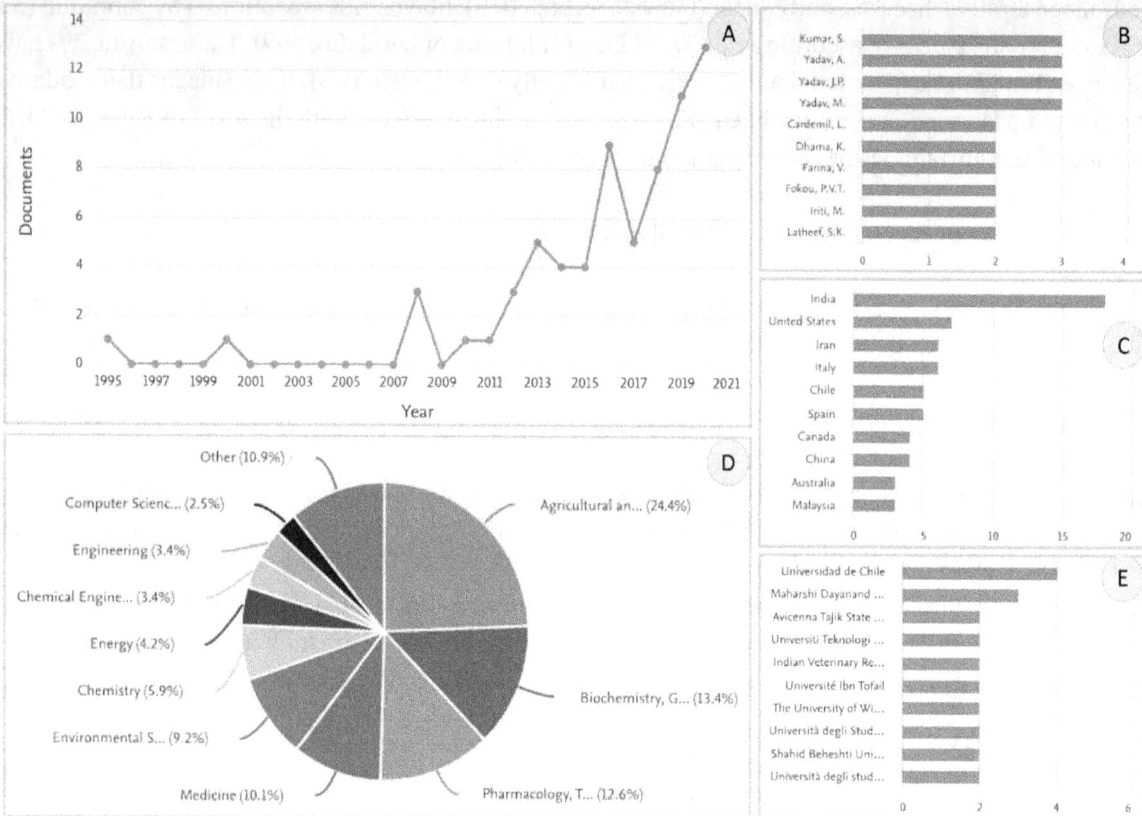


the number of publications in the period between 1995 and 2020 with an exponential increase from 2010 (Figure 7a). Kumar, S., Yadav, A., and Yadav, J.P. are the top three authors in the field (Figure 7b). India, the United States, and Iran are the most influencial in the field (Figure 2c). Agricultural and Biological Sciences, Biochemistry, Genetics and Molecular Biology, and Pharmacology, Toxicology, and Pharmaceutics constitute the main subject areas in the field (Figure 2d) while the Universidad de Chile, the Maharshi Dayanand University, and the Avicenna Tajik State Medical University are the top three affiliations (Figure 2e).

To represent the most relevant publications, the minimum number of occurrences of a key term was 4. Of the 1,621 keywords, 46 meet the threshold. For each of the 46 keywords, the total strength of the co-occurrence links to other keywords was calculated.

The results show three main clusters (Figure 8). The first cluster (Red items) includes key terms such as medicinal plant, aloe vera extract, phytochemicals, antioxidants temperature, growth, etc. while the second cluster represented by green items groups the keywords unclassified drug, drug effect, human, animals, herbaceous agent. However, the third cluster encompasses the blue items including plant extracts, CC, plant leaf, and chemistry (Figure 8).

*Figure 8. Network visualization using keywords "Aloe vera and climate change" and VOSviewer software*

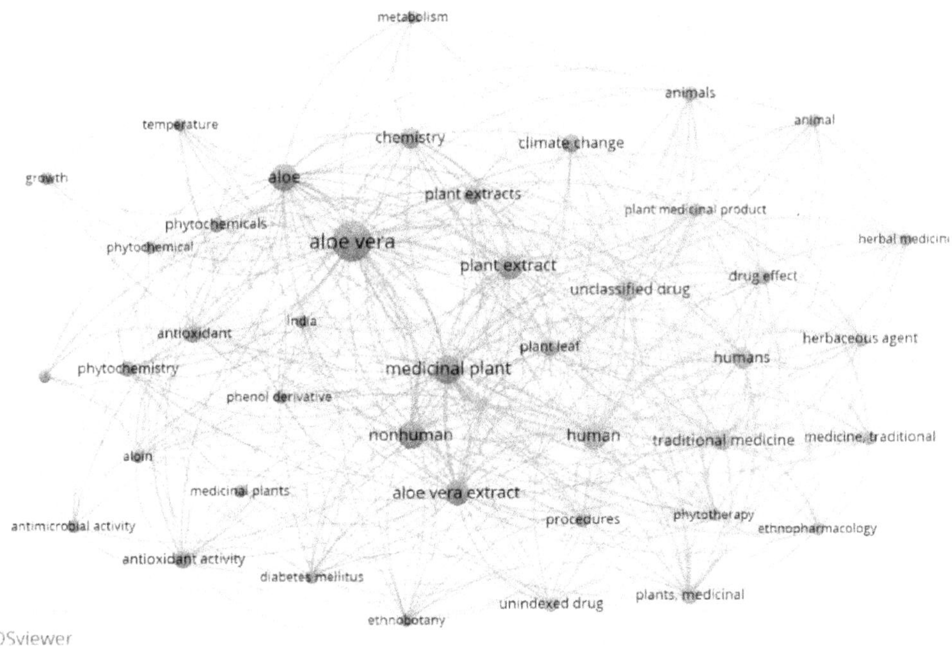


*Figure 9. Trending analysis using keywords "Aloe vera and climate change" and VOSviewer software*

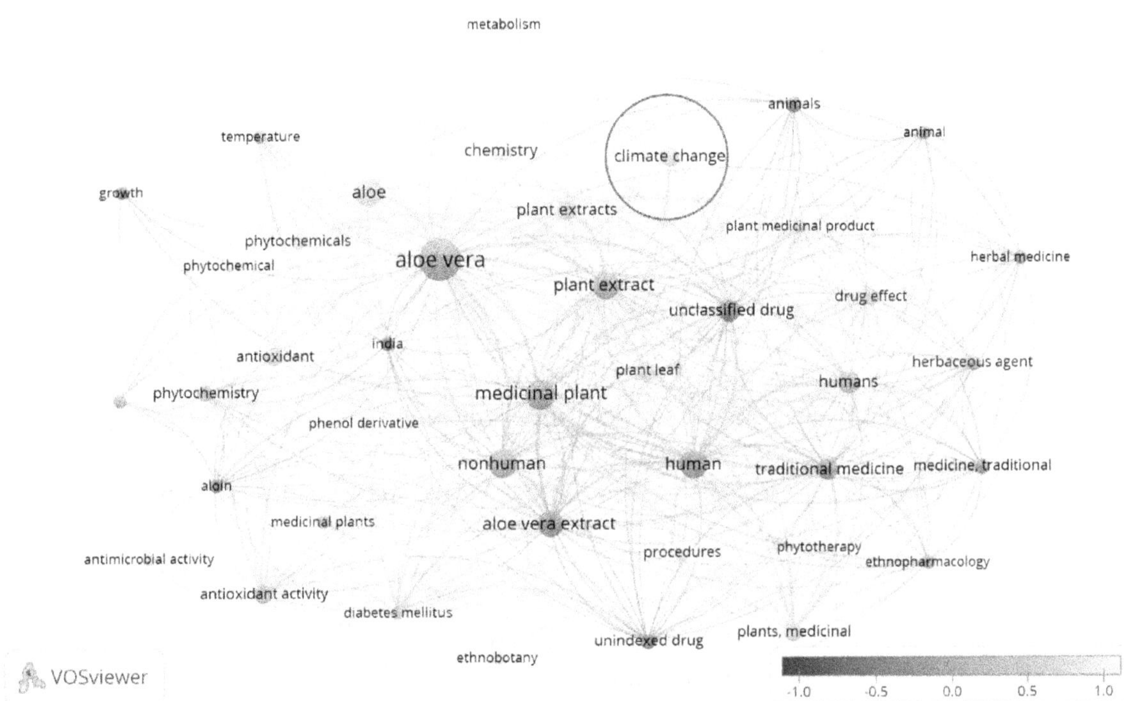

The trending analysis depicted that research is turning toward chemistry, phytotherapy, metabolism, and CC (Figure 9).

As shown in Figure 9, CC was highlighted as a trending topic but not well explored. Studies about the impact of CC on aloe vera plants are infrequent and the majority explored this aspect partially or generally.

CC has a remarkable impact on the life cycle, distribution, and phytochemical composition of plants such as aromatic and medicinal plants (Kumar et al., 2017a). Aloe vera is among the most important medicinal herbs (Benzidia et al., 2019). It grows in hot humid and high-precipitation areas (Manvith & Bidya 2014). In contrast, Kumar et al., (2017a) reported that the plant is also drought resistant and may grow in different climatic conditions. The highest concentrations of phenolic compounds were found in the organs of plants that were native to cold climates, higher altitudes, or semi-arid environments (Kabtni et al., 2020). Particularly, Kumar et al., (2017a) discovered that the highest antioxidant potential is found in the extracts from highland and semi-arid regions while tropical zones had the lowest antioxidant activity.

In Ethiopia, Hussein & Workeneh (2021) modeled the influence of CC on the distribution of Aloe vera species and found that the suitable distribution will become increasingly dispersed in the future depicting that Aloe vera is severely threatened by extinction due to the adverse effects of CC in Ethiopia.

Climate change has worsened, endangering the stability of ecosystems in general, biodiversity, and vegetation growth in particular (Sun et al., 2022). Several studies state the impact of some environmental factors on The aloe vera plant including water and saline stresses, agro-climatic locations, and sun exposure. Severe water stress significantly increased aloin concentration and Total Soluble Solids while reducing leaf yield and plant development (Hazrati et al., 2016). Souguir et al., (2015) explored that:

- The addition of salt to the irrigation water led to an increase in lipid peroxidation as seen by the quantity of malondialdehyde.
- Aloe leaves also demonstrated a notable rise in the concentrations of phenolic compounds in response to salt stress.
- Aloe can be grown in salt-affected soils and watered with salt water, at least at the mild dosage used in the current study.

In fact, environmental factors have the most impact on the ecological niches of some Aloe species such as *A. ballyi, A. penduliflora,* and *A. classenii* (Mkala et al., 2022). Agro-climatic conditions have a significant impact on the phytochemical compounds of plants (Kumar et al., 2017a). Under times of cold stress, plants produce more phytochemicals (Kumar et al., 2017b). The consequences of CC were exacerbated by the vegetation change brought on by increased temperature and precipitation (Zhang et al., 2022). Otherwise, Gao & Zhao (2022) suggested that quantifying the effects of CC on terrestrial ecosystems requires careful observation of the dynamics of vegetation phenology, knowledge of the factors influencing vegetation phenology, and comprehension of the mechanisms underlying phenology changes. Numerous studies have therefore concluded the importance of the CC effect on the dynamics and phytochemical composition of plants including medicinal and aromatic plants such as Aloe vera. The association between this plant and CC was explored by several studies. However, few research works that link with Aloe vera with CC were found including the association between humidity and temperature (Shishegaran et al., 2020) and rainfall (Manvith & Bidya 2014). Abihudi et al., (2020) reported that the most of *Aloe* species are threatened by the effects of CC advancing that they germinated best at moderate temperatures (25 °C–30 °C). According to Salehi et al., (2018), aloe has a range of required temperatures for growth from 4 to 21 °C, yet this cold-tolerant genus can continue to thrive at temperatures as low as

4 °C, specifying that Aloe species can grow up to 61–99 cm tall in ideal climatic conditions. However, Manvith & Bidya (2014) added that Aloe plants can survive and grow in warm tropical climates and cannot withstand frigid temperatures. The intolerance of aloe to low temperatures was advanced also by McKeown (1987) and Cristiano et al., (2016). For example, the aloes *Aloe dichotoma, Aloe pillansii, Aloe ramosissima*, and *Aloe pearsonii* (which all live in harsh regions, where summer temperatures can surpass 50 °C) can resist extremely high temperatures (Cousins & Witkowski, 2012). Manvith & Bidya (2014) disclosed also that Indian Aloe vera farmers frequently struggle with issues like a lack of rain, low groundwater levels because the plant requires a minimum of water. A lack of water and high light intensity reduces production and growth, but they also increase the concentration of many primary and secondary metabolites (Hazrati et al., 2020). Advancing that the main environmental issue limiting agricultural output is water stress, Hazrati et al., (2017) showed also that biochemical and phytochemical substances increased during extreme water stress. This statement was also reported by Mota-Ituarte et al., (2023) adding that water and saline stresses decrease the plant growth and leaves (the yield of fresh biomass and gel) and increase high total solids content. According to Salinas et al., (2016) due to its Crassulaceae acid metabolism, *Aloe barbadensis* Miller (Aloe vera) has a very high tolerance for water constraints. In fact, low soil water potential decreased plant growth rate, leaf number, and leaf weight mostly in leaf growth, demonstrating new leaves' vulnerability to water stress (Hazrati et al., 2017). These authors added that low leaf temperature slows plant and leaf growth rates and increases stomatal resistance. In terms of saline stress, despite the general trend of increasing NaCl, as salinity increased, leaf fresh weight decreased, perhaps as a result of a drop in water content (Murillo-Amador et al., 2015). Aloe vera's long-term salty irrigation raises the plant's concentrations of phenolic chemicals, lipid peroxidation, and hydrogen peroxide (Souguir et al., 2015). Salinas et al., (2016) demonstrated a considerable increase in the total sugars, soluble sugars, oligo and polyfructans, and fructans in Aloe vera plants subjected to water deficiency.

Regarding the effect of the aloe plants on CO2 levels and global warming, a study of Shishegaran et al., (2020) examined the effecacy of Spanish Moss Tillandsia Aerobic, Thailandia Native Fulia, and Aloe Vera on temperature, $CO_2$, and humidity, over an 8-hour period in an environment with high $CO_2$ concentrations. The most effective plant for reducing CO2 levels and global warming is aloe vera. Unfortunatly, it was reported that environmental factors have the most impact on the ecological niches of some Aloe species. In this context, Mkala et al., 2022 advanced that the current range of Aloe species' appropriate habitats is expected to change because to CC disclosing that Aloe species in East Africa will generally suffer from the effects of CC.

## IMPLICATIONS AND LIMITATIONS OF THE STUDY

The potential benefits of aloe vera extract such as antibacterial, anti-fungal, antioxidant, and anti-inflammatory drugs make this plant extraordinary and deserve the name, the "plant of immortality" or the "health longevity plant." Studies on the chemistry of aloe vera, its effectiveness for health, and its socio-economic importance are increasing but are not yet sufficient. Further research and innovation are needed with respect to these areas, in particular in nanomedicine and nanotechnology, which appear to be promising research areas of research.

In regards to the methodological aspect, it is true that there are several studies that have relied solely on the Scopus database alone, such as (Neelam & Sood, 2020; Yuan & Sun, 2020), but there are certainly other quality articles which are indexed in other databases. In addition, only the English language was considered and the grey literature (official and nonofficial reports, guides…) was omitted. Another weakness is that there are several species of aloe vera with different health effects, but the current study targeted the literature on aloe vera in general. This is valuable for general review studies, but for more convenience, future reviews may explore a defined species such as *Aloe Vera Barbandensis* (for more detailed insights). The plant is generally cultivated for its gel. It has many industrial applications, in particular in food supplements, pharmaceuticals, and cosmetics. The antioxidants present in the gel are considered to be the most important compounds which give it many extraordinary properties for health. These substances improve the skin, immunity, inflammation, heart and arteries. The current paper presents a scientometric study exploring the most significant properties of the aloe vera plant, which helps to understand the structure and trend of research.

## CONCLUSION

This article is based on data extracted from a selection of the first 2,000 papers out of a total of 5,111 on aloe vera published worldwide and indexed in the Scopus database during the period 1936-2020. This review presents a bibliographical overview of some aspects of this plant, including the plant itself, its products, its medical importance, and the impact of some stress factors including CC.

The properties of the aloe vera plant are extraordinary, which has long fascinated thousands of researchers around the world primarily in India, the United States, China, and the United Kingdom. Research in the exploration of plant chemistry is growing due to the large number of organic and mineral compounds found in the gel. In general, this study has shown that research on the aloe vera plant is increasing and diversifying. Some of the significant uses of nanotechnology show promise, as mentioned above, in using the gel as a model to adsorb the beneficial compounds. This plant plays also a socio-economic role due to its importance in the creation jobs in agronomic and industrial processes, which deserves a complete study. Furthermore, the effectiveness remains to be clinically proven, which suggests the proliferation of studies on this aspect. CC was mentioned as a trending topic but not well explored in details and the majority of research efforts explored it partially or generally. The review showed that temperature and precipitation are the most important climatic variables that affect the aloe species. However, water and saline stress increase the quality of the gel and reduce the growth and leaf yield. Due to its sensitivity to CC, its socio-economic importance, and its medicine propriety aloe species require immediate conservation action. This research is anticipated to create the necessary groundwork for comprehensively showing how CC affects the plants such as AV.

Funding:

No applicable

**Data Availability Statement:** The data supporting reported results is within the article.

Conflicts of Interest:

The authors declare no conflict of interest.

## REFERENCES

Abihudi, S. A., Venkataramana, P. B., de Boer, H. J., & Treydte, A. C. (2020). Species-specific responses of "Critically Endangered" and "Least Concern" Aloe seed germination to environmental conditions in Tanzania. *Global Ecology and Conservation*, *24*, e01241. doi:10.1016/j.gecco.2020.e01241

Amorati, R., Foti, M. C., & Valgimigli, L. (2013). Antioxidant activity of essential oils. *Journal of Agricultural and Food Chemistry*, *61*(46), 10835–10847. doi:10.1021/jf403496k PMID:24156356

Benzidia, B., Barbouchi, M., Hammouch, H., Belahbib, N., Zouarhi, M., Erramli, H., Ait Daoud, N., Badrane, N., & Hajjaji, N. (2019). Chemical composition and antioxidant activity of tannins extract from green rind of Aloe vera (L.) Burm. F. *Journal of King Saud University. Science*, *31*(4), 1175–1181. doi:10.1016/j.jksus.2018.05.022

Biswas, S. K. (2016). Does the interdependence between oxidative stress and inflammation explain the antioxidant paradox? *Oxidative Medicine and Cellular Longevity*, *2016*, 1–9. doi:10.1155/2016/5698931 PMID:26881031

Chinchilla, N., Carrera, C., Durán, A. G., Macías, M., Torres, A., & Macías, F. A. (2013). Aloe barbadensis: How a miraculous plant becomes reality. *Phytochemistry Reviews*, *12*(4), 581–602. doi:10.100711101-013-9323-3

Choi, Y. M., Shin, D. H., & Kim, C. H. (2016). An optimized methodology to observe internal microstructures of Aloe vera by cryo-scanning electron microscope. *Applied Microscopy*, *46*(2), 76–78. doi:10.9729/AM.2016.46.2.76

Cousins, S. R., & Witkowski, E. T. F. (2012). African aloe ecology: A review. *Journal of Arid Environments*, *85*, 1–17. doi:10.1016/j.jaridenv.2012.03.022

Cristiano, G., Murillo-Amador, B., & De Lucia, B. (2016). Propagation techniques and agronomic requirements for the cultivation of Barbados Aloe (Aloe vera (L.) Burm. F.)—A Review. *Frontiers in Plant Science*, *7*, 1410. doi:10.3389/fpls.2016.01410 PMID:27721816

De Rodrıguez, D. J., Hernández-Castillo, D., Rodrıguez-Garcıa, R., & Angulo-Sánchez, J. L. (2005). Antifungal activity in vitro of Aloe vera pulp and liquid fraction against plant pathogenic fungi. *Industrial Crops and Products*, *21*(1), 81–87. doi:10.1016/j.indcrop.2004.01.002

Debnath, T., Ghosh, M., Lee, Y. M., Nath, N. C. D., Lee, K. G., & Lim, B. O. (2018). Identification of phenolic constituents and antioxidant activity of Aloe barbadensis flower extracts. *Food and Agricultural Immunology*, *29*(1), 27–38. doi:10.1080/09540105.2017.1358254

Elisha, I. L., & Viljoen, A. (2021). Trends in Rooibos Tea (Aspalathus linearis) research (1994–2018): A scientometric assessment. *South African Journal of Botany*, *137*, 159–170. doi:10.1016/j.sajb.2020.10.004

Embuscado, M. E. (2015). Spices and herbs: Natural sources of antioxidants–a mini review. *Journal of Functional Foods*, *18*, 811–819. doi:10.1016/j.jff.2015.03.005

Esua, M. F., & Rauwald, J. W. (2006). Novel bioactive maloyl glucans from Aloe vera gel: Isolation, structure elucidation and in vitro bioassays. *Carbohydrate Research, 341*(3), 355–364. doi:10.1016/j.carres.2005.11.022 PMID:16343466

Gao, X., & Zhao, D. (2022). Impacts of climate change on vegetation phenology over the Great Lakes Region of Central Asia from 1982 to 2014. *The Science of the Total Environment, 845*, 157227. doi:10.1016/j.scitotenv.2022.157227 PMID:35809736

Ginwala, R., Bhavsar, R., Chigbu, D. G. I., Jain, P., & Khan, Z. K. (2019). Potential role of flavonoids in treating chronic inflammatory diseases with a special focus on the anti-inflammatory activity of apigenin. *Antioxidants, 8*(2), 35. doi:10.3390/antiox8020035 PMID:30764536

Grace, O. M. (2011). Current perspectives on the economic botany of the genus Aloe L.(Xanthorrhoeaceae). *South African Journal of Botany, 77*(4), 980-987. doi:10.1016/j.sajb.2011.07.002

Talmadge, J., Chavez, J., Jacobs, L., Munger, C., Chinnah, T., Chow, J. T., & Yates, K. (2004). Fractionation of Aloe vera L. inner gel, purification and molecular profiling of activity. *International Immunopharmacology, 4*(14), 1757-1773. https://doi.org/ doi:10.1016/j.intimp.2004.07.013

Hamman, J. H. (2008). Composition and applications of Aloe vera leaf gel. *Molecules (Basel, Switzerland), 13*(8), 1599–1616. doi:10.3390/molecules13081599 PMID:18794775

Hazrati, S., Tahmasebi, Z., Modares Sanevi, S. A. M., & Mokhtasi, A. (2016). Effect of Water Deficit Stress on Growth, Yield and Aloin Concentration of Aloe vera L. at different harvesting dates. *Journal of Plant Protection Research, 23*(3), 63–80. doi:10.22069/jopp.2017.9304.1886

Hazrati, S., Tahmasebi-Sarvestani, Z., Mokhtassi-Bidgoli, A., Modarres-Sanavy, S. A. M., Mohammadi, H., & Nicola, S. (2017). Effects of zeolite and water stress on growth, yield and chemical compositions of Aloe vera L. *Agricultural Water Management, 181*, 66–72. doi:10.1016/j.agwat.2016.11.026

Hazrati, S., Tahmasebi-Sarvestani, Z., Nicola, S., Kashkooli, A. B., Habibzadeh, F., Mohammadi, H., & Mokhtassi-Bidgoli, A. (2020). Effect of light and water deficiency on growth and concentration of various primary and secondary metabolites of aloe vera l. *Journal of Agricultural Science and Technology, 22*(5), 1343-1358.

Hęś, M., Dziedzic, K., Górecka, D., Jędrusek-Golińska, A., & Gujska, E. (2019). Aloe vera (L.) Webb.: Natural sources of antioxidants–a review. *Plant Foods for Human Nutrition (Dordrecht, Netherlands), 74*(3), 255–265. doi:10.100711130-019-00747-5 PMID:31209704

Hewlings, S. J., & Kalman, D. S. (2017). Curcumin: A review of its effects on human health. *Foods, 6*(10), 92. doi:10.3390/foods6100092 PMID:29065496

Hussein, A., & Workeneh, S. (2021). Modeling the impacts of climate changes on the distribution of Aloe vera species in Ethiopia. *Journal of Earth Science & Climatic Change, 12*(7). Advance online publication. doi:10.4172/2157-7617.1000567

Jain, S., Rathod, N., Nagi, R., Sur, J., Laheji, A., & Gupta, N. (2016). Antibacterial Effect of Aloe Vera Gel against Oral Pathogens: An In-vitro Study. *Journal of Clinical and Diagnostic Research : JCDR, 10*(11), ZC41. doi:10.7860/JCDR/2016/21450.8890 PMID:28050502

Kabtni, S., Sdouga, D., Bettaib Rebey, I., Save, M., Trifi-Farah, N., Fauconnier, M. L., & Marghali, S. (2020). Influence of climate variation on phenolic composition and antioxidant capacity of Medicago minima populations. *Scientific Reports*, *10*(1), 1–15. doi:10.103841598-020-65160-4 PMID:32427946

Kahlon, J. B., Kemp, M. C., Carpenter, R. H., McAnalley, B. H., McDaniel, H. R., & Shannon, W. M. (1991). Inhibition of AIDS virus replication by acemannan in vitro. *Molecular Biotherapy*, *3*(3), 127–135. https://europepmc.org/article/med/1768365 PMID:1768365

Kahramanoğlu, İ., Chen, C., Chen, J., & Wan, C. (2019). Chemical constituents, antimicrobial activity, and food preservative characteristics of Aloe vera Gel. *Agronomy (Basel)*, *9*(12), 831. doi:10.3390/agronomy9120831

Kang, S., Zhao, X., Yue, L., & Liu, L. (2017). Main anthraquinone components in Aloe vera and their inhibitory effects on the formation of advanced glycation end-products. *Journal of Food Processing and Preservation*, *41*(5), e13160. doi:10.1111/jfpp.13160

Karmaoui, A. (2023). The Future of Artificial Intelligence in Agricultural Field: A Bibliometric Analysis. In A. Karmaoui (Ed.), *Intelligent Solutions for Optimizing Agriculture and Tackling Climate Change: Current and Future Dimensions*. doi:10.4018/978-1-6684-4649-2.ch003

Karmaoui, A., El Jaafari, S., Chaachouay, H., & Hajji, L. (2023a). A bibliometric review of geospatial analyses and artificial intelligence literature in agriculture. *GeoJournal*. doi:10.100710708-023-10859-w

Karmaoui, A., Moumane, A., Minucci, G., El Jaafari, S., Chaachouay, H., & Hajji, L. (2023b). Affordable Methods and Techniques to Optimize Farming Productivity in a Context of Climate Change in Arid Areas. In A. Karmaoui (Ed.), *Intelligent Solutions for Optimizing Agriculture and Tackling Climate Change: Current and Future Dimensions*. doi:10.4018/978-1-6684-4649-2.ch006

Keyhanian, S., & Stahl-Biskup, E. (2007). Phenolic constituents in dried flowers of Aloe vera (Aloe barbadensis) and their in vitro antioxidative capacity. *Planta Medica*, *73*(06), 599–602. doi:10.1055-2007-967202 PMID:17520524

Kitture, R., Ghosh, S., More, P. A., Gaware, S., Datar, S., Chopade, B. A., & Kale, S. N. (2015). Curcumin-loaded, self-assembled aloe vera template for superior antioxidant activity and trans-membrane drug release. *Journal of Nanoscience and Nanotechnology*, *15*(6), 4039–4045. doi:10.1166/jnn.2015.10322 PMID:26369010

Kumar, S., Yadav, A., Yadav, M., & Yadav, J. P. (2017a). Effect of climate change on phytochemical diversity, total phenolic content and in vitro antioxidant activity of Aloe vera (L.) Burm. f. *BMC Research Notes*, *10*(1), 1–12. doi:10.118613104-017-2385-3 PMID:28118858

Kumar, S., Yadav, M., Yadav, A., & Yadav, J. P. (2017b). Impact of spatial and climatic conditions on phytochemical diversity and in vitro antioxidant activity of Indian Aloe vera (L.) Burm. f. *South African Journal of Botany*, *111*, 50–59. doi:10.1016/j.sajb.2017.03.012

Lanka, S. (2018). A review on Aloe vera-The wonder medicinal plant. *Journal of Drug Delivery and Therapeutics*, *8*(5-s), 94–99. doi:10.22270/jddt.v8i5-s.1962

Leydesdorff, L., & Rafols, I. (2012). Interactive overlays: A new method for generating global journal maps from Web-of-Science data. Journal of Informetric, 6(2), 318-332. https://doi.org/doi:10.1016/j.joi.2011.11.003

Liontakis, A., & Tzouramani, I. (2016). Economic sustainability of organic Aloe vera farming in Greece under risk and uncertainty. Sustainability, 8(4), 338. https://doi.org/ doi:10.3390u8040338

Liu, J. G., Zhou, Q., Guo, Q., Yang, Z. H., Xie, F., & Han, J. T. (2017). Knowledge diffusion of dynamical network in terms of interaction frequency. *Scientific Reports*, *7*(1), 1–7. doi:10.103841598-017-11057-8 PMID:28883456

Mandal, G., & Das, A. (1980). Structure of the D-galactan isolated from Aloe barbadensis Miller. *Carbohydrate Research*, *86*(2), 247–257. doi:10.1016/S0008-6215(00)85902-9

Manvitha, K., & Bidya, B. (2014). Aloe vera: a wonder plant its history, cultivation and medicinal uses. *Journal of Pharmacognosy and Phytochemistry, 2*(5), 85-88. Available on https://www.phytojournal.com/archives?year=2014&vol=2&issue=5&part=B&ArticleId=249 accessed on Aprtil 04, 2023.

McKeown, E. (1987). Aloe vera. *Cosmet Toiletries*, *102*, 64–65.

Mkala, E. M., Mutinda, E. S., Wanga, V. O., Oulo, M. A., Oluoch, W. A., Waswa, E. N., & Wang, Q. F. (2022). Modeling impacts of climate change on the potential distribution of three endemic Aloe species critically endangered in East Africa. *Ecological Informatics*, *71*, 101765. doi:10.1016/j.ecoinf.2022.101765

Mota-Ituarte, M., Pedroza-Sandoval, A., Minjares-Fuentes, R., Trejo-Calzada, R., Zegbe, J. A., & Quezada-Rivera, J. J. (2023). Water deficit and salinity modify some morphometric, physiological, and productive attributes of Aloe vera (L.). *Botanical Sciences*, *101*(2), 463–475. doi:10.17129/botsci.3175

Murillo-Amador, B., Nieto-Garibay, A., Troyo-Diéguez, E., García-Hernández, J. L., Hernández-Montiel, L., & Valdez-Cepeda, R. D. (2015). Estrés salino moderado en los rasgos fisiológicos y morfológicos del Aloe vera L. *Botanical Sciences*, *93*(3), 639–648. doi:10.17129/botsci.73

Neelam, S., & Sood, S. K. (2020). A scientometric review of global research on smart disaster management. *IEEE Transactions on Engineering Management*, *68*(1), 317–329. doi:10.1109/TEM.2020.2972288

Ni, Y., Turner, D., Yates, K. Á., & Tizard, I. (2004). Isolation and characterization of structural components of Aloe vera L. leaf pulp. *International Immunopharmacology*, *4*(14), 1745–1755. doi:10.1016/j.intimp.2004.07.006 PMID:15531291

Okamura, N., Hine, N., Tateyama, Y., Nakazawa, M., Fujioka, T., Mihashi, K., & Yagi, A. (1998). Five chromones from Aloe vera leaves. *Phytochemistry*, *49*(1), 219–223. doi:10.1016/S0031-9422(97)01071-6

Omidi, M., Fatehinya, A., Farahani, M., Akbari, Z., Shahmoradi, S., & Yazdian, F. (2017). Vashaee, D. Characterization of biomaterials. In *Biomaterials for oral and dental tissue engineering* (pp. 97–115). Woodhead Publishing. doi:10.1016/B978-0-08-100961-1.00007-4

Pahwa, R., Goyal, A., Bansal, P., & Jialal, I. (2020). Chronic inflammation. *StatPearls*.

Patil, R., & Kumar, S. (2020). Bibliometric Survey on Diagnosis of Plant Leaf Diseases Using Artificial Intelligence. *International Journal of Modern Agriculture*, *9*(3), 1111–1131. http://modern-journals.com/index.php/ijma/article/view/316

Pizzorno, J. (2014). Glutathione! *Integrative Medicine (Encinitas, Calif.)*, *13*(1), 8. PMID:26770075

Rajasekaran, S., Sivagnanam, K., & Subramanian, S. (2005). Mineral contents of Aloe vera leaf gel and their role on streptozotocin-induced diabetic rats. *Biological Trace Element Research*, *108*(1), 185–195. doi:10.1385/BTER:108:1-3:185 PMID:16327071

Rosca-Casian, O., Parvu, M., Vlase, L., & Tamas, M. (2007). Antifungal activity of Aloe vera leaves. *Fitoterapia*, *78*(3), 219–222. doi:10.1016/j.fitote.2006.11.008 PMID:17336466

Sabeh, F., Wright, T., & Norton, S. J. (1993). Purification and characterization of a glutathione peroxidase from the Aloe vera plant. *Enzyme & Protein*, *47*(2), 92–98. doi:10.1159/000468662 PMID:8193675

Salehi, B., Albayrak, S., Antolak, H., Kręgiel, D., Pawlikowska, E., Sharifi-Rad, M., Uprety, Y., Tsouh Fokou, P., Yousef, Z., Amiruddin Zakaria, Z., Varoni, E., Sharopov, F., Martins, N., Iriti, M., & Sharifi-Rad, J. (2018). Aloe genus plants: From farm to food applications and phytopharmacotherapy. *International Journal of Molecular Sciences*, *19*(9), 2843. doi:10.3390/ijms19092843 PMID:30235891

Salinas, C., Handford, M., Pauly, M., Dupree, P., & Cardemil, L. (2016). Structural modifications of fructans in Aloe barbadensis Miller (Aloe vera) grown under water stress. *PLoS One*, *11*(7), e0159819. doi:10.1371/journal.pone.0159819 PMID:27454873

Saniasiaya, J., Salim, R., Mohamad, I., & Harun, A. (2017). Antifungal effect of Malaysian Aloe vera leaf extract on selected fungal species of pathogenic otomycosis species in in vitro culture medium. *Oman Medical Journal*, *32*(1), 41–46. doi:10.5001/omj.2017.08 PMID:28042402

Shishegaran, A., Shishegaran, A., Najari, M., Ghotbi, A., & Boushehri, A. N. (2020). Effect of plants on an environment with high carbon dioxide concentration. *Cleaner Engineering and Technology*, *1*, 100002. doi:10.1016/j.clet.2020.100002

Shokri, F., Ziarati, P., & Mousavi, Z. (2016). Removal of Selected Heavy Metals from Pharmaceutical Effluent by Aloe vera L. *Biomedical & Pharmacology Journal*, *9*(2), 705–713. doi:10.13005/bpj/993

Solomou, A. D., Germani, R., & Georgakopoulou, M. (2020). Ecological value, cultivation, utilization and commercialization of Aloe vera in Greece. *JAPS. The Journal of Animal and Plant Sciences*, *30*(4), 1013–1023. doi:10.36899/JAPS.2020.4.0116

Souguir, D., Abd-Alla, H. I., El Ferjani, E., Larbi Khouja, M., & Hachicha, M. (2015). Aloe vera long-term saline irrigation increases contents of hydrogen peroxide, lipid peroxidation and phenolic compounds. *Acta Agriculturæ Scandinavica. Section B, Soil and Plant Science*, *65*(8), 688–696. doi:10.1080/09064710.2015.1049653

Sun, G. Q., Li, L., Li, J., Liu, C., Wu, Y. P., Gao, S., Wang, Z., & Feng, G. L. (2022). Impacts of climate change on vegetation pattern: Mathematical modelling and data analysis. *Physics of Life Reviews*, *43*, 239–270. doi:10.1016/j.plrev.2022.09.005 PMID:36343569

Sun, J., & Yuan, B. Z. (2020). Mapping of the world rice research: A bibliometric analysis of top papers during 2008–2018. [ALIS]. *Annals of Library and Information Studies*, *67*(1), 55–66. http://op.niscair.res.in/index.php/ALIS/article/view/30297

Surjushe, A., Vasani, R., & Saple, D. G. (2008). Aloe vera: A short review. *Indian Journal of Dermatology*, *53*(4), 163. doi:10.4103/0019-5154.44785 PMID:19882025

Tabatabaei-Malazy, O., Atlasi, R., Larijani, B., & Abdollahi, M. (2015). Trends in publication on evidence-based antioxidative herbal medicines in management of diabetic nephropathy. *Journal of Diabetes and Metabolic Disorders*, *15*(1), 1–8. doi:10.118640200-016-0221-2 PMID:26877997

Tanaka, M., Yamamoto, Y., Misawa, E., Nabeshima, K., Saito, M., Yamauchi, K., Abe, F., & Furukawa, F. (2016). Aloe sterol supplementation improves skin elasticity in Japanese men with sunlight-exposed skin: A 12-week double-blind, randomized controlled trial. *Clinical, Cosmetic and Investigational Dermatology*, *9*, 435–442. doi:10.2147/CCID.S118947 PMID:27877061

Van Eck, N. J., & Waltman, L. (2010). Software survey: VOSviewer, a computer program for bibliometric mapping. *Scientometrics*, *84*(2), 523–538. doi:10.100711192-009-0146-3 PMID:20585380

Verrall, B., & Pickering, C. M. (2020). Alpine vegetation in the context of climate change: A global review of past research and future directions. *The Science of the Total Environment*, *141344*, 141344. doi:10.1016/j.scitotenv.2020.141344 PMID:32814293

Wright, C. S. (1936). Aloe vera in the treatment of roentgen ulcers and telangiectasis. *Journal of the American Medical Association*, *106*(16), 1363–1364. doi:10.1001/jama.1936.02770160021006

Yamagata, K. (2019). Effects of Dietary Polyphenols on Chronic Diseases: Epidemiological Data and Molecular Mechanisms of Action. *Frontiers in Natural Product Chemistry*, *5*(5), 55–94. https://www.sciencedirect.com/topics/chemistry/quercetin. doi:10.2174/9789811423772119050005

Yeung, A. W. K., Heinrich, M., & Atanasov, A. G. (2018). Ethnopharmacology—A bibliometric analysis of a field of research meandering between medicine and food science? *Frontiers in Pharmacology*, *9*, 215. doi:10.3389/fphar.2018.00215 PMID:29599720

Yeung, A. W. K., Tzvetkov, N. T., Durazzo, A., Lucarini, M., Souto, E. B., Santini, A., & Atanasov, A. G. (2020). Natural products in diabetes research: Quantitative literature analysis. *Natural Product Research*, 1-15. doi:10.1080/14786419.2020.1821019

Zhang, P., Cai, Y., He, Y., Xie, Y., Zhang, X., & Li, Z. (2022). Changes of vegetational cover and the induced impacts on hydrological processes under climate change for a high-diversity watershed of south China. *Journal of Environmental Management*, *322*, 115963. doi:10.1016/j.jenvman.2022.115963 PMID:36041299

## APPENDIX

*Table 2. Diseases and health problems the most studied in this paper using the aloe vera research through VOSviewer software*

| Keyword | Occurrence | Links | Keyword | Occurrence | Links |
|---|---|---|---|---|---|
| Diabetes mellitus | 151 | 4130 | Abdominal Pain | 33 | 1034 |
| Diarrhea | 81 | 2456 | Jaundice | 30 | 1006 |
| Skin disease | 71 | 2178 | Apoptosis | 48 | 998 |
| Inflammation | 87 | 2112 | Pruritus | 37 | 993 |
| Oxidative stress | 78 | 1806 | Nausea | 34 | 936 |
| Constipation | 59 | 1762 | Skin Injury | 38 | 931 |
| Non insulin dependent diabetes mellitus | 67 | 1732 | Gasterointestinal disease | 30 | 886 |
| Fever | 50 | 1644 | Vomiting | 37 | 1159 |
| Histopathology | 68 | 1621 | Diabetes | 48 | 1074 |
| Hypertension | 49 | 1559 | Liver disease | 32 | 1061 |
| Pathology | 65 | 1541 | Psoriasis | 40 | 854 |
| Asthma | 38 | 1272 | Liver Toxicity | 33 | 773 |
| Pain | 47 | 1257 | Erythema | 33 | 747 |
| Headache | 39 | 1246 | Obesity | 33 | 723 |
| Coughing | 33 | 1197 | Toxicity | 35 | 531 |

Four categories: substances, microbes, techniques, and activities as follows

- Substances: unclassified drug (104, 1649), anti-infective agent (63, 1067), unindexed drug (27, 482), flavonoid (28, 449), antioxidant (27, 449), aazadirachta indica extract (27, 303), antibiotic agent (25, 401), herbaceous agent (26, 383), alkaloid (20, 312), garlic extract (19, 297), anthraquinone derivative (16, 284), garlic (20, 283), essensial oil (17, 217), nanoparticles (20, 208), and metal nanoparticles (17, 165).
- Microbes: staphylococcus aureus (83, 1122), escherichia coli (83, 984), pseudomonas aeruginosa (40, 616), bacteria (30, 380), klebsiella pneumoniae (21, 351), bacterial strain (20, 316), candida albicans (17, 288), enerococcus faecalis (18, 255), and bacillus subtilis (17, 253).
- Techniques and fields: Minimum inhibitory concentration (52, 785), chemistry (36, 655), agar diffusion (23, 345), microbial sensitivity tests (17, 314), antibiotic sensitivity (19, 307), scanning electron microscopy (27, 303), drug isolation (17, 298), antibiotic resistance (19, 289), cytotoxicity (20, 289), drug formulation (19, 282), growth inhibition (17, 276), disk diffusion (test) (19, 263), and zone of inhibition (16, 210)

*Table 3. Occurrence and links of the key words with relationship in research on aloe vera antioxidant activity retrieved using VOSviewer*

| Keyword | Occurrence | Links | Keyword | Occurrence | Links |
|---|---|---|---|---|---|
| Unclassified drug + Unindexed drug | 177+ 68 | 3684+ 1528 | Silymarin | 20 | 465 |
| Ascorbic acid | 75 | 1221 | Scavenger | 20 | 475 |
| Flavonoid + Flavonoids | 64 + 25 | 1427+ 482 | Antiinflamatorry agent | 20 | 503 |
| Phenol Derivative + Phenols | 34 + 24 | 688+ 390 | | | |
| Superoxide Dismutase | 55 | 1263 | Alkaloid | 20 | 505 |
| Alpha tocopherol | 53 | 1047 | Polyphenol | 19 | 393 |
| Catalase | 42 | 1012 | Saponin | 18 | 391 |
| Reactive Oxygene metabolite | 38 | 831 | Nitric oxide | 18 | 429 |
| Glutathione | 38 | 877 | Collagen | 18 | 398 |
| Antiinflamatorry agent | 38 | 906 | Carbohydrate | 18 | 391 |
| | | | Zinc | 17 | 279 |
| Curcumin | 38 | 848 | Vitamin | 17 | 381 |
| Quercetin | 30 | 710 | Free radical scavengers | 17 | 416 |
| Glucose | 30 | 697 | Sunscreen | 16 | 292 |
| Antiinfective Agent | 30 | 680 | Salicylic acid | 16 | 289 |
| Polysaccharide + Polysaccharides | 28 + 16 | 543+ 275 | Anthraquinone derivate | 16 | 393 |
| Malonaldehyde | 27 | 634 | Anthocyanin | 16 | 301 |
| Glutathione Peroxidase | 26 | 578 | Alcohol | 16 | 339 |
| Emodin + Aloe emodin | 25+ 15 | 528+ 318 | Lactate dehydrogenase | 15 | 419 |
| Resveratrol | 22 | 493 | Interleukin 6 | 15 | 361 |
| Free radicals + Free radical | 22 + 18 | 375+ 333 | Carotenoid | 15 | 305 |
| 1,1 diphenyl 2 picrylhydrazyl | 22 | 497 | | | |

- Activities: Wound healing (61, 952), antioxidant activity (83, 984), antineoplastic activity (33, 561), antiviral activity (33, 557), drug efficacy (32, 524), drug effect (30, 509), drug screening (30, 452), immunomodulation (20, 339), drug effects (17, 332), cell proliferation (17, 283), drug mechanism (15, 269), anti-bacterial activity (30, 209), antibacterial (19, 155), antimicrobial (19, 151), and synthesis (chemical) (18, 128).

*Table 4. Occurrence and links of the key words with relationship in research on Aloe vera anti-inflammatory activity retrieved using VOSviewer*

| Keyword | Occurrence | Links | Keyword | Occurrence | Links |
|---|---|---|---|---|---|
| Unclassified drug<br>Unindexed Drug | 135<br>58 | 2521<br>1113 | Anthraquinone derivative | 19 | 385 |
| Antiinflamatory agent<br>Antiinflamatory agents | 70<br>40 | 1376<br>787 | Interleukine 6 | 18 | 368 |
| Herbaceous agent | 68 | 1258 | Ginger extract | 19 | 352 |
| Antioxidant<br>Antioxidants | 53<br>22 | 1092<br>507 | Aloe emodin | 19 | 345 |
| Flavonoid | 38 | 782 | Reactive oxygen metabolite | 18 | 309 |
| Curcumin | 33 | 674 | Emodin | 17 | 302 |
| Antiinfective agent | 25 | 539 | Resveratrol | 15 | 301 |
| Quercetin | 26 | 528 | Immunomodulating agent | 16 | 296 |
| Ascorbic Acid | 27 | 512 | Salicylic Acid | 17 | 286 |
| Alpha tocopherol | 26 | 499 | Indometacin | 17 | 279 |
| Antineoplastic agent | 22 | 489 | Nonsteroid antiinflammatory agent | 15 | 221 |
| Aloin | 24 | 469 | | | |

The research on "Aloe vera antifungal activity" is structured as follows:

- Fungus words: antifungal activity (96, 1715), candida albicans (33, 550), candida (8, 161), aspergillus niger (10, 133), candidiasis (5, 115), penicillium digitatum (5, 59), yeast (5, 72), penicillium (7, 108), aspergillus flavus (7, 95), fluconazole (6, 82), fungicide (5, 45), anti-fugal activity (7, 50).
- Diseases linked to antifungal activity search: skin disease (8, 209), liver toxicity (5, 109), psoriasis (6, 130), acne (5, 101), diabetes mellitus (7, 171), cytotoxicity (5, 92), kkin infection (6, 121), erythema (5, 115), inflammation (7, 157), seborrheic dermatitis (6, 114), burn (10, 220), wound (5, 138), gingivitis (6, 124).
- Plants linked to antifungal activity search: curcuma longa (7, 182), ginger (7, 182), onion (5, 138), curcumin (6, 143), tea, aazadirachta indica (14, 318).
- Bacteria linked to antifungal activity search: pseudomonas aeruginosa (11, 256), bacillus cereus (5, 123), escherichia coli (13, 265), staphylococcus aureus (13, 262),
- Chemical compounds linked to antifungal activity search: unclassified drug (46, 913), antifungal agent (34, 654), herbaceous agent (20, 427), aazadirachta indica extract (13, 270), antioxidant (12, 208), chitosan (9, 104), antiinflamatory agent (7, 192), anti-bacterial agents (7, 180), saponin (6, 132), curcuma longa extract (8, 187), anthraquinone (6, 101), anthraquinone derivate (5, 110), antivirus agent (6, 171), essential oil (7, 128), steroid (5, 101), phenol (5, 99), salicylic acid (5, 88), aloin (9, 170), ginger extract (8, 177), unindexed drug (16, 330), garlic extract (15,322), tetracycline (5, 94), emodin (5, 117), flavonoid (7, 148), glycoside (5, 115), cinnamic acid (5, 116), calotropis procera extract (5, 143), fungicide (5, 45), antibiotic agent (9, 182), antineoplastic agent (6, 163), onion extract (6, 141), immunomodulating agent (5, 124), and anti-infective agents (5, 114).

# Chapter 2
# Carbon Sequestration Strategies in Agricultural Lands:
## Towards a Sustainable Agriculture

**Rachid Aboutayeb**
https://orcid.org/0000-0003-4291-6896
*National Institute of Agricultural Research, Morocco*

**Abdeljalil Boutarfa**
*National Institute of Agricultural Research, Morocco*

**Rachid Mrabet**
https://orcid.org/0000-0003-0311-193X
*National Institute of Agricultural Research, Morocco*

## ABSTRACT

*Carbon sequestration strategies can reduce greenhouse gas emissions, enhance soil fertility, improve water dynamics, and increase soil biodiversity. This chapter investigates such nexuses through the analysis of impacts and benefits mainly from three major strategies: conservation agriculture, agroforestry and organic farming including soil amendments. Moreover, the current study emphasizes the need for integrated approaches to accomplish multiple sustainability objectives in agriculture while simultaneously promoting carbon sequestration. The relationship between carbon sequestration and sustainable agriculture is investigated, and trade-offs between carbon sequestration and other sustainability objectives are outlined. This study emphasizes the policies and governance frameworks that support carbon sequestration, as well as the opportunities and risks associated with their implementation. Overall, carbon sequestration strategies have the potential to provide substantial environmental benefits; however, their implementation requires careful consideration of socioeconomic dimensions.*

DOI: 10.4018/978-1-6684-9838-5.ch002

## INTRODUCTION

The issue of climate change and its repercussions on the environment has garnered significant attention in recent times, hence necessitating the development of strategies aimed at mitigating the levels of carbon dioxide present in the atmosphere. The global ecosystem is significantly affected by climate change,, particularly on carbon sequestration and storage capacity, both of which are global concerns (Hu et al., 2022). The storage of carbon in soil or other natural systems, is an effective method for reducing carbon emissions.

Agricultural lands have a number of challenges that jeopardize their productivity, sustainability, and long-term survival. The depletion of soil fertility is a critical concern for agricultural lands. Soils became mined over time as a result of intensive farming techniques such as monocropping, heavy and intensive tillage, over-grazing and excessive use of chemical fertilizers (Diacono and Montemurro, 2011).

Agriculture intensification also called green revolution is constituting a big burden as well as a significant threat to soils and ecosystems since it strives to maximize crop yields while degrading soil quality, raising greenhouse gas (GHG) emissions, and polluting water resources. Unfortunately, the intensive farming uses excessive amounts of synthetic fertilizers and pesticides which leads to contribute to soil degradation, environmental damage, and biodiversity loss (Thangarajan et al., 2013).

Fortunately, there is a wide variety of land-based mitigation measures that are tested and implemented to improve soil carbon sequestration for sustainable agriculture. Among such measures, the expansion of organic and regenerative agricultural systems (Thangarajan et al., 2013). These measures are considered as multigoal and resilient to multiple insecurities.

The application of organic amendments, such as compost, to the soil has been shown to enhance soil aggregate stability and mitigate the risk of erosion (Annabi et al., 2011),, increase water retention (Hargreaves et al., 2008), provide essential nutrients for plant growth as well as enhancing soil biological activity (Benitez et al., 2003). Meanwhile, the composting process has been found to have several benefits, including trash reduction, soil fertility enhancement (El-Mrini et al., 2022), and the mitigation of greenhouse gas emissions by diverting organic waste away from landfills, besides the soil health enhancement.

Agroecology refers to a comprehensive method of farming that promotes biodiversity and soil health (Wezel et al., 2009). It may include crop rotation, intercropping, agroforestry, and the adoption of natural pest management techniques leading to reducing the use of synthetic phytosanitary products. According to recent studies conducted by Lu et al. (2018) and Ning et al. (2021), it has been observed that ecological restoration activities have made a significant contribution of 56% towards the overall carbon sink in regions dedicated to ecological restoration during the first decade of the twenty-first century..

Conservation agriculture (CA) is a method of farming that aims to reduce soil erosion, improve soil health, and encourage long-term land utilization. CA is based on 3 pillars including minimal tilling, crop rotation, and the use of cover crops which leads to increase soil organic matter (Aboutayeb et al., 2020), enhance water infiltration, and minimize soil erosion (Mrabet et al., 2012).

Although carbon sequestration has promise in mitigating climate change and facilitating sustainable land utilization, several limitations exist that can impede its efficacy. The potential consequences of alterations in land use are widely recognized, including the possibility of unforeseen repercussions, such as an increase of greenhouse gas emissions due to modifications in management practices.Carbon sequestration can be constrained by social factors such as a lack of awareness and understanding of carbon sequestration activities, Economic issues (Fridahl & Lehtveer, 2018) such as the cost of installing carbon sequestration practices (Composting process facilities, No-till seeders, etc).

Ultimately, carbon sequestration is a complicated topic, with various factors affecting its success. To address these limits, a comprehensive approach that takes into account the different dimensions would be required. Regrettably, literature reviews focusing on establishing the link between carbon sequestration strategies and the sustainability of agricultural systems remain less explored. As a result, more research is needed to establish the most effective and sustainable carbon sequestration solutions for agricultural areas, as well as to highlight the most efficient policies and governances promoting these strategies.

The aim of this chapter is to explore current knowledge and research on carbon sequestration strategies in agriculture and their contribution to achieving sustainable agriculture. In other terms, the chapter intends to provide well-evidenced and reasoned key messages concerning contribution of soils to climate change mitigation through regenerative management and carbon sequestration.

## SOIL QUALITY AND CARBON SEQUESTRATION

Soil quality and carbon sequestration are closely related (Lasco et al., 2014). The concept of carbon sequestration pertains to the act of storing carbon in durable carbon sinks. In order to mitigate the effects of climate change, several ecosystems, including forests, soils, and wetlands, have been utilized as carbon sinks (Yu et al., 2022; Epple et al., 2016). Soil represents a significant reservoir of carbon, surpassing the collective carbon content of both the atmosphere and all vegetation, thereby establishing itself as one of the most substantial carbon sinks on Earth (Potter et al., 2003). Therefore, improving soil quality can increase carbon sequestration, helping to mitigate the impacts of climate change.

The term "soil quality" encompasses the various physical, chemical, and biological attributes of the soil that have an impact on its capacity to sustain plant growth and deliver ecosystem services. The quality of soil is determined by several aspects, such as soil structure, nutrient composition, microbial activity, and organic matter content (Mohammadi et al., 2011).

Organic matter, one of the most important soil quality indicators, is particularly important for carbon sequestration in addition to microbial activity and availability of nutrients (Abou-Tammame et al., 2022; Aboutayeb et al., 2020; Bongiorno et al., 2019). According to Trumbore (2000), the preservation of organic matter under undisturbed conditions has the potential to retain carbon and enhance soil quality for extended periods, possibly spanning several centuries.

Various elements, such as soil composition, climatic conditions, land utilization, and management strategies, exert an influence on the process of carbon sequestration and consequently impact the quality of soil (Ramesh et al., 2019). Diverse soil attributes, such as pH levels, soil organic matter content, availability of mineral nutrients, and capacity to retain water, exhibit variations across different areas and climates due to distinct environmental conditions. According to Palm et al. (2014), soils of higher quality exhibit enhanced efficacy in the process of carbon sequestration. High-quality soils are rich in organic matter, which offers plant nutrients and serves as a home for soil microbes that help trap carbon (Kowalska et al., 2020). The inability of low-organic-matter soils to support plant growth and microbial activity reduces their capacity to sequester carbon (Fageria, 2012).

Therefore, it is imperative to prioritize the adoption of measures that improve carbon sequestration and thus soil quality as a fundamental approach to combat climate change. Enhancing soil quality can sustain soil carbon cycle, thereby reducing the effects of climate change and safeguarding soil, considered a precious natural resource, for future generations.

The use of sustainable land management methods that prioritize the enhancement of soil health has the potential to augment carbon sequestration rates and enhance overall soil quality. In light of previous results, it has been shown that Conservation agriculture practices, according to several studies, enhance soil organic matter, thereby augmenting water and nutrient retention, improving soil structure, and increasing soil carbon sequestration capacity (Aboutayeb et al., 2020; Lal, 2015). Agroforestry and reforestation enhance soil quality by augmenting soil organic matter, offering a habitat for soil microorganisms, and mitigating soil erosion. These practices additionally sequester carbon in the above-ground biomass (Stavi and Lal, 2013).

## CARBON SEQUESTRATION STRATEGIES

### Conventional Agriculture

Conventional agricultural practices, including applying synthetic fertilizers, pesticides, and intensive tilling, substantially affect carbon sequestration in agricultural lands. Carbon sequestration involves storage in natural sinks, such as forests, wetlands, and soils (Freibauer et al., 2004). These latter are essential in reducing atmospheric $CO_2$ levels as they absorb and store carbon, mitigating climate change and reducing carbon footprint. In fact, worldwide carbon sequestration is a complex interplay of science, economics, policy, farming, and environment.

Organic matter constitutes a crucial component of carbon sequestration, nevertheless conventional agricultural techniques can reduce its levels (Lal & Kimble, 1997). Intense tillage, which involves regular mechanical disruption of soil, can hasten the breakdown of Organic matter and release carbon dioxide into the atmosphere (Silva-Olaya et al., 2013; Maltas et al., 2018). Moreover, synthetic fertilizers can alter soil ecosystems and inhibit the establishment of soil microbes that aid in carbon sequestration. Pesticides can also destabilize soil ecosystems and impair soil biodiversity, which can have detrimental impacts on carbon sequestration (Dignac et al., 2017).

Powerful reminders that almost all agricultural practices require the disturbance of natural ecosystems, such as forests and grasslands, for crop production which results in the depletion of natural carbon sinks and thus soil carbon content (Jarecki and Lal, 2003). The reduction of natural sinks reduces the ability of different ecosystems to sequester carbon and leads to the release of carbon from biomass and soils.

### Agroecology

Agroecology is an agricultural approach that emphasizes the incorporation of ecological principles and practices (Hecht, 2018). Agro-ecology can also be termed nature or biodiversity-based ecology. Several agroecology approaches, such as conservation agriculture, agroforestry, and the use of agroecological inputs, can improve carbon sequestration (Corbeels et al., 2019). According to Burbi and Olave (2018), the implementation of agroecological systems has the potential to mitigate greenhouse gas (GHG) emissions, specifically carbon dioxide ($CO_2$), by employing various land-use strategies. These strategies encompass landscape diversification, crop rotation, cover crops, intercropping, organic amendments, and the reduction or avoidance of synthetic fertilizers (Kerr et al., 2023). Agroecological methods are

*Table 1. Agroecology and carbon sequestration*

| Practice | Objective | Finding | Reference |
|---|---|---|---|
| Agroecology | Comparison the effectiveness of 3 agroecological systems in carbon sequestration:<br>- Agroforestry<br>-Tree Plantation<br>-Improved farming practices | - Agroforestry: +2.8 t C ha−1 yr−1<br>-Tree Plantation: +1.6 t C ha−1 yr−1<br>-Improved farming practices: -0.9 t C ha−1 yr−1 | Rakotovao et al. 2020. |

significant in decreasing carbon emissions, particularly in small-scale farming situations (Rakotovao, 2017) and are able to enhance agricultural productivity through climate-friendly practices that give priority to the conservation of soil carbon and the sequestration of carbon in vegetation biomass. This presents significant potential for mitigating climate change (Wibbelmann et al., 2013).

Agroforestry is a significant agroecological system (Altieri et al., 2015). The capacity to capture and store atmospheric carbon dioxide (CO2) in vegetation, soils, and biomass products holds substantial potential (Boulbi and Olave, 2018). Parra et al. (2021) found that in the Andean region of Antioquia, silvopastoral systems with a high density of Eucalyptus tereticornis trees (over 500 trees per hectare) can offset 26.6 tonnes of CO2 equivalent per hectare per year.

Agroecology approaches can have a favorable effect on carbon sequestration (Table.1). Conservation agriculture, agroforestry, and the application of agroecological inputs can assist increase the amount of carbon stored in soils, hence decreasing the quantity of carbon dioxide in the atmosphere (Corbeels et al., 2019). The implementation of agroecology methods has the potential to effectively address the issue of climate change and reduce the emission of greenhouse gases. This can be achieved through the promotion of sustainable agriculture practices and the reduction of reliance on synthetic inputs (Adegbeye et al., 2020). It is essential to encourage and transition towards agroecology methods in order to increase carbon sequestration and lessen the negative environmental impacts of conventional agriculture.

Overall, agroecological approaches can also improve small-scale farmers' livelihoods and contribute to advancing food security (Thrupp, 2000). Agroecology can aid in the reduction of hunger and poverty by increasing crop yields while enhancing soil health and sustaining productivity (Altieri, 2009; Siegner et al., 2020) leading to supporting sustainable and diversified farming techniques.

## Conservation Agriculture

Conservation Agriculture, known as CA, comprises three components, namely: no-till, crop rotation, and cover crops. Global-scale adoption of CA is 205 million hectares, which represents 14% of world's arable lands (Kassam et al., 2022). The no-till practice lowers soil disturbance and aids in the retention of more soil organic matter, hence enhancing carbon sequestration in the soil; Du et al. (2017) conducted a comprehensive evaluation of Chinese studies pertaining to soil carbon sequestration. Their findings indicate that slightly more than 50% of the one hundred studies comparing no-till and conventional tillage practices reported a higher level of carbon sequestration in the former. The implementation of crop rotation, such as the combination of cereals and legumes, as well as the utilization of cover crops, has been found to have a positive impact on soil health and fertility. This, in turn, leads to increased plant production and the sequestration of carbon (Patra et al., 2019). The Food and Agriculture Organization (FAO) advocates for the implementation of conservation agriculture as a viable approach to safeguard the

*Table 2. Various components of conservation agriculture and their impact on carbon sequestration*

| Practice | Quantification of Carbon sequestration |
|---|---|
| No-till | According to Swanpeal et al. (2018), the implementation of reduced tillage in clay soil resulted in a higher soil organic carbon (SOC) content of 57.6 t C ha−1, compared to the conventional tillage method which yielded a SOC content of 54.9 t C ha−1.<br>Conversely, in sandy soil, the average SOC content across all treatments was found to be 19.7 t C ha−1. |
| | Zhangliu Du et al. 2017 have found that No-Till practice increased Carbon sequestration by 3.8% comparing to conventional agriculture |
| Rotation | In a study conducted by Venkatesh et al. (2013), it was observed that the implementation of rotation systems involving maize, wheat, mungbean, and pigeonpea-wheat systems resulted in a significant increase of 11% and 10% in total soil organic carbon, respectively. |
| Cover crop | Sanchez et al. 2012 tested the impact of the cover crops implemented in-between perennial tree rows on carbon sequestration and as result this practice increased carbon Sequestration by 1.54 Mg /ha /year |
| | In 2018, Poulton et al., found that the winter cover crop increased carbon sequestered in soil by 0.55 Mg C ha−1 year−1. |
| | The combination of 65% barley and 35% vetch cover crops exhibited a greater capacity for carbon sequestration, resulting in a soil organic carbon increase of almost 1.0 Mg ha−1. (Torres et al., 2021). |
| No-till and rotation | Rice-wheat + residues under zero-till increased SOC at 0.6 m depth by 4.7 t/ha in a seven-year period, whereas rice-wheat under conventional till decreased SOC by 0.9 t C/ha. (Sapkota et al, 2017). |
| | No-till systems with varied crop species boosted SOC stocks by 6–28% (Hok et al. 2015). |
| No-till and cover crop | Reduced-till and no-till increased SOC stock by 6% and 11%, in comparison to conventional tillage, respectively.<br>No-till plus residues raised SOC stock by 12% (Li et al. 2020).<br>No-till and cover crops sequestered 0.85 Mg/ha/year (Sanchez et al, 2012). |

environment and provide sustainable land management, with the aim of mitigating the impacts of climate change (Pisante et al., 2015).It has been demonstrated that CA practices increase carbon sequestration in numerous ways (Table. 2). For instance, the retention of crop residues at the soil surface results in a rise of the soil's organic matter content.. Organic matter is a substantial carbon source that contributes to soil fertility and water retention (Mrabet et al., 2001). Second, reducing soil disturbance during planting and tillage aids in slowing organic matter degradation. This allows organic matter to accumulate in the soil over time, hence boosting carbon sequestration. Finally, crop rotations improve soil health and fertility by improving nutrient availability while reducing insect and disease stress (Reeve et al., 2016). This, in turn, promotes the growth of healthy vegetation capable of carbon sequestration.

As well as promoting sustainable agriculture, conservation agriculture systems may help mitigate climate change. It is a viable approach for carbon sequestration by improving soil quality through cover crops, minimum soil disturbance and crop rotations (Thierfelder and Wall, 2010). These latter lead to promoting soil health and fertility through organic matter enhancement and offer the opportunity to significantly enhance soil carbon sequestration.

## Agroforestry

The practice of systematically integrating tree cultivation into agricultural systems is referred to as agroforestry. According to Feliciano et al. (2018), agroforestry systems contain a wide variety of practices. Some of these practices include crop diversification, long rotation systems for soil conservation, home gardens, boundary plants, enhanced fallows, perennial crops, living fences, and mixed stratum agroforestry. Abbas et al. (2017) posit that the farming community globally finds agroforestry to be a more appealing land use practice compared to the management systems of cropland and forestland. Moreover, trees have the ability to store carbon in both above-ground and below-ground biomass, which helps to reduce the amount of greenhouse gasses that are released into the environment. (Takimoto et al., 2008). Additionally, the presence of trees can enhance soil fertility and augment water retention capabilities (Takimoto et al., 2008; Abbas et al., 2017).

According to Ollinaho and Kroger (2021), Agroforestry, in its various types, seems to be a very positive concept, and very little that is negative has been said about it. Agroforestry offers many benefits, including soil protection, nutrient cycling, and carbon sequestration. Agroforestry systems offer a unique technique of storing carbon due to the presence of trees and other perennial plants. Through photosynthesis, trees and other perennial plants may collect and store a large amount of CO2 from the air. (Dhyani et al., 2020). Furthermore, the implementation of agroforestry systems generates significant benefits in terms of microclimatic enhancement. This, in turn, positively affects the productivity of both trees and intercrops. Additionally, agroforestry systems serve as habitats for soil biodiversity and animals, contribute to the improvement of water quality and nutrient cycling, and ultimately contribute to the overall enhancement of ecosystem functioning on a global scale (Mbow et al., 2014).

Agroforestry is a significant agroecological system (Altieri et al., 2015). The capacity to capture and store atmospheric carbon dioxide (CO2) in vegetation, biomass products, and soils is significant (Boulbi and Olave, 2018). Parra et al., (2021) found that in the Andean region of Antioquia, silvopastoral systems with a high density of Eucalyptus tereticornis trees (over 500 trees per hectare) can offset 26.6 t of CO2 equivalent.ha-1.y-1.

Trees are also essential for healthy soils. They accomplish this by enhancing soil structure and nutrient cycling, which increases carbon sequestration (Hoosbeek et al., 2018).

This system contributes to carbon sequestration by not only storing carbon in trees and soils, but also by lowering greenhouse gas emissions from agricultural activities. When nitrogen-fixing plants are present, agroforestry systems, for example, can reduce fertilizer emissions by reducing the demand for artificial fertilizers (Kim, 2012). Similarly, employing animal manure as a natural fertilizer in agroforestry systems reduces the demand for synthetic fertilizers, lowering emissions associated with their production and transportation (Torres et al., 2017). Agroforestry systems are more effective than standard agricultural approaches at carbon storage, as demonstrated by numerous studies (Jose, 2009; Mbow et al., 2014; Feliciano et al., 2018); it aids in carbon sequestration by storing carbon in trees and soils and reducing greenhouse gas emissions from agricultural activities.

Trees are also essential for healthy soils. They accomplish this by enhancing soil structure and nutrient cycling, which increases carbon sequestration (Hoosbeek et al., 2018).

Agroforestry represents a viable and sustainable alternative to land management that integrates the cultivation of crops or animals with the cultivation of trees and other woody perennials (Nair et al., 2021). Farmers worldwide have long employed a certain practice, which has gained popularity due to its potential to address the effects of climate change through the promotion of carbon sequestration and the prevention of soil erosion..

Agroforestry systems can decrease the need for artificial fertilizers and subsequent fertilizer emissions by utilizing nitrogen-fixing plants (Bentzon-Tarp et al., 2023). The use of organic amendments in agroforestry systems decreases the need for synthetic fertilizers, thereby reducing emissions linked to their production and transportation.

Previous research has demonstrated that agroforestry systems have the potential to enhance soil water retention, mitigate soil erosion, and generate various environmental and socioeconomic advantages

(Mbow et al., 2014; Pavlidis and Tsihrintzis, 2018). According to Tschora & Cherubini (2020), these systems offer additional advantages that strengthen their sustainability. Consequently, they have become an appealing choice for farmers seeking to mitigate the effects of climate change while simultaneously preserving productivity. This appeal originates from their capacity to augment carbon sequestration and promote biodiversity. Wadrom et al., (2017) concluded that agroforesty systems promote synergies in food security, ecosystems and climate change nexus.

## Composting and Soil Amendment

According to Insam & De Bertoldi (2007), composting is a biological process that involves the degradation of organic materials by microorganisms such as bacteria and fungi in an aerobic environment. The process of composting can be broken down into two primary stages: the bio-oxidative stage and the maturation stage.Compost and green manure have the ability to increase the amount of organic matter in the soil, which helps to sequester carbon. According to Alyokhin et al. (2020), natural pest control has the potential to boost soil health and biodiversity while simultaneously reducing the need of synthetic inputs. This, in turn, increases the soil's capacity to store carbon and reduces emissions of greenhouse gases.During the mesophilic stage of composting, which lasts for a few days, bacteria decompose easily accessible organic matter such as fatty acids, carbohydrates, and amino acids, resulting in the production of heat and carbon dioxide (Bernal et al., 2009). Thermophilic bacteria degrade complex organic substances such as lignin and cellulose into simpler ones as the temperature increases. This stage may persist for several weeks, with temperatures reaching 70-80°C. During the maturation stage, compost stabilizes and cools as decomposition is finished by microorganisms such as bacteria and fungi (Mohee et al., 2009; Aboutayeb et al., 2021; El-Mrini and Aboutayeb, 2022).

Organic amendments have been identified as a potential solution for restoring soil quality, increasing plant cover, and enhancing the soil's capacity for carbon sequestration (Soria et al., 2023). A significant alternative for enhancing the long-term storage of soil carbon and boosting the capacity of agricultural systems to moderate and adapt to climate change is the enhancement of soil organic carbon (SOC) by the use of organic inputs. In order to boost soil carbon sequestration and give major co-benefits to rangeland ecosystems and land managers, compost application has been proposed as a novel and potentially fruitful strategy.Moreover, compost application can serve as an effective soil management technique to mitigate greenhouse gas emissions and enhance crop productivity (Jeong et al., 2019). In addition to its role as a soil fertilizer, composting can reduce reliance on certain practices such as chemical fertilizers (which generate greenhouse gases during production), pesticides (which emit pollutants during production), and fuel consumption (by improving soil tilth and workability) (Favoino et al., 2008).

As organic materials are composted, some of the carbon they contain is stored in the resulting compost, resulting in carbon sequestration (El-Mrini et al., 2022). This is due to the carbon being converted into a more stable form that is less susceptible to decomposition (Aboutayeb et al., 2023). Composting also minimizes the quantity of organic waste that is dumped in landfills, where it decomposes anaerobically and emits bad smells and methane, a potent greenhouse gas.

Composting provides various advantages in terms of carbon sequestration (Table. 3), this section will go through the advantages of composting for carbon sequestration:

*Table 3. Effect of composts application on carbon sequestration*

| Practice | Quantification of the carbon sequestration |
|---|---|
| Composting | Manure slurries sequester 14 Mg CO2 ha−1 over a 3-year period, plant waste and Manure-based Compost sequestered 23 Mg CO2e ha−1 over 3 years (DeLonge et al., 2013) |
| | According to Zuo et al. (2023), the application of crop residue and livestock manure in China's croplands between 1980 and 2010 resulted in a significant rise in soil organic carbon levels. Specifically, the soil organic carbon content rose from a range of 29.13–34.54 to 33.51–36.90 Mg C ha−1. This increase corresponds to an average annual growth rate of 113.33 kg C ha−1 yr−1. |
| | The inorganic nutrients and organic nutrients treatments were found to have a mean annual carbon sequestration rate of 0.18 Mg and 0.14 Mg carbon per ha, respectively, in semi-arid tropical soil, according to the findings of Anandakumar et al. (2022). |
| | Application of compost, dry mulching and green manure have increased soil carbon sequestration in soil by 0.49, 0.34 and 0.21 Mg C ha-1 yr-1 respectively (Priori, 2018). |

- Greenhouse Gas Emissions Reduction: By removing organic waste from landfills, composting reduces methane emissions, a powerful greenhouse gas, leading thus to higher amounts of carbon in the end-products of composting process (Favoino and Hogg, 2008).
- Improving Soil Health and Fertility: By providing organic matter, nutrients, and beneficial microorganisms, composting can improve soil health and fertility.

The process of composting soil can improve the structure of the soil, its capacity to store water, and the availability of nutrients, all while increasing the carbon content of the soil. Compost, considered as organic amendment, can also improve soil structure by increasing organic matter content, leading to improve water retention and soil aeration, while minimising soil compaction, and promoting soil biology (Mohammadi et al., 2011).

It has been demonstrated in a number of studies that the application of compost can improve the growth of plants and the yields of crops by supplying the soil with essential nutrients and by improving the structure of the soil (Aboutayeb, 2015; Siedt et al., 2021). Besides, compost is an organic amendment that gradually releases a balanced combination of vital plant nutrients, including nitrogen, phosphorus, potassium, and oligo-elements (El-mrini et al., 2022; Aboutayeb et al., 2023).

Composting reduces landfill waste and associated emissions by diverting organic waste from landfills, where it decomposes anaerobically and emits methane. Composting organic waste instead of sending it to landfills reduces methane emissions and allows for carbon sequestration in the resulting compost (Nordahl et al., 2023).

## Organic Agriculture

The practice of organic farming frequently entails the utilization of natural inputs, such as compost and organic fertilizers, to enhance soil characteristics and thus augment the process of soil carbon sequestration (Leifeld et al., 2010). In the same sense, organic agriculture avoids synthetic fertilizers and pesticides that can disrupt soil ecosystems and impede biological properties of soil that assist in carbon sequestration (Singh et al., 2011).

Organic agriculture practices positively increase carbon sequestration through ecofriendly practices like Composting, cover crops, and crop rotations (Leifeld and Fuhrer, 2010). The aforementioned factors have the potential to enhance the carbon sequestration capacity of soils,, hence lowering atmospheric carbon dioxide levels. By eliminating the use of synthetic inputs and improving soil health, organic agriculture practices can help mitigate climate change and lower greenhouse gas emissions (Niggli et al., 2009). To increase carbon sequestration and decrease the adverse environmental consequences of conventional agriculture, it is vital to encourage farmers to shift towards organic agriculture in order to sustain their agricultural activities.

## SUSTAINABLE AGRICULTURE, POLICIES, AND CARBON SEQUESTRATION: WHAT A RELATIONSHIP?

### Sustainable Agriculture

Sustainable agriculture is a tri-dimensional approach that aims to maintain agricultural productivity over time while reducing environmental impacts and promoting social and economic development. The implementation of sustainable agricultural practices plays a pivotal role in safeguarding the availability of food resources for an expanding worldwide populace, while concurrently mitigating the adverse ecological consequences associated with agricultural activities (Spadaro and Gullino, 2019). Carbon sequestration is closely linked to sustainable agriculture as it can aid in achieving sustainable agricultural goals (Spiertz, 2009).

The increasing recognition of the need of environmentally and socially responsible food production has led to a growing popularity of this farming practice. The objective of sustainable agriculture techniques is to uphold or enhance productivity while concurrently reducing adverse environmental effects, including water pollution, soil degradation, and greenhouse gas emissions (Niggli et al., 2009).Many carbon sequestration mechanisms can help to sustainable agriculture objectives. According to Reicosky (2003), carbon sequestration plays a crucial role in promoting soil health and fertility. It accomplishes this by enhancing soil structure, water retention, nutrient cycling, and ultimately leading to heightened plant productivity and reduced soil erosion. Carbon sequestration has the capacity to not only mitigate greenhouse gas emissions but also effectively extract carbon dioxide from the atmosphere and subsequently sequester it within soil and vegetation (Patil and Kumar, 2017). The provision of habitat for a diverse range of creatures has been found to enhance ecosystem services, including pollination and pest management (Kremen and Miles, 2012).

Carbon sequestration is a crucial component to achieving sustainable agricultural goals by improving soil health, lowering greenhouse gas emissions, and increasing biodiversity. However, trade-offs between carbon sequestration and other sustainable agricultural goals, such as food security and economic sustainability, may exist (Palm et al., 2010). As a result of carbon sequestration strategies, the use of land for food production and the use of land for carbon sequestration may be compromised. This may result in competition for land usage and potential conflicts (Johnson et al., 2014) between objectives focused on mitigating climate change and guaranteeing food security. When implementing carbon sequestration solutions, it is critical to properly balance all these components and consider the local context in order to achieve sustainable agriculture.

Economically, sustainable agriculture can yield economic benefits for farmers since it enhances their profits and reduces input costs by implementing techniques such as crop rotation and integrated pest management. Sustainable farming practices enhance soil health and crop resilience to environmental stresses, thereby mitigating the risk of crop failure and financial losses (Altieri and Nicholls, 2017).

Socially, Sustainable agriculture offers several social benefits for farmers, farm workers, and local communities. Sustainable agriculture may entail agricultural practices that offer employment and revenue for rural communities (Vanclay, 2004). Sustainable agriculture can enhance food security by providing locally accessible, healthy and nutritious food.

## Policy and Governance for Promoting Carbon Sequestration in Agriculture

Carbon sequestration in agriculture is supported by a number of policies and governance structures at various levels (Lal et al., 2015). These programs attempt to boost carbon storage, and improve agricultural sustainability through supporting methods like agroforestry, conservation agriculture, and soil amendments, these initiatives can help with carbon sequestration in agriculture.

Diverse policies and governance structures at various local, national, and regional scales facilitate agricultural carbon sequestration. Terdoo et al. (2014) state that the latter objectives are to increase agricultural resilience to climate change and decrease greenhouse gas emissions.

Carbon pricing is among the best approaches to encourage carbon sequestration in agriculture. Setting a price on carbon emissions provides farmers with a financial incentive to cut their emissions and boost carbon sequestration. Governments can enact a carbon tax system to encourage farmers to adopt measures that lower GHG emissions and support carbon sequestration (Gerber et al., 2010). Besides, financial incentives to farmers could be offered by Governments to farmers that use carbon-sequestering techniques like conservation tillage, cover crops, and agroforestry (Buck and Palumbo-Compton, 2022).

Furthermore, several governments, such as Brazil's Low-Carbon Agricultural Plan (Marengo et al., 2022) and the Building Blocks for Climate-Smart Agriculture (Bazzana et al., 2022), have conceived national programs to enhance carbon sequestration in agriculture. However, putting these regulations and frameworks in place can present both obstacles and possibilities at different sizes. National-level policy coordination across sectors and government levels can be challenging. It is critical to analyze the local context and customize policies to the individual demands of stakeholders and ecosystems (Scherr et al., 2012) in order to effectively promote carbon sequestration in agriculture. Balancing the needs of stakeholders, including farmers, consumers, and environmental groups, may pose challenges. Opportunities exist at regional and local levels to tailor policies to the unique requirements of local communities and ecosystems (Schleyer et al., 2015).

Numerous global initiatives have been undertaken to tackle the issue of climate change and its repercussions on vulnerable sectors like agriculture. The 4 per Thousand initiative and the AAA initiatives exemplify the dedication of governments, organizations, and societies to comprehensively tackle climate change.

The 4 per Thousand initiative, which was introduced during the 21$^{st}$ edition of Climate Change Conference (COP21) in Paris in 2015, signifies a substantial endeavor aimed at tackling climate change by means of sustainable land management techniques and the sequestration of carbon in agricultural lands (Chambers et al., 2016). The principal objective of the endeavor is to attain a yearly augmentation of 0.4% in the inventories of soil organic carbon (SOC), equivalent to an augmentation of 4 parts per thousand. This objective has the potential to effectively reduce greenhouse gas emissions and improve agricultural productivity. The 4 per Thousand initiative acknowledges the significant contribution of agriculture to the mitigation of climate change and seeks to promote the implementation of various practices, including agroforestry, conservation agriculture, cover cropping, and enhanced nutrient management (Soussana et al., 2019). The initiative aims to enhance soil fertility, mitigate erosion, improve water retention capacity, and promote biodiversity through the promotion of these practices. Furthermore, the initiative places significant emphasis on the significance of collaboration, knowledge dissemination, and capacity development in order to facilitate the extensive adoption of these environmentally friendly techniques for managing land (Keesstra et al., 2016).

The AAA (Adaptation of African Agriculture) initiatives, which were introduced in 2011, primarily concentrate on the unique obstacles encountered by African nations in the process of adapting their agricultural systems to the effects of climate change (Zougmoré et al., 2018). The AAA initiatives prioritize the implementation of climate-smart agriculture as a means to improve agricultural productivity, ensure food security, and enhance rural livelihoods in Africa. The proposed approach encompasses the integration of sustainable practices and technologies with the aim of augmenting the resilience of farming communities that are particularly susceptible to vulnerabilities (Mrabet and Moussadek, 2022). The initiatives place a strong emphasis on implementing policy reforms, enhancing capacity-building efforts, promoting knowledge-sharing, and providing financial support. These measures aim to facilitate the transition towards agricultural systems that exhibit resilience to the adverse effects of climate change.. The AAA initiatives seek to empower smallholder farmers and promote inclusive, gender-responsive approaches through the facilitation of climate-resilient technology development and dissemination, establishment of climate information systems, and provision of support for farmer-led adaptation strategies (Mollier et al., 2017). These initiatives acknowledge the necessity for solutions tailored to specific regions and the significance of cooperation among various parties involved in order to tackle the wide-ranging issues posed by climate change in African agriculture.

## CONCLUSION

Carbon sequestration strategies in agriculture have the potential to boost soil health and fertility, lower greenhouse gas emissions, and enhance biodiversity. In this literature review, we discussed different carbon sequestration strategies in agriculture, including conservation agriculture practices, agroforestry, soil amendments, and land management. We also explored the relationship between carbon sequestration and sustainable agriculture, as well as the policies and governance frameworks that promote carbon sequestration in agriculture.

The main outcomes of this chapter are that carbon sequestration strategies can enhance sustainability in agriculture, but there may be trade-offs between carbon sequestration and other sustainability goals, such as food security and economic viability. Additionally, there are various policies and governance frameworks that support carbon sequestration in agriculture, but implementing these policies and frameworks can face challenges and opportunities at different scales.

## REFERENCES

Abou-Tammame, D., Zouhri, A., Boutarfa, A., Fathi, J., & Aboutayeb, R. (2022). The effect of purified wastewater on the physicochemical properties of agricultural soils in Chaouia in Morocco. *Journal of Ecological Engineering*, *23*(1), 34–42. doi:10.12911/22998993/143868

Aboutayeb, R. (2015). *Valorisation des fumiers de volailles de chair: Compostage, épandage et étude des effets des fumiers de volailles et leurs composts sur les propriétés physicochimiques du sol sous cultures de maïs fourrager et de menthe verte*. [Thèse de doctorat, Université Hassan I].

Aboutayeb, R., El-Mrini, S., Zouhri, A., & Azim, K. (2023). Effect of carbon to nitrogen ratio and aeration rate on phosphorus and exchangeable cation contents and their leaching in the soil during olive pomace and turkey manure co-composting. *Journal of Engineering and Applied Sciences (Asian Research Publishing Network), 70*(1), 12.

Aboutayeb, R., El-Mrini, S., Zouhri, A., Idrissi, O., & Azim, K. (2021). Hygienization assessment during heap co-composting of Turkey manure and olive mill pomace. *Eurasian Journal of Soil Science, 10*(4), 332–342. doi:10.18393/ejss.972062

Aboutayeb, R., El Yousfi, B., & El Gharras, O. (2020). Impact of No-Till on physicochemical properties of Vertisols in Chaouia region of Morocco. *Eurasian Journal of Soil Science, 9*(2), 119–125. doi:10.18393/ejss.663502

Adegbeye, M. J., Reddy, P. R. K., Obaisi, A. I., Elghandour, M. M. M. Y., Oyebamiji, K. J., Salem, A. Z. M., & Camacho-Díaz, L. M. (2020). Sustainable agriculture options for production, greenhouse gasses and pollution alleviation, and nutrient recycling in emerging and transitional nations-An overview. *Journal of Cleaner Production, 242*, 118319. doi:10.1016/j.jclepro.2019.118319

Altieri, M. A. (2009). Agroecology, small farms, and food sovereignty. *Monthly Review (New York, N.Y.), 61*(3), 102–113. doi:10.14452/MR-061-03-2009-07_8

Altieri, M. A., & Nicholls, C. I. (2017). The adaptation and mitigation potential of traditional agriculture in a changing climate. *Climatic Change, 140*(1), 33–45. doi:10.100710584-013-0909-y

Altieri, M. A., Nicholls, C. I., Henao, A., & Lana, M. A. (2015). Agroecology and the design of climate change1resilient farming systems. *Agronomy for Sustainable Development, 35*(3), 869–890. doi:10.100713593-015-0285-2

Alyokhin, A., Nault, B., & Brown, B. (2020). Soil conservation practices for insect pest management in highly disturbed agroecosystems–a review. *Entomologia Experimentalis et Applicata, 168*(1), 7–27. doi:10.1111/eea.12863

Anandakumar, S., Bakhoum, N., Chinnadurai, C., Malarkodi, M., Arulmozhiselvan, K., Karthikeyan, S., & Balachandar, D. (2022). Impact of long-term nutrient management on sequestration and dynamics of soil organic carbon in a semi-arid tropical Alfisol of India. *Applied Soil Ecology, 177*, 104549. doi:10.1016/j.apsoil.2022.104549

Annabi, M., Le Bissonnais, Y., Le Villio–Poitrenaud, M., & Houot, S. (2011). Improvement of soil aggregate stability by repeated applications of organic amendments to a cultivated silty loam soil. *Agriculture, Ecosystems & Environment, 144*(1), 382–389. doi:10.1016/j.agee.2011.07.005

Bazzana, D., Foltz, J., & Zhang, Y. (2022). Impact of climate smart agriculture on food security: An agent-based analysis. *Food Policy, 111*, 102304. doi:10.1016/j.foodpol.2022.102304

Benitez, C., Tejada, M., & Gonzalez, J. (2003). Kinetics of the mineralization of nitrogen in a pig slurry compost applied to soils. *Composites Science and Technology, 11*, 72–80.

Bentzon-Tarp, A., Helgadóttir, D., Van den Meersche, K., Gay, F., Priemé, A., Roupsard, O., Mages, C., & Elberling, B. (2023). Spatial-temporal variations of nitrous oxide emissions in coffee agroforestry systems in Costa Rica. *Agriculture, Ecosystems & Environment, 343*, 108257. doi:10.1016/j.agee.2022.108257

Bernal, M. P., Alburquerque, J. A., & Moral, R. (2009). Composting of animal manures and chemical criteria for compost maturity assessment. A review. *Bioresource Technology, 100*(22), 5444–5453. doi:10.1016/j.biortech.2008.11.027

Bongiorno, G., Bünemann, E. K., Oguejiofor, C. U., Meier, J., Gort, G., Comans, R., Mäder, P., Brussaard, L., & de Goede, R. (2019). Sensitivity of labile carbon fractions to tillage and organic matter management and their potential as comprehensive soil quality indicators across pedoclimatic conditions in Europe. *Ecological Indicators, 99*, 38–50. doi:10.1016/j.ecolind.2018.12.008

Buck, H. J., & Palumbo-Compton, A. (2022). Soil carbon sequestration as a climate strategy: What do farmers think? *Biogeochemistry, 161*(1), 59–70. doi:10.100710533-022-00948-2

Burbi, S., & Olave, R. J. (2018). Supporting farmers in the transition to agroecology to promote carbon sequestration from silvopastoral systems. *International Journal of Agricultural Extension, 6*(3), 17–27.

Chambers, A., Lal, R., & Paustian, K. (2016). Soil carbon sequestration potential of US croplands and grasslands: Implementing the 4 per Thousand Initiative. *Journal of Soil and Water Conservation, 71*(3), 68A–74A. doi:10.2489/jswc.71.3.68A

Corbeels, M., Cardinael, R., Naudin, K., Guibert, H., & Torquebiau, E. (2019). The 4 per 1000 goal and soil carbon storage under agroforestry and conservation agriculture systems in sub-Saharan Africa. *Soil & Tillage Research, 188*, 16–26. doi:10.1016/j.still.2018.02.015

Corbeels, M., Cardinael, R., Naudin, K., Guibert, H., & Torquebiau, E. (2019). The 4 per 1000 goal and soil carbon storage under agroforestry and conservation agriculture systems in sub-Saharan Africa. *Soil & Tillage Research, 188*, 16–26. doi:10.1016/j.still.2018.02.015

DeLonge, M. S., Ryals, R., & Silver, W. L. (2013). A lifecycle model to evaluate carbon sequestration potential and greenhouse gas dynamics of managed grasslands. *Ecosystems (New York, N.Y.), 16*(6), 962–979. doi:10.100710021-013-9660-5

Dhyani, S. K., Ram, A., Newaj, R., Handa, A. K., & Dev, I. (2020). Agroforestry for carbon sequestration in tropical India. *Carbon management in tropical and sub-tropical terrestrial systems*, 313-331.

Diacono, M., & Montemurro, F. (2011). Long-term effects of organic amendments on soil fertility. *Sustainable agriculture volume 2*, 761-786.

Dignac, M. F., Derrien, D., Barré, P., Barot, S., Cécillon, L., Chenu, C., Chevallier, T., Freschet, G. T., Garnier, P., Guenet, B., Hedde, M., Klumpp, K., Lashermes, G., Maron, P.-A., Nunan, N., Roumet, C., & Basile-Doelsch, I. (2017). Increasing soil carbon storage: Mechanisms, effects of agricultural practices and proxies. A review. *Agronomy for Sustainable Development, 37*(2), 1–27. doi:10.100713593-017-0421-2

Du, Z., Angers, D. A., Ren, T., Zhang, Q., & Li, G. (2017). The effect of no-till on organic C storage in Chinese soils should not be overemphasized: A meta-analysis. *Agriculture, Ecosystems & Environment, 236*, 1–11. doi:10.1016/j.agee.2016.11.007

Du, Z. L., Zhao, J. K., Wang, Y. D., & Zhang, Q. Z. (2017). Biochar addition drives soil aggregation and carbon sequestration in aggregate fractions from an intensive agricultural system. *Journal of Soils and Sediments*, *17*(3), 581–589. doi:10.100711368-015-1349-2

El-mrini S, Aboutayeb R (2022) Co-composting of olive mill pomace and turkey manure: theoretical risk analysis towards a HACCP plan model. *Environ Eng Manag J 21*(12),1987–2001 doi:.2022.176. doi:10.30638/eemj

El-mrini, S., Aboutayeb, R., & Zouhri, A. (2022). Effect of initial C/N ratio and turning frequency on quality of final compost of turkey manure and olive pomace. *Journal of Engineering and Applied Sciences (Asian Research Publishing Network)*, *69*(1), 1–20.

Epple, C., Rangel, García, Jenkins, S., & Guth, M, M. (2016). Managing ecosystems in the context of climate change mitigation: A review of current knowledge and recommendations to support ecosystem-based mitigation actions that look beyond terrestrial forests. *CBD Technical Series*, *86*, 5.

Fageria, N. K. (2012). Role of soil organic matter in maintaining sustainability of cropping systems. *Communications in Soil Science and Plant Analysis*, *43*(16), 2063–2113. doi:10.1080/00103624.2012.697234

Favoino, E., & Hogg, D. (2008). The potential role of compost in reducing greenhouse gases. *Waste Management & Research*, *26*(1), 61–69. doi:10.1177/0734242X08088584

Favoino, E., & Hogg, D. (2008). The potential role of compost in reducing greenhouse gases. *Waste Management & Research*, *26*(1), 61–69. doi:10.1177/0734242X08088584

Freibauer, A., Rounsevell, M. D., Smith, P., & Verhagen, J. (2004). Carbon sequestration in the agricultural soils of Europe. *Geoderma*, *122*(1), 1–23. doi:10.1016/j.geoderma.2004.01.021

Fridahl, M., & Lehtveer, M. (2018). Bioenergy with carbon capture and storage (BECCS): Global potential, investment preferences, and deployment barriers. *Energy Research & Social Science*, *42*, 155–165. doi:10.1016/j.erss.2018.03.019

Gerber, P., Key, N., Portet, F., & Steinfeld, H. (2010). Policy options in addressing livestock's contribution to climate change. *Animal*, *4*(3), 393–406. doi:10.1017/S1751731110000133

González-Sánchez, E. J., Ordóñez-Fernández, R., Carbonell-Bojollo, R., Veroz-González, O., & Gil-Ribes, J. A. (2012). Meta-analysis on atmospheric carbon capture in Spain through the use of conservation agriculture. *Soil & Tillage Research*, *122*, 52–60. doi:10.1016/j.still.2012.03.001

Hargreaves, J. C., Adl, M. S., & Warman, P. R. (2008). A review of the use of composted municipal solid waste in agriculture. *Agriculture, Ecosystems & Environment*, *123*(1-3), 1–14. doi:10.1016/j.agee.2007.07.004

Hecht, S. B. (2018). The evolution of agroecological thought. In *Agroecology* (pp. 1–19). CRC Press. doi:10.1201/9780429495465-1

Hok, L., de Moraes Sá, J. C., Boulakia, S., Reyes, M., Leng, V., Kong, R., Tivet, F. E., Briedis, C., Hartman, D., Ferreira, L. A., Magno, T., & Pheav, S. (2015). Short-term conservation agriculture and biomass-C input impacts on soil C dynamics in a savanna ecosystem in Cambodia. *Agriculture, Ecosystems & Environment*, *214*, 54–67. doi:10.1016/j.agee.2015.08.013

Hoosbeek, M. R., Remme, R. P., & Rusch, G. M. (2018). Trees enhance soil carbon sequestration and nutrient cycling in a silvopastoral system in south-western Nicaragua. *Agroforestry Systems*, *92*, 263–273.

Hu, Y., Zhang, Q., Hu, S., Xiao, G., Chen, X., Wang, J., Qi, Y., Zhang, L., & Han, L. (2022). Research progress and prospects of ecosystem carbon sequestration under climate change (1992–2022). *Ecological Indicators*, *145*, 109656. doi:10.1016/j.ecolind.2022.109656

Insam, H., & De Bertoldi, M. (2007). Microbiology of the composting process. *Waste management series*, *8*, 25-48.

Jarecki, M. K., & Lal, R. (2003). Crop management for soil carbon sequestration. *Critical Reviews in Plant Sciences*, *22*(6), 471–502. doi:10.1080/713608318

Jeong, S. T., Cho, S. R., Lee, J. G., Kim, P. J., & Kim, G. W. (2019). Composting and compost application: Trade-off between greenhouse gas emission and soil carbon sequestration in whole rice cropping system. *Journal of Cleaner Production*, *212*, 1132–1142. doi:10.1016/j.jclepro.2018.12.011

Johnson, J. A., Runge, C. F., Senauer, B., Foley, J., & Polasky, S. (2014). Global agriculture and carbon trade-offs. *Proceedings of the National Academy of Sciences of the United States of America*, *111*(34), 12342–12347. doi:10.1073/pnas.1412835111

Jose, S. (2009). Agroforestry for ecosystem services and environmental benefits: An overview. *Agroforestry Systems*, *76*(1), 1–10. doi:10.100710457-009-9229-7

Kassam, A., Friedrich, T., & Derpsch, R. (2022). Successful experiences and lessons from conservation agriculture worldwide. *Agronomy (Basel)*, *12*(4), 769. doi:10.3390/agronomy12040769

Keesstra, S. D., Bouma, J., Wallinga, J., Tittonell, P., Smith, P., Cerdà, A., & Fresco, L. O. (2016). Forum paper: The significance of soils and soil science towards realization of the UN sustainable development goals (SDGS). *Soil Discussions*, *2016*, 1–28.

Kerr, R. B., Postigo, J. C., Smith, P., Cowie, A., Singh, P. K., Rivera-Ferre, M., & Neufeldt, H. (2023). Agroecology as a transformative approach to tackle climatic, food, and ecosystemic crises. *Current Opinion in Environmental Sustainability*, *62*, 101275. doi:10.1016/j.cosust.2023.101275

Kim, D. G. (2012). Estimation of net gain of soil carbon in a nitrogen-fixing tree and crop intercropping system in sub-Saharan Africa: Results from re-examining a study. *Agroforestry Systems*, *86*(2), 175–184. doi:10.100710457-011-9477-1

Kowalska, A., Grobelak, A., Almås, Å. R., & Singh, B. R. (2020). Effect of biowastes on soil remediation, plant productivity and soil organic carbon sequestration: A review. *Energies*, *13*(21), 5813. doi:10.3390/en13215813

Kremen, C., & Miles, A. (2012). Ecosystem services in biologically diversified versus conventional farming systems: Benefits, externalities, and trade-offs. *Ecology and Society*, *17*(4), art40. doi:10.5751/ES-05035-170440

Kutos, S., Stricker, E., Cooper, A., Ryals, R., Creque, J., Machmuller, M., Kroegar, M., & Silver, W. L. (2023). Compost amendment to enhance carbon sequestration in rangelands. *Journal of Soil and Water Conservation*, *78*(2), 163–177. doi:10.2489/jswc.2023.00072

Lal, R. (2004). Soil carbon sequestration to mitigate climate change. *Geoderma*, *123*(1-2), 1–22. doi:10.1016/j.geoderma.2004.01.032

Lal, R. (2015). Sequestering carbon and increasing productivity by conservation agriculture. *Journal of Soil and Water Conservation*, *70*(3), 55A–62A. doi:10.2489/jswc.70.3.55A

Lal, R., & Kimble, J. M. (1997). Conservation tillage for carbon sequestration. *Nutrient Cycling in Agroecosystems*, *49*(1/3), 243–253. doi:10.1023/A:1009794514742

Lal, R., Negassa, W., & Lorenz, K. (2015). Carbon sequestration in soil. *Current Opinion in Environmental Sustainability*, *15*, 79–86. doi:10.1016/j.cosust.2015.09.002

Lasco, R. D., Delfino, R. J. P., & Espaldon, M. L. O. (2014). Agroforestry systems: Helping smallholders adapt to climate risks while mitigating climate change. *Wiley Interdisciplinary Reviews: Climate Change*, *5*(6), 825–833. doi:10.1002/wcc.301

Leifeld, J., & Fuhrer, J. (2010). Organic farming and soil carbon sequestration: What do we really know about the benefits? *Ambio*, *39*(8), 585–599. doi:10.100713280-010-0082-8

Li, Y., Li, Z., Chang, S. X., Cui, S., Jagadamma, S., Zhang, Q., & Cai, Y. (2020). Residue retention promotes soil carbon accumulation in minimum tillage systems: Implications for conservation agriculture. *The Science of the Total Environment*, *740*, 140147. doi:10.1016/j.scitotenv.2020.140147

Maltas, A., Kebli, H., Oberholzer, H. R., Weisskopf, P., & Sinaj, S. (2018). The effects of organic and mineral fertilizers on carbon sequestration, soil properties, and crop yields from a long-term field experiment under a Swiss conventional farming system. *Land Degradation & Development*, *29*(4), 926–938. doi:10.1002/ldr.2913

Marengo, J. A., Galdos, M. V., Challinor, A., Cunha, A. P., Marin, F. R., Vianna, M. D. S., Alvala, R. C. S., Alves, L. M., Moraes, O. L., & Bender, F. (2022). Drought in Northeast Brazil: A review of agricultural and policy adaptation options for food security. *Climate Resilience and Sustainability*, *1*(1), e17. doi:10.1002/cli2.17

Mohammadi, K., Heidari, G., Khalesro, S., & Sohrabi, Y. (2011). Soil management, microorganisms and organic matter interactions: A review. *African Journal of Biotechnology*, *10*(86), 19840.

Mohee, R., Jumnoodoo, V., Sobratee, N., Mudhoo, A., & Unmar, G. (2009). Assessing the suitability of the composting process in treating contaminating pesticides and pathogenic wastes. *Compost II. Dynamic Soil, Dynamic Plant*, *3*, 103–114.

Mollier, L., Seyler, F., Chotte, J. L., & Ringler, C. (2017). *End hunger, achieve food security and improved nutrition and promote sustainable agriculture: SDG 2. A Guide to SDG Interactions: From Science to Implementation*. ICSU.

Mrabet, R. (2023). Sustainable agriculture for food and nutritional security. In *Sustainable Agriculture and the Environment* (pp. 25–90). Academic Press. doi:10.1016/B978-0-323-90500-8.00013-0

Mrabet, R., & Moussadek, R. (2022). Development of climate smart agriculture in Africa. In Conservation Agriculture in Africa: Climate Smart Agricultural Development (pp. 17-65). GB: CABI. doi:10.1079/9781789245745.0002

Mrabet, R., Moussadek, R., Fadlaoui, A., & Van Ranst, E. (2012). Conservation agriculture in dry areas of Morocco. *Field Crops Research, 132*, 84–94. doi:10.1016/j.fcr.2011.11.017

Mrabet, R., Saber, N., El-Brahli, A., Lahlou, S., & Bessam, F. (2001). Total, particulate organic matter and structural stability of a Calcixeroll soil under different wheat rotations and tillage systems in a semi-arid area of Morocco. *Soil & Tillage Research, 57*(4), 225–235. doi:10.1016/S0167-1987(00)00180-X

Nair, P. R., Kumar, B. M., Nair, V. D., Nair, P. R., Kumar, B. M., & Nair, V. D. (2021). Definition and concepts of agroforestry. *An Introduction to Agroforestry: Four Decades of Scientific Developments*, 21-28.

Niggli, U., Fließbach, A., Hepperly, P., & Scialabba, N. (2009). Low greenhouse gas agriculture: Mitigation and adaptation potential of sustainable farming systems. *Ökologie und Landbau, 141*, 32–33.

Niggli, U., Fließbach, A., Hepperly, P., & Scialabba, N. (2009). Low greenhouse gas agriculture: Mitigation and adaptation potential of sustainable farming systems. *Ökologie und Landbau, 141*, 32–33.

Ning, K., Chen, J., Li, Z., Liu, C., Nie, X., Liu, Y., Wang, L., & Hu, X. (2021). Land use change induced by the implementation of ecological restoration Programs increases future terrestrial ecosystem carbon sequestration in red soil hilly region of China. *Ecological Indicators, 133*, 108409. doi:10.1016/j.ecolind.2021.108409

Oladele, O., & Braimoh, A. (2014). Potential of agricultural land management activities for increased soil carbon sequestration in Africa–a review. *Applied Ecology and Environmental Research, 12*(3), 741–751. doi:10.15666/aeer/1203_741751

Palm, C., Blanco-Canqui, H., DeClerck, F., Gatere, L., & Grace, P. (2014). Conservation agriculture and ecosystem services: An overview. *Agriculture, Ecosystems & Environment, 187*, 87–105. doi:10.1016/j.agee.2013.10.010

Palm, C. A., Smukler, S. M., Sullivan, C. C., Mutuo, P. K., Nyadzi, G. I., & Walsh, M. G. (2010). Identifying potential synergies and trade-offs for meeting food security and climate change objectives in sub-Saharan Africa. *Proceedings of the National Academy of Sciences of the United States of America, 107*(46), 19661–19666. doi:10.1073/pnas.0912248107

Parra, A. S., de Figueiredo, E. B., de Bordonal, R. O., Moitinho, M. R., Teixeira, D. D. B., & La Scala, N. Jr. (2019). Greenhouse gas emissions in conversion from extensive pasture to other agricultural systems in the Andean region of Colombia. *Environment, Development and Sustainability, 21*(1), 249–262. doi:10.100710668-017-0034-6

Patil, P., & Kumar, A. K. (2017). Biological carbon sequestration through fruit crops (perennial crops-natural "sponges" for absorbing carbon dioxide from atmosphere). *Plant Archives, 17*(2), 1041–1046.

Patra, S., Julich, S., Feger, K. H., Jat, M. L., Sharma, P. C., & Schwärzel, K. (2019). Effect of conservation agriculture on stratification of soil organic matter under cereal-based cropping systems. *Archives of Agronomy and Soil Science, 65*(14), 2013–2028. doi:10.1080/03650340.2019.1588462

Pavlidis, G., & Tsihrintzis, V. A. (2018). Environmental benefits and control of pollution to surface water and groundwater by agroforestry systems: A review. *Water Resources Management, 32*(1), 1–29. doi:10.100711269-017-1805-4

Pisante, M., Stagnari, F., Acutis, M., Bindi, M., Brilli, L., Di Stefano, V., & Carozzi, M. (2015). Conservation agriculture and climate change. *Conservation agriculture*, 579-620.

Potter, C., Klooster, S., Myneni, R., Genovese, V., Tan, P. N., & Kumar, V. (2003). Continental-scale comparisons of terrestrial carbon sinks estimated from satellite data and ecosystem modeling 1982–1998. *Global and Planetary Change*, *39*(3-4), 201–213. doi:10.1016/j.gloplacha.2003.07.001

Poulton, P., Johnston, J., Macdonald, A., White, R., & Powlson, D. (2018). Major limitations to achieving "4 per 1000" increases in soil organic carbon stock in temperate regions: Evidence from long-term experiments at Rothamsted Research, United Kingdom. *Global Change Biology*, *24*(6), 2563–2584. doi:10.1111/gcb.14066

Priori, S. (2018). Effect of organic treatments on soil carbon and nitrogen dynamics in vineyard. *EQA*, *31*, 1–10.

Rakotovao, H. N. (2017). *Impacts environnementaux de l'adoption des pratiques agroécologiques: empreinte car1bone d'exploitation agricole et stocks de carbone, Région Itasy, Madagascar*. [Thèse de Doctorat en Sciences Agro1nomiques, Université d'Antananarivo].

Rakotovao, N. H., Ramifehiarivo, N., Grinand, C., Razakamanarivo, H. R., Albrecht, A., & Razafimbelo, T. M. (2020). *Effects of agroecological practices on soil organic carbon sequestration using synchronic and diachronic approaches in Madagascar.*

Ramesh, T., Bolan, N. S., Kirkham, M. B., Wijesekara, H., Kanchikerimath, M., Rao, C. S., & Freeman, O. W. II. (2019). Soil organic carbon dynamics: Impact of land use changes and management practices: A review. *Advances in Agronomy*, *156*, 1–107. doi:10.1016/bs.agron.2019.02.001

Reicosky, D. C. (2003). Conservation agriculture: Global environmental benefits of soil carbon management. *Conservation Agriculture: Environment, Farmers Experiences, Innovations, Socio-economy, Policy*, 3-12.

Repullo-Ruiberriz de Torres, M. A., Moreno-García, M., Ordóñez-Fernández, R., Rodríguez-Lizana, A., Carceles Rodriguez, B., García-Tejero, I. F., Durán Zuazo, V. H., & Carbonell-Bojollo, R. M. (2021). Cover crop contributions to improve the soil nitrogen and carbon sequestration in almond orchards (SW Spain). *Agronomy (Basel)*, *11*(2), 387. doi:10.3390/agronomy11020387

Sapkota, T. B., Jat, R. K., Singh, R. G., Jat, M. L., Stirling, C. M., Jat, M. K., Bijarniya, D., Kumar, M., Yadvinder-Singh, Saharawat, Y. S., & Gupta, R. K. (2017). Soil organic carbon changes after seven years of conservation agriculture in a rice–wheat system of the eastern Indo-Gangetic Plains. *Soil Use and Management*, *33*(1), 81–89. doi:10.1111um.12331

Scherr, S. J., Shames, S., & Friedman, R. (2012). From climate-smart agriculture to climate-smart landscapes. *Agriculture & Food Security*, *1*(1), 1–15. doi:10.1186/2048-7010-1-12

Schleyer, C., Görg, C., Hauck, J., & Winkler, K. J. (2015). Opportunities and challenges for mainstreaming the ecosystem services concept in the multi-level policy-making within the EU. *Ecosystem Services*, *16*, 174–181. doi:10.1016/j.ecoser.2015.10.014

Siedt, M., Schäffer, A., Smith, K. E., Nabel, M., Roß-Nickoll, M., & van Dongen, J. T. (2021). Comparing straw, compost, and biochar regarding their suitability as agricultural soil amendments to affect soil structure, nutrient leaching, microbial communities, and the fate of pesticides. *The Science of the Total Environment, 751*, 141607. doi:10.1016/j.scitotenv.2020.141607

Siegner, A. B., Acey, C., & Sowerwine, J. (2020). Producing urban agroecology in the East Bay: From soil health to community empowerment. *Agroecology and Sustainable Food Systems, 44*(5), 566–593. doi:10.1080/21683565.2019.1690615

Silva-Olaya, A. M., Cerri, C. E. P., La Scala, N. Jr, Dias, C. T. S., & Cerri, C. C. (2013). Carbon dioxide emissions under different soil tillage systems in mechanically harvested sugarcane. *Environmental Research Letters, 8*(1), 015014. doi:10.1088/1748-9326/8/1/015014

Silva-Parra, A., Trujillo-González, J. M., & Brevik, E. C. (2021). Greenhouse gas balance and mitigation potential of agricultural systems in Colombia: A systematic analysis. *Greenhouse Gases: Science and Technology, 11*(3), 554–572. doi:10.1002/ghg.2066

Singh, J. S., Pandey, V. C., & Singh, D. P. (2011). Efficient soil microorganisms: A new dimension for sustainable agriculture and environmental development. *Agriculture, Ecosystems & Environment, 140*(3-4), 339–353. doi:10.1016/j.agee.2011.01.017

Soria, R., Rodríguez-Berbel, N., Sánchez-Cañete, E. P., Villafuerte, A. B., Ortega, R., & Miralles, I. (2023). Organic amendments from recycled waste promote short-term carbon sequestration of restored soils in drylands. *Journal of Environmental Management, 327*, 116873. doi:10.1016/j.jenvman.2022.116873

Soussana, J. F., Lutfalla, S., Ehrhardt, F., Rosenstock, T., Lamanna, C., Havlík, P., Richards, M., Wollenberg, E. L., Chotte, J.-L., Torquebiau, E., Ciais, P., Smith, P., & Lal, R. (2019). Matching policy and science: Rationale for the '4 per 1000-soils for food security and climate' initiative. *Soil & Tillage Research, 188*, 3–15. doi:10.1016/j.still.2017.12.002

Spadaro, D., & Gullino, M. L. (2019). Sustainable management of plant diseases. *Innovations in Sustainable Agriculture*, 337-359.

Spiertz, J. H. J. (2009). Nitrogen, sustainable agriculture and food security: a review. *Sustainable agriculture*, 635-651.

Stavi, I., & Lal, R. (2013). Agroforestry and biochar to offset climate change: A review. *Agronomy for Sustainable Development, 33*(1), 81–96. doi:10.100713593-012-0081-1

Swanepoel, C. M., Rötter, R. P., Van der Laan, M., Annandale, J. G., Beukes, D. J., du Preez, C. C., Swanepoel, L. H., van der Merwe, A., & Hoffmann, M. P. (2018). The benefits of conservation agriculture on soil organic carbon and yield in southern Africa are site-specific. *Soil & Tillage Research, 183*, 72–82. doi:10.1016/j.still.2018.05.016

Takimoto, A., Nair, P. R., & Nair, V. D. (2008). Carbon stock and sequestration potential of traditional and improved agroforestry systems in the West African Sahel. *Agriculture, Ecosystems & Environment, 125*(1-4), 159–166. doi:10.1016/j.agee.2007.12.010

Tautges, N. E., Chiartas, J. L., Gaudin, A. C., O'Geen, A. T., Herrera, I., & Scow, K. M. (2019). Deep soil inventories reveal that impacts of cover crops and compost on soil carbon sequestration iffer in surface and subsurface soils. *Global Change Biology*, *25*(11), 3753–3766. doi:10.1111/gcb.14762

Terdoo, F., & Adekola, O. (2014). Assessing the role of climate-smart agriculture in combating climate change, desertification and improving rural livelihood in Northern Nigeria. *African Journal of Agricultural Research*, *9*(15), 1180–1191. doi:10.5897/AJAR2013.7665

Thierfelder, C., & Wall, P. C. (2010). Rotation in conservation agriculture systems of Zambia: Effects on soil quality and water relations. *Experimental Agriculture*, *46*(3), 309–325. doi:10.1017/S001447971000030X

Thrupp, L. A. (2000). Linking agricultural biodiversity and food security: The valuable role of agrobiodiversity for sustainable agriculture. *International Affairs*, *76*(2), 265–281. doi:10.1111/1468-2346.00133

Trumbore, S. (2000). Age of soil organic matter and soil respiration: Radiocarbon constraints on belowground C dynamics. *Ecological Applications*, *10*(2), 399–411. doi:10.1890/1051-0761(2000)010[0399:AOSOMA]2.0.CO;2

Tschora, H., & Cherubini, F. (2020). Co-benefits and trade-offs of agroforestry for climate change mitigation and other sustainability goals in West Africa. *Global Ecology and Conservation*, *22*, e00919. doi:10.1016/j.gecco.2020.e00919

Vanclay, F. (2004). Social principles for agricultural extension to assist in the promotion of natural resource management. *Australian Journal of Experimental Agriculture*, *44*(3), 213–222. doi:10.1071/EA02139

Venkatesh, M. S., Hazra, K. K., Ghosh, P. K., Praharaj, C. S., & Kumar, N. (2013). Long-term effect of pulses and nutrient management on soil carbon sequestration in Indo-Gangetic plains of India. *Canadian Journal of Soil Science*, *93*(1), 127–136. doi:10.4141/cjss2012-072

Wezel, A., Bellon, S., Doré, T., Francis, C., Vallod, D., & David, C. (2009). Agroecology as a science, a movement and a practice. A review. *Agronomy for Sustainable Development*, *29*(4), 503–515. doi:10.1051/agro/2009004

Wibbelmann, M., Schmutz, U., Wright, J., Udall, D., Rayns, F., Kneafsey, M., & Turner, M. L. (2013). Mainstreaming agroecology: Implications for global food and farming systems.

Yu, H. Y., Kim, S. H., & Kim, J. G. (2022). Carbon sequestration potential in montane wetlands of Korea. *Global Ecology and Conservation*, *37*, e02166. doi:10.1016/j.gecco.2022.e02166

Zougmoré, R. B., Partey, S. T., Ouédraogo, M., Torquebiau, E., & Campbell, B. M. (2018). Facing climate variability in sub-Saharan Africa: Analysis of climate-smart agriculture opportunities to manage climate-related risks. *Cahiers Agricultures (TSI)*, *27*(3), 1-9..

Zuo, W., Gu, B., Zou, X., Peng, K., Shan, Y., Yi, S., Shan, Y., Gu, C., & Bai, Y. (2023). Soil organic carbon sequestration in croplands can make remarkable contributions to China's carbon neutrality. Journal of Cleaner Production, 382, 135268

# Chapter 3
# Agrochemicals and Climate Change

**Saira Shafiq**
*Department of Botany, University of Agriculture, Faisalabad, Pakistan*

**Muhammad Zia Ul Haq**
https://orcid.org/0000-0001-9914-3367
*Department of Agronomy, University of Agriculture, Faisalabad, Pakistan*

**Amina Shahbaz**
*Department of Agronomy, University of Agriculture, Faisalabad, Pakistan*

**Sadaf Shafique**
*Department of Chemistry, Government Municipal Graduate College, Faisalabad, Pakistan*

**Maryam Riaz**
*Department of Botany, University of Agriculture, Faisalabad, Pakistan*

**Muhammad Tayyab Hanif**
*Department of Agronomy, University of Agriculture, Faisalabad, Pakistan*

**Gulam Jilani**
*Department of Agronomy, University of Agriculture, Faisalabad, Pakistan*

**Hina Ali**
*Department of Botany, University of Agriculture, Faisalabad, Pakistan*

**Wardha Sarfaraz**
*Department of Botany, University of Agriculture, Faisalabad, Pakistan*

**Syed Abbas Raza Naqvi**
*Department of Agronomy, University of Agriculture, Faisalabad, Pakistan*

**Gul Zaib Hassan**
*Department of Agronomy, University of Agriculture, Faisalabad, Pakistan*

## ABSTRACT

*This chapter offers a comprehensive overview of agrochemicals in modern agriculture. Agrochemicals, including fertilizers, herbicides, insecticides, and fungicides, are vital for boosting crop yields, managing pests, and enhancing farm productivity. While they have contributed to addressing global food security challenges since the Green Revolution, their improper and excessive use raises concerns of environmental contamination, soil degradation, and harm to nontarget organisms. Climate change adds further complexities to agrochemical use. To ensure long-term sustainability, integrated pest management, precision agriculture, and eco-friendly alternatives should be embraced. Responsible agrochemical use must be prioritized by farmers and policymakers, considering potential environmental and health impacts. By balancing the benefits and risks, we can work towards a more sustainable future in agriculture, ensuring food availability while safeguarding the environment.*

DOI: 10.4018/978-1-6684-9838-5.ch003

*Figure 1. Classification of agrochemicals*

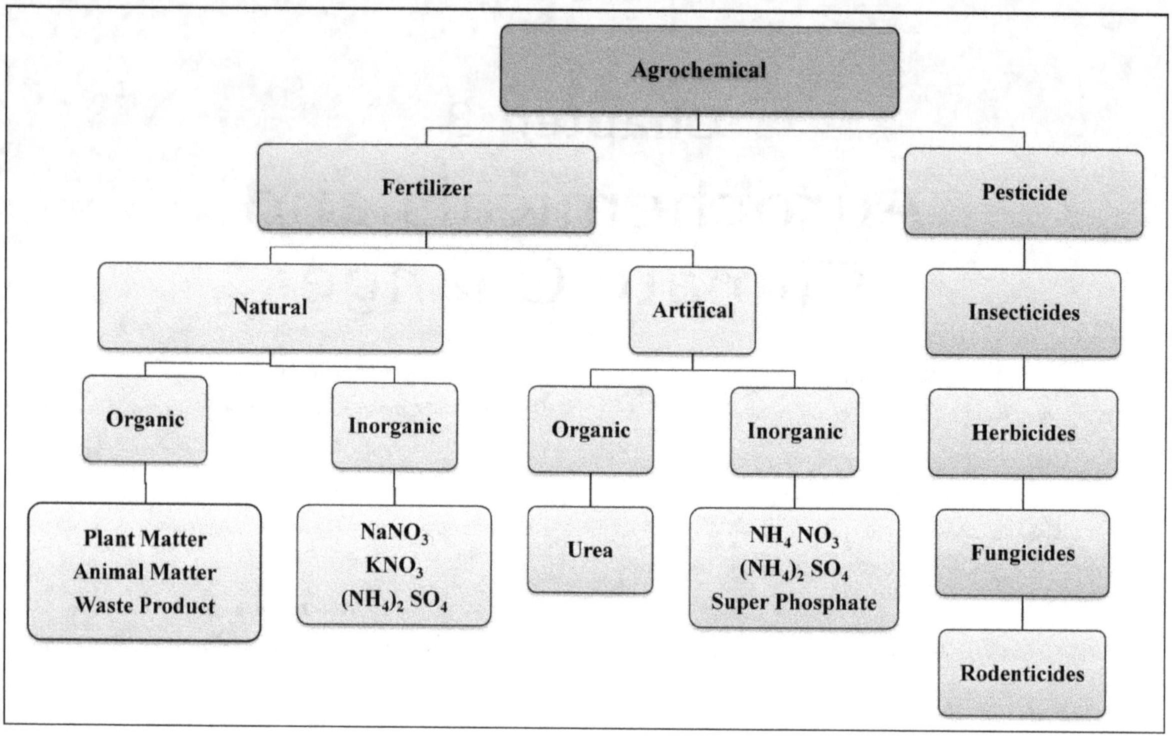


## INTRODUCTION

Agrochemicals are chemical products comprised of fertilizers, plant-protection chemicals or pesticides, and plant-growth hormones used in agriculture. Agrochemicals are substances created and applied in agricultural practices to enhance crop quality, yield, and overall farm production (Roberts et al., 2007). Figure 1 presents the general classification of agrochemicals.

Agrochemicals demonstrate potential advantages in enhancing agricultural production systems' financial viability and addressing food security issues (Figure 2). They are held accountable for the dramatic increase in food production that has been experienced, particularly since the 1930s, when the Green Revolution began (Conway 2019). Agrochemicals are essential for modern agriculture because they provide nutrients, protect crops from pests and diseases, and promote efficient farming practices. 2,4-D, the first herbicide or weed killer, was invented in 1945 and served as a precursor for glyphosate in the 1980s and atrazine in the 1970s. The US government passed the Federal Insecticide, Fungicide, and Rodenticide Act (FIFRA) in 1947 (Devi et al., 2022).

### Relationship Between Agrochemicals and the Water-Soil-Plant-Animal Nexus

The water–soil–plant–animal nexus is a conceptual framework that highlights the interconnectedness of water resources, soil health, plant growth, and animal well-being within agroecosystems (Lal et al., 2017). It emphasizes the intricate relationships and feedback loops between these components, recognizing that changes in one aspect can influence others. This nexus underscores the need for integrated

*Figure 2. Advantages of agrochemicals*

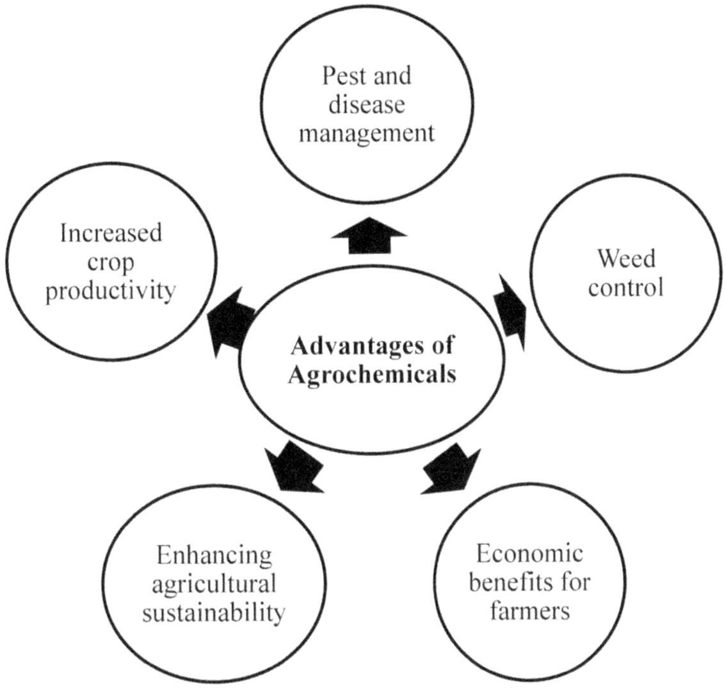


management strategies that consider holistic interactions between water availability, soil quality, plant growth, and animal health. Addressing challenges and optimizing practices within this nexus is essential for achieving sustainable agriculture, as it recognizes the intricate dependencies and trade-offs among various elements of agroecosystems.

Agrochemicals and the Water-Soil-Plant-Animal Nexus: The interconnected relationship between agrochemicals and the complex dynamics that influence agricultural and environmental outcomes. While agrochemicals, such as fertilizers and pesticides, can positively impact agricultural productivity by promoting plant growth and pest control, their improper use can lead to negative consequences (Gagic et al., 2017). The water-soil-plant-animal components are intrinsically linked, and the impact of agrochemicals on one part can cascade through the system, as depicted in Figure 3.

## Agrochemicals and Water

The relationship between agrochemicals and water is critical to understanding the potential impacts of agricultural practices on water resources. One significant concern is agrochemical runoff, which arises from the excessive application of these substances. This can occur during periods of rainfall or irrigation, leading to the contamination of nearby water bodies (Pretty & Bharucha, 2015). Agrochemicals, particularly fertilizers, can leach through the soil and reach groundwater, potentially harming aquatic ecosystems (Kaur & Sinha, 2019). This leads to nutrient waste, soil degradation, and health problems. Proper fertilizer application timing, precise methods, slow-release fertilizers, and vegetation buffers are recommended to prevent this.

*Figure 3. The water-soil-food nexus*

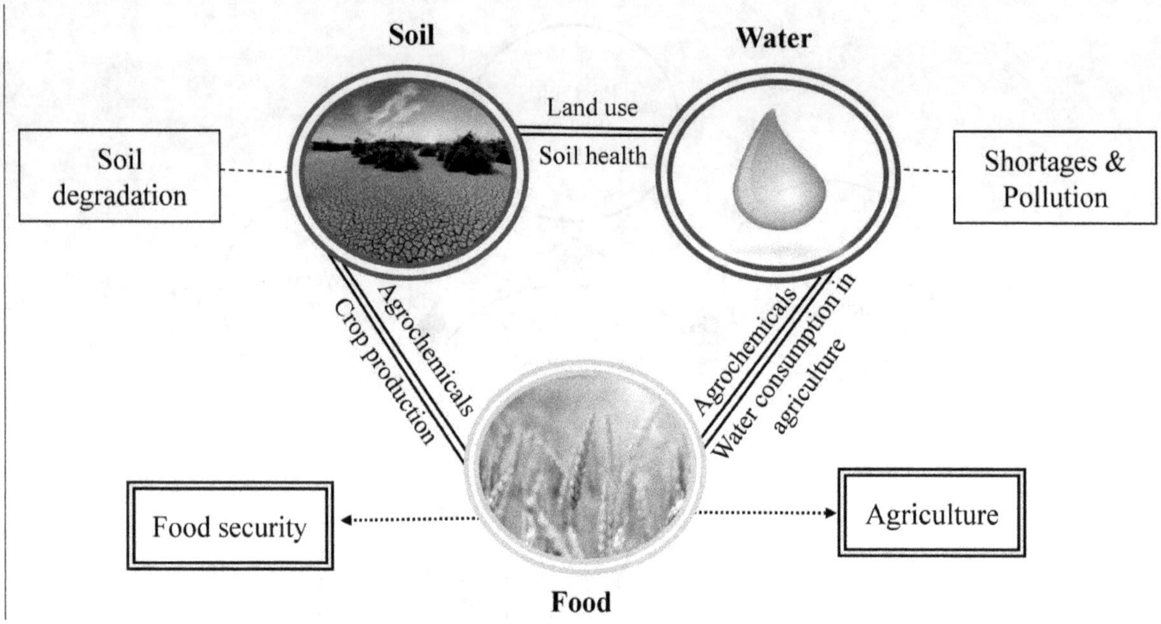


## Agrochemicals and Soil

While fertilizers enhance plant growth, improper use can disrupt soil nutrient balance, leading to degradation and reduced fertility (Tiemann et al., 2015). Agrochemicals, including pesticides, can affect soil microbial communities. Some pesticides may harm beneficial soil organisms, such as earthworms and soil microbes, disrupting important ecological processes like nutrient cycling (Pretty & Bharucha, 2015).

## Agrochemicals, Plants, and Animals

Properly applied fertilizers enhance crop yields and food production, but excessive use can lead to ecological imbalances (Tiemann et al., 2015). Pesticides control pests, but indiscriminate use can harm non-target organisms, affecting biodiversity and resilience (Serrão et al., 2022). Pesticides can change the microbiomes of a wide range of creatures, from insects to mammals, by affecting a number of parameters of the animal microbiome, including the taxonomic composition of bacteria, bacterial biodiversity, and bacterial ratios. Animal immunity is reduced by microbiome changes brought on by pesticides (Syromyatnikov et al., 2020).

# Context of Climate Change and its Influence on the Agrochemical-Water-Soil- Plant-Animal Nexus

Climate change affects the agrochemical-water-soil-plant-animal nexus in many ways. Climate change affects each component as follows:

## Agrochemicals

Climate change can alter agrochemicals' efficacy and usage patterns, including pesticides and fertilizers. Changes in temperature, precipitation patterns, and pest dynamics can influence the effectiveness of pesticide applications and increase the risk of resistance development (Grace et al., 2019). Additionally, shifts in rainfall patterns can impact the leaching and runoff of agrochemicals, potentially leading to water contamination and ecological imbalances.

## Water

Climate change affects water availability and quality, which directly impacts agricultural systems. Changes in precipitation patterns, including droughts and floods, can lead to water scarcity or excess. Increased evapotranspiration rates due to higher temperatures can intensify water stress for plants and animals (Amanullah et al., 2020). Additionally, altered rainfall patterns can affect soil moisture levels, influencing crop growth, nutrient uptake, and productivity (Falkenmark & Rockström, 2006).

## Soil

Climate change can have profound effects on soil properties and processes. Increased temperatures can accelerate soil organic matter decomposition and nutrient cycling, affecting soil fertility. Changes in precipitation patterns can lead to soil erosion and degradation, altering soil structure and nutrient availability (Keesstra et al., 2018). Furthermore, extreme weather conditions like floods and droughts may worsen soil erosion and nutrient leakage.

## Plants

Climate change affects plant development, phenology, and distribution. High temperatures can harm crop development, yields, and nutrition. Heat stress during the reproductive phase increases with warming, causing sterility, lower yields, and crop failure (Sinha et al. 2021). Changes in precipitation can stress or flood plants, affecting growth, nutrient uptake, and pest and disease susceptibility. Plant physiology and nutritional content are also affected by $CO_2$ levels.

## Animals

Climate change poses challenges to animal agriculture and wildlife. Heat stress can affect animal performance, reproductive efficiency, and health, reducing productivity. Altered rainfall patterns can impact forage availability, affecting grazing systems and food availability for wildlife (Godde et al., 2021). Changes in temperature and precipitation can also influence the spread of diseases and pests in animal populations (Fouque & Reeder, 2019).

# TYPES AND USAGES OF AGROCHEMICALS

The various types and usages of agrochemicals, gaining insight into their diverse applications and their challenges in agricultural landscapes.

## Insecticide

Insecticides are chemicals or biological agents that are used to control insects. Different formulations and delivery methods (sprays, baits, slow-release diffusion) of natural or synthetic insecticides (sprays, baits, slow-release diffusion) are used to control vermin. In recent years, biotechnology has even inserted bacterial DNA coding for insecticidal proteins into various crop plants, thereby eliminating pests that feed on these plants (Kumar et al., 2020). Over one million insect species are phytophagous, and approximately 700 are responsible for most crop injury in the field and during storage (Ware & Whitacre, 2004).

### Mechanism of Action of Insecticides

Insecticides are classified into four categories based on their mode of action:

***2.1.1.1 Nerve and muscle target:*** Insecticides harm nerves and muscles. Allethrin and DDT are well-studied pyrethroids. Sodium channel inhibitors were found years ago (Rodrigues et al., 2013), but commercial application is new. Abamectin blocks chlorine channels. Avermectin B1a, a derivative, blocks intraneuronal stimulation of excitatory motor neurons like muscimol and piperazine, GABAergic agonists. Nicotine stimulates, depolarizes, and paralyzes nicotinic receptors (Moreira et al., 2012).

***2.1.1.2 Energy production targets:*** MeBr is odorless and colorless at normal temperature and pressure. Methanol and hydrogen bromide combine to generate fumigant methyl bromide. Methyl bromide cytotoxically alkylates DNA and proteins. First-generation pesticides include arsenic-based insecticides, cryolite, sodium fluoride, sulfates, and carbonates. They halt eating, paralyze, and kill insects by infiltrating their alimentary canal, trachea, and exoskeleton. Rotenone is more toxic to mammals than pyrethrins, yet its breakdown leaves no residue (Aguiar-Menezes, 2005).

***2.1.1.3 Growth and development targets:*** Insects have an evolutionary advantage because their chitinous exoskeleton protects living tissues from the environment. Chitin production alters many biochemical processes, including the chitin synthase enzyme's final polymerization. Insect cuticles are brittle without chitin. Failure to produce a proper cuticle affects ecdysis (Rezende-Teixeira et al., 2022). Terpenoids, or juvenile hormone analogs, directly affect molting and metamorphosis, which are principally governed by the interaction of two hormones, ecdysone (molting hormone) and juvenile hormone (Jindra et al., 2021).

***2.1.1.4 Miscellaneous (biological insecticide):*** Biological insecticides are derived from natural sources, including animals, plants, microbes, and transgenic organisms (Nazir et al., 2019). Biological control is a natural phenomenon by which natural predators manage crops. Its use has expanded substantially and is now integral to integrated pest management.

## Fungicides

Fungicides are chemicals utilized to control and treat fungal infections in plants. They have gained widespread usage in agriculture in recent decades due to their effectiveness, ease of use, and affordability (Vasseghian et al., 2022). Fungal infections can significantly reduce agricultural yields, estimated at around 20% globally (Almeida et al., 2019). Consequently, fungicides have become the primary method for managing fungal diseases (Xia et al., 2006).

Disease control is crucial for agricultural production across various crops, as it directly impacts yield and quality, thereby influencing economic outcomes. Fungicides serve three main purposes:

- Preventing infections during the establishment and growth of grain crops
- Increasing cereal production and eliminating defects
- Enhancing shelf life and quality of harvested plants.

While fungicides have been employed to mitigate mycotoxin contamination in wheat affected by Fusarium head blight, their effectiveness in regulating mycotoxin production associated with other diseases remains limited (Forrer et al., 2014).

### Classification of Fungicides

Protective and non-systemic fungicides are widespread. Protective fungicides protect leaves and stems from a wide range of fungi. Systemic fungicides can be taken by plants and transferred to different tissues where they kill fungus (Baibakova et al., 2019). These chemicals can efficiently treat fungal infections; however, they target certain phases of the fungal life cycle, which may allow resistant strains to survive (Hahn, 2014).

*2.2.1.1* **Non-systemic (contact) fungicides:** This fungicide prevents fungus and fungal spores from growing in plant tissues. Non-systemic fungicides often reside on the plant's surface and may be phytotoxic if swallowed. Zineb, polycarbonate, copper oxychloride, sulfur, Bordeaux mixture, and other contact fungicides cannot heal plants. Non-systemic insecticides can be washed off before harvest, making them safe (Baibakova et al., 2019).

***2.2.1.2 Systemic fungicides:*** Systemic fungicides are absorbed by plants and transferred to infections. They can destroy the fungus even after the mycelia reaches the plant tissue parenchyma, stopping the infection (Baibakova et al., 2019). Systemic fungicides penetrating the plant and travel a short distance (Łozowicka et al., 2020). Dicarboximide fungicides are local-systemic (González et al., 2017). Locally systemic fungicides can flow over leaves but not into the plant.

### Uses of Fungicides

Fungicides make up less than 10% of the total weight of pesticides applied in the United States, but they account for 80% of the total fungicide use in North America. Dithiocarbamates, chloronitriles, demethylation inhibitors (DMIs), and strobilurins account for approximately 65%, 13%, 7%, and 6%, respectively, of the total synthetic fungicide mass used in the United States. These projections are based on provisional 2016 data from the United States Geological Survey (USGS, 2016). Table 1 represents the global pesticide use.

## Herbicides

Herbicides have greatly enhanced crop yields and weed control. Since the Green Revolution, farmers have preferred pesticides for weed management due to their lower labor cost, greater efficacy, and faster results

*Table 1. International pesticide use*

| Nations | Active Ingredients (*metric tons*) |
|---|---|
| Canada | 30000 |
| USA | 248000 |
| China | 240000 |
| Brazil | 57000 |
| Chile | 7000 |
| Colombia | 20000 |
| Costa Rica | 10000 |
| Ecuador | 14000 |
| Mexico | 36000 |
| Paraguay | 3000 |
| Venezuela | 2700 |
| India | 72000 |
| Korea | 26000 |
| Malaysia | 40000 |
| Thailand | 36000 |
| Vietnam | 2000 |

(Hasanuzzaman et al., 2020). These pesticides reduce weed populations but harm nontargeted crops, the environment, and animals. Herbicides are from several chemical families. Several pose considerable risks to human health and the environment when used excessively or without proper information (Drakvik et al., 2020). Herbicide use has had positive and bad effects, making evaluation difficult (Kniss, 2017).

Herbicides pollute air, water, and soil; thus, researchers worldwide are concerned about their long-term repercussions. Weed resistance is a major obstacle to the development of weed-specific herbicides (Peterson et al., 2018). By selecting weed species, the oversupply of herbicides with a single mechanism of action has rapidly created herbicide-resistant plant populations. Over 500 herbicide resistance incidences have been recorded worldwide, involving 167 herbicides and 300 plant species (Qu et al., 2021; Heap, 2023).

## Global Trends in Herbicide Usage in Agriculture

Herbicides are chemicals that kill weeds (Poudel et al., 2020). Herbicides are used worldwide to reduce weed competition and improve crop health (Peerzada et al., 2019). These precisely control weed populations, maximizing plant growth and agricultural productivity. Large-scale farming operations in North America employ herbicides, especially in row crops like maize, soybeans, and cotton (Peterson et al., 2018).

Herbicide use, especially those with active components that may harm the environment and human health, is closely monitored in Europe (Grillo et al., 2021). This has led to stronger rules and a trend toward mechanical cultivation, crop rotation, and integrated weed management. Commercial agriculture

*Table 2. List of herbicides with their site of action*

| Herbicides | Affected Process |
|---|---|
| Aryloxyphenoxypropionates, Cyclohexanediones, Phenylpyrazoline | Acetyl-CoA carboxylase |
| Imidazolinones, Pyrimidinylthiobenzoates, Sulfonylaminocarbonyltriazolinone, Sulfonylureas, Triazolopyrimidines | Acetohydroxyacid synthase |
| Amides, Benzothiadiazinones, Nitriles, Phenyl-carbamates, Pyridazinones | Photosystem II |
| Bipyridylium | Photosystem I |
| Diphenylether, N-phenylphthalimide, Oxadiazole | Protoporphyrinogen oxidase |
| Pyridiazinone, Pyridinone | Phytoene dehydrogenase |
| Callistemone, Isoxazole, Triketone | 4-hydroxyphenylpyruvate dioxygenase |
| Triazole | Carotenoid synthesis |
| Benzofurane, Phosphorodithioate | Lipid synthesis inhibition |
| Dinitrophenols | Membrane disruption |
| Carbamates | Dihydropteroate synthase |
| Phosphinic acids | Glutamine synthetase |
| Glycines | 5-enolpyruvylshikimate-3-phosphate synthase |
| Isoxazolidinone | 1-deoxy-D-xyulose 5-phosphate synthase |

*Table 3. List various herbicides as per their site of action and Herbicide Resistance Action Committee group number*

| HRAC Group | Site of Action | Example herbicide |
|---|---|---|
| 2 | Inhibition of Acetolactate Synthase | Chlorsulfuron |
| 5 | PSII inhibitors - Serine 264 Binders | Chlorotoluron |
| 9 | Inhibition of Enolpyruvyl Shikimate Phosphate Synthase | Glyphosate |
| 1 | Inhibition of Acetyl CoA Carboxylase | Sethoxydim |
| 4 | Auxin Mimics | 2,4-D |
| 22 | PS I Electron Diversion | Paraquat |
| 14 | Inhibition of Protoporphyrinogen Oxidase | Oxyfluorfen |
| 15 | Very Long-Chain Fatty Acid Synthesis inhibitors | Butachlor |
| 3 | Inhibition of Microtubule Assembly 2 | Trifluralin |
| 34 | Inhibition of Lycopene Cyclase | Amitrole |
| 10 | Inhibition of Glutamine Synthetase | Glufosinate-ammonium |
| 6 | PSII inhibitors - Histidine 215 Binders | Bromoxynil |
| 1 | Phytoene Desaturase inhibitors | Diflufenican |
| 27 | Inhibition of Hydroxyphenyl Pyruvate Dioxygenase | Isoxaflutole |
| 29 | Inhibition of Cellulose Synthesis | Dichlobenil |
| 13 | Inhibition of Microtubule Assembly | Clomazone |
| 23 | Inhibition of Microtubule Organization | Propham |

and modern farming have increased herbicide use in developing regions like Asia (Belete et al., 2020). Herbicides boost productivity and efficiency in rice, wheat, sugarcane, and vegetable crops.

Table 2 and 3 show many herbicides and their action site (Heap, 2023).

## Fertilizers

### Nitrogen-Based Fertilizers

As fertilizer, Nitrogen (N), the most essential crop nutrient, degrades the ecosystem and releases greenhouse gases. Ammonia volatilization and nitrous oxide emissions pollute water supplies and harm humans (Figure 4). These issues reduce N uptake and crop productivity, raising fertilization costs and farmers' worries. EEFs address these difficulties. These fertilizers regulate N release, enhancing plant absorption and usage, reducing losses, and increasing agricultural productivity. Coating, method of action on distinct N forms, and development processes distinguish EEFs. Chemical, biological, and nanotechnological improvements, including coating and slow-release agents, reduce N transformation rates and $NH_3$, $N_2O$, and NO losses (Dimkpa et al., 2020). Plant N absorption rises, and leaching, runoff, and denitrification decrease. Efficient N fertilizer use supports food, agriculture, and the environment. Figure 5 describes nitrogen fertilizer efficiency and environmental benefits.

*Figure 4. Strategies to enhance the efficiency of nitrogenous fertilizers and their positive impacts on ecosystem*

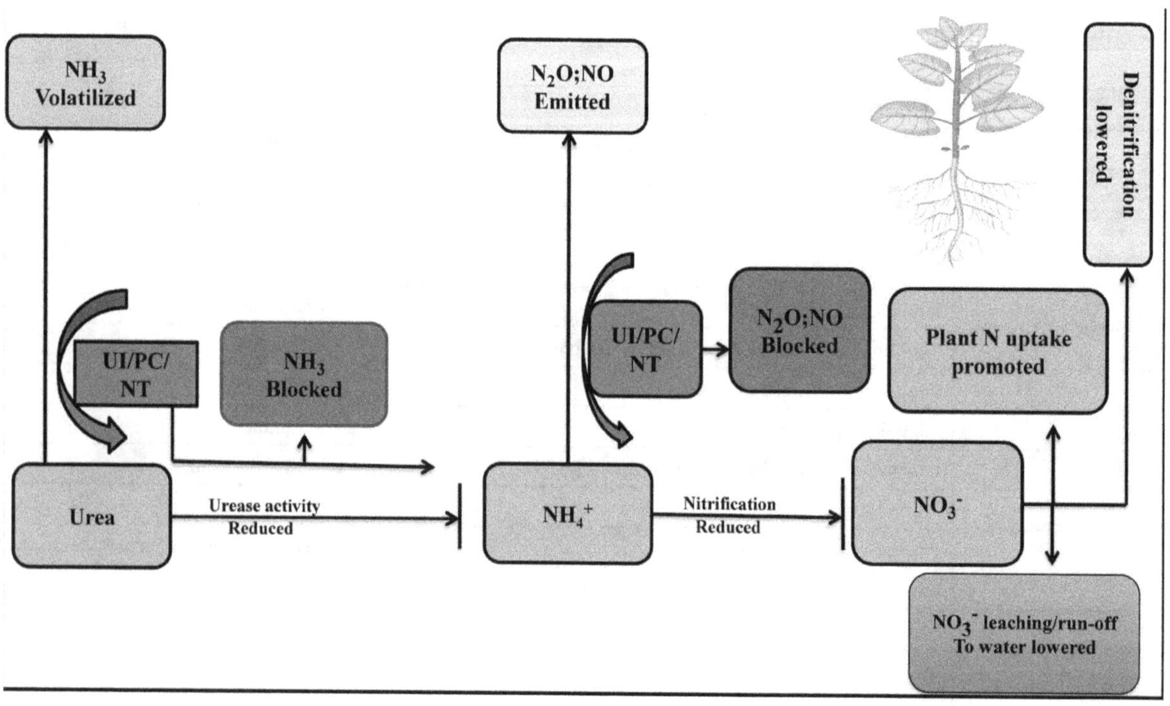

*Figure 5. Fate of urea in soil and environment*

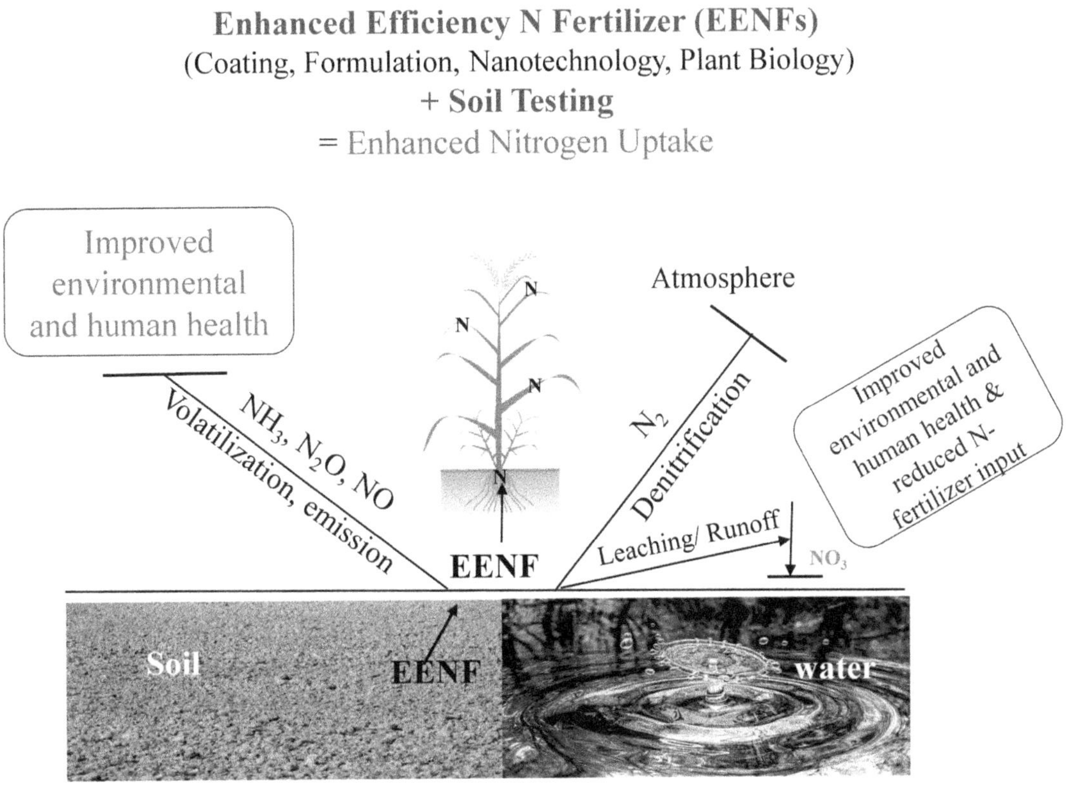


## Phosphate-Based Fertilizers

Agriculture relies on phosphorus (P) from phosphate fertilizers. MAP, DAP, and TSP are phosphate-rock fertilizers (Gong et al., 2022). They boost root growth, flowering, and yield. To avoid phosphorus runoff, proper control is needed. Dosing, timing, and location optimization decrease environmental impact. Phosphorus supplies and fertilizer efficiency are continuous topics. Controlled-release formulations and organic waste phosphorus recovery are sustainable. Global food production relies on phosphate-based fertilizers, which must be used responsibly for sustainable agriculture and environmental protection (Timofeeva, Galyamova, & Sedykh, 2022).

## Potassium-Based Fertilizers

Potassium fertilizers are needed for plant growth. These fertilizers boost agricultural productivity, quality, and health (Maingi & Mbuvi, 2020). They activate enzymes, regulate osmotic equilibrium, and help plants transport nutrients. Potassium fertilizers include KCl, $K_2SO_4$, and $KNO_3$. These fertilizers replenish soil potassium to supply plants. Applying potassium-based fertilizers properly prevents nutrient imbalances and pollution. Farmers must consider fertilizer kind, crop needs, soil conditions, and application rates (Yermiyahu et al., 2020). Carefully using potassium fertilizers can maximize benefits and reduce environmental impacts.

## ENVIRONMENTAL FATE OF AGROCHEMICALS

Agrochemicals migrate and change in the environment—persistence, transit, transformation, and bioaccumulation (Li, 2023). Agrochemicals can last long, be delivered through multiple channels, undergo chemical transformations, and accumulate in organisms. These compounds can harm ecosystems and non-target creatures (Côa et al., 2020). Understanding their environmental fate helps establish sustainable agricultural techniques and reduce their influence. Integrated pest control, precision agriculture, and organic farming are promoted to reduce pesticide use and assure agricultural sustainability.

### Mobility of Agrochemicals

Agrochemical mobility is needed to assess pesticide and fertilizer environmental movement (Avellan et al., 2021). Climate, soil, application, and chemicals affect mobility. Some agrochemicals quickly flow into groundwater, whereas others bond to soil particles. Runoff can carry agrochemicals into waterways. Rates are regulated. Agrochemical mobility management promotes water and agricultural sustainability. Insecticide reactivity in soil may affect surface and groundwater quality. Global irrigation and drinking water depend on groundwater (Kumari & Rai, 2020). Pollutants and pesticides are becoming concerns. Groundwater lacks oxygen, slowing pesticide breakdown. Pesticides from treated fields and rubbish dumps harm groundwater. Pesticides contaminate EU soil (Burri et al., 2019).

### Persistence of Agrochemicals

Agrochemical persistence is pesticides' environmental lifespan. Chemicals, environment, and degradation processes affect persistence (Schäffer et al., 2022). Some agrochemicals accumulate and damage. Chemical, biological, and physical processes degrade agrochemicals (Warra & Prasad, 2020). Active component persistence, toxicity, crop characteristics, rates, application methods, and soil variables determine pesticide dangers (Jepson et al., 2020). "Persistence" defines a pesticide's biological activity in the environment. Persistence has perks and downsides. Pollution, public safety, and non-target plants suffer (Mehdizadeh et al., 2021).

### Degradation

Organic content, pH, moisture, sunlight intensity, microbial population, and oxygen availability affect agrochemical breakdown (Tudi et al., 2021). These factors impede fipronil breakdown in anaerobic rice soil, causing metabolite accumulation. Organic content, pH, moisture, sunshine intensity, microbial population, and oxygen availability affect fipronil degradation and metabolite synthesis. Flooded anaerobic rice soil quickly absorbs fipronil sulphide, accumulating metabolite (Yadav et al., 2023). Bioassays measure herbicide bioavailability. Moderate or high-persistence herbicides can spread (Cuevas et al., 2018). Herbicide persistence depends on chemical composition, application rate, degradation rate, water solubility, soil characteristics, temperature, and other parameters (Hossain, 2015).

### Agrochemical Leaching and Runoff Potential

Pesticide leaching from treated fields pollutes groundwater by colloidal particle co-transport or preferential flow and disperses (Aliste et al., 2021). Soil pesticide contamination worries the EU. Pesticides and synthetic fertilizers in modern agriculture pollute land, air, and water and threaten humans and animals (Alengebawy et al., 2021). Pesticides can enter rivers, lakes, and groundwater through surface runoff, soil leaching, accidental leakage, and subsurface runoff. Rain quickly removes trash cover pesticides (De Buyck et al., 2021).

## ECOLOGICAL IMPACTS OF AGROCHEMICALS

Pesticide overuse in agriculture can harm crops' quality and health (Cooper & Dobson, 2007). Pesticides in food can harm consumers (Leyk et al., 2009). Even low-level pesticide exposure over time can cause asthma (Hoppin et al., 2002). Farming affects the environment, and farmers are vital (Tuc et al., 2007). Unsustainable farming and environmental negligence can degrade the environment (Hou & Wu, 2010). Sustainable pesticide use protects human health and the environment. Pesticides are vital in agriculture; however, their proper use is crucial. Labels offer guidance on active ingredients, usage, and safety. Safe use involves understanding labels, following instructions, wearing protective gear, choosing the right application method, respecting intervals, storing and disposing properly, preventing drift, and maintaining records. Responsible practices ensure effective pest management without harming humans or the environment (Beck et al., 2021).

### Effects on Soil Health and Fertility

Herbicide overuse can harm soil health and fertility, especially soil microorganisms. Enzymatic activity, cellular structure, protein synthesis, and nutrient cycling are crucial for plant growth, but herbicides can disturb these. Compared to no-till, conventional tillage can worsen herbicide damage to soil microorganisms (Kremer & Means, 2009).

Disruption of beneficial insects' population

Pesticides eliminate predators of crop-damaging pests, increasing hazardous bugs. Chemical seed treatments can affect weed-eating insects and soil weed variety. However, reducing herbicide use has increased the Staphylinidae, Neuroptera, and Diptera populations (Sánchez-Bayo, 2021). Fewer chemicals, such as weed killers, bug sprays, and fungal control, enhance Heteroptera, beetles, butterflies, and sawflies. Ground beetles and spiders may suffer little (Frampton & Dorne, 2007).

#### Effects on Non-Target Organisms

Agrochemicals harm humans, animals, aquatic life. Organochlorine insecticides are discussed (Mart, 2018). Many studies show that these pollutants harm juvenile fish. They alter male and female fish hormones, reproductive systems, immunological systems, lipids, and metabolism (Yilmaz et al., 2020). Pesticides and pollutants cause long-term health problems (Yan et al., 2022). Pesticides in crops and food chains damage cellular processes and gene activity, raising the risk of chronic illness. Reducing pesticide exposure is crucial in a society plagued by chronic diseases (Garrigou et al., 2020).

### Implications for Biodiversity and Ecosystem Services

Too many pesticides and fertilizers can affect the environment and living things, causing weed and pest resistance and unclean soil and water. This can harm humans, animals, and the environment (Chandler et al., 2020). Common pesticide research on plants and animals in controlled environments lacks farm animal species monitoring. On the other hand, the French 500 ENI program investigates agriculture in specific areas throughout time, collecting data on land usage and plant and animal variety. This initiative explores field edges, which are important for biodiversity (Wang & Huang, 2021).

## AGROCHEMICALS AND CLIMATE CHANGE INTERACTIONS

### Effects on Agrochemical Application and Effectiveness

Agrochemical use and efficacy are affected by climate change. Increased rainfall and extreme weather events can affect pesticide application by changing spraying liquid reach and timing (Mukherjee et al., 2023). Fall soil moisture deficits or humid soil moisture might affect agricultural operations and need pesticide application schedule adjustments. Pesticide losses can be affected by extreme weather events and seasonal rainfall (Blenkinsop et al., 2008). Moisture and soil temperature affect pesticide-degrading microorganisms (Barriuso et al., 2008). Climate change can increase pesticide use as temperatures rise. Higher temperatures may demand more insecticides and applications (Noyes et al., 2009). Climate change can modify pesticide application and effectiveness, requiring time, dose, and product type adjustments to mitigate adverse impacts and guarantee sustainable pest management. It significantly impacts pest populations in agriculture due to shifts in temperature, precipitation, and extreme weather events. Rising temperatures accelerate pest development, alter phenology, and expand pest ranges. Adapting agrochemical use is essential to manage these changing dynamics. Precision applications using technology like drones and GPS-guided equipment can minimize agrochemical waste.

### Altered Precipitation Patterns and Agrochemical Runoff

Precipitation affects agricultural yields, quality, and pests (Reilly et al., 2003). Greater precipitation during the growing season increases crop yields, but excessive rainfall can damage young plants and lower harvests. Early maturing and water-efficient crop cultivars may thrive (Eid et al., 2007). Precipitation affects plant pesticide uptake and transport, and arid circumstances reduce transpiration, which limits uptake (Keikotlhaile, 2011). Air currents deflect insecticide from the treated area (De Schampheleire et al. 2007). Farmers may monitor wind speed and apply pesticides at the desired wind speed independent of climatic change, reducing drift.

### Extreme Weather Events and Their Influence on Agrochemicals

Climate change increases herbicide use over insecticides and fungicides. This shows herbicide-resistant weeds and decreased herbicide efficacy (Delcour et al., 2015). As temperatures rise, soil and vegetation evaporate pesticides and fertilizers, polluting the air. Milder weather increases organochlorine pesticide evaporation in the US and Korea. Climate change increases pesticide escape risk. Three main causes are:

- Early pesticide application can promote longer growing seasons by enabling more time for breakdown.
- Crop rotation, land use, and new crop varieties affect pesticide use.
- Higher moisture and temperature increase insect pressure and weed populations, increasing pesticide consumption.

## MITIGATION AND SUSTAINABLE ALTERNATIVES

### Strategies for Integrated Pest Management

Integrated Pest Management (IPM) reduces pest populations and chemical pesticide use while limiting environmental effects (Deguine et al., 2021). It uses biological control, cultural practices, and monitoring to provide environmentally and economically sustainable pest management solutions. Biological pest control uses parasites, predators, microbes, and sex traps (Hajek & Eilenberg, 2018). These bio-agents control agricultural pests and weeds without becoming pests. Cultural measures like changing planting and harvesting dates might interrupt crop-pest synchronization and reduce pest damage (Bashyal et al., 2022). Diversified agro-systems and crop rotation disturb pest life cycles and reduce pest populations. A more balanced ecology promotes natural pest management (Lebrazi & Fikri-Benbrahim, 2022). IPM is a holistic approach to pest control that seeks to manage pests economically and environmentally sustainably (Bell & Bennett, 2022). By integrating various pest control methods and considering economic factors, IPM offers potential economic benefits alongside ecological advantages.

### Precision Agriculture and Targeted Agrochemical Use

Precision agriculture with tailored pesticide use can replace conventional farming practices economically and environmentally (Kumar et al., 2022). Site-specific management and modern technology allow precision agriculture to apply pesticides and other inputs. This method improves crop yield, resource use, and environmental impact (Balafoutis et al., 2017). Precision agriculture reduces pesticide use by targeting pests and improving application rates. Tailored input application reduces fertilizer, pesticide, water, and greenhouse gas losses. Precision agriculture reduces fuel use and greenhouse gas emissions. Controlled traffic and machine guiding systems reduce agricultural overlap and fuel use by 25% (Jensen et al., 2012). Variable rate nutrient delivery cuts nitrous oxide emissions by up to 34%. Precision agriculture optimizes resource allocation and reduces waste by targeting inputs to crop needs and field conditions. Precision agriculture is a transformative force in modern farming practices, providing a potent solution to enhancing crop yields while mitigating the impact of agrochemicals (Javaid et al., 2022). The diverse array of examples, ranging from variable-rate pesticide application to robotic weeding and sensor-integrated irrigation, collectively substantiate that precision agriculture is not just a theoretical concept but a tangible and impactful reality.

## POLICY, REGULATIONS, AND EDUCATION

### National and International Regulations on Agrochemical Use

Pesticides' safe and responsible use in agriculture depends on national and international agrochemical laws (Gomes et al., 2020). Pesticides boost crop yields and conserve resources. Developing countries, especially tropical ones, use pesticides to meet global demand for off-season fruits and vegetables (Akter et al., 2018). Herbicides are the most widely utilized insecticides worldwide, mirroring North America and Europe. However, pesticides predominate in Vietnam. Due to a lack of foreign funding, developing nations adopt older, cheaper, and more hazardous pesticidal agents that can be made locally. Inadequate control and the use of hazardous agrochemicals can cause pesticide intoxication and health problems (Negatu et al., 2021).

### Best Management Practices for Agrochemical Application

Agrochemical best management practices limit pesticide risks and environmental impact. Agrochemicals can be safely used with proper storage, awareness campaigns, and sustainable use (WHO, 2010). Agriculture uses pesticides indiscriminately despite their pest control and agricultural productivity benefits (Singh, 2009). Agricultural workers (50% of the worldwide workforce) face the largest pesticide risks. Unsafe pesticide application causes occupational pesticide poisoning in low- and middle-income countries. Unregulated pesticide application in small cans with many holes pollutes the environment and harms health (WHO, 2010).

Legislation resolves pesticide issues rapidly. Underdeveloped nations' governments don't understand these compounds. They import insecticides to increase crop productivity and export. These nations have the simplest pesticide laws. Due to limited government infrastructure, some countries exclusively employ multinational corporate labeling, application rates, consumption patterns, safety data sheets, and toxicity studies.

### Public Awareness and Education Regarding Agrochemical Impacts

Efforts to promote public awareness and education about agrochemical impacts and sustainable agricultural practices require comprehensive strategies involving governments and organizations. Ensuring effective education for farmers is vital to enhancing food production while minimizing the negative consequences of long-term pesticide treatments. Addressing gaps in extension services and knowledge dissemination is crucial for successful outcomes.

***Strengthening extension services***: Governments and agricultural organizations play a pivotal role in strengthening extension services to provide farmers with the necessary knowledge and tools for sustainable agricultural practices. These services bridge the gap between research institutions and farmers, facilitating the transfer of cutting-edge techniques and methodologies.

***Improving support infrastructure:*** Governments should focus on enhancing the support infrastructure for extension service representatives. This includes addressing salaries, lodging, and transportation concerns and ensuring that these professionals are motivated and equipped to carry out their roles effectively.

***Promoting interdepartmental collaboration***: To create a holistic approach to education and awareness, governments can establish interdepartmental collaboration between agricultural, environmental, and health departments. This collaboration can facilitate the sharing of expertise, resources, and information, leading to comprehensive strategies for sustainable agricultural practices.

***Utilizing technology:*** Leveraging technology is essential for reaching a wider audience. Governments and organizations can develop digital platforms, mobile applications, and online resources that provide farmers with information, tutorials, and interactive modules on sustainable practices.

***Empowering farmer communities:*** Engaging farmers themselves is crucial. Governments and organizations can promote the formation of farmer cooperatives and self-help groups. These platforms encourage peer learning, where experienced farmers share their knowledge and success stories, effectively disseminating information at the grassroots level.

***Implementing incentive programs:*** Incentive programs that reward farmers for adopting sustainable practices can be effective motivators. Governments can introduce schemes that offer financial benefits, tax exemptions, or preferential access to resources to farmers who embrace sustainable agricultural practices.

***Promoting research-extension linkages:*** Creating stronger links between research institutions and extension services can ensure that the latest advancements in agricultural science are effectively communicated to farmers. This collaboration can lead to practical, science-based solutions for pest management and sustainable practices.

## INCENTIVES FOR SUSTAINABLE AGRICULTURAL PRACTICES

Environmental stewardship and farming system sustainability depend on sustainable agricultural incentives (Piñeiro et al., 2020). Crop rotation, diversification, cover cropping, and reduced tillage improve farm productivity and sustainability (Wheaton & Kulshreshtha, 2017). These methods improve resource use, climate resilience, and environmental protection. Incentives can promote sustainable habits. Market-based incentives like changing market prices or creating new markets allow flexibility in promoting desirable behaviors (Kiem & Austin, 2013). Market-based initiatives may have unforeseen repercussions.

***Regulatory incentives:*** accompanied by information sharing, capacity building, and technical assistance, can effectively drive sustainable agricultural practices. Effective enforcement, supervision, and monitoring are essential for the success of regulatory measures (Schirmer et al., 2012).

***Cross-compliance incentives***: such as payments for ecosystem services (PES) or agri-environment payments, aim to overcome barriers hindering the adoption of sustainable practices, such as upfront costs or limited access to financial markets (Wheaton & Kulshreshtha, 2017). These incentives compensate farmers for providing ecological services like water quality improvement, forestry, soil erosion control, and air pollution reduction (DeBoe, 2020).

***Compulsory incentives:*** such as regulatory measures, provide certainty and impose sanctions for non-compliance. Certification and market-based incentives depend on farmers' choices (Schirmer et al., 2012). Sustainable agriculture incentives promote environmental care and farming system sustainability. Crop rotation, variety, cover crops, and no-till or reduced tillage can help farmers improve production and environmental sustainability. Sustainable agriculture techniques conserve ecosystems, utilize natural resources efficiently, and promote climate resilience (Wheaton & Kulshreshtha, 2017).

## FUTURE PERSPECTIVES AND RESEARCH NEEDS

Advances in agrochemical technology and formulations

Advances in agrochemical technology and formulations are crucial in enhancing agricultural productivity, reducing environmental impacts, and ensuring food security. Several perspectives and research requirements should be considered in this field.

- Development of environmentally friendly formulations
- Precision and targeted delivery systems

### Climate-Smart Agricultural Practices and Innovations

Climate-smart agricultural practices and innovations are essential to address climate change and ensure sustainable food production. Accessible technologies must be developed for all farmers regardless of economic status or location. These practices aim to minimize greenhouse gas emissions while maintaining productivity. Soil management, crop yield, and carbon storage relationships provide valuable insights into effective mitigation strategies. Advanced technologies, such as precision farming, remote sensing, and data analytics, optimize resource utilization and reduce environmental footprint (Sishodia et al., 2020). Further research should focus on climate-resilient crop varieties and innovative breeding methods. Integrated pest management, agroforestry, and sustainable soil practices can contribute to climate change mitigation. Knowledge sharing and capacity building are crucial for widespread adoption (Sultan & El-Qassem, 2021). Continuous research, innovation, and policy support have unlocked the potential of climate-smart agriculture for a food-secure and resilient future.

### Long-Term Monitoring and Assessment of Agrochemical Impacts

Long-term monitoring and assessing agrochemical effects are crucial for understanding their effects on the environment, human health, and agriculture. Several key perspectives and research needs should be considered to enhance our understanding and address these challenges. These include developing comprehensive monitoring systems, conducting long-term studies, promoting multidisciplinary collaboration, improving risk assessment methodologies, researching alternative agricultural practices, implementing effective policies and regulations, and increasing public awareness and education. By addressing these needs, we can minimize the negative impacts of agrochemicals and work towards sustainable agricultural systems in the future.

### Emerging Technologies and Research Areas in Agrochemical Development

Advancements in agrochemical technology and formulations are pivotal for sustainable agriculture, ensuring productivity while minimizing environmental repercussions and maintaining food security. Key research areas include the development of environmentally friendly formulations and precision/targeted delivery systems. These innovations aim to enhance efficacy while reducing ecological impact.

## Future Agrochemical Technology and Research Perspectives

Research in agrochemical development should focus on creating solutions aligned with sustainability goals. This involves designing products with reduced environmental impact, minimized residues, and enhanced precision. Innovations like nanotechnology-based delivery systems, bio-based pesticides, and natural-product-derived agrochemicals hold promise for the future of sustainable agriculture.

## Importance of International Collaboration and Knowledge Sharing

Addressing the intricate challenges posed by agrochemicals and climate change demands international cooperation. Sharing knowledge, best practices, and research outcomes across borders is crucial for developing holistic solutions. Collaborative efforts can accelerate the adoption of sustainable practices, promote technology transfer, and create a united front against global agricultural threats. In conclusion, the future of agrochemicals and sustainable agriculture lies in integrating emerging technologies, climate-smart practices, and international collaboration. Research needs to encompass advancements in agrochemical formulations, climate-resilient strategies, and comprehensive monitoring systems.

# CONCLUSION

This chapter comprehensively overviews agrochemicals and their significance in modern agriculture. Agrochemicals, including fertilizers, herbicides, insecticides, and fungicides, boost crop yield, manage pest diseases, and improve overall farm productivity. Thus, the significance of agrochemicals in agriculture cannot be overstated. These have been instrumental in addressing global food security challenges, particularly since the Green Revolution. Agrochemicals provide essential nutrients to crops, protect them from pests and diseases, and enable farmers to adopt efficient farming practices, ultimately increasing their food production. However, agrochemical use has raised concerns and challenges. Their improper and excessive use can result in environmental contamination, water and soil degradation, and adverse effects on non-target organisms. The delicate balance of the water-soil-plant-animal nexus can be disrupted, affecting ecosystem health and posing environmental and human well-being risks. Moreover, this chapter highlights the influence of climate change on the agrochemical-water-soil-plant-animal nexus. Changing climatic conditions, such as rising temperatures, altered precipitation patterns, and increased frequency of extreme weather events, add new complexities and challenges to agrochemical use. To ensure the long-term sustainability of agrochemical use, integrated pest management approaches, precision agriculture techniques, and the development of eco-friendly alternatives should be embraced. Farmers and policymakers must prioritize responsible agrochemical use, considering potential environmental and health impacts. By adopting these strategies, we can minimize reliance on agrochemicals, reduce environmental risks, and foster a more sustainable and resilient agricultural sector. Agrochemicals are pivotal for agricultural development, food production, and global food security. Precision agriculture, integrated pest management, organic farming, and bio-based solutions have emerged as potential harmonizers. Converging at the junction of policy, regulation, and education, our exploration underscores the role of global collaboration, knowledge dissemination, and awareness in steering towards sustainable agricultural practices. Through case studies, the challenges and victories of agrochemical use illuminate the transformative potential through cross-regional collaboration. Envisioning this horizon, we anticipate advancements in agrochemical technology, climate-resilient practices, and robust monitoring frameworks. The pursuit of sustainability necessitates a fusion of innovation and stewardship, propelling researchers and practitioners to navigate the intricate nexus between agrochemicals, climate change, and agriculture.

## REFERENCES

Adeleke, B. S., & Babalola, O. O. (2021). The endosphere microbial communities, a great promise in agriculture. *International Microbiology*, *24*(1), 1–17. doi:10.100710123-020-00140-2 PMID:32737846

Aguiar-Menezes, E. L. (2005). *Inseticidas botânicos: seus princípios ativos, modo de ação e uso agrícola*. Embrapa Agrobiologia.

Akter, M., Fan, L., Rahman, M. M., Geissen, V., & Ritsema, C. J. (2018). Vegetable farmers' behaviour and knowledge related to pesticide use and related health problems: A case study from Bangladesh. *Journal of Cleaner Production*, *200*, 122–133. doi:10.1016/j.jclepro.2018.07.130

Alengebawy, A., Abdelkhalek, S. T., Qureshi, S. R., & Wang, M.-Q. (2021). Heavy metals and pesticides toxicity in agricultural soil and plants: Ecological risks and human health implications. *Toxics*, *9*(3), 42. doi:10.3390/toxics9030042 PMID:33668829

Aliste, M., Pérez-Lucas, G., Garrido, I., Fenoll, J., & Navarro, S. (2021). Mobility of insecticide residues and main intermediates in a clay-loam soil, and impact of leachate components on their photocatalytic degradation. *Chemosphere*, *274*, 129965. doi:10.1016/j.chemosphere.2021.129965 PMID:33979936

Almeida, F., Rodrigues, M. L., & Coelho, C. (2019). The still underestimated problem of fungal diseases worldwide. *Frontiers in Microbiology*, *10*, 214. doi:10.3389/fmicb.2019.00214 PMID:30809213

Amanullah, Khalid, S., Imran, Khan, H. A., Arif, M., Altawaha, A. R., Parmar, B. (2020). Effects of climate change on irrigation water quality. *Environment, climate, plant and vegetation growth*, 123-132.

Avellan, A., Yun, J., Morais, B. P., Clement, E. T., Rodrigues, S. M., & Lowry, G. V. (2021). Critical review: Role of inorganic nanoparticle properties on their foliar uptake and in planta translocation. *Environmental Science & Technology*, *55*(20), 13417–13431. doi:10.1021/acs.est.1c00178 PMID:33988374

Baibakova, E. V., Nefedjeva, E. E., Suska-Malawska, M., Wilk, M., Sevriukova, G. A., & Zheltobriukhov, V. F. (2019). Modern fungicides: Mechanisms of action, fungal resistance and phytotoxic effects. *Annual Research & Review in Biology*, 1–16. doi:10.9734/arrb/2019/v32i330083

Balafoutis, A., Beck, B., Fountas, S., Vangeyte, J., Van der Wal, T., Soto, I., & Eory, V. (2017). Precision agriculture technologies positively contributing to GHG emissions mitigation, farm productivity and economics. *Sustainability (Basel)*, *9*(8), 1339. doi:10.3390u9081339

Barriuso, E., Benoit, P., & Dubus, I. G. (2008). Formation of pesticide nonextractable (bound) residues in soil: Magnitude, controlling factors and reversibility. *Environmental Science & Technology*, *42*(6), 1845–1854. doi:10.1021/es7021736 PMID:18409603

Bashyal, S., Poudel, D., & Gautam, B. (2022). A Review On Cultural Practice As An Effective Pest Management Approach Under Integrated Pest Management. [TAEC]. *Tropical Agroecosystems, 3*(1), 34–40. doi:10.26480/taec.01.2022.34.40

Beck, L. L., Young, K. R., Sutherland, C. A., & Smallidge, S. T. (2021). *Pest Management Overview.* College of Agricultural, Consumer and Environmental Sciences.

Belete, T., Wato, T., & Bonga, E. (2020). Challenges and Prospects in the Adoption of Modern Agricultural Inputs in Gimbo District, Kaffa Region, Southern Ethiopia. *Public Policy and Administration Research, 10*, 1–13.

Bell, R., & Bennett, A. (2022). The pros of integrated pest management. *Farmer's Weekly, 2022*(22019), 46–48.

Blenkinsop, S., Fowler, H., Dubus, I. G., Nolan, B. T., & Hollis, J. (2008). Developing climatic scenarios for pesticide fate modelling in Europe. *Environmental Pollution, 154*(2), 219–231. doi:10.1016/j.envpol.2007.10.021 PMID:18045756

Burri, N. M., Weatherl, R., Moeck, C., & Schirmer, M. (2019). A review of threats to groundwater quality in the anthropocene. *The Science of the Total Environment, 684*, 136–154. doi:10.1016/j.scitotenv.2019.05.236 PMID:31153063

Chandler, A. J., Drummond, F. A., Collins, J. A., Lund, J., & Alnajjar, G. (2020). Exposure of the common Eastern bumble bee, Bombus impatiens (Cresson), to sub-lethal doses of acetamiprid and propiconazole in wild blueberry. *Journal of Agricultural and Urban Entomology, 36*(1), 1–23. doi:10.3954/1523-5475-36.1.1

Côa, F., Bortolozzo, L. S., Petry, R., Da Silva, G. H., Martins, C. H., de Medeiros, A. M., Delite, F. S. (2020). Environmental toxicity of nanopesticides against non-target organisms: the state of the art. *Nanopesticides: From research and development to mechanisms of action and sustainable use in agriculture*, 227-279.

Conway, G. (2019). *The doubly green revolution: food for all in the twenty-first century.* Cornell University Press.

Cooper, J., & Dobson, H. (2007). The benefits of pesticides to mankind and the environment. *Crop Protection (Guildford, Surrey), 26*(9), 1337–1348. doi:10.1016/j.cropro.2007.03.022

Costa, L. G., Giordano, G., Guizzetti, M., & Vitalone, A. (2008). Neurotoxicity of pesticides: A brief review. *Frontiers in Bioscience, 13*(4), 1240–1249. doi:10.2741/2758 PMID:17981626

Cuevas, N., Martins, M., & Costa, P. M. (2018). Risk assessment of pesticides in estuaries: A review addressing the persistence of an old problem in complex environments. *Ecotoxicology (London, England), 27*(7), 1008–1018. doi:10.100710646-018-1910-z PMID:29450674

De Buyck, P.-J., Van Hulle, S. W., Dumoulin, A., & Rousseau, D. P. (2021). Roof runoff contamination: A review on pollutant nature, material leaching and deposition. *Reviews in Environmental Science and Biotechnology, 20*(2), 549–606. doi:10.100711157-021-09567-z

De Schampheleire, M., Spanoghe, P., Brusselman, E., & Sonck, S. (2007). Risk assessment of pesticide spray drift damage in Belgium. *Crop Protection (Guildford, Surrey)*, *26*(4), 602–611. doi:10.1016/j.cropro.2006.05.013

DeBoe, G. (2020). *Economic and environmental sustainability performance of environmental policies in agriculture*.

Deguine, J.-P., Aubertot, J.-N., Flor, R. J., Lescourret, F., Wyckhuys, K. A., & Ratnadass, A. (2021). Integrated pest management: Good intentions, hard realities. A review. *Agronomy for Sustainable Development*, *41*(3), 38. doi:10.100713593-021-00689-w

Delcour, I., Spanoghe, P., & Uyttendaele, M. (2015). Literature review: Impact of climate change on pesticide use. *Food Research International*, *68*, 7–15. doi:10.1016/j.foodres.2014.09.030

Devi, P. I., Manjula, M., & Bhavani, R. (2022). Agrochemicals, environment, and human health. *Annual Review of Environment and Resources*, *47*(1), 399–421. doi:10.1146/annurev-environ-120920-111015

Dimkpa, C. O., Fugice, J., Singh, U., & Lewis, T. D. (2020). Development of fertilizers for enhanced nitrogen use efficiency–Trends and perspectives. *The Science of the Total Environment*, *731*, 139113. doi:10.1016/j.scitotenv.2020.139113 PMID:32438083

Drakvik, E., Altenburger, R., Aoki, Y., Backhaus, T., Bahadori, T., Barouki, R., & Bennekou, S. H. (2020). Statement on advancing the assessment of chemical mixtures and their risks for human health and the environment. *Environment International*, *134*, 105267. doi:10.1016/j.envint.2019.105267 PMID:31704565

Eid, H. M., El-Marsafawy, S. M., & Ouda, S. A. (2007). Assessing the economic impacts of climate change on agriculture in Egypt: a Ricardian approach. *World Bank Policy Research Working Paper*(4293).

Falkenmark, M., & Rockström, J. (2006). *The new blue and green water paradigm: Breaking new ground for water resources planning and management* (Vol. 132). American Society of Civil Engineers.

Forrer, H.-R., Musa, T., Schwab, F., Jenny, E., Bucheli, T. D., Wettstein, F. E., & Vogelgsang, S. (2014). Fusarium head blight control and prevention of mycotoxin contamination in wheat with botanicals and tannic acid. *Toxins*, *6*(3), 830–849. doi:10.3390/toxins6030830 PMID:24577585

Fouque, F., & Reeder, J. C. (2019). Impact of past and on-going changes on climate and weather on vector-borne diseases transmission: A look at the evidence. *Infectious Diseases of Poverty*, *8*(03), 1–9. doi:10.118640249-019-0565-1 PMID:31196187

Frampton, G. K., & Dorne, J. L. C. (2007). The effects on terrestrial invertebrates of reducing pesticide inputs in arable crop edges: A meta-analysis. *Journal of Applied Ecology*, *44*(2), 362–373. doi:10.1111/j.1365-2664.2007.01277.x

Gagic, V., Kleijn, D., Báldi, A., Boros, G., Jørgensen, H. B., Elek, Z., & Kovács-Hostyánszki, A. (2017). Combined effects of agrochemicals and ecosystem services on crop yield across Europe. *Ecology Letters*, *20*(11), 1427–1436. doi:10.1111/ele.12850 PMID:28901046

Garrigou, A., Laurent, C., Berthet, A., Colosio, C., Jas, N., Daubas-Letourneux, V., & Baldi, I. (2020). Critical review of the role of PPE in the prevention of risks related to agricultural pesticide use. *Safety Science*, *123*, 104527. doi:10.1016/j.ssci.2019.104527

Gavrilescu, M. (2021). Water, soil, and plants interactions in a threatened environment. *Water (Basel), 13*(19), 2746. doi:10.3390/w13192746

Godde, C. M., Mason-D'Croz, D., Mayberry, D., Thornton, P. K., & Herrero, M. (2021). Impacts of climate change on the livestock food supply chain; a review of the evidence. *Global Food Security, 28*, 100488. doi:10.1016/j.gfs.2020.100488 PMID:33738188

Gomes, H. O., Menezes, J. M. C., da Costa, J. G. M., Coutinho, H. D. M., Teixeira, R. N. P., & do Nascimento, R. F. (2020). A socio-environmental perspective on pesticide use and food production. *Ecotoxicology and Environmental Safety*, *197*, 110627. doi:10.1016/j.ecoenv.2020.110627 PMID:32302856

Gong, H., Meng, F., Wang, G., Hartmann, T. E., Feng, G., Wu, J., & Zhang, F. (2022). Toward the sustainable use of mineral phosphorus fertilizers for crop production in China: From primary resource demand to final agricultural use. *The Science of the Total Environment, 804*, 150183. doi:10.1016/j.scitotenv.2021.150183 PMID:34520915

González, M., Caetano, P., & Sánchez, M. (2017). Testing systemic fungicides for control of Phytophthora oak root disease. *Forest Pathology*, *47*(4), e12343. doi:10.1111/efp.12343

Grace, M. A., Achick, T.-F. E., Bonghan, B. E., Bih, M. E., Ngo, N. V., Ajeck, M. J., & Ntungwen, F. C. (2019). An overview of the impact of climate change on pathogens, pest of crops on sustainable food biosecurity. *International Journal of Ecotoxicology and Ecobiology*, *4*(4), 114–119. doi:10.11648/j.ijee.20190404.15

Grillo, R., Fraceto, L. F., Amorim, M. J., Scott-Fordsmand, J. J., Schoonjans, R., & Chaudhry, Q. (2021). Ecotoxicological and regulatory aspects of environmental sustainability of nanopesticides. *Journal of Hazardous Materials*, *404*, 124148. doi:10.1016/j.jhazmat.2020.124148 PMID:33059255

Hahn, M. (2014). The rising threat of fungicide resistance in plant pathogenic fungi: Botrytis as a case study. *Journal of Chemical Biology*, *7*(4), 133–141. doi:10.100712154-014-0113-1 PMID:25320647

Hajek, A. E., & Eilenberg, J. (2018). *Natural enemies: an introduction to biological control.* Cambridge University Press. doi:10.1017/9781107280267

Hasanuzzaman, M., Mohsin, S. M., Bhuyan, M. B., Bhuiyan, T. F., Anee, T. I., Masud, A. A. C., & Nahar, K. (2020). *Phytotoxicity, environmental and health hazards of herbicides: Challenges and ways forward Agrochemicals detection, treatment and remediation*. Elsevier.

Heap, I. (2023). *The international survey of herbicide resistant weeds.* Online. Internet.

Hoppin, J. A., Umbach, D. M., London, S. J., Alavanja, M. C., & Sandler, D. P. (2002). Chemical predictors of wheeze among farmer pesticide applicators in the Agricultural Health Study. *American Journal of Respiratory and Critical Care Medicine*, *165*(5), 683–689. doi:10.1164/ajrccm.165.5.2106074 PMID:11874814

Hossain, M. (2015). Recent perspective of herbicide: Review of demand and adoption in world agriculture. *Journal of the Bangladesh Agricultural University, 13*(452-2016-35850), 13-24.

Hou, B., & Wu, L. (2010). Safety impact and farmer awareness of pesticide residues. *Food and Agricultural Immunology*, *21*(3), 191–200. doi:10.1080/09540105.2010.484858

Hsu, Y.-T., Shen, T.-C., & Hwang, S.-Y. (2009). Soil fertility management and pest responses: A comparison of organic and synthetic fertilization. *Journal of Economic Entomology*, *102*(1), 160–169. doi:10.1603/029.102.0123 PMID:19253632

IFOAM. (international movement of organic agriculture movements). (2008). The world of organic agriculture - statistics and emerging trends 2008. International federation of organic agriculture movements -ifoam, Bonn, Germany.

IFOAM. (2010). *Definition of organic agriculture*. IFOAM. http://www.ifoam.org/growingorganic/definitions/doa/index.html

Javaid, M., Haleem, A., Singh, R. P., & Suman, R. (2022). Enhancing smart farming through the applications of Agriculture 4.0 technologies. *International Journal of Intelligent Networks*, *3*, 150–164. doi:10.1016/j.ijin.2022.09.004

Jensen, H. G., Jacobsen, L.-B., Pedersen, S. M., & Tavella, E. (2012). Socioeconomic impact of widespread adoption of precision farming and controlled traffic systems in Denmark. *Precision Agriculture*, *13*(6), 661–677. doi:10.100711119-012-9276-3

Jepson, P. C., Murray, K., Bach, O., Bonilla, M. A., & Neumeister, L. (2020). Selection of pesticides to reduce human and environmental health risks: A global guideline and minimum pesticides list. *The Lancet. Planetary Health*, *4*(2), 56–63. doi:10.1016/S2542-5196(19)30266-9 PMID:32112748

Jindra, M., Tumova, S., Milacek, M., & Bittova, L. (2021). A decade with the juvenile hormone receptor [): Elsevier.]. *Advances in Insect Physiology*, *60*, 37–85. doi:10.1016/bs.aiip.2021.03.001

Kaur, T., & Sinha, A. K. (2019). Pesticides in agricultural run offs affecting water resources: A study of Punjab (India). *Agricultural Sciences*, *10*(10), 1381–1395. doi:10.4236/as.2019.1010101

Kaya, A., & Doganlar, Z. B. (2016). Exogenous jasmonic acid induces stress tolerance in tobacco (Nicotiana tabacum) exposed to imazapic. *Ecotoxicology and Environmental Safety*, *124*, 470–479. doi:10.1016/j.ecoenv.2015.11.026 PMID:26629659

Keesstra, S., Mol, G., De Leeuw, J., Okx, J., Molenaar, C., De Cleen, M., & Visser, S. (2018). Soil-related sustainable development goals: Four concepts to make land degradation neutrality and restoration work. *Land (Basel)*, *7*(4), 133. doi:10.3390/land7040133

Keikotlhaile, B. M. (2011). *Influence of the processing factors on pesticide residues in fruits and vegetables and its application in consumer risk assessment*. Ghent University.

Kiem, A. S., & Austin, E. K. (2013). Drought and the future of rural communities: Opportunities and challenges for climate change adaptation in regional Victoria, Australia. *Global Environmental Change*, *23*(5), 1307–1316. doi:10.1016/j.gloenvcha.2013.06.003

Kniss, A. R. (2017). Long-term trends in the intensity and relative toxicity of herbicide use. *Nature Communications*, *8*(1), 1–7. doi:10.1038/ncomms14865 PMID:28393866

Kremer, R. J., & Means, N. E. (2009). Glyphosate and glyphosate-resistant crop interactions with rhizosphere microorganisms. *European Journal of Agronomy*, *31*(3), 153–161. doi:10.1016/j.eja.2009.06.004

Kumar, D., Punetha, A., Verma, P. P., & Padalia, R. (2022). Micronutrient based approach to increase yield and quality of essential oil in aromatic crops. *Journal of Applied Research on Medicinal and Aromatic Plants*, *26*, 100361. doi:10.1016/j.jarmap.2021.100361

Kumar, K., Gambhir, G., Dass, A., Tripathi, A. K., Singh, A., Jha, A. K., & Rakshit, S. (2020). Genetically modified crops: Current status and future prospects. *Planta*, *251*(4), 1–27. doi:10.100700425-020-03372-8 PMID:32236850

Kumar, R., Kumar, N., Rajput, V. D., Mandzhieva, S., Minkina, T., Saharan, B. S., & Duhan, J. S. (2022). Advances in biopolymeric nanopesticides: A new eco-friendly/eco-protective perspective in precision agriculture. *Nanomaterials (Basel, Switzerland)*, *12*(22), 3964. doi:10.3390/nano12223964 PMID:36432250

Kumari, M., & Rai, S. (2020). Hydrogeochemical evaluation of groundwater quality for drinking and irrigation purposes using water quality index in semi arid region of India. *Journal of the Geological Society of India*, *95*(2), 159–168. doi:10.100712594-020-1405-4

Lal, R., Mohtar, R. H., Assi, A. T., Ray, R., Baybil, H., & Jahn, M. (2017). Soil as a basic nexus tool: Soils at the center of the food–energy–water nexus. *Current Sustainable/Renewable. Energy Reports*, *4*, 117–129.

Lebrazi, S., & Fikri-Benbrahim, K. (2022). *Potential of tree legumes in agroforestry systems and soil conservation Advances in Legumes for Sustainable Intensification*. Elsevier.

Leyk, S., Binder, C. R., & Nuckols, J. R. (2009). Spatial modeling of personalized exposure dynamics: The case of pesticide use in small-scale agricultural production landscapes of the developing world. *International Journal of Health Geographics*, *8*(1), 1–16. doi:10.1186/1476-072X-8-17 PMID:19331690

Li, Z. (2023). Modeling Environmental Fate, Transport, and Transformation of Pesticides: First-Order Kinetic Models for Regional and Global Applications. *Reviews of Environmental Contamination and Toxicology*, *261*(1), 14. doi:10.100744169-023-00040-2

Łozowicka, B., Kaczyński, P., Mojsak, P., Rusiłowska, J., Beknazarova, Z., Ilyasova, G., & Absatarova, D. (2020). Systemic and non-systemic pesticides in apples from Kazakhstan and their impact on human health. *Journal of Food Composition and Analysis*, *90*, 103494. doi:10.1016/j.jfca.2020.103494

Maingi, F. M., & Mbuvi, H. M. (2020). The Effect of potassium dosage on selected growth parameters and yield response modeling on potatoes grown in Molo, Kenya. *Journal of Scientific Agriculture, 4*, 0101-0107.

Mart, M. (2018). *Pesticides, a Love Story: America's Enduring Embrace of Dangerous Chemicals*. University Press of Kansas.

Mehdizadeh, M., Mushtaq, W., Siddiqui, S. A., Ayadi, S., Kaur, P., Yeboah, S., & Tampubolon, K. (2021). Herbicide residues in agroecosystems: Fate, detection, and effect on non-target plants. *Reviews in Agricultural Science*, *9*, 157–167. doi:10.7831/ras.9.0_157

Moreira, M. F., Mansur, J. F., & Mansur, J. (2012). Resistência e inseticidas: estratégias, desafios e perspectivas no controle de insetos. Resistência e Inseticidas: Estratégias, Desafios e Perspectivas no Controle de Insetos. 1ed. Rio de Janeiro: INCT-EM, 1-23.

Mukherjee, S., Keswani, K., Nath, P., & Paul, S. (2023). A study on the correlation of anthropogenic activities with climate change. *Environmental Quality Management*, tqem.21984. doi:10.1002/tqem.21984

Nazir, T., Khan, S., & Qiu, D. (2019). Biological control of insect pest. *Pests Control and Acarology*, *21*.

Negatu, B., Dugassa, S., & Mekonnen, Y. (2021). Environmental and health risks of pesticide use in Ethiopia. *Journal of Health & Pollution*, *11*(30), 210601. doi:10.5696/2156-9614-11.30.210601 PMID:34267988

Noyes, P. D., McElwee, M. K., Miller, H. D., Clark, B. W., Van Tiem, L. A., Walcott, K. C., & Levin, E. D. (2009). The toxicology of climate change: Environmental contaminants in a warming world. *Environment International*, *35*(6), 971–986. doi:10.1016/j.envint.2009.02.006 PMID:19375165

Peerzada, A. M., Bukhari, S. A. H., Dawood, M., Nawaz, A., Ahmad, S., & Adkins, S. (2019). *Weed management for healthy crop production. Agronomic Crops* (Vol. 2). Management Practices.

Pérez-Lucas, G., Gambín, M., & Navarro, S. (2020). Leaching behaviour appraisal of eight persistent herbicides on a loam soil amended with different composted organic wastes using screening indices. *Journal of Environmental Management*, *273*, 111179. doi:10.1016/j.jenvman.2020.111179 PMID:32771853

Peterson, C. A., Eviner, V. T., & Gaudin, A. C. (2018). Ways forward for resilience research in agroecosystems. *Agricultural Systems*, *162*, 19–27. doi:10.1016/j.agsy.2018.01.011

Peterson, M. A., Collavo, A., Ovejero, R., Shivrain, V., & Walsh, M. J. (2018). The challenge of herbicide resistance around the world: A current summary. *Pest Management Science*, *74*(10), 2246–2259. doi:10.1002/ps.4821 PMID:29222931

Piñeiro, V., Arias, J., Dürr, J., Elverdin, P., Ibáñez, A. M., Kinengyere, A., & Prager, S. D. (2020). A scoping review on incentives for adoption of sustainable agricultural practices and their outcomes. *Nature Sustainability*, *3*(10), 809–820. doi:10.103841893-020-00617-y

Poudel, S., Poudel, B., Acharya, B., & Poudel, P. (2020). Pesticide use and its impacts on human health and environment. *Environ Ecosyst Sci*, *4*(1), 47–51. doi:10.26480/ees.01.2020.47.51

Pretty, J., & Pervez Bharucha, Z. (2015). Integrated pest management for sustainable intensification of agriculture in Asia and Africa. *Insects*, *6*(1), 152–182. doi:10.3390/insects6010152 PMID:26463073

Qu, R. Y., He, B., Yang, J. F., Lin, H. Y., Yang, W. C., Wu, Q. Y., & Yang, G. F. (2021). Where are the new herbicides? *Pest Management Science*, *77*(6), 2620–2625. doi:10.1002/ps.6285 PMID:33460493

Reilly, J., Tubiello, F., McCarl, B., Abler, D., Darwin, R., Fuglie, K., & Jones, J. (2003). US agriculture and climate change: New results. *Climatic Change*, *57*(1/2), 43–67. doi:10.1023/A:1022103315424

Rezende-Teixeira, P., Dusi, R. G., Jimenez, P. C., Espindola, L. S., & Costa-Lotufo, L. V. (2022). What can we learn from commercial insecticides? Efficacy, toxicity, environmental impacts, and future developments. *Environmental Pollution*, *300*, 118983. doi:10.1016/j.envpol.2022.118983 PMID:35151812

Roberts, T. R., Hutson, D. H., Lee, P. W., Nicholls, P. H., Plimmer, J. R., Roberts, M. C., & Croucher, L. (2007). *Metabolic pathways of agrochemicals: part 1: herbicides and plant growth regulators*. Royal Society of Chemistry.

Rodrigues, C. R. F., Silva, E. N., Ferreira-Silva, S. L., Voigt, E. L., Viégas, R. A., & Silveira, J. A. G. (2013). High $K^+$ supply avoids $Na^+$ toxicity and improves photosynthesis by allowing favorable $K^+$: $Na^+$ ratios through the inhibition of $Na^+$ uptake and transport to the shoots of Jatropha curcas plants. *Journal of Plant Nutrition and Soil Science*, *176*(2), 157–164. doi:10.1002/jpln.201200230

Sánchez-Bayo, F. (2021). Indirect effect of pesticides on insects and other arthropods. *Toxics*, *9*(8), 177. doi:10.3390/toxics9080177 PMID:34437495

Schäffer, A., Fenner, K., Wang, Z., & Scheringer, M. (2022). To be or not to be degraded: In defense of persistence assessment of chemicals. *Environmental Science. Processes & Impacts*, *24*(8), 1104–1109. doi:10.1039/D2EM00213B PMID:35822690

Schirmer, J., Dovers, S., & Clayton, H. (2012). Informing conservation policy design through an examination of landholder preferences: A case study of scattered tree conservation in Australia. *Biological Conservation*, *153*, 51–63. doi:10.1016/j.biocon.2012.04.014

Serrão, J. E., Plata-Rueda, A., Martínez, L. C., & Zanuncio, J. C. (2022). Side-effects of pesticides on non-target insects in agriculture: A mini-review. *Naturwissenschaften*, *109*(2), 17. doi:10.100700114-022-01788-8 PMID:35138481

Shaji, H., Chandran, V., & Mathew, L. (2021). *Organic fertilizers as a route to controlled release of nutrients Controlled release fertilizers for sustainable agriculture*. Elsevier.

Singh, A. (2009). A study on farmers' knowledge, perception and intensity of pesticide use in vegetable cultivation in Western Uttar Pradesh. *Pusa AgriScience*, *32*, 63–69.

Sinha, R., Fritschi, F. B., Zandalinas, S. I., & Mittler, R. (2021). The impact of stress combination on reproductive processes in crops. *Plant Science*, *311*, 111007. doi:10.1016/j.plantsci.2021.111007 PMID:34482910

Sishodia, R. P., Ray, R. L., & Singh, S. K. (2020). Applications of remote sensing in precision agriculture: A review. *Remote Sensing (Basel)*, *12*(19), 3136. doi:10.3390/rs12193136

Sultan, S., & El–Qassem, A. (2021). Future prospects for sustainable agricultural development. *International Journal of Modern Agriculture and Environment*, *1*(2), 54–82. doi:10.21608/ijmae.2023.215952.1012

Syromyatnikov, M. Y., Isuwa, M. M., Savinkova, O. V., Derevshchikova, M. I., & Popov, V. N. (2020). The effect of pesticides on the microbiome of animals. *Agriculture*, *10*(3), 79. doi:10.3390/agriculture10030079

Szynkowska, M. I., Pawlaczyk, A., & Maćkiewicz, E. (2018). Bioaccumulation and biomagnification of trace elements in the environment. *Recent Advances in Trace Elements*, 251-276.

Taylor, M., Lyons, S. M., Davie-Martin, C. L., Geoghegan, T. S., & Hageman, K. J. (2019). Understanding trends in pesticide volatilization from agricultural fields using the pesticide loss via volatilization model. *Environmental Science & Technology*, *54*(4), 2202–2209. doi:10.1021/acs.est.9b04762 PMID:31858785

Tiemann, L., Grandy, A., Atkinson, E., Marin-Spiotta, E., & McDaniel, M. (2015). Crop rotational diversity enhances belowground communities and functions in an agroecosystem. *Ecology Letters, 18*(8), 761–771. doi:10.1111/ele.12453 PMID:26011743

Timofeeva, A., Galyamova, M., & Sedykh, S. (2022). Prospects for using phosphate-solubilizing microorganisms as natural fertilizers in agriculture. *Plants, 11*(16), 2119. doi:10.3390/plants11162119 PMID:36015422

Tuc, V. P., Wangsuphachart, V., Tasanapradit, P., Fungladda, W., Van Trong, P., & Nhung, N. T. (2007). Impacts of pesticide use on semen characteristics among rice famers in Kienxuong District, Thaibinh Province, Vietnam. *The Southeast Asian Journal of Tropical Medicine and Public Health, 38*(3), 569. PMID:17877235

Tudi, M., Daniel Ruan, H., Wang, L., Lyu, J., Sadler, R., Connell, D., & Phung, D. T. (2021). Agriculture development, pesticide application and its impact on the environment. *International Journal of Environmental Research and Public Health, 18*(3), 1112. doi:10.3390/ijerph18031112 PMID:33513796

USGS. (2016). *USGS Pesticide National Synthesis Project*. USGS. https://water.usgs.gov/nawqa/pnsp/usage/maps/

Vasseghian, Y., Arunkumar, P., Joo, S.-W., Gnanasekaran, L., Kamyab, H., Rajendran, S., & Klemeš, J. J. (2022). Metal-organic framework-enabled pesticides are an emerging tool for sustainable cleaner production and environmental hazard reduction. *Journal of Cleaner Production, 373*, 133966. doi:10.1016/j.jclepro.2022.133966

Wang, T., Li, H., & Huang, Y. (2021). The complex ecological network's resilience of the Wuhan metropolitan area. *Ecological Indicators, 130*, 108101. doi:10.1016/j.ecolind.2021.108101

Ware, G. W., & Whitacre, D. M. (2004). An introduction to insecticides. *The pesticide book, 6.*

Warra, A. A., & Prasad, M. N. V. (2020). *African perspective of chemical usage in agriculture and horticulture—their impact on human health and environment Agrochemicals detection, treatment and remediation*. Elsevier.

Wheaton, E., & Kulshreshtha, S. (2017). Environmental sustainability of agriculture stressed by changing extremes of drought and excess moisture: A conceptual review. *Sustainability (Basel), 9*(6), 970. doi:10.3390u9060970

Woodrow, J. E., Gibson, K. A., & Seiber, J. N. (2019). Pesticides and related toxicants in the atmosphere. *Reviews of Environmental Contamination and Toxicology, 247*, 147–196. PMID:30535549

Wu, L., Jiang, Y., Zhao, F., He, X., Liu, H., & Yu, K. (2020). Increased organic fertilizer application and reduced chemical fertilizer application affect the soil properties and bacterial communities of grape rhizosphere soil. *Scientific Reports, 10*(1), 9568. doi:10.103841598-020-66648-9 PMID:32533037

Xia, X. J., Huang, Y. Y., Wang, L., Huang, L. F., Yu, Y. L., Zhou, Y. H., & Yu, J. Q. (2006). Pesticides-induced depression of photosynthesis was alleviated by 24-epibrassinolide pretreatment in Cucumis sativus L. *Pesticide Biochemistry and Physiology, 86*(1), 42–48. doi:10.1016/j.pestbp.2006.01.005

Yadav, S., Singh, M., & Thakur, L. K. (2023). A Comprehensive Review on the Effect of Sugarcane Trash Ash on the Sorption, Degradation and Leaching of Atrazine and Fipronil in Agricultural Soils. *International Journal of Environment and Climate Change*, *13*(9), 305–323. doi:10.9734/ijecc/2023/v13i92233

Yan, Z., Xiong, C., Liu, H., & Singh, B. K. (2022). Sustainable agricultural practices contribute significantly to One Health. *Journal of Sustainable Agriculture and Environment*, *1*(3), 165–176. doi:10.1002ae2.12019

Yilmaz, B., Terekeci, H., Sandal, S., & Kelestimur, F. (2020). Endocrine disrupting chemicals: Exposure, effects on human health, mechanism of action, models for testing and strategies for prevention. *Reviews in Endocrine & Metabolic Disorders*, *21*(1), 127–147. doi:10.100711154-019-09521-z PMID:31792807

Yuan, P., Wang, J., Pan, Y., Shen, B., & Wu, C. (2019). Review of biochar for the management of contaminated soil: Preparation, application and prospect. *The Science of the Total Environment*, *659*, 473–490. doi:10.1016/j.scitotenv.2018.12.400 PMID:31096377

## KEY TERMS AND DEFINITIONS

**Agrochemicals:** These are chemical products comprised of fertilizers, plant-protection chemicals or pesticides, and plant-growth hormones used in agriculture.

**Bioaccumulation:** The accumulation of substances such as pesticides in living organisms over time as they are ingested through the food chain.

**Climate Resilient Practices:** Farming methods and strategies designed to mitigate and adapt to the impacts of climate change, including altered weather patterns, increased temperatures, and extreme events, while maintaining or enhancing agricultural productivity.

**Ecosystem Services:** Benefits that humans derive from ecosystems, including pollination, water purification, soil fertility, and climate regulation.

**Green Revolution:** A period of rapid agricultural innovation and technology adoption that began in the mid-20$^{th}$ century, characterized by the development and widespread use of high-yielding crop varieties, improved irrigation methods, and agrochemicals to boost global food production.

**Integrated Pest Management (IPM):** A holistic approach to pest control that combines various strategies, including biological control, cultural practices, and chemical intervention, to manage pests in a balanced and sustainable manner.

**Persistence:** The ability of agrochemicals to remain active in the environment for an extended period, affecting the soil, water, and organisms.

**Precision Agriculture:** A modern farming approach that uses technology, such as GPS and remote sensing, to gather data and tailor farming practices, such as irrigation and fertilization, to specific areas within a field and optimize resource use and crop production.

**Precision Agriculture:** A technology-driven approach to farming that uses data and information to optimize the use of resources, reduce input waste, and increase productivity while minimizing environmental impact.

**The water–soil–plant–animal nexus:** It is a conceptual framework that highlights the interconnectedness of water resources, soil health, plant growth, and animal well-being within agroecosystems.

# Chapter 4
# Weather-Based Crop Prediction Using Big Data Analytics

**L. Gowri**
*SASTRA University, India*

**S. Pradeepa**
*SASTRA University, India*

**N. Sasikaladevi**
*SASTRA University, India*

## ABSTRACT

*Weather forecasting is an important and indispensable procedure in peoples' day-to-day lives, it evaluates the alteration happening in the current condition of the atmosphere. The major goal of this project is to create a weather-based crop prediction system to predict the crops based on weather forecasting using big data analytics. In this project, the authors gather and analyze data based on temperature, rainfall, soil, seed, crop production, humidity, and wind speed to assist farmers in improving agricultural yields. The data is first preprocessed in a Python environment, and then the MapReduce framework is used to further analyze and process the massive amount of data. Second, based on MapReduce findings, k-means clustering algorithm is used to produce a mean result on the data in terms of accuracy. After that, the authors examine the link between crops, rainfall, temperature, soil, and seed type using bar graphs and scatter plots.*

## INTRODUCTION

Climate change is causing weather patterns to become increasingly brutal and unpredictable (Funk et al.,2015), which can make it challenging for farmers to raise crops. Weather affects productivity in agriculture and other primary sector operations. This is a significant issue since it increases global food instability and hunger (Shen et al., 2019). Agriculture is the backbone of the economy. Most of the farmers are not getting the expected outcomes of crop production. Weather forecasting is a dominant

DOI: 10.4018/978-1-6684-9838-5.ch004

and essential procedure in people's day-to-day lives, it evaluates the changes happening in the current condition of the atmosphere. Because the climate is changing at such a rapid rate these days, traditional weather forecasting systems are becoming less productive and taking more time. These weather forecasts have a huge impact on people's lives and the economy of the country (Gavahi et al., 2021). Farmers and agribusiness are suffering nationwide from a lack of crop production because of abrupt weather changes. The agriculture industry employs more than 40% of the nation's workforce (Girish et al., 2018) and is directly or indirectly responsible for more than 50% of the nation's economic output (Mila et al.,2019). A more accurate and prioritized forecasting system must be developed using the massive amount of data that has been gathered in agriculture rather than the archaic traditional forecasting techniques that use charts (Kafy et al., 2023). Agriculture generates enormous amounts of data each year. Big data will aid in problem-solving and improve industrial understanding (Osman et al.,201). Examining huge data sets with a range of data kinds is known as big data analysis. Big data analytics is the method of analyzing big data to bring out hidden patterns and relevant information that can yield good results (Gavahi et al.,2021). When estimating crop yields, the impact of the climate might be regarded as a top priority. To understand how climate influences agriculture as a factor, there has been a lot of research done, however most of these studies call for detailed and sophisticated information that is not readily available. As a result, statistics are collected using estimates, which may be detrimental. To make up for the availability of data, it is therefore vital to strengthen the approach technique to balance the data, and basics of stream lit for web app development. This chapter aims to improve agricultural decisions by analyzing weather data to forecast crop yield. And also considering how meteorological conditions affect growth and development of the agricultural area.

## RELATED WORK

The weather prediction can be used to enhance productivity in agricultural practices. Farmers today experience a loss for their product when they receive a higher yield from their crop but the market price for that commodity is lower. Farmers run the risk of both market price and production issues, particularly when cultivating new crops. The usage of machine learning technologies helps to solve these issues. A subset of data mining called predictive analysis forecasts probability and trends for the future. The agricultural system analyst still difficult to predict crop yield accurately using environmental, meteorological, and soil fertility parameters (Ramkar et al., 2018). The most often used factors in agricultural production prediction features are temperature, rainfall, and soil type (Sayao et al., 2020). The forecast will enable farmers to decide which crop is most suited for a given rainfall and crop price range (Zeferino et al.,2020). The goal of this strategy is to raise the crop's average yield percentage based on rainfall and temperature values.

(Shukla et al.,2019) discusses several biological remedies for treating or preventing early blight in the potato crop, such as seed dressing and bio formulation. (Chandrasekara et al., 2016) talks about the health benefits of starchy crops like common potatoes, sweet potatoes, and explains its nutritional and even some medicinal benefits. (Fones et al., 2020) reviews the parameters influencing the spread of pathogens and disease control techniques, quoting with a few historic examples like the Irish potato famine. (Gold et al., 2020) explains about various types of late blight diseases found in potato plants across the world and a few instantaneous disease-detection methods used in the industry.

(Osman et al.,2017) build a model that takes weather and climate into account before a certain crop is planted, evaluates them, and plots a correlation between them. To provide both quality and quantity information regarding the state of the crop, as well as the potential for early warnings, allowing for prompt intervention. (Kouadio et al.,2018) suggests that the ability of an Extreme Learning Machine (ELM) model to analyze the characteristics of soil fertility and produce a precise estimation of Robusta coffee output was evaluated. The effectiveness of 18 distinct ELM-based models was assessed using single and multiple combinations of the predictor variables based on the soil organic matter. (Zhong et al., 2018) predicting agricultural yield, a unique hierarchical machine learning approach is proposed. The decision models' estimates from the prediction model were accurate, with a median absolute error of 235 kg/ha.

(Girish et al., 2018) determine a crop's viability for a specific soil based on its basic resources. To employ a variety of predictive algorithms whose precision may be evaluated. The suggested approach will provide the best analysis utilizing a prediction system, and farmers can use it to predict crop diseases and increase crop yields in their agricultural. (Taherei Ghazvinei et al., 2018) propose the current work is to develop an integrated model that uses ELM to forecast the final growth rate of sugarcane. The author demonstrates improved generalization capacity and, the ELM's competence for additional work on improving the prediction model for sugarcane growth was authorized with encouraging outcomes

## WEATHER-BASED CROP PREDICTION ARCHITECTURE.

The objective is to give farmers knowledge about anticipated crop performance, assisting them in making decisions about planting and harvesting. The first step is to analyze the data with the help of MapReduce. To build a Python recommender system that provides the best results based on the local conditions and the season. Next, K means clustering will be used to determine the typical yield of a specific set of crops per unit area. This approach focuses its decisions on variables like temperature, rainfall, wind speed, humidity, soil type, and seed shown in Figure 1. The first stage involves collecting and preprocessing raw data in a Google Colab using Python. Then, using the previously treated data as input, the data is processed using Hadoop's MapReduce framework. to propose, prior to building a recommendation system, combining all the map-reduced datasets for the different parameters into a single, "super" dataset. A Python-created user interface that allows users to enter the necessary data and receive the results from the system displays the results from the recommendation system function. After that, a K-means clustering strategy will be used. To study the relationship between production per area, crop details, and other pertinent and crucial parameters, several 3D graphs and scatter plots would be created.

- Data Collection
- Data Pre-processing
- Hadoop Installation
- Map Reduce
- Recommendation system
- GUI
- Visualization, K-means clustering

*Figure 1. Step wise flow chart of proposed methodology*

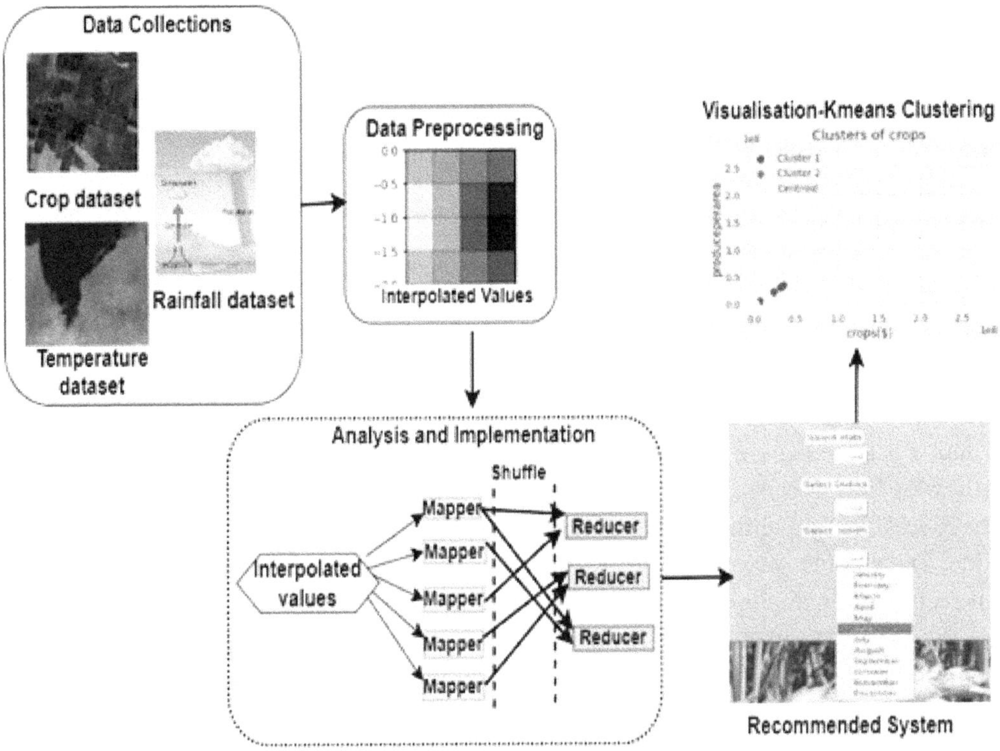


## Data Collection

The input data were gathered from dataset repositories like Kaggle. The Datasets are sourced from numerous websites. The crop dataset, rainfall dataset, and temperature datasets for Chennai, Kolkata, Mumbai, and Delhi are all included. The agricultural production dataset file has most of the states with their respective districts. Also, it includes details on one hundred and twenty-five different crops during the six seasons of Kharif, Autumn, the full year, Rabi, Summer, and Winter from the years 2000 to 2014, as well as information on each crop's production and the area where it was sown. The daily average temperatures of the cities from 1995 to 2020 are contained in the files Chennai.csv, Mumbai.csv, Delhi.csv, and Kolkata.csv. Indian rainfall from 1901 to 2015. The average monthly rainfall for various subdivisions nationwide is included in the dataset.

## Data Preprocessing

The gathered datasets were aggregated and cleansed in this instance. Our data sets are uploaded to the Colab notebook. Link to Colab notebook. Then the Pandas-data frame is utilized for the removal of unnecessary entries while keeping the only necessary ones which are crucial for us. For our calculations, we used the NumPy and SciPy libraries. The anticipated value for the missing item is determined via interpolation. Data frame name is the name of the dataset being used, and dataframename. interpolate(method='linear',direction='forward',inplace=True') does a linear interpolation in the forward direction. The outliers were then found and removed from the datasets using the IQR method.

IQR (Inter Quartile Range)

IQR=Q3-Q1 yields the interquartile range. where Q1 denotes the first quartile or 25th percentile. third quartile, or the seventy-five percentile We find the 25th and 75th percentiles, and since there are only three variances, we use 1.5 as the factor.

Q1=dataframename[columnname].quantile(0.25) Q3=dataframename[columnname].quantile(0.75)

IQR=Q3-Q1 output=dataframename×[~((dataframename[columnname](Q3+1.5∗IQR)))

## Map Reduce

The dataset is divided into key-value pairs using Map Reduce, which is implemented using pre-processed, cleansed data. First, the region, month, and year are used as keys and the temperatures are considered as values in the Map Function. Next, execute computation to find the regions' monthly average temperatures using the reduce function, utilizing the month, year, and region as keys and the average temperature as the value. The cumulative rainfall during the several agricultural seasons was calculated using Map Reduce in accordance with the dataset. Season, year, region, and crop will all be utilized as keys for the Mapping of the Crop Detailed dataset, while the crop's product and the region in which it is produced will be considered for values. The Reduce function will first calculate the value using the year, season, region, and crop as the keys, and the amount of produce produced per unit of area as the value. All the crops' produce, broken down by area, is included in the crop production dataset. Map reduce is a function that processes input data and emits intermediate key-value pairs.

Mapper Class:

step 1: import Apache Hadoop input and output function
step 2: Initialization:

set input variable =rainfall(ranges)
input_stream = sys.stdin

step 3: Iterate for each input data source

I. parse the extracted input values
II. Apply data preporssing and cleaing
III. Perform the custom mapping of newly generated intermediate key-value pairs.

fields = line.strip().split()

IV. emit intermediate key-value pairs

step 4: Create and run an instance of the Mapper class

mapper = Mapper()

mapper.map()

Reducer Class is responsible for processing the intermediate key-value pairs generated by the mapper class and producing the final output.

step 1: Initialization:

set output variable =rainfall(next_ranges)
output_stream = sys.stdout
Step 2: Reduce Function:
collect all associated intermediate key-value pairs
key, value = line.strip().split('\t', 1)
Apply custom reducing function and produce a final result
Emit the final key_values pair as output.
output = sum(map(int, values))

The Driver Class is responsible for setting up the rainfall, specifying various rainfall parameters, and controlling the overall execution flow.

Step 1: Create a rainfall instance

rainfall = new rainfall(conf)

Step 2: Set input and output format.

rainfall.setInputFormatClass(TextInputFormat.class)
rainfall.setOutputFormatClass(TextOutputFormat.class)

Step 3: Set input and output paths.

FileInputFormat.addInputPath(rainfall, new Path(input_path))
FileOutputFormat.setOutputPath(rainfall, new Path(output_path))

Step 4: Specify the mapper and reducer classes.

rainfall.setMapperClass(MapperClass.class)
rainfall.setReducerClass(ReducerClass.class)

Step 5: Submit the rainfall for execution.

rainfall.waitForCompletion(True)

## GUI (Graphical User Interface) and Recommendation System

The crop production and the rainfall map-reduced datasets are integrated. The manually gathered data, soil pH data, and soil humidity data are integrated to create the final actual dataset. After that we use

*Figure 2. Selecting user input month*

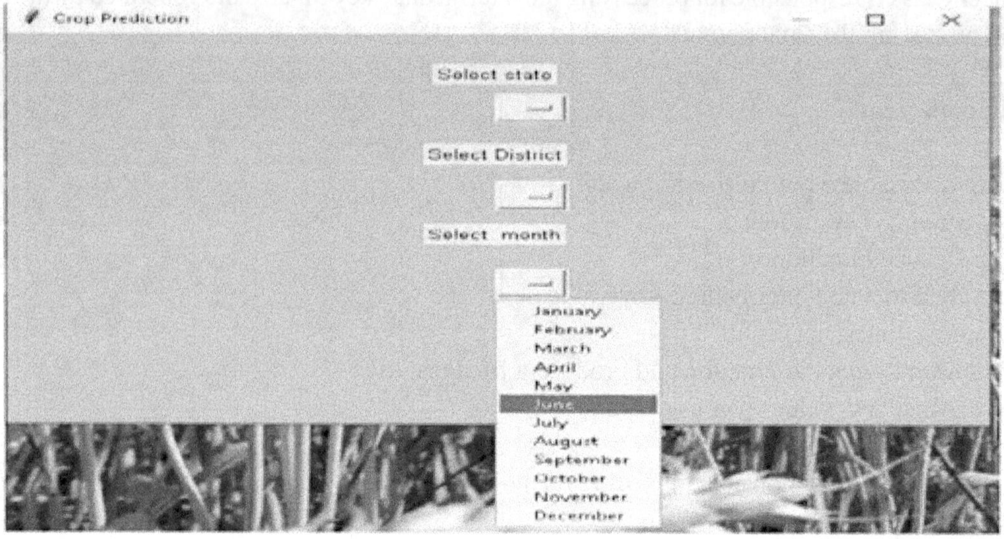


the users' inputs to obtain- state, district, month. Then agricultural seasons are then assigned based on the month that we used as the user's input. Then, depending on the agricultural season, we sort through the data to determine which crops in the district and state with the chosen input had the best yields. The eight best crops yielding the best produce yield for the entire course of year is extracted. crop had delivered exceptional results due to the soil's temperature, rainfall, humidity, and pH levels.

A minimalistic Graphical User Interface is designed in python using Tkinter module for taking inputs from the user and for displaying the results to the user. The user can select the month, state, district from the dropdown menu and the respective output will be displayed in Figure 2.

## EXPERIMENTAL SETUP

### Visualization

The resulting dataset, known as the "super dataset," is used to visualize several bar graphs plotted. Then the scatter plots are plotted for determining the link among the dataset's numerous attributes. Based on the amount of produce produced in a certain region, elbow graphs have been created to determine the clusters number which should be done for that place crops. Where the graph bends dictates how many clusters the dataset should contain. When clustering is used, it is denoted by the K-sign, and the crops plot their production per area close to centroids of clusters. To show the relationship between crops and the other characteristics, various bar graphs are also presented. Crops and their output per area are represented using scatter plots.

## Data Pre-Processing Outputs

Satellite and remote sensing technologies may be used in weather-based crop prediction systems to track the health of crops. Finding observations that significantly differ from much of the data is a critical step in the outlier detection phase of data analysis. It can be very important to identify data points in rainfall monitoring with unusually high or low rainfall levels. Detecting extreme weather conditions or significant temperature changes can help early warning systems and catastrophe management. Based on anticipated weather, this information aids farmers in creating planting, harvesting, and irrigation schedules. Farmers can modify their irrigation and fertilization operations by receiving information on the amount of soil moisture, nutrient shortages, and crop stress.

Analyzing temperature data that includes both minimum and maximum values is crucial in identifying various anomalies. By considering the entire range of temperatures, we can identify not only extreme high and low values but also patterns that might indicate other anomalies. These anomalies could include sudden drops or rises in temperature, fluctuations that are inconsistent with the expected range, and irregular cycles of temperature change Figure 3.

A crucial aspect of data analysis involves outliers. It is important to evaluate the impact of extreme values on the dataset before analysis. When identifying outliers, the temperature dataset instance is compared to the rest of the data to find the most distinctive cases that are isolated from most observed data. The results of removing temperature anomalies are presented in Figure 4.

A scatter plot is a graphical representation that uses a set of points to illustrate the relationship between two variables. This is an ideal way to analyze the correlation between the yearly rainfall deviation of the dataset. The scatter plot for the annual rainfall dataset is given in Figure 5.

The identification of outliers through visual analysis provides insights into the overall patterns and trends in the rainfall data. Figure 6 depicts the results of removing outliers from annual rainfall data.

Using temperature and rainfall data, crop yield was predicted annually. Figure 7 shows crop value deviation, which cannot forecast final evaluation.

Figure 8 shows the production data after the outliers have been removed, which looks like a line and provides insight into the crop cultivation patterns.

## Recommended System Results

Crop production has several facets and accomplishes a variety of goals, from ensuring fundamental needs are met to promoting economic development and environmental sustainability. It is crucial for satisfying the nutritional requirements of the expanding world population and tackling problems with resource management, food security, and climate change. To assist farmers in determining whether a particular crop is acceptable for a given rainfall and crop price range by using big data analysis. The sample results are shown in the Figure 9for the district of Darjeeling.

The elbow method can be used to locate the number cluster. The Darjeeling region dataset has two clusters, as shown in Figure 10.

Figure 11 shows the output of K-Means clustering between crop area and crops. The recalculated centroids, reflecting the mean of instances in each cluster, become the new centers of their respective clusters after all crop instances are added.

*Figure 3. Before removing the outliers for Chennai city*

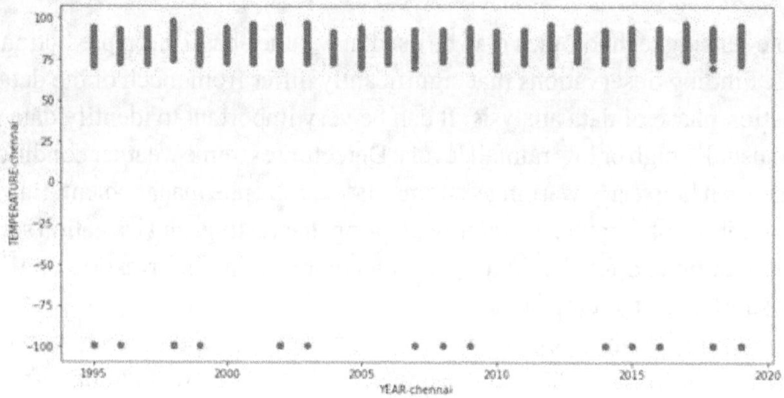


*Figure 4. After removing the outliers for Chennai city*

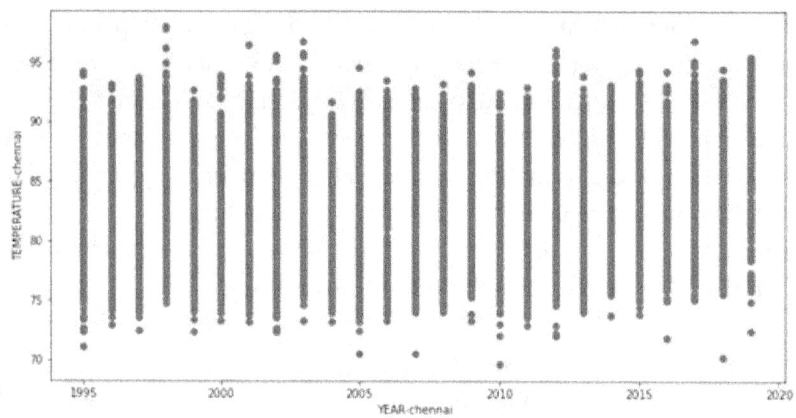


*Figure 5. Before removing the outliers for year vs. annual rainfall*

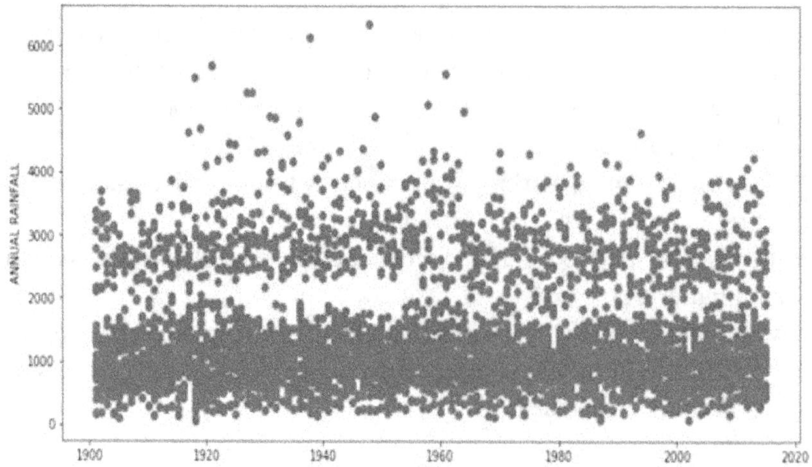

*Figure 6. After removing the outliers for year vs. annual rainfall*

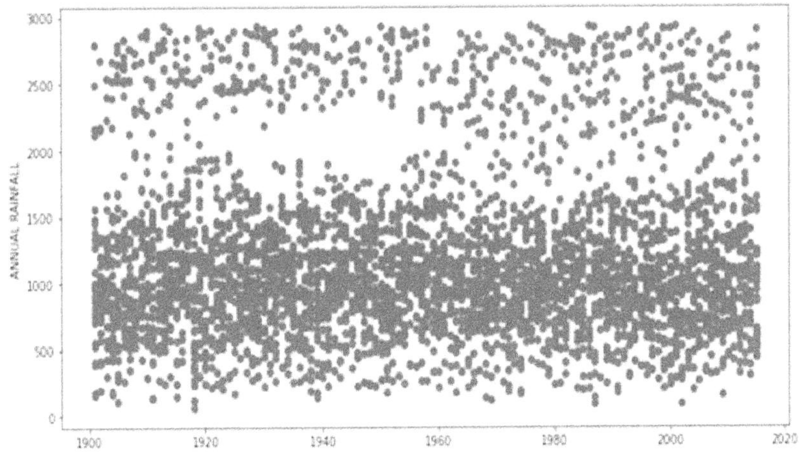


*Figure 7. Before removing the outliers for year vs. production*

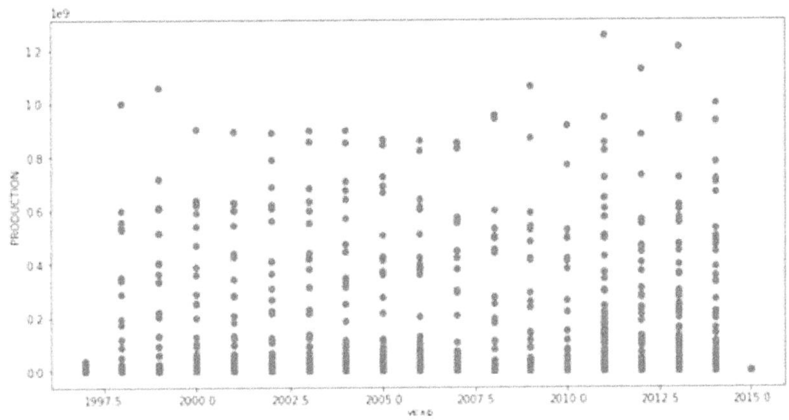


*Figure 8. After removing the outliers for year vs. production*

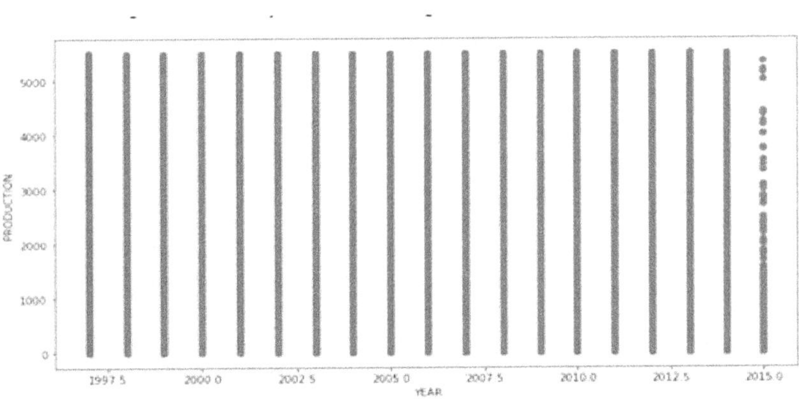

*Figure 9. Total crop production per area for Darjeeling region*

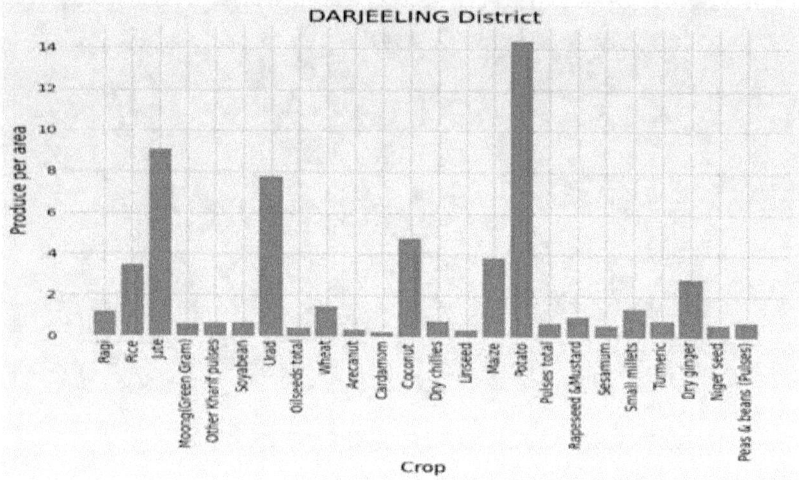


*Figure 10. Elbow graph for Darjeeling region*

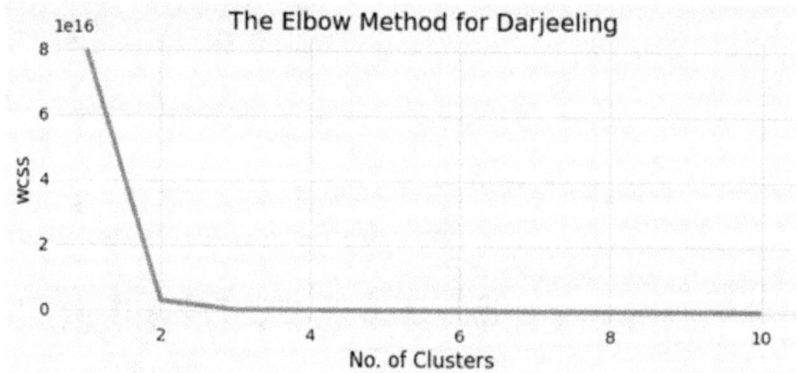


*Figure 11. Output of k-means clustering for Darjeeling region*

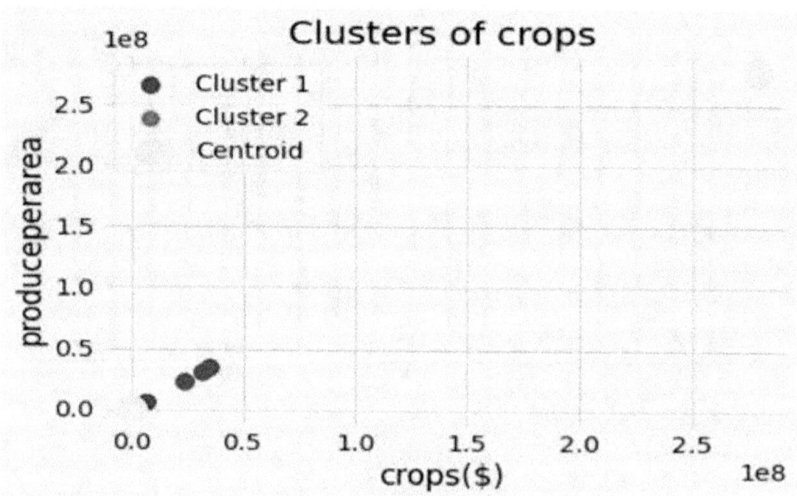

*Figure 12. Seasonal wise total production*

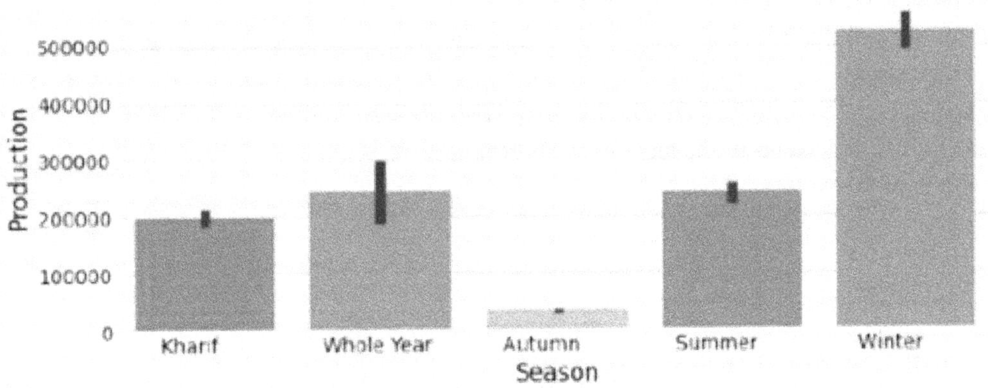


*Figure 13. Year wise total production*

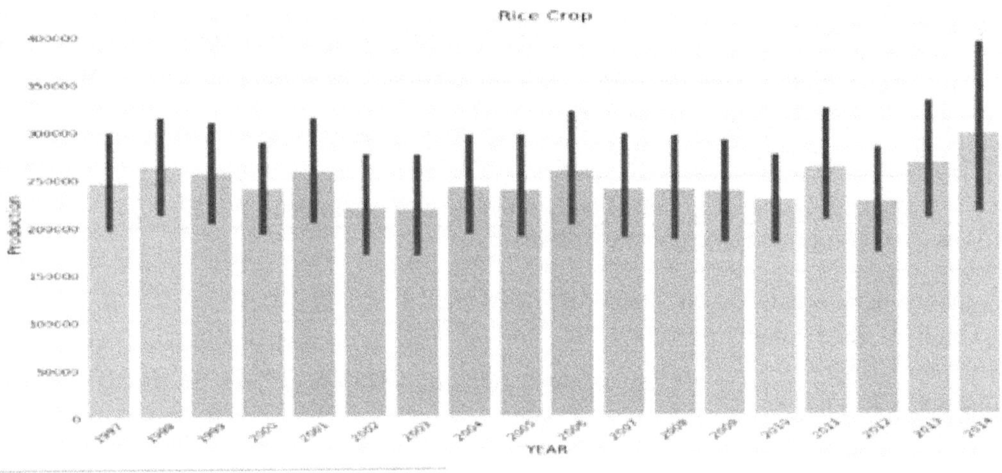


Climate and weather patterns can have a wide range of characteristics, which can influence changes in crop productivity. Therefore, it is necessary to learn about the seasons, which can differ depending on location. Figure 12 shows the total production of the Darjeeling region, categorized by season.

Figure 13 shows the total production of the Darjeeling region, categorized by season.

## CONCLUSION

To find patterns and trends, big data analytics can be utilized to evaluate huge quantities of weather and crop data. In this work, an effective weather-based crop prediction system is developed to anticipate crops based on weather forecasting and big data analytics. Visualizations, graphs, and K-Means clustering are used to calculate the average agricultural yield per area for crops. The suggested study offers a technique for recommending crops that uses MapReduce and K-means clustering to produce results that are computationally efficient. The recommender system can also be developed to provide early warning of diseases that might harm the crops in a specific season and to suggest fertilizers and other nutrients connected to the soil that are required to provide the best output yield with high production. To acquire a more precise result, the work can be further enhanced by integrating additional elements like humidity, wind speed, etc. The work can then be improved by creating a publicly accessible online application that will be highly beneficial for farmers in predicting the crops in a certain place and month to acquire the highest yield of crops.

## REFERENCES

Bose, D. (2017). Big data analytics in Agriculture. *preparation of papers for IEEE Transactions and Journals*. IEEE.

Chandrasekara, A., & Josheph Kumar, T. (2016). Roots and tuber crops as functional foods: A review on phytochemical constituents and their potential health benefits. *International Journal of Food Sciences, 2016*, 2016. doi:10.1155/2016/3631647 PMID:27127779

Fan, W., Chong, C., Xiaoling, G., Hua, Y., & Juyun, W. (2015, December). Prediction of crop yield using big data. In *2015 8th International Symposium on Computational Intelligence and Design (ISCID)* (Vol. 1, pp. 255-260). IEEE. 10.1109/ISCID.2015.191

Fones, H. N., Bebber, D. P., Chaloner, T. M., Kay, W. T., Steinberg, G., & Gurr, S. J. (2020). Threats to global food security from emerging fungal and oomycete crop pathogens. *Nature Food, 1*(6), 332–342. doi:10.103843016-020-0075-0 PMID:37128085

Funk, C., Peterson, P., Landsfeld, M., Pedreros, D., Verdin, J., Shukla, S., Husak, G., Rowland, J., Harrison, L., Hoell, A., & Michaelsen, J. (2015). The climate hazards infrared precipitation with stations—A new environmental record for monitoring extremes. *Scientific Data, 2*(1), 1–21. doi:10.1038data.2015.66 PMID:26646728

Gavahi, K., Abbaszadeh, P., & Moradkhani, H. (2021). DeepYield: A combined convolutional neural network with long short-term memory for crop yield forecasting. *Expert Systems with Applications, 184*, 115511. doi:10.1016/j.eswa.2021.115511

Girish, L., Gangadhar, S., Bharath, T. R., Balaji, K. S., & Abhishek, K. T. (2018). Crop yield and rainfall prediction in Tumakuru district using machine learning. *International Journal for Research in Engineering Application and Management (IJREAM)*, (pp. 61-65). IEEE.

Gold, K. M., Townsend, P. A., Herrmann, I., & Gevens, A. J. (2020). Investigating potato late blight physiological differences across potato cultivars with spectroscopy and machine learning. *Plant Science, 295*, 110316. doi:10.1016/j.plantsci.2019.110316 PMID:32534618

Kafy, A. A., Bakshi, A., Saha, M., Al Faisal, A., Almulhim, A. I., Rahaman, Z. A., & Mohammad, P. (2023). Assessment and prediction of index based agricultural drought vulnerability using machine learning algorithms. *The Science of the Total Environment, 867*, 161394. doi:10.1016/j.scitotenv.2023.161394 PMID:36634773

Kouadio, L., Deo, R. C., Byrareddy, V., Adamowski, J. F., Mushtaq, S., & Phuong Nguyen, V. (2018). Artificial intelligence approach for the prediction of Robusta coffee yield using soil fertility properties. *Computers and Electronics in Agriculture*, *155*, 324–338. doi:10.1016/j.compag.2018.10.014

Kushwaha, A. K., & Bhattacharya, S. (2015). Crop yield prediction using Argo Algorithm in Hadoop. [IJCSITS]. *International Journal of Computer Science and Information Technology & Security*, *5*(2), 271–274.

Mila, F. A., & Parvin, M. (2019). Forecasting area, production and yield of onion in Bangladesh by using ARIMA model. *Asian Journal of Agricultural Extension. Economia e Sociologia*, *37*(4), 1–12.

Priya, R., Ramesh, D., & Khosla, E. (2018, September). Crop prediction on the region belts of India: A Naïve Bayes MapReduce precision agricultural model. In 2018 international conference on advances in computing, communications, and informatics (ICACCI) (pp. 99-104). IEEE. doi:10.1109/ICACCI.2018.8554948

Ramya, M. G. P. G., Balaji, C., & Girish, L. (2015). Environment change prediction to adapt climate-smart agriculture using big data analytics. [IJARCET]. *International Journal of Advanced Research in Computer Engineering and Technology*, *4*(5).

Reddy, P. C., & Babu, A. S. (2017, February). Survey on weather prediction using big data analytics. In *2017 Second international conference on electrical, computer and communication technologies (ICECCT)* (pp. 1-6). IEEE.

Sayão, V. M., dos Santos, N. V., de Sousa Mendes, W., Marques, K. P., Safanelli, J. L., Poppiel, R. R., & Demattê, J. A. (2020). Land use/land cover changes and bare soil surface temperature monitoring in southeast Brazil. *Geoderma Regional*, *22*, e00313. doi:10.1016/j.geodrs.2020.e00313

Shen, R., Huang, A., Li, B., & Guo, J. (2019). Construction of a drought monitoring model using deep learning based on multi-source remote sensing data. *International Journal of Applied Earth Observation and Geoinformation*, *79*, 48–57. doi:10.1016/j.jag.2019.03.006

Shukla, A., & Ratan, V. (2019). Management of Early Blight of Potato by Using Different Bioagents as Tuber Dressing and its Effect on Germination and Growth. *International Journal of Current Microbiology and Applied Sciences*, *8*(06), 2019. http://www. ijcmas. com. doi:10.20546/ijcmas.2019.806.233

Taherei Ghazvinei, P., Hassanpour Darvishi, H., Mosavi, A., Yusof, K. B. W., Alizamir, M., Shamshirband, S., & Chau, K. W. (2018). Sugarcane growth prediction based on meteorological parameters using extreme learning machine and artificial neural network. *Engineering Applications of Computational Fluid Mechanics*, *12*(1), 738–749. doi:10.1080/19942060.2018.1526119

Vijaya Baskar, P. S., Sreemathi, R., & Keertanaa, E. (2017, March). Crop prediction using predictive analytics. In *2017 International Conference on Computation of Power, Energy Information and Communication (ICCPEIC)* (pp. 370-373). IEEE. 10.1109/ICCPEIC.2017.8290395

Zeferino, L. B., de Souza, L. F. T., do Amaral, C. H., Fernandes Filho, E. I., & de Oliveira, T. S. (2020). Does environmental data increase the accuracy of land use and land cover classification? *International Journal of Applied Earth Observation and Geoinformation*, *91*, 102128. doi:10.1016/j.jag.2020.102128

Zhong, H., Li, X., Lobell, D., Ermon, S., & Brandeau, M. L. (2018). Hierarchical modelling of seed variety yields and decision making for future planting plans. *Environment Systems & Decisions*, *38*(4), 458–470. doi:10.100710669-018-9695-4

## KEY TERMS AND DEFINITIONS

**Correlation Analysis:** Correlation analysis is a statistical approach used in feature selection to assess the strength of each feature's linear connection with the target variable.

**Crop Yield Prediction:** It is critical in national, regional, and local decision-making since it is dependent on soil, climatic, environmental, and agricultural factors.

**Elbow method:** The elbow method is a strategy for determining the number of clusters (k) to employ in a k-means clustering algorithm.

**Hadoop:** Hadoop is a free and open-source software framework that allows for the distributed processing and storage of huge datasets across computer clusters by utilising basic programming methods.

**KMean Clustering:** K-Means clustering is an unsupervised machine learning approach that uses characteristics to group comparable data points together.

**MapReduce algorithm:** MapReduce is a programming concept and related implementation for processing and producing large data sets on a cluster using a parallel, distributed algorithm.

**Recommendation system:** A recommendation system is an algorithm that suggests crops that are most appropriate to a certain person.

# Chapter 5
# Crop Improvement Strategies and Principles of Selective Breeding

**Surjit Singha**
https://orcid.org/0000-0002-5730-8677
*Kristu Jayanti College (Autonomous), India*

**Ranjit Singha**
https://orcid.org/0000-0002-3541-8752
*Christ University, India*

## ABSTRACT

*Introducing new crop varieties, crop improvement, and selective breeding enable contemporary agriculture to alleviate food poverty, increase crop production, and enhance product quality. This chapter covers selective breeding, genetically modified crops, genetic resource conservation, and upcoming technologies. Selective breeding discusses heritability, gene frequency, dominance, and epistasis to predict genetic progress. Agricultural innovation involves genetic variety and variation to develop robust and high-performing crop types. Mass selection, pedigree selection, and recurrent selection for trait enhancement are investigated. Marker-assisted and genomic selection are tested for breeding speed and accuracy. The chapter covers genetically modified crops, agricultural improvement, and genetic resource conservation for sustainable agriculture. Gene editing, synthetic biology, and machine learning may boost agricultural yields. According to the abstract, crop development and selective breeding are essential for food security, sustainability, climate change, and agricultural demands.*

## INTRODUCTION

Crop improvement is pivotal in ensuring food security, advancing agricultural sustainability, and fostering robust crop varieties. Various strategies aim to tackle the challenges posed by climate change and meet the growing global food demand by enhancing key traits such as yield, disease resistance, environmental

DOI: 10.4018/978-1-6684-9838-5.ch005

adaptability, and nutrient content. Among these strategies, selective breeding is a fundamental approach for improving agriculture. It involves intentionally selecting and propagating desirable traits within plants, ultimately yielding improved varieties that cater to the needs of both growers and consumers. Crop enhancement encompasses a process dedicated to refining and bolstering the genetic makeup of crops to elevate their productivity, quality, and resilience against biotic and abiotic stressors. In addition to traditional breeding methods, modern techniques such as marker-assisted selection and genetic engineering have emerged as valuable tools in the realm of crop improvement, broadening the scope of possibilities (Zakari et al., 2022; Maharjan & Maharjan, 2018; Adenle et al., 2015). The significance of crop improvement must be considered, especially given the burgeoning global population and the challenges posed by climate change. Furthermore, these enhancement techniques have substantially contributed to producing high-quality agricultural products, increasing farmers' profitability and offering consumers improved nutritional choices (Yang et al., 2021; Kumar et al., 2020).

This chapter probes into the core principles of selective breeding and explores a wide range of crop enhancement strategies, encompassing both traditional and modern approaches within agriculture. The examination includes an in-depth look at the genetic underpinnings of selective breeding, shedding light on the importance of gene diversity and genetic variation in the context of agricultural enhancement. It draws comparisons between traditional breeding techniques, such as mass selection, pedigree selection, and recurrent selection, and their contemporary counterparts, such as marker-assisted and genomic selection. Additionally, we delve into genetically modified crops and their contributions to advancing agriculture, conserving genetic resources, and introducing innovative crop improvement technologies. It contemplates the future of crop enhancement and its pivotal role in securing global food security and promoting sustainability in agriculture.

Crop improvement dates back thousands of years to the earliest days of agriculture when primitive societies began cultivating vegetation for food. Unknowingly, these early cultivators set the framework for future crop enhancement efforts by observing and selecting seeds from plants with desirable characteristics. As societies shifted from hunter-gatherer to agricultural lifestyles, they began consciously propagating plants with desired traits, resulting in the domestication of essential cereal species such as wheat, barley, rice, maize, and legumes (Waqas et al., 2019; Kumar et al., 2020). As plant breeders undertook controlled hybridization experiments during the 18th and 19th centuries, crop improvement made significant progress. This method enabled the combination of desirable characteristics from various parent plants, resulting in offspring with enhanced qualities. Notably, hybridization was instrumental in developing high-yielding maize varieties (Cai et al., 2022; Wu et al., 2021). Late in the 19th century, Gregor Mendel's seminal work on inheritance laws laid the scientific groundwork for understanding how characteristics are transmitted from one generation to the next. Together with the comprehension of quantitative trait inheritance and the function of multiple genes in complex traits, Mendel's discoveries paved the way for more sophisticated breeding strategies (Xu et al., 2022; Godwin et al., 2019). With the introduction of the Green Revolution in the middle of the 20th century, crop improvement reached a significant milestone. This period was marked by the widespread adoption of crop varieties with high yields, synthetic fertilizers, and pesticides. Breeders concentrated on developing semi-dwarf wheat and rice varieties with enhanced fertilizer response and disease resistance. (Baweja et al., 2020; Nelson et al., 2019) The Green Revolution substantially increased global food production, mitigating famine and sustaining population growth. The introduction of biotechnology and genetic engineering in the late 20th century ushered in a new era of crop improvement. Scientists can now transfer DNA from one organism to another, allowing the introduction of specific traits into crops. Genetic engineering techniques resulted in the development of genetically modified (GM) crops with characteristics such as insect resistance, herbicide tolerance, and increased nutritional value. Numerous nations have embraced genetically modified crops, increasing productivity and decreasing pesticide use (Bhattacharya et al., 2021; Klümmer & Qaim, 2014). Recent decades have witnessed extraordinary advances in genomics and molecular biology, accelerating efforts to enhance crop yields. Thanks to advances in genomic data availability, DNA sequencing technologies, and molecular markers, breeders can now pinpoint and select specific genes or genomic regions associated with desired traits. These innovations have facilitated marker-assisted selection (MAS) and genomic selection (GS), enabling the replication of complex characteristics with greater precision and efficiency (Yang et al., 2021; Wishart et al., 2020). From the initial stages of observation and natural selection to the modern era of genomics and biotechnology, the evolution of crop improvement marks an unbroken journey of progress. These methodologies have made substantial contributions to both global food security and the promotion of sustainable agriculture. Selective breeding, a cornerstone of crop enhancement, involves identifying plant genetic diversity and selecting and propagating favourable traits. This process aims to create crop varieties that align with the ever-evolving needs of agriculture.

## GENETIC PRINCIPLES OF SELECTIVE BREEDING

Heritability serves as a metric that gauges the degree to which genetic factors shape the presentation of a particular trait. It quantitatively assesses the percentage of variability in features across a population that can be attributed to individual genetic distinctions. When heritability is low, environmental factors play a more substantial role in influencing how a trait manifests itself. In selective breeding, estimating heritability holds great significance as it aids in forecasting how features will respond to selective pressures. This knowledge gives breeders insights into the potential for genetic enhancements (Rasheed et al., 2023; Robette et al., 2022).

In contrast, gene frequency refers to the proportion of a particular allele within a population. It reveals the prevalence and distribution of a specific gene variant within a gene pool. Gene frequencies can vary between generations due to natural selection, genetic drift, and migration, among other factors. Understanding these variations in gene frequencies is crucial for selective breeding because it permits breeders to monitor and manipulate the presence of desired alleles in populations (Axelsson et al., 2021; Ergon et al., 2019). The term "dominance" in genetics refers to the interaction between alleles at a specific gene locus. In instances of dominance, the dominant allele in a heterozygous individual conceals the effect of the recessive allele. This masking results in the dominant trait being expressed. The distinction is essential in steering breeding strategies within crop enhancement programmes, where breeders may seek to enhance or suppress specific characteristics based on the dominance relationships of the underlying genes (Nguyen et al., 2021; Hufnagel et al., 2020).

The interaction between multiple gene loci is called "epistasis," a crucial genetic concept. In epistasis, the effect of alleles at one locus depends on the alleles present at another locus. These interactions can either promote or suppress the expression of a specific trait. Understanding epistasis is essential to selective breeding because it enables breeders to account for intricate genetic interactions. Breeders can develop more effective breeding strategies that simultaneously target multiple gene loci to accomplish their desired outcomes by contemplating epistatic relationships (Zhang et al., 2023; Johnson et al., 2023).

Crop enhancement is based on genetic diversity, encompassing various genetic traits and variability within populations or species. It is essential for enabling plants to adapt to changing environmental conditions, resist diseases and parasites, and manifest desirable characteristics. Crop breeders and agricultural scientists face significant challenges in light of Karmaoui et al.'s (2019) findings that predict water scarcity in the Middle Draa Valley due to climate change-induced factors such as increased temperatures, decreased precipitation, and increased water demand. Developing drought-resistant and adaptable crop varieties must take precedence to combat the impending water shortage. This objective can be achieved by selecting drought tolerance, efficient water use, and increased yield in environments with limited water. Utilizing the principles of selective breeding, which involve carefully selecting and crossing plants with desirable characteristics, can develop crop varieties that are optimally adapted to water-limited ecosystems. Incorporating genetic diversity associated with water-related traits from wild crop relatives and other germplasm sources is essential for enhancing crop water use efficiency and resiliency. Breeders can contribute significantly to sustainable agricultural production in the face of water scarcity caused by climate change if they take these steps (Azra et al., 2022; Quezada-Martinez et al., 202). Utilizing genetic diversity, crop improvement programmes cultivate superior cultivars. Breeders accomplish this by selecting and crossing individuals with desirable characteristics, ultimately combining advantageous alleles and producing progeny with enhanced traits. In addition to protecting against the depletion of genetic resources, genetic diversity reduces vulnerability to biotic and abiotic stresses. In a review by Hirel et al. (2007), the significance of integrated research programmes emphasizing genetic, physiological, and agronomic aspects to improve nitrogen use efficiency (NUE) in main crop species is highlighted. Investigating the regulatory mechanisms governing nitrogen uptake, assimilation, and recycling is necessary to understand how plants utilize nitrogen. This information is crucial for reducing excessive fertilizer use while maintaining adequate crop yields. To understand NUE thoroughly, it is crucial to conduct comprehensive studies encompassing whole-plant physiology and the analysis of genes, proteins, and metabolites. In addition, developing crop simulation models and diagnostic tools based on physiological or molecular markers can facilitate sustainable fertilizer management practices. Considering economic and environmental constraints, a comprehensive strategy is required to improve NUE and develop crop varieties suited to reduced fertilizer inputs. Breeders can select individuals with desired traits and incorporate them into breeding programmes, increasing crop productivity and sustainability (Ye et al., 2022; Masclaux-Daubresse et al., 2010).

Quantitative genetics investigates the transmission and analysis of complex characteristics with continuous variation. In contrast to traits governed by a single gene, quantitative traits result from the interaction between multiple genes and environmental factors. These characteristics include yield, height, disease resistance, and quality. Selective breeding for quantitative traits relies on statistical methods to estimate genetic parameters, such as heritability and genetic correlations, for estimating heritability and genetic correlations. These parameters aid breeders in assessing the reproductive potential of individuals and choosing the most compatible progenitors. In addition, techniques such as recurrent selection, population enhancement, and hybrid breeding are used to improve quantitative attributes in plants (Cockerton et al., 2021; Worku et al., 2021). Quantitative genetics uses marker-assisted selection (MAS) and genomic selection (GS). MAS utilizes molecular markers associated with desired characteristics to select individuals with those markers.

In contrast, GS uses genomic information to predict the reproductive value of individuals, thereby eliminating the need for extensive phenotypic data. These instruments allow for a more efficient and accurate selection of individuals with desired traits, thereby accelerating crop enhancement (Budhlakoti et al., 2022; Boopathi, 2020;).A comprehensive understanding of the genetic principles underlying selective breeding, the function of gene diversity, and the applications of quantitative genetics provides a solid foundation for crop improvement strategies. Breeders can skillfully manipulate genetic factors, amplify desirable traits, and cultivate enhanced crop varieties to address the challenges of food security, sustainability, and resilience in the context of shifting agricultural demands and climate change (Calicioglu et al., 2019). In their review, Grattapaglia et al. (2018) argue that incorporating genomics into tree genetics has transformative potential. Lengthy reproduction cycles, complex trait inheritance, and emergent environmental factors hinder conventional forest tree breeding techniques.

Nonetheless, the intersection of high-throughput genomics and quantitative genetics has ushered in two innovative paradigms: genomic selection (GS) and the practical use of genomic relationship matrices. GS harnesses genome-wide markers to deduce intricate phenotypes, enabling accelerated breeding cycles, heightened selection precision, and enhanced breeding value estimation. Simultaneously, tracking pedigree variation and genomic relationship matrices opens fresh avenues for assessing genetic parameters and refining breeding strategies. The cost-effectiveness of genotyping further promotes the widespread adoption of GS in tree reproduction.

Future research endeavours in GS should prioritize enhancing prediction models, integrating functional genomics data to improve precision, and incorporating multi-environmental data to forecast the performance of genetic materials under scenarios influenced by climate change. As we witness the accumulation of phenotypic and genome-wide data, coupled with advancements in computational prediction, new opportunities emerge for advancing genomics in tree breeding. This discovery is consistent with the fundamentals of quantitative genetics and the function of selective breeding. As discussed in the review, incorporating genomics data into tree breeding is consistent with quantitative genetics principles. Quantitative genetics focuses on the inheritance and analysis of complex, continuously variable traits, making it an ideal framework for tree breeding characterized by complex features and inheritance patterns. Genomic selection (GS) and realized genomic relationship matrices are prominent examples of applying quantitative genetics in tree breeding. GS is a quantitative genetic method for estimating reproductive values that utilize genome-wide markers to predict complex phenotypes. In the meantime, realized genomic relationship matrices represent a significant advance in genetic parameter estimation, a crucial aspect of quantitative genetics (Serpico, 2020). Incorporating genomics data, a formidable quantitative genetics instrument, into tree breeding can increase the precision and efficacy of selective breeding programmes.

## TRADITIONAL BREEDING STRATEGIES

Mass selection represents one of the earliest reproductive techniques, concentrating on selecting individuals solely based on their observable phenotypic traits without considering their lineage. This strategy entails developing a population of plants or animals and identifying the most preferable individuals based on visual characteristics or performance metrics. These selected individuals serve as the mating stock for the subsequent generation. Mass selection applies to self-pollinating plants and species with high levels of homozygosity, making it a relatively simple technique. In contrast, pedigree selection involves meticulously tracing the ancestry of individuals and selecting superior candidates based on their documented lineage and historical performance records. This method assumes that pertinent characteristics are inherited and can be transmitted to progeny. In their review, advantageous features are from generation to generation. This technique is frequently used in open-pollinated crop and livestock reproduction (Sabadin et al., 2022).

Recurrent selection, on the other hand, seeks to improve populations by targeting multiple characteristics over multiple generations. This technique includes three essential phases: selection, interbreeding, and reselection. Initially, appealing individuals are selected from a population. These individuals are then interbred to introduce genetic diversity and produce individuals with diverse genetic backgrounds. To identify individuals with enhanced trait performance, subsequent selection cycles are utilized. Breeders follow a series of steps throughout the reproductive cycle to develop improved varieties. It typically entails defining the program's goals, such as increasing yield, disease resistance, or qualitative characteristics, and evaluating available genetic resources to identify potential progenitors with desirable traits. Controlled cross-pollination between carefully selected parent plants enables the combination of desired features and the generation of genetic diversity in progeny. Based on the evaluation of their offspring's desired characteristics, parents for the next generation are chosen based on the exceptional trait performance of individuals. These selected individuals are subjected to additional testing in field trials or controlled environments to assess their efficacy and resilience under varying conditions. They are then propagated via spores or vegetative means to generate sufficient improved planting material. Finally, the improved cultivars are made available to farmers or cultivators, and their adoption and efficacy under real-world conditions are closely monitored. Evaluation methods may be based on phenotypic attributes, genetic markers, or a combination. This comprehensive approach requires careful consideration of the reproductive objectives associated with the desired characteristics, the selection criteria, and the genetic basis underlying the reproduction process (Singh et al., 2021).

Traditional breeding is a versatile and efficient method that permits the fusion of diverse genetic materials, creating populations or varieties with a broad genetic base. This method enables breeders to simultaneously target and select multiple desirable traits, thereby enhancing complex features and facilitating the development of well-rounded crop varieties. The adaptability of traditional breeding is a remarkable characteristic, as it can be tailored to meet the specific requirements of various crop species and diverse breeding objectives, making it a universally applicable technique across all agricultural systems. Researchers and farmers can cultivate crops that address current farming challenges and pave the way for a more sustainable and resilient future in agriculture by leveraging the power of traditional breeding (Crossa et al., 2021). Conventional breeding methods, particularly for perennial commodities, frequently entail multiple breeding cycles and lengthy periods. These methods rely mainly on phenotypic characteristics for selection, which are susceptible to environmental influences and may not completely capture the underlying genetic variations. In addition, traditional breeding needs more precision to selectively target specific genes or alleles, which can be problematic for traits governed by complex genetic mechanisms. Continuous selection within a limited reservoir of genetic resources can also reduce genetic diversity, making crops more susceptible to pathogens, parasites, and environmental stresses. Traditional breeding has made significant contributions to crop improvement over the centuries. However, its limitations become apparent when multiple genes control traits or when precise control of specific genetic variations is required. To address these limitations, modern breeding techniques, such as molecular breeding and genetic engineering, have been developed to complement traditional methods and accelerate the breeding process by providing more precise control and the capacity to address complex genetic challenges (Acquaah, 2015).

## MODERN BREEDING STRATEGIES

Marker-assisted selection (MAS) is a reproductive method that identifies genes or attributes of interest associated with molecular markers. These markers function as genetic signposts, enabling breeders to select individuals based on their marker profiles instead of relying solely on observable traits. MAS accelerates the process of reproduction by facilitating the early and accurate selection of individuals carrying the desired genes or alleles (Boopathi, 2020). Genomic Selection (GS) utilizes high-throughput genotyping and statistical models to predict an individual's reproductive value based on genomic information. GS necessitates the development of prediction models that establish links between genetic markers and phenotypic performance. Without the need for conventional phenotypic evaluations, these models are then used to estimate the genetic potential of individuals. GS has the potential to substantially improve reproductive efficacy, especially for characteristics with complex genetic bases.

The three major phases of the MAS procedure are marker discovery, marker validation, and marker-assisted reproduction. Techniques such as DNA sequencing, genotyping arrays, and next-generation sequencing are utilized for marker discovery. After identifying markers, they are validated across diverse genetic backgrounds and populations. Individuals are selected based on their marker profiles during marker-assisted breeding to increase the frequency of desired genes or alleles in the reproductive population (Montesinos-López et al., 2012).On the other hand, Genomic Selection consists of four primary phases: genotyping, phenotyping, model training, and prediction. Numerous genetic markers dispersed throughout the genome are captured using high-density genotyping. Collecting phenotypic data from training populations enables the development of statistical models that establish the associations between genetic characteristics and phenotypic traits (Sanchéz & Akdemir, 2021). The trained models are then used to predict the reproductive value of individuals solely based on their genotypic information. Genomic selection can be applied to many traits and species, including those with complex genetic architectures.

### Case Study 1: Drought-Tolerant Maize Varieties

In regions of India plagued by escalating water scarcity and erratic rainfall patterns, the spectre of drought loomed large, casting a shadow over agricultural productivity. This case study illuminates the development and widespread adoption of drought-tolerant maize varieties, achieved through meticulous selective breeding efforts.

The backdrop was a region grappling with increasing water scarcity and unpredictable rainfall patterns, precipitating substantial crop losses due to recurrent drought. In response to this dire predicament, a cohort of agricultural scientists embarked on a relentless breeding program designed to engineer maize varieties fortified with heightened drought resistance.

The breeding program unfolded systematically, driven by the judicious selection of maize plants demonstrating superior drought-resistant attributes. The initial phase involved the ethical evaluation of a diverse array of maize germplasm, focusing on their innate drought resistance. Those individuals bearing coveted traits such as robust root systems, efficient water utilization, and resilience against withering were singled out for further consideration.

The chosen maize specimens were then subjected to rigorous trials under simulated drought-like stress conditions. Following painstaking scrutiny, plants that exhibited exceptional performance in the face of drought-induced duress were earmarked to proceed to the subsequent reproductive cycle.

Integrating advanced techniques, including the adept deployment of molecular markers, played a pivotal role in expediting the selection of desirable traits linked to drought tolerance. This technological integration ushered in a phase of more precise and productive decision-making in reproduction.

After undergoing multiple iterations of the breeding cycle, an impressive array of maize varieties endowed with drought resistance emerged, including but not limited to Kaveri K 40, Dhanshakti, Prakash, Vivek Hybrid 9, and Ganga 5. These cultivars showcased heightened resistance to water deficiency, maintained elevated yields even in drought-ridden conditions, and demonstrated reduced vulnerability to withering.

The newly developed maize varieties, fortified against drought, were disseminated widely amongst farmers inhabiting drought-prone regions. Extension services and comprehensive training programs were diligently implemented to ensure optimal utilization, educating farmers on the manifold advantages of these varieties and the appropriate cultivation methodologies.

Farmers reaped tangible benefits, experiencing a significant reduction in yield losses during drought episodes, thereby bolstering food security and securing a stable income.

The augmented water-use efficiency inherent to the drought-tolerant varieties ushered in optimized water management practices, lessening the reliance on irrigation.

Farmers adapted to the shifting climate dynamics by cultivating drought-tolerant maize varieties, effectively mitigating the detrimental impact of drought on agricultural output.

Economic gains accrued to farmers via the higher yields and enhanced marketability of the drought-tolerant maize varieties, uplifting their living standards and fostering rural development.

The successful development of drought-tolerant maize varieties through the intricate mechanisms of selective breeding is a compelling testament to the potency of innovative breeding principles and crop enhancement strategies in enhancing agriculture's resilience to climate change-induced challenges.

## Case Study 2: Salt-Tolerant Rice Varieties

The pervasive issue of salinity, resulting from elevated salt levels in both soil and water, has presented a formidable obstacle to rice cultivation in the coastal regions of India. This case study explores the development and widespread adoption of salt-tolerant rice varieties.

The escalating salinization of agricultural lands in coastal regions brought about an alarming decline in rice yields. In response to this pressing problem, scientists launched an ambitious breeding programme to develop rice strains with enhanced salinity tolerance. The breeding initiative centred on identifying and curating salt-tolerant rice germplasm, such as CSR 10, Pokkali, Nonabokra, Luna Sankh, and FL478. Various landraces and species of wild rice were evaluated comprehensively, with an emphasis on their ability to flourish in the presence of elevated salinity levels. Exceptional individuals demonstrating promise were evaluated rigorously through controlled salt stress experiments. Those rice plants with reduced growth inhibition, efficient salt exclusion mechanisms, and the ability to maintain productivity despite salinity stress were designated for the subsequent phases of the breeding cycle. Advanced techniques, such as marker-assisted selection, were utilized dexterously to expedite the identification and transfer of salt-tolerant genes into elite rice varieties.

Multiple generations of the breeding programme produced various salt-tolerant rice varieties. When cultivated under saline conditions, these varieties displayed extraordinary growth enhancements, increased yield stability, and reduced yield losses. Salt-tolerant rice varieties were disseminated to farmers living in salt-affected littoral regions to facilitate their adoption and appropriate cultivation techniques.

This dissemination was supplemented by extension services, farmer training initiatives, and on-field demonstrations intended to promote the optimal use of these varieties.

Modern breeding techniques revolutionize the selection of individuals by concentrating on their genetic makeup, allowing for the precise targeting of genes or alleles linked to desired traits. It improves reproduction efficiency and significantly reduces the time required to cultivate improved varieties. Marker-assisted selection is crucial in incorporating specific target genes or alleles from wild or exotic germplasm sources into elite breeding lines or varieties. It increases genetic diversity, thereby enhancing the adaptability of crops to dynamic environmental changes and evolving threats posed by pests and diseases (Chloupek & Hrstková, 2005). Genomic selection provides more accurate estimates of reproductive values, especially for traits influenced by many genes. It navigates obstacles such as low heritability, genotype-by-environment interactions, and phenotypic evaluation complexities with skill (Bhat et al., 2016). Modern breeding techniques based on the analysis of genomic data enable the identification of advantageous gene combinations, which contributes to the optimization of the selection process, the maximization of genetic advancements, and the accelerated development of enhanced crop varieties.

Modern breeding techniques necessitate significant investments in genotyping technologies, data analysis instruments, and computational resources. (2018) Dagnew et al. These costs can be prohibitive, particularly for small breeding programmes or environments with limited resources. Modern breeding techniques are most effective when applied to traits with identifiable genetic markers or those influenced by a small number of central genes. However, their ability to predict complex traits governed by polygenic inheritance or influenced by complex environmental interactions may be limited (Parihar & Shiwani, 2021). Modern breeding methods rely heavily on the availability of extensive genotypic data, particularly genomic selection. Despite this, a strong emphasis on specific genetic markers or genomic regions may obscure the overall genetic diversity of a reproductive population. It may result in a loss of genetic diversity and adaptive traits (Hasan et al., 2021). Modern breeding techniques, specifically genetic engineering, exacerbate ethical and regulatory concerns about genetically modified organisms (GMOs) and their potential effects on human health and the environment. Strict regulations and obstacles to public acceptability may impede the widespread adoption of these modern breeding techniques (Akhatar et al., 2021). Modern breeding techniques should not be viewed as replacements for traditional breeding strategies but as complementary tools. It is possible to improve the overall efficiency and efficacy of breeding efforts by integrating modern and conventional breeding techniques.

## GENETICALLY MODIFIED CROPS

Genetically modified crops, often genetically engineered or transgenic, are plants that have undergone genetic modifications using recombinant DNA technology to introduce specific traits or attributes. These alterations can encompass improvements in yield, nutrient content, resistance to pests, diseases, herbicides, and the capacity to withstand environmental stresses. Genetically modified organisms (GMOs) are pivotal in crop advancement by equipping breeders with innovative tools and avenues to confront agricultural challenges while satisfying the growing global demand for food production. A noteworthy discovery by Jacobsen and Schouten in 2008 underscores the value of cisgenesis. This genetic modification approach involves transferring genes from closely related plant species or the same species (cisgenesis) without introducing foreign genetic material. This technique proves particularly valuable in crops like potatoes, where conventional breeding methods require assistance in combining multiple desirable traits and incorporating characteristics from wild species. Pre-breeding techniques often entail inadvertent transmission of undesirable alien alleles, which cisgenesis helps mitigate. Although biotechnology permits the introduction of transgenes, strict regulations and opposition to genetically modified (GM) crops prevent significant corporations from adopting them.

In contrast, cisgenesis allows for the transmission of genes within the existing gene pool, preventing the introduction of exogenous genes. The authors propose a systematic method for distinguishing cisgenic plants from GMOs, thereby paving the way for their incorporation into conventional breeding methods. The creation of disease-resistant cisgenic potatoes and pears illustrates the potential of cisgenesis to accelerate breeding processes and make them more accessible to small businesses and developing nations. These findings highlight the importance of genetically modified crops, including cisgenic crops, as potent tools for crop improvement, addressing obstacles in conventional breeding and expanding opportunities for advanced breeding technologies (Koul, 2022; Gheysen & Custers, 2017; Hou et al., 2014).

Genetic engineering entails introducing or manipulating specific genes to modify the genetic composition of an organism. Typically, the procedure consists of the following steps: Scientists isolate and identify the genes responsible for desirable characteristics or traits in a donor organism of the same or a different species. Several techniques introduce the identified gene into the cereal plant's genome, including Agrobacterium-mediated transformation and gene gun delivery. Once integrated, the gene becomes a plant DNA component and expresses the desired characteristic or protein. The effective insertion and expression of the gene determines the identification and selection of transgenic plants. Then, these plants can be utilized in breeding programmes to transmit the trait to elite cultivars or varieties. Developing cultivars with improved disease resistance is one of the many applications of genetic engineering in crop improvement (Kumar et al., 2020; Ye et al., 2022). In their research, Grant et al. (2008) investigated the pivotal role of plant breeding in mitigating trace element deficiencies and toxicity in cereals. Variations in the absorption and distribution of vital and non-vital trace elements among different crop cultivars and varieties profoundly impact the quality of food produced. It is possible to develop crop cultivars with reduced levels of toxic trace elements, such as cadmium (Cd), by implementing selection programs and integrating genetic traits associated with lower concentrations of these harmful elements.

In the case of durum wheat, a single dominant gene displaying high heritability governs the low Cd concentration trait. Incorporating this low-Cd allele into breeding efforts can substantially decrease grain Cd levels, ensuring they remain within international safety limits without compromising essential economic characteristics. While the initial costs associated with Cd selection in breeding programs may be substantial, they tend to decrease as more breeding lines exhibit this valuable trait. The production of low-Cd crop cultivars contributes significantly to food safety by reducing the risk of Cd exposure through the human diet. Genetic modifications play a pivotal role in augmenting crop yields, making a substantial contribution to meeting the demands of a growing global population and ensuring food security. Genetically modified crops, designed with inherent resistance to pests or diseases, offer the advantage of reducing the reliance on chemical pesticides and fungicides. This, in turn, leads to decreased environmental harm and enhanced safety for farmers (Paul et al., 2017; Napier et al., 2019).

Furthermore, certain genetically modified crops can be engineered to confer environmental benefits, such as diminished pesticide usage, preservation of soil and water resources, and reduced emissions of greenhouse gases (Kumar and colleagues, 2020; Freedman, 2013). Genetic engineering can also enhance plants' nutritional composition, effectively addressing specific dietary deficiencies and promoting human health (Smyth, 2020; Warwick et al., 2009). The introduction of genetically modified crops carries the risk of unintended ecological consequences, including the possible development of resistance in targeted pests and potential impacts on non-target organisms. Legitimate concerns exist regarding the control of agricultural biotechnology, which is often concentrated in the hands of a few large corporations. These concerns encompass the potential for market monopolies and the potential adverse effects of genetically modified crops on small-scale producers in developing countries (Jha et al., 2020; Klümper & Qaim, 2014). Additionally, the public's scepticism and opposition to genetically modified crops have been exacerbated by the perceived lack of long-term studies on their effects on human health and the environment. This absence of comprehensive research has contributed to apprehensions surrounding the adoption of these crops.

The risk of gene flow and cross-pollination between genetically modified crops and related natural or non-genetically modified plants represents a potential hazard, as it could spread transgenes and genetic contamination of natural ecosystems. There are scientific, environmental, social, and ethical aspects to the debate surrounding genetically modified crops. Decisions regarding cultivating and using genetically modified products are based on specific contexts and risk-benefit analyses (Zhang et al., 2023; Abushal et al., 2021). Essential aspects of addressing genetically modified crops include extensive research, robust regulation, and open public dialogue. Additional research is required to assess the long-term effects of genetically modified crops on ecosystems, biodiversity, human health, and socioeconomic factors. Developing and deploying genetically modified products must consider safety, environmental impacts, socioeconomic equity, and public acceptance to ensure this technology's ethical and sustainable use for crop enhancement.

## CONSERVATION AND UTILIZATION OF GENETIC RESOURCES

Preserving genetic resources is crucial for advancing crop enhancement and ensuring agriculture's long-term sustainability. These genetic resources comprise various plant genetic materials, including wild cereal relatives, landraces, and traditional varieties, all harbour invaluable traits and genetic diversity. The conservation of these resources provides numerous significant benefits. First, genetic resources are an essential source of genetic diversity for breeding programmes, enabling the development of improved crop varieties with traits such as increased yield, disease resistance, stress tolerance, and improved nutritional content (Datta, 2013; Gong & Liu, 2012). This genetic diversity provides breeders with the materials to address climate change, parasites, and diseases effectively. Secondly, cereal species' genetic diversity provides the capacity for adaptation and resilience in changing environmental conditions. The genes within conserved genetic resources confer resistance to biotic and abiotic stresses, empowering plants to surmount obstacles and maintain productivity as environmental conditions change. Thirdly, conserved genetic resources capture crop species' rich evolutionary history and latent potential (Salgotra & Chauhan, 2023; Burgarella et al., 2019). By conserving genetic diversity, opportunities arise to investigate novel traits and characteristics that may have future applications in breeding programmes, thereby expanding the agricultural toolbox. Preserving traditional cultivars and landraces by conserving genetic resources safeguards agricultural and cultural heritage. It acknowledges the rights and contributions of local communities and upholds the global diversity of agricultural systems, preserving the indigenous traditions and knowledge associated with these commodities.

Conservation efforts, including both in-situ and ex-situ strategies, are crucial in preserving genetic resources. In situ conservation focuses on protecting genetic resources within their natural habitats, prioritizing the preservation of ecosystems and the biodiversity they support. This strategy employs protected areas, national parks, and community-managed conservation zones to protect crop-wild relatives and their habitats. Conservation in situ enables natural populations to preserve their dynamic evolutionary processes and ecological interactions. Ex-situ conservation, on the other hand, entails collecting, storing, and managing genetic resources outside their natural habitats (Hawkes, 1991). This method includes seed banks, field gene banks, tissue culture collections, and cryopreservation. Ex situ conservation ensures the long-term preservation of genetic material and improves its accessibility for research, breeding, and crop improvement efforts. Genetic resources from these conservation efforts play a crucial role in plant breeding programmes. They are used to introduce novel traits, broaden the genetic base, and increase crop varieties' genetic diversity. Crossbreeding, hybridization, and introgression transfer desired genes from conserved genetic resources to elite cultivars. Beyond traditional breeding, genetic resources find applications in biotechnology, genomics, and the development of innovative agricultural practices (Galluzzi et al., 2020; European Commission, 2020).

Kashyap et al. (2022) emphasize a significant finding highlighting crop wild relatives' substantial potential (CWRs) in enhancing cultivated commodities. CWRs offer solutions to challenges posed by abiotic and biotic stresses by enhancing stress tolerance and agronomic traits through introgression breeding. Ultimately, incorporating genetic diversity from natural relatives is the key to sustainable crop enhancement, resulting in increased food production and enhanced resistance to the effects of climate change.

Conservation is essential to protecting genetic resources, employing both in-situ and ex-situ techniques. In situ conservation focuses on preserving genetic resources within their natural habitats, emphasizing protecting ecosystems and biodiversity. This strategy relies on establishing protected areas, national parks, and community-managed conservation zones to protect crop-untamed relatives and their native environments. The dynamic evolutionary processes and ecological interactions that characterize natural populations are preserved by in situ conservation.

Ex-situ conservation, on the other hand, entails collecting, storing, and administering genetic resources outside of their natural habitats. This exhaustive strategy incorporates various techniques, including seed banks, field gene banks, tissue culture collections, and cryopreservation. Ex-situ conservation is crucial for the long-term preservation of genetic material and its accessibility for research, breeding, and crop improvement efforts.

Genetic resources derived from conservation efforts are invaluable in plant breeding programmes. They are utilized to introduce novel traits, broaden the genetic base, and increase crop varieties' genetic diversity. Crossbreeding, hybridization, and introgression are crucial for transferring desirable genes from conserved genetic resources to elite cultivars. In addition, genetic resources have applications in biotechnology, genomics, and the development of innovative agricultural practices that go beyond traditional breeding.

## FUTURE DIRECTIONS IN CROP IMPROVEMENT

Genome editing technologies, such as CRISPR-Cas9, have emerged as precise instruments for altering the DNA sequence of an organism. These technologies enable the targeted modification or deletion of specific genes, thereby introducing desired characteristics or improving existing ones. The implementation of genome editing could have a revolutionary effect on crop improvement. It can expedite reproduction and address crop yield, quality, and stress tolerance challenges. The introduction of high-throughput phenotyping technologies has revolutionized the rapid and accurate evaluation of plant traits on a large scale. Advanced imaging, sensor systems, and remote sensing techniques allow for the efficient collection of data on plant growth, development, stress responses, and productivity. High-throughput phenotyping enables the precise identification and selection of superior individuals in breeding programmes, ultimately resulting in more refined and efficient crop enhancement strategies. Omics technologies, which include genomics, transcriptomics, proteomics, and metabolomics, provide exhaustive insights into the molecular mechanisms regulating diverse plant characteristics and responses. These technologies enable researchers to identify essential genes, decipher regulatory networks, and decode metabolic pathways linked to desirable traits. Breeders can improve the accuracy and efficacy of trait selection and accelerate the development of enhanced crop cultivars by incorporating omics data into their breeding programmes. Breeding informatics, which employs data analytics, computational modelling, and artificial intelligence (AI) methods, is crucial in expediting the breeding process. AI algorithms are proficient at analyzing vast datasets to predict phenotypic performance and identify genotype-phenotype associations.

Climate change's profound influence on global food security and efforts to enhance crop productivity is undeniable. The evolving climate substantially threatens crop yields, primarily due to rising temperatures, fluctuating precipitation patterns, and increased extreme weather events. Beyond disturbing agroecosystems, these climate shifts also disrupt the dynamics of pests and diseases, further endangering food production. In this evolving landscape, our strategies for enhancing crops must adapt, encompassing measures that address the challenges posed by climate change.

A crucial element in this adaptation involves prioritizing the development of cereal varieties equipped with heightened tolerance to various climate-related stressors, including heat, drought, flooding, salinity, and insect and disease pressures. Moreover, diversifying the genetic pool of cereal cultivars by integrating distinctive adaptive traits sourced from crop wild relatives and landraces offers the potential to alleviate the negative impacts of climate change. To bolster these endeavours, it becomes imperative to integrate climate data and predictive modelling into breeding programs. This ensures that the selection of desirable traits and target environments aligns with anticipated climate conditions, facilitating a more effective response to the challenges posed by climate change.

The future of crop improvement has enormous implications for global food security and sustainability. Agricultural systems are under increasing pressure to simultaneously increase food production and reduce their environmental footprint as the global population continues to rise. Crop improvement emerges as a crucial strategy for resolving these obstacles. These endeavours are pivotal in creating crop varieties that exhibit high yields while possessing enhanced nutritional content and resilience to the challenges posed by climate change. Such crop development initiatives offer a sustainable means to fulfil the growing global food demand by increasing crop productivity, nutritional quality, and stress tolerance. Sustainability in agriculture requires striking a delicate equilibrium between productivity, environmental preservation, and socioeconomic factors. Crop enhancement plays a pivotal role by reducing reliance on chemical inputs, conserving water resources, enhancing soil health, and reducing greenhouse gas emissions. It significantly contributes to the long-term viability of agricultural systems by producing produce with increased resource-use efficiency and reduced environmental impacts.

Moreover, crop improvement is essential in the face of emergent hazards such as climate change, evolving pests, diseases, and changing market demands. By incorporating traits that impart resistance, tolerance, and adaptability into crop varieties, we can increase their ability to withstand dynamic threats and changing environmental conditions. Therefore, the assurance of stable and reliable food production is strengthened.

Accessibly improved crop varieties, particularly for small-scale producers, have the potential to increase their productivity and income, providing them with the means to prosper. Moreover, crop enhancement contributes considerably to rural development and poverty reduction by strengthening agricultural value chains and establishing economic opportunities for rural communities.

## DISCUSSION

Traditional breeding techniques, which include mass selection, pedigree selection, and recurrent selection, provide an exhaustive overview of established crop improvement techniques. These tried-and-true strategies have played a crucial role in crop modification, substantially contributing to the creation of enhanced crop varieties. This chapter is devoted to elucidating the advantages and disadvantages of each breeding strategy, allowing readers to determine their applicability in various breeding situations. Traditional breeding techniques, such as mass and pedigree selection, are notable for their simplicity and accessibility. These techniques do not require cutting-edge technology or an in-depth knowledge of genetics. They can be implemented effectively by farmers or breeders with limited resources, making them applicable in various agricultural contexts. In addition, traditional breeding techniques enhance single-gene traits, making them ideal for self-pollinating crops or species with a high degree of homozygosity.

Nevertheless, traditional reproductive methods have limitations. Selection based on phenotypic traits may hinder the evolution of complex traits governed by multiple genes. Knowledge regarding the genetic composition of selected individuals can result in predictable outcomes and limited control over trait transmission. In addition, traditional breeding techniques can be time-consuming, as multiple generations of selection and evaluation are required to achieve the desired improvement. Recent research by Moumane et al. (2022) sheds light on changes in the oasis of Ternata, revealing substantial changes in land use and cover. It includes a worrisome increase in desertification and a concurrent decrease in cultivated land. By analyzing Landsat time-series data from 1991 to 2021, the study uncovered a significant increase in decertified regions (+168.09%) and a decline in cultivated lands by -29.6%. The research established a significant correlation between these land use changes and the amount of water retained in the upland reservoir, highlighting the profound influence of water availability on land degradation. Human activities, such as the expansion of groundwater-dependent agricultural areas, and natural phenomena, such as drought, contribute to desertification.

To address these land use issues, crop enhancement strategies must prioritize the development of varieties adapted to arid and desert-like conditions. It involves selecting characteristics such as drought tolerance, heat tolerance, and water use efficiency to ensure crop productivity and resilience in water-limited and degraded environments. Crop breeders can actively contribute to sustainable agriculture in the Ternata Oasis by implementing selective breeding principles and incorporating genetic diversity from natural relatives or other germplasm sources, thereby mitigating the effects of land degradation on food production.

Despite their limitations, traditional breeding techniques have played a crucial role in crop enhancement and remain the foundation of modern breeding techniques. Integration of conventional and modern breeding techniques, capitalizing on the simplicity and accessibility of traditional breeding combined with the precision and efficacy of modern methods such as marker-assisted selection, would have been a notable addition to this discussion.

This chapter summarises the significance of crop enhancement strategies and principles of selective breeding in assuring food security and sustainability. It emphasizes the role of emerging technologies, the significance of genetic resource conservation, and the future evolution potential of the field. While presenting an exhaustive overview, a more critical discussion of challenges and limitations could further enrich the reader's comprehension of the intricate landscape of crop enhancement and its implications.

## CONCLUSION

Crop enhancement methods and selective breeding represent the cornerstone of contemporary agriculture. The development of fresh, enhanced crop varieties is pivotal in addressing food insecurity, elevating crop yields, and enhancing the overall quality of agricultural products. Incorporating modern breeding technologies, including genetic engineering and marker-assisted selection, has further revolutionized and simplified the process of crop improvement.

The landscape of crop enhancement is dynamic, and the future holds exciting potential for creating novel and innovative technologies. Emerging approaches like gene editing, synthetic biology, and machine learning are poised to propel crop enhancement forward, effectively addressing contemporary agriculture's pressing challenges, including climate change and environmental sustainability.

Efficiently preserving and harnessing genetic resources are crucial in pursuing robust and sustainable agriculture. The preservation and sustainable stewardship of these genetic resources serve as a protective measure to uphold the vital genetic diversity required for future endeavours in crop improvement.

Crop enhancement strategies and the fundamental principles of selective breeding are paramount in the global mission to ensure food security and foster sustainable agriculture. Looking forward, the twin objectives of conserving genetic resources and innovating new technologies will stand as pivotal strategies in meeting the escalating demands for food production in future years.

## REFERENCES

Abushal, L. T., Salama, M., Essa, M. M., & Qoronfleh, M. W. (2021). Agricultural biotechnology: Revealing insights about ethical concerns. *Journal of Biosciences*, *46*(3), 81. doi:10.100712038-021-00203-0 PMID:34423783

Acquaah, G. (2015). Conventional plant breeding principles and techniques. In Springer eBooks (pp. 115–158). doi:10.1007/978-3-319-22521-0_5

Adenle, A. A., Azadi, H., & Arbiol, J. (2015). Global assessment of technological innovation for climate change adaptation and mitigation in the developing world. *Journal of Environmental Management*, *161*, 261–275. doi:10.1016/j.jenvman.2015.05.040 PMID:26189184

Akhatar, J., Kaur, H., & Kumar, H. (2022). Conventional Plant Breeding to Modern Biotechnological Approaches in Crop Improvement. In Technologies in Plant Biotechnology and Breeding of Field Crops (pp. 1–21). doi:10.1007/978-981-16-5767-2_1

Axelsson, E., Ljungvall, I., Bhoumik, P., Conn, L. B., Murén, E., Ohlsson, Å., Olsen, L. H., Engdahl, K., Hagman, R., Hanson, J., Kryvokhyzha, D., Pettersson, M. E., Grenet, O., Moggs, J. G., Del Rio-Espinola, A., Epe, C., Taillon, B. E., Tawari, N. R., Mane, S., & Lindblad-Toh, K. (2021). The genetic consequences of dog breed formation—Accumulation of deleterious genetic variation and fixation of mutations associated with myxomatous mitral valve disease in cavalier King Charles spaniels. *PLOS Genetics*, *17*(9), e1009726. doi:10.1371/journal.pgen.1009726 PMID:34473707

Azra, M. N., Okomoda, V. T., & Ikhwanuddin, M. (2022). Breeding technology as a tool for sustainable aquaculture production and ecosystem services. *Frontiers in Marine Science*, *9*, 679529. doi:10.3389/fmars.2022.679529

Baweja, P., Kumar, S., & Kumar, G. (2020). Fertilizers and pesticides: Their impact on soil health and environment. In Soil Biology (pp. 265–285). doi:10.1007/978-3-030-44364-1_15

Bhat, J. A., Ali, S., Salgotra, R. K., Mir, Z. A., Dutta, S., Jadon, V., Tyagi, A., Mushtaq, M., Jain, N., Singh, P. K., Singh, G. P., & Prabhu, K. V. (2016). Genomic selection in the era of next-generation sequencing for complex traits in plant breeding. *Frontiers in Genetics*, *7*. doi:10.3389/fgene.2016.00221 PMID:28083016

Bhattacharya, A., Parkhi, V., & Char, B. (2021). Genome editing for crop improvement: A perspective from India. *In Vitro Cellular & Developmental Biology. Plant*, *57*(4), 565–573. doi:10.100711627-021-10184-2 PMID:34075289

Boopathi, N. M. (2020). Marker-Assisted Selection (MAS). In Springer eBooks (pp. 343–388). doi:10.1007/978-981-15-2949-8_9

Budhlakoti, N., Kushwaha, A. K., Rai, A., Chaturvedi, K. K., Kumar, A., Pradhan, A. K., Kumar, U., Kumar, R. R., Juliana, P., Mishra, D. C., & Kumar, S. (2022). Genomic selection: a tool for accelerating molecular breeding efficiency for developing Climate-Resilient crops. Frontiers in Genetics. doi:10.3389/fgene.2022.832153

Burgarella, C., Barnaud, A., Kane, N. A., Jankowski, F., Scarcelli, N., Billot, C., Vigouroux, Y., & Berthouly-Salazar, C. (2019). Adaptive introgression: An untapped evolutionary mechanism for crop adaptation. *Frontiers in Plant Science*, *10*, 4. doi:10.3389/fpls.2019.00004 PMID:30774638

Cai, D., Zhang, Z., Zhao, L., Liu, J., & Chen, H. (2022). A novel hybrid seed production technology based on a unilateral cross-incompatibility gene in maize. *Science China. Life Sciences*, *66*(3), 595–601. doi:10.100711427-022-2191-7 PMID:36190647

Calicioglu, Ö., Flammini, A., Bracco, S., Bellú, L. G., & Sims, R. (2019). The Future Challenges of Food and Agriculture: An Integrated Analysis of Trends and Solutions. *Sustainability (Basel)*, *11*(1), 222. doi:10.3390u11010222

Chloupek, O., & Hrstková, P. (2005). Adaptation of crops to the environment. *Theoretical and Applied Genetics*, *111*(7), 1316–1321. doi:10.100700122-005-0060-x PMID:16187121

Cockerton, H. M., Karlström, A., Johnson, A. W., Li, B., Stavridou, E., Hopson, K., Whitehouse, A. B., & Harrison, R. J. (2021). Genomic informed breeding strategies for strawberry yield and fruit quality traits. *Frontiers in Plant Science*, *12*, 724847. doi:10.3389/fpls.2021.724847 PMID:34675948

Crossa, J., Fritsche-Neto, R., Montesinos-Lopez, O. A., Costa-Neto, G., Dreisigacker, S., Montesinos-Lopez, A., & Bentley, A. R. (2021). The modern plant breeding Triangle: Optimizing genomics, phenomics, and environmental data. *Frontiers in Plant Science*, *12*, 651480. doi:10.3389/fpls.2021.651480 PMID:33936136

Dagnew, Y., Urge, M., Tadesse, Y., & Gizaw, S. (2018). Conservation-based breeding program designed for genetic improvement in Gumz sheep in the western lowlands of Ethiopia. *Agriculture & Food Security*, *7*(1), 34. doi:10.118640066-018-0173-5

Datta, A. (2013). Genetic engineering for improving the quality and productivity of crops. *Agriculture & Food Security*, *2*(1), 15. doi:10.1186/2048-7010-2-15

Ergon, Å., Skøt, L., Sæther, V. E., & Rognli, O. A. (2019). Allele Frequency Changes Provide Evidence for Selection and Identification of Candidate Loci for Survival in Red Clover (Trifolium pratense L.). *Frontiers in Plant Science*, *10*, 718. doi:10.3389/fpls.2019.00718 PMID:31244867

European Commission. (2020). *Genetic resources and breeding*. Research and Innovation. https://research-and-innovation.ec.europa.eu/research-area/agriculture-forestry-and-rural-areas/genetic-resources-and-breeding_en

Freedman, D. H. (2013). Are engineered foods evil? *Scientific American*, *309*(3), 80–89. https://www.jstor.org/stable/26017991. doi:10.1038cientificamerican0913-80 PMID:24003560

Galluzzi, G., Seyoum, A., Halewood, M., Noriega, I. L., & Welch, E. W. (2020). Genetic resources' role in climate change breeding: The case of public breeding programmes in eighteen developing countries. *Plants*, *9*(9), 1129. doi:10.3390/plants9091129 PMID:32878309

Gheysen, G., & Custers, R. (2017). Why Organic Farming Should Embrace Co-Existence with Cisgenic Late Blight–Resistant Potato. *Sustainability (Basel)*, *9*(2), 172. doi:10.3390u9020172

Godwin, I. D., Rutkoski, J., Varshney, R. K., & Hickey, L. T. (2019). Technological perspectives for plant breeding. *Theoretical and Applied Genetics*, *132*(3), 555–557. doi:10.100700122-019-03321-4 PMID:30888430

Gong, X., & Liu, J. (2012). Genetic transformation and genes for resistance to abiotic and biotic stresses in Citrus and its related genera. *Plant Cell, Tissue and Organ Culture*, *113*(2), 137–147. doi:10.100711240-012-0267-x

Grant, C. A., Clarke, J. M., Duguid, S., & Chaney, R. L. (2008). Selection and breeding of plant cultivars to minimize cadmium accumulation. *The Science of the Total Environment*, *390*(2-3), 301–310. doi:10.1016/j.scitotenv.2007.10.038 PMID:18036635

Grattapaglia, D., Silva-Junior, O. B., Resende, R. T., Cappa, E. P., Müller, B. S. F., Tan, B., Isik, F., Wang, T., & El-Kassaby, Y. A. (2018). Quantitative Genetics and Genomics Converge to Accelerate Forest Tree Breeding. *Frontiers in Plant Science*, *9*, 1693. doi:10.3389/fpls.2018.01693 PMID:30524463

Hasan, N., Choudhary, S., Naaz, N., Sharma, N., & Laskar, R. A. (2021). Recent advancements in molecular marker-assisted selection and applications in plant breeding programmes. *Journal of Genetic Engineering and Biotechnology*, *19*(1), 128. doi:10.118643141-021-00231-1 PMID:34448979

Hawkes, J. G. (1991). 1. The importance of genetic resources in plant breeding. *Biological Journal of the Linnean Society. Linnean Society of London*, *43*(1), 3–10. doi:10.1111/j.1095-8312.1991.tb00578.x

Hirel, B., Le Gouis, J., Ney, B., & Gallais, A. (2007). The challenge of improving nitrogen use efficiency in crop plants: Towards a more central role for genetic variability and quantitative genetics within integrated approaches. *Journal of Experimental Botany*, *58*(9), 2369–2387. doi:10.1093/jxb/erm097 PMID:17556767

Hou, H., Atlihan, N., & Lu, Z. (2014). New biotechnology enhances the application of cisgenesis in plant breeding. *Frontiers in Plant Science*, *5*. doi:10.3389/fpls.2014.00389 PMID:25157261

Hufnagel, J., Reckling, M., & Ewert, F. (2020). Diverse approaches to crop diversification in agricultural research. A review. *Agronomy for Sustainable Development*, *40*(2), 14. doi:10.100713593-020-00617-4

Jacobsen, E., & Schouten, H. J. (2008). Cisgenesis, a new tool for traditional plant breeding, should be exempted from regulating genetically modified organisms step-by-step. *Potato Research*, *51*(1), 75–88. https://link.springer.com/article/10.1007/s11540-008-9097-y. doi:10.100711540-008-9097-y

Jha, A. K., Chakraborty, S., Kumari, K., & Bauddh, K. (2020). Ecological Consequences of Genetically Modified Crops on Soil Biodiversity. In Ecological and Practical Applications for Sustainable Agriculture. doi:10.1007/978-981-15-3372-3_5

Johnson, M. S., Reddy, G., & Desai, M. M. (2023). Epistasis and evolution: Recent advances and an outlook for prediction. *BMC Biology*, *21*(1), 120. doi:10.118612915-023-01585-3 PMID:37226182

Karmaoui, A., Minucci, G., Messouli, M., Khebiza, M. Y., Ifaadassan, I., & Babqiqi, A. (2019). Climate change impacts the water supply system of the Middle Draa Valley in South Morocco. In *Climate change, food security and natural resource management* (pp. 163–178). Springer. doi:10.1007/978-3-319-97091-2_8

Kashyap, A., Garg, P., Tanwar, K., Sharma, J., Gupta, N. C., Ha, P. T. T., & Rao, M. (2022). Strategies for utilization of crop wild relatives in plant breeding programs. *Theoretical and Applied Genetics*, *135*(12), 4151–4167. https://link.springer.com/article/10.1007/s00122-022-04220-x. doi:10.100700122-022-04220-x PMID:36136128

Klümper, W., & Qaim, M. (2014). A Meta-Analysis of the impacts of Genetically Modified crops. *PLoS One*, *9*(11), e111629. doi:10.1371/journal.pone.0111629 PMID:25365303

Koul, B. (2022). Cisgenics and Crop Improvement. In Cisgenics and Transgenics (pp. 107–129). doi:10.1007/978-981-19-2119-3_3

Kumar, K. S., Gambhir, G., Dass, A., Tripathi, A. K., Singh, A., Jha, A., Yadava, P., Choudhary, M., & Rakshit, S. (2020). Genetically modified crops: Current status and prospects. *Planta*, *251*(4), 91. doi:10.100700425-020-03372-8 PMID:32236850

Maharjan, S. K., & Maharjan, K. L. (2018). Roles and contributions of community seed banks in climate adaptation in Nepal. *Development in Practice*, *28*(2), 292–302. doi:10.1080/09614524.2018.1418838

Masclaux-Daubresse, C., Daniel-Vedele, F., Dechorgnat, J., Chardon, F., Gaufichon, L., & Suzuki, A. (2010). Nitrogen uptake, assimilation and remobilization in plants: Challenges for sustainable and productive agriculture. *Annals of Botany, 105*(7), 1141–1157. doi:10.1093/aob/mcq028 PMID:20299346

Montesinos-López, O. A., Montesinos-López, A., Pérez-Rodríguez, P., Barrón-López, J. A., Martini, J. W. R., Fajardo-Flores, S., Gaytán-Lugo, L. S., Santana-Mancilla, P. C., & Crossa, J. (2021). A review of deep learning applications for genomic selection. *BMC Genomics, 22*(1). doi:10.1186/s12864-020-07319-x

Moumane, A., Al Karkouri, J., Benmansour, A., El Ghazali, F. E., Fico, J., Karmaoui, A., & Batchi, M. (2022). Monitoring long-term land use, land cover change, and desertification in the Ternata Oasis, Middle Draa Valley, Morocco. *Remote Sensing Applications: Society and Environment, 26*, 100745. doi:10.1016/j.rsase.2022.100745

Najafi, E., Devineni, N., Khanbilvardi, R., & Kogan, F. (2018). Understanding the changes in global crop yields through changes in climate and technology. *Earth's Future, 6*(3), 410–427. doi:10.1002/2017EF000690

Napier, J. A., Haslam, R. P., Tsalavouta, M., & Sayanova, O. (2019). The challenges of delivering genetically modified crops with nutritional enhancement traits. *Nature Plants, 5*(6), 563–567. doi:10.103841477-019-0430-z PMID:31160704

Nelson, A. R. L. E., Ravichandran, K., & Antony, U. (2019). The impact of the Green Revolution on indigenous crops of India. *Journal of Ethnic Foods, 6*(1), 8. doi:10.118642779-019-0011-9

Nguyen, H., Chen, Z., Fries, A., Berlin, M., Hallingbäck, H. R., & Wu, H. X. (2021). Effect of additive, dominant and epistatic variances on breeding and deployment strategy in Norway spruce. *Forestry, 95*(3), 416–427. doi:10.1093/forestry/cpab052

Parihar, A. (2022). *Shiwani*. Molecular Breeding and Marker-Assisted Selection for Crop Improvement. In Plant Genomics for Sustainable Agriculture. doi:10.1007/978-981-16-6974-3_6

Paul, M. J., Oszvald, M., Jesus, C., Rajulu, C., & Griffiths, C. A. (2017). Increasing crop yield and resilience with trehalose 6-phosphate: Targeting a feast–famine mechanism in cereals for better source–sink optimization. *Journal of Experimental Botany, 68*(16), 4455–4462. doi:10.1093/jxb/erx083 PMID:28981769

Quezada-Martinez, D., Nyarko, C. P. A., Schiessl, S., & Mason, A. S. (2021). Using wild relatives and related species to build climate resilience in Brassica crops. *Theoretical and Applied Genetics, 134*(6), 1711–1728. doi:10.100700122-021-03793-3 PMID:33730183

Rasheed, A., Ilyas, M., Khan, T. N., Mahmood, A., Riaz, U., Chattha, M. B., Kashgry, N. A. T. A., Binothman, N., Hassan, M. U., Wu, Z., & Qari, S. H. (2023). Study of genetic variability, heritability, and genetic advance for yield-related traits in tomato (Solanum lycopersicon MILL.). *Frontiers in Genetics, 13*, 1030309. doi:10.3389/fgene.2022.1030309 PMID:36685955

Robette, N., Genin, E., & Clerget-Darpoux, F. (2022). Heritability: What is the point? What is it not for? A human genetics perspective. *Genetica, 150*(3–4), 199–208. doi:10.100710709-022-00149-7 PMID:35092541

Sabadin, F., DoVale, J. C., Platten, J. D., & Fritsche-Neto, R. (2022). Optimizing self-pollinated crop breeding employing genomic selection: From schemes to updating training sets. *Frontiers in Plant Science, 13*, 935885. doi:10.3389/fpls.2022.935885 PMID:36275547

Salgotra, R. K., & Chauhan, B. S. (2023). Genetic diversity, conservation, and utilization of plant genetic resources. *Genes, 14*(1), 174. doi:10.3390/genes14010174 PMID:36672915

Sanchéz, J. I., & Akdemir, D. (2021). Training set optimization for sparse phenotyping in Genomic Selection: A Conceptual Overview. *Frontiers in Plant Science, 12*, 715910. doi:10.3389/fpls.2021.715910 PMID:34589099

Serpico, D. (2020). Beyond quantitative and qualitative traits: Three telling cases in the life sciences. *Biology & Philosophy, 35*(3), 34. doi:10.100710539-020-09750-6

Singh, H., Raigar, O. P., & Chahota, R. K. (2021). Estimation of genetic diversity and its exploitation in plant breeding. *Botanical Review, 88*(3), 413–435. doi:10.100712229-021-09274-y

Smyth, S. J. (2020). Regulatory barriers to improving global food security. *Global Food Security*. doi:10.1016/j.gfs.2020.100440

Waqas, M., Korres, N. E., Khan, M. D., Nizami, A., Deeba, F., Ali, I., & Hussain, H. (2019). Advances in the concept and methods of seed priming. In Springer eBooks (pp. 11–41). doi:10.1007/978-981-13-8625-1_2

Warwick, S. I., Beckie, H. J., & Hall, L. M. (2009). Gene flow, invasiveness, and ecological impact of genetically modified crops. *Annals of the New York Academy of Sciences, 1168*(1), 72–99. doi:10.1111/j.1749-6632.2009.04576.x PMID:19566704

Wishart, D. S., Bartok, B., Oler, E., Liang, K. Y. H., Budinski, Z., Berjanskii, M., Guo, A., Cao, X., & Wilson, M. (2020). MarkerDB: An online database of molecular biomarkers. *Nucleic Acids Research, 49*(D1), D1259–D1267. doi:10.1093/nar/gkaa1067 PMID:33245771

Worku, D., Gowane, G. R., Kumar, R., Joshi, P. K., Gupta, I. D., & Verma, A. (2021). Estimation of genetic parameters for production and reproductive traits in Indian Karan-Fries cattle using multi-trait Bayesian approach. *Tropical Animal Health and Production, 53*(3), 369. doi:10.100711250-021-02806-z PMID:34169379

Wu, X., Liu, Y., Zhang, Y., & Gu, R. (2021). Advances in research on the mechanism of heterosis in plants. Frontiers in Plant Science, p. 12. doi:10.3389/fpls.2021.745726

Xu, Q., Yang, C., & Pei, Y. (2022). Editorial: Genetic Pleiotropy in Complex Traits and Diseases. *Frontiers in Genetics, 13*, 897383. doi:10.3389/fgene.2022.897383 PMID:35547251

Yang, Y., Saand, M. A., Huang, L., Abdelaal, W. B., Zhang, J., Wu, Y., Li, J., Sirohi, M. H., & Wang, F. (2021). Applications of Multi-Omics Technologies for crop improvement. *Frontiers in Plant Science, 12*, 563953. doi:10.3389/fpls.2021.563953 PMID:34539683

Ye, J., Tian, W., & Jin, C. (2022). Nitrogen in plants: From nutrition to the modulation of abiotic stress adaptation. *Stress Biology, 2*(1), 4. doi:10.100744154-021-00030-1 PMID:37676383

Ye, R., Yang, X., & Rao, Y. (2022). Genetic engineering technologies for improving crop yield and quality. *Agronomy (Basel), 12*(4), 759. doi:10.3390/agronomy12040759

Zakari, S., Ibro, G., Moussa, B., & Abdoulaye, T. (2022). Adaptation Strategies to Climate Change and Impacts on Household Income and Food Security: Evidence from Sahelian Region of Niger. *Sustainability (Basel), 14*(5), 2847. doi:10.3390u14052847

Zhang, G., Bi, Z., Jiang, J., Lu, J., Li, K., Dan, B., Wang, X., Zhao, X., Li, M., Zhao, X., Wang, W., Xu, J., Li, Z., Zhang, F., & Shi, Y. (2023). Genome-wide association and epistasis studies reveal the genetic basis of saline-alkali tolerance at the germination stage in rice. *Frontiers in Plant Science, 14*, 1170641. doi:10.3389/fpls.2023.1170641 PMID:37251777

Zhang, L., Liu, L., Fang, Z., Shen, W., Dai, Y., Jia, R., Liang, J., & Liu, B. (2023). Fitness changes in wild soybeans are caused by gene flow from genetically modified soybeans. *BMC Plant Biology, 23*(1), 424. doi:10.118612870-023-04398-2 PMID:37710180

## KEY TERMS AND DEFINITIONS

**Breeding:** The controlled reproduction of organisms to produce desired traits and characteristics, often applied to plants or animals in agriculture.

**Climate Change:** Long-term shifts in global or regional climate patterns, often attributed to human activities, impacting weather, ecosystems, and agricultural practices.

**Crop Enhancement:** Improving the quality and yield of crops through various methods, including breeding, genetic modification, and advanced agricultural practices.

**Genetic Diversity:** The variety of genes within a population critical for adaptability and resilience to environmental changes, diseases, and other challenges.

**Genetically Modified Commodities:** Agricultural products altered through genetic engineering to enhance specific traits, such as resistance to pests or improved nutritional content.

**Heritability:** The extent to which genetic factors contribute to observable traits in offspring, influencing the transmission of characteristics from one generation to the next.

**Selective Breeding:** The intentional mating of organisms with desired traits to perpetuate those traits in subsequent generations, shaping the genetic makeup of a population.

# Chapter 6
# Enhancing Crop Resilience in the Face of a Changing Climate:
## Strategies for Sustainable Agricultural Production

**Asad Azeem**
https://orcid.org/0000-0001-6550-3571
*Department of Plant Breeding and Genetics, University of Agriculture, Faisalabad, Pakistan*

**Sami Ul-Allah**
*College of Agriculture, University of Layyah, Pakistan*

**Sana Khan**
*Department of Plant Breeding and Genetics, University of Agriculture, Faisalabad, Pakistan*

**Muhammad Umar Draz Dullu**
*Department of Plant Breeding and Genetics, PMAS-Arid Agriculture University, Rawalpindi, Pakistan*

**Sonia Azeem**
*Department of Zoology, Government College University Faisalabad, Layyah, Pakistan*

**Muhammad Sikandar Zaman**
*Department of Plant Breeding and Genetics, University of Agriculture, Faisalabad, Pakistan*

**Muhammad Zia Ul Haq**
https://orcid.org/0000-0001-9914-3367
*Department of Agronomy, University of Agriculture, Faisalabad, Pakistan*

## ABSTRACT

*Crops can adapt to shifting environmental conditions due to the diversity of traits and characteristics that genetic diversity gives. The capacity to adapt is a fundamental component of crop resilience. Breeding and natural selection are necessary for developing hardy crop types. Breeders can create more resilient crops by subjecting them to various stresses and identifying individuals with favorable features. Agroecological approaches and sustainable farming can improve crop resilience. Crop rotation, intercropping, and agroforestry increase agroecosystem resilience and promote biodiversity. Developing crop varieties with improved resilience to extreme weather events, such as droughts, floods, and heat waves, is essential to addressing the difficulties posed by climate change. Crop resilience can be improved by utilizing agroecological methods, genetic variety, and adaptive ability. The productivity and adaptability of our agricultural systems in the face of shifting environmental conditions can be increased by creating resilient crop types.*

DOI: 10.4018/978-1-6684-9838-5.ch006

## INTRODUCTION

Crop resilience is the capacity of crops to tolerate and recover from a variety of challenges, including drought, pests, diseases, extremely warm or cold temperatures, and various environmental circumstances that might impact their development and, ultimately, yield (Gojon et al., 2023). Resilient crops can maintain productivity even in challenging situations, supporting food sufficiency and sustainable farming (Seppelt et al., 2022). Considering climate change, crop resilience is crucial because shifting weather patterns and an upsurge in the rate of utmost weather incidence can make crop production extremely difficult. Therefore, one of the most important methods for ensuring the availability of food and alleviating the impacts of climate disruption on agriculture is the development and cultivation of resilient crop genotypes. Crop variety resilience is influenced by several variables, including genetic and environmental influences (Kansiime & Mastenbroek, 2016). Environmental circumstances, including water shortage, pests, diseases, harsh heat waves, and other challenges can severely impact crop growth and productivity. For example, drought stress can lower crop output and quality, whereas pests and diseases can seriously harm and destroy crops. Especially during crucial growth stages, extreme temperatures, such as heatwaves or cold snaps, can affect crop growth and production. Crop resilience is critically affected by genetic variables as well (Liliane & Charles, 2020). Crop types can tolerate stressors and recover from them thanks to traits like drought tolerance, pest and disease resistance, and heat tolerance. These characteristics are inherited and can also be acquired by crop breeding or modern genetic modification.

An essential tactic for guaranteeing food security and tackling the impacts of climate variability on agriculture is the development and cultivation of robust crop types (Raza et al., 2019). To create resilient crop types, a variety of procedures can be used, such as traditional breeding methods, modern genome editing techniques, and crop diversification. In traditional breeding methods, crops with desirable traits are chosen and crossed to create offspring with the desired attributes. For example, by looking for plants that use less water or can continue growing in water-limited settings, breeders can choose drought-tolerant cultivars. Similarly, by looking for plants that naturally resist pests or illnesses, breeders can choose for pest and disease resistance. Numerous hardy crop types have been developed over the course of many years using conventional breeding methods (Spaldon et al., 2015). These methods, nevertheless, are prolonged and are dependent on the availability of breeding stock or crop wild relatives (CWR) with desired qualities. Utilizing contemporary science, genetic engineering entails adding or changing genes to crops to give desired qualities.

In genetic engineering, it is possible to change the genetic makeup of crops or to introduce genes for high-temperature tolerance, pest, and disease resistance, or drought tolerance into crops to create strains that are more resilient to stresses. The safety of transgenic organisms and the possibility of genetic tainting of non-GMO crops are two ethical issues that are raised by genetic engineering (Azeem et al., 2023). These issues should be addressed by adopting specific measures and protocols for biosafety and biosecurity in GM plant engineering (Azeem et al., 2023). Crop diversity application plays a significant role in growing several crops with various traits in the same location. Intercropping is also practiced overcoming the effect of various environmental stressors during the whole life span of the crop; by using this methodology, different crop genotypes can lessen the effect of biotic and abiotic stressors by providing a mixed environment that makes it harder for pests and diseases to spread (Han et al., 2023). Agricultural diversity can increase robustness by preserving agricultural genetic variation. The genetic variety that can be derived from diversity can be used to create novel crop types with improved resilience (McCouch & Rieseberg, 2023).

Drought-tolerant crop varieties can tolerate prolonged droughts without loss of yield (Kumar et al., 2023). Crops that can withstand drought include millet, sorghum, and corn. These crops were created using traditional breeding methods, with breeders looking for plants with deep roots, low water requirements, or the capacity to continue growing in water-scarce environments. Pest-resistant crop varieties require less pesticide treatment and cause less crop loss because they are resistant to particular pests or illnesses. For example, certain cotton strains have undergone genetic engineering to create a toxin toxic to specific pests, hence lowering the requirement for pesticide treatment (Fernandez-Cornejo et al., 2014).

Crop cultivars, known as "heat-tolerant," may endure high temperatures without suffering substantial losses (Adhikari et al., 2015). For instance, some rice varieties with improved heat tolerance have been created using traditional breeding methods. This enables them to continue growth and yield even at high temperatures. Like this, certain wheat strains have undergone genetic engineering in order to develop an HSP that aids in their tolerance to heat stress. Therefore, considering climate shifts and other environmental challenges, crop variety resilience is essential for sustainable farming and food sufficiency. Resilient crop types can help reduce risks and guarantee dependable food production for current and future generations by utilizing genetic diversity and cutting-edge breeding procedures (Davis et al., 2021). Therefore, the use of the techniques mentioned above, and methods played a widespread role in increasing the hardiness of crop types and enabling their acceptance in global agricultural systems.

## FACTORS AFFECTING CROP RESILIENCE

Climate change threatens rural lives and global food security like never before. Global warming causes changes in precipitation trends, droughts, floods, and other severe weather (Raman et al., 2019). Cultivars that are more tolerant to heat, drought, and salinity stress are needed to sustain and improve agricultural output and adapt to climate change. To improve the lives of the millions of people who depend on agriculture, farming dangers must be reduced, and productivity and profitability increased. However, abiotic stresses like high temperature, cold, and water shortage can alter plant metabolic processes, molecular and metabolic systems, signal transduction pathways, and cumulative source-sink interactions for adaptation (Maheswari et al., 2015). Droughts cause plants to close their stomata to conserve water (Hsu et al., 2021), whereas hot weather causes them to open to enhance transpiration and cool their leaves. Abiotic and biotic factors impacting plants and their coping strategies are shown in Figure 1.

### Environmental Factors

Climate change is expected to harm soil nutrients, global food supply, food quality, pests, disease, and insect infestations. The negative effects of climate change on crop production are likely to worsen as global temperatures rise (Raman et al., 2019).

#### Drought Stress (DS)

Drought and other environmental stressors impede plant growth and maturation (Shao et al. 2008; Tátrai et al. 2016). Water shortages occur when roots receive insufficient water or evapotranspiration rises (Anjum et al., 2017). Recent years have seen worsening water restrictions for dry region crops. Yields can drop or disappear. Drought stress reduces crop productivity by 21–40% globally (Daryanto et al.

*Figure 1. Abiotic and biotic factors affecting plants and plant strategies to cope with these (Adopted from Sharma et al., 2022).*

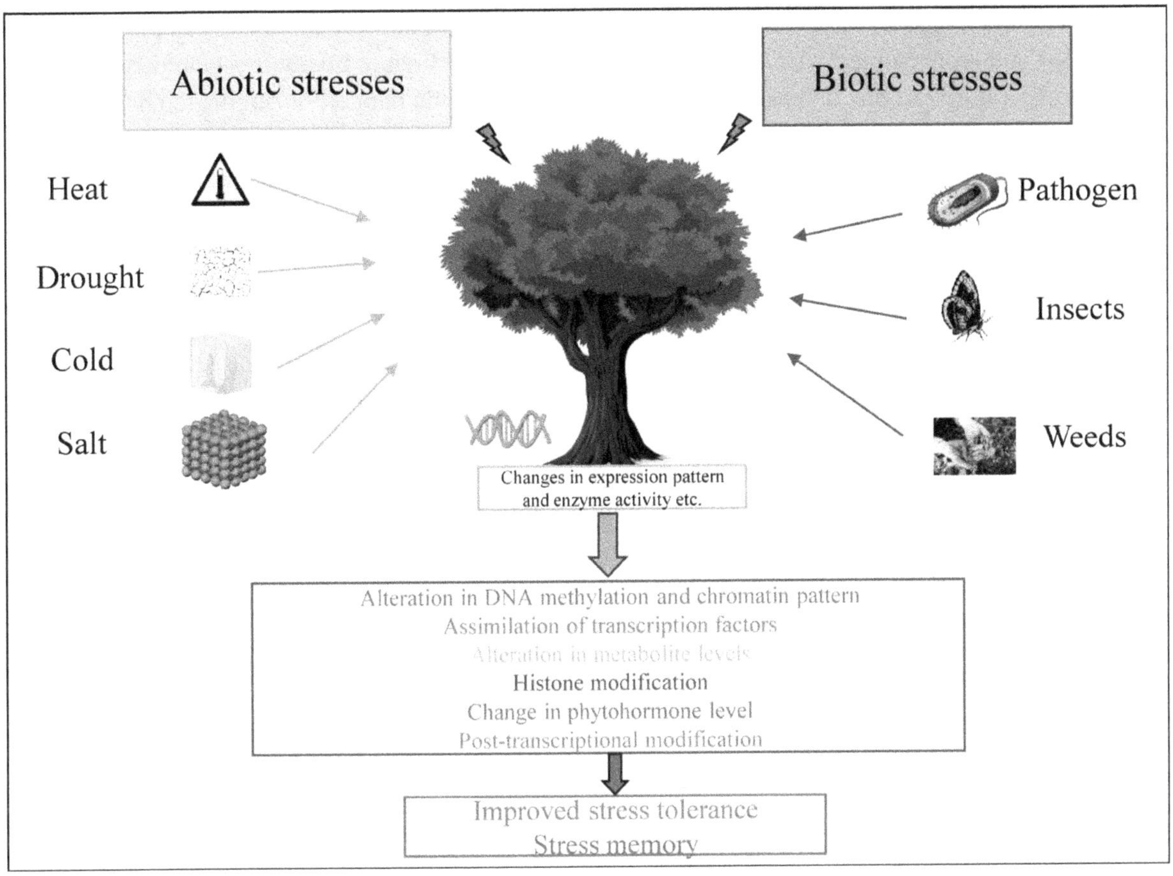


2016). Signal transduction routes relating environmental external cues to plant replies often entail intracellular changes including Ca2+ concentration, ROS accumulation, and cytosolic K+. Many plasma membrane and tonoplast proteins monitor intracellular messengers and respond to stress (Rosero et al., 2020). However, effective drought management and the best agricultural practices can mitigate the negative effects of climatic shifts on crop productivity caused by extreme weather events (Ray et al., 2020). Tangu (2014) discovered that drought-stressed olive trees had significantly reduced leaf volume, size, and area. Ray et al. (2020) found that soybean stem diameter and citrus seedling height decreased. Abiotic factors including salt and dehydration suppressed older rosette leaves via ABA signaling. However, GH3.12, a signaling element connected to the protective phytohormone SA, helped younger rosette leaves overcome this inhibition. (Berens et al., 2019). In response to HS, DS, and their combined stress conditions, Arabidopsis plants consistently elevated 29 transcripts. Many heat shock proteins and ABA and ethylene signaling pathway transcripts were enhanced (Rivero et al., 2022).

## Heat Stress (HS)

Main environmental factor affecting plant growth, development, and production is temperature. Heat stress (HS) and reduced plant yields can result from prolonged high temperatures (Janni et al., 2020). HS is a temperature rise that damages plants irreversibly. It usually happens when the temperature rises 10-15°C above the threshold (Saeed et al., 2023). HS reduces photosynthesis, increases respiration, and decreases plant development (Prasad et al., 2017). Heat stress damages plant roots, which support the plant, absorb nutrients, and water and distribute them (Valdés-López et al., 2016). This disrupts pollination, flowering, root maturity, and growth (Sehgal et al., 2017; Cho, 2018). Wheat florets can be severely damaged by high temperatures for 2–5 days during the reproductive stage (beginning of heading). Wheat's reproductive system can collapse at 35°C every day. (Janni et al., 2020). Increasing the high-temperature time at this stage had similar effects on peas, lentils, and chickpeas (Janni et al., 2020).

Kaushal et al. (2016) examined the threshold temperatures for many crop species and found critical temperatures for vegetative and generative growth, as well as literature-based losses from HS. For instance, high temperatures might slow or stop seed germination. High temperatures affect photosynthesis, respiration, water transport, membrane integrity, ROS and antioxidant production, solute buildup, and regulation. Plants can also adapt to extreme temperatures by activating stress-responsive genes through several pathways (Maheswari et al., 2015). Plant respiration, transpiration, membrane integrity, and photosynthesis are disturbed by HS. Due to PSII-mediated electron transport disruption and thylakoid membrane fluidity changes, this interference destroys PSII light-harvesting complexes and chlorophyll (Raman et al., 2019).

## Cold Stress (CS)

Major food crops are extremely vulnerable to low temperatures, such as rice and maize. When exposed to temperatures below 10°C, crops might experience large yield decreases or possibly fail to produce since these temperatures have a considerable harmful effect on their growth and development. A cold wave or stress is defined as a period when the temperature falls below 5°C for three or more days in a row in areas where the average temperature is at least 10°C. Alternatively, in regions where the average temperature is below 10°C, a temperature drop of more than three days below 3°C is regarded as a cold wave. (Venkateswarlu et al., 2011). When exposed to cold stress above freezing, plants that are acclimated to warm climates sometimes show indications of harm. The loss of leaf growth, withering, chlorosis, and necrosis symptoms that occur in response to cold or chilling stress. When a plant is subjected to chilling stress, the initial quick response results in primary injury; however, this harm is easily repaired if the temperature is increased to non-freezing levels (Maheswari et al., 2015).

At the cellular level, cold stress produces a variety of physiological variability, such as cell/tissue dehydration, cell water crystallization, increased water viscosity, production of free radicals, changed electrolyte levels, decreased membrane conductivity, and the closing of hydroactive stomata. Thirty single nucleotide polymorphisms (SNPs) associated with cold stress during seedling growth have been discovered in sorghum, which is linked to anthocyanin production and plant glucose metabolism. Extensive study has been undertaken on DREB1 and DREB2 in rice and Arabidopsis, indicating their regulatory role in reaction to cold stress (Choudhary et al., 2022).

## Salinity

Salinity inhibits plant growth through osmotic, ionic/hormonal, and other metabolic pathways (Suprasanna, 2020). In many locations, irrigation extends the growing season and enhances crop yield. Salt (NaCl) builds up in damp soils and poisons most crops, especially leaves. Rising soil salinity threatens crop output in 40% of irrigated areas worldwide (Serres et al., 2019). Abiotic factors like salt decrease crop productivity worldwide. Salinization of fertile land would likely harm agricultural productivity in various Indian nations. High salinity causes hyper-ionic and hyperosmotic stress, which kills plants (Maheswari et al., 2015).

The ratio of evaporation to precipitation determines land salinity, which raises salt concentration. Salinity rises due to seawater intrusion. Salinity in clay soils causes sodicity when discharge transfers soluble salts into the subsoil. Na+ ions remain bound to negatively charged clay particles. Over-irrigating agricultural land in arid regions causes severe salinization and water loss from transpiration and evaporation. High soil salt content (Na+) reduces soil porosity, aeration, and water conductivity. Due to standing water's ability to absorb salts from the topsoil, rice can withstand salty soils better than other crops. Many mapping experiments have attempted to reorganize rice salinity resilience QTLs on various chromosomes. A substantial QTL determines almost 40% of salt uptake variance on chromosome 1, known as "SALTOL." Marker-assisted breeding produced multiple SALTOL QTL lines (Maheswari et al., 2015).

## Pests

Climate change may impact pest and pathogen life cycles, development, and dissemination. Increased rainfall will affect insecticide use and its ability to control illnesses and pest insects like the diamondback moth and aphids (Raman et al., 2019). Bt insecticidal proteins have changed agricultural biotechnology by genetically modifying crop plants. Due to poor field pest management, several insect species have developed tolerance to these insecticidal crystal proteins. Some Bt strains produce different vegetative insecticidal proteins (VIPs) in culture media during vegetative development (Shad et al., 2021). SIPS controls coleopteran insects, but Vip3 targets chewing insects (Chakroun et al., 2016). By amino acid order, VIP toxins are categorized into four classes. VIP toxins differ greatly from cry toxins in sequence homology (Chakroun et al., 2016). Vip1 and Vip2 fight insects simultaneously (Ribeiro et al., 2017). Parasporins, crystal proteins with unknown targets, are released by some strains (Shad et al., 2021). Vip3 protein shows no sequence homology with any cry toxin and has many lepidopteran actions (Shad et al., 2021). None of the crystal poisons are structurally like Vip3Aa (Crickmore et al., 2016). With cry protein fusion proteins, cotton and maize have transformed the VIP3a gene individually and together. These fusion proteins improved transgenic plants' insect defense. They modified maize protein Vip3Aa20 and cotton protein Vip3Aa19 separately (Chen et al., 2017). Thus, coordinated efforts are needed to develop climate-ready cultivars that resist current and emerging pests.

## Diseases

Climate change could affect the lifecycle stages, ratios of growth, and geographic dissemination of pathogens and pests. The effectiveness and persistence of mutually qualitative (R) and quantitative resistance (QR) genes for serious ailments may also be impacted by climate shifts. Changes in temperature may affect the severity and spread of diseases like clubroot, which is brought on by *Plasmodiphora brassicae*

(Raman et al., 2019). At least 16 R genes and a number of QR genes for canola have been reported to provide tolerance to Leptosphaeria maculans, which causes blackleg tolerance (Kumar et al. 2018). Two R genes for blackleg resistance, Rlm2 and LepR3, have also been duplicated in canola more recently (Larkan et al., 2015). The Rlm6 gene for resistance to blackleg has been shown by Onaga et al. (2017) to be unsuccessful at giving resistance at high temperatures. Temperature and the length of leaf wetness have an impact on the gene's expression. High temperatures were found to speed up pathogen multiplication and amplify disease symptoms in rice plants that had been exposed to the fungus *Magnaporthe oryzae*, which produces rice blasts (Onaga et al., 2017). Increased temperatures in Arabidopsis caused FLAGELLIN SENSING 2 to express less, which lowered the plant's defenses against *Pseudomonas syringae* pv.

Furthermore, it was discovered that exposure to a pathogen in conjunction with increased temperatures may impact an organism's capacity to withstand both the disease and the heat. For instance, a total of 29 transcripts were consistently increased in Arabidopsis plants in response to heat, drought, and their combined stress. These transcripts include other heat shock proteins in addition to those linked to ABA and ethylene signaling (Rivero et al., 2022)

## Genetic Factors

Novel genetic variability production, germplasm collection screening for resistance characteristics, and genome-wide analysis are priorities in stress biology research (Suprasanna et al., 2018). Recent research suggests using wild, untapped germplasm to uncover evolutionarily stable genes that may respond better to stress. A comparison of GmSALT3 haplotypes in salt-resilient and salt-sensitive soybean accessions found that the resistant ones were more selected in salty environments than the subtle ones, which were chosen in less saline places (Suprasanna, 2020). According to Coolen et al. (2019), Arabidopsis plants show genetic variability in response to combinatorial cues like herbivory or DS on Botrytis cinerea resilience. Several candidate genes have been shown to be important unidentified contributors in plants' ability to respond to environmental stimuli. Over 30% of rice-growing areas lose yield due to plant submersion, however rice is very flood-tolerant. SUBMERGENCE 1 (SUB1), a flooding-tolerant rice landrace, encodes genes for ethylene-responsive transcription factors, including SUB1A-1. The flood-resistant rice variety SUBMERGENCE 1 (SUB1) has genes encoding ethylene-responsive transcription factors, notably SUB1A-1. The transcription factor SUB1A stops genes from oxidizing photo assimilated leaf sugars and starches, which are used for growth and development. This mechanism prevents excessive water intake, helping the plant cope with flooding (Serres et al., 2019). Marker-assisted breeding has extended submersion tolerance in high-yield cultivars by a week. The SUB1A-1 gene increases plant yields in non-submergence situations (Dar et al., 2018).

Plants' HKT1 gene subfamily prevents leaf sodium buildup. Most salt is removed from xylem by HKT1 in the roots. Xylem transports water and nutrients from roots to plants as a vascular conduit. At major quantitative trait loci, several HKT1 alleles improve salt resilience in wheat, rice, and allied crops. Rice with the DEEPER ROOTING (DRO1) gene has deep roots in rice fields, increasing output under water shortages (Serres et al., 2019). In wheat, TaLBD16-D limits the number of lateral roots, but in rice, which contains an auxin transporter, OsAUX1 controls their start (Shoaib et al., 2022).

*Figure 2. Conventional and next-generation breeding strategies to enhance resilience in crops (Adopted from Xiong et al., 2022).*
*(a). Difference between traditional and contemporary crop breeding cycles. (b) The essential stages of a conventional plant breeding program. Breeders initiate the cycle by generating a pool of germplasm and selecting field trial sites that are representative of their intended target environments. Genotypes are evaluated in multi-year, multi-location trials. Selection relies on the performance of the genotype across environments. (c) Crucial stages of a contemporary breeding programme. Breeders create a pool of germplasm using genomic instruments. Genotypes are genotyped employing high-throughput genotyping systems and phenotyped in multi-environment trials aimed at generating trait data that collectively defines end-use traits. To examine the GxE and promote the development of a prediction model, environments are characterised. On the basis of genomic estimated breeding values derived from the prognosis across all potential genotypes (phenotyped and unphenotyped) in all environments (test or untested), selections are made. (d) Element of envirotyping and its connection to climate change.*

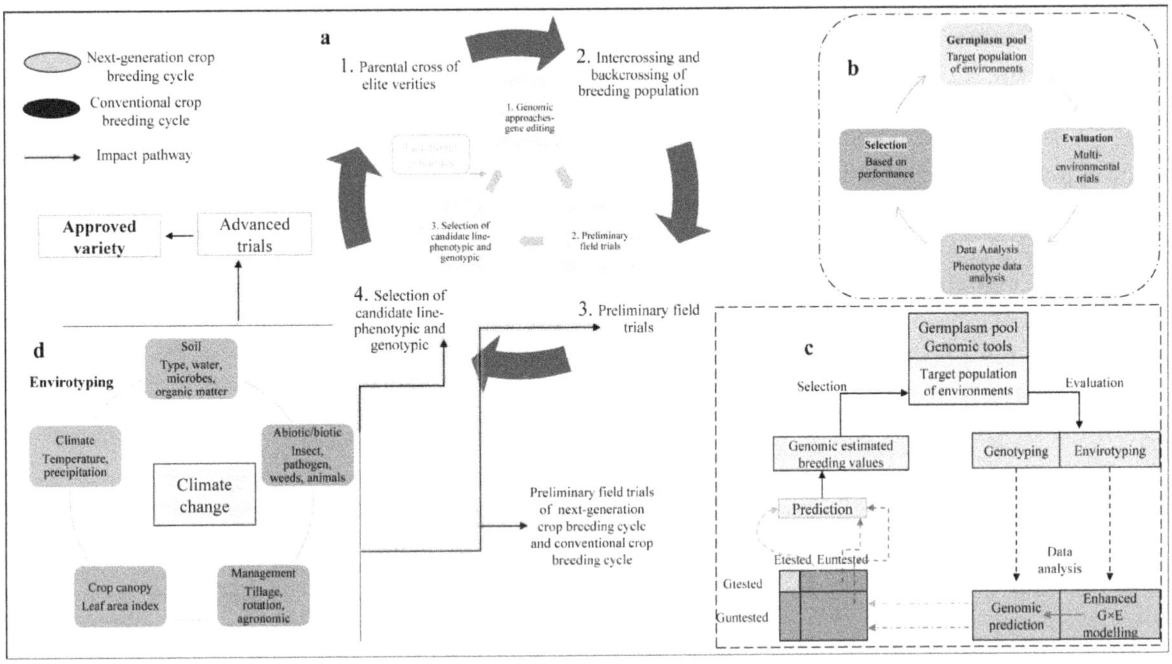


## STRATEGIES FOR DEVELOPING RESILIENT CROP VARIETIES

It is crucial to breed crops with a varied array of resistance genes and produce a diversity of crop types to reduce the potential of diseases to overcome resistance. Enhancing crop resistance to environmental (abiotic) and pathogen (biotic) stress is crucial in view of the catastrophic predictions of climate change as well as the increasing frequency of extreme weather events. This development is essential for providing food security for a population that is expanding globally (Bailey-Serres et al., 2019). Figure 2 summarizes the conventional and next-generation breeding strategies to enhance resilience in crops.

### Conventional Breeding Techniques

To lessen or reverse the effects of climate shift on crop productivity, there have been intensified attempts to find new genetic resources and valuable features. Traditional breeding methods have not been effective in addressing abiotic stress problems. To develop a resilient crop, it is critical to identify a candidate gene linked to temperature stress. The growth and vegetative components are both impacted by this abiotic stress, thus affecting the productivity of the crop (Parmar et al. 2017). Future crops can be developed that are temperature-resistant by thoroughly examining the association between genotype and phenotype (Saeed et al., 2023).

Sunflowers with endurance to drought stress have been successfully bred using *Helianthus anomalous*, a diploid annual species of hybrid derivation that is indigenous to vigorous desert dunes. For interspecific backcrossing in rice, *Oryza glaberrima* has been utilized to increase *Oryza sativa*'s drought resilience (Rosero et al., 2020). The response of crop species to abiotic stress, such as HS, is a complex quantitative feature controlled by numerous genes recognized as QTLs (Quantitative feature Loci), which are dispersed across the genome and function in concert. Historically, novel genetic variants and possible sources of tolerance have been found via QTL mapping and introduced into breeding programs (Rosero et al., 2020).

Several genomic regions in wheat that are related to heat resilience have been identified using QTL analysis, frequently in conjunction with genome-wide association Study (GWAS) and genotyping-by-sequencing (GBS). On multiple chromosomes, there are significant QTL clusters linked to heat and drought tolerance. Specific alleles from the associated key genomic regions have been identified through GWAS of sorghum seedlings grown in high temperatures. These particular areas can be utilized to develop new sorghum cultivars with high-temperature resistance (Chopra et al., 2017). One example of a genetically engineered feature that has spread rather quickly is pest resistance, which is mediated by specific *Bacillus thuringiensis* Cry proteins in cotton.

## Genetic Engineering

Genetic modifications have transformed trait change in plants through genetic engineering. Genetics has acquired encouraging information that can be used to greatly mitigate a variability of biotic and abiotic challenges, including salt, DS, HS, and cold. The development of stress-resistant crop cultivars can be greatly assisted by the detection of stress-associated transcription factors (TFs). In genetically modified crops, these TFs can control the phenotypes of genes associated with diverse stressors (Reynolds et al., 2015). To resist biotic and abiotic stresses, a large number of genetically engineered plants have been created. Compared to natural plants, these genetically modified plants exhibit substantial tolerance to climate change.

Numerous transcription factors (TFs), including the AP2/ERFBP group, are acknowledged as being plant-specific TFs. Numerous plant growth pathways are controlled by this family of AP2/ERFBP TFs, which also play a role in biotic and abiotic stress reactions (Raza et al., 2019). Their similarities and numbers categorize the four sub-groups of AP2/EREBP TFs. The subgroups are made up of the transcription factors ERF (ethylene response factor), DREB (dehydration-responsive element-binding protein), AP2 (Apetala 2), and RAV (related to ABI3/VP1) proteins. Considering their significance in determining how plants react to biotic and abiotic factors, two important subfamilies, DREB and ERF have been extensively studied (Raza et al., 2019). The DREB transcription factors exhibit remarkable regulatory capabilities in various situations characterized by water deficit and cold stress. Numerous plant species, including Arabidopsis, wheat, tomato, rice, and maize, have been studied for DREB transcription factors (Mizoi et al., 2012).

Numerous studies have absorbed the controlling mechanism of DREB1 in rice and Arabidopsis under cold stress, whereas DREB2 reacts to DS, salinity, and high-temperature stresses. Transgenic Arabidopsis with augmented resistance to salinity, drought, and cold conditions was developed by over-expressing the DREB1 TFs (Raza et al., 2019). According to Chen et al. (2007), plants that had been genetically engineered to overexpress the soybean GmDREB2 gene were noticeably salt- and drought-tolerant. Mallikarjuna et al. (2011) reported that the OsDREB2A gene was successfully inserted into genetically modified rice, improving resilience to salinity and drought conditions. The TaPIMP1 gene expression was used to create transgenic wheat, which exhibited substantial tolerance to the fungus *Bipolaris sorokiniana* and water shortage. Microarray analysis revealed TaPIMP1's overexpression. Another experiment examined the role of TaPIMP1 TF in controlling the response of transgenic tobacco to biotic and abiotic stresses. The *Ralstonia solanacearum* infection, excessive salinity, and drought conditions did not harm transgenic tobacco plants. Transgenic tobacco demonstrated resistance to *Ralstonia solanacearum*, salt, and DS (Raza et al., 2019). In field tests, effective powdery mildew tolerance was accomplished by transferring the wheat Pm3e tolerant gene to a vulnerable wheat genotype (Bailey Serres et al., 2019).

## Genome Editing

Model crops have been the main focus of genome editing applications for creating crops with increased resilience to temperature or other abiotic challenges. SSNs are used in genome editing to introduce DSBs in proximity to the target location. A certain nucleic acid sequence can be specifically targeted by single-stranded nucleases (SSNs) for binding and cleavage (Zaidi et al., 2020). Compared to other novel plant breeding techniques (NPBTs), the CRISPR/Cas system has significantly increased in popularity as the best and most innovative SSN. Despite being first proven for plant genome editing in 2013, its uses in plants have swiftly increased. In several commercially significant crops, such as rice, wheat, maize, cotton, soybean, and brassicas, the use of CRISPR has made it possible to introduce essential agricultural traits, including herbicide resistance, cold tolerance, and heat tolerance (Zaidi et al., 2020). The next-generation technique for genome editing is known as CRISPR (Saeed et al., 2023). SDNs were commonly utilized by scientists and private companies to increase the resilience of vegetables to abiotic stressors.

Vegetables' genomes frequently undergo desirable alterations using the CRISPR technology (Adli 2018; Dangol et al. 2021). Tomato plants with modifications were developed by Yu et al. (2019). The organisms' capacity to withstand heat stress was increased as a result of the modification to the mitogen-activated protein kinase 3 (MAPK3) gene. Due to lower ROS levels, they showed less membrane damage. They observed increased antioxidant enzyme activity and reduced leaf wilting in reaction to stress (Yu et al., 2019). Two transcription factors (TFs) known as BZR1 and BZR2 regulate brassinazole signaling. Yin and his coworkers use CRISPR/cas9 to eradicate tomato plants. They alter BZR1, making it recognized to be vulnerable to HS.

HSP and ROS were observed to be present in higher amounts in plants after being exposed to high temperatures. Moreover, HSP and ROS were downregulated in mutants. Nevertheless, $H_2O_2$ generation was similarly lowered in mutants (Yin et al. 2018). Brassica napus was altered using CRISPR technology, which increased seed yield by knocking down the gene CLVTA3. By targeting the TaGW2 gene, which has the potential to limit the seed size to rise, a similar approach was used to enhance the size of wheat seeds (Wang et al., 2018). Using the CRISPR/Cas9 technology, wheat has been grown with minimal gluten levels without any gene alteration (León et al., 2018). The ability to use gene editing tools like ZFNs, TALENs, and CRISPRs to improve the genetics of the crop has been made possible by advances in sympathetic the genetic basis of adaptive and productive characters in canola as well as its response to environmental stresses. The CRISPR/Cas9 system makes use of the guide RNA's specificity to a particular sequence and the endonuclease activity of Cas9 to produce precise genetic changes and, consequently, the desired gene function. Editing genes related to DS in tomato, wheat, maize, and rice has been done using this technique (Raman et al., 2019).

### Use Of Crop Diversity

Several processes of existence and co-adaptation in plants established by natural selection can be utilized by taking advantage of beneficial genes derived from natural inter- and intraspecific diversity (Rosero et al., 2020). CWRs have played a vital role in adapting crop varieties to evolving climatic conditions due to the preservation of certain genes by farmers in landraces or their presence within these wild relatives. Crop improvement efforts are becoming increasingly hampered by the minimal genetic diversity present in current cultivars (Mammadov et al., 2018). The benefits of several crops derived through genetic introgression have been widely recognized. A transcriptome investigation of rice revealed functional enrichment in oxidoreductase and lyase activities in drought-resistant varieties in comparison to cultivated varieties. It takes a significant amount of time, money, and labour to integrate genetic variety from wild species into cultivated cultivars (Zhang et al., 2017). Currently, genes for crops grown in areas where water scarcity is a serious problem are taken from wild relatives. In these regions, analyzing the attributes of locally adapted crop varieties and their wild relatives can reveal characteristics that can be used to develop cultivars and boost their durability and adaptation to such harsh climatic conditions. Using gene sources from wild relatives (WR) for crops that grow in areas where water shortage is a serious problem; for that, the characterization of the WR and landraces enables the discovery of features that can be utilized and incorporated in enhanced genotypes to provide higher adaptation and resistance to such restricted eco-friendly condition. Cultivated landraces, heirloom varieties, and crop relatives in the wild all contain valuable genetic variety for boosting crop resilience. These seed banks provide access to genetic material that can be utilized to develop tolerance to biotic and abiotic challenges, but more funding for high-throughput genotyping and phenotyping is required to speed up downstream breeding, gene discovery, and mapping (Serres et al., 2019).

## RESILIENT CROP VARIETIES

### Drought-Tolerant Crops

The biochemistry, physiology, and molecular aspects of plant responses to various abiotic stressors have undergone significant advancements. Many adaptive features, including Initial robustness, short duration, osmotic balance adjustment, aging of leaves, and prolonged appearance of verdure, have been researched and exploited to improve drought resistance. Plants that practice "stay green" tend to be more resilient to post-flowering senescence brought on by dryness. To develop high-yielding crop genotypes that are suitable for rural households and have improved resilience against prolonged monsoon delays and drought conditions, several ICAR institutes and state agricultural colleges are working together. They have successfully introduced a variety of drought-tolerant varieties of significant food, vegetable, and horticultural crops, including those with shorter growth cycles (Maheswari et al., 2015).

Some crops, such as pomegranates, sorghum, cassava, millet, and sweet potatoes, have a great capacity to withstand drought. In contrast, others, such as sugarcane, bananas, citrus, cotton, and rice, have a low tolerance to the condition (Ray et al., 2020). When compared to its domesticated relative, the wild emmer wheat demonstrates extraordinary drought resistance (Rosero et al., 2020). For instance, maize, the cereal crop with the highest global production, is vulnerable to drought stress, especially during the flowering period. Genetic modification attempts to increase drought tolerance have decreased the yield characteristics of the altered cultivars. However, scientists have achieved significant breakthroughs by locating materials that exhibit yield enhancement comparable to non-genetically modified counterparts when grown in mutually mild and harsh drought conditions (Nuccio et al., 2015).

## Pest-Resistant Crops

To fulfil the needs of the globe's population in 2050, food production is predicted to need to rise by at least 50% over 2005 levels (Fernández et al., 2022). Disturbingly, a hotter climate is predicted to change many insect pests' traits crucial to agriculture, resulting in yield losses that are predicted to rise by 10% to 25% (Deutsch et al., 2018). The first genetically engineered plant with insect resistance was tobacco, which expressed a cowpea trypsin inhibitor. The 'NewLeaf' potato, which was the initial genetically modified crop to be sold and contains the Cry3A protein from *Bacillus thuringiensis* (Bt) to give insect resistance, was commercially released by The Monsanto Company in 1995 (Fernández et al., 2022).

As an illustration, altering the rice plant's CYP71A cytochrome P450 gene, which is responsible for the conversion of tryptamine into serotonin, led to the development of insect-tolerant plants that had lower levels of serotonin but displayed a higher buildup of salicylic acid (Lu et al., 2018). Mao et al. (2007) introduced a method for RNA-mediated gene silencing known as HIGS (host-induced gene silencing) to target a particular insect gene responsible for producing a cytochrome P450 enzyme capable of detoxifying the poisonous plant chemical gossypol. This method was employed to control the lepidopteran insect known as the cotton bollworm, *Helicoverpa armigera* and presented a method for HIGS to target an insect gene that encodes a cytochrome P450 that purifies the poisonous plant metabolite, gossypol, in order to regulate the lepidopteran pest cotton bollworm, *Helicoverpa armigera*. Recently, the gene responsible for creating a phenolic glucoside malonyltransferase enzyme, which is involved in the detoxification of phenolic glycosides, was targeted by double-stranded RNA to make tomato plants completely immune to the tobacco whitefly, *Bemisia tabaci* (Fernández et al., 2022). For instance, three genes, Cry1Ac, Cry2A, and Gna (snowdrop lectin), were put into indica rice in order to target three main pests: the brown planthopper, *Nilaparvata lugens*, the yellow stemborer, *Scirpophaga incertulas*, and the rice leaf folder, *Cnaphalocrocis medinalis*. For instance, resistance to the potato aphid Macrosiphum was provided by transgenic production of Mi-1.2, an NBS-LRR protein from wild tomatoes in a vulnerable marketable genotype. Although NBS-LRR R genes are well known for being specific to particular biotypes, when the Mi-1.2 gene was introduced into eggplant, it failed to confer resistance to the biotypes of the same species that consume eggplants (Fernández et al., 2022).

## Heat-Tolerant Crops

When plants are visible to HS during crop growth, heat stress reduces agriculture productivity. Plant growth and economic yield are affected by HS-induced morphological, physiological, and biochemical changes. The prolonged exposure of plants to temperatures above a critical threshold that permanently damages their growth and development is called "heat stress." A brief temperature surge of 10–15°C is called heat shock. Genetically developed thermotolerant agricultural varieties reduce the detrimental effects of high temperatures. The use of heat-resistant cultivars and mitigation and adaptation methods can reduce productivity losses. Heat tolerance and yield loss depend on crop and development stage. Land races and WR from climates similar to future agriculturally productive locations can be used to develop heat-tolerant crop varieties. Some successful heat-tolerant plant breeding has used conventional breeding methods (Maheswari et al., 2015).

Developing heat-tolerant crops requires a comprehensive strategy combining conventional and molecular breeding. Identifying heat tolerance genes, cloning, and marker-assisted selection can help solve this problem (Maheswari et al., 2015). When exposed to high temperatures, transgenic cotton plants overexpressing HSP AtHSP101 in their pollen had higher pollen germination and tube growth. In greenhouse and field testing on transgenic plants, reproductive organs' heat tolerance increased, reducing yield losses from high temperatures (Burke and Chen, 2015).

The most widely utilized genome editing method, CRISPR/Cas9, has been used to confer HS resistance to approximately 20 crop species, including rice, wheat, cotton, maize, sorghum, and potato. HS-tolerant traits from line N22 have been incorporated into additional rice varieties to make them climate-ready (Ye et al., 2015). Tomatoes are prone to heat because continuous mild heat reduces pollen viability. The Asian Vegetable Research and Development Centre has found 39 tolerant lines after years of investigating natural genetic diversity. Equinox and Sun Leaper were used in their tomatoes. An example of "Amelia," a tomato variety engineered to withstand tropical heat (Janni et al., 2020).

## BENEFITS OF RESILIENT CROP VARIETIES

Food security, agricultural sustainability, and climate change mitigation depend on adaptable crop types (Davis et al., 2021). In drought-prone, flood-prone, or excessively hot or cold places, robust crop kinds are essential for food security. Producing crop varieties that can withstand and recover from these shocks allows farmers to maintain a food supply. Resilient crop types reduce dependence on pesticides, fertilizers, and water, improving agricultural sustainability. By producing pest- and drought-tolerant crops, farmers can reduce pollution and water use (Qaim, 2020). RCVs are crucial for addressing climate change's effects on agriculture. Farmers need crop varieties that can withstand weather pressures. By breeding crop kinds that are more resilient to biotic and abiotic stresses, breeders can reduce the effects of climate change on agriculture.

### Ensuring Food Security

Population growth and climate change are becoming increasingly obvious, making food supply one of the biggest global issues (Wijerathna-Yapa & Pathirana, 2022). Resilient crop varieties improve food security by increasing production, reducing crop failure, increasing crop diversity, improving nutrition, and reducing agriculture's environmental effect. Robust crop kinds boost productivity. Resilient crop varieties sometimes outperform conventional cultivars in adverse conditions (Brodt et al., 2011). This can assist meet the growing food demand by producing more food on the same land. Drought-resistant plants can withstand long droughts, improving crop productivity and reducing crop failure. This is especially important in drought-prone areas. A resilient crop variety can help reduce crop failure and farmer income loss. These genotypes resist pests and diseases, reducing crop failure and supporting farmers (Fand et al., 2012). It may be easier to maintain food production and availability during severe times. Resilient crop types boost agricultural variety. These changes can encourage crop diversity by lowering monoculture farming, which increases crop diseases and pests. Crop diversity reduces the risk of crop failures due to single-crop diseases or pests and ensures a more reliable food supply.

## Enhancing Agricultural Sustainability

RCVs can also boost agricultural sustainability (Umesha et al., 2018). Farming must be profitable and sustainable while maintaining its natural resources. Resilient crops improve agriculture's sustainability in several ways. Hardy crops help reduce agriculture's environmental impact. These types may require less water and pesticides, reducing agriculture's environmental impact and climate change (El-ssawy et al., 2023). Resilient crop types reduce the need for new land, preserving natural ecosystems and reducing agriculture's impact on biodiversity. RCVs promote sustainable agriculture. These cultivars often work with conservation agriculture, which reduces soil disturbance, protects soil cover, and improves crop rotation. Creating crop kinds resistant to these methods can boost their use and reduce agriculture's environmental impact. Thirdly, resilient crop cultivars reduce chemical inputs (Pathirana & Carimi, 2022). Engineered insecticides and fungicides that resist biotic stress may reduce their use. This can reduce chemical inputs' environmental and health risks and agriculture's environmental impact. Fourth, vigorous crops improve soil fertility and health. These cultivars can be cultivated for deeper roots and increased nutrient uptake to improve soil fertility and health. This maintains long-term soil production and reduces fertilizer use. Fifth, robust crops support agro-biodiversity. These variants can help preserve plant genetic resources and ecosystem diversity by increasing crop variety. Thus, pests, diseases, and environmental stressors are less likely to destroy crops, ensuring agriculture's long-term sustainability. Thus, strong crop types boost agricultural sustainability. RCVs promote robust crop varieties to ensure agriculture's long-term viability and reduce its environmental impact (Wang et al., 2022).

## Addressing the Impacts of Climate Change on Agriculture

One of the biggest benefits of resilient crop varieties is their ability to withstand climate change (Altieri et al., 2015). Temperature, rainfall, and severe weather events are already causing crop failures and production losses in agriculture. Resilient crops can mitigate these consequences in several ways. First, DS, HS, and flooding-tolerant crop types can be developed. In conditions where traditional varieties would perish, they can grow and yield. This can provide food security and prevent crop failures despite changing weather. Creating resilient crop varieties with increased genetic variability reduces crop failure due to extreme weather or other environmental challenges.

Modern genetic engineering or conventional breeding can produce this genetic variant. In either situation, the goal is to improve crops that can adapt to changing environmental conditions. Thirdly, resilient crop kinds may reduce agriculture's environmental impact and climate change. RCVs can reduce agricultural yield-related greenhouse gas emissions by reducing fertilizer and pesticide use (Balafoutis et al., 2017). Through land expansion reduction, resilient crop kinds can help preserve natural ecosystems and reduce agriculture's impact on biodiversity. Resilient crop kinds can also help develop sustainable agricultural strategies to reduce and adjust greenhouse gas emissions. In conservation agriculture, resilient crop types limit soil disturbance, maintain soil cover, and stimulate crop rotation (Shrestha et al., 2020). These strategies can improve soil health, reduce chemical inputs, and sequester carbon. Therefore, robust crop kinds can reduce farmers' climate change vulnerability. Giving farmers robust crop kinds can help them adapt to changing weather and reduce crop failure. This can boost rural food security and reduce poverty.

## CHALLENGES AND LIMITATIONS

Strong crop varieties have many benefits, but they must overcome some challenges to reach their full potential (Mekonnen et al., 2022). These constraints can be categorized. Genetic diversity is a major crop variety resiliency concern. Selective breeding of modern crop kinds limits genetic variation, which is necessary for crop resilience. The organism may not be able to resist pests and diseases or adapt to environmental changes. Cost and access are crop variety resilience's biggest challenges (Cramer, 2019). Developing robust crop kinds is expensive and time-consuming, requiring extensive research. These varieties may be unavailable to smallholder farmers, who require their benefits, due to their high production and distribution costs. Policy and regulatory constraints reduce agricultural variety robustness (Dimnwobi et al., 2023). Due to complex and contentious genetically modified agricultural regulations, introducing these commodities to the market can be difficult. This may limit hardy crop choices and postpone future crop introductions. Many undeveloped nations lack resources and infrastructure, limiting agricultural adaptation. Lack of transportation, storage, and water makes it hard to develop and distribute robust crop types in rural areas. Climate change substantially reduces crop resilience. It might be difficult to predict which crop kinds would survive rising temperatures and shifting rainfall patterns. Thus, improving resilient crop genotypes that can adapt to future climate change may be difficult.

## FUTURE RESEARCH DIRECTIONS

Genomic analysis should be used to uncover resilience-related genetic markers to enable concentrated breeding for more resilient crops. Furthermore, breeding programs using cutting-edge methodologies and considering several environmental stressors must include a diversity of resilience features. Crop and climate modelling can accurately predict climatic conditions and their consequences on crops, driving resilience policies. Investigate hybridization between farmed and wild plants to increase genetic variation and resistance. Researching agricultural communities' socioeconomic resilience and environmentally friendly crop management is crucial. Climate-smart crop insurance, risk management, and international collaboration and knowledge sharing are needed to increase resilience and food security in a changing climate.

## CONCLUSION

Crop types must be resilient to climate change and other environmental stressors to ensure global food security and agricultural sustainability. RCV crops are more resistant to pests, diseases, extreme weather, and environmental changes. Robust crop types boost yields, nutrition, climate change sensitivity, and agricultural sustainability. In addition, robust crop types can improve smallholder farmers' livelihoods, minimize agriculture's environmental effect, and alleviate food insecurity. Producing and promoting robust crop kinds is difficult due to many restrictions. These include lack of genetic variety, access and cost issues, legal limits, poor infrastructure and resource availability, and climate change. These concerns can be addressed by increasing genetic diversity, crop variety availability and affordability, resource and infrastructure investments, and climate-smart agriculture. These methods can promote crop variety resilience and agriculture's long-term viability. Crop variety resilience is crucial to agricultural sustainability and global food security. We can help future generations get healthy, sustainable food by developing and adopting resilient agricultural varieties.

## REFERENCES

Adhikari, U., Nejadhashemi, A. P., & Woznicki, S. A. (2015). Climate change and eastern Africa: A review of impact on major crops. *Food and Energy Security*, *4*(2), 110–132. doi:10.1002/fes3.61

Adli, M. (2018). The CRISPR tool kit for genome editing and beyond. *Nature Communications*, *9*(1), 1911. doi:10.103841467-018-04252-2 PMID:29765029

Altieri, M. A., Nicholls, C. I., Henao, A., & Lana, M. A. (2015). Agroecology and the design of climate change-resilient farming systems. *Agronomy for Sustainable Development*, *35*(3), 869–890. doi:10.100713593-015-0285-2

Anjum, S. A., Ashraf, U., Zohaib, A., Tanveer, M., Naeem, M., Ali, I., Tabassum, T., & Nazir, U. (2017). Growth and development responses of crop plants under drought stress: A review. *Zemdirbyste*, *104*(3), 267–276. doi:10.13080/z-a.2017.104.034

Azeem, A., Ul-Allah, S., Sher, A., Ijaz, M., Sattar, A., Ahmad, W., & Qayyum, A. (2023). Biosafety and Biosecurity in Genetically Modified Crops. In Sustainable Agriculture in the Era of the OMICs Revolution (pp. 501-510). doi:10.1007/978-3-031-15568-0_23

Bailey-Serres, J., Parker, J. E., Ainsworth, E. A., Oldroyd, G. E., & Schroeder, J. I. (2019). Genetic strategies for improving crop yields. *Nature*, *575*(7781), 109–118. doi:10.103841586-019-1679-0 PMID:31695205

Balafoutis, A., Beck, B., Fountas, S., Vangeyte, J., Van der Wal, T., Soto, I., Gómez-Barbero, M., Barnes, A., & Eory, V. (2017). Precision agriculture technologies positively contributing to GHG emissions mitigation, farm productivity and economics. *Sustainability (Basel)*, *9*(8), 1339. doi:10.3390u9081339

Brodt, S., Six, J., Feenstra, G., Ingels, C., & Campbell, D. (2011). Sustainable agriculture. *Nat. Educ. Knowl*, *3*(1). https://charlesscd.com/wp-content/uploads/2019/03/agriculture-and-the-environment-2019-Pennsylvania-Envirothon.pdf

Burke, J. J., & Chen, J. (2015). Enhancement of reproductive heat tolerance in plants. *PLoS One*, *10*(4), e0122933. doi:10.1371/journal.pone.0122933 PMID:25849955

Chakroun, M., Banyuls, N., Bel, Y., Escriche, B., & Ferré, J. (2016). Bacterial vegetative insecticidal proteins (Vip) from entomopathogenic bacteria. *Microbiology and Molecular Biology Reviews*, *80*(2), 329–350. doi:10.1128/MMBR.00060-15 PMID:26935135

Chen, M., Wang, Q. Y., Cheng, X. G., Xu, Z. S., Li, L. C., Ye, X. G., Xu, Z. S., Li, L. C., Ye, X. G., Xia, L. Q., & Ma, Y. Z. (2007). GmDREB2, a soybean DRE-binding transcription factor, conferred drought and high-salt tolerance in transgenic plants. *Biochemical and Biophysical Research Communications*, *353*(2), 299–305. doi:10.1016/j.bbrc.2006.12.027 PMID:17178106

Chen, W. B., Lu, G. Q., Cheng, H. M., Liu, C. X., Xiao, Y. T., Xu, C., Shen, Z. C., & Wu, K. M. (2017). Transgenic cotton coexpressing Vip3A and Cry1Ac has a broad insecticidal spectrum against lepidopteran pests. *Journal of Invertebrate Pathology*, *149*, 59–65. doi:10.1016/j.jip.2017.08.001 PMID:28782511

Cho, R. (2018). Agriculture and Climate: How Climate Change will Alter Our Food, State of the Planet. *Earth Institute, Colombia University*. https://blogs. ei. columbia. edu/2018/07/25/climate-change-food-agriculture.

Chopra, R., Burow, G., Burke, J. J., Gladman, N., & Xin, Z. (2017). Genome-wide association analysis of seedling traits in diverse Sorghum germplasm under thermal stress. *BMC Plant Biology*, *17*(1), 1–15. doi:10.118612870-016-0966-2 PMID:28086798

Choudhary, J. R., Tripathi, A., Kaldate, R., Rana, M., Mehta, S., Ahlawat, J., & Wani, S. H. (2022). Breeding efforts for crop productivity in abiotic stress environment. In *Augmenting crop productivity in stress environment* (pp. 63–103). Springer Nature Singapore. doi:10.1007/978-981-16-6361-1_5

Coolen, S., Van Pelt, J. A., Van Wees, S. C., & Pieterse, C. M. (2019). Mining the natural genetic variation in Arabidopsis thaliana for adaptation to sequential abiotic and biotic stresses. *Planta*, *249*(4), 1087–1105. doi:10.100700425-018-3065-9 PMID:30547240

Cramer, L. K. (2019). Access to early generation seed: Obstacles for delivery of climate-smart varieties. *The climate-smart agriculture papers: Investigating the business of a productive, resilient and low emission future*, 87-98. doi:10.1007/978-3-319-92798-5_8

Crickmore, N., Zeigler, D. R., Schnepf, E., Van Rie, J., Lereclus, D., Baum, J., Bravo, A., & Dean, D. H. (2016). *Bacillus thuringiensis toxin nomenclature*.

Dar, M. H., Zaidi, N. W., Waza, S. A., Verulkar, S. B., Ahmed, T., Singh, P. K., Roy, S. K. B., Chaudhary, B., Yadav, R., Islam, M. M., Iftekharuddaula, K. M., Roy, J. K., Kathiresan, R. M., Singh, B. N., Singh, U. S., & Ismail, A. M. (2018). No yield penalty under favorable conditions paving the way for successful adoption of flood tolerant rice. *Scientific Reports*, *8*(1), 1–7. doi:10.103841598-018-27648-y PMID:29915310

Daryanto, S., Wang, L., & Jacinthe, P. A. (2016). Global synthesis of drought effects on maize and wheat production. *PLoS One*, *11*(5), e0156362. doi:10.1371/journal.pone.0156362 PMID:27223810

Das Dangol, S., Hashmi, M. H., Saeed, F., Yel, I., Öztürk, A., & Bakhsh, A. (2021). Utilizing RNA-Based Approaches to Understand Plant-Insect Interactions. *RNA-Based Technologies for Functional Genomics in Plants*, 393-428. doi:10.1007/978-3-030-64994-4_17

Davis, K. F., Downs, S., & Gephart, J. A. (2021). Towards food supply chain resilience to environmental shocks. *Nature Food*, *2*(1), 54–65. doi:10.103843016-020-00196-3 PMID:37117650

Deutsch, C. A., Tewksbury, J. J., Tigchelaar, M., Battisti, D. S., Merrill, S. C., Huey, R. B., & Naylor, R. L. (2018). Increase in crop losses to insect pests in a warming climate. *Science, 361*(6405), 916–919. doi:10.1126cience.aat3466 PMID:30166490

Dimnwobi, S. K., Okere, K. I., Onuoha, F. C., & Ekesiobi, C. (2023). Energy poverty, environmental degradation and agricultural productivity in Sub-Saharan Africa. *International Journal of Sustainable Development and World Ecology, 30*(4), 428–444. doi:10.1080/13504509.2022.2158957

El-ssawy, W., Darwish, A., & Hassanien, A. E. (2023). The Role of Artificial Intelligence in Water Management in Agriculture for Climate Change Impacts. In The Power of Data: Driving Climate Change with Data Science and Artificial Intelligence Innovations (pp. 221-238). Springer. doi:10.1007/978-3-031-22456-0_13

Fand, B. B., Kamble, A. L., & Kumar, M. (2012). Will climate change pose serious threat to crop pest management: A critical review. *International journal of scientific and Research publications, 2*(11), 1-14. https://www.academia.edu/download/40340132/ijsrp-nov-2012-print.pdf#page=52

Fernandez-Cornejo, J., Wechsler, S., Livingston, M., & Mitchell, L. (2014). Genetically engineered crops in the United States. *SSRN*, (162). Advance online publication. doi:10.2139srn.2503388

Gojon, A., Nussaume, L., Luu, D. T., Murchie, E. H., Baekelandt, A., Rodrigues Saltenis, V. L., Cohan, J., Desnos, T., Inzé, D., Ferguson, J. N., Guiderdonni, E., Krapp, A., Klein Lankhorst, R., Maurel, C., Rouached, H., Parry, M. A. J., Pribil, M., Scharff, L. B., & Nacry, P. (2023). Approaches and determinants to sustainably improve crop production. *Food and Energy Security, 12*(1), e369. doi:10.1002/fes3.369

Han, F., Guo, R., Hussain, S., Guo, S., Cai, T., Zhang, P., Jia, Z., Naseer, M. A., Saqib, M., Chen, X., & Ren, X. (2023). Rotation of planting strips and reduction in nitrogen fertilizer application can reduce nitrogen loss and optimize its balance in maize–peanut intercropping. *European Journal of Agronomy, 143*, 126707. doi:10.1016/j.eja.2022.126707

Hsu, P. K., Dubeaux, G., Takahashi, Y., & Schroeder, J. I. (2021). Signaling mechanisms in abscisic acid-mediated stomatal closure. *The Plant Journal, 105*(2), 307–321. doi:10.1111/tpj.15067 PMID:33145840

Janda, M., Lamparová, L., Zubíková, A., Burketová, L., Martinec, J., & Krčková, Z. (2019). Temporary heat stress suppresses PAMP-triggered immunity and resistance to bacteria in Arabidopsis thaliana. *Molecular Plant Pathology, 20*(7), 1005–1012. https://bsppjournals.onlinelibrary.wiley.com/doi/pdfdirect/10.1111/mpp.12799. doi:10.1111/mpp.12799 PMID:30924595

Janni, M., Gullì, M., Maestri, E., Marmiroli, M., Valliyodan, B., Nguyen, H. T., & Marmiroli, N. (2020). Molecular and genetic bases of heat stress responses in crop plants and breeding for increased resilience and productivity. *Journal of Experimental Botany, 71*(13), 3780–3802. doi:10.1093/jxb/eraa034 PMID:31970395

Kansiime, M. K., & Mastenbroek, A. (2016). Enhancing resilience of farmer seed system to climate-induced stresses: Insights from a case study in West Nile region, Uganda. *Journal of Rural Studies, 47*, 220–230. doi:10.1016/j.jrurstud.2016.08.004

Kaushal, N., Bhandari, K., Siddique, K. H., & Nayyar, H. (2016). Food crops face rising temperatures: An overview of responses, adaptive mechanisms, and approaches to improve heat tolerance. *Cogent Food & Agriculture*, *2*(1), 1134380. doi:10.1080/23311932.2015.1134380

Kumar, A., Sengar, R., Pathak, R. K., & Singh, A. K. (2023). Integrated approaches to develop drought-tolerant rice: Demand of era for global food security. *Journal of Plant Growth Regulation*, *42*(1), 96–120. doi:10.100700344-021-10561-6

Kumar, V., Paillard, S., Fopa-Fomeju, B., Falentin, C., Deniot, G., Baron, C., Vallée, P., & Delourme, R. (2018). Multi-year linkage and association mapping confirm the high number of genomic regions involved in oilseed rape quantitative resistance to blackleg. *Theoretical and Applied Genetics*, *131*(8), 1627–1643. doi:10.100700122-018-3103-9 PMID:29728747

Larkan, N. J., Ma, L., & Borhan, M. H. (2015). The *Brassica napus* receptor-like protein RLM 2 is encoded by a second allele of the L ep R 3/R lm2 blackleg resistance locus. *Plant Biotechnology Journal*, *13*(7), 983–992. https://onlinelibrary.wiley.com/doi/pdf/10.1111/pbi.12341. doi:10.1111/pbi.12341 PMID:25644479

Liliane, T. N., & Charles, M. S. (2020). Factors affecting yield of crops. *Agronomy-climate change & food security*, *9*. doi:10.1186/s12870-016-0877-2

Lu, H. P., Luo, T., Fu, H. W., Wang, L., Tan, Y. Y., Huang, J. Z., Wang, Q., Ye, G., Gatehouse, A. M. R., Lou, Y., & Shu, Q. Y. (2018). Resistance of rice to insect pests mediated by suppression of serotonin biosynthesis. *Nature Plants*, *4*(6), 338–344. doi:10.103841477-018-0152-7 PMID:29735983

Maheswari, M., Sarkar, B., Vanaja, M., Rao, M. S., Rao, C. S., Venkateswarlu, B., & Sikka, A. K. (2015). *Climate resilient crop varieties for sustainable food production under aberrant weather conditions*. Krishi. https://krishi.icar.gov.in/jspui/bitstream/123456789/2586/1/CCRV_11th%20August%202015.pdf

Mallikarjuna, G., Mallikarjuna, K., Reddy, M. K., & Kaul, T. (2011). Expression of OsDREB2A transcription factor confers enhanced dehydration and salt stress tolerance in rice (Oryza sativa L.). *Biotechnology Letters*, *33*(8), 1689–1697. doi:10.100710529-011-0620-x PMID:21528404

Mammadov, J., Buyyarapu, R., Guttikonda, S. K., Parliament, K., Abdurakhmonov, I. Y., & Kumpatla, S. P. (2018). Wild relatives of maize, rice, cotton, and soybean: Treasure troves for tolerance to biotic and abiotic stresses. *Frontiers in Plant Science*, *9*, 886. doi:10.3389/fpls.2018.00886 PMID:30002665

Mao, Y. B., Cai, W. J., Wang, J. W., Hong, G. J., Tao, X. Y., Wang, L. J., & Chen, X. Y. (2007). Silencing a cotton bollworm P450 monooxygenase gene by plant-mediated RNAi impairs larval tolerance of gossypol. *Nature Biotechnology*, *25*(11), 1307–1313. doi:10.1038/nbt1352 PMID:17982444

Mateos Fernández, R., Petek, M., Gerasymenko, I., Juteršek, M., Baebler, Š., Kallam, K., & Patron, N. J. (2022). Insect pest management in the age of synthetic biology. *Plant Biotechnology Journal*, *20*(1), 25–36. https://onlinelibrary.wiley.com/doi/pdf/10.1111/pbi.13685. doi:10.1111/pbi.13685 PMID:34416790

McCouch, S. R., & Rieseberg, L. H. (2023). Harnessing crop diversity. In *National Acad Sciences, 120*. https://www.pnas.org/doi/full/10.1073/pnas.2221410120

Mekonnen, T. W., Gerrano, A. S., Mbuma, N. W., & Labuschagne, M. T. (2022). Breeding of vegetable cowpea for nutrition and climate resilience in Sub-Saharan Africa: Progress, opportunities, and challenges. *Plants*, *11*(12), 1583. doi:10.3390/plants11121583 PMID:35736733

Mizoi, J., Shinozaki, K., & Yamaguchi-Shinozaki, K. (2012). AP2/ERF family transcription factors in plant abiotic stress responses. *Biochimica et Biophysica Acta (BBA)-. Biochimica et Biophysica Acta. Gene Regulatory Mechanisms*, *1819*(2), 86–96. doi:10.1016/j.bbagrm.2011.08.004 PMID:21867785

Nuccio, M. L., Wu, J., Mowers, R., Zhou, H. P., Meghji, M., Primavesi, L. F., Paul, M. J., Chen, X., Gao, Y., Haque, E., & Lagrimini, L. M. (2015). Expression of trehalose-6-phosphate phosphatase in maize ears improves yield in well-watered and drought conditions. *Nature Biotechnology*, *33*(8), 862–869. doi:10.1038/nbt.3277 PMID:26473199

Onaga, G., Wydra, K. D., Koopmann, B., Séré, Y., & von Tiedemann, A. (2016). Elevated temperature increases in planta expression levels of virulence related genes in *Magnaporthe oryzae* and compromises resistance in *Oryza sativa* cv. Nipponbare. *Functional Plant Biology*, *44*(3), 358–371. doi:10.1071/FP16151 PMID:32480570

Parmar, N., Singh, K. H., Sharma, D., Singh, L., Kumar, P., Nanjundan, J., & Thakur, A. K. (2017). Genetic engineering strategies for biotic and abiotic stress tolerance and quality enhancement in horticultural crops: a comprehensive review. *3 Biotech*, *7*, 1-35. doi:10.1007/s13205-017-0870-y

Pathirana, R., & Carimi, F. (2022). Management and utilization of plant genetic resources for a sustainable agriculture. *Plants*, *11*(15), 2038. doi:10.3390/plants11152038 PMID:35956515

Prasad, P. V., Bheemanahalli, R., & Jagadish, S. K. (2017). Field crops and the fear of heat stress—Opportunities, challenges and future directions. *Field Crops Research*, *200*, 114–121. doi:10.1016/j.fcr.2016.09.024

Qaim, M. (2020). Role of new plant breeding technologies for food security and sustainable agricultural development. *Applied Economic Perspectives and Policy*, *42*(2), 129–150. doi:10.1002/aepp.13044

Raman, H., Uppal, R. K., & Raman, R. (2019). Genetic solutions to improve resilience of canola to climate change. *Genomic designing of climate-smart oilseed crops*, pp. 75-131. Springer. https://link.springer.com/chapter/10.1007/978-3-319-93536-2_2

Ray, R. L., Ampim, P. A., & Gao, M. (2020). Crop protection under drought stress. *Crop protection under changing climate*. 145-170. doi:10.1007/978-3-030-46111-9_6

Raza, A., Razzaq, A., Mehmood, S. S., Zou, X., Zhang, X., Lv, Y., & Xu, J. (2019). Impact of climate change on crops adaptation and strategies to tackle its outcome: A review. *Plants*, *8*(2), 34. doi:10.3390/plants8020034 PMID:30704089

Raza, A., Razzaq, A., Mehmood, S. S., Zou, X., Zhang, X., Lv, Y., & Xu, J. (2019). Impact of climate change on crops adaptation and strategies to tackle its outcome: A review. *Plants*, *8*(2), 34. doi:10.3390/plants8020034 PMID:30704089

Reynolds, M., Tattaris, M., Cossani, C. M., Ellis, M., Yamaguchi-Shinozaki, K., & Pierre, C. S. (2015). Exploring genetic resources to increase adaptation of wheat to climate change. In *Advances in Wheat Genetics: From Genome to Field: Proceedings of the 12th International Wheat Genetics Symposium* (pp. 355-368). Springer Japan. https://library.oapen.org/bitstream/handle/20.500.12657/27833/1002172.pdf?sequence=1#page=336

Ribeiro, B. M., Martins, É. S., de Souza Aguiar, R. W., & Corrêa, R. F. T. (2017). Expression of Bacillus thuringiensis toxins in insect cells. *Bacillus thuringiensis and Lysinibacillus sphaericus: Characterization and use in the field of biocontrol*, 99-110. Springer. doi:10.1007/978-3-319-56678-8_8

Rivero, R. M., Mittler, R., Blumwald, E., & Zandalinas, S. I. (2022). Developing climate-resilient crops: Improving plant tolerance to stress combination. *The Plant Journal*, *109*(2), 373–389. doi:10.1111/tpj.15483 PMID:34482588

Rosero, A., Berdugo-Cely, J. A., Šamajová, O., Šamaj, J., & Cerkal, R. (2020). A dual strategy of breeding for drought tolerance and introducing drought-tolerant, underutilized crops into production systems to enhance their resilience to water deficiency. *Plants*, *9*(10), 1263. doi:10.3390/plants9101263 PMID:32987964

Saeed, F., Chaudhry, U. K., Raza, A., Charagh, S., Bakhsh, A., Bohra, A., Ali, S., Chitikineni, A., Saeed, Y., Visser, R. G. F., Siddique, K. H. M., & Varshney, R. K. (2023). Developing future heat-resilient vegetable crops. *Functional & Integrative Genomics*, *23*(1), 47. doi:10.100710142-023-00967-8 PMID:36692535

Sánchez-León, S., Gil-Humanes, J., Ozuna, C. V., Giménez, M. J., Sousa, C., Voytas, D. F., & Barro, F. (2018). Low-gluten, nontransgenic wheat engineered with CRISPR/Cas9. *Plant Biotechnology Journal*, *16*(4), 902–910. doi:10.1111/pbi.12837 PMID:28921815

Sehgal, A., Sita, K., Kumar, J., Kumar, S., Singh, S., Siddique, K. H., & Nayyar, H. (2017). Effects of drought, heat and their interaction on the growth, yield and photosynthetic function of lentil (Lens culinaris Medikus) genotypes varying in heat and drought sensitivity. *Frontiers in Plant Science*, *8*, 1776. doi:10.3389/fpls.2017.01776 PMID:29089954

Seppelt, R., Klotz, S., Peiter, E., & Volk, M. (2022). Agriculture and food security under a changing climate: An underestimated challenge. *iScience*, *105551*(12), 105551. https://www.cell.com/iscience/pdf/S2589-0042(22)01823-5.pdf. doi:10.1016/j.isci.2022.105551 PMID:36458255

Shad, M., Yasmeen, A., Azam, S., Bakhsh, A., Latif, A., Shahid, N., & Shahid, A. A. (2021). Enhancing the resilience of transgenic cotton for insect resistance. *Molecular Biology Reports*, 1–9. doi:10.1007/978-3-319-90318-7_8 PMID:34839448

Shao, H. B., Chu, L. Y., Shao, M. A., Jaleel, C. A., & Hong-mei, M. (2008). Higher plant antioxidants and redox signaling under environmental stresses. *Comptes Rendus Biologies*, *331*(6), 433–441. doi:10.1016/j.crvi.2008.03.011 PMID:18510996

Sharma, M., Kumar, P., Verma, V., Sharma, R., Bhargava, B., & Irfan, M. (2022). Understanding plant stress memory response for abiotic stress resilience: Molecular insights and prospects. *Plant Physiology and Biochemistry*, *179*, 10–24. doi:10.1016/j.plaphy.2022.03.004 PMID:35305363

Shoaib, M., Banerjee, B. P., Hayden, M., & Kant, S. (2022). Roots' drought adaptive traits in crop improvement. *Plants*, *11*(17), 2256. doi:10.3390/plants11172256 PMID:36079644

Shrestha, J., Subedi, S., Timsina, K. P., Chaudhary, A., Kandel, M., & Tripathi, S. (2020). Conservation agriculture as an approach towards sustainable crop production: A review. *Farming and Management, 5*(1), 7-15. doi:10.31830/2456-8724.2020.002

Spaldon, S., Samnotra, R., & Chopra, S. (2015). Climate resilient technologies to meet the challenges in vegetable production. *International Research on Current and Academic Review, 3*(2), 28-47. https://www.researchgate.net/publication/277664533

Suprasanna, P. (2020). Plant abiotic stress tolerance: Insights into resilience build-up. *Journal of Biosciences*, *45*(1), 1–8. doi:10.100712038-020-00088-5 PMID:33097677

Suprasanna, P., Ghuge, S. A., Patade, V. Y., Mirajkar, S. J., & Nikalje, G. C. (2018). Genomic roadmaps for augmenting salinity stress tolerance in crop plants. *Salinity Responses and Tolerance in Plants, Volume 2: Exploring RNAi, Genome Editing and Systems Biology*, 189-216. doi:10.1007/978-3-319-90318-7_8

Tangu, N. A. (2014). Effects on plant morphology of drought in olive. *Türk Tarım ve Doğa Bilimleri Dergisi, 1*(Özel Sayı-1), 900-904. https://dergipark.org.tr/en/pub/turkjans/issue/13310/160838

Tátrai, Z. A., Sanoubar, R., Pluhár, Z., Mancarella, S., Orsini, F., & Gianquinto, G. (2016). Morphological and physiological plant responses to drought stress in Thymus citriodorus. *International Journal of Agronomy*, *2016*, 1–8. doi:10.1155/2016/4165750

Valdés-López, O., Batek, J., Gomez-Hernandez, N., Nguyen, C. T., Isidra-Arellano, M. C., Zhang, N., Joshi, T., Xu, D., Hixson, K. K., Weitz, K. K., Aldrich, J. T., Paša-Tolić, L., & Stacey, G. (2016). Soybean roots grown under heat stress show global changes in their transcriptional and proteomic profiles. *Frontiers in Plant Science*, *7*, 517. https://www.frontiersin.org/articles/10.3389/fpls.2016.00517/full. doi:10.3389/fpls.2016.00517 PMID:27200004

Venkateswarlu, B., Singh, A.K., Prasad, Y.G., Ravindra C., G, Srinivasa, R. C., Rao, K.V., Ramana, D.B.V. and Rao, V.U.M. (2011). *District level contingency plans for weather aberration in India.* Central Research Institute for Dryland Agriculture, Natural Resource Management Division.

Wang, D., Saleh, N. B., Byro, A., Zepp, R., Sahle-Demessie, E., Luxton, T. P., Ho, K. T., Burgess, R. M., Flury, M., White, J. C., & Su, C. (2022). Nano-enabled pesticides for sustainable agriculture and global food security. *Nature Nanotechnology*, *17*(4), 347–360. https://www.ncbi.nlm.nih.gov/pmc/articles/PMC9774002/. doi:10.103841565-022-01082-8 PMID:35332293

Wang, W., Pan, Q., He, F., Akhunova, A., Chao, S., Trick, H., & Akhunov, E. (2018). Transgenerational CRISPR-Cas9 activity facilitates multiplex gene editing in allopolyploid wheat. *The CRISPR Journal*, *1*(1), 65–74. https://www.liebertpub.com/doi/full/10.1089/crispr.2017.0010?_ga=2.19975736.1449706263.1533772800-181300196.1533772800. doi:10.1089/crispr.2017.0010 PMID:30627700

Wijerathna-Yapa, A., & Pathirana, R. (2022). Sustainable Agro-Food Systems for Addressing Climate Change and Food Security. *Agriculture*, *12*(10), 1554. https://www.mdpi.com/2077-0472/12/10/1554/pdf. doi:10.3390/agriculture12101554

Xiong, W., Reynolds, M., & Xu, Y. (2022). Climate change challenges plant breeding. *Current Opinion in Plant Biology*, *70*, 102308. doi:10.1016/j.pbi.2022.102308 PMID:36279790

Ye, C., Tenorio, F. A., Argayoso, M. A., Laza, M. A., Koh, H. J., Redoña, E. D., Jagadish, K. S. V., & Gregorio, G. B. (2015). Identifying and confirming quantitative trait loci associated with heat tolerance at flowering stage in different rice populations. *BMC Genetics*, *16*(1), 1–10. https://link.springer.com/article/10.1186/s12863-015-0199-7. doi:10.118612863-015-0199-7 PMID:25895682

Yin, Y., Qin, K., Song, X., Zhang, Q., Zhou, Y., Xia, X., & Yu, J. (2018). BZR1 transcription factor regulates heat stress tolerance through FERONIA receptor-like kinase-mediated reactive oxygen species signaling in tomato. *Plant & Cell Physiology*, *59*(11), 2239–2254. https://academic.oup.com/pcp/article/59/11/2239/5071972. doi:10.1093/pcp/pcy146 PMID:30107607

Yu, W., Wang, L., Zhao, R., Sheng, J., Zhang, S., Li, R., & Shen, L. (2019). Knockout of SlMAPK3 enhances tolerance to heat stress involving ROS homeostasis in tomato plants. *BMC Plant Biology*, *19*(1), 1–13. https://bmcplantbiol.biomedcentral.com/articles/10.1186/s12870-019-1939-z. doi:10.118612870-019-1939-z PMID:31412779

Zaidi, S. S. E. A., Mahas, A., Vanderschuren, H., & Mahfouz, M. M. (2020). Engineering crops of the future: CRISPR approaches to develop climate-resilient and disease-resistant plants. *Genome Biology*, *21*(1), 1–19. https://genomebiology.biomedcentral.com/articles/10.1186/s13059-020-02204-y. doi:10.118613059-020-02204-y PMID:33256828

Zhang, H., Mittal, N., Leamy, L. J., Barazani, O., & Song, B. H. (2017). Back into the wild—Apply untapped genetic diversity of wild relatives for crop improvement. *Evolutionary Applications*, *10*(1), 5–24. https://onlinelibrary.wiley.com/doi/pdfdirect/10.1111/eva.12434. doi:10.1111/eva.12434 PMID:28035232

## KEY TERMS AND DEFINITIONS

**Biodiversity Hotspots:** Areas with extraordinarily high levels of species diversity and endemism are known as biodiversity hotspots.

**Crop resilience:** The capacity of a crop variety to withstand and adapt to adverse effects of climate change.

**Crop varieties:** These are distinct cultivated plants from a single species that display particular traits or qualities.

**Genetic Diversity:** The variation of genetic material within a species or group is referred to as genetic diversity.

**Selective Breeding:** It is commonly referred to as artificial selection, is the process of choosing particular plants with desirable qualities for reproduction and passing on those traits to the next generations.

# Chapter 7
# Impacts and Mitigation Strategies of Abiotic and Biotic Stresses in Olive Orchards in the Era of Climate Change

**Rachid Azenzem**
https://orcid.org/0000-0003-0666-3211
*National Institute of Agricultural Research, Morocco*

**Tayeb Koussa**
*Faculty of Sciences, Chouaïb Doukkali University, Morocco*

**Mohamed Najib Alfeddy**
*National Institute of Agricultural Research, Morocco*

**Jalal Kassout**
https://orcid.org/0000-0002-3702-2751
*National Institue of Agricultural Research, Morocco*

## ABSTRACT

*Climate change is the most critical issue threatening human survival. Droughts, heat waves, and floods are examples of extreme events that pose a threat to global ecosystems and future food security. Such changes have a significant impact on the biotic and abiotic components of agroecosystems. The olive tree (Olea europaea L.) is of considerable economic importance in many countries worldwide. Nowadays, its cultivation is facing new challenges related to climate change, such as environmental stresses, pest outbreaks and increased incidence of pathogens. Therefore, various abiotic stresses, such as heat, drought, flooding, and salt, are hostile to the growth and development of olive trees, resulting in significant yield losses. Global climate variations can make trees vulnerable, promote the spread of pests and pathogens, accelerate their development, and weaken or eliminate their competitors and natural enemies. To address this situation, adaptable management techniques and changes in agricultural practices are needed to mitigate the effects of biotic and abiotic stresses.*

DOI: 10.4018/978-1-6684-9838-5.ch007

## INTRODUCTION

Plants are subject to biotic and abiotic stress factors that are often unfavorable to their growth and development processes, reducing their productivity and the quality of their products, which can lead to considerable economic losses. Hence, major abiotic stress factors, including drought, salinity, low and high temperatures and waterlogging, pose risks to global agricultural production and food supply. Similarly, epidemics of biotic stress factors such as diseases and herbivore attacks represent major and growing threat to primary productivity, global food security and biodiversity loss in many sensitive regions of the planet.

The olive is the most representative tree of the Mediterranean region, and is of considerable economic, social and ecological importance. The proliferation of olive groves with the development of human civilization can be explained by the many uses of olive trees, both domestic and wild, as a source of food, wood and livestock feed. Olive-growing covers some 10.3 million hectares worldwide, over 95% of which is in the Mediterranean basin, mainly in Spain (25%), Tunisia (12%), Italy (11%), Morocco (11%), Turkey (9%) and Greece (8%) (FAOSTAT, 2021). Recent decades have seen an expansion of olive-growing areas and an increase in world production, both attributed to greater public awareness of the nutritional benefits of olive oil and table olives. Therefore, rising global demand is putting considerable pressure on arable land to increase production, and pushing olive growers to improve their productivity in the face of new constraints linked to climate change.

According to the latest contribution by Working Group II to the IPCC's Sixth Assessment Report, all regions of the planet are already suffering the effects of climate change, and humanity is expecting even more drastic and powerful effects in the near future (IPCC, 2022). Depending on future trends in GHG concentrations, which are themselves sensitive to social and technical developments, the observed warming trends are likely to persist and intensify throughout the 21st century (IPCC, 2021; Zittis et al., 2019). According to several studies (Cramer et al., 2018; Lelieveld et al., 2012; Waha et al., 2017), these significant changes in the environment and climate could have a negative impact on a number of industries and socio-economic activities, including water resource management and agriculture, human health, energy demand and production, transport, ecosystems, biodiversity, forest fires, and many others.

The Mediterranean region is particularly vulnerable to climate change, with an increased frequency of severe weather events (Zittis et al., 2022). Droughts, heat waves and intense rainfall are just some of the phenomena that pose a major and additional challenge for agriculture in this part of the world. The impact of these occurrences can be observed through changes in atmospheric $CO_2$ concentration, temperature and water resource availability, which affect plant growth and productivity, the ability to conduct agricultural operations and the geographical distribution of crops (Brito et al., 2019). Hence, the combined impact of prolonged periods of drought and higher temperatures will have a significant impact on the main Mediterranean crops, such as olives, as well as on the viability of olive orchards (Fraga et al., 2020). Furthermore, elements that affect the months preceding the flowering season, such as weather conditions, management practices or diseases, are reflected in the annual production of flowers and pollen (Ribeiro et al., 2008). In the past, environmentally induced alterations were often poorly understood and went unnoticed, particularly in the case of plants such as olive, which can thrive for years with little or no maintenance. However, growers are now aware of the importance of abiotic factors in determining high yields and good quality. In addition, abiotic stresses can affect the level of resistance to parasitic and non-parasitic diseases (Graniti et al., 2011).

*Figure 1. Schematic representation of the different causes of biotic and abiotic stresses likely to affect olive cultivation in the context of climate change*

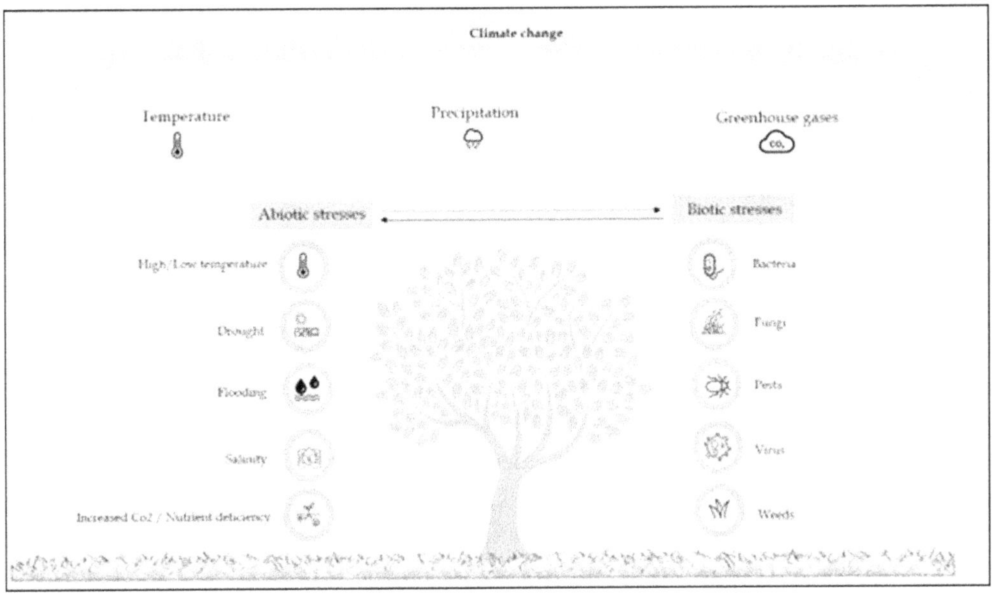


The interaction of plants and biotic stressors with the environment is a key factor in plant disease and pest epidemics (Teshome et al., 2020). Hence, weather conditions have a major influence on plant health and the dispersal of pests and diseases, although other factors, including changes in the distribution of host crops, agricultural intensification and farming practices, can also have a significant impact on the severity of damage (West et al., 2012). Therefore, changes in temperature, $CO_2$ concentrations, precipitation, and drought may have an impact on the yield losses caused by pests and diseases, as well as on the ongoing co-evolution of plants and their enemies (Elad and Pertot, 2014). In fact, climate change can exacerbate the damage caused by biotic stresses in a number of ways, including affecting the distribution and structure of pest and pathogen populations, host-pathogen/pest interactions and vector physiology, as well as promoting the emergence of new species likely to erode host plant resistance (Linnakoski et al., 2019). In this context, the collection of organisms potentially harmful to olive trees could be affected by the effects of climate change, with direct impacts on their biology and indirect impacts linked to their environment, host plants and natural enemies (Azenzem et al., 2023). This could lead to further difficulties for olive growers, who will have to reduce the climatic vulnerability of their groves and adopt adaptation measures to mitigate the negative effects of these increasingly frequent climatic hazards.

In this chapter, we examine the impact of the main abiotic stresses on olive production in the context of climate change. We then emphasise some of the possible effects of global climate variations on olive pest and disease epidemics. Finally, we highlight a recent trend towards adaptable management techniques and changes in agricultural practices that are needed to mitigate the effects of biotic and abiotic stresses on olive cultivation in a changing environment.

## OLIVE MAJOR ABIOTIC STRESSES UNDER CLIMATE CHANGE

Plant growth, crop production, and ecosystem sustainability are significantly influenced by abiotic stress factors, including drought, extreme temperatures, inadequate or excessive water, high salinity, heavy metals, and ultraviolet (UV) radiation. Such stresses pose a serious threat to both natural and agricultural ecosystems (Xu et al., 2019). Olive trees (*Olea europaea* L.) are widely found in Mediterranean agroecosystems and extensively cultivated in various regions worldwide (FAO, 2010). The increasing economic value of olive products, particularly olive oil, has led to changes in olive cultivation practices, shifting from traditional low-density and low-input systems to high-density and high-input systems. Consequently, this transition has brought about new challenging conditions, resulting in abiotic stress factors. The impact of these stressors on olive trees can significantly affect both yield production and quality. Nevertheless, olives exhibit a remarkable ability to thrive under diverse environmental conditions (Kassout et al., 2019), especially the Mediterranean climate characterized by mild and wet winters and hot and dry summers (Driouech et al., 2010). Olive adaptability is associated with changes in its morphological characteristics and physiological mechanisms, which occur in response to various abiotic stress factors (Terral et al., 2004; Connor and Fereres, 2005; Fernández, 2014; Kassout et al., 2019, 2021). Generally, abiotic stress in olive trees can induce alterations in the anatomical and morphological traits of roots, stems, and leaves, consequently affecting physiological and molecular mechanisms (Terral et al., 2004; Fernández, 2014; Kassout et al., 2021). These changes are likely to have adverse effects on olive growth, development, and overall performance (Fernández et al., 1991; Rossi et al., 2013; Kassout et al., 2019, 2021, 2022).

## IMPACT AND RESPONSES OF OLIVE TO DROUGHT

Drought stress is a complex form of stress that induces alterations in the physiological, morphological, ecological, biochemical, and molecular traits of plants (Choat et al., 2018; Kassout et al., 2019; Yang et al., 2021). It has a significant negative impact on crop production and yield performance (Fang et al., 2010; Cohen et al., 2021; Ibba et al., 2023). Drought stress leads to a substantial decrease in plant water content, resulting in the loss of turgor pressure and a drastic reduction in leaf water potential ($\Psi$) as well as cell division and expansion (Farooq et al., 2012). Consequently, there is a significant decrease in leaf size and number, and even leaf abscission, leading to a reduction in the number of branches and shoots (Farooq et al., 2012). Additionally, the photosynthetic capacity of the entire canopy is limited (Taiz and Zeiger, 2006). Prolonged drought stress causes a decline in xylem water potential within the hydraulic system, increasing vulnerability to cavitation and embolism (Terral et al., 2004; Fernández, 2014; Choat et al., 2018) and disrupting the overall water transport within the plant. However, plant species have developed diverse and intricate morphological, anatomical, physiological, and biochemical mechanisms to mitigate the impacts of drought stress (Michaletz et al., 2015; Brodribb et al., 2020; Kassout et al., 2021). For example, Kassout et al. (2019) found strong correlations between various leaf traits (such as dry matter content, surface area, and stomatal density) in Moroccan wild olive populations and ecological parameters associated with aridity. They demonstrated that mechanistic traits, including stomatal density and leaf water content, play a crucial role in the response of wild olive to the climate variations in the Mediterranean region (Kassout et al., 2019). Hence, the sclerophylly characteristic of olive leaves plays a vital role in maintaining favorable gas exchange and leaf water content, enabling trees to endure water deficit conditions (Marchi et al., 2008). Under drought conditions, olive leaves typically exhibit increased thickness and higher leaf tissue density (Kassout et al., 2019), which can be considered a mechanism contributing to enhanced stress tolerance due to improved resistance to desiccation damage (Mediavilla et al., 2012).

Olive leaves possess stomata on the lower surface (abaxial surface), which is concealed by numerous trichomes, making them hypostomatous species and effectively reducing water loss (Pallioti et al., 1994). Stomatal regulation serves as a crucial mechanism for withstanding aridity and is influenced by both hydraulic and hormonal signals (Torres-Ruiz et al., 2014). The response of stomatal conductance (gs) to water deficit in olive trees is strongly linked to leaf hydraulic conductance (Hernandez-Santana et al., 2016), as well as hydraulic failure in roots and stems (Torres-Ruiz et al., 2014). Consequently, under severe drought conditions, olive water use efficiency significantly decreases (Bacelar et al., 2007). In fact, as drought stress intensifies, stomatal conductance (gs) declines, leading to reduced photosynthetic rates in olive trees (Fernández, 2014; Fernández et al., 1991). Hormonal regulation, particularly abscisic acid (ABA), plays a role in promoting stomatal closure through direct actions on guard cells and indirect hydraulic action by reducing leaf water permeability (Pantin et al., 2013). Notably, a significant correlation has been observed between ABA levels and leaf water potential (Ψleaf) in olive trees (Kitsaki and Drossopoulos, 2005). Additionally, in the drought-tolerant chemlali cultivar, lower levels of ABA were observed compared to the drought-sensitive "Chetoui" cultivar (Guerfal et al., 2009). Furthermore, the regulation of the antioxidant system plays a pivotal role in controlling oxidative stress induced by reactive oxygen species (ROS). Drought stress promotes the production of ROS, which can act as signaling molecules or cause oxidative stress in plants (Mattos and Moretti, 2015). Consequently, drought stress leads to a decrease in the activities of certain antioxidant enzymes, including ascorbate peroxidase, catalase, and superoxide dismutase, as well as non-enzymatic antioxidant mechanisms such as the accumulation of phenolic compounds, tocopherols, carotenoids, ascorbate, and glutathione (Bacelar et al., 2006; Petridis et al., 2012; Abdallah et al., 2017).

According to Kassout et al. (2021), increasing aridity and alterations in available water resources play a significant role in controlling and modulating the sap conduction performance of wild olive trees. The study revealed a strong correlation between decreasing vessel lumen area (SVS) and increasing aridity. Interestingly, the individuals growing under lower water availability appeared to produce wood with higher hydraulic conductivity at an earlier stage of branch diameter growth. Consequently, the inability to develop larger vessels during the growth unit's development could potentially lead to branch mortality in olive trees and, consequently, affect the persistence of the population (Kassout et al., 2021). This finding aligns with the report by Terral et al. (2004), who noted that vessel size in olive trees is heavily influenced by seasonal or annual variations in precipitation. Additionally, the anatomical response becomes even more pronounced in irrigated trees, as demonstrated by Terral and Durand (2006) in cultivated olives. Therefore, the intraspecific variation in wood anatomy of O. europaea can be considered a dynamic compromise between efficient water transport through the formation of large vessels and safe sap conduction through narrower, numerous, and interconnected vessels that limit the risks of cavitation, embolism, and provide structural support (Terral et al., 2004; Kassout et al., 2021). Furthermore, Torres-Ruiz et al. (2013) demonstrated that olives have the ability to tolerate a low stem water potential (Ψstem) approaching -5 MPa with minimal risk of xylem embolism. Moreover, under drought conditions, olive trees exhibit a strong capacity for leaf and stem osmotic adjustments, enabling them to decrease osmotic potential and establish a soil-plant gradient to extract water from the soil even at low water potentials (Dichio et al., 2006; Sofo et al., 2008).

## IMPACT AND RESPONSES OF OLIVE TO SALINITY

Soil salinity is widely recognized as the most limiting stress factor for the growth and development of plant species (Zhu, 2002). With increasing temperatures and decreasing precipitation rates, soil properties are significantly affected (Gelybó et al., 2018), thereby creating more challenging conditions for both wild and cultivated species. The impact of salt stress is particularly evident in crop species, leading to a significant reduction in productivity and overall performance (Connor and Fereres, 2005; Isayenkov, 2012). The elevated concentration of sodium (Na+) and chlorine (Cl-) ions, rather than calcium (Ca2+) and sulfate (SO42-) ions, contributes to the development of soil salinity. Additionally, salt stress in plants triggers oxidative stress due to the generation of reactive oxygen species (ROS) (Isayenkov, 2012). Furthermore, a correlation has been observed between the reduction in osmotic potential between the roots and soil solutions and the uptake of Na+ and Cl- ions (Abbasi et al., 2016).

The olive tree exhibits moderate resistance to salinity, as demonstrated in previous studies (Rugini and Fedeli, 1990; Gucci and Tattini, 2010). Early investigations on olive tolerance to salinity revealed that when electrical conductivity (EC) exceeded 4 to 6 dS m-1, a 10% reduction in olive production was observed (Bernstein, 1965). Subsequently, Maas and Hoffman (1977) highlighted that olive production is significantly impacted by increasing salinity beyond an ECs of 4 to 6 dS m-1. Consequently, an elevation in NaCl concentration was associated with a decrease in olive growth and total leaf area (Vigo et al., 2005), primarily due to a notable reduction in the photosynthetic capacity of olive trees (Vigo et al., 2002). Nevertheless, salt tolerance appears to vary among olive cultivars (Chartzoulakis et al., 2002, 2006; Vigo et al., 2002). For instance, Chartzoulakis et al. (2002, 2006) found that the Kalomon cultivar exhibited resistance, while Megaritiki and Kothreiki were considered salt-tolerant. On the other hand, Mastoidis, Amphissis, and Koroneiki demonstrated lower tolerance. Similarly, the tolerance of the Arbequina cultivar to salinity decreased depending on the duration of exposure and concentration (Aragüés et al., 2005). Notable differences in response to salinity were also observed between the Frantoio and Leccino cultivars (Tattini et al., 1992). It is important to mention that the maximum tolerated concentration of NaCl in olive trees was estimated to be around 137 mM, while a significant decline in production was observed above 30 mM (Rugini and Fedeli, 1990; Gucci and Tattini, 2010). Apart from reductions in productivity and growth performance, salt stress in olive trees is characterized by apparent foliar chlorosis and necrosis, desiccation of flowers, ovary abortion, and under severe conditions, necrosis of stem tips and roots (Gucci et al., 2003). However, while fruit production is negatively affected by salinity stress, it does not significantly alter the olive oil content (Gucci et al., 2003; Chartzoulakis, 2005). Ultimately, salt tolerance in olive trees is associated with their capacity to exclude and retain Na+ and Cl- ions in the roots, involving complex physiological and morphological adjustments from organ to whole-plant levels (Chartzoulakis, 2005).

## IMPACT AND RESPONSES OF OLIVE TO WATERLOGGING

Waterlogging refers to the saturation of soil pores with excessive water, resulting in the inhibition of root respiration (Sasidharan et al., 2017). It disrupts gas exchange between the soil and atmosphere, thereby blocking root activity (Armstrong and Drew, 2002). In addition, the accumulation of toxic substances such as H2S, N2, Mn+2, Fe+2, along with reactive oxygen species (ROS), creates a more stressful environment for plant growth and development during waterlogging (Tian et al., 2019; Zhou et al., 2020). Prolonged waterlogging leads to hypoxic conditions that can ultimately result in plant death (Fukao et al., 2019). While olive trees, like many fruit trees, are highly susceptible to waterlogging and soil hypoxia, the degree of susceptibility varies depending on the cultivar (Hassan and Seif, 1990). For instance, Hassan and Seif (1990) demonstrated that the Mission cultivar died after 30 days of waterlogging, whereas the Kalamon cultivar withstood the stress for a period of 60 days. Furthermore, Aragüés et al. (2010) found that the combined effect of salinity and waterlogging significantly influenced the survival and growth of the Arbequina cultivar. Waterlogging represents a complex stress factor that affects various physiological processes throughout the entire plant, potentially explaining the previously observed negative impact on olive growth and performance (Pan et al., 2020).

## IMPACT OF CLIMATE CHANGE ON OLIVE INSECT PESTS

Like other crops, the olive tree is vulnerable to numerous pests and diseases that prevent it from achieving its production performance. The group of these enemies comprises over 255 species, including insect pests, mites, nematodes and pathogens (Bueno and Jones, 2002). The effects of ongoing climate change are likely to affect the development and distribution of pests, diseases, weeds and other biotic variables, such as the natural enemies. Thus, increasing extreme weather events, changing rainfall patterns, rising temperatures and increasing CO2 concentrations can profoundly alter the distribution and intensity of infestations in olive groves, through direct effects on the life cycle of crop pests and pathogens, and indirect effects on hosts and natural enemies. In fact, the environmental factors may have an impact on biological activities such as development, reproduction and survival rates of insect pests. They can lead to an increase in their geographical range, improved winter survival rates, increased voltinism and shorter generation times, altered synchrony between plants and pests, altered interspecific interactions, an increased risk of species invasions and secondary pest outbreaks, and a reduction in the effectiveness of biological control (Wang et al., 2022).

It is estimated that 15% of olive production in the Mediterranean basin is lost every year to insect-related yield losses. In the absence of any sanitary intervention, a number of insect pests can cause economically significant damage in olive orchards. Olive fly (*Bactrocera oleae*), olive psyllid (*Euphyllura olivinea*), olive moth (*Prays oleae*) and black scale (*Saissetia oleae*) are among the main pests of concern to olive growers. In certain circumstances, other pests such as the bark beetle (*Hylesinus oleiperda*), the otiorhynchus (*Otiorhynchus cribricolis*) and the oleander scale (*Aspidiotus nerii*) are considered secondary and can inflict varying degrees of damage. As mentioned before, climate variability directly affects insect pest populations in terms of growth rates, overwintering success, migration patterns and habitat expansion, while indirectly affecting trophic interactions with their environment.

### Impact of Rising Temperature

Due to their poikilothermic nature, temperature has a significant impact on insect growth, development and multiplication. Their metabolic rate approximately doubles when the temperature rises by 10°C (Ma et al., 2021). Consequently, an increase in temperature can alter their geographic range, population size, generation time, survival and fecundity (Harvey et al., 2020). Hence, insect pests are expected to migrate from tropical and subtropical regions to temperate regions at higher altitudes as temperatures rise, due to changes in the development environments of their host plants (Shrestha, 2019). They will also be able to colonize new sites that were previously unsuitable for them. In this way, to understand the effects of climate change on the distribution of *Olea europaea* and *Bactrocera oleae*, a number of temperature-based predictive models have been studied. (Gutierrez et al., 2009) examined the distribution and abundance of both species in the ecological zones of Arizona–California (AZ–CA) and Italy under three climate change scenarios and observed weather conditions using a physiologically-based demographic model (PBDM). They reported that in AZ-CA, global warming is expected to increase the olive fly's range northwards and in coastal areas, while reducing it in southern areas. They also found that in Italy, the olive fly's range is expanding northwards as the climate warms. The same model was used by Ponti et al. (2014) to assess the ecological and economic effects of an estimated 1.8°C climate warming on olives and olive fly, throughout the Mediterranean basin. The analysis suggests that, as conditions improve for both species, fly populations in Europe should increase inland and at higher altitudes. Fly populations should decline in warmer parts of the basin as temperatures approach or exceed upper thermal limits. Likewise, temperature and phenology of olive trees, as well as olive moth, are strongly correlated both within and between years (Armendáriz et al., 2007). Indeed, a study by Mansour et al. (2017) in olive groves in southwest Morocco reported that *Prays oleae* adult population densities are high along the coast, where temperatures are moderate and humidity levels high. By contrast, in inland regions, where the environment is warmer and less humid, emergence is later and densities are lower. Furthermore, the second generation of the black scale *Saissetia oleaea* is generally only observed in coastal regions, where humidity is higher and temperatures more temperate than in continental areas of Mediterranean countries.

In addition to the globalization and liberalization of agricultural trade, global warming has increased opportunities for the introduction of exotic species, creating habitats previously unsuitable for their establishment and encouraging the introduction of vulnerable cultivars or crops (Gregory et al., 2009). For instance, *Euphyllura olivina* is expanding its geographical range, creating risks in areas previously spared. (Azenzem et al., 2023). A recent study by Hougardy et al (2020) revealed that *E. olivina* has been spreading slowly in California since it was first reported in 2007. If it ever reaches the olive-growing regions of this state, where the climate is very similar to that of southeastern Spain, the psyllid could become a serious economic pest. Moreover, *E. olivina* has also been imported in the Netherlands, although there has been no sign of the species becoming established (Den Bieman, 2019). More recently, the psyllid has also been reported for the first time in the eastern Palearctic region of South Korea following its introduction in 2020, although its establishment in this country has not yet been proven (Cho et al., 2021).

Insect development is accelerated by higher temperatures in certain favorable ranges. Insects are thus able to produce a greater number of generations each year, ultimately causing greater damage to crops. Furthermore, according to climate change forecasts, warmer temperatures will accelerate development, reducing the time needed to reach reproductive maturity and shortening life cycles, leading to a sharp increase in pest populations (Bale et al., 2002). In that respect, *B. oleae* can generate one to four generations in the Mediterranean region, depending on temperature and regional factors (such as altitude and distance from the sea) (Marchi et al., 2016). According to studies by Tsitsipis (1980) and Fletcher (1989) on the growth of olive fly larvae under laboratory conditions at constant temperature, the lower temperature threshold for olive fly larvae is between 10 and 12.5°C, and the upper threshold between 30 and 32°C. Conversely, high summer temperatures favor *B. oleae* mortality and reduce its parasitic activity, as the adults' physiological processes cease around 35°C (Gutierrez et al., 2009). In addition, Wang et al. (2009) showed that eggs laid inside an olive fruit had a mortality rate of over 50% when exposed to 35°C, and that no eggs survived when exposed to 38°C for 10 days. The risk of infestation decreases as olive trees are planted further from the shore, as the olive fly prefers coastal regions with cool summers and warm winters (Volpi et al., 2020). In the same way, the biology of *E. olivina* is also closely linked to that of its host plant and to temperature. These two variables influence the explosion or decline of its population. According to numerous studies, *E. olivina* is fatally affected by high temperatures, and its average lifespan decreases when the temperature is above or close to 30°C (Chermiti, 1989; Tajnari, 2001; Ksantini et al.,2002; Ksantini, 2003). A temperature of 32 degrees Celsius brings egg-laying to a complete halt, while a temperature of 35 degrees or more causes 100% of mature females to lose their developing oocytes and mature eggs, placing them in reproductive diapause (Ksantini, 2003). On the other hand, the damage caused by *Saissetia oleae* varies regularly according to the abundance of its populations in the same year and from one year to the next, under the effect of difficult climatic conditions, such as high temperatures (Mendel et al. 1984). In fact, first-stage scale mortality can approach 80% at 30°C and 30% relative humidity.

Insect pests entering winter diapause are likely to experience the greatest changes in their thermal environment (Bale and Hayward, 2010). At warmer temperatures, increased metabolic activity shortens the insects' diapause period and reduces their winter mortality (Hahn and Denlinger, 2007). Consequently, global warming would lead to early infestations, an increase in the insect population and, ultimately, further damage to olive orchards.

In agroecosystems, temperature extremes and fluctuations have an effect on the biology of insect pests and their natural enemies, affecting characteristics such as generation time, sex ratio, lifespan, fecundity, activity, distribution and survival (Chidawanyika et al. 2012). Temperature changes will affect the timing of daytime activity of various insect species, altering interactions between insect pests and their natural enemies and upsetting the balance of biological management (Thomson et al., 2010). In a study conducted in two Egyptian localities, Abd El-Salam et al. (2019) reported that the balance between olive fly larvae and their parasitoid, *Psyttalia concolor*, was disturbed by unfavorable fluctuations in temperature and humidity. They revealed that temperature was a key factor in predicting *P. concolor* survival compared to relative humidity.

In addition, insect species contribute to crop losses by transmitting pathogens as vectors. Recent research has shown that the Philaenus spumarius L. stink bug is responsible for spreading the Xylella fastidiosa bacterium, the main cause of rapid olive decline syndrome (Cornara et al., 2017). Climate change may increase the incidence of insect-borne plant diseases, as insect vectors expand their range and reproduce more rapidly (Petzoldt and Seaman, 2006).

The pest outbreak is the result of the availability of host plants, accelerated metabolism and altered voltinism. They can also be caused by unpredictable changes in the population of insect pests and their natural enemies. Higher temperatures have been shown to facilitate insect herbivory by improving host plant nutritional quality and increasing photosynthesis outbreaks (Visser and Both, 2005). In temperate regions, pest epidemics are likely to become more frequent as a result of global warming and more frequent droughts (Logan et al., 2003). Recent outbreaks of the gall pest *Dasineura oleae* in a number of Mediterranean countries have led to this pest being considered a serious threat to olive groves (Picchi et al., 2022). Several authors have put forward hypotheses on the importance of climate change in this phenomenon (Simoglou et al., 2012). In fact, changes in climatic parameters can affect pest abundance, the success of parasitizm activity and the tritrophic interaction between *D. oleae*, its parasitoids and the olive tree (Simoglou et al., 2012 ; War et al., 2016). Therefore, pest outbreaks can pose additional problems for the technical management of olive groves, particularly for smallholders.

## Impact of Rising $CO_2$ Levels

High levels of greenhouse gas (GHG) emissions, notably carbon dioxide (CO2), methane, nitrous oxide, etc., are widely recognized as the main cause of climate change (NASA, 2022). In less than 200 years and over the last 15 years, human activity has increased the level of CO2 in the atmosphere by 50% and 10% respectively (Montzka, 2022). An increase in atmospheric CO2 levels is likely to have an impact on plant physiology by stimulating photosynthetic activity, leading to better growth and higher plant yields (Gornall et al., 2010). In turn, this phenomenon will have an indirect impact on insects by modifying the quantity and quality of foliage and plants. The majority of current research focuses on the indirect impacts of increased atmospheric CO2 on insects through altered host characteristics (Jactel et al., 2019). In general, elevated atmospheric CO2 increases carbohydrate content and decreases nitrogen content, increasing the C/N ratio and C-based defences that are detrimental to defoliating herbivores (Robinson et al., 2012 ; Boullis et al., 2015), with notable variations between host species and insect functional groups. Therefore, increased CO2 concentration encourages certain pest groups to consume more plants (Bezemer et al., 1998), as nitrogen is essential for insect development, which can lead to higher levels of plant damage. For instance, when nitrogen levels fall, foliage feeders such as caterpillars, miners and chewers frequently increase their consumption rate by compensatory feeding (DeLucia et al., 2008 ; Hamilton et al., 2005). Moreover, thrips and other whole-cell feeders show an increase in population size (Bezemer et al., 1998). While, phloem-feeding insect pests, such as whiteflies and aphids, show a dual response: faster population growth rates and lower population densities (Sutherst et al., 2011).

## Impact of Change in Precipitation Patterns

Changes in precipitation patterns affect the frequency and duration of droughts, as well as the seasonal averages, intensity and timing of individual rainfall events (Kahaduwa and Rajapakse 2021). All these aspects can have considerable effects on insect populations. Hence, drought has an impact on the performance of herbivores by altering the nutritional value of the tissues they feed on. Dry environments can offer ideal conditions for the growth and development of certain herbivorous insects. Other insect species are attracted to plants that suffer from drought. For example, bark beetles (Scolytidae) can detect ultrasonic acoustic emissions produced by the rupture or cavitation of water columns in the xylem when plants lose moisture during the transpiration process. In addition, drought-stressed plants produce fewer secondary metabolites, which have a defensive function, making them more vulnerable to insect attack (Yihdego, 2019).

Heavy rainfall can lead to prolonged water stagnation and flooding, which has a direct impact on insect species that hibernate in the soil. Furthermore, flooding and heavy rain can wash away insect eggs and larvae, as well as small pests such as aphids, mites, jassids, whiteflies and others (Pathak et al., 2012).

# IMPACT OF CLIMATE CHANGE ON OLIVE DISEASE PATHOGENS

In agricultural and forestry systems, plant diseases caused by fungi, bacteria and viruses are major biotic constraints on production. Climatic factors play an important role in their spatial and temporal distribution, reproduction and spread (Graniti et al., 2011). Changes in environmental conditions are strongly associated with the incidence and severity of plant diseases, and play a major role in differences in the

*Figure 2. Possible impacts of climate change on insect pests in olive groves*

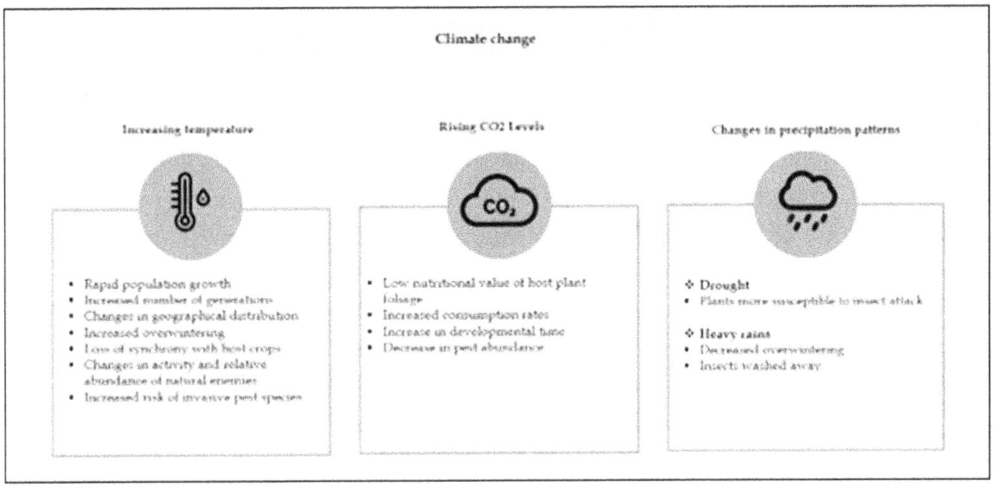


level of losses caused by these diseases. Therefore, the expected changes associated with global warming (increasing temperature, changing rainfall patterns, and rising CO2 concentrations) are likely to have an impact on all components of the disease triangle, including the host, pathogen, weather conditions and their interactions.

In addition to the predicted implications of climate change for the olive tree, its production is expanding outside the regions where it has been practiced since antiquity, and traditional cultivation methods are being modified to increase yields and profits. These changes could alter the spatial distribution of pathogens and encourage the emergence of new diseases. In fact, numerous bacterial, fungal, viral and phytoplasma diseases affect olive trees, but only a small number are likely to cause economically significant damage. The group of major diseases likely to cause serious economic losses in olive orchards includes olive leaf spot, verticillium wilt, cercosporiosis, root rots, olive quick decline syndrome and olive knot. The occurence and severity of these diseases are strongly influenced by temperature, atmospheric carbon dioxide concentrations, relative humidity and rainfall.

## Impact of Elevated Temperatures

Changes in temperature linked to climate change can have an impact on host plant physiology and resistance, as well as on the growth stage, development rate and pathogenicity of infectious pathogens. Hence, an increase in temperature combined with sufficient soil moisture can promote evapotranspiration, creating a humid microclimate under the plant canopy and perhaps increasing the frequency of diseases that thrive in hot, humid environments (Mcelrone et al., 2005). Moreover, pathogens may follow their host if climate change leads to a progressive modification of growing sites. Thus, the geographical expansion of olive growing, due to global warming or other factors, and the spread of non-native cultivars may lead to the appearance of diseases of minor importance in traditional growing areas, encouraging the development of many dormant infections and causing the emergence of new diseases. For instance, olive knot caused by *Pseudomonas savastanoi pv. savastanoi* was reported as a new disease in Australia in 2003 (Hall et al., 2004) and in 2008 in Nepal (Balestra et al., 2009), where olive cultivation is relatively recent. On a local scale, the spread of verticillium wilt in Spain is linked to the expansion of olive growing and the modification of cultivation techniques to increase yields (Jimenez-Dıaz, 2011). Furthermore, temperature can also affect plant resistance to disease, which can have an impact on both disease incidence and severity. Examples include the relationship between temperature and susceptibility to rust in wheat and oats, and the greater resistance of certain forage species (Coakley et al., 1999). The reason why plants are more vulnerable to infection at high temperatures is not well understood at the molecular level (Cohen et Leach, 2020). But high temperatures can weaken plant defenses, leading to increased pathogen infections (Cheng et al., 2019). Hence, salicylic acid, an essential plant defense hormone, is produced less frequently in Arabidopsis at elevated temperatures because key immune transcription factors such as CBP60g are less activated (Castroverde et Dina, 2021). Transcription factors of the CBP60g family are widely conserved in plants (Zheng et al., 2022), and knowing how they regulate heat-sensitive immunity in plants may help us better understand how global warming affects plant diseases.

The life cycle of a pathogen comprises several crucial stages that are directly affected by temperature changes, including infection, reproduction, dispersal and survival between seasons. Hence, rising temperature may be partly responsible for the recent resurgence of outbreaks of olive fruit anthracnose caused by the *Colletothricum* species in many olive groves in southern Italy. Warmer autumn temperatures also favor the pathogen's ability to overwinter on mummified drupes, which do not fall to the ground and serve as inoculum for the following season, shortening the incubation period for drupe infections and increasing the number of secondary infection cycles (Graniti et al., 2011). Moreover, in southern United Kingdom, where the first British olive groves were planted benefiting from climate change, cercosporiosis of olive trees caused by *Pseudocercospora cladosporioides* was first reported in 2009 (Henricot et al., 2009). Both occurrences were perceived as a harbinger of warmer weather in this part of the world, enabling olive trees to survive further north europe than they did 30 years ago.

Temperature is known to affect the distribution of fast-growing opportunistic fungi and animal pests, as well as the distribution of soil microbial communities (Tedersoo et al., 2014 ; Oliverio et al., 2017 ; Bebber et al., 2013) . Some soil-borne diseases may become more frequent as soil temperatures rise (Delgado-Baquerizo et al., 2020). A study by Lopez-Escudero and Blanco-Lopez (2007) on Verticillium wilt of olive trees showed that the incidence of the disease correlates with the density of the initial soil inoculum. But symptom manifestation in relation to inoculum density varies and is strongly influenced by ambient and soil factors (DeVay et al., 1974). Therefore, Verticillium wilt development and fungal proliferation are strongly influenced by soil temperature. The ideal temperature for in vitro development of *V. dahliae* isolates from cotton varies from 24 to 27 °C in pathotype D and from 21 to 24 °C in pathotype ND, according to Bejarano-Alca'zar et al. (1996). Xu et al. (2012) found that the best growth temperature for *V. dahliae* isolates of pathotypes D and ND was 25 °C although isolates of pathotype D can adapt effectively to high temperatures and infect cotton severely at temperatures ranging from 25 to 30 °C (Jimenez-Dıaz Et al., 1998). In olive plants, Calderon et al. (2014) reported that models predicted that soil temperatures between 16 and 24 uC for cv. Picual and 20 to 24 °C for cv. Arbequina favored infection by the defoliating *V. dahliae* pathotype. Soil temperatures between 16 and 20 °C were found to be ideal for infection by the non-defoliating pathotype. The development of Verticillium wilt in olive trees is favored by air and soil temperatures close to the optimal growth zone of *V. dahliae* (Navas-Cortes et al., 2008; Levin et al., 2003). Moderate spring air temperatures in Mediterranean-type regions favor the intensity of Verticillium wilt attacks, but hot summers prevent the disease from spreading further

(Jimenez-Dı´az et al., 2012). Recently, Requena-Mullor and al. (2020) found that isothermal conditions had a negative effect on occurrence, but this effect was modulated by rainfall seasonality and watering. They also concluded that, as *V. dahliae* thrives under mild temperature conditions, rainy periods must coincide with warm periods to provide optimal conditions for pathogen development.

The prevalence of vector-borne diseases will also change with climate change. Climate has a significant impact on vector growth and dispersal. Geographical distribution, increased overwintering, variations in population growth rates, increased number of generations, longer development seasons, variations in the timing of crop and pest phenology and variations in interspecific interactions are all possible effects of the changes. Given that insects have short life cycles, are mobile, have a high reproductive potential and are physiologically sensitive to temperature, even small-scale climate change will have a significant impact on vector distribution and abundance. Hemipteran insects that feed on mesophyll, phloem or xylem sap, such as psyllids, leafhoppers and spittlebugs, are the main vectors for transmitting bacterial pathogens to plants. The vector-borne *Xylella fastidiosa* bacterium represents a global threat to agriculture, horticulture, forestry and unmanaged environments, with around 600 host species in over 85 families of wild and cultivated plant species (EFSA, 2022). The " Olive Quick Decline Syndrome" (OQDS) first recognized in October 2013 in southern Italy (Salento peninsula, located in the Puglia region) was linked to the quarantine bacterium *Xylella fastidiosa* (Xf) subsp. *pauca* and, since its initial appearance, has been confirmed in various Mediterranean regions of Europe (Saponari et al., 2013). Thousands of olive trees have perished as a result of recent OQDS epidemics and, depending on cultivar resistance and the effectiveness of disease control, the economic impact on the European olive sector over a 50-year period could reach 5.2 billion euros (5.6 billion US dollars) (Schneider et al., 2020). The meadow spittlebug *Philaenus spumarius* L. is the only epidemiologically significant vector of Xf currently identified in European outbreaks. It has a large geographic range that includes Europe, Asia, the Middle East, North Africa, North America, and New Zealand (Cornara et al. 2018). This insect appears to be sensitive to humidity and temperature, with cool, moist conditions being the best for its local and long-term proliferation. For example, it has been hypothesized that a local drop in *P. spumarius* populations on the Californian coast in recent decades was due to rising summer temperatures and drought. According to a recent study by Godefroid et al. (2021), models predict that many regions of Western Europe and the Mediterranean basin, including all areas where severe Xf has occurred to date, are climatically very well adapted to this vector. In contrast, the warmer, drier regions of the Mediterranean basin are likely to be less suitable for *P. spumarius*. According to the same authors, models also predict that by 2040-2060, important agricultural areas in the southern Mediterranean region could experience a dramatic decline in climatic suitability for *P. spumarius*. Moreover, they reported that the current spatial pattern of Xf epidemics in Europe can be explained in part by the climatic tolerances of *P. spumarius*, and areas predicted to be highly favorable to the bacterium, but not optimal for this vector, still seem to be free of serious Xf epidemics. A more recent study by Farigoule et al. (2022) found a statistically significant positive association between the frequency of *X. fastidiosa*-positive insect vectors and temperature, based on a four-year study in different climatic regions on the French island of Corsica. They found that milder winters correlated with higher insect frequency. They found that the risk of *X. fastidiosa* epidemics will increase in the future, using climate estimates up to the year 2100. They concluded that the fraction of vectors carrying the pathogen is likely to increase, but the environment will remain favorable to the bacterium and its main vector, which may move to higher altitudes.

## Impact of Increased CO2 Concentrations

The functioning of various ecosystems and the carbon cycle are likely to be strongly affected by rising CO2 levels in the atmosphere. The variable incidence of disease in environments with high concentrations of carbon dioxide implies CO2 sensitivity in both hosts and pathogens. In general, rising CO2 levels lead to an increase in photosynthesis, leaf area, plant height, total biomass (shoots and roots), crop yield, sugar and starch content, water use efficiency, growth and yield (Heagle et al. 1979). Consequently, an increase in biomass can modify the microenvironment and the risk of infection. In general, higher plant density tends to lengthen the period during which leaves are wet and regulate temperature, increasing the likelihood of leaf infection by pathogens (Yáñez-López et al., 2012). Thus, many foliar infections can benefit from a greater availability of host tissue and a more humid microenvironment due to denser plant growth. Under such circumstances, diseases such as olive leaf spot (OLS), caused by *Venturia oleaginea*, can find conditions that greatly favor infection, including lack of canopy ventilation, increased humidity and longer wetting periods. In addition to the direct effects of increased CO2 on plant development, it can also have indirect effects, such as reducing the expression of induced resistance. Plant immune responses and hormone levels are influenced by atmospheric CO2, which can alter the interaction between plants and pathogens. As an illustration, higher basal expression of jasmonic acid-sensitive genes under increased CO2 enhanced resistance to the necrotrophic leaf pathogen *Botrytis cinerea*, but decreased resistance to the hemi-biotrophic leaf pathogen *Pseudomonas syringae pv. Tomato (*Zhou et al., 2019).

In addition, there can be direct effects of elevated levels of CO2 on the growth of pathogens. For example, *Colletotrichum gloeosporioides*, the agent responsible for the anthracnose disease that affects many plants, including olive trees, grows faster in *Stylosanthes scabra* at high CO2 levels (Chakraborty and Datta, 2003). High CO2 concentrations promote pathogen fecundity, accelerating the reproduction of organisms in response to climate change. Therefore, the combination of higher fecundity and a suitable environment in larger canopies will provide more opportunities for infection. Thus, increasing population size and number of generations in favorable microclimates would increase the likelihood of more harmful pathotypes emerging more rapidly in populations of sexually reproduced pathogens with high genetic diversity (Sutherst et al. 1996).

## Impact of Changing Precipitation Patterns

The increase in the interannual variance of precipitation caused by climate change is the main cause of calamities such as floods and drought (De Jong et al., 2018). In addition, rainfall variability has a negative impact on the economic well-being of mankind, and this effect is exacerbated when a country's economy is primarily based on agricultural operations (Fahad and Wang, 2018). As variations in precipitation, relative humidity and soil moisture are among the main factors influencing the incidence and severity of plant diseases, climate-induced changes in these variables are likely to have an impact on the occurrence of plant disease epidemics in the future (Romero et al., 2022). With regard to olive diseases, a study by Obanor et al. (2008) showed that free water is necessary for *Spilocaea oleagina* (causing Olive Leaf Spot) conidial germination on leaves incubated at 20°C for 48 hours, and that high relative humidity alone is not sufficient to induce its germination. They reported that conidial germination requires initial leaf wetness of around 6 hours, followed by 100% relative humidity. Recent research indicates that, in general, an increase in relative humidity can increase the prevalence of fungal diseases (Romero et al., 2022). On the other hand, for soil-borne pathogens, many of which cause plant wilt diseases, soil moisture is more important than air humidity. In olive orchards, Verticillium wilt incidence, severity and inoculum density are generally higher in wetter soils (Tsror, 2011; Lopez-Escudero and Blanco-Lopez, 2005; Perez-Rodrıguez et al., 2016).

Drought has a wide range of effects on pathogen infection rates and disease severity (Wakelin et al., 2018). A side effect of water stress is stomatal closure and reduced photosynthesis (Daszkowska-Golec and Szarejko, 2013), which often leads to inhibition of leaf growth, as well as changes in shoot architecture and root/shoot ratio. In this way, drought can hinder plant growth or the production of defense material, which favors the spread of infection. In contrast, a number of diseases are less severe when humidity is low. Furthermore, new diseases capable of tolerating extreme environmental conditions and benefiting from adjustments in plant physiology in response to stress could potentially develop as a result of drought (Rai et al., 2021). Necrotrophs, in general, accelerate drought-induced tree mortality by depleting tree resources as a result of repair and compartmentalization processes, while diseases caused by biotrophs should be less severe under drought conditions due to the close relationship between pathogen performance and tree nutritional status. However, because they deplete the carbohydrate reserves essential to plants' ability to withstand drought, biotrophs should have more serious effects on stressed trees if they manage to infiltrate (Oliva et al., 2014).

Given that drought can have a positive or negative effect on pathogen infection, it's important to discuss how these two types of stress can coexist under field conditions. Some studies have shown that drought stress increases susceptibility to bacterial pathogens (Choi e al., 2013), while other research has shown that it promotes resistance to bacterial pathogens (Ramegowda et al., 2013 ; Gupta et al., 2016). For instance, vessel occlusion caused by the commensal endophyte *Xylella fastidiosa* (Xf) *subsp. pauca* (Cardinale et al., 2018), which spreads from the infection site to colonize the xylem, causes drought stress symptoms in olive trees. Mediterranean olive orchards are regularly confronted with abiotic challenges, in particular water scarcity, which can exacerbate these symptoms. As the two stress factors appear to overlap, potentially creating a detrimental synergy, the interaction between plant and water balance is particularly complex in environments sensitive to Xf infections and water stress. De Pascali et al. (2022) assessed how the two stress factors overlap, evaluating carbohydrate content and expression profiles of carbohydrate metabolism-related genes in two olive cvs (Xf-susceptible Cellina di Nardò and Xf-resistant Leccino) and reporting transcriptional dynamics induced by Xf infection, drought or combined stress (drought/Xf). They found that the response to drought stress induced beneficial effects for *Xylella fastidiosa* resistance in cv Leccino.

## STRATEGIES TO MITIGATE ABIOTIC STRESSES IN OLIVE ORCHARDS UNDER CLIMATE CHANGE

The occurrence of rapid climate change has intensified a multitude of abiotic stresses affecting plant crops, resulting in profound alterations to their functioning and eco-physiological mechanisms (Ma et al., 2020; Kassout et al., 2021). The escalating pace of global warming carries dire consequences for plant growth, crop productivity, and nutritional value, thereby endangering sustainable agricultural production on a global scale (Wheeler and Von Braun, 2013). The combined effects of intensive anthropogenic activities and ongoing climate change have further deteriorated the quality of agricultural land (Fedoron et al., 2010). In their natural habitat, plants typically confront an amalgamation of diverse environmental stressors, including drought, salinity, extreme temperatures, heavy metals, and UV radiation, thus intensifying the severity of combined stresses they experience (Lamers et al., 2020; Xu et al., 2019). However, not all species possess the inherent capacity to adapt and manifest plastic responses that enable their growth in such challenging ecological conditions (Cipollini, 2004).

The adverse impacts of temperature extremes and limited water availability stand as prominent abiotic stressors that exert a profound negative influence on the growth and productivity of crop plants. With impending climate change, projections from the Intergovernmental Panel on Climate Change indicate that by the year 2050, the average global temperature is anticipated to increase by 2-5°C or potentially even more. Such a climatic shift poses considerable challenges to agricultural systems. Of particular concern is the prevalence of drought stress, a major driver of yield fluctuations across diverse crop ecosystems. To counteract the detrimental effects of abiotic stress, plants have evolved a range of mechanisms encompassing avoidance, adaptation, and acclimation strategies (Cramer et al., 2011). Additionally, in their quest to manage stress conditions, plants possess an array of tolerance mechanisms. These encompass the involvement of ion transporters, osmoprotectants, proteins, as well as factors associated with transcriptional control and signaling cascades that are activated to counterbalance the biochemical and molecular changes induced by abiotic stressors (Mathivanan, 2021; Hadiarto and Tran, 2011; Waqas et al., 2019; Pan et al., 2019).

The ability of plants to respond to stress stimuli, generate appropriate signaling, and initiate relevant biochemical and physiological adaptations holds critical significance for their growth and survival (Zhang et al., 2022). Various key factors underpin this process, encompassing the biochemical, physiological, and molecular mechanisms of plants. Furthermore, innovative approaches, such as harnessing the potential of microbiomes (Latef et al., 2014; Lau, et al., 2022) and the utilization of nanotechnology (Silva et al., 2022), have been employed to combat abiotic stresses and ensure sustainable crop productivity. For instance, the enhancement of various physiological processes in plants, such as olive trees, and the subsequent improvement of plant tolerance to stresses, have been attributed to the symbiotic association between plants and multiple strains of Arbuscular mycorrhizal fungi (AMF), including *Glomus* spp, *Rhizophagus* spp, and *Funneliformis* spp (Oliveira et al., 2022). AMF have the ability to establish symbiotic relationships with plants and are widely recognized as a prevalent form of biostimulant employed in agricultural practices (Du Jardin, 2015). Consequently, the utilization of seaweed and plant extracts has garnered substantial attention in both research and commercial sectors (Cristofano et al., 2021). Seaweed, which has been utilized in agriculture since ancient times, possesses remarkable beneficial effects on crop plants (Franzoni et al., 2022). On the other hand, certain plant extracts are known to harbor a wealth of bioactive compounds that can activate diverse physiological processes in plants, thereby bolstering their overall performance (Zulfiqar et al., 2020).

## STRATEGIES TO MITIGATE BIOTIC STRESSES IN OLIVE ORCHARDS IN THE CONTEXT OF CLIMATE CHANGE

There are urgent and serious concerns about how climate change will affect productivity, pest and pathogen populations and their control methods. Therefore, the spatial and temporal distribution of pests and diseases, as well as control strategies, will have to evolve in line with potential climate change scenarios. Hence, climate change would mainly affect the effectiveness of chemical and biological control agents and/or host resistance. In fact, conventional strategies to control diseases and pests rely mainly on the use of fungicides, but the dynamics of residues of these chemicals in foliage can be altered by changes in temperature and variable rainfall (Delcour et al., 2015) . In the worst-case scenario, a number of crops could require more pesticide spray treatments or higher application rates, increasing farmers' expenses, consumer prices and the risk of fungicide resistance developing (Juroszek and von Tiedemann 2011). Similarly, as biocontrol agent populations are sensitive to environmental changes and extreme conditions, the application of these agents in the field may encounter additional problems. Furthermore, and as a consequence of climate change, the accelerated evolution of insects and pathogens due to increased fecundity, a greater number of generations per season or a microclimate more conducive to development would pose the greatest threat to the longevity of host resistance. Consequently, changes in the Earth's climate have highlighted the need for new, environmentally-friendly approaches to pest and disease control, for sustainable agricultural production that guarantees food security. In this way, it will be strategically essential to assess the effectiveness of current cultural, genetic, biological and chemical management techniques under changing climatic conditions, and to research new tools and techniques to cope with the anticipated changes in olive orchards. The use of resistant cultivars, the adaptation of planting dates, the use of diversified species in cropping systems, the use of reliable tools to predict epidemics and the application of integrated pest management strategies are examples of current preventive crop protection measures that could become more crucial in the future. The aim is to adopt crop protection technologies that are available, effective and adaptable to altered conditions, while prioritizing monitoring and surveillance to guard against any unexpected incidents that may result from climate change.

## Improving Monitoring Capacity

Currently, disease and pest surveillance is mainly coordinated at regional and national levels, with significant differences between developing and developed countries (Carvajal-Yepes et al., 2019). These differences and the lack of global coordination make it difficult to respond rapidly to new or rapidly spreading epidemics. It is therefore necessary to set up an international system for exchanging information between regions, including essential data on insects, invasive alien species, diseases and ecological conditions, including meteorological information. It is also essential to strengthen international cooperation, including that of national, regional and global organizations. This system could be modeled and extended by linking existing regional surveillance systems to national and regional centers, and would enable real-time monitoring and rapid response to reduce the risk of new or rapidly spreading international diseases and pests. As an illustration, concern about the possible spread of *Xylella fastidiosa* in Europe has been heightened by the extraordinary loss of olive trees (Olmo et al., 2021). Compulsory surveys were therefore carried out in EU member states, revealing that the bacterium had also spread to Corsica, Provence, the Alps and the Côte d'Azur in France, as well as Alicante and the Balearic Islands in Spain (Denancé et al., 2017). Multiple Xf subspecies and genotypes have been found in Italy, France and Spain, reinforcing the threat to Europe and the Mediterranean basin (Landa et al., 2020). Countries in the region where there are as yet no active outbreaks need to monitor their crops regularly, as early detection of the disease could help avoid massive losses caused by Xf (El Handi et al., 2022). Monitoring and using information on climate forecasts and the risks associated with biotic stress factors can also help growers prevent pest and disease problems by taking special preventive precautions (Heeb et al., 2019).

## Adapted Integrated Pest Management (IPM) Methods

Integrated Pest Management (IPM) is a comprehensive approach to pest and disease management that uses all methods while minimizing the use of chemical agents to keep pest numbers below economic thresholds (Wang et al., 2022). IPM strategies are mainly developed by farmers and scientists to reduce negative environmental impacts while maximizing crop yields and financial benefits (FAO, 2023). IPM stimulates natural pest and disease control mechanisms while promoting the growth of a healthy crop with the least possible disruption to agroecosystems (FAO, 2020). The procedure includes a number of management processes during which decisions, assessments and control measures must be taken in order to correctly implement the chosen approach. Hence, an economic or intervention threshold is first established, i.e. a level at which all relevant variables indicate that appropriate management measures must be taken to keep pest populations under control before economic yield losses occur. Regular field monitoring is also a fundamental part of this approach, to detect the appearance and spread of pests and diseases, and to identify natural enemies. Therefore, early discovery often enables the pest problem to be controlled by inexpensive spot treatments, or by eliminating infested or infected pests and plant material. As the first line of pest control, IPM programs aim to manage crops in such a way as to prevent pests from becoming a threat. This involves applying cultural techniques such as selecting pest-resistant cultivars and planting pest-free rootstocks. These control methods present little or no risk to humans or the environment, and can be highly effective and economical. When monitoring, identification and action thresholds indicate that pest control treatment is necessary and preventive methods are no longer available or effective, IPM programs evaluate the appropriate control strategy based on efficacy and risk.

Selecting desired traits for future needs is the best method of adapting crops to climate change. Breeding initiatives for trees and agricultural plants can encourage genetic variety, stress tolerance and disease resistance. These must be combined with other breeding objectives, including yield, fruit quality and appropriate shelf life. The selection of resistant cultivars has long been used to combat olive diseases, and breeding programs have included the selection of cultivars and wild olives resistant to *Verticillium dahliae*, due to the lack of effective chemical control of Verticillium wilt. Selection is still ongoing, and hundreds of cultivars have so far been tested in different parts of the world. Only three cultivars - 'Empeltre', 'Frantoio' and 'Changlot Real' - have demonstrated a high level of resistance to Verticillium wilt. However, these cultivars have considerable agronomic drawbacks which have discouraged their use in affected orchards (Valverde et al., 2023).

To reduce the impact of agricultural pests and diseases on crops in a changing environment, it is necessary to modify farming practices and adopt adaptive management strategies (Thomson et al., 2010). The term "cultural control" refers to the use of sound agronomic methods to prevent or minimize disease and pest infestations and damage. Vulnerability to pest outbreaks can be reduced by planting several varieties, at different times of the year, and increasing habitat diversity in edges to encourage the number of natural enemies. Other strategies are used to minimize the impact of pests and diseases at farm level, such as mulching, raised beds and shelters to preserve soil moisture, protecting crops from heavy rain, heat and flooding, and preventing soil deterioration.

The use of macrobial agents (predators or parasitoids) or microbial agents (bacteria, viruses, fungi or nematodes) against insect pests, mites and diseases is an essential part of integrated pest management (Kenis et al., 2017). In certain situations, natural enemies such as parasitic wasps and predatory arthropods can be very effective in significantly reducing pest populations (Hajek and Eilenberg, 2018). Common methods of controlling endemic pests include regular releases of commercially available natural enemies, and protecting natural enemy populations by providing refugia or avoiding actions that harm them. In the Mediterranean region, biological control of olive fly *Bactrocera oleae* is mainly carried out by the chalcidoidea ectoparasitoids *Eupelmus urozonus* (Eupelmidae), *Pnigalio mediterraneus* (Eulophidae), *Eurytoma martellii* (Eurytomidae) and *Cyrtoptyx latipes* (Pteromalidae), and by the endoparasitoid ichneumonoidea *Psyttalia concolor* (Braconidae) (Arambourg, 1986; Pinheiro et al., 2017). According to Garantonakis et al. (2017), most research has focused on the braconid *Psyttalia concolor*. Although this parasitoid has become established in a few southern regions (Miranda et al., 2008), it has not been able to produce significant numbers to exert sufficient control over the fly.

Chemical control is also an important element of pest and disease management techniques, but it is necessary to understand how global warming affects the efficacy of many synthetic insecticides, their longevity and the emergence of pesticide resistance in pathogens and pest populations.

## CONCLUSION

Climate change threatens food security through its effects on food crops and animal feed, both in terms of quality and quantity. Droughts, heat waves and floods are examples of increasingly frequent extreme weather events that pose risks to agriculture by affecting plant development and production. Traditional, intensive olive-growing systems are facing climate change in the Mediterranean region, where most of the world's olive oil is produced and which is particularly sensitive to climatic disturbances. The olive tree, a xerophilous species that has evolved to thrive in the harsh Mediterranean environment, faces a number of biotic and abiotic stresses, resulting in considerable economic losses every year. Climate change is likely to considerably exacerbate the negative effects of these stresses and increase crop losses. Indeed, despite being considered highly resistant and tolerant to various abiotic stresses, olive trees can be affected by adverse climatic conditions, such as lack of water, heat waves, salinity and flooding, and their protective mechanisms consume a significant amount of energy, hampering their growth and productivity. These abiotic variables can also modify the way the olive tree, as a host, adapts to its pests and diseases, further increasing the risk of epidemics. Climate change further increases the risk of these outbreaks by affecting the evolution of biotic stress factors and the introduction of new pathogenic strains and invasive insect pests. Possible effects on these factors include changes in distribution and abundance, phenology and changes in certain aspects of population dynamics, as well as the emergence of new species. Furthermore, the combination of biotic and abiotic stress factors is likely to further increase yield losses in olive groves in a changing climate.

In conclusion and in order to ensure the sustainability of olive-growing ecosystems, preventive, mitigation and adaptation measures are essential to reduce projected increases in biotic stressors risk in olive orchards. In addition, the use of effective monitoring can help improve predictive power and provide more appropriate pest and disease management advice under a climate change scenario. To preserve our olive heritage, a global surveillance strategy is also needed to constantly monitor and predict the emergence of the main pests and diseases, in order to reduce the risk of rapid international spread of species. The use of new technological (digital, satellite) and biotechnological (genomics) tools will make it possible to set up such a system and enable early detection and forecasting of outbreaks.

## REFERENCES

Abbasi, H., M. Jamil, A. Haq, S. Ali, R. Ahmad, and Z. Malik. (2016). Salt stress manifestation on plants, mechanism of salt tolerance and potassium role in alleviating it: a review. *Zemdirbyste- Agricul. 103*, 229– 238. doi:10.13080/z-a.2016.103.030

Abdallah, M. B., Methenni, K., Nouairi, I., Zarrouk, M., & Youssef, N. B. (2017). Drought priming improves subsequent more severe drought in a drought- sensitive cultivar of olive cv. Chétoui. *Scientia Horticulturae, 221*, 43–52. doi:10.1016/j.scienta.2017.04.021 PMID:28713194

Abdel Latef, A. A. H., & Miransari, M. (2014). The role of arbuscular mycorrhizal fungi in alleviation of salt stress. *Use of Microbes for the Alleviation of Soil Stresses: Volume 2: Alleviation of Soil Stress by PGPR and Mycorrhizal Fungi*, 23-38.

Aragüés, R., Puy, J., & Isidoro, D. (2004). Vegetative growth response of young olive trees (*Olea europaea* L., cv. Arbequina) to soil salinity and waterlogging. *Plant and Soil, 258*(1), 69–80. doi:10.1023/B:PLSO.0000016537.61832.6e

Arambourg, Y. (1986). Traité d'entomologie oléicole (pp. 360-pp). COI.

Armendáriz, A., De La Iglesia, L., Santiago, Y., Campillo, G., Alberte, C., Barroso, L. M., & Sanz, A. P. (2007). Ciclo del prays del olivo (" Prays oleae" Bern.) en Arribes del Duero. *Boletin de Sanidad Vegetal, Plagas, 33*(4), 443–456.

Armstrong, W., & Drew, M. C. (2002). Root growth and metabolism under oxygen defciency. In Y. Waisel, A. Eshel, & U. Kafkaf (Eds.), *Plant roots: the hidden half* (3rd ed., pp. 729–761). Marcel Dekker.

Azenzem, R., Alfeddy, M. N., & Kassout, J. (2023). Climate Change and Insect Pests: Economic Damages and Adaptive Strategies in Mediterranean Olive Groves. In Climate Change and the Economic Importance and Damages of Insects (pp. 14-47). IGI Global.

Bacelar, E. A., Moutinho-Pereira, J. M., Gonçalves, B. C., Ferreira, H. F., & Correia, C. M. (2007). Changes in growth, gas exchange, xylem hydraulic properties and water use effciency of three olive cultivars under contrasting water availability regimes. *Environmental and Experimental Botany, 60*(2), 183–192. doi:10.1016/j.envexpbot.2006.10.003

Bacelar, E. A., Santos, D. L., Moutinho-Pereira, J. M., Gonçalves, B. C., Ferreira, H. F., & Correia, C. M. (2006). Immediate responses and adaptative strategies of three olive cultivars under contrasting water availability regimes: Changes on structure and chemical composition of foliage and oxidative damage. *Plant Science, 170*(3), 596–605. doi:10.1016/j.plantsci.2005.10.014

Bale, J. S., & Hayward, S. A. L. (2010). Insect overwintering in a changing climate. *The Journal of Experimental Biology, 213*(6), 980–994. doi:10.1242/jeb.037911 PMID:20190123

Bale, J. S., Masters, G. J., Hodkinson, I. D., Awmack, C., Bezemer, T. M., Brown, V. K., Butterfield, J., Buse, A., Coulson, J. C., Farrar, J., Good, J. E. G., Harrington, R., Hartley, S., Jones, T. H., Lindroth, R. L., Press, M. C., Symrnioudis, I., Watt, A. D., & Whittaker, J. B. (2002). Herbivory in global climate change research: Direct effects of rising temperature on insect herbivores. *Global Change Biology, 8*(1), 1–16. doi:10.1046/j.1365-2486.2002.00451.x

Balestra, G. M., Lamichhane, J. R., Kshetri, M. B., Mazzaglia, A., & Varvaro, L. (2009). First report of olive knot caused byPseudomonas savastanoipv.savastanoiin Nepal. *Plant Pathology*, *58*(2), 393–393. doi:10.1111/j.1365-3059.2008.02007.x

Bebber, D. P., Holmes, T., & Gurr, S. J. (2014). The global spread of crop pests and pathogens. *Global Ecology and Biogeography*, *23*(12), 1398–1407. doi:10.1111/geb.12214

Bejarano-Alcázar, J., Blanco-López, M. A., Melero-Vara, J. M., & Jiménez-Díaz, R. M. (1996). Etiology, importance, and distribution of Verticillium wilt of cotton in southern Spain. *Plant Disease*, *80*(11), 1233–1238. doi:10.1094/PD-80-1233

Bernstein, L. (1965). Salt tolerance of fruit crops (No. 292). Agriculture Research Service, US Department of Agriculture.

Bezemer, T. M., Jones, T. H., & Knight, K. J. (1998). Long-term effects of elevated CO 2 and temperature on populations of the peach potato aphid Myzus persicae and its parasitoid Aphidius matricariae. *Oecologia*, *116*(1-2), 128–135. doi:10.1007004420050571 PMID:28308517

Boullis, A., Francis, F., & Verheggen, F. J. (2015). Climate change and tritrophic interactions: Will modifications to greenhouse gas emissions increase the vulnerability of herbivorous insects to natural enemies? *Environmental Entomology*, *44*(2), 277–286. doi:10.1093/ee/nvu019 PMID:26313181

Brito, C., Dinis, L. T., Moutinho-Pereira, J., & Correia, C. M. (2019). Drought stress effects and olive tree acclimation under a changing climate. *Plants*, *8*(7), 232. doi:10.3390/plants8070232 PMID:31319621

Brodribb, T. J., Powers, J., Cochard, H., & Choat, B. (2020). Hanging by a thread? Forests and drought. *Science*, *368*(6488), 261–266. doi:10.1126cience.aat7631 PMID:32299945

Bueno, A. M., & Jones, O. (2002). Alternative methods for controlling the olive fly, Bactrocera oleae, involving semiochemicals. *IOBC/WPRS Bulletin*, *25*(9), 147–156.

Cardinale, M., Luvisi, A., Meyer, J. B., Sabella, E., De Bellis, L., Cruz, A. C., Ampatzidis, Y., & Cherubini, P. (2018). Specific fluorescence in situ hybridization (FISH) test to highlight colonization of xylem vessels by Xylella fastidiosa in naturally infected olive trees (Olea europaea L.). *Frontiers in Plant Science*, *9*, 431. doi:10.3389/fpls.2018.00431 PMID:29681910

Carvajal-Yepes, M., Cardwell, K., Nelson, A., Garrett, K. A., Giovani, B., Saunders, D. G., Kamoun, S., Legg, J. P., Verdier, V., Lessel, J., Neher, R. A., Day, R., Pardey, P., Gullino, M. L., Records, A. R., Bextine, B., Leach, J. E., Staiger, S., & Tohme, J. (2019). A global surveillance system for crop diseases. *Science*, *364*(6447), 1237–1239. doi:10.1126cience.aaw1572 PMID:31249049

Castroverde, C. D. M., & Dina, D. (2021). Temperature regulation of plant hormone signaling during stress and development. *Journal of Experimental Botany*, *72*(21), 7436–7458. doi:10.1093/jxb/erab257 PMID:34081133

Chakraborty, S., & Datta, S. (2003). How will plant pathogens adapt to host plant resistance at elevated CO2 under a changing climate? *The New Phytologist*, *159*(3), 733–742. doi:10.1046/j.1469-8137.2003.00842.x PMID:33873600

Chartzoulakis, K., Loupassaki, M., Bertaki, M., & Androulakis, I. (2002). Effects of NaCl salinity on growth, ion content and C02 assimilation rate of six olive cultivars. *Scientia Horticulturae*, *96*(1-4), 235–247. doi:10.1016/S0304-4238(02)00067-5

Chartzoulakis, K., Psarras, G., Vemmos, S., Loupassaki, M., & Bertaki, M. (2006). Response of two olive cultivars to salt stress and potassium supplement. *Journal of Plant Nutrition*, *29*(11), 2063–2078. doi:10.1080/01904160600932682

Chartzoulakis, K. S. (2005). Salinity and olive: growth, salt tolerance, photosynthesis and yield. *Agricultural Water Management, 78*(1– 2), 108– 121.

Cheng, Y. T., Zhang, L., & He, S. Y. (2019). Plant-microbe interactions facing environmental challenge. *Cell Host & Microbe*, *26*(2), 183–192. doi:10.1016/j.chom.2019.07.009 PMID:31415751

Chermiti, B. (1989). Dynamique des populations du psylle de l'olivier E. olivina, en conditions méditerranéennes [Thèse Doctorat Es-Sciences, Université d'AixMarseille, France].

Chidawanyika, F., Mudavanhu, P., & Nyamukondiwa, C. (2012). Biologically based methods for pest management in agriculture under changing climates: Challenges and future directions. *Insects*, *3*(4), 1171–1189. doi:10.3390/insects3041171 PMID:26466733

Cho, G., Lee, J. H., & Choi, D. S. (2021). First interception of olive psyllid Euphyllura olivina (Costa, 1839)(Hemiptera: Psylloidea) in the East Palaearctic region. *Phytoparasitica*, *49*(5), 865–868. doi:10.100712600-021-00924-w

Choat, B., Brodribb, T. J., Brodersen, C. R., Duursma, R. A., López, R., & Medlyn, B. E. (2018). Triggers of tree mortality under drought. *Nature*, *558*(7711), 531–539. doi:10.103841586-018-0240-x PMID:29950621

Choi, H. K., Iandolino, A., da Silva, F. G., & Cook, D. R. (2013). Water deficit modulates the response of Vitis vinifera to the Pierce's disease pathogen Xylella fastidiosa. *Molecular Plant-Microbe Interactions*, *26*(6), 643–657. doi:10.1094/MPMI-09-12-0217-R PMID:23425100

Cipollini, D. (2004). Stretching the limits of plasticity: Can a plant defend against both competitors and herbivores? *Ecology*, *85*(1), 28–37. doi:10.1890/02-0615

Coakley, S. M., Scherm, H., & Chakraborty, S. (1999). Climate change and plant disease management. *Annual Review of Phytopathology*, *37*(1), 399–426. doi:10.1146/annurev.phyto.37.1.399 PMID:11701829

Cohen, I., Zandalinas, S. I., Huck, C., Fritschi, F. B., & Mittler, R. (2021). Meta- Analysis of Drought and Heat Stress Combination Impact on Crop Yield and Yield Components. *Physiologia Plantarum*, *171*(1), 66–76. doi:10.1111/ppl.13203 PMID:32880977

Cohen, S. P., & Leach, J. E. (2020). High temperature-induced plant disease susceptibility: More than the sum of its parts. *Current Opinion in Plant Biology*, *56*, 235–241. doi:10.1016/j.pbi.2020.02.008 PMID:32321671

Connor, D. J., & Fereres, E. (2005). The physiology of adaptation and yield expression in olive. *Horticultural Reviews*, *31*, 155–229.

Cornara, D., Saponari, M., Zeilinger, A. R., de Stradis, A., Boscia, D., Loconsole, G., Bosco, D., Martelli, G. P., Almeida, R. P. P., & Porcelli, F. (2017). Spittlebugs as vectors of Xylella fastidiosa in olive orchards in Italy. *Journal of Pest Science*, *90*(2), 521–530. doi:10.100710340-016-0793-0 PMID:28275326

Cramer, G. R., Urano, K., Delrot, S., Pezzotti, M., & Shinozaki, K. (2011). Effects of abiotic stress on plants: A systems biology perspective. *BMC Plant Biology*, *11*(1), 1–14. doi:10.1186/1471-2229-11-163 PMID:22094046

Cramer, W., Guiot, J., Fader, M., Garrabou, J., Gattuso, J. P., Iglesias, A., Lange, M. A., Lionello, P., Llasat, M. C., Paz, S., Peñuelas, J., Snoussi, M., Toreti, A., Tsimplis, M. N., & Xoplaki, E. (2018). Climate change and interconnected risks to sustainable development in the Mediterranean. *Nature Climate Change*, *8*(11), 972–980. doi:10.103841558-018-0299-2

Cristofano, F., El-Nakhel, C., & Rouphael, Y. (2021). Biostimulant substances for sustainable agriculture: Origin, operating mechanisms and effects on Cucurbits, leafy greens, and nightshade vegetables species. *Biomolecules*, *11*(8), 1103. doi:10.3390/biom11081103 PMID:34439770

Daszkowska-Golec, A., & Szarejko, I. (2013). Open or close the gate–stomata action under the control of phytohormones in drought stress conditions. *Frontiers in Plant Science*, *4*, 138. doi:10.3389/fpls.2013.00138 PMID:23717320

De Jong, P., Tanajura, C. A. S., Sánchez, A. S., Dargaville, R., Kiperstok, A., & Torres, E. A. (2018). Hydroelectric production from Brazil's São Francisco River could cease due to climate change and inter-annual variability. *The Science of the Total Environment*, *634*, 1540–1553. doi:10.1016/j.scitotenv.2018.03.256 PMID:29710652

De Pascali, M., Vergine, M., Negro, C., Greco, D., Vita, F., Sabella, E., De Bellis, L., & Luvisi, A. (2022). Xylella fastidiosa and drought stress in olive trees: A complex relationship mediated by soluble sugars. *Biology (Basel)*, *11*(1), 112. doi:10.3390/biology11010112 PMID:35053110

Delcour, I., Spanoghe, P., & Uyttendaele, M. (2015). Literature review: Impact of climate change on pesticide use. *Food Research International*, *68*, 7–15. doi:10.1016/j.foodres.2014.09.030

Delgado-Baquerizo, M., Guerra, C. A., Cano-Díaz, C., Egidi, E., Wang, J.-T., Eisenhauer, N., Singh, B. K., & Maestre, F. T. (2020). The proportion of soil-borne pathogens increases with warming at the global scale. *Nature Climate Change*, *10*(6), 550–554. doi:10.103841558-020-0759-3

DeLucia, E. H., Casteel, C. L., Nabity, P. D., & O'Neill, B. F. (2008). Insects take a bigger bite out of plants in a warmer, higher carbon dioxide world. *Proceedings of the National Academy of Sciences of the United States of America*, *105*(6), 1781–1782. doi:10.1073/pnas.0712056105 PMID:18256185

Den Bieman, C. F. M., Malenovský, I., Burckhardt, D., & Heijerman, T. (2019). First checklist of the Dutch jumping plant lice since 93 years (Hemiptera: Psylloidea). *Nederlandse Faunistische Mededelingen*, *53*, 55–118.

Denancé, N., Legendre, B., Briand, M., Olivier, V., De Boisseson, C., Poliakoff, F., & Jacques, M. A. (2017). Several subspecies and sequence types are associated with the emergence of Xylella fastidiosa in natural settings in France. *Plant Pathology*, *66*(7), 1054–1064. doi:10.1111/ppa.12695

DeVay, J. E., Forrester, L. L., Garber, R. H., & Butterfield, E. J. (1974). Characteristics and Concentration of Propagules of. *Phytopathology*, *64*(1), 22–29. doi:10.1094/Phyto-64-22

Dichio, B., Xiloyannis, C., Sofo, A., & Montanaro, G. (2006). Osmotic regulation in leaves and roots of olive trees during a water defcit and rewatering. *Tree Physiology*, *26*(2), 179–185. doi:10.1093/treephys/26.2.179 PMID:16356914

Driouech, F., M., Déqué, & E. Sánchez- Gómez. (2010). Weather regimes— Moroccan precipitation link in a regional climate change simulation. *Global and Planetary Change, 72*(1– 2), 1– 10.

Du Jardin, P. (2015). Plant biostimulants: Definition, concept, main categories and regulation. *Scientia Horticulturae*, *196*, 3–14. doi:10.1016/j.scienta.2015.09.021

El Handi, K., Hafidi, M., Sabri, M., Frem, M., El Moujabber, M., Habbadi, K., Haddad, N., Benbouazza, A., Abou Kubaa, R., & Achbani, E. H. (2022). Continuous pest surveillance and monitoring constitute a tool for sustainable agriculture: Case of Xylella fastidiosa in Morocco. *Sustainability (Basel)*, *14*(3), 1485. doi:10.3390u14031485

El-Salam, A., Ezzat, A. M., Salem, S. A. W., Abdel-Rahman, R. S., El-Behery, H. H., & Magd Elden, M. A. (2019). Effects of climatic changes on olive fly, Bactrocera oleae (Rossi) population dynamic with respect to the efficacy of its larval parasitoid in Egyptian olive trees. *Bulletin of the National Research Center*, *43*(1), 1–9. doi:10.118642269-019-0220-9

Elad, Y., & Pertot, I. (2014). Climate Change Impacts on Plant Pathogens and Plant Diseases. *Journal of Crop Improvement*, *28*(1), 99–139. doi:10.1080/15427528.2014.865412

European Food Safety Authority (EFSA), Delbianco, A., Gibin, D., Pasinato, L., & Morelli, M. (2021). Update of the Xylella spp. host plant database–systematic literature search up to 31 December 2020. *EFSA Journal*, *19*(6), e06674. PMID:34188716

Fahad, S., & Wang, J. (2018). Farmers' risk perception, vulnerability, and adaptation to climate change in rural Pakistan. *Land Use Policy*, *79*, 301–309. doi:10.1016/j.landusepol.2018.08.018

Fang, X., Turner, N. C., Yan, G., Li, F., & Siddique, K. H. (2010). Flower numbers, pod production, pollen viability, and pistil function are reduced and flower and pod abortion increased in chickpea (*Cicer arietinum* L.) under terminal drought. *Journal of Experimental Botany*, *61*(2), 335–345. doi:10.1093/jxb/erp307 PMID:19854801

FAO. (2010). *The second report on the state of the World's Plant Genetic Resources for Food and Agriculture*. Commission on Genetic Resources for Food.

FAO. (2020). *NSP – Integrated Pest Management, FAO definition*. FAO. http:// www.fao.org/agriculture/crops/thematic-sitemap/theme/pests/ipm/ en/

FAO. (2023). *How to Practice Integrated Pest Management*. FAO. https://www.fao.org/agriculture/crops/thematic-sitemap/theme/compendium/tools-guidelines/how-to-ipm/en/

FAOSTAT. (2021). Food and Agriculture Organization of the United Nations Statistical Dataset. FAO: Rome, Italy.

Farigoule, P., Chartois, M., Mesmin, X., Lambert, M., Rossi, J. P., Rasplus, J. Y., & Cruaud, A. (2022). Vectors as Sentinels: Rising Temperatures Increase the Risk of Xylella fastidiosa Outbreaks. *Biology (Basel)*, *11*(9), 1299. doi:10.3390/biology11091299 PMID:36138778

Farooq, M., M. Hussain, A. Wahid, & K. H. M. Siddique. (2012). Drought stress in plants: an overview. *Plant responses to drought stress* 1– 33

Fedoroff, N. V., Battisti, D. S., Beachy, R. N., Cooper, P. J., Fischhoff, D. A., Hodges, C. N. et al. (2010). Radically rethinking agriculture for the 21st century. *science, 327*(5967), 833-834.

Fernández, J. E. (2014). Understanding olive adaptation to abiotic stresses as a tool to increase crop performance. *Environmental and Experimental Botany*, *103*, 158–179. doi:10.1016/j.envexpbot.2013.12.003

Fernández, J. E., Moreno, F., Cabrera, F., Arrue, J. L., & Martín-Aranda, J. (1991). Drip irrigation, soil characteristics and the root distribution and root activity of olive trees. *Plant and Soil*, *133*(2), 239–251. doi:10.1007/BF00009196

Fletcher, B. S. (1989). Temperature-development rate relationships of immature stages and adult of tephritid fruit flies. In A. S. Robinson and G. Hooper [eds.], Fruit flies: their biology, natural enemies, and control. Elsevier, Amsterdam.

Fraga, H., Pinto, J. G., Viola, F., & Santos, J. A. (2020). Climate change projections for olive yields in the Mediterranean Basin. *International Journal of Climatology*, *40*(2), 769–781. doi:10.1002/joc.6237

Franzoni, G., Cocetta, G., Prinsi, B., Ferrante, A., & Espen, L. (2022). Biostimulants on crops: Their impact under abiotic stress conditions. *Horticulturae*, *8*(3), 189. doi:10.3390/horticulturae8030189

Fukao, T., Barrera-Figueroa, B. E., Juntawong, P., & Peña-Castro, J. M. (2019). Submergence and waterlogging stress in plants: A review highlighting research opportunities and understudied aspects. *Frontiers in Plant Science*, *10*, 340. doi:10.3389/fpls.2019.00340 PMID:30967888

Garantonakis, N., Varikou, K., & Birouraki, A. (2017). Parasitism of Psytallia concolor (Hymenoptera: Braconidae) on Bactrocera oleae (Diptera: Tephritidae) infesting different olive varieties. *Phytoparasitica*, *45*(4), 461–469. doi:10.100712600-017-0597-7

Gelybó, G., Tóth, E., Farkas, C., Horel, Á., Kása, I., & Bakacsi, Z. (2018). Potential impacts of climate change on soil properties. *Agrokémia és Talajtan*, *67*(1), 121–141. doi:10.1556/0088.2018.67.1.9

Godefroid, M., Morente, M., Schartel, T., Cornara, D., Purcell, A., Gallego, D., & Fereres, A. (2021). Climate tolerances of Philaenus spumarius should be considered in risk assessment of disease outbreaks related to Xylella fastidiosa. *Journal of Pest Science*, 1–14.

Gornall, J., Betts, R., Burke, E., Clark, R., Camp, J., Willett, K., & Wiltshire, A. (2010). Implications of climate change for agricultural productivity in the early twenty-first century. *Philosophical Transactions of the Royal Society of London. Series B, Biological Sciences*, *365*(1554), 2973–2989. doi:10.1098/rstb.2010.0158 PMID:20713397

Graniti, A., Faedda, R., Cacciola, S. O., & di San Lio, G. M. (2011). *Olive diseases in a changing ecosystem.*

Gregory, P. J., Johnson, S. N., Newton, A. C., & Ingram, J. S. (2009). Integrating pests and pathogens into the climate change/food security debate. *Journal of Experimental Botany*, *60*(10), 2827–2838. doi:10.1093/jxb/erp080 PMID:19380424

Gucci, R., Mancuso, S., & Sebastiani, L. (2003). Resistenza agli stress ambientali. In: Fiorino P (ed) Olea— Trattato di Olivicoltura. Edagricole. Edizioni Agricole de Il Sole 24 ORE Edagricole Srl, pp 91– 111.

Gucci, R., & Tattini, M. (2010). Salinity tolerance in olive. In J. Janick (Ed.), *Horticultural Reviews* (Vol. 21, pp. 177–214). John Wiley & Sons, Inc.

Guerfel, M., Baccouri, O., Boujnah, D., Chaïbi, W., & Zarrouk, M. (2009). Impacts of water stress on gas exchange, water relations, chlorophyll content and leaf structure in the two main Tunisian olive (*Olea europaea* L.) cultivars. *Scientia Horticulturae*, *119*(3), 257–263. doi:10.1016/j.scienta.2008.08.006

Gupta, A., Sarkar, A. K., & Senthil-Kumar, M. (2016). Global transcriptional analysis reveals unique and shared responses in Arabidopsis thaliana exposed to combined drought and pathogen stress. *Frontiers in Plant Science*, *7*, 686. doi:10.3389/fpls.2016.00686 PMID:27252712

Gutierrez, A. P., Ponti, L., & Cossu, Q. A. (2009). Effects of climate warming on olive and olive fly (Bactrocera oleae (Gmelin)) in California and Italy. *Climatic Change*, *95*(1), 195–217. doi:10.100710584-008-9528-4

Hadiarto, T., & Tran, L. S. P. (2011). Progress studies of drought-responsive genes in rice. *Plant Cell Reports*, *30*(3), 297–310. doi:10.100700299-010-0956-z PMID:21132431

Hahn, D. A., & Denlinger, D. L. (2007). Meeting the energetic demands of insect diapause: Nutrient storage and utilization. *Journal of Insect Physiology*, *53*(8), 760–773. doi:10.1016/j.jinsphys.2007.03.018 PMID:17532002

Hajek, A. E., & Eilenberg, J. (2018). *Natural enemies: an introduction to biological control*. Cambridge University Press. doi:10.1017/9781107280267

Hall, B. H., Cother, E. J., Whattam, M., Noble, D., Luck, J., & Cartwright, D. (2004). First report of olive knot caused by Pseudomonas savastanoi pv. savastanoi on olives (Olea europaea) in Australia. *Australasian Plant Pathology*, *33*(3), 433–436. doi:10.1071/AP04031

Hamilton, J. G., Dermody, O., Aldea, M., Zangerl, A. R., Rogers, A., Berenbaum, M. R., & Delucia, E. H. (2005). Anthropogenic changes in tropospheric composition increase susceptibility of soybean to insect herbivory. *Environmental Entomology*, *34*(2), 479–485. doi:10.1603/0046-225X-34.2.479

Harvey, J. A., Heinen, R., Gols, R., & Thakur, M. P. (2020). Climate change-mediated temperature extremes and insects: From outbreaks to breakdowns. *Global Change Biology*, *26*(12), 6685–6701. doi:10.1111/gcb.15377 PMID:33006246

Hassan, M. M., & Seif, S. A. (1990). Response of seven olive cultivars to water logging. *Gartenbauwissenschaft*, *55*, 223–225.

Heagle, A. S., Spencer, S., & Letchworth, M. B. (1979). Yield response of winter wheat to chronic doses of ozone. *Canadian Journal of Botany*, *57*(19), 1999–2005. doi:10.1139/b79-250

Heeb, L., Jenner, E., & Cock, M. J. (2019). Climate-smart pest management: Building resilience of farms and landscapes to changing pest threats. *Journal of Pest Science*, *92*(3), 951–969. doi:10.100710340-019-01083-y

Henricot, B., Gorton, C., Denton, J., & Denton, G. (2009). Pseudocercospora cladosporioides, the cause of leaf spot on olive, a pathogen new to the United Kingdom. *Plant Pathology*, *58*(4), 803–803. doi:10.1111/j.1365-3059.2009.02079.x

Hernandez-Santana, V., Rodriguez-Dominguez, C. M., Fernández, J. E., & Diaz-Espejo, A. (2016). Role of leaf hydraulic conductance in the regulation of stomatal conductance in almond and olive in response to water stress. *Tree Physiology*, *36*(6), 725–735. doi:10.1093/treephys/tpv146 PMID:26846979

Hougardy, E., Wang, X., Hogg, B. N., Johnson, M. W., Daane, K. M., & Pickett, C. H. (2020). Current distribution of the olive psyllid, Euphyllura olivina, in California and initial evaluation of the Mediterranean parasitoid Psyllaephagus euphyllurae as a biological control candidate. *Insects*, *11*(3), 146. doi:10.3390/insects11030146 PMID:32110919

Ibba, K., Kassout, J., Boselli, V., Er-Raki, S., Oulbi, S., Mansouri, L. E., Bouizgaren, A., Sikaoui, L., & Hadria, R. (2023). Assessing the impact of deficit irrigation strategies on agronomic and productive parameters of Menara olive cultivar: Implications for operational water management. *Frontiers in Environmental Science*, *11*, 1100552. doi:10.3389/fenvs.2023.1100552

Intergovernmental Panel on Climate Change (IPCC). (2021). Summary for policymakers. In V. Masson-Delmotte, P. Zhai, A. Pirani, S. L. Connors, C. Péan, & S. Berger (Eds.), *Climate change 2021: The physical science basis. Contribution of working group I to the sixth assessment report of the intergovernmental panel on climate change*. Cambridge University Press.

Intergovernmental Panel on Climate Change (IPCC). (2022). *Sixth assessment report*. IPCC. https://www.ipcc.ch/assessment-report/ar6/

Isayenkov, S. V. (2012). Physiological and molecular aspects of salt stress in plants. *Cytology and Genetics*, *46*(5), 302–318. doi:10.3103/S0095452712050040

Jactel, H., Koricheva, J., & Castagneyrol, B. (2019). Responses of forest insect pests to climate change: Not so simple. *Current Opinion in Insect Science*, *35*, 103–108. doi:10.1016/j.cois.2019.07.010 PMID:31454625

Jimenez-Diaz, R. M., Cirulli, M., Bubici, G., del Mar Jimenez-Gasco, M., Antoniou, P. P., & Tjamos, E. C. (2012). Verticillium wilt, a major threat to olive production: Current status and future prospects for its management. *Plant Disease*, *96*(3), 304–329. doi:10.1094/PDIS-06-11-0496 PMID:30727142

Jiménez-Díaz, R. M., Olivares-García, C., Landa, B. B., del Mar Jiménez-Gasco, M., & Navas-Cortés, J. A. (2011). Region-wide analysis of genetic diversity in Verticillium dahliae populations infecting olive in southern Spain and agricultural factors influencing the distribution and prevalence of vegetative compatibility groups and pathotypes. *Phytopathology*, *101*(3), 304–315. doi:10.1094/PHYTO-07-10-0176 PMID:20942654

Jiménez-Díaz, R. M., Tjamos, E. C., & Cirulli, M. (1998). Verticillium wilt of major tree hosts: olive. *A compendium of Verticillium wilts in tree species*, 13-16.

Juroszek, P., & Von Tiedemann, A. (2011). Potential strategies and future requirements for plant disease management under a changing climate. *Plant Pathology*, *60*(1), 100–112. doi:10.1111/j.1365-3059.2010.02410.x

Kahaduwa, A., & Rajapakse, L. (2022). Review of climate change impacts on reservoir hydrology and long-term basin-wide water resources management. *Building Research and Information*, *50*(5), 515–526. doi:10.1080/09613218.2021.1977908

Kassout, J., Ater, M., Ivorra, S., Barbara, H., Limier, B., Ros, J., Girard, V., Paradis, L., & Terral, J. F. (2021). Resisting aridifcation: Adaptation of sap conduction performance in Moroccan wild olive subspecies distributed over an aridity gradient. *Frontiers in Plant Science*, *12*, 1288. doi:10.3389/fpls.2021.663721 PMID:34276726

Kassout, J., Terral, J. F., El Ouahrani, A., Houssni, M., & Ivorra, S. (2022). Species Distribution Based-Modelling Under Climate Change: The Case of Two Native Wild *Olea europaea* Subspecies in Morocco, *O. e.* subsp. *europaea* var. *sylvestris* and *O. e.* subsp. *maroccana*. In *Climate Change in the Mediterranean and Middle Eastern Region* (pp. 21–43). Springer. doi:10.1007/978-3-030-78566-6_2

Kassout, J., Terral, J.-F., Hodgson, J., & Ater, M. (2019). Trait- based plant ecology a flawed tool in climate studies? The leaf traits of wild olive that pattern with climate are not those routinely measured. *PLoS One*, *14*(7), e0219908. doi:10.1371/journal.pone.0219908 PMID:31314789

Kenis, M., Hurley, B. P., Hajek, A. E., & Cock, M. J. (2017). Classical biological control of insect pests of trees: Facts and figures. *Biological Invasions*, *19*(11), 3401–3417. doi:10.100710530-017-1414-4

Kitsaki, C. K., & Drossopoulos, J. B. (2005). Environmental effect on ABA concentration and water potential in olive leaves (*Olea europaea* L. cv "Koroneiki") under non- irrigated feld conditions. *Environmental and Experimental Botany*, *54*(1), 77–89. doi:10.1016/j.envexpbot.2004.06.002

Ksantini, M. (2003). *Contribution à l'étude de la dynamique des populations du psylle de l'olivier Euphylluraolivina Costa (Homoptera, Aphalaridae) et de sa nuisibilité dans la région de Sfax* [Thèse de Doctorat en Sciences biologiques]. Fac. Sc. Sfax.

Ksantini, M., Jardak, T., & BouAïn, A. (2002). Temperature Effect on the Biology of Euphylluraolivina Costa. *Acta Horticulturae*, (586), 827–829. doi:10.17660/ActaHortic.2002.586.179

Lamers, J., Van Der Meer, T., & Testerink, C. (2020). How plants sense and respond to stressful environments. *Plant Physiology*, *182*(4), 1624–1635. doi:10.1104/pp.19.01464 PMID:32132112

Landa, B. B., Castillo, A. I., Giampetruzzi, A., Kahn, A., Román-Écija, M., Velasco-Amo, M. P., Navas-Cortés, J. A., Marco-Noales, E., Barbé, S., Moralejo, E., Coletta-Filho, H. D., Saldarelli, P., Saponari, M., & Almeida, R. P. (2020). Emergence of a plant pathogen in Europe associated with multiple intercontinental introductions. *Applied and Environmental Microbiology*, *86*(3), e01521–e19. doi:10.1128/AEM.01521-19 PMID:31704683

Lau, S. E., Teo, W. F. A., Teoh, E. Y., & Tan, B. C. (2022). Microbiome engineering and plant biostimulants for sustainable crop improvement and mitigation of biotic and abiotic stresses. *Discover Food*, *2*(1), 9. doi:10.100744187-022-00009-5

Lelieveld, J., Hadjinicolaou, P., Kostopoulou, E., Chenoweth, J., El Maayar, M., Giannakopoulos, C., Hannides, C., Lange, M. A., Tanarhte, M., Tyrlis, E., & Xoplaki, E. (2012). Climate change and impacts in the Eastern Mediterranean and the Middle East. *Climatic Change*, *114*(3–4), 667–687. doi:10.100710584-012-0418-4 PMID:25834296

Levin, A. G., Lavee, S., & Tsror, L. (2003). Epidemiology of Verticillium dahliae on olive (cv. Picual) and its effect on yield under saline conditions. *Plant Pathology*, *52*(2), 212–218. doi:10.1046/j.1365-3059.2003.00809.x

Linnakoski, R., Kasanen, R., Dounavi, A., & Forbes, K. M. (2019). Editorial: Forest Health Under Climate Change: Effects on Tree Resilience, and Pest and Pathogen Dynamics. *Frontiers in Plant Science*, *10*, 1157. doi:10.3389/fpls.2019.01157 PMID:31681344

Logan, J. A., Régnière, J., & Powell, J. A. (2003). Assessing the impacts of global warming on forest pest dynamics. *Frontiers in Ecology and the Environment*, *1*(3), 130–137. doi:10.1890/1540-9295(2003)001[0130:ATIOGW]2.0.CO;2

López-Escudero, F. J., & Blanco-López, M. A. (2005). Effects of drip irrigation on population of Verticillium dahliae in olive orchards. *Journal of Phytopathology*, *153*(4), 238–239. doi:10.1111/j.1439-0434.2005.00961.x

López-Escudero, F. J., & Blanco-López, M. A. (2007). Relationship between the inoculum density of Verticillium dahliae and the progress of Verticillium wilt of olive. *Plant Disease*, *91*(11), 1372–1378. doi:10.1094/PDIS-91-11-1372 PMID:30780740

Ma, C. S., Ma, G., & Pincebourde, S. (2021). Survive a warming climate: Insect responses to extreme high temperatures. *Annual Review of Entomology*, *66*(1), 163–184. doi:10.1146/annurev-ento-041520-074454 PMID:32870704

Ma, Y., Dias, M. C., & Freitas, H. (2020). Drought and salinity stress responses and microbe-induced tolerance in plants. *Frontiers in Plant Science*, *11*, 591911. doi:10.3389/fpls.2020.591911 PMID:33281852

Maas, E. V., & Hoffman, G. J. (1977). Crop salt tolerance— Current assessment. *Journal of the Irrigation and Drainage Division*, *103*(2), 115–134. doi:10.1061/JRCEA4.0001137

Mansour, A. A., Ouanaimi, F., Chemseddine, M., & Boumezzough, A. (2017). Study of the flight dynamics of Prays oleae (Lepidoptera: Yponomeutidae) using sexual trapping in olive orchards of Essaouira region, Morocco. *Journal of Entomology and Zoology Studies*, *5*(2), 943–952.

Marchi, S., Guidotti, D., Ricciolini, M., & Petacchi, R. (2016). Towards understanding temporal and spatial dynamics of Bactrocera oleae (Rossi) infestations using decade-long agrometeorological time series. *International Journal of Biometeorology*, *60*(11), 1681–1694. doi:10.100700484-016-1159-2 PMID:27044274

Marchi, S., Tognetti, R., Minnocci, A., Borghi, M., & Sebastiani, L. (2008). Variation in mesophyll anatomy and photosynthetic capacity during leaf development in a deciduous mesophyte fruit tree (*Prunus persica*) and an evergreen sclerophyllous Mediterranean shrub (*Olea europaea*). *Trees (Berlin)*, *22*(4), 559–571. doi:10.100700468-008-0216-9

Mathivanan, S. (2021). Abiotic Stress-Induced Molecular and Physiological Changes and Adaptive Mechanisms in Plants. *Abiotic Stress in Plants*, 315.

Mattos, L., & Moretti, C. (2015). Oxidative stress in plants under drought conditions and the role of different enzymes. *Enzyme Engineering*, *5*, 1–6.

Mcelrone, A. J., Reid, C. D., Hoye, K. A., Hart, E., & Jackson, R. B. (2005). Elevated CO2 reduces disease incidence and severity of a red maple fungal pathogen via changes in host physiology and leaf chemistry. *Global Change Biology*, *11*(10), 1828–1836. doi:10.1111/j.1365-2486.2005.001015.x

Mediavilla, S., Gallardo-López, V., González-Zurdo, P., & Escudero, A. (2012). Patterns of leaf morphology and leaf N content in relation to winter temperatures in three evergreen tree species. *International Journal of Biometeorology*, *56*(5), 915–926. doi:10.100700484-011-0498-2 PMID:21969112

Mendel, Z., Podoler, H., & Rosen, D. (1984). Population dynamics of the mediterranean black scale, Saissetia oleae (Olivier), on citrus in Israel. 4. The natural enemies. *Journal of the Entomological Society of Southern Africa*, *47*(1), 1–21.

Michaletz, S. T., Weiser, M. D., Zhou, J., Kaspari, M., Helliker, B. R., & Enquist, B. J. (2015). Plant Thermoregulation: Energetics, Trait– Environment Interactions, and Carbon Economics. *Trends in Ecology & Evolution*, *30*(12), 714–724. doi:10.1016/j.tree.2015.09.006 PMID:26476814

Miranda, M. A., Miquel, M., Terrassa, J., Melis, N., & Monerris, M. (2008). Parasitism of Bactrocera oleae (Diptera; Tephritidae) by Psyttalia concolor (Hymenoptera; Braconidae) in the Balearic Islands (Spain). *Journal of Applied Entomology*, *132*(9-10), 798–805. doi:10.1111/j.1439-0418.2008.01358.x

Montzka, S. (2022). *The NOAA Annual Greenhouse Gas Index (AGGI).* NOAA Global Monitoring Laboratory Website. https://gml.noaa.gov/aggi/aggi.html

NASA. (2022). Global Climate Change. Carbon dioxide. *Latest Measurement June 2022.* Carbon Dioxide|Vital Signs—Climate Change: Vital Signs of the Planet.

Navas-Cortés, J. A., Landa, B. B., Mercado-Blanco, J., Trapero-Casas, J. L., Rodríguez-Jurado, D., & Jiménez-Díaz, R. M. (2008). Spatiotemporal analysis of spread of infections by Verticillium dahliae pathotypes within a high tree density olive orchard in southern Spain. *Phytopathology*, *98*(2), 167–180. doi:10.1094/PHYTO-98-2-0167 PMID:18943193

Obanor, F. O., Walter, M., Jones, E. E., & Jaspers, M. V. (2008). Effect of temperature, relative humidity, leaf wetness and leaf age on Spilocaea oleagina conidium germination on olive leaves. *European Journal of Plant Pathology*, *120*(3), 211–222. doi:10.100710658-007-9209-6

Oliva, J., Stenlid, J., & Martinez-Vilalta, J. (2014). The effect of fungal pathogens on the water and carbon economy of trees: Implications for drought-induced mortality. *The New Phytologist*, *203*(4), 1028–1035. doi:10.1111/nph.12857 PMID:24824859

Oliveira, T. C., Cabral, J. S. R., Santana, L. R., Tavares, G. G., Santos, L. D. S., Paim, T. P., Müller, C., Silva, F. G., Costa, A. C., Souchie, E. L., & Mendes, G. C. (2022). The arbuscular mycorrhizal fungus Rhizophagus clarus improves physiological tolerance to drought stress in soybean plants. *Scientific Reports*, *12*(1), 9044. doi:10.103841598-022-13059-7 PMID:35641544

Oliverio, A. M., Bradford, M. A., & Fierer, N. (2017). Identifying the microbial taxa that consistently respond to soil warming across time and space. *Global Change Biology*, *23*(5), 2117–2129. doi:10.1111/gcb.13557 PMID:27891711

Olmo, D., Nieto, A., Borràs, D., Montesinos, M., Adrover, F., Pascual, A., Gost, P. A., Quetglas, B., Urbano, A., García, J. D., Velasco-Amo, M. P., Olivares-García, C., Beidas, O., Juan, A., Marco-Noales, E., Gomila, M., Rita, J., Moralejo, E., & Landa, B. B. (2021). Landscape epidemiology of Xylella fastidiosa in the Balearic Islands. *Agronomy (Basel)*, *11*(3), 473. doi:10.3390/agronomy11030473

Pallioti, A., Bongi, G., & Rocchi, P. (1994). Peltate trichomes effects on photosynthetic gas exchange of *Olea europaea* L. leaves. *Plant Physiology*, *13*, 35–44.

Pan, J., Sharif, R., Xu, X., & Chen, X. (2021). Mechanisms of waterlogging tolerance in plants: Research progress and prospects. *Frontiers in Plant Science*, *11*, 627331. doi:10.3389/fpls.2020.627331 PMID:33643336

Pan, T., Sun, X., Liu, Y., Li, H., Deng, G., Lin, H., & Wang, S. (2018). Heat stress alters genome-wide profiles of circular RNAs in Arabidopsis. *Plant Molecular Biology*, *96*(3), 217–229. doi:10.100711103-017-0684-7 PMID:29177640

Pantin, F., Monnet, F., Jannaud, D., Costa, J. M., Renaud, J., Muller, B., Simonneau, T., & Genty, B. (2013). The dual effect of abscisic acid on stomata. *The New Phytologist*, *197*(1), 65–72. doi:10.1111/nph.12013 PMID:23106390

Pathak, H., Aggarwal, P. K., & Singh, S. D. (2012). *Climate change impact, adaptation and mitigation in agriculture: methodology for assessment and applications*. Indian Agricultural Research Institute.

Pérez-Rodríguez, M., Serrano, N., Arquero, O., Orgaz, F., Moral, J., & López-Escudero, F. J. (2016). The effect of short irrigation frequencies on the development of Verticillium wilt in the susceptible olive cultivar 'Picual'under field conditions. *Plant Disease*, *100*(9), 1880–1888. doi:10.1094/PDIS-09-15-1018-RE PMID:30682984

Petridis, A., Therios, I., Samouris, G., Koundouras, S., & Giannakoula, A.. (2012). Effect of water defcit on leaf phenolic composition, gas exchange, oxidative damage and antioxidant activity of four Greek olive (Olea europaea L.) cultivars. Plant physiology and biochemistry, 60, 1– 11.

Petzoldt, C., & Seaman, A. (2006). Climate change effects on insects and pathogens. *Climate change and agriculture: Promoting practical and profitable responses, 3*, 6-16.

Picchi, M. S., Tondini, E., Albertarelli, N., Monteforti, G., & Petacchi, R. (2022). Following the pest outbreak: Preliminary findings on the landscape effect on Dasineura oleae and its parasitoids in central Italy. *Phytoparasitica*, *50*(2), 1–15. doi:10.100712600-021-00962-4

Pinheiro, L. A., Medina, P., Bento, A., Pereira, J. A., & Santos, S. A. P. (2017). Bioecology of Psyttalia concolor (Szépligeti) (Hymenoptera: Braconidae) and management strategies to increase biological control of the olive fruit fly. In S. A. P. Santos (Ed.), *Natural Enemies: Identification, Protection Strategies and Ecological Impacts* (pp. 141–172). Nova Science Publishers.

Ponti, L., Gutierrez, A. P., Ruti, P. M., & Dell'Aquila, A. (2014). Fine-scale ecological and economic assessment of climate change on olive in the Mediterranean Basin reveals winners and losers. *Proceedings of the National Academy of Sciences of the United States of America*, *111*(15), 5598–5603. doi:10.1073/pnas.1314437111 PMID:24706833

Rai, A., Irulappan, V., & Senthil-Kumar, M. (2022). Dry root rot of chickpea: A disease favored by drought. *Plant Disease*, *106*(2), 346–356. doi:10.1094/PDIS-07-21-1410-FE PMID:34649462

Ramegowda, V., Senthil-Kumar, M., Ishiga, Y., Kaundal, A., Udayakumar, M., & Mysore, K. S. (2013). Drought stress acclimation imparts tolerance to Sclerotinia sclerotiorum and Pseudomonas syringae in Nicotiana benthamiana. *International Journal of Molecular Sciences*, *14*(5), 9497–9513. doi:10.3390/ijms14059497 PMID:23644883

Requena-Mullor, J. M., García-Garrido, J. M., García, P. A., & Rodríguez, E. (2020). Climatic drivers of Verticillium dahliae occurrence in Mediterranean olive-growing areas of southern Spain. *PLoS One*, *15*(12), e0232648. doi:10.1371/journal.pone.0232648 PMID:33378350

Ribeiro, H., Cunha, M., & Abreu, I. (2008). Quantitative forecasting of olive yield in Northern Portugal using a bioclimatic model. *Aerobiologia*, *24*(3), 141–150. doi:10.100710453-008-9094-2

Robinson, E. A., Ryan, G. D., & Newman, J. A. (2012). A meta-analytical review of the effects of elevated CO2 on plant–arthropod interactions highlights the importance of interacting environmental and biological variables. *The New Phytologist*, *194*(2), 321–336. doi:10.1111/j.1469-8137.2012.04074.x PMID:22380757

Romero, F., Cazzato, S., Walder, F., Vogelgsang, S., Bender, S. F., & van der Heijden, M. G. (2022). Humidity and high temperature are important for predicting fungal disease outbreaks worldwide. *The New Phytologist*, *234*(5), 1553–1556. doi:10.1111/nph.17340 PMID:33713447

Rossi, L., Sebastiani, L., Tognetti, R., d'Andria, R., Morelli, G., & Cherubini, P. (2013). Tree- ring wood anatomy and stable isotopes show structural and functional adjustments in olive trees under different water availability. *Plant and Soil*, *372*(1), 567–579. doi:10.100711104-013-1759-0

Rugini, E., & Fedeli, E. (1990). Olive (*Olea europaea* L.) as an oilseed crop. In *Legumes and oilseed crops I* (pp. 593–641). Springer. doi:10.1007/978-3-642-74448-8_29

Saponari, M., Boscia, D., Nigro, F., & Martelli, G. P. (2013). Identification of DNA sequences related to Xylella fastidiosa in oleander, almond and olive trees exhibiting leaf scorch symptoms in Apulia (Southern Italy). *Journal of Plant Pathology*, *95*(3).

Sasidharan, R., Bailey-Serres, J., Ashikari, M., Atwell, B. J., Colmer, T. D., Fagerstedt, K., Fukao, T., Geigenberger, P., Hebelstrup, K. H., Hill, R. D., Holdsworth, M. J., Ismail, A. M., Licausi, F., Mustroph, A., Nakazono, M., Pedersen, O., Perata, P., Sauter, M., Shih, M.-C., & Voesenek, L. A. C. J. (2017). Community recommendations on terminology and procedures used in flooding and low oxygen stress research. *The New Phytologist*, *214*(4), 1403–1407. doi:10.1111/nph.14519 PMID:28277605

Schneider, K., Van der Werf, W., Cendoya, M., Mourits, M., Navas-Cortés, J. A., Vicent, A., & Oude Lansink, A. (2020). Impact of Xylella fastidiosa subspecies pauca in European olives. *Proceedings of the National Academy of Sciences of the United States of America, 117*(17), 9250–9259. doi:10.1073/pnas.1912206117 PMID:32284411

Shrestha, S. (2019). Effects of climate change in agricultural insect pest. *Acta Sci. Agric, 3*(12), 74–80. doi:10.31080/ASAG.2019.03.0727

Silva, S., Dias, M. C., & Silva, A. M. S. (2022). Titanium and zinc based nanomaterials in agriculture: A promising approach to deal with (a)biotic stresses? *Toxics, 10*(4), 172. doi:10.3390/toxics10040172 PMID:35448432

Simoglou, K. B., Karataraki, A., Roditakis, N. E., & Roditakis, E. (2012). Euzophera bigella (Zeller) (Lepidoptera: Pyralidae) and Dasineura oleae (F. Low)(Diptera: Cecidomyiidae): emerging olive crop pests in the Mediterranean? *Journal of Pest Science, 85*(2), 169–177. doi:10.100710340-012-0418-1

Sofo, A., Manfreda, S., Fiorentino, M., Dichio, B., & Xiloyannis, C. (2008). The olive tree: A paradigm for drought tolerance in Mediterranean climates. *Hydrology and Earth System Sciences, 12*(1), 293–301. doi:10.5194/hess-12-293-2008

Sutherst, R. W., Constable, F., Finlay, K. J., Harrington, R., Luck, J., & Zalucki, M. P. (2011). Adapting to crop pest and pathogen risks under a changing climate. *Wiley Interdisciplinary Reviews: Climate Change, 2*(2), 220–237. doi:10.1002/wcc.102

Sutherst, R. W., Yonow, T., Chakraborty, S., O'Donnell, C., & White, N. (1996). A generic approach to defining impacts of climate change on pests, weeds and diseases in Australasia.

Taiz, L., and E. Zeiger. (2006). *Fisiologia vegetal* (Vol. 10). Universitat Jaume I

Tajnari, H. (2001). *Étude de la nuisibilité du psylle de l'olivier, Euphyllura olivina Costa*. Symposium sur la Protection Intégrée des Cultures dans la région Méditerranéenne, DPVCTRF, Rabat, Maroc

Tattini, M., Bertoni, P., & Caselli, S. (1992). Genotipic responses of olive plants to sodium chloride. *Journal of Plant Nutrition, 15*(9), 1467–1485. doi:10.1080/01904169209364412

Tedersoo, L., Bahram, M., Põlme, S., Kõljalg, U., Yorou, N. S., Wijesundera, R., & Abarenkov, K. (2014). Global diversity and geography of soil fungi. *science, 346*(6213), 1256688.

Terral, J.-F., Badal, E., Heinz, C., Roiron, P., Thiébault, S., & Figueiral, I. (2004). A hydraulic conductivity model points to post- Neogene survival of the Mediterranean Olive in riparian habitat. *Ecology, 85*(11), 3158–3165. doi:10.1890/03-3081

Teshome, D. T., Zharare, G. E., & Naidoo, S. (2020). The Threat of the Combined Effect of Biotic and Abiotic Stress Factors in Forestry Under a Changing Climate. *Frontiers in Plant Science, 11*, 601009. doi:10.3389/fpls.2020.601009 PMID:33329666

Thomson, L. J., Macfadyen, S., & Hoffmann, A. A. (2010). Predicting the effects of climate change on natural enemies of agricultural pests. *Biological Control, 52*(3), 296–306. doi:10.1016/j.biocontrol.2009.01.022

Tian, L., Li, J., Bi, W., Zuo, S., Li, L., Li, W., & Sun, L. (2019). Effects of waterlogging stress at different growth stages on the photosynthetic characteristics and grain yield of spring maize (*Zea mays* L.) Under feld conditions. *Agricultural Water Management*, *218*, 250–258. doi:10.1016/j.agwat.2019.03.054

Torres-Ruiz, J. M., Cochard, H., Mayr, S., Beikircher, B., Diaz-Espejo, A., Rodriguez-Dominguez, C. M., Badel, E., & Fernández, J. E. (2014). Vulnerability to cavitation in *Olea europaea* current—year shoots: Further evidence of an open-vessel artifact associated with centrifuge and air-injection techniques. *Physiologia Plantarum*, *152*(3), 465–474. doi:10.1111/ppl.12185 PMID:24611594

Torres-Ruiz, J. M., Díaz-Espejo, A., Perez-Martin, A., & Hernandez-Santana, V. (2013). Loss of hydraulic functioning at leaf, stem and root level and its role in the stomatal behaviour during drought in olive trees. *Acta Horticulturae*, (991), 333–339. doi:10.17660/ActaHortic.2013.991.41

Tsitsipis, J. A. (1980). Effect of constant temperatures on larval and pupal development of olive fruit flies reared on artificial diet. *Environmental Entomology*, *9*(6), 764–768. doi:10.1093/ee/9.6.764

Tsror, L. (2011). Epidemiology and control of Verticillium wilt on olive. *Israel Journal of Plant Sciences*, *59*(1), 59–69. doi:10.1560/IJPS.59.1.59

Valverde, P., Barranco, D., López-Escudero, F. J., Díez, C. M., & Trapero, C. (2023). Efficiency of breeding olives for resistance to Verticillium wilt. *Frontiers in Plant Science*, *14*, 1149570. doi:10.3389/fpls.2023.1149570 PMID:36909426

Vigo, C., Therios, I., & Bosabalidis, A. (2005). Plant growth, nutrient concentration and leaf anatomy in olive plants irrigated with diluted seawater. *Journal of Plant Nutrition*, *28*(6), 101–102. doi:10.1081/PLN-200058897

Vigo, C., Therios, I., Patakas, A., Karatassou, A., & Nastou, A. (2002). Changes in photosynthetic parameters and nutrient distribution of olive plants (*Olea europaea* L.) cultivar 'Chondrolia Chalkidikis' under NaCl, Na2SO4 and KCl salinities. *Agrochimica*, *46*, 33–46.

Visser, M. E., & Both, C. (2005). Shifts in phenology due to global climate change: The need for a yardstick. *Proceedings. Biological Sciences*, *272*(1581), 2561–2569. doi:10.1098/rspb.2005.3356 PMID:16321776

Volpi, I., Guidotti, D., Mammini, M., Petacchi, R., & Marchi, S. (2020). Managing complex datasets to predict Bactrocera oleae infestation at the regional scale. *Computers and Electronics in Agriculture*, *179*, 105867. doi:10.1016/j.compag.2020.105867

Waha, K., Krummenauer, L., Adams, S., Aich, V., Baarsch, F., Coumou, D., Fader, M., Hoff, H., Jobbins, G., Marcus, R., Mengel, M., Otto, I. M., Perrette, M., Rocha, M., Robinson, A., & Schleussner, C.-F. (2017). Climate change impacts in the Middle East and Northern Africa (MENA) region and their implications for vulnerable population groups. *Regional Environmental Change*, *17*(6), 1623–1638. doi:10.100710113-017-1144-2

Wakelin, S. A., Gomez-Gallego, M., Jones, E., Smaill, S., Lear, G., & Lambie, S. (2018). Climate change induced drought impacts on plant diseases in New Zealand. *Australasian Plant Pathology*, *47*(1), 101–114. doi:10.100713313-018-0541-4

Wang, B. X., Hof, A. R., & Ma, C. S. (2022). Impacts of climate change on crop production, pests and pathogens of wheat and rice. *Frontiers of Agricultural Science and Engineering*, *9*(1), 4–18. doi:10.15302/J-FASE-2021432

Wang, X. G., Johnson, M. W., Daane, K. M., & Nadel, H. (2009). High summer temperatures affect the survival and reproduction of olive fruit fly (Diptera: Tephritidae). *Environmental Entomology*, *38*(5), 1496–1504. doi:10.1603/022.038.0518 PMID:19825305

Waqas, M. A., Kaya, C., Riaz, A., Farooq, M., Nawaz, I., Wilkes, A., & Li, Y. (2019). Potential mechanisms of abiotic stress tolerance in crop plants induced by thiourea. *Frontiers in Plant Science*, *10*, 1336. doi:10.3389/fpls.2019.01336 PMID:31736993

War, A. R., Taggar, G. K., War, M. Y., & Hussain, B. (2016). Impact of climate change on insect pests, plant chemical ecology, tritrophic interactions and food production. *International Journal of Clinical and Biological Sciences*, *1*(02), 16–29.

West, J. S., Townsend, J. A., Stevens, M., & Fitt, B. D. L. (2012). Comparative biology of different plant pathogens to estimate effects of climate change on crop diseases in Europe. *European Journal of Plant Pathology*, *133*(1), 315–331. doi:10.100710658-011-9932-x

Wheeler, T., & Von Braun, J. (2013). Climate change impacts on global food security. *Science*, *341*(6145), 508–513. doi:10.1126cience.1239402 PMID:23908229

Xu, C., McDowell, N. G., Fisher, R. A., Wei, L., Sevanto, S., Christoffersen, B. O., Weng, E., & Middleton, R. S. (2019). Increasing impacts of extreme droughts on vegetation productivity under climate change. *Nature Climate Change*, *9*(12), 948–953. doi:10.103841558-019-0630-6

Xu, F., Yang, L., Zhang, J., Guo, X., Zhang, X., & Li, G. (2012). Effect of temperature on conidial germination, mycelial growth and aggressiveness of the defoliating and nondefoliating pathotypes of Verticillium dahliae from cotton in China. *Phytoparasitica*, *40*(4), 319–327. doi:10.100712600-012-0232-6

Yáñez-López, R., Torres-Pacheco, I., Guevara-González, R. G., Hernández-Zul, M. I., Quijano-Carranza, J. A., & Rico-García, E. (2012). The effect of climate change on plant diseases. *African Journal of Biotechnology*, *11*(10), 2417–2428. doi:10.5897/AJB10.2442

Yang, Y., Kang, L., Zhao, J., Qi, N., Li, R., Wen, Z., Kassout, J., Peng, C., Lin, G., & Zheng, H. (2021). Quantifying Leaf Trait Covariations and Their Relationships with Plant Adaptation Strategies along an Aridity Gradient. *Biology (Basel)*, *10*(10), 1066. doi:10.3390/biology10101066 PMID:34681167

Yihdego, Y., Salem, H. S., & Muhammed, H. H. (2019). Agricultural pest management policies during drought: Case studies in Australia and the state of Palestine. *Natural Hazards Review*, *20*(1), 05018010. doi:10.1061/(ASCE)NH.1527-6996.0000312

Zhang, H., Zhu, J., Gong, Z., & Zhu, J. K. (2022). Abiotic stress responses in plants. *Nature Reviews. Genetics*, *23*(2), 104–119. doi:10.103841576-021-00413-0 PMID:34561623

Zheng, Q., Majsec, K., & Katagiri, F. (2022). Pathogen-driven coevolution across the CBP60 plant immune regulator subfamilies confers resilience on the regulator module. *The New Phytologist*, *233*(1), 479–495. doi:10.1111/nph.17769 PMID:34610150

Zhou, W., Chen, F., Meng, Y., Chandrasekaran, U., Luo, X., Yang, W., & Shu, K. (2020). Plant waterlogging/ flooding stress responses: From seed germination to maturation. *Plant Physiology and Biochemistry*, *148*, 228–236. doi:10.1016/j.plaphy.2020.01.020 PMID:31981875

Zhou, Y., Van Leeuwen, S. K., Pieterse, C. M., Bakker, P. A., & Van Wees, S. C. (2019). Effect of atmospheric CO 2 on plant defense against leaf and root pathogens of Arabidopsis. *European Journal of Plant Pathology*, *154*(1), 31–42. doi:10.100710658-019-01706-1

Zhu, J. K. (2002). Salt and drought stress signal transduction in plants. *Annual Review of Plant Biology*, *53*(1), 247–273. doi:10.1146/annurev.arplant.53.091401.143329 PMID:12221975

Zittis, G., Almazroui, M., Alpert, P., Ciais, P., Cramer, W., Dahdal, Y., et al. (2022). Climate change and weather extremes in the Eastern Mediterranean and Middle East. *Reviews of Geophysics, 60*, e2021RG000762. doi:10.1029/2021RG000762

Zittis, G., Hadjinicolaou, P., Klangidou, M., Proestos, Y., & Lelieveld, J. (2019). A multi-model, multi-scenario, and multi-domain analysis of regional climate projections for the Mediterranean. *Regional Environmental Change*, *19*(8), 2621–2635. doi:10.100710113-019-01565-w

Zulfiqar, F., Casadesús, A., Brockman, H., & Munné-Bosch, S. (2020). An overview of plant-based natural biostimulants for sustainable horticulture with a particular focus on moringa leaf extracts. *Plant Science*, *295*, 110194. doi:10.1016/j.plantsci.2019.110194 PMID:32534612

## KEY TERMS AND DEFINITIONS

**Abiotic Stress:** Nonliving environmental factors (such as drought, extreme cold or heat, high winds) that can have harmful effects on plants.

**Biological Control:** Management of plant diseases and pests using bio-control agents.

**Biotic stress:** A general term for the adverse effect on plants caused by living organisms, including viruses, fungi, parasites, bacteria, insects, weeds and competing plants.

**Crop loss:** The extent to which yield of crop is reduced by the damage caused by pest infestation.

**Cultural control:** Method of pest control by means of skillful combination of agronomic practices

**Disease Incidence:** Number of plant unit infected expressed as proportion or percentage of diseased entities within a sampling unit.

**Disease Severity:** Intensity of disease on individual plants or their parts

**Economic Pest Threshold or Intervention Threshold:** The density of a pest at which control measure should be initiated to prevent increasing population.

**Epidemiology:** Science dealing with the study of factors affecting the outbreak and spread of an infectious disease

**Poikilotherm:** An organism whose internal temperature varies with the temperature of its surroundings.

# Chapter 8
# Bridging the Gap Between Climate Change and Plant Biology:
## Enhancing Resilience Through Genetic Diversity and Agroecological Practices

**Asad Azeem**
https://orcid.org/0000-0001-6550-3571
*Department of Plant Breeding and Genetics, University of Agriculture, Faisalabad, Pakistan*

**Muhammad Zia Ul Haq**
https://orcid.org/0000-0001-9914-3367
*Department of Agronomy, University of Agriculture, Faisalabad, Pakistan*

**Haroon Ur Rasheed**
*Institute of Horticultural Sciences, University of Agriculture, Faisalabad, Pakistan*

**Safeer Uddin**
*University of Agriculture Faisalabad, Okara, Pakistan*

**Muhammad Umar Draz Dullu**
*Department of Plant Breeding and Genetics, PMAS-Arid Agriculture University. Rawalpindi, Pakistan*

**Sonia Azeem**
*Department of Zoology, Government College University Faisalabad, Layyah, Pakistan*

**Muhammad Sikandar Zaman**
*Department of Plant Breeding and Genetics, University of Agriculture, Faisalabad, Pakistan*

## ABSTRACT

*The inherent genetic diversity of crops equips them with a remarkable ability to adapt to shifting environmental conditions, a crucial aspect of crop resilience. This adaptability stems from the vast array of traits and characteristics encoded within their genomes, enabling them to respond to diverse environmental challenges. By subjecting crops to stress conditions, breeders can identify individuals with desirable traits that confer enhanced resilience. Agroecological approaches and sustainable farming practices also contribute significantly to crop resilience. Crop rotation, intercropping, and agroforestry systems promote biodiversity and enhance the resilience of agroecosystems. In the face of climate change, developing crop varieties with improved resilience to extreme weather events is paramount. By harnessing agroecological principles, genetic diversity, and adaptive capacity, we can bolster crop resilience, ensuring the productivity and adaptability of our agricultural systems in a changing environment.*

DOI: 10.4018/978-1-6684-9838-5.ch008

## INTRODUCTION

Climate change has become the most burning issue around the globe, which has a major effect on plants (Ahmed et al., 2022). Various consequences are changing how plants grow, reproduce, and survive as temperatures rise and weather patterns become more unpredictable. The effects of climate change on plants and their implications for ecosystems and human society are discussed. Environmental fluctuation is one of the most important elements influencing plant growth and development (Abobatta, 2023). Plants are enduring a variety of consequences because of human activities like burning fossil fuels and deforestation that are contributing to the continued rise in global temperatures. The fact that plants are flowering earlier in the spring due to the warmer temperatures is one of the most notable effects. However, it can result in conflicts with pollinators and other creatures that depend on plants for food and habitat, upsetting ecosystems and possibly contributing to a reduction in biodiversity (Gora et al., 2023). At the same time, rising temperatures are accelerating plant growth and increasing biomass production.

Even though it could seem like a good result, it might be detrimental. Faster-growing trees, for instance, may shade out other species and lower biodiversity in forests. Rapid growth in agricultural systems can cause crops to be less nutritious and of worse quality, and it can also increase the pressure from pests and diseases. The global pattern of precipitation is changing as a result of climate change, in addition to changes in temperature resulting in drought and desertification in some areas (Usman et al., 2023). On the other hand, increased rainfall results in flooding (Shehzad, 2023). Either of these consequences may significantly impact plants. Increased rainfall may cause waterlogging in some regions, which can harm roots and slow growth in plants (Kaur et al., 2020). In agricultural systems, too much rain can also cause soil erosion and nutrient loss, which lowers crop output.

On the other hand, a drought can cause plants to wilt and die, lowering production and possibly creating a food shortfall. As dry conditions make it simpler for fires to spread, it can also increase the risk of wildfires (Richardson et al., 2022). The increased frequency of extreme weather phenomena like hurricanes, cyclones, and storms affects plants as well. These occurrences can potentially harm vegetation widely, uprooting trees and harming crops, among other effects. They may also result in floods, which might drown plants and erode soil. Ecosystems and human society are significantly impacted by how plants respond to climate change. Ecological services, including pollination, carbon sequestration, and water purification, may be jeopardized as a result of ecological disruption. On other species and the overall health of ecosystems, this may have a domino effect.

In addition, how climate change affects plants may directly affect human culture. For instance, alterations in crop yields and quality may cause food shortages and price increases, potentially worsening food insecurity in vulnerable people. Homes and infrastructure may sustain damage in areas with greater floods, resulting in monetary losses and forced evictions (Randell et al., 2022). Despite the fact that climate change has had a large negative influence on plants, there may be ways to adapt and mitigate the effects. For instance, some plants may be able to change their geographic range or develop new features to adapt to changing environments. Crop rotation, soil conservation, and water management are other agricultural strategies that can help lessen the effects of climate change on crops (Smith & Olesen, 2010). At the same time, tackling the underlying causes of climate change and lessening its effects on plants and ecosystems requires cutting greenhouse gas emissions and moving toward a low-carbon economy. Therefore, there is a dire need to make a cohesive worldwide effort that incorporates governmental regulations, technical advancement, and individual initiatives to tackle prevailing climate fluctuation. As a result, the effects of climate change on plants globally are profound, affecting their ability to grow, reproduce, and survive. These effects have important ramifications for ecosystems and human society, underscoring the pressing need to deal with the primary causes of climate change.

## CLIMATE CHANGE AND AGRICULTURE

The wide-ranging effects of climate change on agriculture present serious difficulties for global food supply systems (Behnassi et al., 2022). Several of the main impacts of climate change on agriculture are listed below: Temperature increases may have an impact on crop growth and development. Increased temperatures can cause heat stress, lower crop yields, and have an impact on food's nutritional value (Choukri et al., 2022). In addition, adjustments to the length and duration of the growing seasons may cause scheduling issues for planting and harvesting, which could have an impact on crop yield and overall agricultural output. Precipitation patterns can alter due to climate change, which can affect rainfall levels, frequency, and distribution. In many locations, this may lead to an increase in droughts, floods, and water scarcity. Lower agricultural output can result from crop yield reductions, changes in soil moisture levels, and reduced availability of water for irrigation (Slamini et al., 2022). A changing climate may affect the amount of water available for agriculture through the diminished snowpack, glacier melt, and altered rainfall patterns.

Water scarcity, particularly in areas that largely rely on irrigation, can limit agricultural growth and production potential. Under the influence of shifting climatic circumstances, competition for water supplies among various industries, such as agriculture, industry, and urban regions, might become more intense. Pest and disease distribution, population dynamics, and behavior can all be impacted by climate change. Invasive pests and plant diseases can proliferate as a result of rising temperatures, altered precipitation patterns, and shifting seasonal cycles. Due to the need to modify their pest management techniques, farmers may end up using more pesticides and incurring higher production costs. Climate change has the potential to accelerate processes that lead to soil deterioration, such as salinization, nitrogen depletion, and erosion (Kumar et al., 2023). Increased soil erosion may result from climate change-related heavy rain and wind events, especially on sloped agricultural lands. In areas that are already in danger of desertification, soil deterioration can decrease crop output, worsen soil health, and raise desertification risks. Crop pests and weeds can modify their life cycles and behaviors as a result of climate change. Increased infestations of some pests and weeds, as well as agricultural losses, can result from warmer temperatures and changing precipitation patterns (Abobatta, 2023).

This may call for modifications to weed control and pest management plans. The geographic distribution and adaptability of some crops can change as a result of variations in temperature and precipitation patterns. While other places might become more suited to various crop kinds, some regions might see decreased yield or perhaps become unfit for traditional crops. Farmers may need to modify their cropping strategies in response to such changes, which could result in alterations to agricultural practices and land use patterns. A combination of adaptation strategies, such as the creation of climate-resilient crop varieties, improved water management procedures, sustainable soil management practices, and the encouragement of diverse and resilient farming systems, are needed to address the effects of climate change on agriculture (Pound et al., 2018). Long-term sustainability also depends on reducing greenhouse gas emissions and the carbon footprint of agriculture.

## EFFECTS OF CLIMATE CHANGE ON PLANTS

The effects of climate change on plants, both agricultural crops and wild vegetation, are extensive and complicated. Numerous factors, including variations in temperature and precipitation patterns, can influence how plants grow and develop (Agnihotri & Mitra, 2023). For instance, greater temperatures can speed up photosynthesis, which will boost plant growth and productivity. Since greater temperatures also result in higher evapotranspiration rates, which causes water stress in plants, this effect might be constrained by the availability of water. Decreased crop productivity and greater pest and disease pressure can result from changes in precipitation patterns, which can also increase rainfall variability and severe droughts. Phenological changes can greatly impact agricultural and environmental systems (Brown et al., 2012). Changes in fruiting, for instance, can have an impact on animals and human communities that depend on these crops for food and living.

In contrast, changes in flowering patterns can have an impact on pollination and seed generation. The risk of pests and diseases in plants, including cultivated crops and wild vegetation, is also rising as a result of climate change. Insect and other pest populations may grow as a result of warming temperatures, which might harm plants and lower crop output. Changes in precipitation patterns may foster the development of fungi that can harm both agricultural crops and wild plants (Singh et al., 2023). A general overview of the consequences of climate change on plant physiology, biochemistry, and morphology is depicted in Figure 1. Therefore, the effects of climate change on plants are beautifully described below:

### Impact on Plant Growth and Development

Plant growth and development are being significantly impacted by climate change, which has ramifications for agricultural productivity and food security. The effects of climate change on field crops will be discussed along with suggested solutions. One of the most important elements influencing plant growth and development is temperature. Field crops are enduring a variety of repercussions as global temperatures continue to rise as a result of human activities like burning fossil fuels and deforestation (Kumari, 2021). Since greater temperatures also result in higher evapotranspiration rates, which causes water stress in plants, this effect might be constrained by the availability of water. Temperature fluctuations can have indirect impacts on crop phenology and flowering patterns, in addition to their direct effects on plant growth and production. Warmer temperatures, for instance, may lead crops to flower sooner in the growing season, which may be advantageous for particular crops like rice or wheat. However, early flowering may result in decreased yields for other crops, such as soybeans, because of a mismatch with pollinators or pest pressure. With increased rainfall unpredictability and drought circumstances brought on by climate change, precipitation patterns are crucial for crop growth and development. Increased rainfall in some areas may cause waterlogging and soil erosion, which would lower agricultural yields and productivity (Wolka et al., 2018).

On the other hand, drought conditions can cause crop wilting and mortality, which can have a large negative impact on farmers' yields and finances. The risk of pests and diseases in field crops is rising as a result of climate change, in addition to its direct effects on crop growth and development. For instance, rising temperatures might result in a rise in populations of pests like mites and aphids, which can harm crops and lower harvests. Changes in precipitation patterns can also foster the development of fungal diseases like Fusarium head blight in cereals like wheat (Moraes et al., 2023). The effects of climate change on field crops significantly impact agribusiness production and food security. The effects of

*Figure 1. Consequences of climate change and its impacts on plants*
*(Adopted from Patni et al. 2022;). Part of the diagram is obtained from Pixabay.*

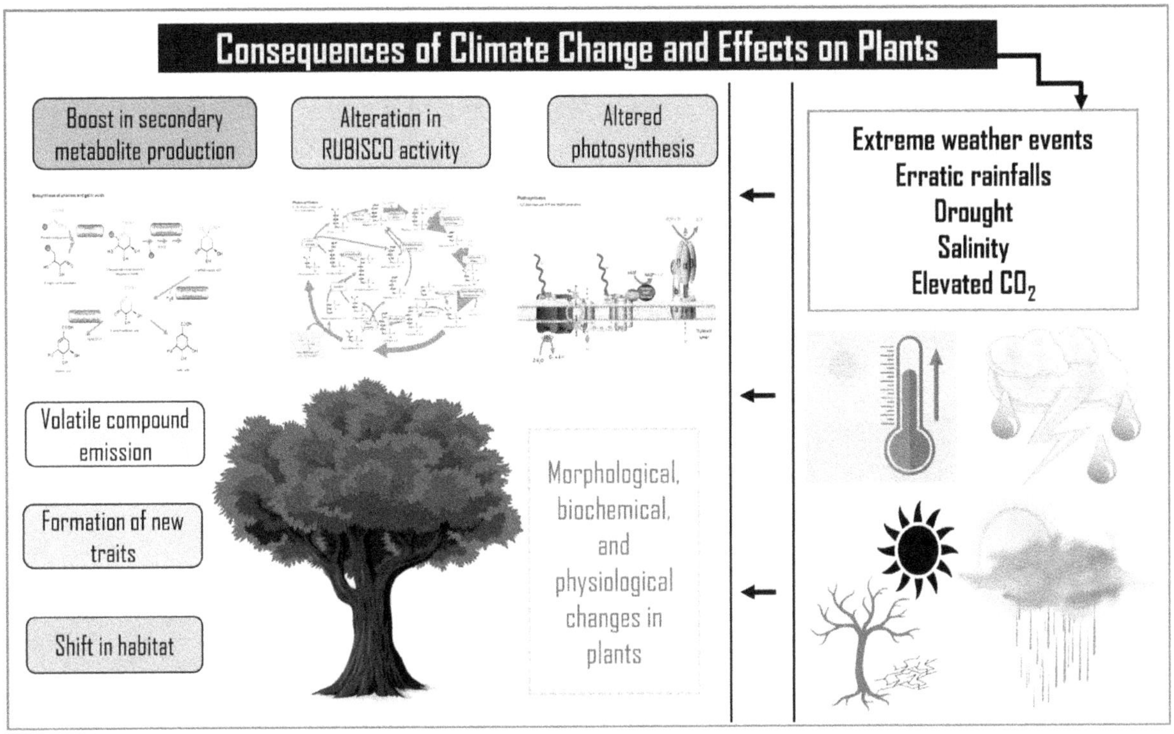


climate change may exacerbate crop output constraints in areas where they already exist due to things like water scarcity and increasing food insecurity. For instance, by 2050, climate change is expected to lower maize yields in some parts of Sub-Saharan Africa by up to 30%, which might result in food shortages and hunger (Kotir, 2011).

## Changes in Plant Phenology and Reproduction

Climate change profoundly impacts plant phenology and reproduction, like blooming, fruiting, and seed generation (Volk et al., 2023). Figure 2 is a depiction of damage caused by heat and drought stress on the plant reproductive stage. These alterations, which have complicated implications on plant growth and reproductive performance, can be brought on by changes in temperature and precipitation patterns. The change in flowering dates is one of the most important effects of climate change on plant phenology. Plants may start to bloom earlier in the year when temperatures rise, which may have an impact on pollination and seed output. Plant phenology can be impacted by variations in precipitation patterns, with drought conditions delaying flowering and fruiting. The productivity of crops and environmental dynamics can both be significantly impacted by phenological changes (Inouye, 2022). For instance, if the timing of flowering does not coincide with the availability of crucial resources like nutrients, water, or light, it may cause mismatches with pollinators or result in reduced seed output.

On the other hand, a delayed flowering period could result in decreased crop yields or a higher susceptibility to pests and diseases (Ul-Allah et al., 2021). In addition to altering phenology, climate

*Figure 2. Effect of heat and drought stress on plant reproductive ability*
*(Adopted from Gray & Brady, 2016). Part of the diagram is obtained from brgfx on Freepik.*

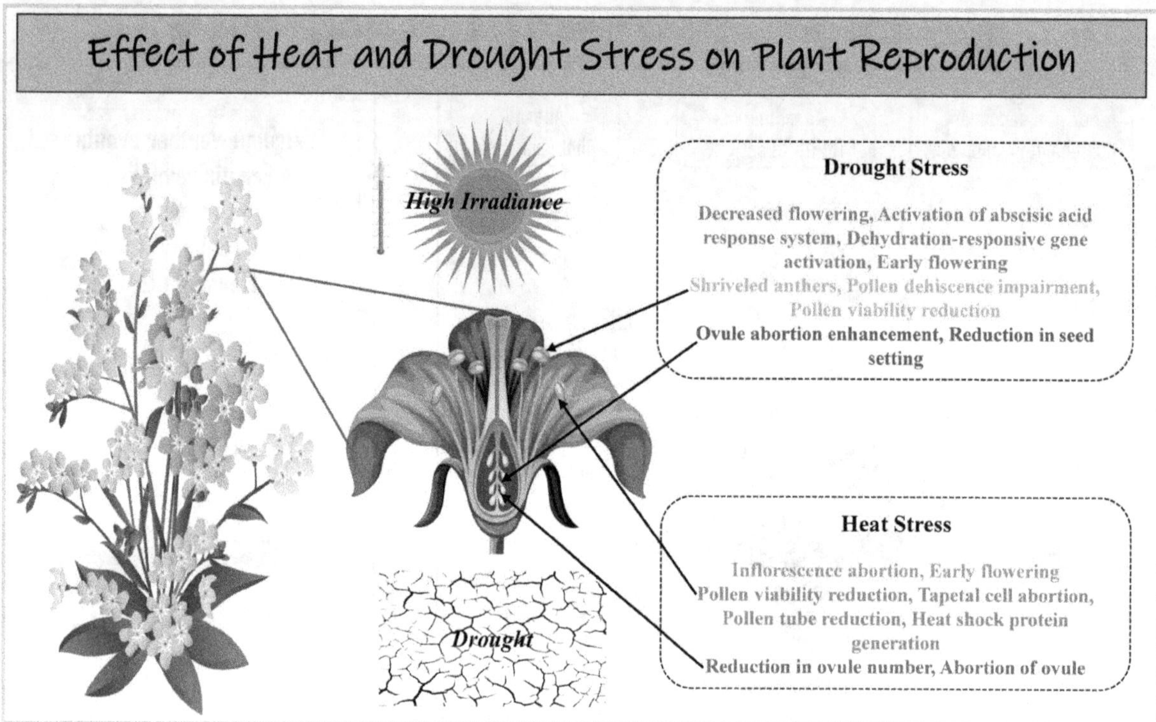


change can also have an impact on plant reproduction by altering how resources are distributed. As temperatures rise, changes in seed size, germination rates, and seedling vigor may result from plants devoting more energy to growth and reproduction. These changes may significantly impact the dynamics of ecosystems, agricultural output, and plant fitness. Therefore, there are numerous and complex effects of climate change on plant phenology and reproduction, with consequences for both agricultural and natural ecosystems. In order to create effective adaptation and mitigation policies that can assist in promoting sustainable development and help people become more resilient to climate change, it is essential to understand these effects.

## Damage to Plant Tissues

Plant tissues are susceptible to climate fluctuation (Volk et al., 2023). The temperature rises, modifications to precipitation patterns, and severe weather can all physically harm plants and stunt their development and productivity. Through heat stress, climate change can harm plant tissues in one of the most obvious ways. High temperatures have been linked to decreased growth and productivity because they can harm cell membranes, slow down photosynthesis reactive oxygen species generation, and speed up respiration (Hasanuzzaman et al., 2013). Extreme heat events have the potential to kill plant tissues in some circumstances. Changes in precipitation patterns can also harm the tissues of plants. For instance, a protracted drought can make plants stressed out, which can cause wilting, scorched leaves, and even death. Like soil erosion, nutrient leaching, and root damage, high rainfall events can also negatively impact plant growth and output. Hurricanes, tornadoes, and hailstorms are only a few examples of extreme weather events linked to climate change that can physically harm plant tissues. Strong winds have the potential to uproot trees, break branches, and hurt flowers and foliage. Floods can drown plants and result in soil erosion, while hailstorms can shred the leaves and blossoms off of plants. As a result, climatic change may harm plant tissues, resulting in decreased growth and yield output.

## Reduced Photosynthesis Rates

The rate of photosynthesis in plants can be significantly impacted by climate change, which can, therefore, have an impact on plant growth, productivity, and carbon sequestration (Elbasiouny et al., 2022). Plant photosynthesis rates can be impacted by changes in precipitation patterns, rising carbon dioxide ($CO_2$) levels, and temperature rise. Heat stress is one of the most direct ways that climate change can lower photosynthetic rates. Cell membranes may be harmed by high temperatures, which may interfere with the operation of the photosynthetic apparatus within plant cells (Sharma et al., 2020). Reduced photosynthetic rates can also result in decreased growth and output. Changes in precipitation patterns can also impact the rate of photosynthesis in plants. Long-term dryness can cause plants to experience water stress, which can lower photosynthetic rates and induce leaf withering and leaf drop. The soil can get saturated during periods of severe rain, which limits the oxygen available to plant roots and can consequently slow photosynthetic rates. Increased rates of plant photosynthesis can be impacted by elevated atmospheric $CO_2$, which is a significant contributor to climate change. Some plant species may experience greater growth and productivity as a result of higher $CO_2$ levels stimulating photosynthetic rates. However, other elements, like the availability of water and nutrients, frequently affect this effect. Elevated $CO_2$ levels in other plant species may have little to no impact on photosynthesis rates, or they may even cause photosynthesis rates to decrease if there are insufficient amounts of other resources (Maschler et al., 2022). Climate change so causes a reduction in photosynthesis, which in turn results in a reduction in plant yield. For the purpose of creating effective adaptation and mitigation methods that can aid in promoting sustainable development and enhancing climate change resilience, it is essential to comprehend these effects.

## Plant Mortality

Plant mortality is significantly impacted by climate change, and this can have a large influence on ecosystems as a whole. The primary ways that climate change influences plant mortality are listed below. Heat stress in plants is a potential consequence of climate change's rising temperatures. A plant's tolerance for high temperatures can be exceeded, which can cause physiological stress, reactive oxygen species production, stunted development, and, ultimately, plant death (Azeem et al., 2023). Heat waves can be highly harmful, especially to plants that are sensitive to heat. Traditional precipitation patterns may be disrupted by climate change, resulting in more frequent and severe droughts or heavy rainfall events. Water scarcity brought on by droughts, particularly in arid and semi-arid areas, can influence plant viability. Rainfall that is too heavy can kill plants by causing waterlogging, root damage, and increased susceptibility to diseases. The distribution and number of diseases and pests may alter as a result of climate change. Increased plant mortality can result from the spread of some pests and pathogens that are favoured by warmer temperatures and shifting rainfall patterns (Trębicki et al., 2017).

Additionally, invasive species can prosper amid climate change and harm native plant species. Plants may have trouble adjusting to their current environments as the climate changes. Due to physical obstacles or inadequate dispersal mechanisms, some plants can be unable to migrate or disperse to more favourable areas. Range shifts may emerge from this when plants become less fit and more susceptible to death since their existing environments are no longer suitable for them. Critical plant life cycle stages like flowering and leaf emergence might change in timing as a result of climate change. Lack of synchronization between a plant's phenology and the changing climate cues can affect both the reproductive success and death of the plant (Ovaskainen et al., 2013). Plants may suffer severe mortality, for instance, if flowers bloom before pollinators appear or if frost develops after bud bursts. Dryer conditions can be a result of climate change, which raises the danger and intensity of wildfires. Particularly in habitats that are prone to fire, intense flames can result in widespread plant mortality. Wildfire-related vegetation loss can have a long-lasting impact on the dynamics of plant groups and ecosystems (Bladon et al., 2014). It's crucial to remember that the precise effects of climate change on plant mortality can fluctuate based on the type of plant, where it is located, and the local environmental circumstances overall. However, climate change poses serious risks to plant life and can have a domino effect on ecosystems, affecting things like changes in biodiversity and food supply.

## IMPACTS OF CLIMATE CHANGE ON ECOSYSTEMS

Ecosystems are being significantly impacted by climate change worldwide. Some of the main consequences are as follows: Biodiversity is in danger due to climate change since it disrupts ecosystems and species' habitats. Many species have particular needs in terms of temperature and climate, and as these conditions change, some may find it difficult to adapt or move to new habitats (Bladon et al., 2014). Local extinctions and a loss of biodiversity generally may result from this. Modifying habitats, climate change, altered precipitation patterns, and sea level rise have an impact on ecosystems. For instance, as ocean temperatures rise, coral reefs become more and more susceptible to bleaching, which causes a loss of habitat for many marine species (Asch et al., 2019). The plants and animals that rely on forests, wetlands, and other ecosystems are impacted by changes in temperature and precipitation. Many species are compelled to move their geographic ranges to follow favourable conditions when temperatures change.

As a result, interactions between species may change because they may run across new rivals, predators, or prey. Such changes have a ripple effect across the food web and have the potential to upset ecosystems' fragile balance. The timing of biological activities, including flowering, migration, and reproduction, is impacted by climate change. Mismatches between species that depend on one another, such as plants and pollinators or predators and prey, can result from changes in phenology. The timing of biological activities, including flowering, migration, and reproduction, is impacted by climate change (Asch et al., 2019). These disturbances may significantly impact the functioning of ecosystems and the survival of species. Existing challenges to species and ecosystems are exacerbated by climate change, raising the possibility of extinction. For instance, melting ice is causing some Arctic animals, such as polar bears and Arctic foxes, to lose their habitats on the sea ice, which makes it more difficult for them to hunt and reproduce. The extinction of keystone species or species with special ecological responsibilities may significantly impact the stability and resilience of ecosystems. Primary production in terrestrial and marine ecosystems is impacted by climate change. The development of plants and algae, which serve as the foundation of food webs, can be impacted by variations in temperature, precipitation, and the availability of nutrients.

Additionally, changes in soil carbon storage or an increase in wildfires can affect the carbon cycle, which can have an impact on ecosystems' overall carbon balance as well as climate feedback loops. Ocean acidification results from the oceans' increased uptake of atmospheric carbon dioxide. This acidification may have a deleterious effect on marine life, especially on creatures like coral reefs, shellfish, and some planktonic species that develop calcium carbonate skeletons or shells (Haigh et al., 2015). The destruction or loss of these ecosystems impacts the productivity and biodiversity of marine ecosystems. Among the effects of climate change on ecosystems, these are only a few. It is critical to undertake conservation strategies to save ecosystems and the species that depend on them, reduce greenhouse gas emissions, and preserve habitats.

## Changes In Biodiversity

The distribution of species, the size of populations, and ecological interactions are all affected negatively by climate change. Several of the main effects of climate change on biodiversity are listed below:

1. **Range Shifts:** Species are forced to relocate to better habitats as temperatures and climate conditions change. As a result, animals migrate to higher elevations or latitudes or seek refuge in milder microclimates, shifting their geographic range (Kumar et al., 2021). However, not all species can change or spread out quickly enough, which causes extinctions in specific areas. As a result, ecosystems' general species composition and distribution are changed.
2. **Loss of Habitat:** Critical habitats may be lost or degraded as a result of climate change. Ecosystems like forests, wetlands, and coral reefs are susceptible to damage or extinction due to rising temperatures, shifting rainfall patterns, and an increase in the frequency of severe weather (such as droughts and floods). Many species depend on these environments for food and refuge, and their extinction could have a ripple effect on biodiversity (Roy et al., 2023).
3. **Phenological Mismatches:** Natural phenological processes, including flowering, breeding, and migration, can be thrown off by climate change. Climate change may result in mismatched interactions between species if each species depends on distinct cues (such as temperature or day duration) to synchronize different stages of its life cycle (Vázquez et al., 2017). For instance, if a plant blooms sooner in the spring because of warmer temperatures but its pollinator migrates at a different period, this may result in less successful pollination and have an effect on plant reproduction.
4. **Species Interactions:** Species' interactions and relationships with one another can change due to climate change. Predator-prey dynamics, resource competition, and mutualistic interactions (such as those between plants and pollinators), for instance, can all be impacted by changes in temperature and precipitation patterns. While some species may gain from climate change, others may suffer detrimental consequences. The equilibrium and operation of ecosystems may be disturbed by these changes.
5. **Increased Extinction Risk:** Climate change provides an additional concern for species already under strain from other factors like habitat loss or overexploitation. Particularly susceptible are species with restricted climatic ranges or distinct habitat needs (Bhat et al., 2023). The risk of extinction increases as suitable habitats become harder to find or are fragmented, making it harder for species to migrate or adapt.

Coordinated efforts are needed to reduce greenhouse gas emissions, safeguard and restore habitats, advance sustainable land and water management strategies, and advance conservation and restoration projects in order to address climate change and its effects on biodiversity. By preserving biodiversity, we can increase ecosystems' adaptability and lessen the impact of climate change on both natural systems and societal structures.

## Impacts on Plant-Pollinator Interactions

The interactions between plants and their pollinators, which are crucial to pollination and the reproductive success of many plant species, are significantly impacted by climate change. Following are a few significant effects of climate change on plant-pollinator interactions:

1. **Phenological Mismatches:** The timing of plant flowering and the activity of pollinators can be thrown off by climate change. Phenological mismatches may occur if plants flower earlier or later than expected due to changing climate conditions while their pollinators continue to bloom simultaneously. As a result, population levels and biodiversity may be impacted, as well as decreased pollination success and reproductive failure for both pollinators and plants (Bhat et al., 2023).
2. **Changes in Geographic Range:** The geographic ranges of plant species and the pollinators that visit them can alter as a result of climate change (Kolanowska, 2023). In reaction to shifting temperatures and precipitation patterns, some species may migrate to higher latitudes or altitudes. Local plant-pollinator relationships may change if plant and pollinator species do not shift at the same rate or in the same direction, disrupting their historical coexistence and mutual dependency.
3. **Habitat Loss and Fragmentation:** Pollinators and plants may both suffer as a result of habitat loss and degradation brought on by climate change. Reduced connectivity due to habitat fragmentation makes it more difficult for pollinators to access and travel between suitable feeding and breeding areas (Jiao et al., 2023). This can reduce the number of pollinator species and their capacity to properly pollinate plant species, particularly those with broad home ranges or particular habitat needs.
4. **Indirect Effects on Mutualistic Interactions:** The distribution and quantity of various organisms that interact with plants and pollinators can alter as a result of climate change (Guimaraes Jr et al., 2017). For instance, variations in temperature and precipitation may have an impact on the frequency of diseases, parasites, and predators of both plants and pollinators. Such indirect effects may further impact the well-being and survival of pollinators and the plants they interact with.

In the face of climate change, efforts must be made in conservation and management to preserve and improve plant-pollinator interactions. The maintenance of floral diversity, the promotion of native plant species, the protection and restoration of habitats, the use of climate-smart land management techniques, and the preservation and restoration of habitats can all support resilient plant-pollinator networks. Understanding the unique reactions of various plant and pollinator species to climate change can also help to guide focused conservation measures and facilitate the adaptation of plant-pollinator interactions in a changing environment.

## Impact on Ecosystem Function

Ecosystem function and the activities that take place within ecosystems and contribute to their general structure, stability, and productivity are significantly impacted by climate change. Some of the main impacts of climate change on ecosystem function are listed below:

1. **Primary Productivity:** The primary productivity of ecosystems is the rate at which plants and algae use photosynthesis to turn sunlight into organic matter, which can be impacted by climate change. Changes in $CO_2$ levels, precipitation patterns, and temperature can all have an impact on the timing and length of the growing season as well as the development of plants (Cavicchioli et al., 2019). In some places, rising temperatures and $CO_2$ levels can initially promote plant growth, but protracted droughts or heat waves can reduce output. Changes in primary production may affect how energy moves through food webs and how readily available resources are to other organisms.
2. **Nutrient Cycling:** Ecosystem processes that involve nutrient cycling may be impacted by climate change. Increased temperatures, for instance, may hasten the decomposition of organic waste and hasten the release of nutrients from the soil. Nutrient imbalances in ecosystems may result from variations in precipitation patterns' effects on nutrient leaching and discharge. These changes in nutrient availability and cycling may have a cascade impact on the make-up and operation of microbial and plant communities.
3. **Water Availability and Hydrological Processes:** Modifications in climatic patterns, including different rainfall patterns and higher evaporation rates, can affect the amount of water that is available in ecosystems (Asif et al., 2023). The distribution and abundance of plant species can be impacted by droughts and reduced water availability, which can modify the composition of the vegetation and the structure of the ecosystem. The operation of aquatic ecosystems is also influenced by the availability of water, which affects the amount and distribution of aquatic creatures, the growth of algae, and the quality of the water.
4. **Carbon Sequestration:** Ecosystems are essential to carbon sequestration because they remove carbon dioxide from the atmosphere, thereby reducing climate change (Daba & Dejene, 2018). However, the ability of ecosystems to store carbon can be impacted by climate change. For instance, hotter temperatures and droughts can raise the danger of wildfires, which release carbon dioxide from the atmosphere's reserves. Changes in vegetation production and composition can also impact the quantity of carbon stored in ecosystems.
5. **Ecosystem Resilience and Adaptation:** Ecosystems may become less resilient as a result of climate change, making them more susceptible to perturbations and less able to recover (Mumby et al., 2014). Ecosystems with low biodiversity or unique habitat needs may be especially vulnerable. However, as they attempt to adjust to shifting climatic conditions, some ecosystems may display adaptive responses, such as changes in species composition or phenology.

The reduction of greenhouse gas emissions, the preservation and restoration of habitats, and the promotion of sustainable land and water management techniques are all necessary to address climate change and its effects on ecosystem function. For ecosystem services like carbon sequestration, water purification, and the provision of food and resources for human communities to continue, ecosystem function must be preserved and improved.

## CLIMATE CHANGE AND IMPLEMENTATION OF SUSTAINABLE AGRICULTURAL PRACTICES

Global agricultural systems are facing serious difficulties owing to changing climate, yet resilient agriculture methods can help to lessen these effects. Following are a few ways that sustainable farming methods help to combat climate change:

### Climate-Smart Crop Selection

A key element of sustainable agriculture operations in the face of climate change is the choice of climate-smart crops. It entails finding and growing crop types that are tolerant of shifting climatic circumstances while reducing agriculture's negative environmental effects. Farmers may increase production, increase food security, and lower greenhouse gas emissions by implementing climate-smart crop selection practices (Dutta et al., 2022). The following are some crucial elements of applying climate-smart crop selection: Temperatures rise as a result of climate change, which also alters rainfall patterns and increases the frequency of extreme weather occurrences. It is crucial to choose crop kinds that are suited to these shifting environmental factors. Resilient crops may withstand extreme temperatures, droughts, flooding, or pests while still producing steady yields. In order to lessen the vulnerability of agricultural systems to climate change, a variety of crops should be planted. Farmers can reduce the risks associated with climatic unpredictability by choosing crops with various growth cycles and environmental requirements (Hajimirzajan et al., 2021).

Crop variety also boosts ecosystem services, enhances soil health, and lowers pest and disease burdens. Climate-smart agriculture heavily relies on breeding initiatives to create new crop kinds. Breeding for characteristics including disease resistance, nutritional efficiency, and tolerance to heat and drought are the main research areas. Creating crop types that are more tolerant to climate change can be accelerated through genetic enhancement methods like biotechnology and gene editing (Munaweera et al., 2022). The emphasis on agroecological principles, which encourage the integration of ecological processes into farming systems, is a key component of climate-smart crop selection. Organic farming, conservation agriculture, and agroforestry are agroecological approaches that increase biodiversity, boost soil health, and lessen reliance on artificial inputs. For the crop selection process to be successfully implemented, farmers must be involved (Garibaldi et al., 2017). Farmers are in a unique position to identify good crop types because they have the practical experience and significant traditional knowledge. Participatory methods promote knowledge exchange, experimentation, and the preservation of regional crop diversity. In light of climate change, choosing crops that are climate-smart is a realistic way to implement sustainable agriculture methods. Farmers can increase their adaptability, productivity, and sustainability in a changing environment by choosing and growing resilient crop types, diversifying agricultural systems, and adhering to agroecological principles.

### Soil Management

Implementing sustainable agriculture methods in the context of climate change requires careful consideration of soil management. By raising crop resilience, increasing nutrient cycling, improving water infiltration, and sequestering carbon, healthy soils are essential for reducing climate change and preparing for it (Lal et al., 2011). Here are some crucial components of soil management as an application of environmentally friendly farming methods in the face of climate change: No-till and minimal tillage techniques cause the least amount of soil disturbance and keep crop leftovers on the soil's surface. These techniques lessen the effects of high rainfall and boost water availability during dry seasons by preventing soil erosion, enhancing soil structure, and increasing water infiltration. Planting crops known as "cover crops" is done expressly to maintain and improve the soil while it is fallow or in between harvests of cash crops. Cover crops contribute to improved soil structure, decreased weed growth, increased organic matter content, decreased soil erosion, and increased soil's ability to retain water (Kaye & Quemada, 2017).

Additionally, they support biodiversity, nitrogen cycling, and carbon sequestration. Sustainable soil management requires the application of effective nutrient management techniques. The efficiency of nutrient utilization is increased, nutrient runoff is minimized, and techniques like balanced fertilizer application, precision farming, and the use of organic fertilizers decrease the risk of water pollution. Accurate soil testing and analysis aid in identifying the precise nutrient needs of crops, enabling farmers to use fertilizers sparingly. Increasing the amount of organic matter in soils is essential for climate-smart agriculture (Scherr et al., 2012). The structure, water-holding capabilities, and nutritional availability of the soil are all improved by organic matter. Crop rotation, using compost or manure, and other techniques assist in boosting soil fertility, increasing organic matter content, and encouraging carbon sequestration. Because erosion causes soil degradation and the loss of priceless topsoil, it is essential to prevent soil erosion for sustainable soil management.

Utilizing erosion control techniques like contour ploughing, terracing, windbreaks, and buffer strips can successfully reduce soil erosion brought on by strong winds or rains, preserving the health and fertility of the soil. Agroforestry and windbreak systems that incorporate trees and other woody plants into agricultural landscapes have many positive effects on soil management (Pantera et al., 2021). Trees and bushes aid in soil stabilization, wind erosion reduction, shade production, nutrient cycling enhancement, and biodiversity promotion. Their canopies regulate temperature and lessen evaporation, while their root systems enhance soil structure and water infiltration. Farmers who regularly monitor and evaluate their soils can better comprehend the condition and quality of those soils. Making educated judgments about soil management strategies requires access to useful information from soil testing for nutrient levels, pH, organic matter content, and soil structure analyses. Monitoring soil moisture levels can aid in water resource conservation and irrigation scheduling optimization. It is crucial to educate and teach farmers about sustainable soil management techniques. The transmission of knowledge, the promotion of best practices, and the empowerment of farmers to embrace sustainable soil management practices are made possible via extension services, farmer field schools, and knowledge-sharing platforms.

## Water Management

Implementing sustainable farming techniques in the face of climate change requires careful water management. It is crucial to maximize water usage efficiency, conserve water resources, and lessen the effects of drought and water scarcity since changing weather patterns, rising temperatures, and altered rainfall distributions affect water availability (Iglesias & Garrote, 2015). The following are some crucial components of water management as an application of climate change-friendly agriculture techniques: Adopting effective irrigation techniques is essential for long-term water management. By delivering water directly to the root zone of crops, technologies, including drip irrigation, precision sprinklers, and soil moisture monitors, assist in reducing water loss due to evaporation and runoff. These methods increase crop productivity, minimize stress on freshwater resources, and optimize water use. Sustainable agriculture can be considerably aided by adopting water conservation measures. Mulching, cover crops, and conservation tillage are methods that assist in keeping the soil moist, lower evaporation, and decrease water loss. Conservation methods also improve the water-holding capacity of agricultural areas, minimize erosion, and promote the health of the soil (Ahuchaogu et al., 2022). It is critical to evaluate crop water needs based on regional climate, crop characteristics, and soil types. Farmers can use this knowledge to identify the best irrigation schedules and quantities needed to suit the unique requirements of diverse crops at various growth stages. Water waste is reduced, and total water use efficiency is increased with proper irrigation scheduling. In order to use rainwater for agriculture, it must be captured and stored. When it rains a lot, strategies like building rooftop rainwater harvesting systems, farm ponds, or reservoirs can help gather rainwater. Then, during dry spells, harvested rainwater can be used for irrigation, minimizing reliance on freshwater sources. It is crucial to select crop kinds that are suitable to the local water supply.

Crops that are drought-tolerant and water-efficient can endure dry spells and still generate respectable yields (Patrick, 2015). Crop selection should take into account elements, including the crop's water needs, the nature of its root system, and its compatibility with the local climate. Putting in place technologies for water reuse and recycling can help agriculture make the best use of its water supplies. Sustainable water management strategies include methods like recycling water from manufacturing sites, catching and reusing agricultural runoff, and using treated wastewater for irrigation (Ahuchaogu et al., 2022). The reused water is treated and monitored properly to guarantee that it satisfies quality standards and to reduce any possible dangers. When managing water, an integrated strategy must take into account every stage of the water cycle, from sourcing to disposal. This method entails managing watershed-level water resources, encouraging water-wise farming techniques, coordinating water consumption across various industries, and implementing comprehensive plans to address water-related issues. Farmers may adapt to shifting patterns of water availability, lower water-related hazards, increase agricultural production, and contribute to overall sustainability in the face of climate change by putting these water management methods into practice.

## Agrobiodiversity Conservation

Implementing sustainable farming practices in the face of climate change requires preserving agrobiodiversity on a fundamental level. The variety and variability of organisms, including plants, animals, and microorganisms, necessary for agricultural productivity is referred to as agrobiodiversity. Agrobiodiversity preservation strengthens ecosystem functions, supports food security, and keeps farming systems flexible and adaptable (Erenso & Andemo, 2022). Key components of agrobiodiversity preservation as a means of putting into practice climate change-affected sustainable agriculture methods are as follows: To practice climate-smart agriculture, crop diversification needs to be encouraged. In order to reduce the risks brought on by climate change, it is beneficial to grow a diversity of crops with various genetic features and adaptability abilities (Awiti et al., 2022). Diverse crops are better able to withstand shifting temperatures and precipitation patterns, fend off pests and diseases, and sustain consistent yields in a variety of environments. To maintain agrobiodiversity, traditional and native crop types must be conserved. These regionally adapted types have developed over many generations and have special qualities like insect or drought tolerance. The cultivation of these plants helps to preserve genetic variety, cultural history, and food independence. A key tactic for preserving plant genetic resources is the establishment of seed and gene banks. Numerous plant species, particularly rare and endangered ones, are preserved and protected in these repositories.

Additionally, seed banks help farmers, scientists, and breeders exchange and distribute seeds, fostering agricultural diversity and resilience. Agroforestry systems sustainably combine trees, crops, and livestock. By providing a habitat for a variety of plant and animal species, these systems support the conservation of biodiversity (Montagnini & del Fierro, 2022). Agroforestry practices also increase soil fertility, lessen erosion, increase water retention, and sequester carbon, all of which help in reducing greenhouse gas emissions and preparing for climate change. The use of integrated pest management techniques reduces the need for chemical pesticides and encourages organic pest control. In this strategy, various methods are combined, including crop rotation, biological control agents, habitat modification, and resistant crop varieties. Biodiversity is safeguarded by limiting chemical inputs, including helpful insects, birds, and soil creatures. Bees, butterflies, and birds are examples of pollinators that play a crucial role in agricultural environments by promoting pollination and guaranteeing crop output (Sabbahi, 2022). Farmers are given the ability to conserve, share, and improve their seed varieties through supporting farmer-managed seed systems.

Through this method, local seed diversity is encouraged, farmer autonomy is strengthened, and crop adaptation to shifting climatic conditions is encouraged. Networks that share seeds between farmers make it easier to preserve and spread traditional crop kinds. It is crucial to raise awareness of the value of agrobiodiversity conservation among farmers, communities, and politicians. It is possible to spread awareness of the importance of preserving traditional knowledge and agrobiodiversity-related practices through educational programs, workshops, and campaigns. Farmers can increase resilience by using these agrobiodiversity conservation techniques.

## FUTURE RESEARCH DIRECTIONS

To address pressing issues, future research on "Climate Change and its Impacts on Plants" should take an interdisciplinary approach. For precise forecasts of changes in plant distribution, growth patterns, and phenology under various climate scenarios, predictive modeling incorporating climate data, $CO_2$ levels, and land-use changes is essential. Understanding the genetic basis of plant resilience through genomic research is essential for guiding targeted breeding efforts to create climate-resilient cultivars.

For food security in the face of changing climates, integrated sustainable agriculture policies that take into account climate-resilient crops and contemporary technologies are essential. Research on the connections between plants and their microbiomes as well as the use of remote sensing to keep track of plant health are crucial. Additionally, by including communities in citizen science projects, data collection on plant responses to climate change can be expanded while raising awareness. By focusing on these areas of study, we may better comprehend the complex interplay between plants and climate, which will help us create plans to lessen and accommodate the effects of climate change on plant life.

## CONCLUSION

Plants and ecosystems all across the world are threatened significantly by climate change. Temperatures have risen, precipitation patterns have changed, and extreme weather events have become more frequent due to the atmosphere's rising greenhouse gas concentrations. These changes severely impact the development, distribution, and general ecological balance of plants. The changing of phenological patterns in plants is one of the most obvious effects of climate change. Early flowering and leafing of plants affect pollinators and upend fundamental ecological interactions. Moreover, drought stress, decreased output, and greater susceptibility to diseases and pests might result from rising temperatures and shifting rainfall patterns. When coupled with habitat fragmentation, these elements represent major threats to plant biodiversity and may result in the extinction of numerous species. The effects of climate change on plants go beyond individual species. The availability of animal food and shelter might change due to changes in plant distribution, which can have a cascade effect on biodiversity. Furthermore, the loss of plant communities can reduce ecosystems' ability to store carbon, aggravating the issue of greenhouse gas emissions and further accelerating global warming. Because of this, we can work to put into practice effective solutions to prevent and adapt to climate change, ensuring a sustainable future for both plants and people.

## FUNDING

The present book chapter did not receive any grant from the funding agency.

## REFERENCES

Abobatta, W. F. (2023). The influence of climate change on interactions between environmental stresses and plants. In *Plant Stress Mitigators* (pp. 425–434). Elsevier. doi:10.1016/B978-0-323-89871-3.00021-5

Agnihotri, P., & Mitra, A. K. (2023). *Understanding the Impact of Global Climate Change on Abiotic Stress in Plants and the Supportive Role of PGPR.*

Ahmed, M., Hayat, R., Ahmad, M., Ul-Hassan, M., Kheir, A. M., Ul-Hassan, F., Ur-Rehman, M. H., Shaheen, F. A., Raza, M. A., & Ahmad, S. (2022). Impact of climate change on dryland agricultural systems: A review of current status, potentials, and further work need. *International Journal of Plant Production, 16*(3), 341–363. doi:10.100742106-022-00197-1 PMID:35614974

Ahuchaogu, I. I., Ehiomogue, P. O., & Udoumoh, U. I. (2022). Effects of soil and water conservation measures on the environment: A review. *Poljoprivredna tehnika, 47*(3), 42-55.

Asch, R. G., Stock, C. A., & Sarmiento, J. L. (2019). Climate change impacts on mismatches between phytoplankton blooms and fish spawning phenology. *Global Change Biology, 25*(8), 2544–2559. doi:10.1111/gcb.14650 PMID:31152499

Asif, Z., Chen, Z., Sadiq, R., & Zhu, Y. (2023). Climate Change Impacts on Water Resources and Sustainable Water Management Strategies in North America. *Water Resources Management, 37*(6-7), 1–16. doi:10.100711269-023-03474-4

Awiti, H. A., Gido, E. O., & Obare, G. A. (2022). Smallholder Farmers Climate-Smart Crop Diversification Cost Structure: Empirical Evidence From Western Kenya. *Frontiers in Sustainable Food Systems, 6*, 86. doi:10.3389/fsufs.2022.842987

Azeem, A., Abbas, N., Azeem, S., Iqbal, Z., & Ul-Allah, S. (2023). Physiological and Molecular Mechanism of Nanoparticles Induced Tolerance in Plants. In Emerging Contaminants and Plants: Interactions, Adaptations and Remediation Technologies (pp. 233-248). Springer. doi:10.1007/978-3-031-22269-6_9

Behnassi, M., Baig, M. B., Sraïri, M. T., Alsheikh, A. A., & Risheh, A. W. A. A. (2022). Food Security and Climate-Climate Smart Food Systems—An Introduction. In Food Security and Climate-Smart Food Systems: Building Resilience for the Global South (pp. 1-13). Springer.

Bhat, I. A., Fayaz, M., Rafiq, S., Guleria, K., Qadir, J., Wani, T. A., & Kaloo, Z. A. (2023). Predicting potential distribution and range dynamics of Aquilegia fragrans under climate change: Insights from ensemble species distribution modelling. *Environmental Monitoring and Assessment*, *195*(5), 623. doi:10.100710661-023-11245-2 PMID:37115430

Bladon, K. D., Emelko, M. B., Silins, U., & Stone, M. (2014). Wildfire and the future of water supply. In: ACS Publications. doi:10.1021/es500130g

Brown, M., De Beurs, K., & Marshall, M. (2012). Global phenological response to climate change in crop areas using satellite remote sensing of vegetation, humidity and temperature over 26 years. *Remote Sensing of Environment*, *126*, 174–183. doi:10.1016/j.rse.2012.08.009

Cavicchioli, R., Ripple, W. J., Timmis, K. N., Azam, F., Bakken, L. R., Baylis, M., Behrenfeld, M. J., Boetius, A., Boyd, P. W., Classen, A. T., Crowther, T. W., Danovaro, R., Foreman, C. M., Huisman, J., Hutchins, D. A., Jansson, J. K., Karl, D. M., Koskella, B., Mark Welch, D. B., & Webster, N. S. (2019). Scientists' warning to humanity: Microorganisms and climate change. *Nature Reviews. Microbiology*, *17*(9), 569–586. doi:10.103841579-019-0222-5 PMID:31213707

Choukri, H., El Haddad, N., Aloui, K., Hejjaoui, K., El-Baouchi, A., Smouni, A., Thavarajah, D., Maalouf, F., & Kumar, S. (2022). Effect of High Temperature Stress During the Reproductive Stage on Grain Yield and Nutritional Quality of Lentil (Lens culinaris Medikus). *Frontiers in Nutrition*, *9*, 725. doi:10.3389/fnut.2022.857469 PMID:35495922

Daba, M. H., & Dejene, S. W. (2018). The role of biodiversity and ecosystem services in carbon sequestration and its implication for climate change mitigation. *Environmental Sciences and Natural Resources*, *11*(2), 1–10.

Dutta, A., Trivedi, A., Nath, C. P., Gupta, D. S., & Hazra, K. K. (2022). A comprehensive review on grain legumes as climate-smart crops: Challenges and prospects. *Environmental Challenges*, *7*, 100479. doi:10.1016/j.envc.2022.100479

Elbasiouny, H., El-Ramady, H., Elbehiry, F., Rajput, V. D., Minkina, T., & Mandzhieva, S. (2022). Plant nutrition under climate change and soil carbon sequestration. *Sustainability (Basel)*, *14*(2), 914. doi:10.3390u14020914

Erenso, F., & Andemo, A. (2022). Vulnerability of Agrobiodiversity and Agroforestry Settings to Climate Change in Gedeo Zone, Ethiopia. *International Journal of Agronomy*, *2022*, 2022. doi:10.1155/2022/8738482

Garibaldi, L. A., Gemmill-Herren, B., D'Annolfo, R., Graeub, B. E., Cunningham, S. A., & Breeze, T. D. (2017). Farming approaches for greater biodiversity, livelihoods, and food security. *Trends in Ecology & Evolution*, *32*(1), 68–80. doi:10.1016/j.tree.2016.10.001 PMID:27793463

Gora, M. M., Mir, M., Mehraj, S., Sharma, R., Hussain, A., Nazim, N., & Khan, I. (2023). Biodiversity for Sustainable Development. In biodiversity (pp. 369-379). CRC Press.

Gray, S. B., & Brady, S. M. (2016). Plant developmental responses to climate change. *Developmental Biology*, *419*(1), 64–77. doi:10.1016/j.ydbio.2016.07.023 PMID:27521050

Guimaraes, P. R. Jr, Pires, M. M., Jordano, P., Bascompte, J., & Thompson, J. N. (2017). Indirect effects drive coevolution in mutualistic networks. *Nature*, *550*(7677), 511–514. doi:10.1038/nature24273 PMID:29045396

Haigh, R., Ianson, D., Holt, C. A., Neate, H. E., & Edwards, A. M. (2015). Effects of ocean acidification on temperate coastal marine ecosystems and fisheries in the Northeast Pacific. *PLoS One*, *10*(2), e0117533. doi:10.1371/journal.pone.0117533 PMID:25671596

Hajimirzajan, A., Vahdat, M., Sadegheih, A., Shadkam, E., & El Bilali, H. (2021). An integrated strategic framework for large-scale crop planning: Sustainable climate-smart crop planning and agri-food supply chain management. *Sustainable Production and Consumption*, *26*, 709–732. doi:10.1016/j.spc.2020.12.016

Hasanuzzaman, M., Nahar, K., & Fujita, M. (2013). Extreme temperature responses, oxidative stress and antioxidant defense in plants. *Abiotic stress-plant responses and applications in agriculture, 13*, 169-205.

Iglesias, A., & Garrote, L. (2015). Adaptation strategies for agricultural water management under climate change in Europe. *Agricultural Water Management*, *155*, 113–124. doi:10.1016/j.agwat.2015.03.014

Inouye, D. W. (2022). Climate change and phenology. *Wiley Interdisciplinary Reviews: Climate Change*, *13*(3), e764. doi:10.1002/wcc.764

Jiao, L., Zhang, P., Zhang, H., Shao, D., Yang, W., & Sun, T. (2023). Density-dependent dispersal strategy of pollinator moderates the adverse effect of habitat loss on plant reproduction: An integrated model based on pollinators' behavioural response. *Functional Ecology*, *37*(3), 732–747. doi:10.1111/1365-2435.14252

Kaur, G., Singh, G., Motavalli, P. P., Nelson, K. A., Orlowski, J. M., & Golden, B. R. (2020). Impacts and management strategies for crop production in waterlogged or flooded soils: A review. *Agronomy Journal*, *112*(3), 1475–1501. doi:10.1002/agj2.20093

Kaye, J. P., & Quemada, M. (2017). Using cover crops to mitigate and adapt to climate change. A review. *Agronomy for Sustainable Development*, *37*(1), 1–17. doi:10.100713593-016-0410-x

Kolanowska, M. (2023). Loss of fungal symbionts and changes in pollinator availability caused by climate change will affect the distribution and survival chances of myco-heterotrophic orchid species. *Scientific Reports*, *13*(1), 6848. doi:10.103841598-023-33856-y PMID:37100884

Kotir, J. H. (2011). Climate change and variability in Sub-Saharan Africa: A review of current and future trends and impacts on agriculture and food security. *Environment, Development and Sustainability*, *13*(3), 587–605. doi:10.100710668-010-9278-0

Kumar, N. H., Murali, M., Girish, H., Chandrashekar, S., Amruthesh, K., Sreenivasa, M., & Jagannath, S. (2021). Impact of climate change on biodiversity and shift in major biomes. In *Global Climate Change* (pp. 33–44). Elsevier. doi:10.1016/B978-0-12-822928-6.00007-1

Kumar, S., David Raj, A., Kalambukattu, J. G., & Chatterjee, U. (2023). Climate Change Impact on Land Degradation and Soil Erosion in Hilly and Mountainous Landscape: Sustainability Issues and Adaptation Strategies. In Ecological Footprints of Climate Change: Adaptive Approaches and Sustainability (pp. 119-155). Springer.

Kumari, S. (2021). Chapter-5 Global Warming: Causes, Effects and Solutions. *Emerging Trends in Agriculture Sciences*, 67.

Lal, R., Delgado, J., Groffman, P., Millar, N., Dell, C., & Rotz, A. (2011). Management to mitigate and adapt to climate change. *Journal of Soil and Water Conservation*, *66*(4), 276–285. doi:10.2489/jswc.66.4.276

Maschler, J., Bialic-Murphy, L., Wan, J., Andresen, L. C., Zohner, C. M., Reich, P. B., Lüscher, A., Schneider, M. K., Müller, C., Moser, G., Dukes, J. S., Schmidt, I. K., Bilton, M. C., Zhu, K., & Crowther, T. W. (2022). Links across ecological scales: Plant biomass responses to elevated CO2. *Global Change Biology*, *28*(21), 6115–6134. doi:10.1111/gcb.16351 PMID:36069191

Montagnini, F., & del Fierro, S. (2022). Functions of agroforestry systems as biodiversity islands in productive landscapes. In Biodiversity Islands: Strategies for Conservation in Human-Dominated Environments (pp. 89-116). Springer. doi:10.1007/978-3-030-92234-4_4

Moraes, W. B., Madden, L. V., Baik, B.-K., Gillespie, J., & Paul, P. A. (2023). Environmental Conditions After Fusarium Head Blight Visual Symptom Development Affect Contamination of Wheat Grain with Deoxynivalenol and Deoxynivalenol-3-Glucoside. *Phytopathology*, *113*(2), 206–224. doi:10.1094/PHYTO-06-22-0199-R PMID:36131392

Mumby, P. J., Chollett, I., Bozec, Y.-M., & Wolff, N. H. (2014). Ecological resilience, robustness and vulnerability: How do these concepts benefit ecosystem management? *Current Opinion in Environmental Sustainability*, *7*, 22–27. doi:10.1016/j.cosust.2013.11.021

Munaweera, T., Jayawardana, N., Rajaratnam, R., & Dissanayake, N. (2022). Modern plant biotechnology as a strategy in addressing climate change and attaining food security. *Agriculture & Food Security*, *11*(1), 1–28. doi:10.118640066-022-00369-2

Ovaskainen, O., Skorokhodova, S., Yakovleva, M., Sukhov, A., Kutenkov, A., Kutenkova, N., Shcherbakov, A., Meyke, E., & Delgado, M. M. (2013). Community-level phenological response to climate change. *Proceedings of the National Academy of Sciences of the United States of America*, *110*(33), 13434–13439. doi:10.1073/pnas.1305533110 PMID:23901098

Pantera, A., Mosquera-Losada, M., Herzog, F., & Den Herder, M. (2021). Agroforestry and the environment. *Agroforestry Systems*, *95*(5), 767–774. doi:10.100710457-021-00640-8

Patni, B., Bhattacharyya, M., kumari, A., & Purohit, V. K. (2021). Alarming influence of climate change and compromising quality of medicinal plants. *Plant Physiology Reports*, 1-10.

Patrick, R. (2015). When the well runs dry: The slow train wreck of global water scarcity. *Journal - American Water Works Association, 107*(3), 65–76. doi:10.5942/jawwa.2015.107.0042

Pound, B., Lamboll, R., Croxton, S., Gupta, N., & Bahadur, A. V. (2018). *Climate-Resilient Agriculture in South Asia: An analytical framework and insights from practice*. Action on Climate Today.

Randell, H., Gray, C., & Shayo, E. H. (2022). Climatic conditions and household food security: Evidence from Tanzania. *Food Policy, 112*, 102362. doi:10.1016/j.foodpol.2022.102362 PMID:37064798

Richardson, D., Black, A. S., Irving, D., Matear, R. J., Monselesan, D. P., Risbey, J. S., Squire, D. T., & Tozer, C. R. (2022). Global increase in wildfire potential from compound fire weather and drought. *NPJ Climate and Atmospheric Science, 5*(1), 23. doi:10.103841612-022-00248-4

Roy, P., Pal, S. C., Chakrabortty, R., Chowdhuri, I., Saha, A., & Shit, M. (2023). Effects of climate change and sea-level rise on coastal habitat: Vulnerability assessment, adaptation strategies and policy recommendations. *Journal of Environmental Management, 330*, 117187. doi:10.1016/j.jenvman.2022.117187 PMID:36610196

Sabbahi, R. (2022). Effects of Climate Change on Insect Pollinators and Implications for Food Security—Evidence and Recommended Actions. In The Food Security, Biodiversity, and Climate Nexus (pp. 143-163). Springer.

Scherr, S. J., Shames, S., & Friedman, R. (2012). From climate-smart agriculture to climate-smart landscapes. *Agriculture & Food Security, 1*(1), 1–15. doi:10.1186/2048-7010-1-12

Sharma, A., Kumar, V., Shahzad, B., Ramakrishnan, M., Singh Sidhu, G. P., Bali, A. S., Handa, N., Kapoor, D., Yadav, P., Khanna, K., Bakshi, P., Rehman, A., Kohli, S. K., Khan, E. A., Parihar, R. D., Yuan, H., Thukral, A. K., Bhardwaj, R., & Zheng, B. (2020). Photosynthetic response of plants under different abiotic stresses: A review. *Journal of Plant Growth Regulation, 39*(2), 509–531. doi:10.100700344-019-10018-x

Shehzad, K. (2023). Extreme flood in Pakistan: Is Pakistan paying the cost of climate change? A short communication. *The Science of the Total Environment, 880*, 162973. doi:10.1016/j.scitotenv.2023.162973 PMID:36958563

Singh, B. K., Delgado-Baquerizo, M., Egidi, E., Guirado, E., Leach, J. E., Liu, H., & Trivedi, P. (2023). Climate change impacts on plant pathogens, food security and paths forward. *Nature Reviews. Microbiology, 21*(10), 1–17. doi:10.103841579-023-00900-7 PMID:37131070

Slamini, M., Sbaa, M., Arabi, M., & Darmous, A. (2022). Review on Partial Root-zone Drying irrigation: Impact on crop yield, soil and water pollution. *Agricultural Water Management, 271*, 107807. doi:10.1016/j.agwat.2022.107807

Smith, P., & Olesen, J. E. (2010). Synergies between the mitigation of, and adaptation to, climate change in agriculture. *Journal of Agricultural Science, 148*(5), 543–552. doi:10.1017/S0021859610000341

Trębicki, P., Dáder, B., Vassiliadis, S., & Fereres, A. (2017). Insect–plant–pathogen interactions as shaped by future climate: Effects on biology, distribution, and implications for agriculture. *Insect Science, 24*(6), 975–989. doi:10.1111/1744-7917.12531 PMID:28843026

Ul-Allah, S., Azeem, A., Sher, A., Ijaz, M., Sattar, A., Saleem, M. A., Bibi, M., Abbas, N., & Hussain, M. (2021). Assessment of genetic variability and direct-indirect contribution of post-anthesis traits to the grain yield in bread wheat (Triticum aestivum) at different sowing dates. *International Journal of Agriculture and Biology*, *26*(1), 193–200. doi:10.17957/IJAB/15.1824

Usman, S., Umar, N., & Abdulhamid, A. (2023). Climate Change and Drought in the Dryland Areas of Nigeria. In Climate Change Impacts on Nigeria: Environment and Sustainable Development (pp. 361-378). Springer. doi:10.1007/978-3-031-21007-5_19

Vázquez, D. P., Gianoli, E., Morris, W. F., & Bozinovic, F. (2017). Ecological and evolutionary impacts of changing climatic variability. *Biological Reviews of the Cambridge Philosophical Society*, *92*(1), 22–42. doi:10.1111/brv.12216 PMID:26290132

Volk, G. M., Carver, D., Irish, B. M., Marek, L., Frances, A., Greene, S., Khoury, C. K., Bamberg, J., del Rio, A., Warburton, M. L., & Bretting, P. K. (2023). Safeguarding plant genetic resources in the United States during global climate change. *Crop Science*, *63*(4), 2274–2296. doi:10.1002/csc2.21003

Wolka, K., Mulder, J., & Biazin, B. (2018). Effects of soil and water conservation techniques on crop yield, runoff and soil loss in Sub-Saharan Africa: A review. *Agricultural Water Management*, *207*, 67–79. doi:10.1016/j.agwat.2018.05.016

## KEY TERMS AND DEFINITIONS

**Biotic Interactions:** It include interactions between living things, including plants and other creatures like pathogens, pests, and symbiotic organisms.

**Climate Change:** Changes in the Earth's climate that are significant and persistent.

**Global Warming:** The term "global warming" refers to the gradual rise in Earth's average surface temperature brought on by human activity, particularly the buildup of greenhouse gases in the atmosphere that trap heat and have a warming impact.

**Phenological Shifts:** Phenological shifts are modifications in the timing of biological events in plants, such as flowering, fruiting, and leaf budding, as a result of alterations in climatic circumstances, particularly changes in temperature and precipitation patterns.

# Chapter 9
# The Missing Component:
## Intraspecific Traits Variation Enhances Wild Fruit Tree Tolerance to Climate Change

**Jalal Kassout**
https://orcid.org/0000-0002-3702-2751
*National Institute of Agricultural Research, Morocco*

**Khalil Kadaoui**
https://orcid.org/0000-0001-5306-6861
*Faculty of Sciences of Tétouan, University of Abdelmalek Essaâdi, Morocco*

**Younes Hmimsa**
*Polydisciplinary Faculty of Larache, University of Abdelmalek Essaadi, Morocco*

**Salama El Fatehi**
*Polydisciplinary Faculty of Larache, University of Abdelmalek Essaadi, Morocco*

**Soufian Chakkour**
*Faculty of Sciences of Tétouan, University of Abdelmalek Essaâdi, Morocco*

**Mhammad Houssni**
*Faculty of Sciences of Tétouan, University of Abdelmalek Essaâdi, Morocco*

**Sara Oulbi**
*National Institute of Agricultural Research, Morocco*

**Rachid Azenzem**
https://orcid.org/0000-0003-0666-3211
*National Institute of Agricultural Research, Morocco*

**Abdelouahab Sahli**
*Faculty of Sciences of Tétouan, University of Abdelmalek Essaâdi, Morocco*

**Mohammed Ater**
*Faculty of Sciences of Tétouan, University of Abdelmalek Essaâdi, Morocco*

## ABSTRACT

*Plants contend with a myriad of environmental stressors that exert constraints on their performances, posing considerable concerns regarding the enduring persistence and resilience of plant populations over the long term. This apprehension is especially pertinent for Mediterranean species, exemplified by the wild olive and the carob tree. Nevertheless, a multitude of studies have illustrated that plant species exhibit the capability to endure adverse environmental conditions through the deployment of various morpho-physiological and biochemical traits that support physiological processes. Given the pivotal role of wild fruit trees as invaluable genetic resources, a comprehensive understanding of their potential responses to evolving environmental conditions becomes imperative for the development of apt strategies for resource preservation and breeding programs. In the context of this chapter, our hypothesis posits that intraspecific trait variability emerges as a central player in augmenting the resilience of these emblematic wild fruit trees in the face of climate change.*

DOI: 10.4018/978-1-6684-9838-5.ch009

## INTRODUCTION

Climate change stands as an imminent and pervasive threat, casting a looming shadow over the world's natural ecosystems. The phenomenon is underpinned by the rapid alteration of our planet's climate, primarily attributed to the relentless activities of humankind. The implications of this drastic environmental shift are far-reaching, encompassing the long-term survival prospects and adaptive capabilities of countless plant species. Scientific discourse (Allen et al., 2010; Vellend et al., 2017; Pörtner et al., 2022) has consistently sounded the alarm, emphasizing the urgency of our collective response to this existential crisis. Of particular concern is the Mediterranean region, where the effects of climate change and anthropogenic pressures have already begun to materialize. The authoritative voices of scientific research (Polade et al., 2014; Matesanz and Valladares, 2014; Oulbi et al., 2023), have provided a stark portrait of this ongoing transformation. Within this complex web of changes lies the looming challenge posed to the Mediterranean plant species, each uniquely adapted to the region's distinctive ecological tapestry.

Mediterranean plant species, including iconic representatives such as the olive tree and carob tree, are at the frontline of this ecological battle. Their significance extends beyond mere botanical curiosity, as they hold pivotal roles within both natural forests and agroecosystems. These trees, iconic symbols of the Mediterranean landscape, are characterized by their intricate biogeographical requirements, making them highly sensitive barometers of environmental change. The ongoing investigations, such as those by Kassout et al. (2021, 2022a, b, 2023b), have illuminated the extent of their vulnerability to the relentless onslaught of climate change. The dynamic interplay between climate change and its impacts on the natural world is an issue of paramount importance. Scientific research has consistently underscored the urgency of this situation, with a particular focus on the Mediterranean region and its unique plant species. The olive and carob trees, emblematic of this region, are emblematic of the broader challenge we face in preserving the intricate balance of our planet's ecosystems in the face of rapid climate change and human-induced pressures. Thus, their study not only provides insight into their own survival but also serves as a bellwether for the broader health of our planet's natural systems. Inextricably linked to these shifts are the terrestrial dynamics of energy and water flow. The pioneering work of van Bodegom et al. (2014) and Forzieri et al. (2020) delves into the cascading consequences. The disruptions in plant distribution, brought about by climate change, have a profound impact on the flow of energy and water across these landscapes. This, in turn, reverberates throughout the entire terrestrial ecosystem, altering the very essence of how these environments function. Moreover, such climatic variations could have a significant effects on insect pest populations and their dynamics (Azenzem et al., 2023).

The existing plant life in the Mediterranean region comprises an intricate blend of taxa originating from diverse biogeographical regions and having distinct evolutionary histories. This diverse flora encompasses over 48,000 species of flowering plants, with approximately 50% of them being endemic to the region (Blondel et al., 2010). Consequently, Mediterranean plant communities are characterized by the prevalence of stress-tolerant evergreen trees, shrubs, and semideciduous shrubs. These plant groups often exhibit shared morphological, anatomical, and phenological traits, which have commonly been attributed to an evolutionary convergence driven by the unique Mediterranean climate (Davis et al., 1996; Blondel et al., 2010). Among the most ubiquitous and best-valuable Mediterranean plant species, the wild olive (*Olea europaea* subsp. *europaea* var. *sylvestris*) and the carob tree (*Ceratonia siliqua*) have been considered as key component of the Mediterranean vegetation (Terral et al., 2004; Baumel et al., 2018; Kassout et al., 2021, 2022b). In fact, both trees has been exploited around the Mediterranean region since antiquity (Zohary et al., 2012; Kaniewski et al., 2012), and linked to a significant and high cultural and economic importance (Zohary, 2002; Besnard et al., 2018). Wild olive and carob tree are long-lived, evergreen and thermophilous Mediterranean fruit trees (Quézel and Médail, 2003), and both characterized by a high level of genetic diversity (Besnard et al., 2018; Baumel et al., 2022). In addition, both species represents an important component of natural habitats of the Mediterranean region and its traditional agroecosystems (Besnard et al., 2018; Viruel et al., 2018). However, the predicted extreme warming and drying trends in the Mediterranean region compared to other climatic ecoregions (Cramer et al., 2018), is likely to challenge the sustainability of olive and carob based naturel and agroecosystems. Consequently, understanding the factors that underlie the dramatic differences in the ecological strategies employed by the olive and carob trees to deal with the challenging environmental conditions of the Mediterranean region is an urgent matter.

The Mediterranean region, often celebrated as a remarkable cradle of biodiversity, is a tapestry woven from a diverse array of plant life. These botanical communities, originating from varied biogeographical origins and shaped by distinct evolutionary histories, combine to form a complex and intricate mosaic. Within this rich tapestry, an astonishing diversity of over 48,000 species of flowering plants thrives, with nearly half of them being unique to this vibrant region (Blondel et al., 2010). As a consequence of this remarkable diversity, Mediterranean plant communities bear the unmistakable imprint of the region's climatic and environmental nuances, presenting a dynamic tableau of life. One hallmark of these Mediterranean plant communities is the resilience embodied by the prevalence of stress-tolerant evergreen trees, shrubs, and semideciduous shrubs. These botanical groups, while hailing from different lineages, remarkably share an array of morphological, anatomical, and phenological traits. These shared characteristics have been ascribed to a phenomenon known as evolutionary convergence, a process driven by the region's distinctive Mediterranean climate (Davis et al., 1996; Blondel et al., 2010). This convergence reflects nature's exquisite ability to mold life forms in response to the prevailing environmental challenges, forging a harmony amid diversity. Within this harmonious tapestry, two botanical luminaries, the wild olive (*Olea europaea* subsp. *europaea* var. *sylvestris*) and the carob tree (*Ceratonia siliqua*), rise to prominence. These venerable species, deeply intertwined with the Mediterranean's history, culture, and economy, have played pivotal roles in the lives of the region's inhabitants for millennia (Zohary et al., 2012; Kaniewski et al., 2012; Zohary, 2002; Besnard et al., 2018). They are revered as symbols of sustainability and tenacity, having thrived in this challenging environment. Both the wild olive and the carob tree are remarkable in their own right. They are long-lived sentinels, their evergreen foliage a testament to their resilience in the face of Mediterranean climatic extremes. Thriving in thermophilous enclaves, they have weathered countless seasons and stood as witnesses to the region's evolving natural and cultural landscapes (Quézel and Médail, 2003). Furthermore, these botanical icons exhibit a rich tapestry of genetic diversity, further underscoring their ecological significance (Besnard et al., 2018; Baumel et al., 2022). But amidst this reverence and admiration lies a daunting challenge. The Mediterranean region is on the cusp of a climatic transformation, marked by extreme warming and drying trends that far outpace those seen in other climatic ecoregions (Cramer et al., 2018). The sustainability of the intricate nature-based ecosystems and traditional agroecosystems built around the wild olive and carob tree now hangs in the balance. It is in the face of this impending crisis that we recognize the urgency of comprehending the enigmatic strategies employed by these two botanical sentinels to navigate the ever-shifting environmental conditions of the Mediterranean. The tale of the wild olive and carob tree is not merely a story of survival but a lesson in adaptation and resilience. As we confront the escalating challenges of climate change, it becomes increasingly imperative to unravel the factors that delineate the distinct ecological strategies employed by these botanical stalwarts. The harmonious blend of life within the Mediterranean's bosom may hold the keys to resilience and adaptation in an ever-changing world. It is an endeavor that goes beyond scientific curiosity; it is an urgent quest to safeguard the ecological heritage of a region that has nurtured life for millennia.

In the realm of ecological research, a powerful lens through which we can dissect and comprehend the intricate strategies of plant species under varying environmental circumstances is that of plant functional traits. These traits encompass an array of characteristics, ranging from structural and chemical properties to physiological and phenological attributes of plants and their various parts. Collectively, these traits wield a profound influence over a plant's performance as it navigates the complex gradients of its environment (Violle et al., 2007). The application of plant functional traits as a research tool has burgeoned, gaining prominence in our quest to decipher the strategies employed by plant species when faced with contrasting environmental conditions. It is through the prism of these traits that we can discern the adaptive mechanisms that underlie the ability of plants to thrive in diverse ecological niches (Terral et al., 2004; Kassout et al., 2021, 2022b, 2023b). In essence, these traits serve as the fingerprints of plants, revealing their secrets of survival and success. Within this scientific backdrop, our primary objective in this chapter takes center stage. We embark on a quantitative journey, driven by the quest to deepen our understanding of the intricate mechanisms that mold the ecological strategies and functional responses of two iconic denizens of the Mediterranean landscape: the wild olive and the carob fruit tree. These botanical stalwarts have long graced the region with their presence, weathering the tests of time and nature, their roots entwined with the fabric of the Mediterranean's intricate ecosystems. Our mission is clear: to subject these venerable botanical guardians to rigorous scrutiny under the lens of plant functional traits. By doing so, we seek to shed light on the intricate ballet of adaptations and responses that allow these trees to flourish under disparate environmental conditions. Through quantitative tests and meticulous observation, we aim to unlock the secrets of how the wild olive and carob tree thrive amidst the challenges presented by their Mediterranean home, which is now grappling with the ominous specter of climate change. As we delve deeper into this scientific exploration, we anticipate uncovering the hidden facets of these botanical treasures, discovering the subtle but profound ways in which their traits shape their strategies and responses. It is our hope that the knowledge gleaned from this endeavor will not only enrich our understanding of these iconic species but also provide valuable insights into the broader mechanisms governing the ecological resilience of plant life in the face of an ever-changing world. Plant functional traits stand as a formidable tool in our pursuit of ecological understanding. Through their meticulous examination, we embark on a scientific voyage, striving to unravel the intricacies of the wild olive and carob tree's ecological strategies and functional responses in the Mediterranean mosaic. In doing so, we cast light upon the resilient tapestry of life that thrives amidst the challenges of a shifting environment.

## FROM TRAITS TO FUNCTIONAL RESPONSES TO ENVIRONMENTAL CHANGES

Plants, those fascinating living organisms that anchor ecosystems and shape the world around us, exhibit an astonishing array of variations in their physical and physiological attributes. These attributes, aptly termed functional traits by Violle et al. (2007), serve as the compass by which plants navigate the complex tapestry of environmental conditions. They are the keys to understanding how plants respond, adapt, and thrive in diverse ecological settings (Violle et al., 2012; Shipley et al., 2016; Garland et al., 2022). The intrigue of functional traits lies in their power to elucidate the intricate trade-offs that underpin plant fitness, species distribution patterns, the assembly of ecological communities, and the very processes that govern ecosystems (Wright et al., 2004; Siefert et al., 2015; Diaz et al., 2016). These traits form the language through which plants communicate their strategies for survival and success in the natural world. At the heart of the study of functional traits lies a revolutionary approach that enables us to classify plant species into functional groups based on the unique combination of traits they possess. These groups, often referred to as syndromes, represent consistent associations between traits, akin to a plant's ecological strategy in response to its environment (Grime, 2001; Lavorel and Garnier, 2002; Kattge et al., 2011; Garnier et al., 2016; Pierce et al., 2013; Diaz et al., 2016). In essence, it's a language that allows us to decode how plants adapt to their surroundings. Delving deeper into the realm of functional traits, meta-analyses have unveiled fundamental dimensions of plant functional strategies linked to resource utilization and environmental conditions (Wright et al., 2004; Diaz et al., 2004, 2016; Pierce et al., 2017). At the heart of these strategies are two prominent ecological theories that have shaped our understanding of the intricate interplay between form and function in plants (Diaz et al., 2016). The first theory, known as the 'leaf economics spectrum - LES,' unveils a pivotal trade-off that defines plant ecology along a spectrum. This trade-off, etched into the very leaves of plants, distinguishes them along an ecological axis. At one end, we find species geared for rapid nutrient acquisition in resource-rich environments. They boast expansive leaf areas, high photosynthesis rates, and a specific leaf area (SLA) that allows them to efficiently harvest nutrients. On the other end, we encounter conservative species, masters of resource conservation in demanding conditions. They flaunt high values of leaf dry mass content (LDMC), with dense leaves that conserve precious resources (Reich et al., 1992; Wright et al., 2004). The second theory peers into the realm of plant size, unraveling the intricacies of growth rate and competitive prowess, especially under favorable conditions (Diaz et al., 2004, 2016). Plant size, a reflection of a plant's potential dominance and competitive ability, plays a symphony in the ecological orchestra, shaping the dynamics of plant communities and their interactions (Westoby et al., 2002). Functional traits are the keys to understanding the trade-offs, strategies, and ecological intricacies that underpin the magnificent tapestry of plant life on Earth. Through their study, we unveil the hidden secrets of the green world and gain insight into the dynamic forces that shape our natural world.

While studying functional traits among species is useful for assessing the main functional and ecological strategies of plant species worldwide (Diaz et al., 2016; Pierce et al., 2017), there is still a lack of understanding regarding trait variation and coordination within individual species. In fact, our previous understanding of trait variation was primarily based on the assumption that intraspecific trait variation within species could be largely disregarded (Albert et al., 2010; Shipley et al., 2016). However, recent studies have shown that intraspecific trait variability (ITV) is indeed important (Albert et al., 2010, 2011; Messier et al., 2010; Siefert et al., 2015; Kassout et al., 2019, 2021, 2022b), with significant consequences for ecological processes (Laforest-Lapoint et al., 2014), as well as plants' responses to environmental conditions (Violle et al., 2012; Messier et al., 2010; Moran et al., 2015; Kupller et al., 2020; Kassout et al., 2021, 2022b). For instance, Siefert et al. (2015) revealed that ITV accounts for 25% of trait variation at the global scale. Similarly, Fajardo and Piper (2011) found substantial ITV (23%) in *Nothofagus pumilio* in Chile when compared with interspecific studies. Additionally, Kassout et al. (2019) identified significant ITV in key resource use traits (such as SLA and LDMC) within a single species, the emblematic wild olive tree (*Olea europaea* subsp. *europaea* var. *sylvestris*), across its geographical range in Morocco. In fact, ITV is expected to be particularly important for widely distributed species, as suggested by Siefert et al. (2015) and Fajardo and Siefert (2018), at local geographical scales according to Messier et al. (2017), or under extreme environments, as observed by Fajardo and Piper (2011) and

Thomas et al. (2020). Consequently, numerous studies have demonstrated that ITV varies in response to climatic factors and the extent of stressful conditions (Fajardo and Piper, 2011; Kuppler et al., 2020; Kassout et al., 2021). Moreover, ITV within populations and within individuals of single species is expected to be a key mechanism for coping with environmental variations (Moran et al., 2015; Messier et al., 2017; Kassout et al., 2019, 2022b). However, local micro-environmental conditions can also play a crucial role in influencing the expression and magnitude of trait variation, including ITV (Messier et al., 2017; Fajardo and Siefert, 2018; Kassout et al., 2019, 2021). Therefore, understanding how ITV, and thus ecological strategies, vary within single species across different environments is crucial for highlighting the evolutionary and ecological mechanisms driving plant community adaptation and for comprehending future responses to ongoing climate change.

## INTRASPECIFIC TRAITS VARIABILITY OF MEDITERRANEAN WILD FRUIT TREES: METHODOLOGICAL APPROACH

In this chapter will focus upon climate and, in particular, wild olive (*Olea europaea* subsp. *europaea* var. *sylvestris*) and carob tree (*Ceratonia siliqua*), growing along a climatic gradient in Mediterranean Morocco. Both species and region are of climatic relevance and interest (Kassout et al., 2021, 2022a,b, 2023b). Specifically, intraspecific variation in trait expression will be explored between and within several populations of both studied species from climatically-contrasted habitats. Thus, the traits to be presented fall into two groups. The first group comprises leaf functional traits, attributes that identify the 'worldwide leaf economics spectrum' (Wright et al., 2004; Diaz et al., 2016), measured and sampled from wild olive populations. These traits are essentially those routinely measured in 'functional' studies. The second group includes seed-related traits sampled from carob populations, previously demonstrated to be linked to the success of flowering plants in their environment and to the persistence, dispersal, and expansion of plant species (Saatkamp et al., 2014; Beckman et al., 2020).

To assess intraspecific variation in both studied species, several populations of spontaneous trees growing under natural conditions and distributed along a latitudinal gradient from the North to South region of Morocco were investigated (Figure 1). Therefore, eighteen population of carob tree, and nineteen population of wild olive tree were sampled. The field context includes a wide range of habitats and vegetation types across different bioclimatic stages, from sub-humid to semi-arid (Kassout et al., 2019, 2022b, 2023b).

For the carob tree, five healthy and mature specimens were systematically chosen from each population, as outlined in the works by Kassout et al. (2022b, 2023b). Subsequently, 1 kg of healthy pods was harvested from four cardinal exposures (north, south, east, and west) of each selected tree. Within this collection, 25 pods were randomly chosen, and from each of these, 20 seeds were selected at random for subsequent trait measurements. These measurements included seed area (SA), width (SW), length (SL), and aspect ratio (SAR). The entire dataset comprised measurements from 1,740 seeds. In the case of the wild olive tree, ten healthy oleaster trees were randomly identified in each population. From each tree, thirty healthy and mature leaves were randomly collected from the sun-exposed (south) side of the upper canopy. These leaves were promptly transported to the laboratory under cool and hydrated conditions, adhering to established protocols (Cornelissen et al., 2003). In the laboratory, various traits, including stomatal density (DS), leaf water content (LWC), leaf area (LA), leaf aspect ratio (LAR), specific leaf area (SLA), and leaf dry matter content (LDMC), were meticulously measured following the methods

*Figure 1. Geographical distribution of wild olive and carob populations in Morocco*

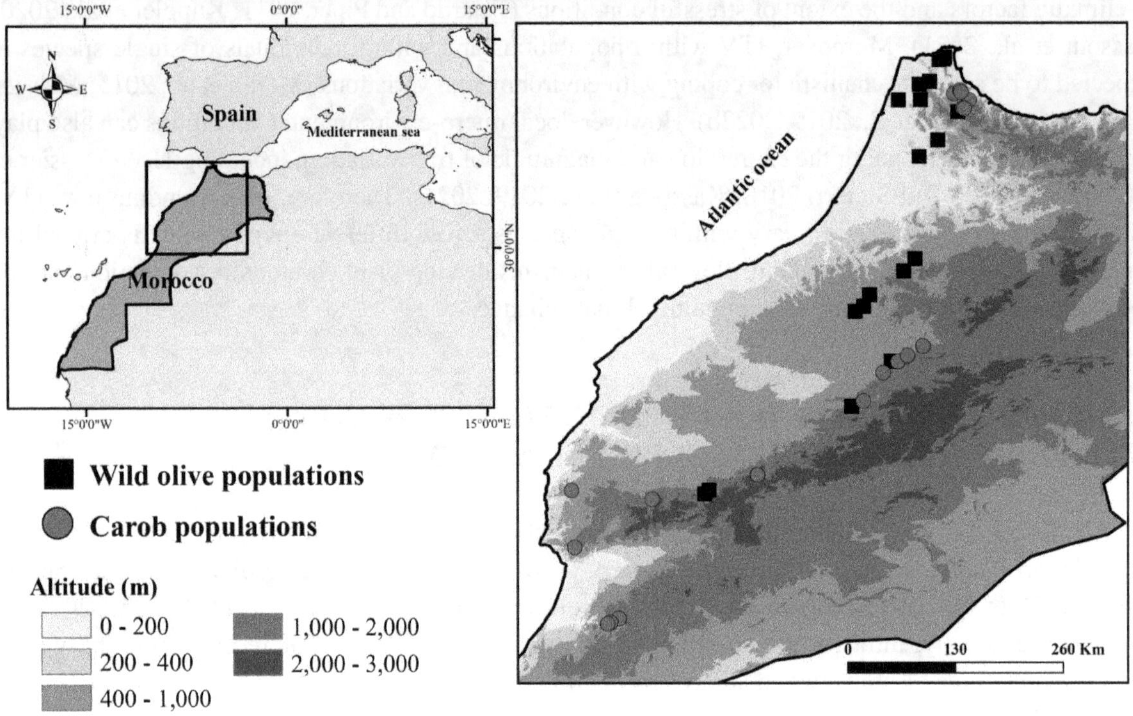


described by Kassout et al. (2019). To assess the extent and significance of intraspecific variation in the measured traits for both species, spanning from the intra-individual to the inter-population level, a variance decomposition procedure was employed, following the methodology outlined in Kassout et al. (2019, 2022b, 2023b). Specifically, the lme() function in the nlme package (Pinheiro et al., 2021) was utilized to fit a general linear model using the restricted maximum likelihood method (RMEL) across the hierarchical levels under consideration. Subsequently, the varcomp() function in the ape package (Paradis et al., 2004) was employed to extract the variance contributions at each hierarchical level. All statistical analyses were performed using the R software (R Core Team, 2020).

## THE ECOLOGICAL IMPORTANCE OF INTRASPECIFIC VARIATION: IMPORTANCE AND IMPLICATIONS FOR MITIGATING CLIMATE CHANGE EFFECTS

The variability in plant functional traits arises from a combination of evolutionary processes and environmental factors operating at different scales (Des Roches et al., 2017). Traditionally, trait-based approaches have assumed that within-species trait variation is insignificant compared to between-species variation, and that species can be characterized by average trait values regardless of their genetic or environmental contexts. However, it is now recognized that within-species variation is crucial for species to adapt to changing environments (Albert et al., 2011; Moore et al., 2020; Kuppler et al., 2020). Furthermore, numerous studies have emphasized the significance of within-species trait variation across environmental gradients (Fajardo and Siefert, 2018; Kassout et al., 2019, 2021, 2022b, 2023b; Kuppler et al., 2020). For example, within-species trait variability not only enhances the average plant responses to environmental changes (Kuppler et al., 2020) but also enables species to establish over a broader range of environmental conditions (Fajardo and Siefert, 2018). Therefore, comprehending the ecological implications of within-species variation is vital for understanding the effects of climate change on plant species.

The assessment of within-species trait variation has been conducted for two species under study, namely the wild olive and carob. This quantification takes into account three distinct levels of spatial and organizational complexity. The first level is inter-population variation, which captures differences observed between different populations. The second level, intra-population variation, focuses on variation among trees within the same population. Lastly, the third level, intra-individual variation, examines variation within organs of the same individual. Specifically, for the intra-individual level, the leaves of the wild olive and the seeds of the carob tree are the organs under consideration. In this chapter, our analysis focused on examining the distribution and organization of trait variance across three ecological scales within two wild fruit trees in Morocco. The patterns observed in the variance components of the key plant functional traits yielded valuable insights. The results highlight two crucial points that we wish to emphasize: (a) the significance of intraspecific variation was found to vary according to the specific scale under investigation. Its relative contribution was deemed important within the studied levels. (b) the inclusion of intraspecific variability in trait-based studies investigating the impact of climate change on plant responses was demonstrated to be of great importance. Overall, these findings shed light on the intricate relationship between trait variance, ecological scales, and the implications for studying plant responses to climate change.

First, when considering the carob tree (*Ceratonia siliqua*), the distribution of variance across different hierarchical levels (population, tree, and seed) was examined for various measurements, including seed area (SA), seed width (SW), seed length (SL), and seed aspect ratio (SAR). At the population level, SA, SW, and SL exhibited a higher proportion of variance compared to other levels. Specifically, they accounted for 58.82%, 46.55%, and 48.52% of the total variance among the studied populations (Figure 2). Interestingly, these traits also demonstrated a considerable amount of variance at the tree level. The range of variance at the tree level was observed to be 29.95% for SA and 34% for SL (Figure 2). On the other hand, SAR displayed higher values at the tree level, contributing to 43.29% of the total variance. Notably, a smaller portion of the variance was observed within seeds of the same tree, referred to as the "seed level." For SA, this intra-individual variance accounted for 4.4% of the total variability. Secondly, the wild olive traits exhibited significant intraspecific variations (ITV) both among and within populations. Consequently, when partitioning the variance, distinct patterns of trait variations emerged depending on the ecological level under investigation (Figure 3). For instance, LWC and DS exhibited substantial variation at the population level, accounting for 79.9% and 74% of the total variance, respectively. On the other hand, LA and LAR had higher variance ratios at the intra-population level, encompassing 43.40% to 42.80% of the total variance. Thus, SLA and LDMC displayed substantial variations, with only 48.20% and 48% of the total variance observed at the tree level, respectively. Notably, LAR, LA, SLA, and LDMC also demonstrated significant variations at the population level. Additionally, SLA and LDMC exhibited non-negligible variability at the population level (Figure 3).

Based on the findings obtained from the analysis of carob seed traits, the considerable magnitude of variability observed across different ecological levels indicates a high degree of phenotypic variability and significant intraspecific variation (Albert et al., 2010; Kassout et al., 2019, 2023b). The phenotypic

*Figure 2. Estimated percentage variance across hierarchical levels (population:tree:seed) patterns differently for wild olive leaf traits*

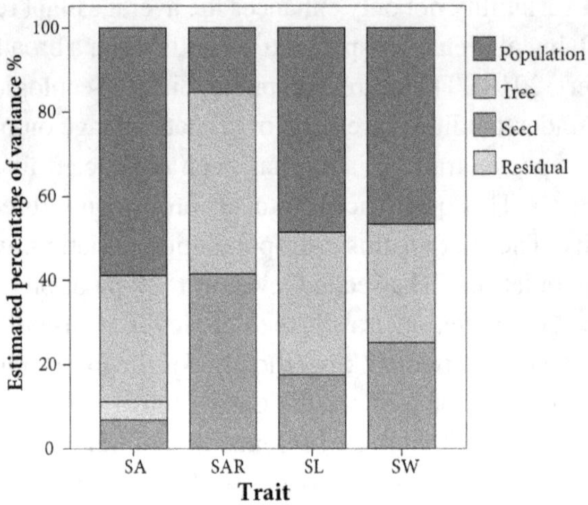


variation observed in the seeds suggests that carob populations respond differently to various biotic and/or abiotic factors within their respective environments (Sultan, 2000). The heterogeneous and dynamic environmental conditions along the sampling gradient likely influenced the expression and extent of the observed variation (Vázquez et al., 2017; Kassout et al., 2019). Consequently, the notable relationship between trait variability and geographic distance at the inter-population level reflects the impact of divergent environmental conditions among carob populations on seed trait variability. The population genetic structure of Moroccan carob trees also reveals distinct differentiation between populations from the South and those from the North region (Baumel et al., 2021), which could potentially explain the observed pattern of seed variation. Baumel et al. (2021) identified four main regions as ancestral areas in the evolutionary history of carob, including South Morocco and North Morocco with the Iberian Peninsula as major regions. These findings, coupled with the strong climatic gradient along the North–Center–South transect, likely contribute to shaping the variability of carob seeds. Additionally, the clear differentiation of carob populations based on their geographic origins aligns with the concept of Mediterranean biogeographic refugia in Morocco (Medail and Diadema, 2009), which has likely played a crucial role in the diversification of plants, including thermophilous species like the carob and the olive tree. The findings of this study reveal the presence and expression of intraspecific variations across multiple levels, ranging from geographical to individual scales (Evangelista et al., 2019). Additionally, these results suggest that certain seed traits, such as seed area and seed width, exhibit greater variability in response to large-scale environmental factors, such as climate. Conversely, other traits, like aspect ratio, appear to be more sensitive to variations in local-scale conditions present at the individual tree level (Messier et al., 2010).

Regarding the wild olive tree (*Olea europaea* subsp. *europaea* var. *sylvestris*), the results obtained from our study unveil significant intraspecific trait variations (ITV) both among and within populations, in addition to variations along environmental gradients (Figure 3). These findings underscore the consistency of intraspecific trait variation within the species (Fajardo and Piper, 2011; Kassout et al., 2019, 2021, 2022b) and its substantial contribution to overall trait variation (Kuppler et al., 2020).

*Figure 3. Estimated percentage variance across hierarchical levels (population:tree:leaf) patterns differently for carob seed traits*

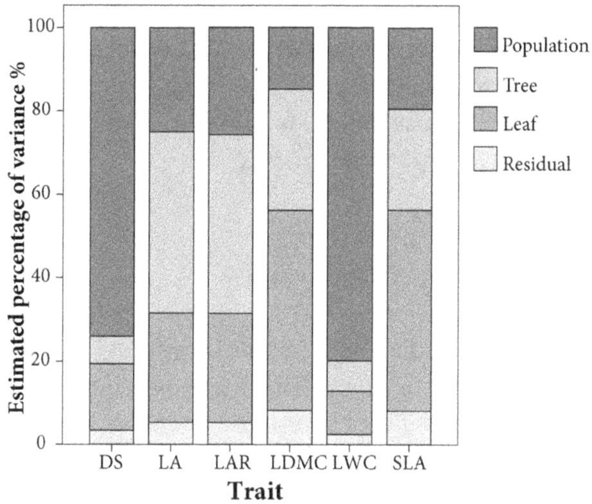


Exploring ITV in wild olive trees offers valuable insights into the underlying mechanisms governing plant functional responses to challenging conditions. Our findings suggest that variations in climate (inter-population level) and local conditions (intra-population and intra-individual levels) account for a significant proportion of trait variability (Figure 3). For instance, traits like leaf water content (LWC), and stomatal density (DS) demonstrate substantial variability in response to climatic variations along the sampling transect. These traits are linked to gas exchange and water use strategies in plants (Pérez-Harguindeguy et al., 2013), and thus, their variability suggests that wild olive trees employ adaptive responses to cope with stressful conditions, such as aridity, in order to maintain physiological functions (Eggli and Nyffeler, 2009; Wang et al., 2011, 2015). On the other hand, leaf area (LA), leaf aspect ratio (LAR), specific leaf area (SLA), and leaf dry matter content (LDMC) exhibit significant variations at the intra-population level and among individual trees, indicating that intraspecific variations in these traits are more sensitive to local-scale variations in environmental conditions (Messier et al., 2010; Kassout et al., 2021, 2022a). As a slow-growing species with considerable genetic variability (Besnard et al., 2018), the wild olive tree experiences diverse types of stressful conditions. Therefore, the variations expressed at each ecological level in our study are influenced by factors such as genotype, phenotypic plasticity (Albert et al., 2010), and potential differences in local conditions like water availability and changes in land cover (Kassout et al., 2021, 2023a).

These findings contribute significantly to the growing body of evidence indicating that intraspecific trait variation remains consistent within species (Fajardo and Piper 2011; Kassout et al., 2019, 2023b; Kuppler et al., 2020) and plays a crucial role in the overall trait variation (Albert et al. 2010; Violle et al., 2012). The observed intraspecific variation can stem from either genetic variability or phenotypic plasticity influenced by variations in environmental conditions (Violle et al. 2012). The wild olive and the carob tree, being slow-growing species (Batlle and Tous 1997; Kassout et al., 2021), are subjected to diverse and potentially stressful environmental conditions. Consequently, they exhibit substantial genetic variability (Besnard et al., 2018; Viruel et al., 2019), which likely contributes to the observed intraspecific variation. Thus, the variation expressed across the studied ecological scales is likely influenced by a combination of genotype, phenotypic plasticity (Geber and Grifen, 2003; Albert et al., 2010), and potential differences in phenology, land use, and the historical context of the study area. It is essential to note that while various studies underscore the role of genetic factors and phenotypic plasticity (Jung et al., 2014; Siefert et al., 2015; Kuppler et al., 2020), several investigations have pointed to the predominant influence of environmental factors in shaping intraspecific variability (Kassout et al., 2021, 2023b). This emphasizes the intricate interplay between genetic and environmental determinants in governing the observed trait variations within species.

## CONCLUSION

Despite the considerable number of studies on functional traits and stressful conditions, there is still a lack of understanding regarding the response of plant traits to climate change. The current findings contributes to fill this gap by investigating the expression, extent, and shift in leaf and seed size-related traits of the native *O. e.* subsp *e.* var *sylvestris* and *C. siliqua* along an aridity gradient within their distribution area. The study also identifies how the studied traits respond to changes in climatic conditions found between the studied populations, shedding light on the main functional responses to resist stressful conditions. The findings also highlight the substantial intraspecific variation in both species, indicating a high level of resilience to climate change, specifically to aridity and aridification. Furthermore, the study raises questions about the role of trait-based approaches in explaining the driving factors behind plant adaptive strategies under stressful conditions.

## REFERENCES

Albert, C. H., Grassein, F., Schurr, F. M., Vieilledent, G., & Violle, C. (2011). When and how should intraspecific variability be considered in trait-based plant ecology? *Perspectives in Plant Ecology, Evolution and Systematics*, *13*(3), 217–225. doi:10.1016/j.ppees.2011.04.003

Albert, C. H., Thuiller, W., Yoccoz, N. G., Douzet, R., Aubert, S., & Lavorel, S. (2010). A multi-trait approach reveals the structure and the relative importance of intra-vs. interspecific variability in plant traits. *Functional Ecology*, *24*(6), 1192–1201. doi:10.1111/j.1365-2435.2010.01727.x

Allen, C. D., Macalady, A. K., Chenchouni, H., Bachelet, D., McDowell, N., Vennetier, M., Kitzberger, T., Rigling, A., Breshears, D. D., Hogg, E. H. T., Gonzalez, P., Fensham, R., Zhang, Z., Castro, J., Demidova, N., Lim, J.-H., Allard, G., Running, S. W., Semerci, A., & Cobb, N. (2010). A global overview of drought and heat-induced tree mortality reveals emerging climate change risks for forests. *Forest Ecology and Management*, *259*(4), 660–684. doi:10.1016/j.foreco.2009.09.001

Azenzem, R., Al Feddy, M.-N., & Kassout, J. (2023). Climate Change and Insect Pests: Economic Damages and Adaptive Strategies in Mediterranean Olive Groves. In Climate Change and the Economic Importance and Damages of Insects (pp. 14-47). IGI Global.

Baumel, A., Mirleau, P., Viruel, J., Bou Dagher Kharrat, M., La Malfa, S., Ouahmane, L., Diadema, K., Moakhar, M., Sanguin, H., & Médail, F. (2018). Assessment of plant species diversity associated with the carob tree (Ceratonia siliqua, Fabaceae) at the Mediterranean scale. *Plant Ecology and Evolution*, *151*(2), 185–193. doi:10.5091/plecevo.2018.1423

Baumel, A., Nieto Feliner, G., Médail, F., La Malfa, S., Di Guardo, M., Bou Dagher Kharrat, M., Lakhal-Mirleau, F., Frelon, V., Ouahmane, L., Diadema, K., Sanguin, H., & Viruel, J. (2022). Genome-wide footprints in the carob tree (Ceratonia siliqua) unveil a new domestication pattern of a fruit tree in the Mediterranean. *Molecular Ecology*, *31*(15), 4095–4111. doi:10.1111/mec.16563 PMID:35691023

Beckman, N. G., Aslan, C. E., Rogers, H. S., Kogan, O., Bronstein, J. L., Bullock, J. M., Hartig, F., HilleRisLambers, J., Zhou, Y., Zurell, D., Brodie, J. F., Bruna, E. M., Cantrell, R. S., Decker, R. R., Efiom, E., Fricke, E. C., Gurski, K., Hastings, A., Johnson, J. S., & Zambrano, J. (2020). Advancing an interdisciplinary framework to study seed dispersal ecology. *AoB Plants*, *12*(2), plz048. doi:10.1093/aobpla/plz048 PMID:32346468

Besnard, G., Terral, J. F., & Cornille, A. (2018). On the origins and domestication of the olive: A review and perspectives. *Annals of Botany*, *121*(3), 385–403. doi:10.1093/aob/mcx145 PMID:29293871

Blondel, J., Aronson, J., Bodiou, J.-Y., & Boeu, G. (2010). *The Mediterranean region: biological diversity through time and space*. Oxford University Press.

Bréda, N., Huc, R., Granier, A., & Dreyer, E. (2006). Temperate forest trees and stands under severe drought: A review of ecophysiological responses, adaptation processes and long-term consequences. *Annals of Forest Science*, *63*(6), 625–644. doi:10.1051/forest:2006042

Choat, B., Jansen, S., Brodribb, T. J., Cochard, H., Delzon, S., Bhaskar, R., Bucci, S. J., Feild, T. S., Gleason, S. M., Hacke, U. G., Jacobsen, A. L., Lens, F., Maherali, H., Martínez-Vilalta, J., Mayr, S., Mencuccini, M., Mitchell, P. J., Nardini, A., Pittermann, J., & Zanne, A. E. (2012). Global convergence in the vulnerability of forests to drought. *Nature*, *491*(7426), 752–755. doi:10.1038/nature11688 PMID:23172141

Cramer, W., Guiot, J., Fader, M., Garrabou, J., Gattuso, J. P., Iglesias, A., Lange, M. A., Lionello, P., Llasat, M. C., Paz, S., Peñuelas, J., Snoussi, M., Toreti, A., Tsimplis, M. N., & Xoplaki, E. (2018). Climate change and interconnected risks to sustainable development in the Mediterranean. *Nature Climate Change*, *8*(11), 972–980. doi:10.103841558-018-0299-2

Davis, G. W., Richardson, D. M., Keeley, J. E., & Hobbs, R. J. (1996). Mediterranean-type ecosystems: the influence of biodiversity on their functioning. *Functional roles of biodiversity: a global perspective.*, 151-183.

Des Roches, S., Post, D. M., Turley, N. E., Bailey, J. K., Hendry, A. P., Kinnison, M. T., Schweitzer, J. A., & Palkovacs, E. P. (2018). The ecological importance of intraspecific variation. *Nature Ecology & Evolution*, *2*(1), 57–64. doi:10.103841559-017-0402-5 PMID:29203921

Diaz, S., Hodgson, J. G., Thompson, K., Cabido, M., Cornelissen, J. H., Jalili, A., Montserrat-Martí, G., Grime, J. P., Zarrinkamar, F., Asri, Y., Band, S. R., Basconcelo, S., Castro-Díez, P., Funes, G., Hamzehee, B., Khoshnevi, M., Pérez-Harguindeguy, N., Pérez-Rontomé, M. C., Shirvany, F. A., & Zak, M. R. (2004). The plant traits that drive ecosystems: Evidence from three continents. *Journal of Vegetation Science*, *15*(3), 295–304. doi:10.1111/j.1654-1103.2004.tb02266.x

Díaz, S., Kattge, J., Cornelissen, J. H., Wright, I. J., Lavorel, S., Dray, S., Reu, B., Kleyer, M., Wirth, C., Colin Prentice, I., Garnier, E., Bönisch, G., Westoby, M., Poorter, H., Reich, P. B., Moles, A. T., Dickie, J., Gillison, A. N., Zanne, A. E., & Gorné, L. D. (2016). The global spectrum of plant form and function. *Nature*, *529*(7585), 167–171. doi:10.1038/nature16489 PMID:26700811

Driouech, F., Déqué, M., & Sánchez-Gómez, E. (2010). Weather regimes—Moroccan precipitation link in a regional climate change simulation. *Global and Planetary Change*, *72*(1-2), 1–10. doi:10.1016/j.gloplacha.2010.03.004

Eggli, U., & Nyffeler, R. (2009). Living under temporarily arid conditions-succulence as an adaptive strategy. *Bradleya*, *2009*(27), 13–36. doi:10.25223/brad.n27.2009.a10

Evangelista, C., Olden, J. D., Lecerf, A., & Cucherousset, J. (2019). Scale-dependent patterns of intraspecific trait variations in two globally invasive species. *Oecologia*, *189*(4), 1083–1094. doi:10.100700442-019-04374-4 PMID:30824999

Fajardo, A., & Piper, F. I. (2011). Intraspecific trait variation and covariation in a widespread tree species (Nothofagus pumilio) in southern Chile. *The New Phytologist*, *189*(1), 259–271. doi:10.1111/j.1469-8137.2010.03468.x PMID:21039558

Fajardo, A., & Siefert, A. (2016). Phenological variation of leaf functional traits within species. *Oecologia*, *180*(4), 951–959. doi:10.100700442-016-3545-1 PMID:26796408

Fajardo, A., & Siefert, A. (2018). *Intraspecific trait variation and the leaf economics spectrum across resource gradients and levels of organization*.

Forzieri, G., Miralles, D. G., Ciais, P., Alkama, R., Ryu, Y., Duveiller, G., Zhang, K., Robertson, E., Kautz, M., Martens, B., Jiang, C., Arneth, A., Georgievski, G., Li, W., Ceccherini, G., Anthoni, P., Lawrence, P., Wiltshire, A., Pongratz, J., & Cescatti, A. (2020). Increased control of vegetation on global terrestrial energy fluxes. *Nature Climate Change*, *10*(4), 356–362. doi:10.103841558-020-0717-0

Garland, T. Jr, Downs, C. J., & Ives, A. R. (2022). Trade-offs (and constraints) in organismal biology. *Physiological and Biochemical Zoology*, *95*(1), 82–112. doi:10.1086/717897 PMID:34905443

Garnier, E., Navas, M. L., & Grigulis, K. (2016). *Plant functional diversity: organism traits, community structure, and ecosystem properties*. Oxford University Press.

Geber, M. A., & Griffen, L. R. (2003). Inheritance and natural selection on functional traits. *International Journal of Plant Sciences*, *164*(S3), S21–S42. doi:10.1086/368233

Grime, J. P. (2001). *Plant strategies, vegetation processes, and ecosystem properties* (2nd ed.). Wiley.

Jung, V., Albert, C. H., Violle, C., Kunstler, G., Loucougaray, G., & Spiegelberger, T. (2014). Intraspecific trait variability mediates the response of subalpine grassland communities to extreme drought events. *Journal of Ecology*, *102*(1), 45–53. doi:10.1111/1365-2745.12177

Kaniewski, D., Van Campo, E., Boiy, T., Terral, J. F., Khadari, B., & Besnard, G. (2012). Primary domestication and early uses of the emblematic olive tree: Palaeobotanical, historical and molecular evidence from the Middle East. *Biological Reviews of the Cambridge Philosophical Society*, *87*(4), 885–899. doi:10.1111/j.1469-185X.2012.00229.x PMID:22512893

Kassout, J., Ater, M., Ivorra, S., Barbara, H., Limier, B., Ros, J., Girard, V., Paradis, L., & Terral, J.-F. (2021). Resisting aridification: Adaptation of sap conduction performance in Moroccan wild olive subspecies distributed over an aridity gradient. *Frontiers in Plant Science*, *12*, 663721. doi:10.3389/fpls.2021.663721 PMID:34276726

Kassout, J., Hmimsa, Y., El Fatehi, S., El Ouahrani, A., Kadaoui, K., Chakkour, S., Ariza-Mateos, D., Palacios-Rodríguez, G., Navarro-Cerrillo, R., & Ater, M. (2022a). Image analysis of Moroccan carob seeds (*Ceratonia siliqua* L.) revealed substantial intraspecific variations depending on climate and geographic origin. *Ecological Processes*, *11*(1), 1–14. doi:10.118613717-022-00378-w

Kassout, J., Hmimsa, Y., Fatehi, S. E., Kadaoui, K., Houssni, M., Chakkour, S., Sahli, A., El Chami, M. A., Ariza-Mateos, D., Palacios-Rodríguez, G., Navarro-Cerrillo, R. M., & Ater, M. (2023b). Aridity Gradients Shape Intraspecific Variability of Morphological Traits in Native *Ceratonia siliqua* L. of Morocco. *Plants*, *12*(19), 3447. doi:10.3390/plants12193447 PMID:37836187

Kassout, J., Terral, J. F., Azenzem, R., El Ouahrani, A., Sahli, A., Houssni, M., & Ater, M. (2023a). *Olea europaea* and Stressful Conditions. In *Medicinal Plant Responses to Stressful Conditions* (pp. 243–272). CRC Press. doi:10.1201/9781003242963-13

Kassout, J., Terral, J. F., El Ouahrani, A., Houssni, M., Ivorra, S., Kadaoui, K., et al. (2022b). Species distribution based-modelling under climate change: the case of two native wild *Olea europaea* subspecies in Morocco, *O. e.* subsp. *europaea* var. *sylvestris* and *O. e.* subsp. *maroccana. Climate change in the mediterranean and middle eastern region*, 21-43.

Kassout, J., Terral, J. F., Hodgson, J. G., & Ater, M. (2019). Trait-based plant ecology a flawed tool in climate studies? The leaf traits of wild olive that pattern with climate are not those routinely measured. *PLoS One*, *14*(7), e0219908. doi:10.1371/journal.pone.0219908 PMID:31314789

Kattge, J., Diaz, S., Lavorel, S., Prentice, I. C., Leadley, P., Bönisch, G., Garnier, E., Westoby, M., Reich, P. B., Wright, I. J., Cornelissen, J. H. C., Violle, C., Harrison, S. P., Van BODEGOM, P. M., Reichstein, M., Enquist, B. J., Soudzilovskaia, N. A., Ackerly, D. D., Anand, M., & Wirth, C. (2011). TRY–a global database of plant traits. *Global Change Biology*, *17*(9), 2905–2935. doi:10.1111/j.1365-2486.2011.02451.x

Kuppler, J., Albert, C. H., Ames, G. M., Armbruster, W. S., Boenisch, G., Boucher, F. C., Campbell, D. R., Carneiro, L. T., Chacón-Madrigal, E., Enquist, B. J., Fonseca, C. R., Gómez, J. M., Guisan, A., Higuchi, P., Karger, D. N., Kattge, J., Kleyer, M., Kraft, N. J. B., Larue-Kontić, A.-A. C., & Junker, R. R. (2020). Global gradients in intraspecific variation in vegetative and floral traits are partially associated with climate and species richness. *Global Ecology and Biogeography*, *29*(6), 992–1007. doi:10.1111/geb.13077

Laforest-Lapointe, I., Martínez-Vilalta, J., & Retana, J. (2014). Intraspecific variability in functional traits matters: Case study of Scots pine. *Oecologia*, *175*(4), 1337–1348. doi:10.100700442-014-2967-x PMID:24850418

Lavorel, S., & Garnier, E. (2002). Predicting changes in community composition and ecosystem functioning from plant traits: Revisiting the Holy Grail. *Functional Ecology*, *16*(5), 545–556. doi:10.1046/j.1365-2435.2002.00664.x

Lenoir, J., Gégout, J. C., Marquet, P. A., de Ruffray, P., & Brisse, H. (2008). A significant upward shift in plant species optimum elevation during the 20th century. *science, 320*(5884), 1768-1771.

Lionello, P., & Scarascia, L. (2018). The relation between climate change in the Mediterranean region and global warming. *Regional Environmental Change*, *18*(5), 1481–1493. doi:10.100710113-018-1290-1

Lloret, F., de la Riva, E. G., Pérez-Ramos, I. M., Marañón, T., Saura-Mas, S., Díaz-Delgado, R., & Villar, R. (2016). Climatic events inducing die-off in Mediterranean shrublands: Are species' responses related to their functional traits? *Oecologia*, *180*(4), 961–973. doi:10.100700442-016-3550-4 PMID:26801493

Matesanz, S., & Valladares, F. (2014). Ecological and evolutionary responses of Mediterranean plants to global change. *Environmental and Experimental Botany*, *103*, 53–67. doi:10.1016/j.envexpbot.2013.09.004

Médail, F., & Diadema, K. (2009). Glacial refugia influence plant diversity patterns in the Mediterranean Basin. *Journal of Biogeography*, *36*(7), 1333–1345. doi:10.1111/j.1365-2699.2008.02051.x

Messier, J., McGill, B. J., Enquist, B. J., & Lechowicz, M. J. (2017). Trait variation and integration across scales: Is the leaf economic spectrum present at local scales? *Ecography*, *40*(6), 685–697. doi:10.1111/ecog.02006

Messier, J., McGill, B. J., & Lechowicz, M. J. (2010). How do traits vary across ecological scales? A case for trait-based ecology. *Ecology Letters*, *13*(7), 838–848. doi:10.1111/j.1461-0248.2010.01476.x PMID:20482582

Moore, T. E., Jones, C. S., Chong, C., & Schlichting, C. D. (2020). Impact of rainfall seasonality on intraspecific trait variation in a shrub from a Mediterranean climate. *Functional Ecology*, *34*(4), 865–876. doi:10.1111/1365-2435.13533

Moran, E. V., Hartig, F., & Bell, D. M. (2016). Intraspecific trait variation across scales: Implications for understanding global change responses. *Global Change Biology*, *22*(1), 137–150. doi:10.1111/gcb.13000 PMID:26061811

Oulbi, S. Kassout, J., Bouizgaren, A., El Khatri, N., Sikaoui, L., & Hadria, R. (2023). Assessment of Carbon Balance in Olive Cultivation Systems in Morocco: Toward an Intelligent Solution in Climate Change Mitigation. In Intelligent Solutions for Optimizing Agriculture and Tackling Climate Change: Current and Future Dimensions (pp. 96-114). IGI Global.

Paradis, E., Claude, J., & Strimmer, K. (2004). APE: Analyses of phylogenetics and evolution in R language. *Bioinformatics (Oxford, England)*, *20*(2), 289–290. doi:10.1093/bioinformatics/btg412 PMID:14734327

Perez-Harguindeguy, N., Diaz, S., Garnier, E., Lavorel, S., Poorter, H., Jaureguiberry, P., Bret-Harte, M. S., Cornwell, W. K., Craine, J. M., Gurvich, D. E., Urcelay, C., Veneklaas, E. J., Reich, P. B., Poorter, L., Wright, I. J., Ray, P., Enrico, L., Pausas, J. G., de Vos, A. C., & Cornelissen, J. H. C. (2013). New handbook for standardised measurement of plant functional traits worldwide. *Australian Journal of Botany*, *61*(3), 167–234. doi:10.1071/BT12225

Pierce, S., Negreiros, D., Cerabolini, B. E., Kattge, J., Díaz, S., Kleyer, M., Shipley, B., Wright, S. J., Soudzilovskaia, N. A., Onipchenko, V. G., van Bodegom, P. M., Frenette-Dussault, C., Weiher, E., Pinho, B. X., Cornelissen, J. H. C., Grime, J. P., Thompson, K., Hunt, R., Wilson, P. J., & Tampucci, D. (2017). A global method for calculating plant CSR ecological strategies applied across biomes world-wide. *Functional Ecology*, *31*(2), 444–457. doi:10.1111/1365-2435.12722

Pinheiro, J., Bates, D., DebRoy, S., & Sarkar, D. (2021). Nonlinear mixed-effects models. *R package version, 3*, 1-153.

Polade, S. D., Pierce, D. W., Cayan, D. R., Gershunov, A., & Dettinger, M. D. (2014). The key role of dry days in changing regional climate and precipitation regimes. *Scientific Reports*, *4*(1), 4364. doi:10.1038rep04364 PMID:24621567

Pörtner, H. O., Roberts, D. C., Poloczanska, E. S., Mintenbeck, K., & Tignor, M. (2022). IPCC, 2022: Summary for policymakers.

Quézel, P., & Médail, F. (2003). *Ecologie et biogéographie des forêts du bassin méditerranéen*. Elsevier.

R Core Team. (2020). *R: A language and environment for statistical computing*. R.

Reich, P. B., Walters, M. B., & Ellsworth, D. S. (1992). Leaf life-span in relation to leaf, plant, and stand characteristics among diverse ecosystems. *Ecological Monographs*, *62*(3), 365–392. doi:10.2307/2937116

Saatkamp, A., Poschlod, P., & Venable, D. L. (2014). The functional role of soil seed banks in natural communities. In *Seeds: the ecology of regeneration in plant communities* (pp. 263–295). CABI. doi:10.1079/9781780641836.0263

Shipley, B., De Bello, F., Cornelissen, J. H. C., Laliberté, E., Laughlin, D. C., & Reich, P. B. (2016). Reinforcing loose foundation stones in trait-based plant ecology. *Oecologia*, *180*(4), 923–931. doi:10.100700442-016-3549-x PMID:26796410

Siefert, A., Violle, C., Chalmandrier, L., Albert, C. H., Taudiere, A., & Fajardo, A. (2015). A global meta-analysis of the relative extent of intraspecific trait variation in plant communities. *Ecology Letters*, *18*(12), 1406–1419. doi:10.1111/ele.12508 PMID:26415616

Sultan, S. E. (2000). Phenotypic plasticity for plant development, function and life history. *Trends in Plant Science*, *5*(12), 537–542. doi:10.1016/S1360-1385(00)01797-0 PMID:11120476

Terral, J. F., Badal, E., Heinz, C., Roiron, P., Thiebault, S., & Figueiral, I. (2004). A hydraulic conductivity model points to post-neogene survival of the mediterranean olive. *Ecology*, *85*(11), 3158–3165. doi:10.1890/03-3081

Thomas, H. J., Bjorkman, A. D., Myers-Smith, I. H., Elmendorf, S. C., Kattge, J., Diaz, S., Vellend, M., Blok, D., Cornelissen, J. H. C., Forbes, B. C., Henry, G. H. R., Hollister, R. D., Normand, S., Prevéy, J. S., Rixen, C., Schaepman-Strub, G., Wilmking, M., Wipf, S., Cornwell, W. K., & de Vries, F. T. (2020). Global plant trait relationships extend to the climatic extremes of the tundra biome. *Nature Communications, 11*(1), 1351. doi:10.103841467-020-15014-4 PMID:32165619

Van Bodegom, P. M., Douma, J. C., & Verheijen, L. M. (2014). A fully traits-based approach to modeling global vegetation distribution. *Proceedings of the National Academy of Sciences of the United States of America, 111*(38), 13733–13738. doi:10.1073/pnas.1304551110 PMID:25225413

Vázquez, D. P., Gianoli, E., Morris, W. F., & Bozinovic, F. (2017). Ecological and evolutionary impacts of changing climatic variability. *Biological Reviews of the Cambridge Philosophical Society, 92*(1), 22–42. doi:10.1111/brv.12216 PMID:26290132

Vellend, M., Baeten, L., Becker-Scarpitta, A., Boucher-Lalonde, V., McCune, J. L., Messier, J., Myers-Smith, I. H., & Sax, D. F. (2017). Plant biodiversity change across scales during the Anthropocene. *Annual Review of Plant Biology, 68*(1), 563–586. doi:10.1146/annurev-arplant-042916-040949 PMID:28125286

Wang, R., Huang, W., Chen, L., Ma, L., Guo, C., & Liu, X. (2011). Anatomical and physiological plasticity in Leymus chinensis (Poaceae) along large-scale longitudinal gradient in northeast China. *PLoS One, 6*(11), e26209. doi:10.1371/journal.pone.0026209 PMID:22087223

Wang, R., Yu, G., He, N., Wang, Q., Zhao, N., Xu, Z., & Ge, J. (2015). Latitudinal variation of leaf stomatal traits from species to community level in forests: Linkage with ecosystem productivity. *Scientific Reports, 5*(1), 14454. doi:10.1038rep14454 PMID:26403303

Westoby, M., Falster, D. S., Moles, A. T., Vesk, P. A., & Wright, I. J. (2002). Plant ecological strategies: Some leading dimensions of variation between species. *Annual Review of Ecology and Systematics, 33*(1), 125–159. doi:10.1146/annurev.ecolsys.33.010802.150452

Wright, I. J., Reich, P. B., Westoby, M., Ackerly, D. D., Baruch, Z., Bongers, F., Cavender-Bares, J., Chapin, T., Cornelissen, J. H. C., Diemer, M., Flexas, J., Garnier, E., Groom, P. K., Gulias, J., Hikosaka, K., Lamont, B. B., Lee, T., Lee, W., Lusk, C., & Villar, R. (2004). The worldwide leaf economics spectrum. *Nature, 428*(6985), 821–827. doi:10.1038/nature02403 PMID:15103368

Zohary, D. (2002). Domestication of the carob (*Ceratonia siliqua* L.). *Israel Journal of Plant Sciences, 50*(supl), 141-145.

Zohary, D., Hopf, M., & Weiss, E. (2012). *Domestication of Plants in the Old World: The origin and spread of domesticated plants in Southwest Asia, Europe, and the Mediterranean Basin.* Oxford University Press. doi:10.1093/acprof:osobl/9780199549061.001.0001

# Section 2

# Water and Soil in the Context of Climate Change

# Chapter 10
# A Case Study From Kanpur Nagar, Uttar Pradesh for Assessing the Economic Value of Water and Its Determinants

**Ajay Kumar Singh**
https://orcid.org/0000-0003-0429-0925
*Department of Humanities and Social Sciences, Graphic Era University (Deemed), Dehradun, India*

**Santosh Singh**
*Department of English, School of Liberal Arts and Management, DIT University, Dehradun, India*

## ABSTRACT

*The chapter applied a willingness to pay ability (WPA) approach to estimate the economic value of drinking water (EVM) using primary data. The sample of 350 households are randomly taken from four wards of Kanpur Nagar, Uttar Pradesh (India). Simple OLS, probit, logit, and mixed-effect logit regression models are applied to examine the effect of social- economic variables on WPA of households to receive better water supply from suppliers. It found that the municipality is charging a very low price to provide water supply of 30 Kilolitre/household per month. However, implicit cost (IC) of water/household is found higher as compared to WPA. IC/household appeared high due to consumption of unsafe drinking water, and incidence of various waterborne health issues and diseases. Family size, occupation, and annual income showed a positive impact on WPA of households.*

## INTRODUCTION

Production of goods is not possible without water, rivers, lakes, forest and air and other) (Streimikiene, 2015; Cosgrove & Loucks, 2015; Jamel & Derbali, 2016; Kapur, 2016; Singh et al., 2018). People or human receive different types of goods and services from nature which is known as ecosystem services (Jamel & Derbali, 2016; Singh, 2020). Ecosystem services provide several benefits to regulate climate

DOI: 10.4018/978-1-6684-9838-5.ch010

change, reduce pollution, conserve water resources, maintain soil quality and wind speed (Zho et al., 2021). Hence, natural and ecosystem services is crucial determinant of economic and economic development (Golubev, 1983; Jamel & Derbali, 2016; Goswami & Bisht, 2017; Singh et al., 2018; Singh, 2020; Singh et al., 2020; Singh et al., 2021). Water is a most essential component of ecosystem services. Water is an essential component of socio-economic development (Golubev, 1983; Chowdhury & Behera, 2021). It is an essential resource to meet the numerous requirements of human livelihood (Borgoyary, 2013). Water is essential public goods that meets nutritional requirement of human body. Production of agricultural and industrial sector is not possible without water (Singh & Jyoti, 2021b). It helps to generate energy and electricity for people and industries. Water meets the irrigation requirement of the agricultural sector which ensure the food security of all (Singh & Sharma, 2018a; Jyoti & Singh, 2020; Kumar & Singh, 2023). Hence, the ecosystem of water produces food, clean drinking water, entertainment, and recreate products and various species of fishes in the rivers (Pour & Kalashami, 2012; Cosgrove & Loucks, 2015; Kumar et al., 2017). Water and natural resources are vital for humanity (MEA, 2005; Chowdhury & Behera, 2021). So, all countries should give priority to preserve natural resources to increase sustainable development (Singh et al., 2018; Getnet et al., 2022).

The developing countries are using an overwhelming quantity of natural resources in production activities (Wang et al., 2010; Singh et al., 2023a). Therefore, natural and water resources are consistently decreasing (Singh et al., 2018). Also, high population growth, urbanization and industrialization produce negative impacts on water resources sustainability (Golubev, 1983; Tundisi, 2008; Schnoor, 2010; Borgoyary, 2013; Chen et al., 2013; Rios et al., 2018; Upadhyaya et al., 2023). It is projected that world population would be 9 billion by 2050s (Singh, 2017), and it would cause to increase more requirements of water (Arfanuzzaman & Rahman, 2017). Borgoyary (2013) also observed that water demand would increase by 65% by 2030s. On the other hand, around 1/3rd population of the world do not have secure sources of water to complete their basic wants (Rios et al., 2018). UNESCO report is also found that around 2 billion do not have source of safe drinking water world-wide.[1] So, to maintain water resources sustainability is seemed as a challenge for policy makers. While most developing countries could not maintain water resources sustainability due to financial restrictions of people and high dependency of production sectors on water resources (Wang et al., 2010; Borgoyary, 2013; Biswas & Venkatachalam, 2015; Arfanuzzaman & Rahman, 2017; Singh & Jyoti, 2019). The agricultural sector cannot grow without irrigation water and water resources sustainability (Singh & Jyoti, 2021b). Agricultural sustainability is also negatively impacting in presence of climate change and water scarcity (Singh et al., 2022a). Furthermore, safe and consistent water supply is remaining a serious concern in most economies (Va´squez et al., 2009). Water scarcity is high in developing countries due to extensive stress of population, and urbanization and industrialization (Wang et al., 2010; Arfanuzzaman & Rahman, 2017; Singh & Jyoti, 2021b).

Water is an essential input for agricultural sector (Singh & Jyoti, 2021a). Thus, agricultural production and its allied activities would be in alarming position due to water scarcity in largely agrarian countries (Arfanuzzaman & Rahman, 2017). Accordingly, food, livelihood security and income of common people may be in dangerous position due to water scarcity. Water scarcity may be responsible for food insecurity. Also, water scarcity has a negative impact on energy and its sources (Wang et al., 2010; Cosgrove & Loucks, 2015). Hence, the developing countries should formulate specific policies to increase water resources sustainability to ensure the future of all. Accordingly, valuation of ecosystem services helps to frame effective water management policies to protect ecosystem services of water (Chowdhury & Behera, 2021).

At present availability of water is a serious problem due to human driven activities and climate change across countries (Borgoyary, 2013; Kumar, 2015a,b). At present around 0.8 billion people have low accessibility of drinking water across the globe (Katko & Hukka, 2015). About 6 – 8 million people of the world die every year due to unavailability of safe drinking water (Katko & Hukka, 2015). Also, water recharge in the ground is declining due to climate change especially increasing rainfall pattern or variability in different seasons (Kumar et al., 2015b; Vallgårda, 2018; Singh & Jyoti, 2021a; Singh et al., 2023c). The world would need 40% additional water to meet water demand of people in coming 40 years. Moreover, water supplied has dropped below 1,700 $m^3$ per capita in most countries (Goswami & Bisht, 2017).

## Water Related Challenges and Issues in India

India has a high disparity in socio-economic structure, natural resources and infrastructure development across states (Kumar et al., 2016). It also has a high variation in poverty, food security, hunger, literacy rate and industrial development (Kumar & Sharma 2013). India needs more water to sustain the growth of agricultural sector (Samui et al., 2014). India uses large quantities of ground water in the production sector (Samui et al., 2014). Surface and groundwater are depleting due to urbanization and industrial growth in India. So, food grain production may be decreased by 20% due to water scarcity in India. In India, Punjab, Haryana, Tamil Nadu and Gujarat have a high dependency on subsurface water (Borgoyary, 2013; Jyoti et al., 2023). Also, there is high diversity in water resources in Indian states. The groundwater level is in a critical stage in most Indian states. Crop production in the hilly area of India is also susceptible due to water scarcity (Singh et al., 2023c). Therefore, over utilization of water in production activities will hamper the agricultural productivity in India. Moreover, largely populated cities like Delhi, Indore, Agra and Mumbai of India are facing several problems due to water shortages (Ghosh, 2021). Monsoon rainfall is also shifting and changing due to incidence of climate change. Therefore, India's water resources are facing numerous challenges due to high dependency of its on-monsoon rainfall. Unpredictability of drought and flood are also caused to increase vulnerability for water resources sustainability in India.

In India, around 163 million people have low accessibility of water, and around 21% of communicable diseases in people happened due to unsafe drinking water.[2]_The *per capita water availability has declined from 1816 cubic meter in 2011 to 1544 cubic meter in 2001 in India.*[3] Water pollution is also increasing due to industrialization and excessive use of fertilizer in cultivation. Hence, rising water pollution is a big challenge in India. There is no equal distribution and allocation of water in rural and urban areas, and across India states (Upadhyaya et al., 2023). It is also expected that water demand will increase rapidly due to unplanned growth of urban cities (Ghosh, 2021). While, as more Indian population lives in rural areas which have low accessibility of clean water. India is in extremely poor position in water availability as compared to other agrarian and developing economies (Wang et al., 2010). Also, water availability is stressed in rural areas due to extreme weather events and climate change. India has limited clean water bodies which cannot meet the water requirements of present population (Mathur & Thakur, 2003). Poor people are also incapable to access clean and safe drinking water to maintain their health, food and nutritional security in India (Wang et al., 2010). Around 33% and 8% households use handpump and tube well/borehole, to receive water respectively in India (Figure: 1). While, around 62% urban Indian households are using tap water from treated sources to meet drinking water (Figure: 2). India has also a high dependency on groundwater to meet the requirement of water in rural areas (Figure:

*Figure 1. Main sources of drinking water of households in India*
**Source:** *Census (Government of India), 2011.*

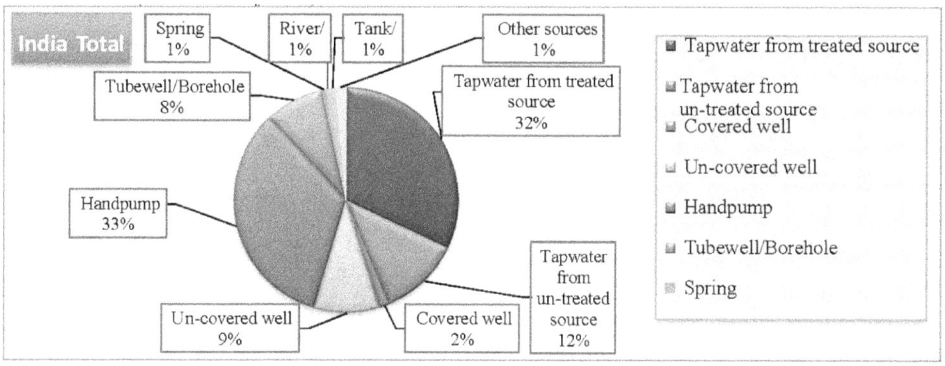


*Figure 2. Main sources of drinking water of households in urban India*
**Source:** *Census (Government of India), 2011.*

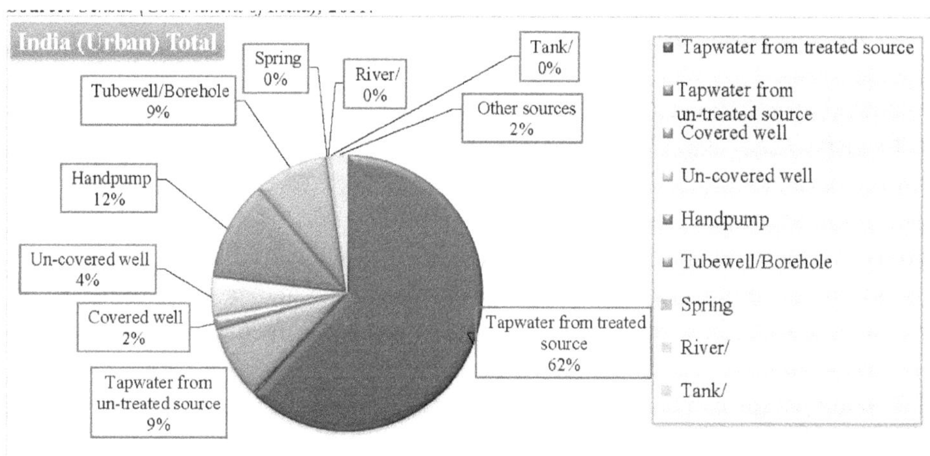


*Figure 3. Main sources of drinking water of households in rural India*
**Source:** *Census (Government of India), 2011.*

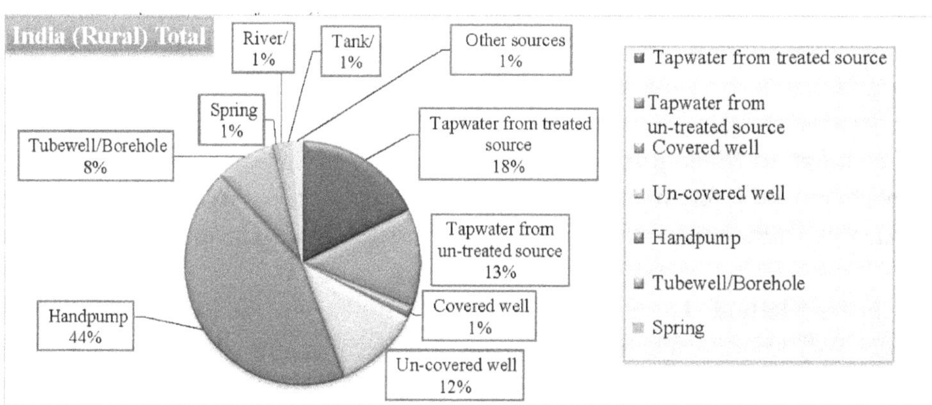

3). Therefore, national policy makers and development thinkers should implement effective policies to provide clean water to all Indians.

### Valuation of Economic Value of Water in India (EVW)

Economic value of any resources or ecosystem services exists when its supply is relative scarce as compared to its demand. India has more water scarcity as compared to other countries. India has the lowest quality of water in the world. The water resources are degrading due to multiple reasons in India (Singh et al., 2016; Singh, 2020). So, it is vital to investigate the economic value of water (EVW) to increase water resources sustainability. It would also be beneficial to understand the view of people towards water resources sustainability. Water is a public goods and everyone has an equal right to use it. Implications of EVW would be positive to provide safe and clean water as per the needs of households in rural and urban India.

## REVIEW OF LITERATURE

### Methods for Measuring the EVW of Ecosystem Services

Following methods can be applied to examine the economic value of water and other ecosystem services:

**Market Price Method:** This method is appropriate to estimate the monetary value for those ecosystem services which have an open market. Similar to general goods, the ecosystem services can be bought and sold in the commercial market, e.g., wild animals, birds, etc.

**Productivity Method:** The method is viable to predict the financial value of those ecosystem services which are used to produce other products that can be sold in the market. This technique is rational to observe the economic value of collected raw materials, woods and medicinal plants from forestry, and sand and products from rivers.

**Hedonic Pricing Method:** This method is rational to estimate the monetary value of those ecosystem services which have an impact on price of other goods in the market. For instance, availability of medicinal plants in forest area determines the price of domestic medicine. Also, price of milk and meat depend on availability of fodder for livestock.

**Travel Cost Method:** It is effective to guess the monetary value of those services which increase the entertainment of people. It includes willingness to pay ability of individual for visiting a specific tourist place, e.g., park, religious place, zoo, hill station, etc.

**Contingent Choice Method:** This method includes the awareness of people that how much amount they are willing to pay to trade-offs among the various ecosystem services. For example, for which ecosystem services the people are ready to pay more and less amount.

**Contingent Valuation Method (CVM):** It comprises a hypothetical situation and it increases the ability of people to pay for receiving a specific quantity of services (Reddy, 1999). The monetary value of drinking water, irrigation water, street cleaning and street light, can be observed through mentioned method.

**Benefit Transfer Method:** The technique is valid to estimate the monetary value of those services which have a diversity in original site and study site. For instance, the economic value of recreational fishing may not be similar in different locations.

## Economic Value of Water (EVW) and Its Factors Affecting

It is stated earlier that also ecosystem services have monetary value. Thus, existing researchers estimated the monetary value of drinking water, irrigation and non-irrigation water in the agricultural sector, and other ecosystem services across countries using various methods (Bjornlund & McKay, 1998; Reddy, 1999; Dasgupta, 2004; Siew et al., 2015). Reddy (1999) estimated the EVM of drinking water in Rajasthan (India) using a CVM. It also perceived inter-village variation in water use and its affecting factors. It concluded that government policies are responsible for increasing the water shortage and environmental degradation. Dasgupta (2004) also observed the EVW in Delhi (India) using CVM method. Dutta et al. (2005) applied WTP technique to estimate the EVW in Delhi (India). Dutta and Tiwari (2005) also considered CVM approach for measuring the willingness to pay ability of households for safe drinking water in Delhi (India). Using the similar method, Majumdar and Gupta (2009) observe the monetary value of water in Calcutta, West Bengal (India). It also used probit model to measure the impact of demographic and social-economic factors on willingness to pay ability of households. Devi et al. (2009) concluded that water pricing systems depend on political, financial and socio-economic profile of people. Jala and Nandagiri (2015) observed the EVW in Pilikula Lake, Magalore (India) using CVM method. It highlighted that age; education income and geographical location have an important impact on willingness to pay ability (WPA). While, the empirical findings of Francis et al. (2015) claimed that it is difficult to identify the maintenance costs of water. Ramsey et al. (2017) examined the residential water use and its conservation practices adopted by people in Jaipur (India). It is concluded that water conservation behaviour and technologies were positively correlated with high income, education level and water resources. Chowdhury and Behera (2021)'s study also argued that female members, animal and lower social castes are found conscious to avail services from water bodies and tanks.

Somanathan & Ravindranath (2006) estimated the monetary value of water in irrigated and non-irrigated land using price difference method in Andhra Pradesh and Karnataka. It found a difference between the EVW for irrigated and non-irrigated land. While Chandrasekaran et al. (2009) found insignificant impact of water use system on sustainability of resources in Tamil Nadu. It claimed that pricing of irrigated water is influenced by education, family size, age and water requirement of consumers. Karthikeyan (2010) reported that pricing of irrigated water is determined by education level, family size and area under rice crop. Barton and Taron (2010) estimated the monetary value of irrigated water in Karnataka (India). Veettil et al. (2011) examined the price variation of irrigated area in farming in Krishna River basin in India. Venkatachalam and Narayanamoorthy (2012) also noted variation in value of water on irrigated and non-irrigated land in Tamil Nadu. The previous studies also mentioned that price of agricultural products show a crucial impact on value of irrigated water (Bjornlund & McKay, 1998). Similar opinion is also provided by Rodgers & Hellegers (2005) that argued that value of water depends on the monetary value of irrigated crops in Indonesia. It is also observed that there is regional diversity in WPA of households to avail drinking water services from public and private water suppliers across India (Majumdar & Gupta, 2009; Jala & Nandagiri, 2015; Ramsey et al., 2017; Upadhyaya et al., 2023).

For other countries, Va´squez et al. (2009) also considered the CVM method for valuation of EVW in Mexico. Wang et al. (2010) estimated the EVW using CVM technique in economically feasible households in China. Pour and Kalashami (2012) also applied the mentioned method to predict the value of water in Iran. Khan et al. (2014) reported that WPA of beneficiary to avail water depends on income, water requirement and family size in Bangladesh. Using a descriptive and correlation coefficient analytical technique, Blum et al. (2014) highlighted that education level, water treatment and health status showed a significant impact on WTP of respondents to acquire water in rural Kenya. Jianjun et al. (2016) also noted a substantial impact of gender and income on EVW in China. Water sources and quantity also have a significant impact on WPA of households (Dhungana & Baral, 2016). Vallgårda (2018) claimed that EVW has multiple values and its estimation is worth more.

## Research Gap and Objectives of this Chapter

Alternative cost method, choice experiment method, Hedonic pricing method, opportunity cost method, profit function method, market price method, damage cost method, selling to pay method, meta-analysis, value transfer method, scaling up techniques, contingent valuation method, choice experiment method and market value of direct use method and utility function method can be applied to predict the monetary value of water (EVM) and other natural resources in different countries using (MEA, 2005; Farraro et al., 2011; Chiabai et al., 2011; Pour & Kalashami, 2012; Ninan & Inoue, 2013; Biswas & Venkatachalam, 2015; Ghosh & Bandyopadhyay, 2017). Notwithstanding, limited studies could estimate the EVW and implicit cost of water in developing countries like India. There is a significant diversity in market value of water (MVW) and EVW which is not observed by existing researchers in India. Moreover, EVW must include the full cost of environmental degradation. While, existing studies could not provide the rational and consistent technique to estimate EVW in India. Thereby, all people can have an economic and social accessibility of clean water. Accordingly, limited studies could examine the factors affecting economic value of drinking water in India. Appropriate estimation of EVW would be beneficial for policy makers to decide the price of ecosystem services that are scare. Most specifically, academician is searching the answer for following research questions:

- What must be appropriate EVW for different segments of the community?
- How and why EVW varies across households?
- How does EVW depend on social and economic strictures of households?
- What are the basic determinants of WPA of households to avail better water supply?
- What must be policy initiatives to increase water resources sustainability?

According to above-mentioned research questions, this chapter achieve following objectives:

- To estimate the EVW based on a primary survey of households.
- To assess the implications of social -economic activities on EVW.
- To identify the factors affecting WPA of households to receive better supply of water.

# RESEARCH METHODS AND MATERIALS

## Overview of Study Area

The primary data was collected from Kanpur district of Uttar Pradesh (UP), India. UP is the most populated state and counts 16.50% of the population of India (Census, GoI, 2011). The state has 17.51% youth population of India (Census, GoI, 2011). Around 77.63% of the people lives in rural areas and involved in the agricultural and its allied sectors in Uttar Pradesh (Census, GoI, 2011). Administratively the state

*Figure 4. Main sources of drinking water of households in UP*
**Source:** *Census (Government of India), 2011.*

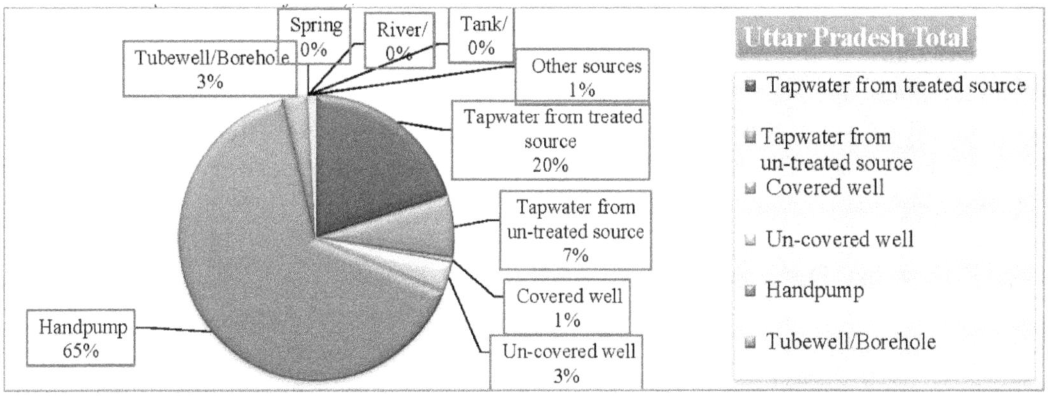


*Figure 5. Main sources of drinking water of households in urban UP*
**Source:** *Census (Government of India), 2011.*

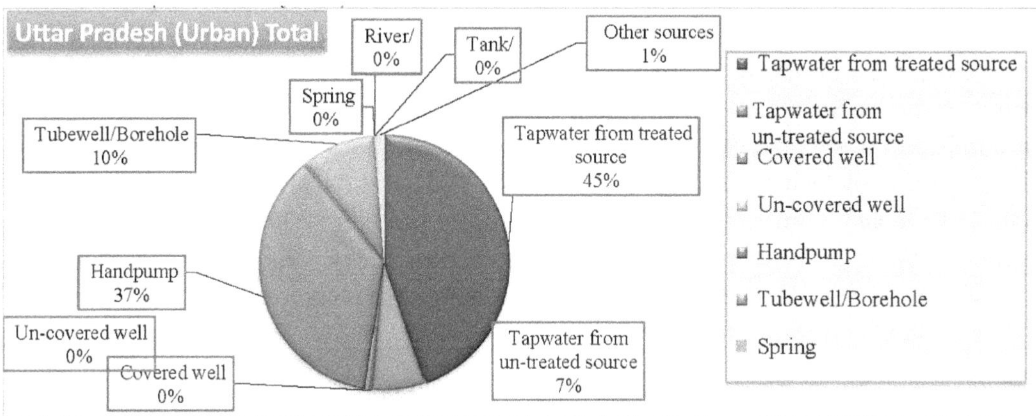


*Figure 6. Main sources of drinking water of households in rural UP*
**Source:** *Census (Government of India), 2011.*

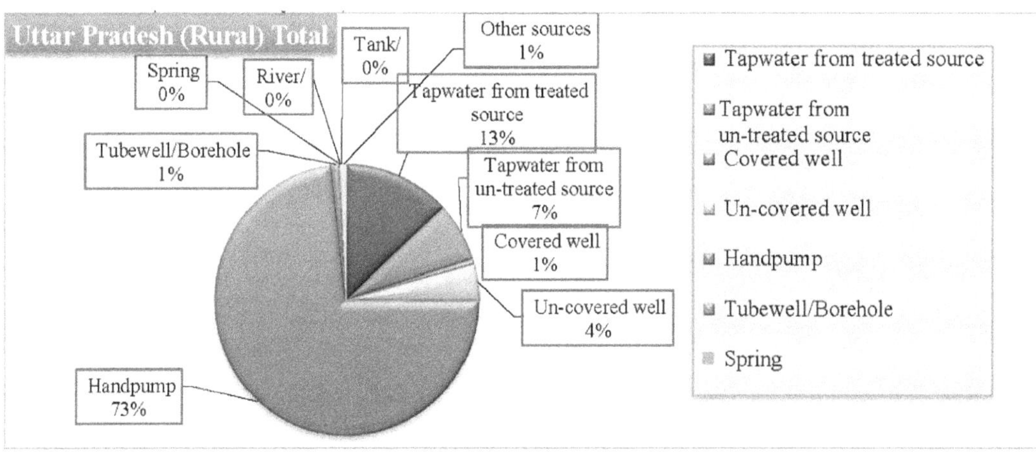

is divided in 18 divisions, 71 districts, 658 blocks and 106,773 villages (Census, GoI, 2011). The state is a highly agriculturally intensive of India. Uttar Pradesh has a dominant position in livestock, milk and agricultural production in India. The state has 12.31% main workers and 17.75% marginal workers of India. The agriculture sector of this state provides jobs to 17.52% marginal agricultural labour of India.

Uttar Pradesh has high poverty, infant mortality rate, income inequality, and low literacy rate among the Indian states (Kumar & Sharma, 2013). The most community of this states use water from hand-pumps. Also, there is significant diversity in the different sources of water in rural and urban area of Uttar Pradesh (Figure: 4, Figure: 5, Figure: 6).

## Source of Primary Data and Sample Size

The descriptive and empirical analysis of this chapter is based on a primary survey of randomly selected 350 respondents from Kanpur Nagar, UP. The chapter is proposed to know the opinion of people towards their willingness to pay ability (WPA) to avail and safe water supply from the municipality. Kanpur Nagar is an industrial city and it has around 5500 small and large industries. Kanpur Nagar is categorized into 6 zones administratively by Kanpur Jal Nigam. The main sources of drinking water of households in Kanpur Nagar are given in Figure: 7. The primary survey of selected households was conducted in the 2nd Zone of Kanpur Nagar. While, 4 wards from this zone were selected randomly for field survey. Face to-face interview of 350 households was conducted using a structural questionnaire. The survey was conducted during November 2022 - March 2023. However, only 311 households could provide the answers on all questions which help to achieve prescribed objectives of this chapter. The questionnaire has 61 different questions which are divided into 4 sections. The questionnaire includes diverse information related to socio-economic structure, educational profile, expenditure on health and water, expenditure on physical assets, physical resources, income-generating resources, family size, existence of diseases in various seasons, annual income, occupation, gender, number of income earners and water availability.

*Figure 7. Main sources of drinking water of households in Kanpur Nagar*
**Source:** *Census (Government of India), 2011.*

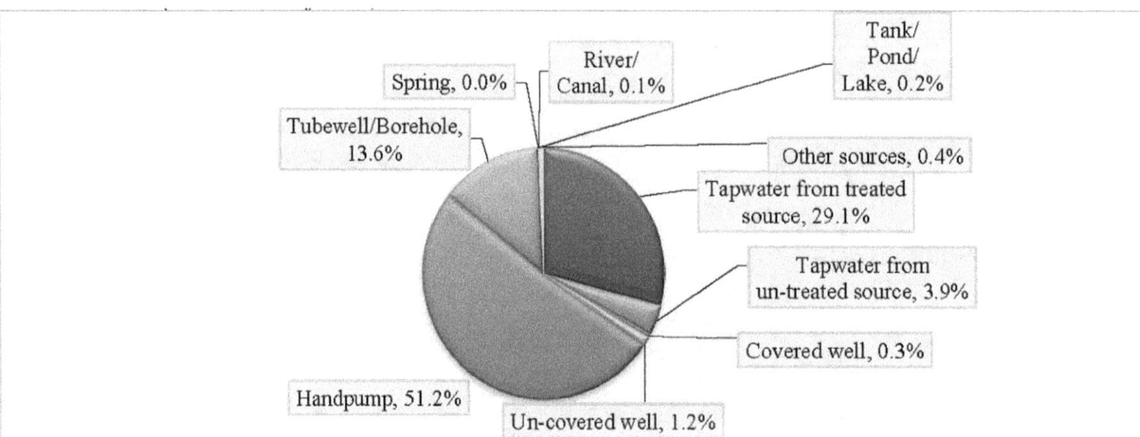

## Mathematical Approach to Estimate Economic Value of Water (EVW)

The simple descriptive method is considered to observe the economic value of water (EVW). It applied willingness to pay ability (WPA) of respondents to determine the EVW. WPA is the maximum amount that an individual is ready to pay to achieve a certain quantity of goods and services from a buyer. Thereupon, average values of all respondents are considered as EVM. WPA approach has a high viability and consistency to estimate the economic value of ecosystem services (Getnet et al., 2022). Getnet et al. (2022) observed the willingness to pay decisions of farmers to achieve the required irrigated water in south Gonder Zone, Ethiopia. This chapter also used the WPA approach to measure the EVW (Dutta et al., 2005; Majumdar & Gupta, 2009; Blum et al., 2014; Jala & Nandagiri, 2015). While, WPA of households is estimated under non-parametric approach (Blum et al., 2014). The WPA is used as:

$$Mean\,WPA = \frac{\sum_{i=1}^{n} \mathrm{WTPi}}{n} \tag{1}$$

Here, *WPA* is the amount that a specific household ($i$) wish to pay for certain quantity of water in a year and $n$ is the number of respondents in equation (1).

## Formulation of Econometric Models

Reddy (1999) estimated the factor affecting per capita water use in Rajasthan (India) using a linear regression model. Dutta et al. (2005) also used specified model to assess the probability of factors which affect the EVW. Somanathan & Ravindranath (2006); Karthikeyan (2010) also considered a linear regression model to examine the influence of soil type on the economic value of irrigated and non-irrigated water. Khan et al. (2014) applied the Bivariate Probit model also to determine the factor affecting WTP of water in Bangladesh. Chandrasekaran et al. (2009) also used a logit model to assess the WPA of farmers affecting factors in Tamil Nadu (India). Using a bivariate probit model, Getnet et al. (2022) also analyzed the determinants of WPA of farmers to receive irrigated water in Ethiopia. Singh et al. (2023b) also preferred logit and probit models to understand the implications of social-economic indicators on climate change adaptation strategies in Gujarat, India. The present chapter also used WPA of beneficiary as a dependent variable (*DV*) and characteristics of selected households are used as independent variables (*IVs*) in the empirical analysis (Siew et al., 2015; Jala & Nandagiri, 2015). Ordinary least square estimation, logit model, probit model and mixed effect logit model using maximum likelihood estimation are used to assess the implication of undertaken variables on WPA of households in this chapter (Majumdar & Gupta, 2009). Ordinary least square estimation model is used in the following form:

$$(AWPPP)_i = \alpha_0 + \alpha_1 (OCSR)_i + \alpha_2 (TFSR)_i + \alpha_3 (SDW)_i + \alpha_4 (MEWBD)_i + \alpha_5 (MWSC)_i + \alpha_6 (AAPP)_i + \alpha_7 (TWBU)_i + \alpha_8 (EWTPY)_i + \alpha_9 (PB)_i + \alpha_{10} (QWS)_i + \alpha_{11} (UGW)_i + \alpha_{12} (AAWC)_i + \alpha_{13} (GEN)_i + U_i \tag{2}$$

Here, *AWPPP* is willingness to pay ability (WPA) (Yes = 1, No = 0); *OCSR* is occupation of household (Salaried = 1, Self-employed = 0); *TFSR* is type of family (Joint = 1, Single = 0); *SDW* is source of drinking water (Public supply = 1, Owned borewell = 0); *MEWBD* is medical expenses (in INR/

year); *MWSC* is municipal water supply connection (Yes = 1, No = 0); *AAPP* is actual amount paid for drinking water (in INR/year); *TWBU* is water treatment before using it (Yes = 1, No = 0); *EWTPY* is expenditure on water treatment (in INR/year); *pb* is paid amount as bribe (Yes = 1, No = 0); *QWS* is quantity of water suitability (Yes = 1, No = 0); *UGW* is using groundwater (Yes = 1, No = 0); *AAWC* is awareness of respondents towards water conservation (Yes = 1, No = 0); *GEN* is gender of household's head (Male = 1, Female = 0); $\alpha_0$ is the constant coefficient; $\alpha_1 \ldots \alpha_{13}$ are the coefficients of associated *IV*s; and $U_i$ is residual-term in equation (2). Previous studies also applied logit, probit and mixed logit models to examine the impact of social-economic factors on WPA of people to acquire water from the suppliers (Dutta & Tiwari, 2005; Chandrasekaran et al., 2009; Barton & Taron, 2010; Majumdar & Gupta, 2009). Logit model, probit model and mixed effect logit models are also used as:

$$Pr\ \{\theta < [\beta_0 + \beta_1\,(OCSR)_i + \beta_2\,(TFSR)_i + \beta_3\,(SDW)_i + \beta_4\,(MEWBD)_i + \beta_5\,(MWSC)_i + \beta_6\,(AAPP)_i + \beta_7\,(TWBU)_i + \beta_8\,(EWTPY)_i + \beta_9\,(PB)_i + \beta_{10}\,(QWS)_i + \beta_{11}\,(UGW)_i + \beta_{12}\,(AAWC)_i + \beta_{13}\,(GEN)_i + Z_i]/\sigma\}^a \tag{3}$$

Here, $Pr\ \{\theta\}$ function of willingness to pay ability (WPA) of respondents (Yes = 1, No = 0); the explanation of other variables are given with equation (2); $\beta_0$ is the constant coefficient; $\beta_1 \ldots \beta_{13}$ are the coefficients of related *IV*s; and $\sigma$ is constant coefficient in equation (3). Willingness to pay ability (WPA) of respondents to avail drinking water also depend on family size, occupation, education level, geographical location and usage of water for various purposes (Dutta & Tiwari, 2005; Khan et al., 2014). So, it is essential to estimate the factors affecting WPA of respondents. For this investigation, actual values of WPA of respondents are used as dependent variables (*DVs*) and stated social-economic factors are treated as *IV*s in regression analysis (Bjornlund & McKay, 1998; Blum et al., 2014). Following functional from of empirical model is developed to determine the association of WPA of respondents with social-economic characteristics:

$$(AAWP)_i = \xi_0 + \alpha_1\,(OCSR)_i + \xi_2\,(TFSR)_i + \xi_3\,(SDW)_i + \xi_4\,(MEWBD)_i + \xi_5\,(MWSC)_i + \xi_6\,(AAPP)_i + \xi_7\,(TWBU)_i + \xi_8\,(EWTPY)_i + \xi_9\,(PB)_i + \xi_{10}\,(QWS)_i + \xi_{11}\,(UGW)_i + \xi_{12}\,(AAWC)_i + \xi_{13}\,(GEN)_i + \tau_i \tag{4}$$

Here, *AAWP* is the actual amount willingness to pay (in INR/year) to achieve better quality of water; $\xi_0$ is constant term; $\xi_1 \ldots \xi_{13}$ are the regression coefficients of associated *IV*s; and $\tau_i$ is the error-term in equation (4). This study includes the data from four wards (i.e., Ward 10, Ward 19, Ward 25 and Ward 59). These wards have a significant diversity in socio-economic characteristics of household and geographical locations. The ward-wise dummy variables are also included in the empirical model to capture the influence of *IV*s on WPA in these wards. The empirical model is applied as:

$$(AAWP)_i = \lambda_0 + \alpha_1\,(OCSR)_i + \lambda_2\,(TFSR)_i + \lambda_3\,(SDW)_i + \lambda_4\,(MEWBD)_i + \lambda_5\,(MWSC)_i + \lambda_6\,(AAPP)_i + \lambda_7\,(TWBU)_i + \lambda_8\,(EWTPY)_i + \lambda_9\,(PB)_i + \xi_{10}\,(QWS)_i + \lambda_{11}\,(UGW)_i + \lambda_{12}\,(AAWC)_i + \lambda_{13}\,(GEN)_i + D_1\,(Ward\ 10) + D_2\,(Ward\ 19) + D_3\,(Ward\ 59) + \eta_I \tag{5}$$

Here, *AAWP* is the actual amount (in INR/year) which a household is willingness to pay for receiving a better quality of drinking water; $\lambda_0$ is constant term; $\lambda_0 \ldots \lambda_{13}$ are the regression coefficients of associated *IV*s; $D_1$, $D_2$ and $D_3$ are regression coefficients of dummy variables for Ward 10, Ward 19 and Ward 25, respectively; and $\eta_i$ is the error-term in equation (5).

## Process of Selection of Appropriate Model

This chapter applied uniform processes to select an appropriate and consistent model. Some variables are excluded in regression analysis due to presence of multicollinearity among them. The variance of inflation factor (VIF) values is estimated to check multicollinearity between the explanatory variables (Sharma & Singh, 2017). The validity of functional form of empirical model is verified by Ramsey RESET test (Singh, 2017). It also used ordinary least square, logit, probit and probit mixed effect models to estimate the coefficients of *IV*s with *DV*. Akaike Information Criterion (AIC) and Schwarz Information Criteria (SIC) is adopted to finalize appropriate model (Singh, 2017). The descriptive and empirical analysis are done by SPSS and STATA statistical software.

# RESULTS AND DISCUSSION

## Basic Characteristics of Sample

The social-economic features of respondents are presented in Table: 1. We have given equal weightage in selection of respondents from each ward. 23.20%, 25.70%, 25.40% and 25.70% households are taken from Ward 10, Ward 19, Ward 25 and Ward 59, respectively. There were 77.80% male and 22.20% female respondents in the sample. As the selection of respondents is based on random sampling techniques. The sample have 44.7% undergraduate, 27.7% post-graduate and 27.6% high-secondary qualified respondents. The sample also included respondents from all communities. 62.6%, 25.1%, 6.1% and 6.10% respondents are from general community, other backward class community, scheduled caste and scheduled tribe community, respectively.

Around 25.1%, and 12.5% respondents are engaged in large and small business sector, and government sector, private and service sector, respectively. Only 6.1% respondents are engaged in agricultural and livestock sector. The sample have 41.8% and 58.2% salaried and 58.2% self-employed workers, correspondingly. The water related amenities and concerns are given in Table: 2. Around 92.3% respondents are using drinking water supplied by the public sector. Only 7.7% respondents are using drinking water from their own bore well. Around 33.8% respondents are receiving water three times in a week; while, 24.4% respondents are receiving water every day from the municipality. Only 16.7% respondents are getting appropriate quality of water. In contrast, 41.8% of respondents perceived a bad smell in water. Therefore, around 53% respondents are using RO, 17.7% are using ceramic candles, and 17.7% respondents are using aqua-guard for water cleaning before using it. Only 11.6% of respondents are boiling water before utilization of it. The respondents are also facing several health issues due to consumption of unsafe and polluted water. Most households are facing allergy, fungal, itching, body pain, tooth pain, eye and ear infection, cold, headache, nausea, acidity, dysentery, abdominal pain, dehydration, loose motion, stomach pain, cough, throat infection, tonsillitis, stone and vomiting due to contaminated water.

*Table 1. Social-economic profile of households*

| Features | Group | No. of Households | Percentage |
|---|---|---|---|
| Geographical location | Ward 10 | 72 | 23.20 |
| | Ward 19 | 80 | 25.70 |
| | Ward 25 | 79 | 25.40 |
| | Ward 59 | 80 | 25.70 |
| Gender | Female | 69 | 22.20 |
| | Male | 242 | 77.80 |
| Age groups (in Year) | 30-40 | 178 | 57.20 |
| | 40-50 | 44 | 14.10 |
| | 50-60 | 89 | 28.60 |
| Educational qualification | High secondary | 86 | 27.60 |
| | Graduate | 139 | 44.70 |
| | Post graduate | 86 | 27.70 |
| Cast-wise distribution | General | 195 | 62.60 |
| | Other backward class | 78 | 25.10 |
| | Schedule Cast | 19 | 6.10 |
| | Schedule Tribe | 19 | 6.10 |
| Income source | Agricultural sector | 19 | 6.10 |
| | Businessmen | 78 | 25.10 |
| | Government servant | 39 | 12.50 |
| | Livestock rearing business | 19 | 6.10 |
| | Private job | 39 | 12.50 |
| | Service sector | 39 | 12.50 |
| | Small businessman | 78 | 25.10 |
| Occupation | Salaried workers | 130 | 41.80 |
| | Self-employed workers | 181 | 58.20 |

**Source:** Author's estimated using primary data.

## Water Related Amenities and Concerns

## WPA, Medical Expenses, Expenditure on Water Treatment, and Quantity of Water

Table: 3 showed the standard deviation (SD) of family size that is noticed 3.026. While, the values of SD for other variables are found extremely high. Thus, the results show that there is found a high diversity in WPA of respondent's accessibility of water, WPA of respondents to receive water supply of 30 kiloliters/ month, medical expenses on water borne diseases, annual payment paid for water, annual expenditure on water treatment and quantity of water required to meet all needs.

*Table 2. Water related concern*

| Characteristics | Group | No. of Households | Percentage |
|---|---|---|---|
| Main sources of drinking water | Bore well | 24 | 7.70 |
| | Public supply | 287 | 92.30 |
| Water supply frequency | Five times a week | 34 | 10.90 |
| | Four times a week | 65 | 20.90 |
| | Once a day | 76 | 24.40 |
| | Thrice a week | 105 | 33.80 |
| | Twice a day | 31 | 10.00 |
| Quality of piped water | Appropriate | 52 | 16.70 |
| | Bad smell | 130 | 41.80 |
| | Poor taste | 25 | 8.00 |
| | Turbidity | 104 | 33.40 |
| Coping mechanism adopted by respondents for cleaning water | Aqua-guard | 55 | 17.70 |
| | boiling | 36 | 11.60 |
| | Ceramic candles | 55 | 17.70 |
| | RO | 165 | 53.10 |

**Source:** Author's estimation based on primary data.

*Table 3. WPA, medical expenses, expenditure on water treatment and quantity of water*

| Characteristics and measurement | Minimum | Maximum | Mean | SD |
|---|---|---|---|---|
| No. of family members | 2 | 15 | 5.99 | 3.026 |
| WPA to access water supply/month (INR/month) | 50 | 200 | 127.97 | 56.415 |
| WPA to receive water supply of 30 kilolitres/month (INR/month) | 200 | 1500 | 793.57 | 393.853 |
| Medical expenses on water borne diseases (INR/year) | 1000 | 111000 | 2.42E4 | 26224.415 |
| Actual annual amount paid for water supply (INR/year) | 1000 | 11798 | 4614.72 | 2316.256 |
| Annual expenditure on water treatment (INR/year) | 500 | 12500 | 5094.21 | 3325.915 |
| Water requires to meet all needs (litters/day) | 200 | 4500 | 1193.25 | 877.415 |

**Source:** Author's estimation based on primary data.

## Payments Made by Respondents for Various Activities in Different Wards

The mean annual payments paid by respondents for various activities are given in Table: 4. The highest and lowest amounts paid by respondents on medical expenses to reduce water borne diseases are reported in wards 25 and ward 59, respectively. The mean annual payments paid by respondents for water supply are observed INR3867, INR4187, INR6356 and INR3995 in Ward 10, Wards 19, Ward 25 and Wards 59, respectively. Thus, there is a high variation in annual payment paid by responding to avail water from municipalities in these wards. Annual expenditure on water treatment paid by respondents is reported INR5369, INR5119, INR4701 and 5210 in ward 10, wards 19, ward 25 and wards 59, respectively. Hence, it is also specified that the quality of drinking water varies among these wards.

*Table 4. Mean annual payments paid by respondents for various activities*

| Wards | Annual medical expenses on water borne diseases (INR/Year) | Annual payment paid by respondents for water supply (INR/Year) | Annual expenditure on water treatment (INR/Year) |
|---|---|---|---|
| Ward 10 | 22694 | 3867 | 5369 |
| Ward 19 | 12825 | 4187 | 5119 |
| Ward 25 | 48203 | 6356 | 4701 |
| Ward 59 | 13175 | 3995 | 5210 |

**Source:** Author's estimation based on primary data.

## Ward-Wise Willingness to Pay Ability (WPA) of Respondents

*WPA* of respondents to avail better quality of water supply from the public sector is given in Table: 5. It infers that only 8.68% respondents are willing to pay an amount of INR200 to avail better quality of drinking water. 6.75% respondents are agreed to pay an amount of INR300 and 9% respondents are ready to pay an amount of INR400 to acquire better quality of water. The results also showed that only 4.18% households are agreed to pay a maximum amount of INR1500 to obtain good quality of water from the suppliers.

## WPA of Respondents to Receive Improved Water Supply

We also include the opinion of respondent's willingness to pay (WPA) to receive better water supply (Table: 6). It showed that around 23%, 26%, 23% and 28% respondents are conscious to pay amount of INR50, INR100, INR150 and INR200, respectively to receive better water supply from the municipality.

## Validity of Estimated EVW

The average value of WPA is considered as EVW in this chapter. It is necessary to check the validity and consistency of EVW. Past researchers recommended that correlation coefficients of estimated value of a specific activity with its associated variables can be used for its validation. In this chapter, the correlation coefficient of EVW with number of family members, expenditure on water treatment, total amount

*Table 5. Willingness to pay ability (WPA) of respondents in each ward*

| | WPA of respondents for better quality water supply per month for 30 kiloliters | | | | | | | | | | | | | |
|---|---|---|---|---|---|---|---|---|---|---|---|---|---|---|
| Range (in INR) | 200 | 300 | 400 | 500 | 600 | 700 | 800 | 900 | 1000 | 1100 | 1200 | 1300 | 1400 | 1500 |
| Ward 10 | 3 | 5 | 8 | 7 | 9 | 1 | 2 | 7 | 0 | 8 | 8 | 3 | 6 | 5 |
| Ward 19 | 6 | 6 | 6 | 7 | 7 | 4 | 6 | 9 | 3 | 4 | 8 | 6 | 5 | 3 |
| Ward 25 | 11 | 5 | 4 | 9 | 8 | 5 | 4 | 7 | 2 | 10 | 1 | 4 | 5 | 4 |
| Ward 59 | 7 | 5 | 10 | 6 | 6 | 9 | 5 | 2 | 6 | 7 | 12 | 1 | 3 | 1 |
| Total | 27 | 21 | 28 | 29 | 30 | 19 | 17 | 25 | 11 | 29 | 29 | 14 | 19 | 13 |
| % of total | 8.68 | 6.75 | 9.00 | 9.32 | 9.65 | 6.11 | 5.47 | 8.04 | 3.54 | 9.32 | 9.32 | 4.50 | 6.11 | 4.18 |

**Source:** Author's estimation based on primary data.

*Table 6. WPA of respondents to receive better water supply*

| | Willingness to Pay Ability of respondents for improved existing water supply per months | | | |
|---|---|---|---|---|
| Range (INR) | 50 | 100 | 150 | 200 |
| Ward 10 | 24 | 14 | 13 | 21 |
| Ward 19 | 11 | 20 | 21 | 28 |
| Ward 25 | 20 | 24 | 18 | 17 |
| Ward 59 | 17 | 22 | 20 | 21 |
| Total | 72 | 80 | 72 | 87 |
| in % | 23 | 26 | 23 | 28 |

**Source:** Author's estimation based on primary data.

*Table 7. Consistency of respondents' view on amount willing to pay*

| *Variables* | *FAMESR* | *EXWATRPY* | *TAWRMAN* | *AMWIPAWA* | *AWPIEWS* |
|---|---|---|---|---|---|
| *FAMESR* | 1 | | | | |
| *EXWATRPY* | -0.044* | 1 | | | |
| *TAWRMAN* | 0.773** | -0.059 | 1 | | |
| *AMWIPAWA* | 0.045* | 0.060 | 0.096 | 1 | |
| *AWPIEWS* | 0.000** | 0.019 | 0.078 | 0.036** | 1 |

**Source:** Author's estimation based on primary data. **Note:** *FAMSER:* Number of family members; *EXWATRPY:* Expenditure on water treatment; *TAWRMAN:* Quantity of water required to meet all needs; *AMWIPAWA:* WPA for current water supply; *AWPIEWS:* WPA for better water supply.

of water required and WPA of households to receive better quality of water supply is estimated for the validation of EVW. The correlation coefficient EVW with family size, expenditure on water treatment, amount of water required and amount willing to pay for better water supply appeared positive and statistically significant (Table: 7). Thus, the statistical value of EVW is found valid. The correlation coefficients of WPA with number of family members, expenditure on water treatment, quantity of water required and WPA for current water supply are also observed statistically significant. It infers that willingness to pay ability increases as a family size and quantity of water requirement increase. Chandrasekaran et al. (2009); Khan et al. (2014) also noted a significant impact of family size on willingness to pay ability of households.

## Cost-Benefit Analysis of Economic Value of Water (EVW)

The Cost-benefit analysis is used to check the reliability of estimated economic value of water (EVW) of 1 Kilolitre and 30 Kilolitre. Implicit cost (IC) of water for each household is estimated using a simple mathematical technique that is used by Majumdar and Gupta (2009). The IC of water is the sum of three components: medical expenses on water borne diseases, expenditure on water treatment and paid payment for current water supply (Table: 8). The results claimed that each household is willing to pay an average amount of INR793.56 for 30 kilolitre water supply/month. While, per household is paying amount of INR2824.61 as implicit cost/month. Thus, there is a need to create awareness among the

*Table 8. Results of cost-benefit analysis*

| Quantity of water | WPA for current water supply of 30 kilolitres (INR/Month) | Total implicit cost (INR/Month) | Actual amount charged by municipality (INR/Month) |
|---|---|---|---|
| EVW for 30 Kilolitre | 793.5691 | 2824.6190 | 177 |
| EVW for 1 Kilolitre | 26.4523 | 94.1540 | 5.90 |

**Source:** Author's estimation based on primary data.

*Table 9. Descriptive statistics of undertaken variables*

| Symbol | Brief Description | Max. value | Min. value | Mean Value | Standard Deviation |
|---|---|---|---|---|---|
| *AWPPP* | Household willing to pay (Yes = 1, No = 0) | 1.00 | 0.00 | 0.60 | 0.49 |
| *AAWP* | Actual amount willingness to pay (INR/Year) | 700 | 13.33 | 169.96 | 126.94 |
| *OCSR* | Occupation of household (Salaried = 1, Self-employed = 0) | 1.00 | 0.00 | 0.42 | 0.49 |
| *TFSR* | Type of family (Joint = 1, Single = 0) | 1.00 | 0.00 | 0.62 | 0.49 |
| *SDW* | Source of drinking water (Public supply = 1, Borewell = 0) | 1.00 | 0.00 | 0.92 | 0.27 |
| *MEWBD* | Medical expenses (INR/year) | 111000 | 1000.00 | 24186.5 | 26224.42 |
| *MWSC* | Municipal water supply connection (Yes = 1, No = 0) | 1.00 | 0.00 | 0.86 | 0.35 |
| *AAPP* | Actual amount paid for drinking water at present (in INR/year) | 11798 | 1000.00 | 4614.72 | 2316.26 |
| *TWBU* | Water treatment before use it (Yes = 1, No = 0) | 1.00 | 0.00 | 0.62 | 0.49 |
| *EWTPY* | Expenditure on water treatment/year (INR/year) | 12500 | 500.00 | 5094.21 | 3325.92 |
| *PB* | Paid bribe (Yes = 1, No = 0) | 1.00 | 0.00 | 0.48 | 0.50 |
| *QWS* | Quantity of water suitable (Yes = 1, No = 0) | 1.00 | 0.00 | 0.50 | 0.50 |
| *UGW* | Usage of groundwater (Yes = 1, No = 0) | 1.00 | 0.00 | 0.33 | 0.47 |
| *AAWC* | Awareness towards water conservation (Yes = 1, No = 0) | 1.00 | 0.00 | 0.67 | 0.47 |
| *GEN* | Gender of household's head (Male = 1, Female = 0) | 1.00 | 0.00 | 0.78 | 0.42 |

**Source:** Author's estimation based on primary data.

households towards the IC of water. On the contrary, the municipality is charging INR177/household for a water supply of 30 Kilolitre/month. The significant differences are also detected for EVW of 1 kilolitre water supply/month.

## Descriptive Statistics of *DV* and *IVs*

The statistical description of variables is presented in Table: 9. The estimated values of SD for of most variables (expect amount of WPA for current water supply, medical expenses, actual amount paid for current water supply, expenditure on water treatment) are found less than 1. Thus, most variables are in the normal form.

## Empirical Results based on OLS Estimation

The impact of socio-economic indicators on WPA of respondents is given in Table: 10. The statistical value of *VIF* indicates that there is no existence of multicollinearity among *IV*s. Also, statistically significant *F* – value showed that the model is fitted well. The $R^2$ value is reported 0.1036. Thus, it proposed that there is a requirement to incorporate more variables in the econometric model to increase the statistical value of $R^2$. The coefficient of occupation with WPA is seen as positive. Thus, it demonstrated that salaried households have a capacity to pay more to avail drinking water supply from municipalities as compared to self-employed households. A salaried worker receives a fixed amount every month and they appeared economically well to pay more amount for water services. Association of type of family with WPA is found negative. It indicates that joint families do not wish to pay more amount for availing the water services from the suppliers in the study area.

It is also noticed a negative coefficient of sources of drinking water with WPA. The results can be interpreted that people think that there is a responsibility of the government to provide them drinking water. The coefficient of medical expenses with WPA of respondents is appeared reported negative. Thus, the estimates indicate that more medical expenses on seasonal diseases disturb economic budget of respondents. The survey results of this research are also claimed that respondents are facing various seasonal diseases due to consumption of unsafe water. Therefore, medical expenditure is caused to reduce WPA of respondents.

*Table 10. Results with simple ordinary least square estimation*

| *AWPPP=DV* | *Reg. Coef.* | *Std. Err.* | *t-value* | *P>\|t\|* | *[95% Conf. Interval]* | |
|---|---|---|---|---|---|---|
| *OCSR* | 0.080879 | 0.0606 | 1.34 | 0.183 | -0.0383 | 0.2001 |
| *TFSR* | -0.2780 | 0.0561 | -4.96 | 0.000 | -0.3882 | -0.1677 |
| *SDW* | -0.1332 | 0.1052 | -1.27 | 0.206 | -0.3401 | 0.0738 |
| *MEWBD* | -3.43E-07 | 1.14E-06 | -0.3 | 0.764 | -2.59E-06 | 1.91E-06 |
| *MWSC* | -0.1420 | 0.0807 | -1.76 | 0.079 | -0.3007 | 0.0168 |
| *AAPP* | -9.65E-06 | 0.0001 | -0.74 | 0.458 | -3.5E-05 | 1.59E-05 |
| *TWBU* | -0.0310 | 0.0574 | -0.54 | 0.590 | -0.1439 | 0.0820 |
| *EWTPY* | 8.89E-06 | 8.25E-06 | 1.08 | 0.282 | -7.35E-06 | 2.51E-05 |
| *PB* | 0.0505 | 0.0641 | 0.79 | 0.431 | -0.0756 | 0.1766 |
| *QWS* | 0.0252 | 0.0542 | 0.46 | 0.643 | -0.0815 | 0.1319 |
| *UGW* | -0.0538 | 0.0708 | -0.76 | 0.448 | -0.1931 | 0.0856 |
| *AAWC* | -0.0139 | 0.0690 | -0.2 | 0.841 | -0.1497 | 0.1219 |
| *GEN* | 0.0223 | 0.0658 | 0.34 | 0.735 | -0.1071 | 0.1518 |
| *Cons. Coef.* | 0.9861 | 0.1724 | 5.72 | 0.000 | 0.6470 | 1.3253 |
| | | | | | | |
| **No. of Obs.** | | 311 | | ***R-squared*** | | 0.1036 |
| ***F*(13 297)** | | 2.64 | | ***Adj. R-squared*** | | 0.0643 |
| ***Prob > F*** | | 0.0017 | | ***VIF*** | | 1.19 |

**Source:** Author's estimation based on primary data.

The coefficient of municipality water supply connection with WPA of respondents is observed negative. It infers that respondents are not seemed keen to pay more money to the municipality. The actual amount for current water supply is negatively associated with WPA. The result shows that respondents are not interested to pay additional amount for existing water supply due to low quality of water. The coefficient of WPA with water treatment is appeared negative. Expenditure on water treatment is also caused to reduce WPA of respondents. The respondents are using Aqua-guard, boiling, ceramic candles and RO for water cleaning. Thus, respondents are receiving unsafe water from the municipality. Low quality of water is also a prime reason which reduced the respondent's intention to increase their WPA.

The coefficient of paid bribe with WPA is found positive. Bribe is a crucial reason which provides incentive for respondents to pay more amount of money for water services. Coefficient of suitable quantity and quality of water with WPA of households is found positive. The respondents are interested to pay an additional amount to receive better quality of water from the municipality. The respondents who are using groundwater do not show their enthusiasm to pay additional amounts to avail the water services. It is observed a negative association of WPA of respondents with usage of groundwater. While, awareness of people towards water conservation has a negative impact on WPA. Hence, most respondents are unconscious about water conservation. The respondents do not wish to pay additional amounts for water services. Male respondents have a more economic capacity to pay significant amount for availing water services as compared to females. Male respondents have more job opportunities, social ability, high working capacity and efficiency as compared to females (Chowdhury & Behera, 2021). Accordingly, the result also showed a positive impact of male respondents on WPA.

## Empirical Result based on Logit, Probit, and Mixed Effect Logit Models

This study observed the probability of factors that show a significant impact on WPA of respondents using logit, probit and mixed effect logit models. The empirical findings based on above-mentioned models are given in Table: 11. The probability function of WPA with occupation of the household appeared positive. The estimates showed that occupation plays a vital role to increase the intention of respondents to pay rational amounts for availing water services from the municipality. The salaried respondents are conscious to pay more amount of money for water services as compared to self-employed. Type of family also showed an undesirable influence on WPA. The respondents belonging to joint family are also not seemed keen to pay an additional amount to receive water.

It is also observed a positive association of source of drinking water with WPA under logit model. While, the results based on probit model and mixed effect logit model showed a negative association of source of drinking water with WPA. Furthermore, the results based on the logit, probit and mixed effect model, showed that medical expenses have a negative influence on WPA. It provides a clarity that more expenditure on health-related issues reduce the economic capacity of the people to pay additional amounts to obtain the desired quantity of water. The results indicate that the municipality is not providing better quality of water to the consumers. The result also stated a negative association of municipal water supply connection with WPA. The association of amount actually paid by respondents and WPA is reported negative. The coefficient of water treatment with WPA is perceived as negative. It is stated earlier that respondents who are using various techniques to sanitize the water are undesirable to pay more money to avail water from the suppliers.

Expenditure on water treatment also show a positive impact on WPA of respondents. The association of paid bribes with WPA is found negative under the logit model. The estimate can be interpreted

*Table 11. Results of willingness to pay based on logit, probit and mixed effect logit model*

| AWPPP=DV | Logit model | | Probit model | | Mixed effect logit model | |
|---|---|---|---|---|---|---|
| AWPPP | Reg. Coef. | P>\|z\| | Reg. Coef. | P>\|z\| | Reg. Coef. | P>\|z\| |
| OCSR | 0.0596 | 0.828 | 0.2230 | 0.186 | 0.1827 | 0.989 |
| TFSR | -0.1324 | 0.61 | -0.7801 | 0 | -0.6406 | 0.96 |
| SDW | 0.0830 | 0.865 | -0.3872 | 0.201 | -0.3312 | 0.989 |
| MEWBD | -8.95E-07 | 0.866 | -8.85E-07 | 0.78 | -6.46E-07 | 0.998 |
| MWSC | -0.2541 | 0.485 | -0.4096 | 0.077 | -0.3360 | 0.986 |
| AAPP | -5.05E-06 | 0.934 | -3.00E-05 | 0.405 | -2.60E-05 | 0.993 |
| TWBU | -0.0496 | 0.852 | -0.0847 | 0.593 | -0.0726 | 0.995 |
| EWTPY | 1.75E-05 | 0.645 | 2.51E-05 | 0.267 | 2.06E-05 | 0.991 |
| PB | -1.1652 | 0.000 | 0.1356 | 0.45 | 0.1097 | 0.994 |
| QWS | -0.0942 | 0.709 | 0.0861 | 0.568 | 0.0786 | 0.995 |
| UGW | -0.1407 | 0.475 | -0.1407 | 0.475 | -0.1060 | 0.995 |
| AAWC | -0.0460 | 0.811 | -0.0460 | 0.811 | -0.0469 | 0.998 |
| GEN | -0.2690 | 0.37 | 0.0639 | 0.725 | 0.0499 | 0.997 |
| Cons. Coef. | 0.2433 | 0.754 | 1.3953 | 0.005 | -16.9079 | 0.689 |

**Source:** Author's estimation based on primary data.

*Table 12. Association of actual amount WPA with IVs*

| AAWP=DV | Reg. Coef. | Std. Err. | t | P>\|t\| | [95% Conf. Interval] | |
|---|---|---|---|---|---|---|
| OCSR | 8.6592 | 14.1998 | 0.61 | 0.542 | -19.2853 | 36.6037 |
| TFSR | -127.3770 | 13.1470 | -9.69 | 0.000 | -153.2500 | -101.504 |
| SDW | 13.9436 | 24.5509 | 0.57 | 0.570 | -34.3715 | 62.2586 |
| MEWBD | -0.0002 | 0.0003 | -0.62 | 0.538 | -0.0007 | 0.0004 |
| MWSC | -0.3111 | 18.8687 | -0.02 | 0.987 | -37.4438 | 36.8218 |
| AAPP | -0.0020 | 0.0031 | -0.63 | 0.527 | -0.0080 | 0.0041 |
| TWBU | -8.5328 | 13.4440 | -0.63 | 0.526 | -34.9899 | 17.9243 |
| EWTPY | 0.0028 | 0.0019 | 1.44 | 0.151 | -0.0010 | 0.0066 |
| PB | 5.9806 | 15.0288 | 0.40 | 0.691 | -23.5955 | 35.5567 |
| QWS | 6.9614 | 12.7048 | 0.55 | 0.584 | -18.0412 | 31.9640 |
| UGW | 6.7404 | 16.5981 | 0.41 | 0.685 | -25.9240 | 39.4047 |
| AAWC | -24.3294 | 16.1841 | -1.50 | 0.134 | -56.1789 | 7.5202 |
| GEN | 10.0500 | 15.4650 | 1.31 | 0.192 | -10.1980 | 50.6750 |
| Con. Coef. | 244.8981 | 39.2687 | 6.24 | 0.000 | 167.6191 | 322.1771 |
| | | | | | | |
| **No. of Obs.** | | 311 | | **R-squared** | | 0.2610 |
| **F(12, 298)** | | 8.77 | | **Adj. R-squared** | | 0.2313 |
| **Prob > F** | | 0.0000 | | **VIF** | | 1.20 |

**Source:** Author's estimation based on primary data.

*Table 13. Association of actual amount WPA with IVs including ward-wise dummies*

| *AAWP=DV* | *Reg. Coef.* | *Std. Err.* | *t* | P>\|t\| | *[95% Conf. Interval]* | |
|---|---|---|---|---|---|---|
| *OCSR* | 8.053 | 14.247 | 0.57 | 0.572 | -19.9870 | 36.0920 |
| *TFSR* | -127.5290 | 13.200 | -9.66 | 0.000 | -153.5070 | -101.5510 |
| *SDW* | 11.0900 | 24.707 | 0.45 | 0.654 | -37.5350 | 59.7150 |
| *MEWBD* | 0.0000 | 0.000 | -0.80 | 0.426 | -0.0010 | 0.0000 |
| *MWSC* | -1.1420 | 18.988 | -0.06 | 0.952 | -38.5130 | 36.2280 |
| *AAPP* | -0.0020 | 0.003 | -0.58 | 0.562 | -0.0080 | 0.0040 |
| *TWBU* | -7.3170 | 13.487 | -0.54 | 0.588 | -33.8600 | 19.2270 |
| *EWTPY* | 0.0030 | 0.002 | 1.47 | 0.142 | -0.0010 | 0.0070 |
| *PB* | 5.8150 | 15.087 | 0.39 | 0.700 | -23.8780 | 35.5070 |
| *QWS* | 6.0670 | 12.735 | 0.48 | 0.634 | -18.9970 | 31.1310 |
| *UGW* | 7.4140 | 16.633 | 0.45 | 0.656 | -25.3200 | 40.1480 |
| *AAWC* | -23.8480 | 16.220 | -1.47 | 0.143 | -55.7700 | 8.0740 |
| *GEN* | 20.2390 | 15.465 | 1.31 | 0.192 | -10.1980 | 50.6750 |
| $WARD_{10}D_1$ | 5.1690 | 20.576 | 0.25 | 0.802 | -35.3250 | 45.6630 |
| $WARD_{19}D_2$ | -1.3850 | 20.952 | -0.07 | 0.947 | -42.6210 | 39.8500 |
| $WARD_{59}D_3$ | -16.4810 | 21.049 | -0.78 | 0.434 | -57.9070 | 24.9450 |
| *Con. Coef.* | 236.5720 | 45.606 | 5.19 | 0.000 | 146.8170 | 326.3260 |
| | | | | | | |
| ***No. of Obs.*** | | 311 | | ***R-squared*** | | 0.2690 |
| ***F*(12, 298)** | | 6.76 | | ***Adj. R-squared*** | | 0.2292 |
| **Prob > F** | | 0.0000 | | ***VIF*** | | 1.38 |

**Source:** Author's estimation based on primary data.

that as respondents pay bribes, therefore, they are undesirable to make more payment to avail the water services. The respondents are keen to increase their WPA for suitable quantity and quality of water supply from the municipality. We also found a positive association of suitable quantity of water with WPA of respondents under probit and mixed effect logit models. Usage of groundwater has a negative impact on WPA of respondents under logit, probit and mixed effect logit models. The estimate implies that respondents who are using groundwater and they are undesired to increase their WPA. The result also indicates that awareness of respondents towards water conservation is negatively associated with WPA. The female dominated households has a significant impact on WPA. The result is similar with the findings of Reddy (1999). Male dominated households have a better economic capacity to pay more amount to avail the water services.

## Empirical Findings Factors Affecting Willingness to Pay Ability of Respondents

The empirical results which assess the coefficients of actual amount willingness to pay ability (WPA) of respondents with *IV*s are presented in Table:12 and Table: 13. For this, we used two simultaneous regressions equations. In the 1st model, we considered only *IV*s, While, in the 2nd model, we included *IV*s with ward-wise dummy variables. The empirical results indicate that type of family, medical expenses, water supply connection, actual amount paid by respondents for water services, expenditure on water treatment, respondent's awareness towards water conservation are negatively associated with WPA of respondents. Hence, it is proposed that policy makers need to focus on aforementioned indicators to avoid the negative impact of these indicators on WPA. We also found a positive impact of occupation, source of drinking water, expenditure on water treatment, paid bribes, suitable quantity of water and usage of groundwater on WPA.

## CONCLUSION

This study implied that willingness to pay ability (WPA) of households depend on water frequency, quantity and quality of water, income, occupation, family size, geographical location, health expenditure and water borne diseases. Thus, municipality need to maintain water frequency and water quality to enhance the WPA of households. Most respondents claimed that they are facing several seasonal diseases due to consumption of unsafe water. Most households, therefore, are spending large share of their income on medical treatment to prevent from water borne diseases in different seasons. Most respondents are using several coping mechanisms such as RO, ceramic candles, boiling and aqua-guard for making water consumable. The average amount of WPA is estimated INR793.57, while, maximum and minimum amount of WPA is observed INR1500 and INR200, respectively. The water users are paying an amount of INR127.97/month for availing better water supply form the municipality. The mean economic value of water (EVW) is found INR793.57/month to acquire water from the municipality.

It is also projected the implicit cost (IC) of water for individual household. IC of water is the linear sum of medical expenses on water borne diseases, expenditure on water treatment and water payment for current water supply. Per household is paying the average amount of INR2825 as implicit cost/year. Hence, IC of water is significantly higher than the actual amount which is paid by the household to acquire water from the supplier. The municipality should improve water quality to increase the WPA of households. The empirical results based on logit, probit and mixed effect probit models showed that occupation, source of drinking water, gender, quality of water and expenditure on water treatment are positively associated with WPA. The type of family, medical expenses, water supply connection, expenditure on water treatment, usage of groundwater and awareness towards water conservation have a negative impact on WPA.

## IMPLICATIONS

It is true that India will face serious water scarcity due to several reasons which are stated in this chapter. Hence, the production activities of all sectors will be disturbed due to water scarcity, low water conservation practices, wastage of water, water pollution, low technological upgradation and low financial capacity of people in India. It cannot be denied that consumption of unsafe drinking water will hamper human health in India. Consequently, water scarcity may be caused to increase food insecurity, livelihood insecurity, health insecurity, and income of rural and urban dwellers in India. India, therefore, needs to improve water conservation practices using technological upgradation to increase water resources sustainability (Ramsey et al., 2017). Otherwise, water-dependent ecosystem services will be distressed due to water scarcity. It is also essential for India to increase awareness among the people toward water conservation practices to meet the water demand in all sectors and to increase water resources sustainability.

Indian policy makers should also treat water as a public good for estimating its economic value. It is also essential to determine water pricing as per availability of water in different locations and social-economic conditions of people in India. India should also create a hydrological model to increase water sustainability. Public private partnerships (PPP) may be positive to provide drinking water facilities to all citizen in India (Jama & Mourad, 2019). Increasing temperature is caused to reduce water resources sustainability and hydrological process (Singh & Jyoti, 2021a). Climate change is also appeared a critical challenge to maintain the water resources sustainability in India. The dwellers in rural and urban areas should use advanced technologies to conserve the rainwater to maintain water resource sustainability. Rainwater must be utilized at greater scale to increase the ground water recharge (Linneck et al., 2015). Hence, India needs to provide and develop proper facilities and infrastructure to store the rainwater (Va´squez et al., 2009).

India should also develop water harvest technologies to increase water resources sustainability (Kumar & Singh, 2023). India also develops and discover digital technologies to conserve and protect water resources. Development institutions must take active steps to maintain the water resources sustainability in India (Reddy, 1999; Jama & Mourad, 2019). There must be effective role of ongoing policies to increase water resources sustainability in India (Reddy, 1999). Appropriate institutional arrangements must be setup (i.e., water markets, define water rights, and proper legal and regulatory instruments) to increase water resources sustainability in India (Borgoyary, 2013). There is also a requirement to increase awareness among the people to maintain the water resources sustainability, and conserve and harvest rainwater for various purposes (Linneck et al., 2015). Communication and community involvement in decision–making process must be implemented to promote water conservation and its sustainability in India (Borgoyary, 2013). Transparency in demand and supply of water help to increase water resources sustainability in India (Fan et al., 2014).

## SCOPE OF FUTURE RESEARCH

Water pricing methods should include various factors such as financial and socio-economic condition of peoples, WTP ability and affordability, technical constraints, geographical location, political factors, climatic factors and demographic change (Devi et al., 2009). Policy makers could not estimate the appropriate EVW in India. The scientific research community must discover and maintain a database to resolve the above-mentioned issues for better implementation of water policies and to evaluate EVW. Climate change and water scarcity affecting factors must be included in estimation of EVW. Further research, therefore, can be considered to estimate the impact of climatic factors on water resources. This chapter could not include the involvement of different income groups of the community in water price estimation. Therefore, existing researchers may also consider this aspect in estimation of EVW in further research. Existing researchers can also increase the large sample size of households from various states of India to validate the empirical finding of this chapter in further study.

## LIMITATIONS

Most specifically, the empirical findings of this chapter are based on a sample of 350 households who are randomly selected from Kanpur Nagar, Uttar Pradesh (India). Although, there is a variation in geographical location, availability of natural resources, availability of water, urbanization, industrialization, agricultural production sector, climatic conditions, infrastructural development and government policies towards social welfare across cities in India. Therefore, the findings of this chapter may not be generalized across Indian states and regions. Accordingly, this study acknowledged the above-mentioned issue as limitations of this chapter. Henceforth, existing researchers can collect the similar data on related variables from different locations of India for validating the descriptive and empirical results of this chapter.

## REFERENCES

Arfanuzzaman, M., & Atiq Rahman, A. (2017). Sustainable water demand management in face of rapid urbanization and ground water depletion for social-ecological resilience building. *Global Ecology and Conservation*, *10*(1), 9–22. doi:10.1016/j.gecco.2017.01.005

Barton, D. N., & Taron, A. (2010). Valuing irrigation water using survey-based methods in the Tungabhadra River Basin, India. *Irrigation and Drainage Systems*, *24*(1), 265–277. doi:10.100710795-010-9098-2

Biswas, D., & Venkatachalam, L. (2015). Farmers' willingness to pay for improved irrigation water – A case study of Malaprabha irrigation project in Karnataka, India. *Water Economics and Policy*, *1*(1), 1–24. doi:10.1142/S2382624X14500040

Bjornlund, H., & McKay, J. (1998). Factors affecting water price in a rural water market: A South Australian experience. *Water Resources Research*, *34*(6), 1563–1570. doi:10.1029/97WR03470

Blum, A. G., Null, C., & Hoffmann, V. (2014). Marketing households water treatment: Willingness to pay results from an experiment in Rural Kenya. *Water (Basel)*, *6*(7), 1873–1886. doi:10.3390/w6071873

Borgoyary, M. (2013). Valuation of water: Options for sustainable development in developing countries. *Water and Development*, *2*(1), 1-18. https://www.eolss.net/sample-chapters/c07/E2-24M-04-01.pdf

Chandrasekaran, K., Devarajulu, S., & Kuppannan, P. (2009). Farmers' willingness to pay for irrigation water: A case of tank irrigation systems in south India. *Water (Basel)*, *1*(1), 5–18. doi:10.3390/w1010005

Chen, M., Zhang, H., Liu, W., & Zhang, W. (2013). The global pattern of urbanization and economic growth: Evidence from the last three decades. *PLoS One*, *9*(8), 1–15. doi:10.1371/journal.pone.0103799 PMID:25099392

Chiabai, A., Travisi, C. M., Markandya, A., Ding, H., & Nunes, P. A. L. D. (2011). Economic assessment of forest ecosystem services losses: Cost of policy inaction. *Environmental and Resource Economics*, *50*(3), 405–445. doi:10.100710640-011-9478-6

Chowdhury, K., & Behera, B. (2021). Economic significance of provisioning ecosystem services of traditional water bodies: Empirical evidence from West Bengal, India. *Resources. Environmental Sustainability*, *5*(1), 1–10. doi:10.1016/j.resenv.2021.100033

Cosgrove, W. J., & Loucks, D. P. (2015). Water management: Current and future challenges and research directions. *Water Resources Research*, *51*(1), 4823–4839. doi:10.1002/2014WR016869

Dasgupta, P., & Dasgupta, R. (2004). Economic value of safe water for infrastructurally disadvamtaged urban household: A case study in Delhi, India. *Water Resources Research*, *40*(11401), 1–10. doi:10.1029/2003WR002461

Devi, J. S., Joseph, B., Karunakaran, K., Anurdha, B., & Devi, K. R. (2009). People's attitudes towards paying for water. *Current Science*, *97*(9), 1296–1302.

Dhungana, A. R., & Baral, B. (2016). Factors affecting willingness to pay for improved water supply system in rural Tanahu, Nepal. *Janapriya Journal of Interdsciplinary Studies*, *5*(1), 1–13. doi:10.3126/jjis.v5i0.17836

Dutta, V., Chander, S., & Srivastava, L. (2005). Public support for water supply improvements: Empirical evidence from unplanned settlements of Delhi, India. *Journal of Environment & Development*, *14*(4), 439–462. doi:10.1177/1070496505281841

Dutta, V., & Tiwari, A. P. (2005). Cost of service and willingness to pay for reliable urban water supply: A study from Delhi, India. *Water Science and Technology: Water Supply*, *5*(6), 135–144. doi:10.2166/ws.2005.0058

Fan, L., Wang, F., Liu, G., Yang, X., & Qin, W. (2014). Public perception of water consumption and its effects on water conservation behavior. *Water (Basel)*, *6*(1), 1771–1784. doi:10.3390/w6061771

Farraro, P. J., Lawlor, K., Mullan, K. L., & Pattanayak, S. K. (2011). Forest figures: Ecosystem services valuation and policy evaluation in developing countries. *Review of Environmental Economics and Policy*, *1*(1), 1–26. doi:10.1093/reep/rer019

Francis, M. R., Nagarajan, G., Sarkar, R., Mohan, V. R., Kang, G., & Balraj, V. (2015). Perception of drinking water safety and factors influencing acceptance and sustainability of a water quality intervention in rural southern India. *BMC Public Health*, *15*(731), 1–9. doi:10.118612889-015-1974-0 PMID:26223687

Getnet, A., Tesfaye, E., Ahmed, Y., & Ahmed, M. (2022). Economic valuation and its determinates of improved irrigation water use; evidence based on South Gondar Zone, Ethiopia. *Cogent Economics & Finance*, *10*(1), 1–22. doi:10.1080/23322039.2022.2090663

Ghosh, N., & Bandyopadhyay, J. (2017). Methods of valuation of water resources: A review. *Sawas Journal*, *1*(1), 19–50.

Ghosh, P. (2021). Water stress and water crisis in large cities of India. In R. K. Mishra, R. B. Singh, & A. Dubey (Eds.), *Sustainable Climate Action and Water Management. Advances in Geographical and Environmental Sciences*. Springer. doi:10.1007/978-981-15-8237-0_11

Golubev, G. N. (1983). Economic activity, water resources and the environment: A challenge for hydrology. *Hydrological Sciences Journal*, *28*(1), 57–75. doi:10.1080/02626668309491143

Goswami, K., & Bisht, P. S. (2017). The role of water resources in socio-economic development. *International Journal for Research in Applied Science and Engineering Technology*, *5*(12), 669–1674.

Houtven, G. L. V., Pattanayak, S. K., Usmani, F., & Yang, J. C. (2017). What are households willing to pay for improved water access? Results from a meta-analysis. *Ecological Economics*, *136*(1), 126–135. doi:10.1016/j.ecolecon.2017.01.023

Jala, & Nandagiri, L. (2015). Evaluation of economic value of Pilikula Lake using travel cost and contingent valuation methods. *Aquatic Procedia*, *4*(1), 1315-1321. doi:10.1016/j.aqpro.2015.02.171

Jama, A. A., & Mourad, K. A. (2019). Water services sustainability: Institutional arrangements and shared responsibilities. *Sustainability (Basel)*, *11*(3), 1–15. doi:10.3390u11030916

Jamel, L., & Derbali, A. (2016). Do energy consumption and economic growth lead to ED? Evidence from Asian economies. *Cogent Economics & Finance*, *4*(11), 1–19. doi:10.1080/23322039.2016.1170653

Jianjun, J., Wenyu, W., Ying, F., & Xiaomin, W. (2016). Measuring the willingness to pay for drinking water quality improvements: Results of a contingent valuation survey in Songzi, China. *Journal of Water and Health*, *14*(3), 504–512. doi:10.2166/wh.2016.247 PMID:27280614

Jyoti, B., & Singh, A. K. (2020). Projected sugarcane yield in different climate change scenarios in Indian states: A state-wise panel data exploration. *International Journal of Food and Agricultural Economics*, *8*(1), 343–365.

Jyoti, B., Singh, A. K., & Ashraf, S. N. (2023). Appropriate technologies and their implications in the agricultural sector. In P. H. Alex Khang (Ed.), *Advanced Technologies and AI-Equipped IoT Applications in High-Tech Agriculture*. IGI Global. doi:10.4018/978-1-6684-9231-4.ch004

Kapur, R. (2016). Natural resources and environmental issues. *Journal of Ecosystem & Ecology*, *6*(2), 1–4.

Karthikeyan, C. (2010). *Economic and social value of irrigation water: Implications for sustainability*. Fourteenth International Water Technology Conference, IWTC 14 2010, Cairo, Energy.

Katko, T. S., & Hukka, J. J. (2015). Social and economic importance of water services in the build environment: Need for more structural thinking. *Procedia Economics and Finance*, *21*(1), 217–223. doi:10.1016/S2212-5671(15)00170-7

Khan, N. I., Brouwer, R., & Yang, H. (2014). Household's willingness to pay for arsenic safe drinking water in Bangladesh. *Journal of Environmental Management*, *143*(1), 151–161. doi:10.1016/j.jenvman.2014.04.018 PMID:24905645

Kumar, A. (2015). *Effect of climatic factors on agricultural productivity and food security in India: An econometric analysis*. [PhD Thesis, IIT Indore, India].

Kumar, A., Ahmad, M. M., & Sharma, P. (2017). Influence of climatic and non-climatic factors on sustainable food security in India: A statistical investigation. *International Journal of Sustainable Agricultural Management and Informatics*, *3*(1), 1–30. doi:10.1504/IJSAMI.2017.082917

Kumar, A., & Sharma, P. (2013). Impact of climate variation on agricultural productivity and food security in rural India. Economics Discussion Papers, No. 2013-43. Publisher: Kiel Institute for the World Economy Germany.

Kumar, A., & Sharma, P. (2014). Climate change and sugarcane productivity in India: An econometric analysis. *Journal of Social and Development Sciences*, *5*(2), 111–122. doi:10.22610/jsds.v5i2.811

Kumar, A., Sharma, P., & Ambrammal, S. K. (2015a). Climatic effects on sugarcane productivity in India: A stochastic production function application. *International Journal of Economics and Business Research*, *10*(2), 179–203. doi:10.1504/IJEBR.2015.070984

Kumar, A., Sharma, P., & Joshi, S. (2015b). Effects of climatic factors on agricultural productivity in India: A state-wise panel data analysis. *International Journal of Basic and Life Sciences*, *3*(1), 48–67.

Kumar, A., Sharma, P., & Joshi, S. (2016). Assessing the impacts of climate change on land productivity in Indian crop agriculture: An evidence from panel data analysis. *Journal of Agricultural Science and Technology*, *18*(1), 1–13. https://jast.modares.ac.ir/article-23-4491-en.pdf

Kumar, A., Singh, B. J., & Sharma, P. (2020). Assessing the climate change impact on rice and wheat production in Uttar Pradesh and Haryana states of India. *Climatic Change*, *6*(21), 74–93.

Kumar, S., & Singh, A. K. (2023). Modeling the effects of climate change on agricultural productivity: Evidence from Himachal Pradesh. *Asia-Pacific Journal of Regional Science*, *7*(2), 1–28. doi:10.100741685-023-00291-w

Linneck, S., Parker, A. H., Smeets, P., Cruddas, P. H., & Jeffrey, P. (2015). *Social and economic factors influencing the 'choice' of water supply in rural Karnataka, India.* 38th WEDC International Conference on Water, Sanitation and Hygiene Beyond 2015: Improving Access and Sustainability, Loughborough University, UK. https://wedc-knowledge.lboro.ac.uk/resources/conference/38/Linneck-2136.pdf

Majumdar, C., & Gupta, G. (2009). Willingness to pay and municipal water pricing in transition: A case study. *Journal of Integrative Environmental Sciences*, *6*(4), 247–260. doi:10.1080/19438150903068224

Mathur, O. P., & Thakur, S. (2003). *Urban water pricing: Setting the stage for reforms*. National Institute of Public Finance and Policy. https://www.nipfp.org.in/publications/reports/1068/

Millennium Ecosystem Assessment (MEA). (2005). *Ecosystems and Human Well-being: Synthesis*. Island Press, Washington DC. https://www.millenniumassessment.org/documents/document.356.aspx.pdf

Ninan, K. N., & Inoue, M. (2013). Valuing forest ecosystem services: Case study of a forest reserve in Japan. *Ecosystem Services*, *5*(1), 78–87. doi:10.1016/j.ecoser.2013.02.006

Pour, M. T., & Kalashami, M. K. (2012). Applying CVM for economic valuation of drinking water in Iran. *International Journal of Agricultural Management and Development*, *2*(3), 209–214.

Ramsey, E., Berglund, E. Z., & Goyal, R. (2017). The impact of demographic factors, beliefs, and social influences on residential water consumption and implications for non-price policies in urban India. *Water (Basel)*, *9*(11), 1–21. doi:10.3390/w9110844

Reddy, V. R. (1999). Quenching the thirst: The cost of water in fragile environments. *Development and Change*, *30*(1), 79–113. doi:10.1111/1467-7660.00108

Rios, P. C. S., Deen, T. A., Nagabhatla, N., & Ayala, G. (2018). Explaining water pricing through a water security lens. *Water (Basel)*, *10*(9), 1–20. doi:10.3390/w10091173

Rodgers, C., & Hellegers, P. J. G. J. (2005). Water pricing and valuation in Indonesia: Case study of the Brantas River Basin. *EPT Discussion Paper 141*. International Food Policy Research Institute, USA. doi:10.22004/ag.econ.58586

Samui, R. P., Kulkarni, P. S., Kamble, M. V., & Vaidya, N. G. (2014). A critical evaluation of sugarcane yield variation as influenced by climatic parameters in Uttar Pradesh and Maharashtra states of India. *Time Journals of Agriculture and Veterinary Sciences*, *2*(1), 63–69.

Schnoor, J. L. (2010). *Water sustainability in a changing world*. National Water Research Institute. https://research.uiowa.edu/sites/research.uiowa.edu/files/files/Clarke Prize Lecture.pdf

Sharma, P., & Singh, A. K. (2017). Association of state-wise food security index with climatic factors in India: Evidence from state-wise panel data. *Journal of Global Agriculture and Ecology*, *6*(3), 196–205.

Siew, M. K., Yacob, M. R., Radam, A., Adamu, A., & Alias, E. F. (2015). Estimating willingness to pay for wetland conservation: A contingent valuation study of Paya Indah Wetland, Selangor Malaysia. *Procedia Environmental Sciences*, *30*(1), 268–272. doi:10.1016/j.proenv.2015.10.048

Singh, A. K. (2017). *Climate Change and Productivity of Food-grain and Cash Crops in India*. LAP LAMBERT Academic Publishing.

Singh, A. K. (2020). Economic valuation of ecosystem services of river and human well-being in developing countries: A research direction. *JOJ Wildlife & Biodiversity*, *3*(1), 1–2.

Singh, A. K., Ahmad, M. M., & Sharma, P. (2017a). Implications of socioeconomic factors on food security in selected economies: An empirical assessment. *Journal of Global Economics. Management and Business Research*, *8*(2), 103–115.

Singh, A. K., Ashraf, S. N., Sharma, S. K., & Kumar, S. (2023b). Factors affecting farm income and climate adaptation strategies in the farming sector of Gujarat state of India. In B. J. Sadhik, J. Adewole, A. Yeneneh, & L. Abidoye (Eds.), *Opportunities and Challenges in Climate-Friendly Clean Water and Energy Technologies* (pp. 122–147). IGI Global. doi:10.4018/978-1-6684-7303-0.ch006

Singh, A. K., & Issac, J. (2018). Impact of climatic and non-climatic factors on sustainable livelihood security in Gujarat state of India: A statistical exploration. *Agriculture and Food Sciences Research*, *5*(1), 30–46. doi:10.20448/journal.512.2018.51.30.46

Singh, A. K., Issac, J., & Narayanan, K. G. S. (2019a). Measurement of environmental sustainability index and its association with socio-economic indicators in selected Asian economies: An empirical investigation. *International Journal of Environment and Sustainable Development*, *18*(1), 58–100. doi:10.1504/IJESD.2019.098641

Singh, A. K., & Jyoti, B. (2019). Measuring the climate variability impact on cash crops farming in India: An empirical investigation. *Agriculture and Food Sciences Research*, *6*(2), 155–165. doi:10.20448/journal.512.2019.62.155.165

Singh, A. K., & Jyoti, B. (2021a). Projected food-grain production and yield in India: An evidence from state-wise panel data investigation during 1977-2014. *Journal of Agricultural Sciences – Sri Lanka*, *16*(1), 108-125. doi:10.4038/jas.v16i1.9188

Singh, A. K., & Jyoti, B. (2021b). Projected productivity of cash crops in different climate change scenarios in India: Use of marginal impact analysis technique. *Finance & Economics Review*, *3*(1), 63–87.

Singh, A. K., Jyoti, B., Kumar, S., & Lenka, S. K. (2021). Assessment of global sustainable development, environmental sustainability, economic development and social development index in selected economies. *International Journal of Sustainable Development and Planning*, *16*(1), 123–138. doi:10.18280/ijsdp.160113

Singh, A. K., & Kumar, S. (2021). Assessing the performance and factors affecting industrial development in Indian states: An empirical analysis. *Journal of Social Economics Research*, *8*(2), 135–154. https://archive.conscientiabeam.com/index.php/35/article/view/1374/1915. doi:10.18488/journal.35.2021.82.135.154

Singh, A. K., Kumar, S., Ashraf, S. N., & Jyoti, B. (2023c). Implications of farmer's adaptation strategies to climate change in agricultural sector of Gujarat: Evidence from farm level data. *Research on World Agricultural Economy*, *3*(1), 42–57. doi:10.36956/rwae.v3i1.498

Singh, A. K., Kumar, S., & Jyoti, B. (2022a). Influence of climate change on agricultural sustainability in India: A state-wise panel data analysis. *Asian Journal of Agriculture*, *6*(1), 15–27. doi:10.13057/asianjagric/g060103

Singh, A. K., Kumar, S., & Jyoti, B. (2022b). Association of sustainable food security with climatic and non-climatic factors in Gujarat state of India: A district-wise panel data investigation. In: Leal Filho, W., Kovaleva, M., Popkova, E. (eds.), Sustainable Agriculture and Food Security. World Sustainability Series (pp. 259-282). Springer Cham, Springer Nature Switzerland. doi:10.1007/978-3-030-98617-9_16

Singh, A. K., Narayanan, K. G. S., & Sharma, P. (2017b). Effect of climatic factors on cash crop farming in India: An application of Cobb-Douglas production function model. *International Journal of Agricultural Resources, Governance and Ecology*, *13*(2), 175–210. doi:10.1504/IJARGE.2017.086452

Singh, A. K., Narayanan, K. G. S., & Sharma, P. (2019b). Measurement of technical efficiency of climatic and non-climatic factors in sugarcane farming in Indian states: Use of stochastic frontier production function approach. *Climatic Change*, *5*(19), 150–166.

Singh, A. K., Sharma, A. K., & Jyoti, B. (2023a). Does economic development have a causal relationship with environmental degradation? Experience from different income group countries. In G. Sart (Ed.), *Considerations on Education for Economic, Social, and Environmental Sustainability* (pp. 300–333). IGI Global. doi:10.4018/978-1-6684-8356-5.ch015

Singh, A. K., & Sharma, P. (2018a). Implications of climatic and non-climatic variables on food security in developing economies: A conceptual review. *MOJ Food Processing & Technology*, *6*(1), 1–12. doi:10.15406/mojfpt.2018.06.00138

Singh, A. K., & Sharma, P. (2018b). Measuring the productivity of food-grain crops in different climate change scenarios in India: An evidence from time series investigation. *Climatic Change*, *4*(16), 661–673.

Singh, A. K., Sharma, P., & Singh, D. K. (2016). Measuring the influence of weather variables on productivity of food-grain crops in India: An application of Just & Pope's production technique. *ABBS Management Business and Entrepreneurship Review*, *7*(2), 29–46. doi:10.23874/amber/2016/v7/i2/121780

Singh, A. K., Singh, B. J., & Negi, V. (2020). Does sustainable development have a causal relationship with environmental development? Evidence from a country-wise panel data analysis. *International Journal of Technology Management & Sustainable Development*, *19*(2), 147–171. doi:10.1386/tmsd_00020_1

Somanathan, E., & Ravindranath, R. (2006). Measuring the marginal value of water and elasticity of demand for water in agriculture. *Economic and Political Weekly*, *6*(1), 2712–2715. doi:10.2307/4418398

Streimikiene, D. (2015). Environmental indicators for the assessment of quality of life. *Intellectual Economics*, *9*(1), 67–79. doi:10.1016/j.intele.2015.10.001

Tundisi, J. G. (2008). Water resources in the future: Problems and solutions. *Estudos Avançados*, *22*(63), 1–8.

Twesige, D., & Mbabazize, P. M. (2013). Relationship between environmental accounting indicators, macroeconomic indicators, and sustainable development indicators in developing countries: Case study Rwanda. *Researchjournali's. Journal of Economics*, *1*(1), 1–11. https://www.jkuat.ac.ke/campuses/kigali/wp-content/uploads/2014/06/my-publication-1.pdf

Upadhyaya, A., Jeet, P., Singh, A. K., & Sundaram, P. K. (2023). Estimation of the economic value of irrigation water in canal and tube well command areas. *$H_2$ Open Journal,* *6*(2), 1 – 9. doi:10.2166/h2oj.2023.011

Va'squez, W. F., Mozumder, P., Herna'ndez-Arce, J., & Berrens, R. P. (2009). Willingness to pay for safe drinking water: Evidence from Parral, Mexico. *Journal of Environmental Management*, *90*(1), 3391–3400. doi:10.1016/j.jenvman.2009.05.009 PMID:19525059

Vallgårda, A. F. (2018). *The value of water – A valuation study of groundwater in Uppsala municipality.* [Degree thesis No 1132, Sveriges lantbruksuniversitet Swedish University of Agricultural Sciences]. http://stud.epsilon.slu.se

Veettil, P. C., Speelman, S., Frija, A., Buysse, J., Mondelaers, K., & van Huylenbroeck, G. (2011). Price sensitivity of farmer's preferences for irrigation water-pricing method: Evidence from choice model analysis in Krishna River basin, India. *Journal of Water Resources Planning and Management*, *1*(1), 205–214. doi:10.1061/(ASCE)WR.1943-5452.0000103

Venkatachalam, L., & Narayanamoorthy, A. (2012). Estimating economic value of irrigation through contingent valuation of method: Results from Bhavani River Basin, Tamil Nadu. *Indian Journal of Agricultural Economics*, *67*(3), 308–315. doi:10.22004/ag.econ.204814

Wang, H., Xie, J., & Li, H. (2010). Water pricing with household surveys: A study of acceptability and willingness to pay in Chongqing, China. *China Economic Review*, *21*(1), 136–149. doi:10.1016/j.chieco.2009.12.001

Zho, X., Wang, J., Su, J., & Sun, W. (2021). Ecosystem service value evaluation method in a complex ecological environment: A case study of Gansu province, China. *PLoS One*, *16*(2), 1–25. doi:10.1371/journal.pone.0240272 PMID:33544729

## ADDITIONAL READINGS

Dolan, F., Lamontagne, J., Link, R., Hejazi, M., Reed, P., & Edmonds, J. (2021). Evaluating the economic impact of water scarcity in a changing world. *Nature Communication*, *12*(1915). doi:10.1038/s41467-021-22194-0

Loureiro, M. L., Loomis, J. B., & Vázquez, M. X. (2009). Economic valuation of environmental damages due to the prestige oil spill in Spain. *Environmental and Resource Economics*, *44*(1), 537–553. doi:10.100710640-009-9300-x

Lowe, B. H., Zimmer, Y., & Oglethorpe, D. R. (2022). Estimating the economic value of green water as approach to foster the virtual green-water trade. *Ecological Indicators*, *136*(1), 1–18. doi:10.1016/j.ecolind.2022.108632

Owusu-Sekyere, E., Jordaan, H., & Chouchane, H. (2017). Evaluation of water footprint and economic water productivities of dairy products of South Africa. *Ecological Indicators*, *83*(1), 32–40. doi:10.1016/j.ecolind.2017.07.041

Rost, S., Gerten, D., Bondeau, A., Lucht, W., Rohwer, J., & Schaphoff, S. (2008). Agricultural green and blue water consumption and its influence on the global water system. *Water Resources Research*, *44*(9), 2007WR006331. Advance online publication. doi:10.1029/2007WR006331

Vanham, D., Mekonnen, M. M., & Hoekstra, A. Y. (2013). The water footprint of the EU for different diets. *Ecological Indicators*, *32*(1), 1–8. doi:10.1016/j.ecolind.2013.02.020

Zhao, D., Hubacek, K., Feng, K., Sun, L., & Liu, J. (2019). Explaining virtual water trade: A spatial-temporal analysis of the comparative advantage of land, labor and water in China. *Water Resources*, *153*(1), 304–314. doi:10.1016/j.watres.2019.01.025 PMID:30738227

## KEY TERMS AND DEFINITIONS

**Climate Change:** Variability or variation in weather parameters from mean in a long-term known as climate change.

**Economic Value of Water:** Monetary value of water which an individual consumer or group of consumers are willing to pay to receive water as per their need.

**Ecosystem Services:** Goods and services that consumer and producers receive from nature known as ecosystem services.

**Implicit Cost of Water:** It is the linear sum of medical expenses on water borne diseases, expenditure on water treatment and water payment made by household for availing current water supply.

**Water Pollution:** Rising contamination in the available water due to rising GHGs emissions, acid rain, fertilizer, chemicals, pesticides and wastage of raw materials.

**Water Scarcity:** Consistent declining in the per capita availability of ground and surface water.

**Willingness to Pay Ability:** The economic ability of individual consumer or person to pay a desire amount to avail a specific quantity of ecosystem services from the supplier or producer.

# Chapter 11
# Strengthening Capacity for Climate Change Adaptation of Moroccan Agriculture:
## Deficit Irrigation Strategies for Resilient Olive Sector

**Khaoula Ibba**
*Cadi Ayyad University, Morocco & National Institute of Agricultural Research, Morocco*

**Jalal Kassout**
https://orcid.org/0000-0002-3702-2751
*National Institute of Agricultural Research, Morocco*

**Salah Er-Raki**
*Cadi Ayyad University, Morocco & Mohammed VI Polytechnic University, Morocco*

**Sara Oulbi**
*National Institute of Agricultural Research, Morocco*

**Loubna El Mansouri**
*Hassan II Institute of Agronomy and Veterinary, Morocco*

**Vladimiro Boselli**
https://orcid.org/0000-0003-2985-6358
*Università degli Studi di Padova, Italy*

**Abdelaziz Bouizgaren**
*National Institute of Agricultural Research, Morocco*

**Lhassane Sikaoui**
*International Olive Oil Council, Spain*

**Rachid Hadria**
https://orcid.org/0000-0001-5779-2764
*National Institute of Agricultural Research, Morocco*

## ABSTRACT

*Agriculture in the Mediterranean region is substantially challenged by climate change, and olive growers in particular are facing difficulties because of its effects. In this regard, the olive sector can adopt deficit irrigation strategies, to achieve sustainable management practices that improve water productivity and save irrigation water. This chapter aims to investigate the potential impacts of deficit irrigation strategies on Menara olive tree. Agronomical parameters were assessed in 2021 and 2022, focusing on trees subjected to regulated (RDI) and sustained (SDI) deficit irrigation. The results showed that the performances of Menara olive trees under controlled water stress are better under RDI strategy compared to SDI one. Moreover, the RDI strategy, particularly under specific treatment, allowed to reduction of sup-*

DOI: 10.4018/978-1-6684-9838-5.ch011

*plied water by 20% and 40% during sensitive and normal periods, respectively. These results highlight the importance of water management strategies and the need to consider their implications on flowering traits, crop yield, and fruit quality over consecutive growing seasons.*

## INTRODUCTION

The Olive tree (*Olea europaea* L.), native to the Mediterranean basin, is an evergreen tree considered highly suitable and well-suited to the Mediterranean environmental conditions (Kassout et al., 2021). Nevertheless, the Mediterranean area is recognized as a climate change trouble spot, experiencing notable trends of warming and drying. Presently, olive groves encounter new challenges and risks, with climate change being one of the most significant (Kassout et al., 2023). In the coming decades, and following recent climate change scenarios, olive growers will confront rising temperatures, drought, and increased occurrences of severe weather incidents, exemplified by extreme heat events, necessitating their attention and adaptation (Fraga et al., 2020a).

Morocco, with its arid and semi-arid climate marked by hot and dry summer conditions, limited and irregular precipitation, is grappling with a considerable water scarcity challenge, reflecting the global issue of declining water resources (Er-Raki et al., 2010). Moreover, crop production in Morocco heavily relies on irrigation due to the limited productivity of rainfed cropping (Taheripour et al., 2020). Precipitation decreases overall, leading to a reduction in soil water availability and resulting water stress can have various adverse effects on olive trees (Arampatzis et al., 2018). Although olive trees are considered resilient to drought conditions, adequate water availability is beneficial for olive cultivation and is considered a vital resource for improving final yields. Considering the pivotal importance of the olive sector within the Mediterranean region, there is an urgent need for sustainable water management practices. This includes adopting efficient irrigation systems and implementing water-saving strategies, such as regulated and sustained deficit irrigation (RDI and SDI), to sustain olive production and enhance the sector's resilience to climate change and water scarcity, particularly in areas facing drastic reductions in available water resources.

Nowadays, many studies have demonstrated that appropriate deficit irrigation strategies can maintain profitable olive productivity (Fernández et al., 2013; Gucci et al., 2019; Ibba et al., 2023). Deficit irrigation is a water management strategy that intentionally applies less water than the crop's full water requirement, aiming to optimize water use efficiency and enhance crop resilience. Deficit irrigation strategies are necessary for optimizing the application of irrigation concerning both oil quality and production (Fernández and Moreno, 1999). These strategies can assist farmers in determining the ideal timing and quantity of required irrigation water. Over recent years, there has been a progressive rise in research related to regulated deficit irrigation, contributing to its enrichment and extensive application across diverse crop species (Guo et al., 2021; Romero-Trigueros et al., 2020; Wu et al., 2020; Yang et al., 2022). This, innovative irrigation approach, known as RDI, has the capability to substantially enhance the adaptive capacity of olive trees in agriculture when facing drought stress and ultimately fulfill the aim of enhancing economic benefits for orchards (Gucci et al., 2019; Ibba et al., 2023; Lima et al., 2019). In this regard, the optimization of regulated and sustained deficit irrigation strategies involves exploring various treatments aimed at achieving an optimal balance between yield and water productivity. Several studies have reported that the RDI strategy involves inducing stress in the pit hardening and the initial stages of oil synthesis (Fereres et al., 2012; Goldhamer, 1999; Gucci et al., 2019). However, maintaining a full water supply during spring and oil synthesis is recommended (Hueso, Camacho, & Gómez-del-Campo, 2021; Moriana et al., 2003). Meanwhile, the SDI strategy entails supplying water throughout the irrigation season at a level below the crop's evapotranspiration (ETc) demand (Moriana et al., 2003). On the other hand, forthcoming climatic changes will have a significant impact on the entire agricultural sector, encompassing the olive tree industry as well (Fraga et al., 2020b; Ponti et al., 2014). Given the present worldwide changes, understanding how olive cultivars adapt and respond to drought stress is critical, specifically in the context of deficit irrigation strategies. According to Fraga et al. (2020b), irrigation has been identified as a potentially impactful strategy for olive tree cultivators to adapt to the challenges posed by climate change. Furthermore, Mairech et al. (2020) suggest that deficit irrigation is currently regarded as a sustainable management approach.

Identifying research gaps is crucial for guiding future research efforts. It can help in assessing the long-term sustainability of deficit irrigation strategies. This involves evaluating the agricultural system's resilience. Indeed, challenges related to deficit irrigation in perennial crops may occur from prolonged stress that extends beyond the present season, resulting in adverse effects in subsequent years (Gucci et al., 2019). In fact, adverse impacts of RDI on other fruit trees have been seen following a span of two to three consecutive seasons with deficit irrigation, with negative impacts on the yield of peach trees observed after three successive years of RDI treatment (Girona et al., 2003). Despite the fact that RDI has been utilized in a number of crops, previous investigation has focused on its effect, but the mechanism of its action remains uncompleted. The research would require a great deal of understanding of cell biology and molecular biology. The mechanism of RDI would become more obvious with thorough research of plant mechanisms. On the other hand, cost is another major issue for promotion. Drip irrigation and water measuring devices are essential since deficit irrigation strategies are dependent on precise water control. The construction and maintenance of this equipment require specific costs, posing challenges for widespread promotion on a large scale. As a consequence, specific RDI application techniques for certain crops are urgently needed, which may serve as a model for the design of RDI equipment and are useful for promotion as indicated by Yang et al. (2022).

The present chapter aims to propose insights into the impact of regulated and sustained deficit irrigation strategies on the locally grown olive cultivar *Menara* in Morocco over two consecutive years (2021 and 2022). It also offers an overview of the suitable adaptation of this variety to drought stress. Moreover, this chapter provides an applicable solution to address the effects of climate change on olive agroecosystems.

## MATERIAL AND METHODS

### Experimental Study Site

This study is based on experience conducted during 2021 and 2022 in a 12-year-old 'Menara' olive orchard, irrigated with a drip irrigation system, located at the Saada research station under the management of the National Institute of Agricultural Research (INRA) in Marrakech, Morocco, with a density of 156 trees/ha. The study area has a semi-arid Mediterranean climate, marked by hot and dry summers and limited rainfall. The cumulative reference evapotranspiration (ETo), was 1295.7 mm and 1316 mm in 2021 and 2022, respectively.

*Table 1. Summary of different regulated and sustained deficit irrigation treatments*

| Irrigation strategy | Treatment | Irrigation% of ETc by phase | |
|---|---|---|---|
| | | Sensitive Phase (SP) | Normal Phase (NP) |
| Control | T0 | 100 | 100 |
| RDI | T1 | 100 | 70 |
| | T2 | 100 | 60 |
| | T3 | 80 | 70 |
| | T4 | 80 | 60 |
| SDI | T5 | 70 | 70 |
| | T6 | 60 | 60 |

## Irrigation Strategies and Methodology

Irrigation treatments were studied according to the sensitivity of phenological periods to water stress, with (SP) 'Sensitive Phase' covering the flowering and oil synthesis phase and (NP) 'Normal Phase' covering the pit hardening phase. In addition to a control regime (FI) fully irrigated: T0 (100% $_{ETc}$), four water regimes were tested for RDI: T1 (SP 100- NP 70% $_{ETc}$), T2 (SP 100- NP 60% $_{ETc}$), T3 (SP 80- NP 70% $_{ETc}$), T4 (SP 80- NP 60% $_{ETc}$). The RDI strategy was designed using knowledge of the olive tree's sensitivity to water stress during various phenological periods. Phenological development was monitored annually from 2021 to 2022 in accordance with the methodology outlined by (Sanz-Cortés et al., 2002). For SDI, two water regimes were evaluated: T5 (70% $_{ETc}$) and T6 (60% $_{ETc}$).

Irrigation treatments were studied according to the sensitivity of phenological periods to water stress, with 'Sensitive Phase' (SP) covering the flowering and oil synthesis phase and 'Normal Phase' (NP) covering the pit hardening phase. The experiment consisted of testing seven deficit irrigation scenarios based on daily crop evapotranspiration (ETc).

In addition to a control regime fully irrigated (FI), T0 (100% $_{ETc}$), four water regimes were tested for RDI and two regimes were tested for SDI. The following table summarizes the seven studied regimes:

## Data Collection

Agronomical parameters were assessed in the seven studied treatments during the experiment, focusing on plants subjected to regulated and sustained deficit irrigation.

The daily irrigation water amounts were estimated by using the standard FAO56 formula for crop evapotranspiration ($ETc = ETo \cdot Kc \cdot Kr$), where the daily reference evapotranspiration ETo was estimated from measured in-situ meteorological data by using Penman-Monteith model (Allen et al., 1998, 2006). The crop coefficient Kc was estimated as 0.65 in April and October; 0.6 in May and June; 0.55 from July to September; and 0.7 in November and December as reported by Orgaz et al. (2007). The reduction coefficient, Kr, was determined using the formula $Kr = 2C/100$, with C representing the percentage of canopy cover. This calculation is applicable to canopies that extend to less than 50% of the ground, representing the degree of plant cover on the orchard floor.

The following parameters were carried out every year and tracked for each of the six selected trees per each treatment, and one-year-old shoots were identified and marked: length of shoots (SL, cm), the number of buds (BUN), fruit number per shoot (FNS), flowering rate (FR, %). The flowering rate (FR) was represented as the proportion of buds that progressed into inflorescences and can be derived from the measured parameters (Amar, 2017). Additionally, to determine fruit yield (FY, kg/tree), twelve trees were employed for each treatment. A set of 100 olives from six replications of each treatment was sampled to measure their fruit weight (FW, g) as detailed by Ibba et al. (2023).

## Data Analysis

To assess the responses of *Menara* cultivar to different deficit irrigation treatments, the principal component analysis method (PCA) was applied. Box plots were used to highlight agronomical trait variability between 2021 and 2022. Moreover, PCA was employed to determine the relationships between the studied agronomical parameters and the consecutive years of deficit irrigation among the irrigation treatments. This approach allowed to effectively demonstrate the variability of functional responses of *Menara* olive variety to deficit irrigation strategies across the different treatments and years. The R software (R Core Team, 2020) was used to conduct these analyses.

# RESULTS AND DISCUSSION

## Variablity of *Menara* Agronomic Parameters Under Deficit Irrigation Treatments

The principal components analysis (PCA) was executed to investigate the inter-relationships between all the studied agronomical traits. The results of this analysis are shown in Figure 1. The PCA axes account for 76.6% of the total variance (see Table 2). The first PCA axis alone represents 55.5% of the total variance and shows a positive correlation with all agronomical variables. This suggests a consistency in

*Figure 1. Plot illustrating the contribution of agronomic parameters studied to the PCA axes*

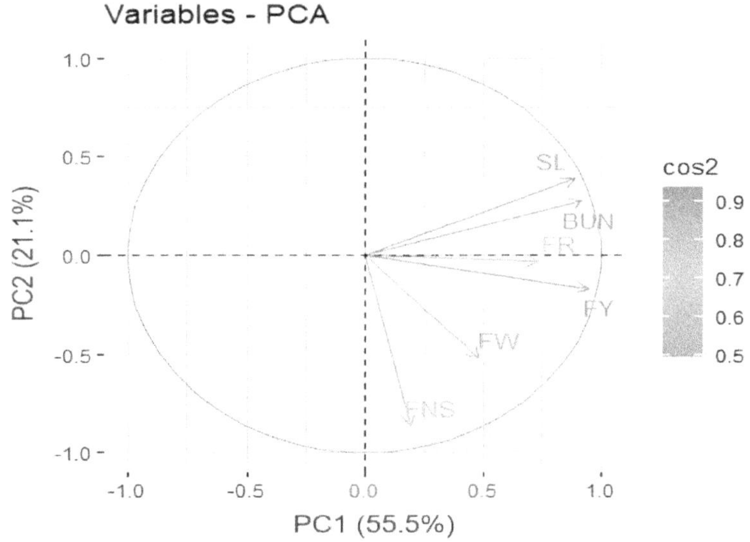

*Table 2. Correlations between agronomical traits and the two first PCA axes identified*

| Variables | PC1 | PC2 |
|---|---|---|
| SL | 0.89 | 0.39 |
| BUN | 0.91 | 0.27 |
| FR | 0.73 | -0.03 |
| FY | 0.95 | -0.03 |
| FW | 0.48 | -0.51 |
| FNS | 0.19 | -0.86 |
| Variance % | 55.5 | 21.1 |
| Cumulative variance % | 55.54 | 76.6 |

*Table 3. Correlation coefficients between the studied variables*

| Variables | SL | BUN | FR | FY | FW | FNS |
|---|---|---|---|---|---|---|
| SL | 1.00 | | | | | |
| BUN | 0.96** | 1.00 | | | | |
| FR | 0.52 | 0.54* | 1.00 | | | |
| FY | 0.75** | 0.82*** | 0.66* | 1.00 | | |
| FW | 0.26 | 0.24 | 0.19 | 0.51 | 1.00 | |
| FNS | -0.13 | 0.02 | 0.17 | 0.32 | 0.24 | 1.00 |

responses to deficit irrigation treatments, which affect all studied agronomical parameters. On the other hand, the second axis (accounting for 21.1% of the total variance) exhibits a positive correlation with shoot length (SL) and bud number (BUN), while demonstrating a negative correlation with flowering rate (FR), fruit yield (FY), fruit weight (FW), and fruit number per shoot (FNS) (Table 2 and Figure 1).

Tree-related variables (shoot length (SL), number of buds (BUN), and fruit yield (FY)) were found highly and positively correlated when all considered, as indicated in Table 3. These findings suggest the presence of relationships among the studied agronomical variables, particularly highlighting the significance of shoot length (SL) and bud number (BUN) as indicators of olive tree growth and development. SL represents the tree's vigor and is involved in photosynthesis, while bud number reflects its potential for flowering and fruit production (Dag et al., 2010). Interestingly, shoot growth may also impact flowering and production, as previously reported by Hueso et al. (2021).

Abbreviation of the studied variables are given in Materials and Methods section.

Within the table, asterisks represent statistical significance levels, with *** $p < 0.001$, ** $p < 0.01$ and * $p < 0.1$. Abbreviation of the studied variables are given in Materials and Methods section.

## Adaptive Responses of Different Irrigation Strategies Over the Two Consecutive Years 2021 and 2022

Figure 2 illustrates the projection of various deficit irrigation strategies for the two consecutive years, 2021 (A) and 2022 (B), on the plane determined by the first two principal components. In 2021, noticeable differences were observed between two groups. The first group consisted solely of the control (T0A), while the second group included all RDI treatments (TA1, T2A, T3A, and T4A) and SDI treatments (T5A and T6A). These differences were clearly manifested by two distinct response patterns to water stress when compared to the control. During the initial year, both the RDI and SDI strategies exhibited generally similar response patterns. However, in 2022, the response of the different irrigation treatments displayed notable variations among the strategies, leading to the identification of three groups of responses. The first group consisted of the control (T0B), the second group included the RDI treatments (T1B, T2B, T3B, and T4B), and the last group comprised the SDI treatments (T5B and T6B). These findings highlight the adaptive nature of the responses to water stress over time and demonstrate the impact of each irrigation strategy on the performance of Menara olive trees in consecutive years.

Observed differences between the studied irrigation treatments over these two consecutive growing seasons could be mainly ascribed to the cumulative impact of deficit irrigation strategies. Over time, Menara olive trees subjected to deficit irrigation, specifically the RDI treatments, demonstrate their ability to adapt and cope with limited water availability. In contrast, the SDI treatments (T5 and T6) had a cumulative water deficit effect, leading to a decline in the agronomic performance of Menara olive trees as reported by Ibba et al. (2023). As the water deficit accumulates over two years, its impact becomes more pronounced, resulting in varying responses between irrigation treatments. The influence of drought on plants can be mainly characterized by cumulative responses over time (Zhao et al., 2020). Additionally, Ivits et al. (2016) also reported that the intensification of drought effects on terrestrial ecosystems can result from cumulative water deficits. These results imply that trees exposed to the RDI treatments exhibited higher adaptation to water stress compared to those of SDI treatments, revealing that water deprivation during critical growth periods resulted in reduced agronomic parameters, as observed in the SDI strategy. Girona et al. (2003), on the other hand, showed adverse carryover effects on peach tree yield after applying the RDI treatment for three consecutive years. Consequently, the challenges associated

*Figure 2. Graphical representation of the PCA showing deficit irrigation treatments over two consecutive years 2021-2022. Black and orange colors stand for years 2021 and 2022, respectively.*

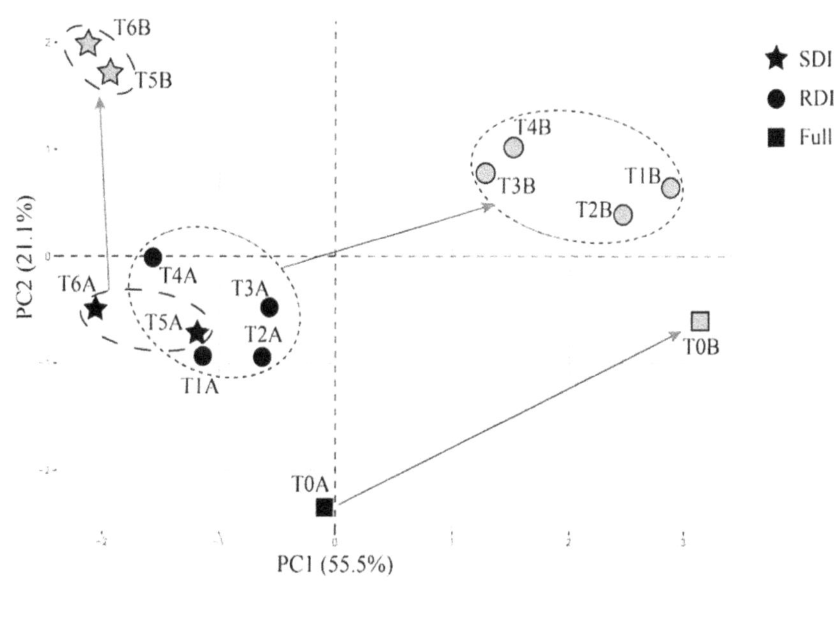

with deficit irrigation in perennial crops arise from enduring stress effects that persist beyond the present growing season and may lead to detrimental consequences in the subsequent years (Gucci et al., 2019).

## Agronomic Traits Responses to Deficit Irrigation Strategies Across 2021 and 2022

The box plots in the figures below display the distribution of the studied agronomic traits under various irrigation treatments for the years 2021 and 2022. In the first year (2021), no significant differences were noted in shoot length (SL) and bud number (BUN) among the deficit irrigation treatments. However, in 2022, a decline in SL and BUN was specifically observed under the SDI strategy compared to the control and RDI strategy, as shown by the treatments T5 and T6 (see Figures 3 and 4). Notably, there exists a strong correlation between these two parameters. Generally, the application of water deficit through the RDI strategy did not exhibit a negative impact on SL and BUN, providing evidence that Menara olive trees subjected to RDI treatments demonstrated important adaptation to water deficit conditions.

Additionally, the analysis of flowering rate (FR) and fruit number per shoot (FNS) further supports these findings, generally confirming that the results of the RDI treatments were significantly higher in comparison to the SDI treatments, as shown in Figures 5 and 6. Water shortage during sensitive periods, according to Rapoport et al. (2012), resulted in a decrease in several flowering traits. The RDI strategy, particularly under the T4 treatment, effectively reduced supplied water by 20% during sensitive periods (SP) and 40% during normal periods (NP). Remarkably, this treatment had minimal impact on the studied agronomic parameters of the Menara olive variety, highlighting its promising potential for sustainable

*Figure 3. Box plot analysis of shoot length (SL, cm) distribution under different irrigation treatments over the years 2021 and 2022; the top and bottom sides of each box represent the quartiles, and the band inside the box represents the median*

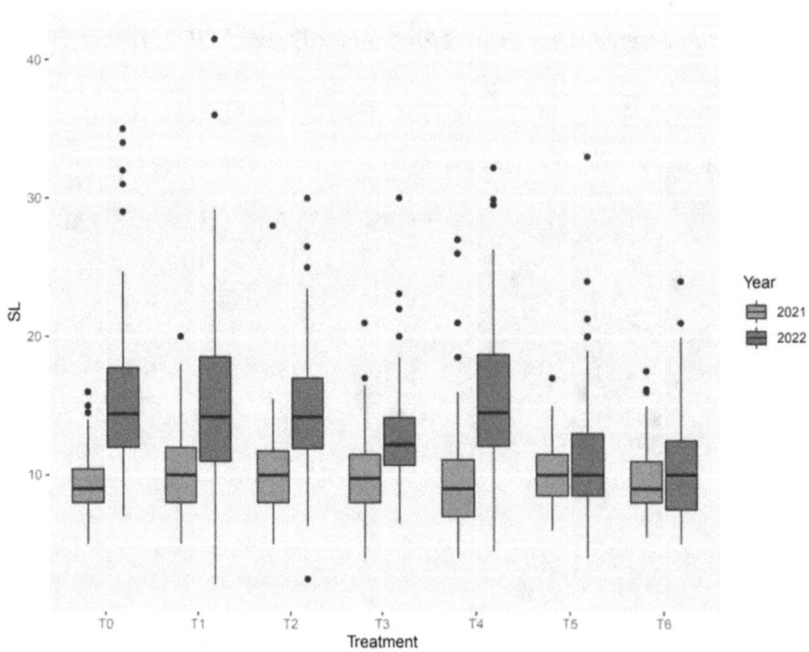

*Figure 4. Same as figure 3, but for the bud number (BUN)*

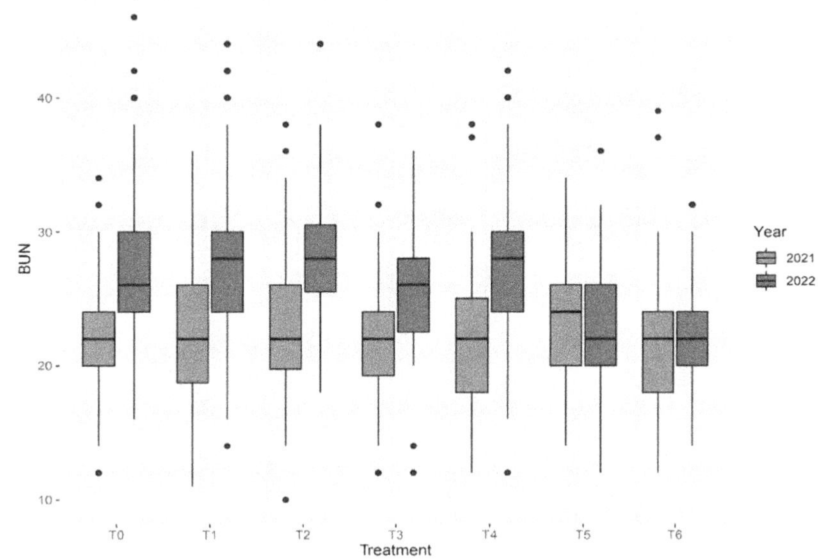


water management in olive orchards. Thus, during this phase when pit hardening occurs, it is feasible to conserve a substantial quantity of irrigation water without negatively impacting production, as demonstrated by Caruso et al. (2014) and Dabbou et al. (2015).

Comparing the RDI and SDI treatments with the fully irrigated control group gives rise to an interesting observation. In 2021, both the RDI and SDI treatments resulted in lower yields (FY). However, in 2022, the trends diverged significantly: fruit yield in treatments T1 and T2 showed substantial increases,

*Figure 5. Same as figure 3 but for the flowering rate (FR,%)*

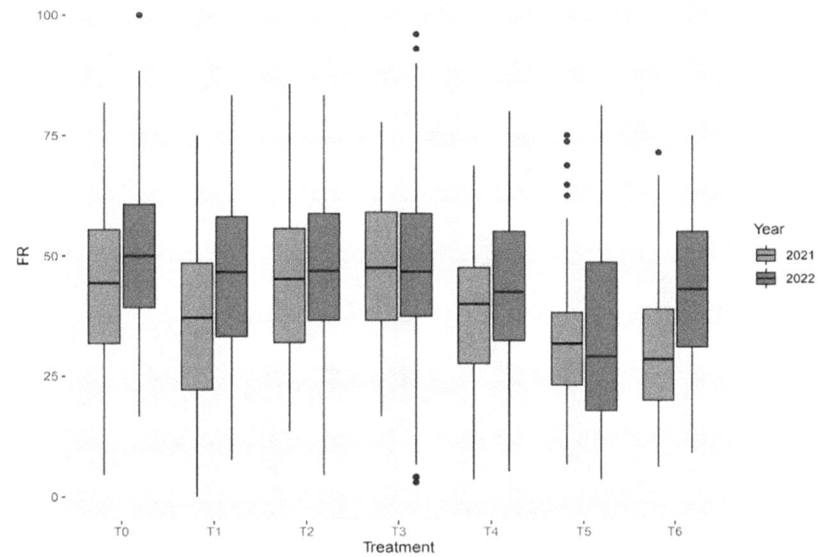

*Figure 6. Same as figure 3 but for the fruit number per shoot (FNS)*

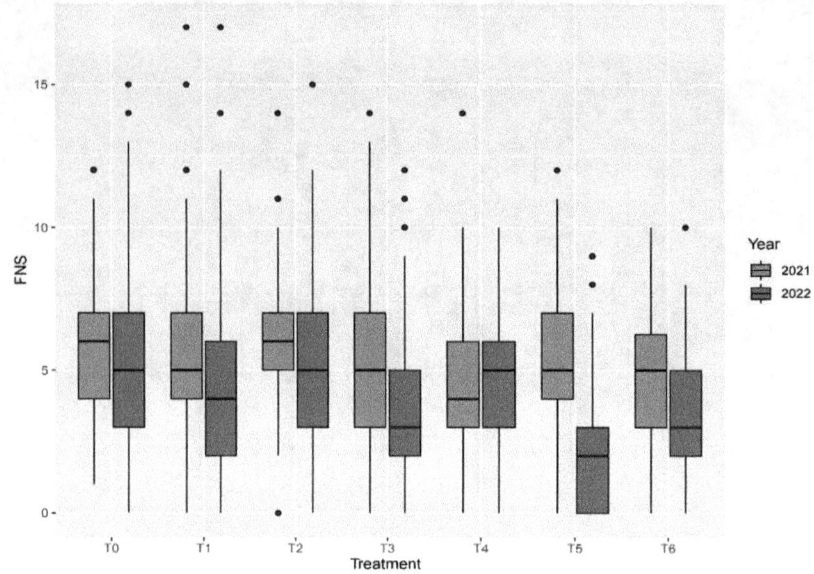


while treatments T5 and T6 showed notable decreases (see Figure 7). Pierantozzi et al. (2020) reported that the RDI treatments applied from fruit set to pit hardening exhibited no significant difference in fruit yield at the end of the third year. The significant water-saving strategy SDI (T6), reducing irrigation by 40% throughout the entire growing season, had a noticeable negative impact on fruit weight in 2022, as shown in Figure 8. The substantial decrease in water availability appeared to influence fruit development, resulting in reduced fruit weight. These findings emphasize the critical role of water management

*Figure 7. Same as figure 3 but for the fruit yield (FY, kg)*

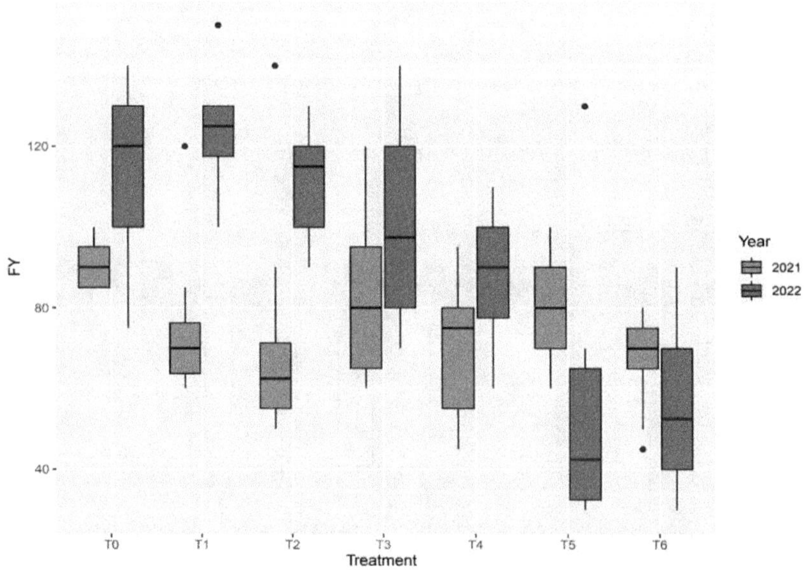

*Figure 8. Same as figure 3 but for the fruit weight (FW, g)*

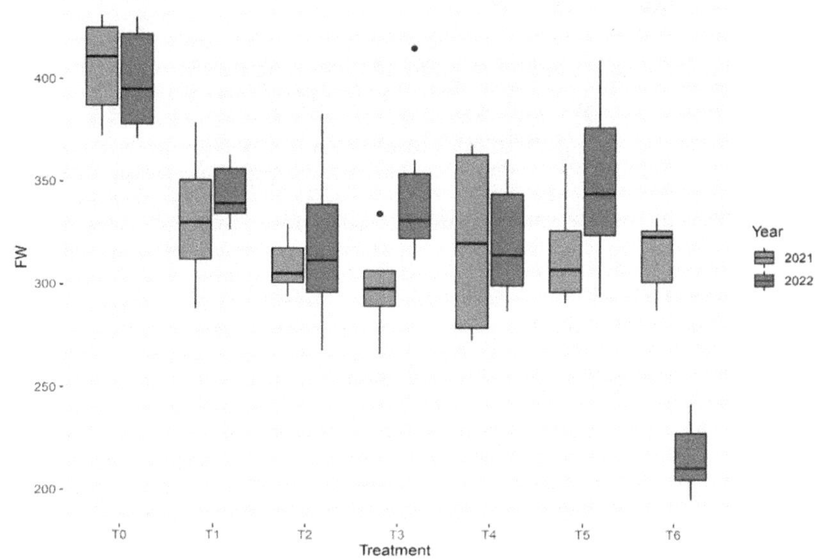


strategies and the need to consider their implications for crop yield and fruit quality over successive growing seasons.

## CONCLUSION

Faced with the grave concerns of climate change and water scarcity, the pursuit of sustainable water management practices in arid regions, notably in Morocco, has become an urgent priority. The need for efficient and sustainable water management cannot be overstated. In our study, we undertook a comprehensive examination of the efficacy of two distinct irrigation strategies: Regulated Deficit Irrigation (RDI) and Sustained Deficit Irrigation (SDI). These strategies primarily differ in terms of their irrigation timing and the quantity of water applied.

Our research endeavors revealed compelling evidence that the adoption of the RDI strategy can yield substantial water savings while maintaining or even enhancing agricultural productivity. Specifically, the key to success appears to lie in reducing plant water supply by 20% during sensitive growth periods and by 40% during normal growth periods. These findings hold significant promise for sustainable agriculture and water conservation practices, particularly in water-scarce regions like Morocco. The implications of our findings extend beyond immediate water conservation benefits. By implementing the RDI approach, agricultural practices can enhance their resilience in response to changes in water availability induced by climate change. This strategic approach not only conserves water but also contributes to the overall sustainability of agricultural systems, ensuring they remain productive and economically viable in a changing climate.

However, while our research provides valuable insights, it is important to acknowledge that it represents a preliminary step in a complex journey. To comprehensively understand how different olive cultivars respond to various irrigation regimes and to establish the long-term adaptability of plants, we advocate for further investigation. The principal constraints of this study relate to the relatively short time of the experiment observation, which could restrict a comprehensive knowledge of the plant's long-term adaptation and production beyond the initial two-year period. Extending this research entails prolonging the duration of the investigation to observe how the plant adapts to the irrigation regimen, enabling a more holistic exploration of its long-term performance. Extended observation periods and studies involving a broader range of olive cultivars are recommended. This holistic approach will help refine and validate the practical implications of the RDI strategy, ensuring that sustainable water management practices can be tailored to the specific needs and conditions of different agricultural contexts. Our research underscores the urgency of adopting sustainable water management practices in arid regions like Morocco. The RDI strategy, with its demonstrated potential for water savings and continued productivity, emerges as a promising avenue for achieving this goal. Yet, the journey towards sustainable agriculture is ongoing, and further research and adaptation are essential to navigate the challenges posed by climate change and water scarcity successfully.

## REFERENCES

Allen, R. G., Pereira, L. S., Raes, D., & Smith, M. (1998). FAO Penman–Monteith equation. *Crop Evapotranspiration: Guidelines for Computing Crop Water Requirements. FAO Irrigation and Drainage Paper, 56*, 17–28.

Allen, R. G., Pruitt, W. O., Wright, J. L., Howell, T. A., Ventura, F., Snyder, R., Itenfisu, D., Steduto, P., Berengena, J., Yrisarry, J. B., Smith, M., Pereira, L. S., Raes, D., Perrier, A., Alves, I., Walter, I., & Elliott, R. (2006). A recommendation on standardized surface resistance for hourly calculation of reference ETo by the FAO56 Penman-Monteith method. *Agricultural Water Management, 81*(1–2), 1–22. doi:10.1016/j.agwat.2005.03.007

Amar, F. B. (2017). Etude comparative de la biologie florale des variétés d ' olive de table (Olea europaea L. ) Meski et Picholine Languedoc dans la région de Sfax (Tunisie) Floral biology comparative study of Meski and Picholine languedoc table olive varieties. Bulletin de l'Institut Scientifique, Rabat. Section Sciences de La Vie, 55–61. doi:10.2166/wcc.2018.224

Arampatzis, G., Hatzigiannakis, E., Pisinaras, V., Kourgialas, N., Psarras, G., Kinigopoulou, V., Panagopoulos, A., & Koubouris, G. (2018). Soil water content and olive tree yield responses to soil management, irrigation, and precipitation in a hilly Mediterranean area. Journal of Water and Climate Change, 9(4), 672–678. https://doi.org/10.2166/wcc.2018.224

Caruso, G., Gucci, R., Urbani, S., Esposto, S., Taticchi, A., Di Maio, I., Selvaggini, R., & Servili, M. (2014). Effect of different irrigation volumes during fruit development on quality of virgin olive oil of cv. Frantoio. *Agricultural Water Management, 134*, 94–103. doi:10.1016/j.agwat.2013.12.003

Dabbou, S., Dabbou, S., Chehab, H., Taticchi, A., Servili, M., & Hammami, M. (2015). Content of Fatty Acids and Phenolics in *Coratina* Olive Oil from Tunisia: Influence of Irrigation and Ripening. *Chemistry & Biodiversity, 12*(3), 397–406. doi:10.1002/cbdv.201400142 PMID:25766913

Dag, A., Bustan, A., Avni, A., Tzipori, I., Lavee, S., & Riov, J. (2010). Timing of fruit removal affects concurrent vegetative growth and subsequent return bloom and yield in olive (Olea europaea L.). *Scientia Horticulturae*, *123*(4), 469–472. doi:10.1016/j.scienta.2009.11.014

Er-Raki, S., Chehbouni, A., Khabba, S., Simonneaux, V., Jarlan, L., Ouldbba, A., Rodriguez, J. C., & Allen, R. (2010). Assessment of reference evapotranspiration methods in semi-arid regions: Can weather forecast data be used as alternate of ground meteorological parameters? *Journal of Arid Environments*, *74*(12), 1587–1596. doi:10.1016/j.jaridenv.2010.07.002

Fereres, E., Goldhamer, D., Sadras, V., & ... (2012). Yield response to water of fruit trees and vines: Guidelines. *FAO Irrigation and Drainage Paper*, *66*, 246–497.

Fernández, J. E., Perez-Martin, A., Torres-Ruiz, J. M., Cuevas, M. V., Rodriguez-Dominguez, C. M., Elsayed-Farag, S., Morales-Sillero, A., García, J. M., Hernandez-Santana, V., & Diaz-Espejo, A. (2013). A regulated deficit irrigation strategy for hedgerow olive orchards with high plant density. *Plant and Soil*, *372*(1–2), 279–295. doi:10.100711104-013-1704-2

Fraga, H., Moriondo, M., Leolini, L., & Santos, J. A. (2020a). Mediterranean Olive Orchards under Climate Change: A Review of Future Impacts and Adaptation Strategies. *Agronomy (Basel)*, *11*(1), 56. doi:10.3390/agronomy11010056

Fraga, H., Pinto, J. G., & Santos, J. A. (2020b). Olive tree irrigation as a climate change adaptation measure in Alentejo, Portugal. *Agricultural Water Management, 237*, 106193. https://doi.org/ doi:10.1016/j.agwat.2020.106193Girona

Goldhamer, D. A. (1999). Regulated deficit irrigation for canning olives. *Acta Horticulturae*, (474), 369–372. doi:10.17660/ActaHortic.1999.474.76

Gucci, R., Caruso, G., Gennai, C., Esposto, S., Urbani, S., & Servili, M. (2019). Fruit growth, yield and oil quality changes induced by deficit irrigation at different stages of olive fruit development. *Agricultural Water Management*, *212*, 88–98. doi:10.1016/j.agwat.2018.08.022

Guo, S.-H., Yang, B.-H., Wang, X.-W., Li, J.-N., Li, S., Yang, X., Ren, R.-H., Fang, Y.-L., Xu, T.-F., Zhang, Z.-W., & Meng, J.-F. (2021). ABA signaling plays a key role in regulated deficit irrigation-driven anthocyanins accumulation in 'Cabernet Sauvignon' grape berries. *Environmental and Experimental Botany*, *181*, 104290. doi:10.1016/j.envexpbot.2020.104290

Hueso, A., Camacho, G., & Gómez-del-Campo, M. (2021). Spring deficit irrigation promotes significant reduction on vegetative growth, flowering, fruit growth and production in hedgerow olive orchards (cv. Arbequina). *Agricultural Water Management*, *248*, 106695. doi:10.1016/j.agwat.2020.106695

Ibba, K., Kassout, J., Boselli, V., Er-Raki, S., Oulbi, S., Mansouri, L. E., Bouizgaren, A., Sikaoui, L., & Hadria, R. (2023). Assessing the impact of deficit irrigation strategies on agronomic and productive parameters of Menara olive cultivar: Implications for operational water management. *Frontiers in Environmental Science*, *11*, 1100552. doi:10.3389/fenvs.2023.1100552

Ivits, E., Horion, S., Erhard, M., & Fensholt, R. (2016). Assessing European ecosystem stability to drought in the vegetation growing season: Ecosystem stability to drought. *Global Ecology and Biogeography*, *25*(9), 1131–1143. doi:10.1111/geb.12472

Kassout, J., Ater, M., Ivorra, S., Barbara, H., Limier, B., Ros, J., Girard, V., Paradis, L., & Terral, J.-F. (2021). Resisting aridification: Adaptation of sap conduction performance in Moroccan wild olive subspecies distributed over an aridity gradient. *Frontiers in Plant Science*, *12*, 1288. doi:10.3389/fpls.2021.663721 PMID:34276726

Lima, F. A., Córcoles, J. I., Tarjuelo, J. M., & Martínez-Romero, A. (2019). Model for management of an on-demand irrigation network based on irrigation scheduling of crops to minimize energy use (Part II): Financial impact of regulated deficit irrigation. *Agricultural Water Management*, *215*, 44–54. doi:10.1016/j.agwat.2019.01.006

Mairech, H., López-Bernal, Á., Moriondo, M., Dibari, C., Regni, L., Proietti, P., Villalobos, F. J., & Testi, L. (2020). Is new olive farming sustainable? A spatial comparison of productive and environmental performances between traditional and new olive orchards with the model OliveCan. *Agricultural Systems*, *181*, 102816. doi:10.1016/j.agsy.2020.102816

Moriana, A., Orgaz, F., Pastor, M., & Fereres, E. (2003). Yield responses of a mature olive orchard to water deficits. *Journal of the American Society for Horticultural Science*, *128*(3), 425–431. doi:10.21273/JASHS.128.3.0425

Orgaz, F., Villalobos, F. J., Testi, L., & Fereres, E. (2007). A model of daily mean canopy conductance for calculating transpiration of olive canopies. *Functional Plant Biology*, *34*(3), 178. doi:10.1071/FP06306 PMID:32689344

Pierantozzi, P., Torres, M., Tivani, M., Contreras, C., Gentili, L., Parera, C., & Maestri, D. (2020). Spring deficit irrigation in olive (cv. Genovesa) growing under arid continental climate: Effects on vegetative growth and productive parameters. *Agricultural Water Management*, *238*, 106212. doi:10.1016/j.agwat.2020.106212

Ponti, L., Gutierrez, A. P., Ruti, P. M., & Dell'Aquila, A. (2014). Fine-scale ecological and economic assessment of climate change on olive in the Mediterranean Basin reveals winners and losers. *Proceedings of the National Academy of Sciences of the United States of America*, *111*(15), 5598–5603. doi:10.1073/pnas.1314437111 PMID:24706833

Rapoport, H. F., Hammami, S. B. M., Martins, P., Pérez-Priego, O., & Orgaz, F. (2012). Influence of water deficits at different times during olive tree inflorescence and flower development. *Environmental and Experimental Botany*, *77*, 227–233. doi:10.1016/j.envexpbot.2011.11.021

Romero-Trigueros, C., Cabañero, J. J., Tortosa, P. A., Gambín, J. M., Maestre-Valero, J. F., & Nicolás, E. N. (2020). Medium-long term effects of saline reclaimed water and regulated deficit irrigation on fruit quality of citrus. *Journal of the Science of Food and Agriculture*, *100*(3), 1350–1357. doi:10.1002/jsfa.10091 PMID:31617215

Sanz-Cortés, F., Martinez-Calvo, J., Badenes, M. L., Bleiholder, H., Hack, H., Llácer, G., & Meier, U. (2002). Phenological growth stages of olive trees (Olea europaea). *Annals of Applied Biology*, *140*(2), 151–157. doi:10.1111/j.1744-7348.2002.tb00167.x

Taheripour, F., Tyner, W. E., Haqiqi, I., & Sajedinia, E. (2020). *Water Scarcity in Morocco*. World Bank., doi:10.1596/33306

Wu, Y., Zhao, Z., Liu, S., Huang, X., & Wang, W. (2020). Does partial root-zone drying have advantages over regulated deficit irrigation in pear orchard under desert climates? *Scientia Horticulturae*, *262*, 109099. doi:10.1016/j.scienta.2019.109099

Yang, B., Fu, P., Lu, J., Ma, F., Sun, X., & Fang, Y. (2022). Regulated deficit irrigation: An effective way to solve the shortage of agricultural water for horticulture. *Stress Biology*, *2*(1), 28. doi:10.100744154-022-00050-5 PMID:37676363

Zhao, A., Yu, Q., Feng, L., Zhang, A., & Pei, T. (2020). Evaluating the cumulative and time-lag effects of drought on grassland vegetation: A case study in the Chinese Loess Plateau. *Journal of Environmental Management*, *261*, 110214. doi:10.1016/j.jenvman.2020.110214 PMID:32148284

# Chapter 12
# Eutrophication: A Bionomical Vision

**Ann Theresa Jose**
https://orcid.org/0009-0002-2590-1430
*Indian Agricultural Research Institute, India*

**S. Syam**
*Indian Agricultural Research Institute, India*

**A. Deepasree**
*N.M. College of Agriculture, India*

**Abshiba Jithu**
*Indian Agricultural Research Institute, India*

**A. Jayakishore Reddy**
*Polytechnic of Agriculture, India*

**Mayurakshi Chanda**
*Indian Agricultural Research Institute, India*

## ABSTRACT

*Major advances in the scientific understanding of eutrophication and the significant effects of eutrophication have been made over the decades. The effects of eutrophication include increased biomass of freshwater phytoplankton and periphyton; shifts in phytoplankton species composition to taxa that may be toxic or inedible, e. g. bloom-forming cyanobacteria; reduced biodiversity of waterbodies; changes in vascular plant production, biomass, and species composition; reduced water clarity; and decreases in the perceived aesthetic value of the water. Relative abundances of major bacterial and archaeal groups indicate high similarities at the phylum- and class-level groups across eutrophicated lakes. The impact of climate change and greenhouse gases emissions regards to affect and effect of eutrophication leads to various studies. The effect of reducing nitrogen inputs alone for controlling eutrophication by controlling inputs of phosphorus.*

DOI: 10.4018/978-1-6684-9838-5.ch012

## INTRODUCTION

Eutrophication is a process that occurs when water bodies become overly enriched with nutrients, particularly nitrogen and phosphorus. These nutrients promote the growth of algae and other aquatic plants, leading to an excessive proliferation known as an algal bloom. The excessive growth of these organisms can have detrimental effects on the ecosystem and the water quality of the affected area. Due to ever increasing human interventions of natural as well as managed ecosystems, eutrophication has become an urgent environmental issue worldwide. The Organization for Economic Cooperation and Development (OECD) (2016) defined eutrophication as 'the enrichment of water with nutrient salts that causes structural changes to the ecosystem, namely, the increase in production of microalgae and aquatic plants, depletion of fish species, general deterioration of water quality and other effects that reduce and preclude use'. The potential sources triggering eutrophication in lakes are complicated and involve various types of factors, which can be classified into natural-driven factors and human intervention factors. Hence, eutrophication can be broadly divided into natural eutrophication and cultural eutrophication (Lin *et al.* 2020). The main difference is that natural eutrophication is a natural process taking place in geological time scale while cultural eutrophication is the process that speeds up natural eutrophication due to human activities. It is a rapid process taking only decades. There can be point and non-point sources of eutrophication. In point, sources nutrient waste travels directly from the source to water. They are relatively easy to regulate. However, in non-point sources, they come from diffused sources and are difficult to regulate. The contribution of non-point sources to eutrophication is much more than point sources (Carpenter *et al.* 1998).

Eutrophication typically occurs as a result of human activities, such as agricultural runoff, sewage discharge, and the use of fertilizers. When these nutrients enter water bodies, they stimulate the rapid growth of algae and other plants. As the algae bloom, they block sunlight from reaching the deeper parts of the water, hindering the growth of underwater plants and disrupting the balance of the ecosystem. Once the algae die and decompose, bacteria consume the organic matter, consuming oxygen in the process. This leads to a depletion of oxygen levels in the water, creating hypoxic or anoxic conditions, which are harmful to many aquatic organisms. The reduced oxygen levels can result in fish kills and the decline of other marine life. Additionally, eutrophication can lead to the formation of harmful algal blooms (HABs), which produce toxins that can be harmful to humans and other animals, imposing a threat to public health (Rathore *et al.* 2016). To combat eutrophication, it is important to reduce nutrient inputs into water bodies. This can be achieved through improved agricultural practices, such as minimizing the use of fertilizers and implementing buffer zones to trap runoff. Wastewater treatment plants can also play a significant role in removing nutrients from sewage before it is discharged into water bodies. Overall, addressing eutrophication requires a combination of measures aimed at reducing nutrient pollution, promoting sustainable land use practices, and raising awareness about the importance of protecting water quality.Top of Form So this chapter will focus on eutrophicaton, its global status, causes, impact and the various management options to combat it.

## EFFECTS OF EUTROPHICATION

Understanding eutrophication is crucial as it poses significant threats to freshwater, marine, and estuarine ecosystems worldwide. By studying its causes, effects, and management strategies, we can develop effective solutions to mitigate its adverse impacts and promote the restoration of impacted water bodies. The effects of eutrophication include increased biomass of freshwater phytoplankton and peri-phyton, shifts in phytoplankton species composition to taxa that may be toxic or inedible (e.g. bloom-forming cyanobacteria), reduced biodiversity of waterbodies, changes in vascular plant production, biomass, and species composition, reduced water clarity, decreases in the perceived aesthetic value of the water body, taste, odor, and water supply filtration problems, possible health risks in water supplies, elevated pH and dissolved oxygen depletion in the water column, shifts in fish species composition towards less desirable species, increased probability of fish kills and bottom water hypoxia (Rathore *et al.* 2016). Eutrophication decreases the diversity and richness of aquatic macro-invertebrates, fish, and aquatic primary producers. The trophic structure of the environment is directly impacted by the shift in biodiversity. Under the coastal regions that receive direct nutrient input from several of the major rivers, the diversity of the flora and fauna is under threat (Petersen *et al.* 2015). Lesser biodiversity results in lesser competition and hence prolific growth of some undesirable groups. The ecological balance of the entire food chain is thus affected.

Therefore, the several significant effects of eutrophication can be explained under the following heads:

**Algal Blooms:** Eutrophication promotes the excessive growth of algae and other aquatic plants, leading to the formation of algal blooms. These blooms can cover the water surface, reducing sunlight penetration and hindering the growth of submerged plants. The dominance of certain algal species can disrupt the balance of the ecosystem and negatively impact other organisms.

**Oxygen Depletion**: As the algae and other plants die and decompose, bacteria break down the organic matter, consuming oxygen in the process. This results in the depletion of dissolved oxygen levels in the water, leading to hypoxic or anoxic conditions. Oxygen depletion can suffocate fish and other aquatic organisms, causing fish kills and the decline of species that rely on well-oxygenated water.

**Loss of Biodiversity**: Eutrophication can lead to a decline in biodiversity. The excessive growth of algae and the subsequent oxygen depletion can harm or kill many aquatic species, including fish, insects, and plants. Additionally, some species may thrive in eutrophic conditions, outcompeting native species and reducing overall biodiversity.

**Harmful Algal Blooms (HABs):** Eutrophication can result in the formation of harmful algal blooms (HABs). Certain species of algae produce toxins that can be harmful to humans and other animals. These toxins can accumulate in shellfish and fish, posing risks to human health if consumed. HABs can also create "dead zones" in water bodies, where oxygen levels are extremely low or absent, making them unsuitable for most forms of life.

**Changes in Water Clarity and Appearance:** Algal blooms caused by eutrophication can make the water appear murky, green, or discolored. The increased turbidity and reduced water clarity can have aesthetic impacts, making recreational activities such as swimming, boating, and fishing less enjoyable.

**Disruption of Food Chains**: Eutrophication can alter the food chains and trophic dynamics in aquatic ecosystems. Changes in nutrient availability and the dominance of certain algal species can impact the availability of food for higher trophic levels, affecting fish, birds, and other organisms that rely on specific prey.

**Economic Impact**: Eutrophication can have economic consequences. The decline or loss of fish populations can affect commercial and recreational fishing industries. HABs can lead to the closure of shellfish beds and fisheries, causing financial losses for fishermen, aquaculture operations, and coastal communities that depend on these resources.

## GLOBAL STATUS OF EUTROPHICATION LEVELS

Eutrophication is a widespread environmental issue affecting various water bodies around the world, including lakes, rivers, estuaries, and coastal areas. The severity of eutrophication can vary significantly from one region to another, depending on factors such as population density, agricultural practices, industrial activities, and wastewater treatment.

According to statistics from the Water Research Commission 54%, 53%, 48%, 41% and 28% of lakes in Asia, Europe, North America, South America and Africa face eutrophication problems, respectively (Nyenje et al., 2010). The number of hypoxic zones in coastal waters has increased dramatically over the past 50 years, rising from about 10 cases in 1960 to over 500 in 2010 covering over 2,40,000 $Km^2$ around the globe (Rabalais *et al.* 2010).

## MICROBIAL COMMUNITIES IN EUTROPHICATION LAKES

Relative abundances of major bacterial and archaeal groups indicate high similarities at the phylum- and class-level groups across eutrophicated lakes. There is a decrease in β- and γ-Proteobacteria, Bacteroidetes, Verrucomicrobia, Cyanobacteria, and Acidobacteria with sediment depth, but a rise in -Proteobacteria, Chloroflexi, Firmicutes, Planctomycetes, Actinobacteria, Armatimonadetes, Acetothermia, and Aminicenantes. Archaea, Pacearchaeota, and Woesearchaeota often decrease with sediment depth, whereas Thermoplasmata, Altiarchaeales, Diapherotrites, and Lokiarchaeota generally rise. Co-occurrence patterns show that the stated Bacteria and Archaea constitute two big, separated surface and subsurface sedimentary clusters. Marine Group I (MGI), Nitrospirae, and Gemmatimonadetes, as well as the bathyarchaeotal MCG-6 and Group C3 subgroups, form smaller clusters. The clustering that was observed at the phylum- and class-level is similar, but less clear, when dominant orders are incorporated into the analysis.

That relative abundances of the orders Nitrospirales (Nitrospirae), Sva0485, Syntrophobacterales, 43F-1404R, NB1-j (all δ-Proteobacteria), vadinHA17 (Bacteroidetes), and Xanthomonadales (γ-Proteobacteria) differ significantly between the three eutrophic lakes and oligotrophic lakes.

## IMPACT OF CLIMATE CHANGE ON EUTROPHICATION

Climate change can have several effects on eutrophication, both directly and indirectly. Here are some ways in which climate change can affect eutrophication:

**Precipitation Patterns**: Climate change can alter precipitation patterns, leading to changes in the distribution and intensity of rainfall. Increased rainfall can result in more runoff from land, carrying excess nutrients from agricultural fields and urban areas into water bodies. This increased nutrient load contributes to eutrophication.

**Temperature Changes**: Rising temperatures associated with climate change can accelerate the rate of nutrient cycling in ecosystems. Warmer temperatures can enhance the growth of algae and bacteria, increasing their metabolic rates and nutrient uptake. This can intensify the eutrophication process and lead to larger algal blooms.

**Changes in Water Circulation**: Climate change can affect water circulation patterns in lakes, rivers, and oceans. Changes in wind patterns and water currents can impact nutrient transport and mixing within water bodies. This can influence the spatial distribution and severity of eutrophication, as nutrients may accumulate in certain areas or be transported to new regions.

**Increased Frequency of Extreme Events**: Climate change is associated with an increased frequency and intensity of extreme weather events, such as heavy rainfall and storms. These events can cause sudden nutrient influxes into water bodies, as runoff carries nutrients from fertilized agricultural fields or urban areas. The sudden nutrient loading can trigger intense algal blooms and worsen eutrophication.

**Sea-Level Rise**: Rising sea levels, a consequence of climate change, can exacerbate eutrophication in coastal areas. Higher sea levels can cause saltwater intrusion into freshwater sources, leading to changes in salinity levels. This can affect the composition and distribution of aquatic species, disrupt nutrient cycling, and contribute to eutrophication dynamics.

It is important to note that the interactions between climate change and eutrophication are complex and can vary depending on regional and local factors. The specific impacts of climate change on eutrophication may differ across different ecosystems and geographical locations. Understanding these interactions is crucial for developing effective strategies to mitigate and adapt to the combined effects of climate change and eutrophication.

## IMPACT OF NUTRIENT LOAD IN EUTROPHICATION

In the study conducted by Padedda et al., (2022) on an ancient Mediterranean natural lake, high nutrient input in combination with a reduced water supply results in a high nutrient concentration, which severely limits the quality of the system. It is obvious that the abundance of nutrients played a major role in the development of eutrophic processes in the lake. It was substantially derived from the high loads. Phosphorus is considered as key nutrient in controlling primary production. The regulation of P export by the sources is most crucial for mitigating eutrophication. The initial priority of mitigating measures should be to restrict exports from the agriculture and animal husbandry sectors.

## CONTRIBUTION OF FRESHWATER EUTROPHICATION TO GREENHOUSE GAS EMISSIONS

Recently, low oxygen content, abundant organic matter and nutrients, dominant primary producers and HABs in eutrophic freshwaters have been found to favor GHG production and emissions. Increased freshwater eutrophication projected in the future is likely to significantly promote aquatic GHG fluxes and strengthen the greenhouse effect; therefore, resulting in serious environmental problems and socio-economic loss. Management organisations must take action in order to prevent freshwater eutrophication, which will enhance water quality and lower GHG emissions. Attempts to reduce nutrient inputs, balance nutrient stoichiometry, and maintain a clear, macrophyte dominated water regime are indispensable. Additionally, in order for management organisations to put into place efficient mitigation measures and safeguard freshwater ecosystems and the Earth's climate, a greater and more in-depth understanding of the future dynamics of freshwater eutrophication and GHG emissions is required. This should begin by quantifying the effects of key influencing factors or indicators, which is a major undertaking and

*Figure 1.*

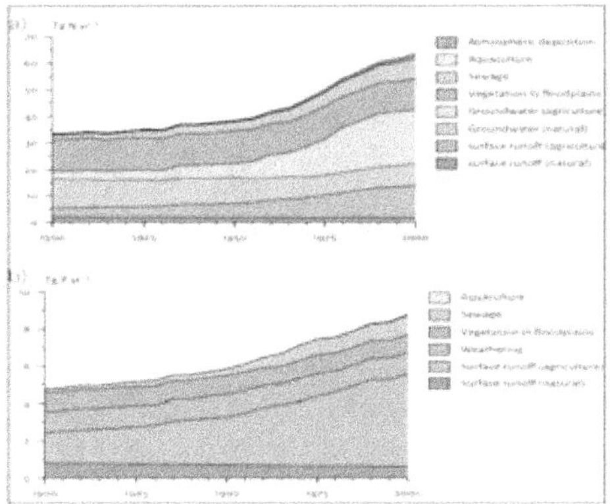

requires close cooperation between freshwater ecologists, meteorologists, modelers, and management agencies (Li et al.,2021).

## MANAGEMENT OF EUTROPHICATION

1. Effect of reducing nitrogen inputs alone for controlling eutrophication

In a 37 year whole lake experiment done by Schindler *et al.* (2008) in lake (5.0 ha), Northwestern Ontario, Canada concluded that eutrophication of lakes cannot be controlled by reducing nitrogen inputs alone. The experimental lake area has been fertilized for 37 years with constant annual inputs of phosphorous and decreasing inputs of nitrogen to test whether controlling nitrogen inputs can control eutrophication (Table 1). Reducing nitrogen inputs increasingly favored nitrogen-fixing cyanobacteria, nitrogen fixation was sufficient to allow biomass to continue to be produced in proportion to phosphorous, and lake remained eutrophic (figure 2). Therefore, this concluded that to reduce eutrophication the focus of management must be to reduce inputs of phosphorus rather than controlling nitrogen inputs alone.

*Table 1. Annual fertilizer additions to lake, 1969-2005*

| Year | N (Kg/Year) | P (Kg/Year) | N:P (by weight) |
|---|---|---|---|
| 1969 | 249 | 20.7 | 12.1 |
| 1970-74 | 308 | 24.8 | 12.4 |
| 1975-82 | 110 | 23.6 | 4.66 |
| 1983 | 110 | 19.8 | 5.54 |
| 1984-89 | 110 | 23.6 | 4.66 |
| 1990-97 | 0 | 23.6 | 0 |
| 1998 | 0 | 31.9 | 0 |
| 1999-2005 | 0 | 24.5 | 0 |

*Figure 2.*

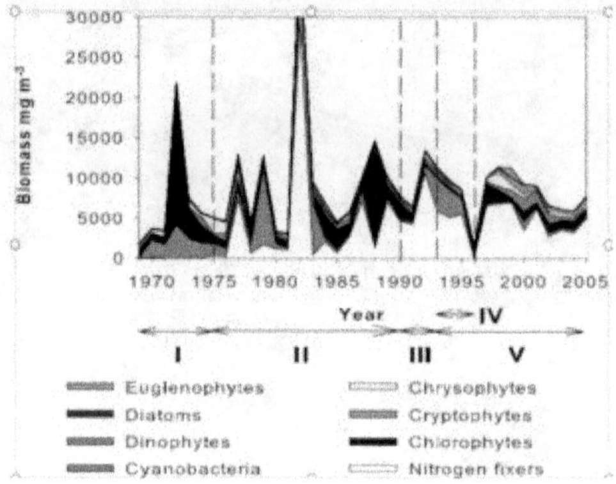


2. Effect of lowering soil test phosphorus (STP) for eutrophication control

In a study by Withers *et al.* (2019) on the intensive farming areas of Europe, The United States, and Australia, it became clear that most soils in Europe and the United States with ≤20 mg kg$^{-1}$ P- Olsen equivalent have a greater chance of meeting the low target Soluble Reactive Phosphorous (SRP) concentrations required for eutrophication control (as shown by the gray shaded area in Figure 3). Threshold Soil Test Phosphorus (STP) concentrations nearer 10 mg kg$^{-1}$ P-Olsen equivalent will lower eutrophication risk further. Additional measures appear necessary on Australian soils due to their lower phosphorus buffering power (despite more extensive weathering which may be due to the type of iron and aluminum oxides and hydroxides present in the soils). The relationship between soil test phosphorus and soluble reactive phosphorus in land runoff suggest that eutrophication control targets would be more achievable if soil test phosphorus concentrations were kept at or below the current recommended.

Threshold values for fertilizer response Figure 3).

*Figure 3.*

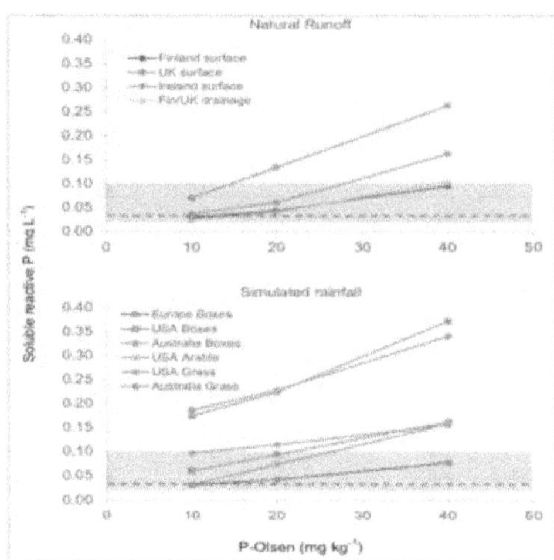

3. Effect of Aluminium pillared bentonite for phosphate removal from aquaculture discharge water

In a study carried out by Kumararaja *et al.* (2019) at ICAR-Central Institute of Brackish Water (ICAR-CIBW) for Aquaculture, Chennai for reducing phosphorus levels of aquaculture discharge water with low cost available bentonite with improved efficiency by pillaring with polyhydroxy aluminum (pillaring with polyhydroxy aluminium is more effective due to the affinity of phosphate anion to the aluminum, improved surface area and porosity). Applied 500g aluminum pillared bentonite to 1.5 m discharge water treatment system. Phosphate concentration was measured at different time intervals and it was observed that phosphate concentration reduced with time intervals. The phosphate level was reduced by 60% within 2 hours of application. Phosphate concentration reduced below the critical level of 0.2 ppm after 6 hours of application. The results indicate that pillared bentonite treatment resulted in more than 80% reduction in phosphate levels of aquaculture discharge water (Figure 4).

The effect of salinity on the phosphorus adsorption capacity of pillared bentonite was studied and it was observed that the phosphorus sorption capacity of pillared bentonite decreased with increasing salinity from 11.85 mg/g at 0 ppt to 8.21, 6.26 and 5.01 at 10, 20 and 30 ppt respectively. Reduction in adsorption arises from competition for adsorption sites from carbonates and bicarbonates.

The phosphate sorption capacity of pillared bentonite improved with rising temperatures from 11.85 mg/g at 25°C to 13.81 and 15.85 mg/g at 35 °C and 45°C respectively (Table 3).

The result demonstrates that the phosphate removal process is endothermic. Increased mobility of phosphate with increasing kinetic energy at higher temperatures results in higher adsorption.

Aluminum pillared bentonite developed in the study has higher phosphate adsorption capacity than modified bentonites (Table 4). Novel pillared bentonite hence could enhance the efficiency of the discharge water treatment system of the aquaculture facility by removing the phosphate and thereby avoiding eutrophication.

*Figure 4.*

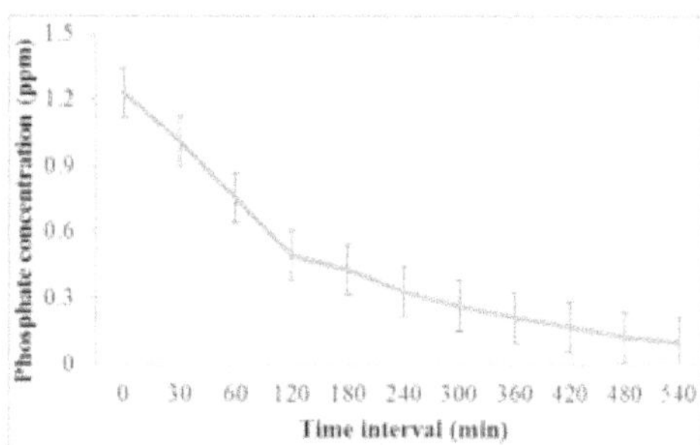


*Table 3. Effect of temperature on phosphorus adsorption capacity of pillared bentonite*

| Temperature (C) | Adsorption capacity (mg/g) |
|---|---|
| 25 | 11.8 |
| 35 | 13.81 |
| 45 | 15.85 |

*Table 4. Phosphate adsorption capacity of the modified bentonites*

| | Adsorption capacity (mg/g) | Reference |
|---|---|---|
| Al-bentonite | 7.1 | |
| Al-CTMAB-Bent | 7.63 | Ma and Zhu, 2006 |
| HyAl-Mt | 10.06 | Zhu and Zhu, 2007 |
| Al-bentonite | 8.98 | Tian et al., 2009 |
| Fe–Al-Bent | 10.5 | Yan et al., 2010 |
| Fe–Al-Bent | 8.33 | Rahni et al., 2016 |
| Al pillared-bentonite | 11.85 | This study |

4. Effect of Phoslock treatment on phosphorus removal in tropical eutrophic reservoir

In a study by Barcante *et al.* (2020) in a hypereutrophic Pampulha reservoir, in Brazil to know the effect of Phoslock (Lanthanum modified bentonite clay developed at Commonwealth Scientific and Industrial Research Organization (CSIRO), Australia) treatment on phosphorus removal in eutrophic water bodies, treatment was carried out from April 2016 to February 2018.

Phoslock is used to control internal phosphorus loading to remove dissolved phosphorus from the water column of eutrophic lakes. The mechanism of action is by binding of lanthanum ion of bentonite clay with phosphate ion from water and it forms LaPO4 (rhabdophane Mineral). It is then precipitated from water column and immobilized on lake sediments.

Lanthanum is effective over a wide pH range in binding phosphorus and shows a very high phosphorus removal over pH 5-9. The phosphorus binding capacity of lanthanum is not affected by anoxia. These characteristics make Phoslock a promising agent in reducing sediment phosphorus release, even though humic substances can reduce their efficiency.

Phoslock is highly effective at both stripping soluble reactive phosphorus from the water column and intercepting phosphorus released from the sediments once settled where it forms a reactive capping.

Average amounts of around 33 to 47 tons of Phoslock were applied every month. Before the treatment, cyanobacteria were the dominant, accounting for more than 80% of total phytoplankton biomass. The positive effect of reduction of cyanobacteria biomass was observed between September 2016 and February 2017 and between December 2017 and June 2018. There was no effect in between. From December 2017, cyanobacteria decreased and biomass remained low and the time represents a period of high biodiversity (Figure 5). A significant increase in phytoplankton species richness other than cyanobacteria was observed in 2018, 21 months after Phoslock treatment (figure 6). Phytoplankton diversity and changes in the structure of the phytoplankton community is the remarkable achievement of Phoslock treatment.

But even at this high rate of Phoslock application, results are not satisfactory as large amount s of nutrients still enter the reservoir by tributaries. So, this concludes that to achieve acceptable results, external nutrient loads should be limited. Limiting nutrient input into aquatic systems is the main strategy to successfully reduce eutrophication.

*Figure 5.*

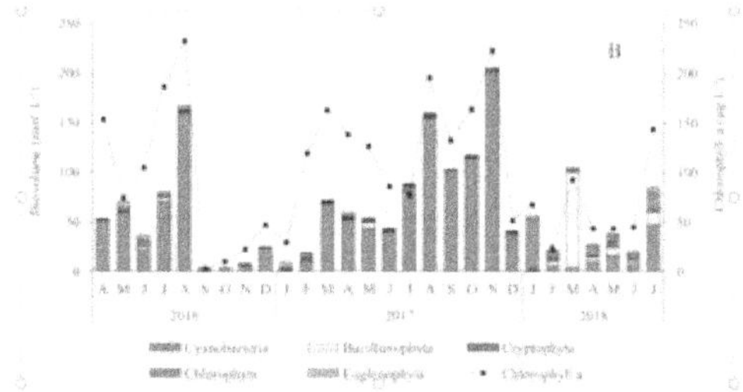

*Figure 6.*

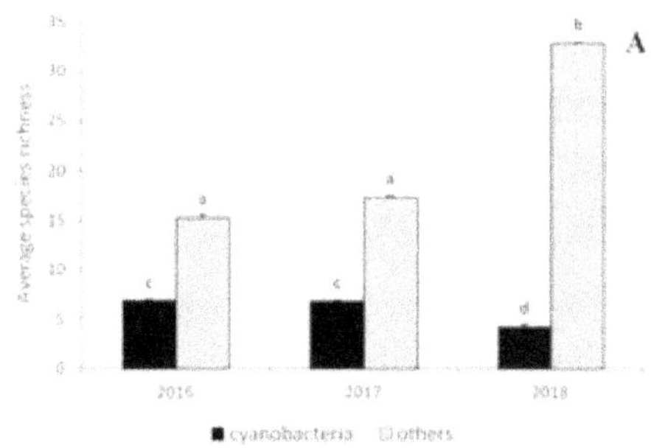


5. Effect of geo-engineering in ponds to control eutrophication

In a study done by Waajen et al. (2016) on the effect of geo-engineering in two urban ponds, Dongen and Eindhoven in the Netherlands, the ponds were separated into six different compartments, and six different treatments were carried out (table 5).

Biomanipulation consisted of fish stock control and the introduction of submerged macrophytes. Original fishes removed and restocking done. Dredging involves the mechanical removal of sediments from water bodies. It is very effective but the cost is very high. An attractive alternative to sediment dredging is the in-situ capping of sediments with an active barrier capable of capturing nutrients released from the pore water and minimizing their release to the water column. A promising active capping agent

*Table 5. List of treatments carried out in Dongen and Eindhoven ponds in Netherlands*

| Sl. No | Treatments |
|---|---|
| 1 | Dredging + biomanipulation (DB) |
| 2 | Dredging + PAC (polyaluminiumchloride) + biomanipulation (DPB) |
| 3 | LMB (lanthanum-modified bentonite) + biomanipulation (LB) |
| 4 | LMB + PAC + biomanipulation (LPB) |
| 5 | Biomanipulation (B) |
| 6 | Control (C). |

*Table 6 (a). Median Secchi depth (m), median turbidity (NTU), and median concentrations of total and cyanobacterial chlorophyll-a, total phosphorus (TP), and soluble reactive phosphorus (SRP) (mg/ L) in compartments of pond Dongen and pond Eindhoven (31 August 2009 - 20 August 2011). Treatments: DB - dredging + biomanipulation, DPB - dredging + PAC + biomanipulation, LB - LMB + biomanipulation, LPB- LMB + PAC + biomanipulation, B - biomanipulation, C - control*

| | Pond Dongen | | | | | |
|---|---|---|---|---|---|---|
| | Compartments | | | | | |
| | DB | DPB | LB | LPB | B | C |
| Secchi depth (m) | 0.85 | 0.70 | 0.58 | 0.50 | 0.55 | 0.25 |
| Turbidity (NTU) | 8.1 | 5.8 | 8.4 | 14.2 | 8.8 | 45.5 |
| Total chlorophyll-a (μg $L^{-1}$) | 5.3 | 3.4 | 5.9 | 9.2 | 5.0 | 268.9 |
| Cyanobacterial chlorophyll-a (μg $L^{-1}$) | 0.0 | 0.0 | 0.0 | 0.0 | 0.0 | 0.0 |
| TP (μg $L^{-1}$) | 39 | 46 | 57 | 71 | 53 | 448 |
| SRP (μg $L^{-1}$) | 14 | 12 | 15 | 14 | 20 | 47 |

is lanthanum-modified bentonite clay (LMB), Phoslock developed by CSIRO, Australia. Poly aluminum chloride is a flocculant.

In pond Dongen, the greatest secchi depth was found in dredged compartments, that is DB and DPB. The secchi depth of LB, LPB and B were twice the control. The lowest turbidity was observed in DPB compartment. Chlorophyll-a concentration were substantially lower than control for all the treated counterparts. Similar is the case with total phosphorus and soluble reactive phosphorus (Table 6).

In pond Eindhoven, dredged and LMB treated compartments gradually became similar with secchi depths after 2 years. Similarly, the turbidity was lowest in four either dredged or LMB treated compartments. The median chlorophyll-a concentration was lower in dredged and LMB treated compartments than in biomanipulated compartment which was lower than in control. LB and LPB compartments had lower total phosphorus concentration and soluble reactive phosphorus concentration (Table 6).

So, from the results we can conclude that a combination of measures reducing internal phosphorus loading with biomanipulation can improve water quality and create and maintain a clear water state in urban ponds without high phytoplankton biomass provided the external phosphorus load is limited. Bio manipulation as a single measure was shown only to be effective for a limited period or was not effective at all. Geo-engineering using LMB to control the internal phosphorus loan can be an alternative to costly dredging. The use of LMB is sufficient and the additional advantage of the flocculant PAC was not shown. The difference between the two experimental sites supports the importance of site-specific system analysis to choose the most promising mitigating measure.

## 6. The use of shellfish for eutrophication control

A study was carried out by Petersen et al. (2015) on the effect of selfish for eutrophication control in Skive Fjord, Denmark i.e., the fundamental idea behind employing extractive cultures to mitigate the consequences of excessive nutrient runoff is to view excess nutrient levels in coastal waters as a resource that can be recycled back to land. Back on land, the mitigation crop can be used for various

*Table 6 (b). Median Secchi depth (m), median turbidity (NTU), and median concentrations of total and cyanobacterial chlorophyll-a, total phosphorus (TP), and soluble reactive phosphorus (SRP) (mg/ L) in compartments of pond Dongen and pond Eindhoven (31 August 2009 - 20 August 2011). Treatments: DB - dredging + biomanipulation, DPB - dredging + PAC + biomanipulation, LB - LMB + biomanipulation, LPB- LMB + PAC + biomanipulation, B - biomanipulation, C - control*

| | Pond Eindhoven | | | | | |
|---|---|---|---|---|---|---|
| | Compartments | | | | | |
| | DB | DPB | LB | LPB | B | C |
| Secchi depth (m) | 1.20 | 1.00 | 1.29 | 1.06 | 0.58 | 0.50 |
| Turbidity (NTU) | 5.0 | 5.3 | 3.7 | 7.0 | 10.6 | 19.2 |
| Total chlorophyll-a (μg $L^{-1}$) | 6.2 | 19.7 | 7.6 | 13.3 | 44.5 | 52.4 |
| Cyanobacterial chlorophyll-a (μg $L^{-1}$) | 0.0 | 1.5 | 0.4 | 2.1 | 2.0 | 6.5 |
| TP (μg $L^{-1}$) | 92 | 73 | 28 | 43 | 90 | 88 |
| SRP (μg $L^{-1}$) | 29 | 22 | 17 | 13 | 19 | 19 |

purposes, e.g. food, feed, etc. By depleting the water of particles, mussels will enhance light penetration and increase water transparency even outside the specific mussel culture unit. Nutrients in the marine environment are being taken up by mussels and returned to land after the harvest of the crop (Figure 7A). Ingestion of particulate matter by the mitigation mussels will result in mussel tissue, excretion of dissolved nutrients, and release of fecal material from the mussels (Figure 7B). Some of the released fecal material will be consumed by associated fauna on the production lines and the remainder will lead to increased sedimentation of organic material below and in the vicinity of culture units. The culture unit corresponds to a removal of 0.6–0.9 t N $ha^{-1}$ $year^{-1}$ and 0.03–0.05 t P $ha^{-1}$ $year^{-1}$. It is more area-efficient nutrient removal than most land-based abatement measures. Even when there is an actual increase in biodeposition below culture units, it was found that the introduction of mitigation cultures in the water column results in a net reduction in total biodeposition on a basin scale. According to model results, the Skive Fjord's increased biodeposition below the culture unit as a result of mussel filtration and the subsequent production of faecal material was balanced out by a decrease in sedimentation outside the

*Figure 7.*

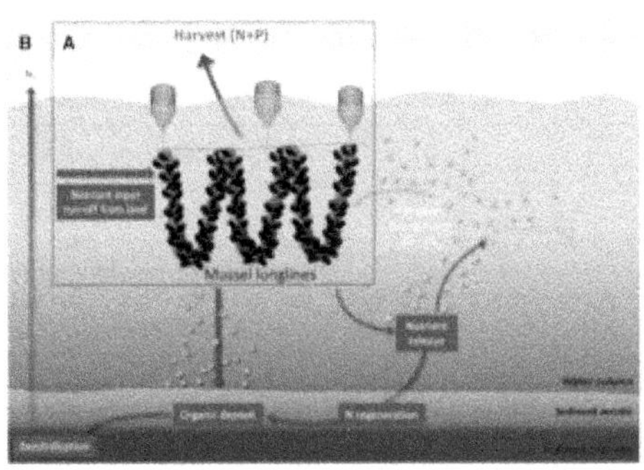

unit as a result of the removal of organic particles from the surrounding waters, and the effects on basin scale resulted in a net decrease in total sedimentation. So basin scale reduction in sedimentation was observed in the long run.

7. Purifying eutrophic river waters with integrated floating island systems

A study was conducted by Zhao *et al.* (2012) on the effect of purifying eutrophic river waters with integrated floating island system in Jiaxing City, China with an integrated floating island system consisted of aquatic vegetation near riverside and massive floating islands with adsorptive films (figure 8). Hydrophytes have been widely applied in floating islands for remediation of surface water and wastewater due to their efficacy in assimilating nutrients and creating favorable conditions for microbial decomposition of organic matter. Aquatic plants also limit algal growth by competing for nutrients and sunlight. The plant biomass must be removed periodically from the water bodies to maintain purification efficiency. Nine aquatic plants were applied to the system (Table 7).

Due to the greater biomass of 2339 g dry biomass/m$^2$, the alligator flag showed the most nitrogen and phosphorus uptake (Figure 9, 10). Hence, the alligator flag was found domain having the highest phytoremediation efficiency.

Rich in protein and microelements, the plant biomass from integrated floating island system can be used for animal feed and the food value could partially offset the whole cost of harvesting. Crude protein, calcium, magnesium, iron and manganese can meet the daily requirements of livestock and poultry at different growth stages according to National Research Council (NRC), China. The accumulation of

*Figure 8.*

*Table 7. Nine aquatic plants applied to integrated treatment system*

| Common name | Scientific name |
|---|---|
| Water hyacinth | *Eichhornia crassipes* (EC) |
| Water lettuce | *Pistia stratiotes* (PS) |
| Water dragon | *Jussiaea reppens* (JR) |
| Pennywort | *Hydrocotyle verticillata* (HV) |
| Frogbit | *Hydrocharis dubia* (HD) |
| Parrot weather | *Myriophyllum aquaticum* (MA) |
| Pickerel weed | *Pontederia cordata* (PC) |
| Canna | *Canna indica* (CI) |
| Alligator flag | *Thalia dealbata* (CP) |

trace mineral elements in the tested plants was in the order Cu > Pb > Cr > As. They were below the standard of NRC and hygienical standard for feeds in China therefore safe as animal feeds (Table 8).

8. Phytoremediation potential of azolla for phosphorus controlled water eutrophication

Akhtar *et al.* (2015) carried out microcosm investigation on the differential potential of azolla for phytoremediation of phosphorus controlled water eutrophication Okayama University, Japan using three different azolla species; *A. japonica*, *A. pinnata* and *A. hybrid.*

In study 1-azolla plants with one-week phosphorus, hunger treatment was exposed to 5-ppm phosphorus. Here the concentration of phosphorus depleted completely within 24 hours of exposure by *A. japonica* and *A. pinnata* compared to control while *A. hybrid* depleted phosphorus within 48 hours (figure 11A).

In study 2 azolla plants with one week phosphorus hunger period exposed to 10 ppm phosphorus. Here the phosphorus concentration depleted from 10 to 2 ppm within 96 hours when exposed to *A. japonica* and *A. pinnata* and remained less than one ppm after 144 hours.

*A. hybrid* depleted phosphorus concentration only from 10 to 4 ppm in 96 hours. It does clearly indicate more removal ability of naturally growing *A. japonica* and *A. pinnata* (figure 11B).

*Figure 9.*

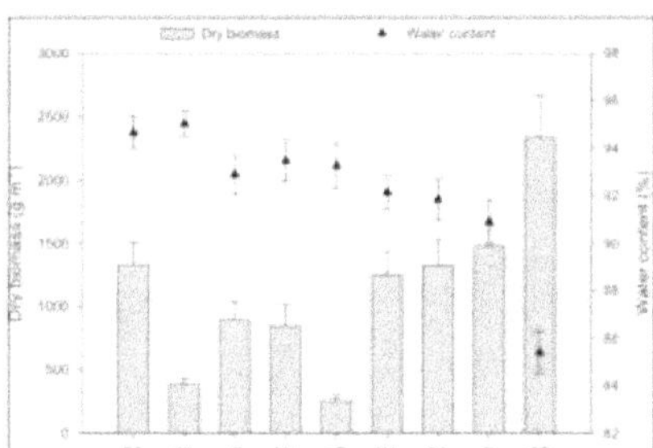

*Figure 10.*

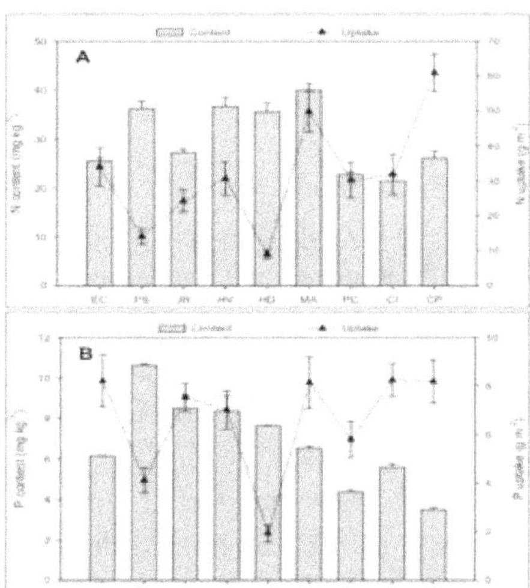

*Figure 11.*

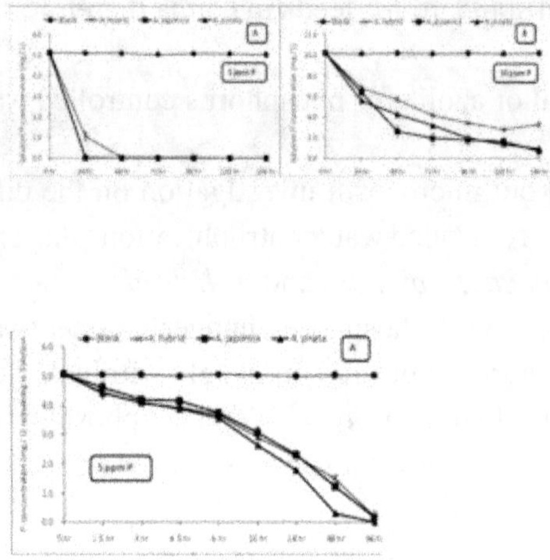

*Figure 12.*

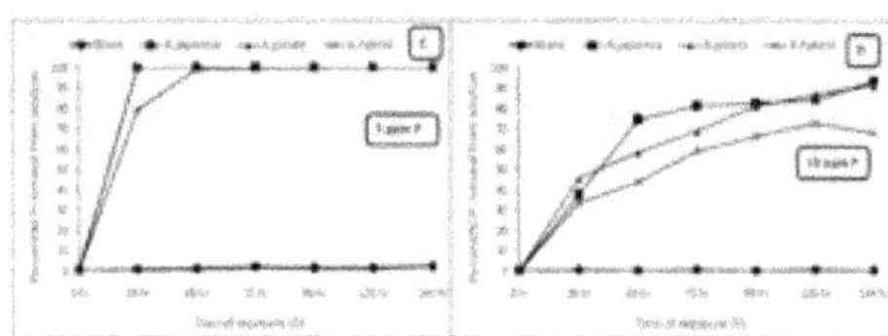

In study 3 azolla plants without the phosphorus hunger treatment was exposed to 5 ppm phosphorus. The results showed that removal of phosphorus in solution exposed to plants without phosphorus hunger was slow compared to solutions exposed to plants with the phosphorus hunger (figure 12).

Do you do faster growth, high tolerance and accumulation ability, free-floating azolla might be the best candidate among macrophytes to combat phosphorus driven eutrophication.

9. Stakeholder-driven hydrologic modeling of agricultural conservation practices in the **Watershed for mitigation of Lake Eutrophication, a case study of Lake Macatawa, Michigan, USA** (Lavorivska et al.,2020)

The study demonstrated that widespread adoption of numerous Best Management Practises (BMPs) such as continuous no-till with high residue, cover crops, filter strips, and conversion of some marginal croplands to perennial grasses across all the watershed's row croplands is necessary to achieve the total phosphorus (TP) reduction of 72% required by the Total Maximum Daily Load (TMDL). Applying combinations of multiple agricultural BMPs on the same fields, for example no-till and residue management, led to greater load reductions. Embracing P-based fertilizer applications at trim take-up rates, debilitating fertilizer inputs to soils with exceptionally high P tests, and applying additional excrement on low-P soils or moving it outside of the watershed are all reasonable options (Spiegal et al., 2020). These practices were found suitable for small agricultural tile-drained watersheds with high legacy soil P levels in the Great Lakes basin.

## 10. Managing nutrient loading to abate eutrophication

According to Padedda et al., (2022), the measures that can be implemented in agricultural areas are improvements in fertilizer management, association with measures to reduce soil erosion, transitioning to species depends on synthetic chemicals and adopting organic farming. Compared with the point sources, the implementation of animal husbandry should be discouraged as they are responsible for the majority of the amount of soluble P that enters the lake system. Measures modifying the P and N paths in the drainage basin by building small wetland systems and restoring large ribbons of riparian vegetation along the tributaries, which are effective channel for supplement runoff within the lake and give other environmental services locally. Nature-Based Solutions (NBS) is an major term which includes biological solutions aims to provide ecosystem benefits in many ways. In agricultural context, the three suitable NBS approaches are: (1) phytoremediation, (2) re-vegetation of drainage ditches, and (3) buffer strips.

Phytoremediation consists of constructed wetlands having significantly lower operating costs than conventional technological systems like wastewater treatment plants. These recent plant based solutions, in addition to guaranteeing good purification yields and being characterized by management simplicity and low construction and management costs, are particularly advantageous to reduce the surfaces required and energy consumption and have a greater regulation capacity compared with conventional activated sludge systems, which allow coping with very significant increases and fluctuations in load. Ditch re-vegetation results in low retention capacity, as it is designed as a free flow wetland, but one of its advantages is that it significantly improves the self-cleaning capacity of rivers.

Over the past 20 years of buffer strip research, it has become clear that different plant configurations, ranging from grass to tree, and shrub topsoil, can significantly reduce surface flow, as well as suspended solids, nitrogen, and phosphates from agricultural runoff. With this approach, even 2–4-m strips are enough to intercept more than 70–80% of nutrients. Other advantages of buffer strips include

*Table 8. Chemical compositions in aboveground tissues of nine aquatic plants growing on floating islands and maximum tolerable levels (MTL) in animal feed. c Hygienically standard for feeds of China (GB 13078-2001). NG Not given*

| Species | Crude protein | Zn (mg/Kg) | Cu (mg/Kg) | Pb (mg/Kg) | Cr (mg/Kg) | As (mg/Kg) |
|---|---|---|---|---|---|---|
| Water hyacinth | 162.5 | 0.17 | 12.61 | 3.18 | 3.29 | 0.46 |
| Water lettuce | 219.4 | 0.08 | 14.01 | 3.93 | 3.47 | 0.88 |
| Water dragon | 167.5 | 0.11 | 13.84 | 2.57 | 2.07 | 0.71 |
| Pennywort | 236.6 | 0.12 | 15.85 | 4.32 | 3.19 | 1.51 |
| Frogbit | 230.2 | 0.08 | 16.74 | 4.8 | 5.83 | 1.78 |
| Parrot weather | 255.4 | 0.09 | 19.74 | 3.32 | 2.27 | 1.04 |
| Pickerel weed | 150.6 | 0.16 | 11.15 | 4.14 | 3.57 | 0.84 |
| Canna | 127.8 | 0.08 | 12.45 | 3.07 | 2.59 | 0.43 |
| Alligator flag | 157 | 0.04 | 5.01 | 3.25 | 1.39 | 0.43 |
| MTL in China [c] | NG | 250 | 25-200 | 5.0-40 | 10-200 | 2.0-10.0 |
| | | | | | | |

the possibility of retaining and treating up to 60% of pesticides used in agriculture and integrating with short-rotation forestry systems for the production of biomass.

## INTEGRATED STRATEGIES TO MANAGE SOIL PHOSPHORUS STATUS FOR EUTROPHICATION CONTROL

In many cultivated areas the soil test P is disproportionately high in the past. Soil P accumulation poses an accelerated risk to freshwater eutrophication and needs to be addressed through drawdown of STP and controls over soil runoff and erosion. STP analysis was originally created to help with agronomic management, but it turned out to be a good way to figure out how much dissolved P might be lost in field runoff. We need to do more to make sure it's used more consistently so it's more sustainable for soil P fertility, especially if it's combined with extra buffering or saturation indices in highly eutrophic catchments.

The meta-analysis across three continents suggests agronomic optimum STP concentrations need to be in the range of 10 to 20 mg $kg^{-1}$ P-Olsen equivalents to match the very challenging eutrophication controls targets set for freshwaters. It will take many years to draw down STP, and it's definitely a long-term mitigation solution, but it goes hand-in-hand with developing more sustainable ways to use P in the food supply chain by helping to end the P cycle through the reuse of P. A better understanding of the environmental benefits of controlling STP could help policy makers work with industry to encourage more voluntary soil testing and compliance with recommended productivity thresholds, or if that doesn't work, push for more regulation as part of more effective and sustainable P management practices. Soil and soil P management are just one of the many steps needed to reduce agricultural P losses and make food production sustainable in the future (Withers et al., 2019).

### Importance of Temperature and Water Residence Time

During low temperature period, nitrogen and phosphorus can be controlled to reduce the eutrophication level, while in moderate and high temperature period, reducing nitrogen may not suitable for controlling eutrophication in some lakes. The nitrogen fixation has a great impact on nitrogen loading, indicating that it provides much nitrogen to water bodies for cyanobacteria growth. Low N:P ratio and high temperature can shift lakes to be dominated by nitrogen fixation. In addition, short lake water residence time can severely inhibit the nitrogen fixation capacities via inhibiting the growth of nitrogen-fixing cyanobacteria, changing the N:P ratio and resuspending solids from sediments. With the increase of temperature, shallow lakes are easy to become nitrogen fixation dominance. Therefore, the management of lake eutrophication should be seasonal and adapted to local conditions (Zhao et al.,2022).

## PUBLIC AWARENESS AND EDUCATION ON EUTROPHICATION

Public awareness and education play a crucial role in addressing the issue of eutrophication. By increasing knowledge and understanding among the general public, individuals can actively contribute to the prevention and management of eutrophication. Public awareness is essential for promoting behavioral changes and fostering a sense of responsibility towards the environment. When individuals understand the causes and consequences of eutrophication, they are more likely to adopt sustainable practices and support initiatives aimed at reducing nutrient pollution. By involving the public, eutrophication management efforts can benefit from increased participation, collaboration, and advocacy.

## POLICY AND REGULATORY MEASURES TO CONTROL EUTROPHICATION

Policy and regulatory measures are essential tools for controlling eutrophication and ensuring sustainable management of water bodies. Governments, regulatory agencies, and international organizations play a crucial role in formulating and implementing policies that address the root causes of eutrophication. This section explores some key policy and regulatory measures commonly employed to control eutrophication.

1. **Nutrient Water Quality Standards**: Establishing water quality standards is a fundamental step in controlling eutrophication. These standards define acceptable nutrient levels in water bodies and set limits on nitrogen and phosphorus concentrations. Monitoring programs are put in place to assess compliance with these standards and take appropriate actions if thresholds are exceeded.
2. **Source Control and Nutrient Management**: Policies focusing on source control and nutrient management aim to regulate and reduce nutrient inputs into water bodies. Some key measures include: a. Agricultural Practices: Implementing best management practices in agriculture, such as precision farming, optimized fertilizer application, cover cropping, and buffer zones to minimize nutrient runoff. b. Urban and Industrial Regulations: Enforcing regulations on wastewater treatment and stormwater management to reduce nutrient discharge from urban areas and industrial activities. c. Phosphorus Detergents and Fertilizer Restrictions: Regulating the use of phosphorus-based detergents and fertilizers to prevent excessive nutrient loading in water bodies.
3. **Wastewater Treatment**: Efficient wastewater treatment is crucial for reducing nutrient loads entering water bodies. Policies related to wastewater treatment include: a. Upgrading Treatment Facilities: Encouraging the upgrade and expansion of wastewater treatment plants to incorporate advanced nutrient removal technologies. b. Effluent Standards: Setting strict effluent standards for wastewater treatment plants, particularly for nitrogen and phosphorus concentrations. c. Reclaimed Water Use: Promoting the use of reclaimed water for irrigation and industrial purposes to reduce the discharge of nutrients into water bodies.
4. **Land-Use Planning and Zoning**: Integrating land-use planning and zoning regulations is an effective approach to control eutrophication. This includes: a. Buffer Zones: Designating buffer zones along water bodies, which are areas with limited human activities to reduce nutrient runoff and promote natural filtration. b. Riparian Protection: Implementing regulations to protect riparian areas and wetlands that act as natural filters and nutrient sinks. c. Green Infrastructure: Incorporating green infrastructure practices, such as green roofs, rain gardens, and permeable pavements, to manage stormwater runoff and reduce nutrient loads.
5. **Environmental Impact Assessments**: Requiring environmental impact assessments for new developments and major projects ensures that potential impacts on water bodies and nutrient loads are considered in decision-making processes. This enables the adoption of mitigation measures and the avoidance of activities that may exacerbate eutrophication.

6. **Collaborative Governance and Stakeholder Engagement**: Promoting collaboration among different stakeholders, including governments, local communities, NGOs, and industries, is crucial for effective eutrophication management. This can be achieved through: a. Multi-Stakeholder Platforms: Establishing platforms for dialogue, coordination, and knowledge exchange among stakeholders involved in eutrophication management. b. Public Participation: Engaging the public in decision-making processes, such as through public consultations, to ensure diverse perspectives and local knowledge are considered in policy development. c. Partnerships: Encouraging partnerships between governmental bodies, research institutions, and non-profit organizations to develop and implement eutrophication control measures.
7. **International Cooperation and Agreements**: Eutrophication is a transboundary issue, and international cooperation is essential for its effective control. Collaborative agreements and frameworks between countries can address shared water bodies and promote joint actions to reduce nutrient pollution. Examples include the Baltic Sea

## CONCLUSION AND PATH AHEAD

A joint management of nitrogen and phosphorus is essential to curb eutrophication rather than either alone. Agronomic optimum soil test phosphorus concentrations in the range of 10 to 20 mg/kg P-Olsen equivalent necessary to match eutrophication controls targets for freshwaters. Aluminum pillared bentonite is efficient in reducing the phosphate level of aquaculture discharge water before releasing into the environment. Mitigation culture using mussels are more area-efficient nutrient removal than most land-based abatement measures. A combination of measures targeting the internal phosphorus load and bio manipulation can be effective in creating and maintaining a clear water state provided the external phosphorus load is limited. Acid-modified sugarcane-bagasse-based adsorbent can be successfully employed to remove microalga biomass from eutrophic water sustainably. Azolla plants are efficient phytoremediation plants for phosphorus removal and can be deployed for phosphorus removal in aquatic systems. Bacterial and archaeal communities rise significantly from the phylum to the order-level and less so from the order- to lower taxonomic levels, indicating that trophic state filtering is highest at the order-level. Thus, trophic state-related variations in OM source distributions and amounts appear to be the primary drivers of long-term changes in bacterial community structure. Archaeal communities, on the other hand, are distinguished by trophic state in newly formed strata but converge in older sediments produced under multiple trophic regimes. Lake sediments are major carbon sinks on a worldwide scale. Although microorganisms play an important role in the fate of organic carbon (OC) in lake sediments, few research have looked at the constraints on lake sedimentary microbial communities. Hence there is importance of a site-specific system analysis to choose the most promising mitigating measures.

Farmers are naturally concerned that omitting phosphorus fertilizer will compromise crop productivity. Future research must therefore provide them with the evidence base to guide the management of STP drawdown so that productivity is not compromised. Development of a nutrient classification system that accounts for productivity as well as the risk of losses to surface waters in various agroecosystem and crops. An additional in-depth study focusing on the parameters that limit phytoextraction strategies and link their findings to hydrology to recommend it to policy makers and farmers. The current state of knowledge about the effects of oysters (on reefs or in aquaculture) on nitrogen dynamics is incomplete in many respects.

## REFERENCES

Akhtar, M. S., Oki, Y., Nakashima, Y., Nishigaki, M., Adachi, T., & Kamigaki, T. (2016). Microcosm investigation on differential potential of free-floating Azolla macrophytes for phytoremediation of P-controlled water eutrophication. *International Journal of Agriculture and Biology*, *18*(1), 204–212. doi:10.17957/IJAB/15.0119

Barçante, B., Nascimento, N. O., Silva, T. F. G., Reis, L. A., & Giani, A. (2020). Cyanobacteria dynamics and phytoplankton species richness as a measure of waterbody recovery: Response to phosphorus removal treatment in a tropical eutrophic reservoir. *Ecological Indicators*, *117*, 106702. doi:10.1016/j.ecolind.2020.106702

Basri, W. N. F., Daud, H., Lam, M. K., Cheng, C. K., Oh, W. D., Tan, W. N., Shaharun, M. S., Yeong, Y. F., Paman, U., Kusakabe, K., Kadir, E., Show, & Lim. (2019). A sugarcane-bagasse-based adsorbent employed for mitigating eutrophication threats and producing biodiesel simultaneously. *Processes (Basel, Switzerland)*, *7*(9), 572. doi:10.3390/pr7090572

Iavorivska, L., Veith, T. L., Cibin, R., Preisendanz, H. E., & Steinman, A. D. (2021). Mitigating Lake Eutrophication through stakeholder-driven hydrologic modeling of agricultural conservation practices: A case study of Lake Macatawa, Michigan. *Journal of Great Lakes Research*, *47*(6), 1710–1725. doi:10.1016/j.jglr.2021.10.001

Kumararaja, P., Suvana, S., Saraswathy, R., Lalitha, N., & Muralidhar, M. (2019). Mitigation of eutrophication through phosphate removal by aluminum pillared bentonite from aquaculture discharge water. *Ocean and Coastal Management*, *182*, 104951. doi:10.1016/j.ocecoaman.2019.104951

Li, Y., Shang, J., Zhang, C., Zhang, W., Niu, L., Wang, L., & Zhang, H. (2021). The role of freshwater eutrophication in greenhouse gas emissions: A review. *The Science of the Total Environment*, *768*, 144582. doi:10.1016/j.scitotenv.2020.144582 PMID:33736331

Lin, S. S., Shen, S. L., Zhou, A., & Lyu, H. M. (2021). Assessment and management of Lake Eutrophication: A case study in Lake Erhai, China. *The Science of the Total Environment*, *751*, 141618. doi:10.1016/j.scitotenv.2020.141618 PMID:33167190

Nyenje, P. M., Foppen, J. W., Uhlenbrook, S., Kulabako, R., & Muwanga, A. (2010). Eutrophication and nutrient release in urban areas of sub-Saharan Africa—A review. *The Science of the Total Environment*, *408*(3), 447–455. doi:10.1016/j.scitotenv.2009.10.020 PMID:19889445

Padedda, B. M., Lugliè, A., Lai, G. G., Giadrossich, F., Satta, C. T., & Pulina, S. (2022). Land-based impact of nutrient loads and eutrophication on an ancient Mediterranean natural lake. *Hydrology*, *9*(1), 7. doi:10.3390/hydrology9010007

Petersen, J. K., Saurel, C., Nielsen, P., & Timmermann, K. (2016). The use of shellfish for eutrophication control. *Aquaculture International*, *24*(3), 857–878. doi:10.100710499-015-9953-0

Rathore, S. S., Chandravanshi, P., Chandravanshi, A., & Jaiswal, K. (2016). Eutrophication: Impacts of excess nutrient inputs on aquatic ecosystem. *IOSR Journal of Agriculture and Veterinary Science*, *09*(10), 89–96. doi:10.9790/2380-0910018996

Schindler, D. W., Hecky, R. E., Findlay, D. L., Stainton, M. P., Parker, B. R., Paterson, M. J., Beaty, K. G., Lyng, M., & Kasian, S. E. M. (2008). Eutrophication of lakes cannot be controlled by reducing nitrogen input: Results of a 37-year whole- ecosystem experiment. *Proceedings of the National Academy of Sciences of the United States of America, 105*(32), 11254–11258. doi:10.1073/pnas.0805108105 PMID:18667696

Waajen, G., van Oosterhout, F., Douglas, G., & Lürling, M. (2016). Geo-engineering experiments in two urban ponds to control eutrophication. *Water Research, 97*, 69–82. doi:10.1016/j.watres.2015.11.070 PMID:26725204

Withers, P. J. A., Vadas, P. A., Uusitalo, R., Forber, K. J., Hart, M., Foy, R. H., Delgado, A., Dougherty, W., Lilja, H., Burkitt, L. L., Rubæk, G. H., Pote, D., Barlow, K., Rothwell, S., & Owens, P. R. (2019). A global perspective on integrated strategies to manage soil phosphorus status for eutrophication control without limiting land productivity. *Journal of Environmental Quality, 48*(5), 1234–1246. doi:10.2134/jeq2019.03.0131 PMID:31589721

Zhao, F., Xi, S., Yang, X., Yang, W., Li, J., Gu, B., & He, Z. (2012). Purifying eutrophic river waters with integrated floating island systems. *Ecological Engineering, 40*, 53–60. doi:10.1016/j.ecoleng.2011.12.012

Zhao, F., Zhan, X., Xu, H., Zhu, G., Zou, W., Zhu, M., Kang, L., Guo, Y., Zhao, X., Wang, Z., & Tang, W. (2022). New insights into eutrophication management: Importance of temperature and water residence time. *Journal of Environmental Sciences (China), 111*, 229–239. doi:10.1016/j.jes.2021.02.033 PMID:34949352

## KEYWORDS AND DEFINITIONS

**Eutrophication:** The infusion of nutrient salts into the water, leading to alterations in the ecosystem's structure, such as a rise in microalgae and aquatic plant growth, a decline in fish populations, a decline in water quality, and various other impacts that limit and prevent its utilization.

**Harmful algal bloom:** It refers to the excessive growth of microalgae in marine or brackish water bodies. These blooms can lead to water discoloration, extensive fish die-offs, the contamination of seafood with harmful toxins, or disruptions to ecosystems and services that humans view as harmful.

**Non-Point Source Pollution:** Non-point sources are described as any sources that fall outside the scope of the official definition of "point source" provided in Section 502(14) of the United States Clean Water Act (Water Quality Act) of 1987.

**Phytoremediatio:** The utilization of plants, either directly or indirectly, for the purpose of cleansing polluted soil or water.

**Point Source:** This refers to any identifiable, restricted, and separate conduit or outlet, which may include, among other things, pipes, ditches, channels, tunnels, wells, specific openings, containers, transport vehicles, concentrated animal feeding operations, or waterborne vessels or other floating structures, from which pollutants are discharged or could potentially be discharged. It's important to note that this definition excludes stormwater runoff from agricultural activities and the water that returns from irrigated agriculture.

**Watershed:** Watershed is an area where rainwater runoff is gathered and directed towards a shared outlet or point.

# Chapter 13
# Composting for a Sustainable Future:
## Turning Waste Into Nutrient-Rich Soil

**Ranjit Singha**
https://orcid.org/0000-0002-3541-8752
*Christ University, India*

**Surjit Singha**
https://orcid.org/0000-0002-5730-8677
*Kristu Jayanti College (Autonomous), India*

## ABSTRACT

*As the worldwide waste management problem has risen, composting has gained popularity. It turns organic trash into nutrient-rich soil for agriculture, landscaping, and environmental rehabilitation. This chapter on composting, "Composting for a Sustainable Future: Turning Waste into Nutrient-Rich Soil," is comprehensive. It covers decomposition science, composting methods, materials, and procedures. Composting's agricultural, landscaping, and environmental remediation benefits are also covered. The chapter also examines composting's role in climate change mitigation, its obstacles, and remedies. Decomposition can help people, businesses, and communities live more sustainably. It urges decomposition and trash reduction and provides information about tools to start composting, a simple yet efficient solution to worldwide waste management.*

## 1. INTRODUCTION

This chapter examines decomposition, the transformation of organic refuse into nutrient-rich soil. Composting is a sustainable and natural method of reprocessing organic materials with environmental, agricultural, and economic benefits. The chapter discusses the simplicity of composting, its scalability, and its potential to alleviate the global waste management crisis. It provides an overview of decomposition principles, techniques, materials, and procedures, focusing on its agricultural, landscaping, and

DOI: 10.4018/978-1-6684-9838-5.ch013

environmental remediation applications. In addition, the chapter emphasizes composting's role in mitigating climate change and discusses potential obstacles and solutions. The chapter's ultimate objective is to encourage individuals, businesses, and communities to embrace composting as a vital component of sustainable waste management, thereby contributing to a more environmentally favourable future. Composting has a long history dating back thousands of years (Wu, 2013). We will travel through time to discover the origins and development of decomposition over the centuries. From ancient civilizations' practices of using organic waste as soil amendments to contemporary scientific understanding and advancements in composting techniques, we will examine how this age-old process has evolved and garnered worldwide recognition. Composting dates back thousands of years and has a fascinating history. Ancient civilizations, such as Egypt, understood the value of decomposition and utilized it to improve soil fertility. Composting was used to recycle crop residues and animal manure as part of traditional agricultural practices in numerous regions, including China (Ju et al., 2005). Sir Albert Howard's influential work from the early 20th century emphasized the importance of decomposition for sustainable agriculture and soil health. The need for sustainable waste management solutions has led to composting programmes to divert organic waste from landfills and produce valuable compost, contributing to the rise of municipal and industrial composting over the past few decades. Decomposition innovations, such as mechanized rotating apparatus and specialized decomposition systems, have increased productivity and scalability. Composting is now widely acknowledged as an essential component of sustainable waste management, with benefits including organic waste recycling, reduced reliance on synthetic fertilizers, enhanced soil health, and mitigation of environmental impacts. Today, composting is crucial in promoting sustainability, circular economy principles, and an eco-friendly approach to waste management and soil fertility improvement. The practice of composting has a long and rich history, dating back thousands of years. Here is a brief overview of the historical background and evolution of composting:

Composting's roots can be traced back to ancient civilizations. Agricultural societies in ancient Egypt recognized the value of humus and utilized it to increase soil fertility. Egyptians practised "pit composting," in which organic refuse was interred in trenches with soil and permitted to decompose. Composting techniques were widespread in conventional agricultural systems worldwide. Farmers in China, for instance, utilized decomposition techniques to recycle crop residues and animal manure, recognizing the value of returning organic matter to the soil. Early in the 20th century, the British scientist Sir Albert Howard, the father of composting, made significant contributions to the comprehension and promotion of decomposition. Howard's research highlighted the significance of decomposition for healthy soil and sustainable agriculture. He advocated using agricultural residues and animal manure in decomposition to enhance soil fertility and productivity (Howard, 1947). Composting techniques have evolved and refined over time. Aerobic composting methods, which rely on oxygen for decomposition, have resulted in quicker and more effective composting processes. Windrow composting, in-vessel composting, and aerated static pile composting have gained popularity for larger-scale composting operations. There has been a growing awareness of the need for sustainable waste management practices in recent decades. Composting has emerged as a viable solution, particularly for municipal and industrial waste. Numerous nations and municipalities have instituted composting programmes to divert organic waste from landfills, reduce greenhouse gas emissions, and generate valuable compost for agricultural and landscaping applications. Composting practices have expanded and become more effective due to technological advances in composting. These include the development of automated turning apparatus, specialized composting receptacles and systems, and biofilters for pollution control in large-scale composting facilities. These innovations have increased the accessibility, scalability, and efficiency of decomposition. Composting has garnered widespread recognition in recent years as an integral component of sustainable waste management strategies. Composting is an attractive option for individuals, businesses, and communities seeking to reduce their ecological footprint due to its ability to recycle organic waste, reduce reliance on synthetic fertilizers, improve soil health, and mitigate environmental impacts.

The historical context and development of decomposition illustrate how this practice has evolved from traditional agricultural techniques into a recognized and valued waste management and soil fertility enhancement strategy. Composting is essential for promoting sustainability, recycling organic waste, and fostering a circular economy. Composting is necessary for both sustainable agriculture and effective waste management. Composting reduces the burden on waste management systems by diverting organic waste from landfills and incinerators. It is a valuable source of organic matter that promotes soil health, agricultural productivity, and sustainable farming practices. Both sustainable agriculture and waste management rely heavily on composting. Compost is a soil amendment abundant in nutrients that increases soil fertility. It enhances soil structure, water retention capacity, and plant nutrient accessibility. Composting replenishes the soil's organic matter, encouraging beneficial microbial activity and nutrient cycling. Farmers can reduce the risk of nutrient discharge and water pollution by reducing their reliance on synthetic fertilizers and increasing their use of compost as a natural fertilizer. Composting is an efficient way to divert organic waste from landfills. Composting organic debris such as food leftovers, yard waste, and agricultural residues reduces methane emissions, a potent greenhouse gas. Composting helps combat climate change by minimizing greenhouse gas emissions and landfill space requirements. Composting closes the nutrient cycle by returning organic waste to the soil. Composting transforms organic waste into a valuable resource rather than a disposable material. By returning nutrients and organic matter to the ground through decomposition, we create a sustainable system that minimizes nutrient depletion and reduces reliance on external inputs. Compost enhances soil's physical, chemical, and biological properties to improve health. It enhances soil structure, enhances water infiltration, and decreases soil erosion.

Healthy soils containing sufficient organic matter foster robust root development, nutrient absorption, and plant growth—composting aids in erosion prevention, watershed protection, and long-term soil productivity. Composting can be used for remediation of the environment. It is used to remediate polluted soils and restore degraded land. In areas affected by industrial activities, mining, or other forms of pollution, compost functions as a soil amendment that helps sequester pollutants, enhance soil quality, and promote plant growth. Composting adheres to the circular economy principles by concluding the organic refuse cycle. It transforms waste into valuable resources that promote sustainable agriculture and improve soil health. Composting is a sustainable approach to resource management in which organic waste is considered a valuable input instead of a disposable waste product. Composting is necessary for sustainable agriculture because it improves soil fertility, reduces the use of synthetic fertilizers, and promotes nutrient cycling. It contributes to efficient waste management by diverting organic waste from landfills, reducing greenhouse gas emissions, and producing valuable resources such as compost abundant in nutrients.

This chapter provides a comprehensive overview of decomposition in the context of the water-soil-plant-animal nexus and the challenges climate change poses to achieve a sustainable future. We will investigate the science of composting, including the biological processes and factors that influence decomposition. In addition, we will discuss various decomposition methods, techniques, and factors that influence the produced compost's quality. By grasping the purpose and scope of this chapter, readers will appreciate the potential of decomposition as a method for converting refuse into nutrient-rich soil and contributing to a more sustainable and resilient agricultural system.

## 2. UNDERSTANDING COMPOSTING

The scientific basis of decomposition in composting derives from the degradation of organic matter by microorganisms, such as bacteria, fungi, and other decomposers. Composting is a biological process involving decomposing organic materials by microorganisms like bacteria, fungi, and actinomycetes. The intricate details of the composting process include the role of aerobic and anaerobic organisms, the decomposition of organic matter, and the transformation of nutrients into forms that plants can utilize. Composting science studies the biological and chemical processes that occur during the decomposition of organic materials. Composting is a complex physical process involving microorganisms like bacteria, fungi, and actinomycetes and larger organisms like earthworms. Diverse metabolic processes enable these organisms to decompose organic matter into simpler compounds. Composting requires four fundamental components: organic matter, moisture, oxygen, and the proper ratio of carbon to nitrogen. The source of nutrients for microbial activity is organic matter, such as food scraps, yard detritus, and agricultural by-products. Moisture is essential to maintain the optimal environment for microbial growth, while oxygen is required for aerobic decomposition.

The ratio of carbon to nitrogen, known as the C/N ratio, is a critical factor influencing the efficiency and speed of the composting process. Microorganisms responsible for decomposition require a well-balanced mixture of carbon-rich ("browns") and nitrogen-rich ("greens") materials to thrive. Common examples of green materials with high nitrogen content include fresh grass clippings with a C/N ratio around 15:1, vegetable scraps with a C/N ratio of approximately 20:1, coffee grounds with a C/N ratio also around 20:1, and manure, such as cow or horse manure, with a C/N ratio of approximately 25:1. On the other hand, brown materials, which are high in carbon, contribute to the overall C/N balance. Examples include dry leaves with a C/N ratio ranging from 40-80:1, straw with a C/N ratio of about 50-100:1, and wood chips with a C/N ratio falling between 100-400:1. These diverse C/N ratios highlight the importance of achieving a harmonious balance between carbon and nitrogen sources in the compost pile. This balance ensures optimal conditions for microbial activity. It determines the resulting compost's overall nutrient content and quality, playing a crucial role in sustainable and effective waste management practices.

It is essential to observe that the optimal C/N ratio for decomposition is typically between 25:1 and 30:1. This proportion provides the ideal environment for microorganisms to decompose organic matter into nutrient-rich compost. Nitrogen-rich substances include fresh grass clippings, food scraps, and manure. Carbon-rich substances include dried leaves, straws, and wood shavings. During decomposition, microorganisms convert complex organic compounds into simpler substances via enzymatic reactions. This process generates heat, causing the compost material's temperature to rise. The heat generated during decomposition is essential for eradicating plant seeds, pathogens, and dangerous bacteria, resulting in a clean and secure product. According to Karmaoui (2023), Insects play a crucial role in the global ecosystem, representing a significant portion of the animal kingdom with over one million documented species. Their abundance and distribution are being impacted by climate change, leading to profound consequences for human well-being. "Climate Change and the Economic Importance and Damages of Insects" explores the intersection of economic entomology and climate change, highlighting the urgent need for research in this critical area. This research is highly relevant to composting for a sustainable future, emphasizing the intricate relationship between climate change, insects, and their economic implications. Insects, including decomposers like beetles and flies, play a vital role in the breakdown of organic matter, such as food waste and plant materials, which form the foundation of composting processes. Climate change-induced shifts in insect populations and their activities can affect the decomposition process and nutrient cycling, impacting the production of nutrient-rich soil. Understanding insects' economic importance and damage in climate change is crucial for environmentalists, scientists, researchers, scholars, practitioners, instructors, and students engaged in composting and sustainable agriculture. By shedding light on the consequences of climate change on insects and their role in various sectors, such as pollination and crop productivity, this reference source provides valuable insights into the challenges and opportunities for sustainable development in the face of climate change impacts. It encourages the integration of scientific findings into composting practices, ensuring the preservation and enhancement of soil fertility, biodiversity, and overall ecosystem resilience.

The phases are mesophilic and thermophilic. Microorganisms degrade readily available organic materials between 68° and 113° Fahrenheit (20° and 45° Celsius) during the mesophilic phase. As the compost pile decomposes, the thermophilic phase, characterized by elevated temperatures (between 122-160°F or 50-70°C), occurs. The thermophilic phase accelerates the breakdown of more robust organic compounds and ensures the destruction of pathogens and plant spores. The compost material must be turned or aerated intermittently throughout the composting process to maintain adequate oxygen levels and promote uniform decomposition. It helps prevent anaerobic conditions, which can lead to noxious aromas and methane production, a potent greenhouse gas. Composting can be accomplished using various techniques suited to multiple dimensions and environmental conditions. Traditional or backyard composting, vermicomposting (composting with worms), windrow composting, static pile composting, and in-vessel composting are all composting techniques. There are advantages, considerations, and applications for each method. Numerous decomposition techniques are tailored to dimensions, space constraints, and objectives. The backyard/On-site Composting method is appropriate for individual households or small-scale applications. Creating a compost pile or utilizing a compost container in the courtyard or garden is involved. Organic waste materials, such as kitchen remnants, yard debris, and leaves, are added to the pile or container and turned or mixed periodically to encourage decomposition.

Seri-composting, also known as the Seri method, originated in the arid regions of Rajasthan, India. It is named after the Seri tribe, whose inventive approach to decomposition in dry and scorching climates inspired its creation. Seri-composting is designed to conserve moisture and facilitate decomposition in compost mounds, making it ideal for arid regions. It entails the underground construction of circular or semi-circular decomposition basins. To reduce nutrient leaching, the ditches are enclosed with desiccated leaves or straws to prevent direct contact between the organic debris and the soil. Seri-composting entails alternating layers of organic waste and moisture-retaining material, such as straw or desiccated leaves, in the composting process. This stratification technique aids in maintaining the required carbon-to-nitrogen ratio and promotes efficient air circulation within the compost pile. It covered the trenches with a layer of soil combined with bovine dung slurry or other microbial inoculants to introduce microorganisms that assist in decomposition. This covering layer provides insulation by retaining moisture and heat within the compost pile, accelerating the decomposition process even in arid conditions. Vermicomposting, earthworms speed up the decomposition process. Organic waste is added to specialized receptacles, vermicomposting systems, and bedding material (e.g., shredded paper coconut coir). Earthworms, typically red wigglers (*Eisenia fetida*), convert organic matter into nutrient-rich worm castings or vermicompost. Vermicomposting is optimal for small-scale or indoor composting and yields high-quality compost.

Composting in windrows is appropriate for larger quantities of organic refuse. It entails constructing long, narrow stacks or aisles (windrows) of organic materials in outdoor areas, such as yard refuse, agricultural residues, and manure. Periodically, machinery or equipment revolves around the windrows to aerate the compost and ensure uniform decomposition. This procedure is commonly employed in commercial composting facilities and requires substantial space. The aerated Static Pile Composting method involves piling or enclosing organic materials in a large mass. The compost mound is aerated by introducing ventilation through implanted perforated pipelines or ducts. Typically, ventilation is regulated to maintain optimal oxygen concentrations for aerobic decomposition. Composting with aerated static piles is efficient and widely used in large-scale composting operations. In-vessel Composting in enclosed containers or structures, such as barrels, silos, or caverns, is known as in-vessel decomposition. The containers offer enhanced temperature, moisture, and odour control. Decomposition in-vessel is ideal for urban areas or facilities with limited space and can speed up decomposition.

Bokashi Composting is a fermentation-based process that employs a specialized inoculant containing beneficial microorganisms, typically cereal or sawdust. In an airtight container, organic detritus, including culinary remnants, is layered with the inoculant. The waste is fermented by microorganisms, which break it down into pre-compost material. Bokashi composting is well-known for indoor composting and is compatible with other techniques. These composting techniques give individuals, communities, and larger-scale enterprises the flexibility and options necessary to manage organic waste and produce nutrient-rich compost effectively. The choice of method is determined by variables such as available space, refuse volume, available resources, and desired composting results. Multiple variables affect the effectiveness of the decomposition procedure.

Temperature, moisture content, aeration, and the carbon-to-nitrogen ratio (C/N ratio) are crucial factors. The rate of decomposition and the elimination of pathogens and vegetation spores largely depend on temperature. Moisture levels influence microbial activity and the decomposition process as a whole. Adequate aeration is essential for supplying oxygen to aerobic microorganisms and preventing anaerobic conditions. The carbon-to-nitrogen ratio influences the decomposition rate and the quality of the resultant compost, as well as the optimal ranges for each factor and their influence on the composting efficacy. Several crucial variables influence the effectiveness and efficiency of the decomposition procedure. The impact of temperature on decomposition rate and microbial activity is essential to the composting process (Li *et al.*, 2023). Microorganisms generate heat as they decompose organic matter during the composting process. Amorim et al. (2023) During the thermophilic phase, the decomposition pile's temperature can typically rise between 50 and 70 ° Celsius (122-160 ° Fahrenheit). Higher temperatures facilitate the decomposition of more resistant materials, eradicate plant spores and pathogens, and speed up composting. However, Barakat *et al.* (2023) excessively high temperatures can harm beneficial microorganisms. Monitoring and maintaining the optimum temperature range is essential for effective decomposition.

Moisture is necessary for composting because it facilitates microbial activity and ensures proper decomposition. Ideally, the compost mound should contain between 40 and 60 per cent moisture. Insufficient moisture can delay decomposition, whereas excessive humidity can result in anaerobic conditions, disagreeable aromas, and the loss of essential nutrients. Maintaining a hydrated, but not saturated, compost pile is always necessary. Adding water or moistening the organic materials can help if the compost pile is overly dry. However, if the compost pile is too saturated, turning or mixing the compost and adding dry materials such as leaves or detritus can restore the moisture balance. Aerobic decomposition, in which microorganisms decompose organic matter in the presence of oxygen, requires a sufficient oxygen supply. Bai *et al.* (2023). Oxygen promotes the growth of aerobic organisms, which are responsible for efficient decomposition and the prevention of anaerobic conditions. Aeration keeps the compost pile aerobic, minimizing unpleasant fragrances and promoting the decomposition of organic material. By regularly mixing or rotating the compost pile, oxygen is introduced, and aeration is maintained. Large-scale decomposition systems commonly use mechanical rotation or forced aeration to maintain sufficient oxygen levels.

The carbon-to-nitrogen ratio is the proportion of carbon-rich (brown) compost materials to nitrogen-rich (green) compost materials (Azis *et al.*, 2023). The optimal C/N ratio for decomposition ranges between 25:1 and 40:1, although minor variations can still yield satisfactory results (Temel, 2023). Carbon-rich substances are energy sources, whereas nitrogen-rich substances supply proteins and promote microbial growth (Khan et al., 2023). A suitable carbon-to-nitrogen ratio ensures optimal microbial activity and effective decomposition (Zhang et al., 2023). If the C/N ratio (excess carbon) is excessively high, decomposition may be sluggish (Yu et al., 2023). If the C/N ratio is too low (excess nitrogen), the pile may emit a noxious fragrance and lose nitrogen through volatilization (Cao et al., 2023). By adjusting the C/N ratio with the addition of suitable materials, a compost pile can be kept in balance (Martínez-Gallardo et al., 2023). By understanding and controlling these factors, composters can create an optimal environment for microbial activity, decomposition, and production of high-quality compost. For efficient decomposition, regular monitoring, adjustments, and maintaining temperature, moisture, aeration, and the C/N ratio are required. Composting materials include kitchen leftovers, yard refuse, agricultural residues, manure, etc. We will discuss the characteristics of various composting materials, including their carbon-to-nitrogen ratios, moisture content, and composting advantages and considerations. Understanding the properties of different decomposition materials is essential for producing a nutrient-rich, well-balanced compost. Composting materials include various organic residues that can be used as composting feedstock. These materials are carbon-rich (brown) or nitrogen-rich (green) substances.

Carbon-rich materials, often called "browns," form the structural foundation of compost piles, contributing essential carbon and promoting aeration. Dried leaves from deciduous trees are a readily available and abundant carbon source, boasting high carbon content and aiding the pile's overall aeration. Straw and hay, typically used as animal bedding, serve as desiccated plant residues rich in carbon, offering a practical repurposing option for decomposition. Wood chips and sawdust, derived from wood, provide a high carbon-to-nitrogen ratio. Beyond their carbon contribution, they enhance aeration and add volume to the compost pile. Shredded paper and cardboard, sourced from paper products like newspapers and office paper, are carbon-rich materials that contribute to the compost pile's structure. These paper products should be sliced or torn into smaller fragments to expedite decomposition. Woody materials such as twigs and small branches also contribute to aeration and structural integrity in the compost pile. While they decompose slowly, they provide a long-lasting carbon source, extending the half-life of carbon in the compost. Incorporating these carbon-rich materials ensures a well-balanced compost, fostering optimal conditions for microbial activity and nutrient-rich end product, ultimately benefiting soil health and plant growth.

Nitrogen-rich materials, commonly called "greens," are crucial in composting, contributing essential nutrients and facilitating rapid decomposition. Kitchen scraps such as eggshells, coffee grounds, tea leaves, and vegetable and fruit trimmings are excellent nitrogen sources. These materials decompose swiftly, infusing the compost pile with vital nutrients. However, sparingly using freshly cut grass clippings is advisable to prevent matting, as they deteriorate rapidly and potentially heat the compost pile. Livestock manure from herbivores like cows, horses, and poultry is another valuable nitrogen source, providing both nitrogen and beneficial microorganisms to enhance decomposition. It is crucial, however, to age or compost the manure separately before incorporating it into the compost pile to mitigate potential pathogen concerns. Green plant trimmings, such as fresh pruning debris, vegetation, and garden residue, are rich in nitrogen and decompose rapidly. These materials contribute to a balanced blend of nutrients, further enhancing the nutrient profile of the compost. In composting, these nitrogen-rich materials are vital in creating a well-balanced and nutrient-dense compost that promotes healthy plant growth and soil fertility.

A successful compost pile requires a balance of carbon-rich and nitrogen-rich materials. The optimal decomposition carbon-to-nitrogen (C/N) ratio is between 25:1 and 40:1 (Sohil & Kichloo, 2023). A mixture of browns and greens assures the correct proportion of nutrients and microbial activity for effective decomposition. In addition to these materials, it is best to avoid including meat, dairy products, oily/greasy substances, and pet refuse in compost, as they can attract vermin or introduce pathogens. Combining carbon-rich and nitrogen-rich materials allows composters to create a diverse and nutrient-rich compost pile that supports healthy microbial activity and produces high-quality compost. Composting is a complex, multi-stage procedure that promotes efficient decomposition and makes superior-quality compost composting. The essential stages comprising this process are the selection and preparation of materials, the stratification or layering of compost ingredients, continuous monitoring and adjustment of decomposition conditions, and organized compost pile administration. Furthermore, effective composting necessitates implementing strategies such as rotating or aerating the compost pile, diligently monitoring temperature and moisture levels, and promptly resolving prevalent challenges that may emerge throughout the decomposition process.

Commencing the composting process entails carefully selecting an appropriate composting method customized to the specific requirements and available area. Composting methods include vermicomposting, in-vessel composting, and windrow composting, among others, in addition to backyard decomposition. After selecting the appropriate manner, it is critical to designate a specific area for composting, considering variables including sunlight exposure, drainage, and accessibility. Following this is the assortment of decomposition materials, emphasizing a balanced mixture of nitrogen-rich "greens" and carbon-rich "browns." Carbon provides energy-sustaining nitrogen to microbial growth, whereas greens contribute carbon. Sheered paper, dried leaves, kitchen leftovers, and grass trimmings are all examples of materials. Composting is an environmentally conscious and sustainable method of waste management that converts organic refuse into a nutrient-rich resource by observing these fundamental steps and considerations.

A systematic procedure is required to produce nutrient-dense compost at home, which commences with the compost mound or bin arrangement. In a compost pile, carbonaceous materials (e.g., dried leaves or straw) and nitrogenaceous materials (e.g., culinary scraps) coexist balanced, with compost pile layers rotating between the two substances. To establish a harmonious composition in a compost receptacle, browns and greens are incorporated directly into one another in layers. It is critical to maintain the ideal moisture level comparable to that of a damp sponge; add water to compensate for arid conditions and desiccated materials such as leaves to prevent oversaturation. Consistent mixing or stirring of the compost is essential for facilitating aeration and ensuring constant decomposition. This prevents compaction by distributing oxygen and hydration; it can be accomplished with a pitchfork or by transferring substances between containers. The compost is monitored by observing temperature, humidity, and odour. Preferred conditions for the pile during the thermophilic phase are increased temperature (from 122-160°F or 50-70°C), moisture retention, and the emission of an earthy odour. Maintenance of the compost mound becomes necessary as the decomposition process progresses. This entails adding supplements, adjusting moisture levels, and turning the pile frequently to promote decomposition. Weeks to months may pass between applying a particular method and the prevailing environmental conditions as compost matures. Compost that has reached maturity possesses a dark, granular consistency and emits a fragrant odour. The final compost product functions as an organic soil amendment and fertilizer, bolstering soil structure, augmenting nutrient concentration, and fostering holistic plant well-being. Adhering to the prescribed procedures and ensuring optimal environmental conditions, transforming organic waste into compost emerges as a viable and enduring waste management approach, concurrently promoting soil health.

## 3. COMPOSTING AT HOME

The practice of residential composting presents an extensive array of benefits. Through waste reduction within one's household, organic substances such as by-products of vegetables, fruits, and yard refuse are diverted from landfills, effectively mitigating the quantity of waste intended for disposal. The soil amendment produced from decomposition is abundant in nutrients and organic matter and contains beneficial microorganisms. By improving the soil's structure, fertility, and ability to retain moisture, this compost promotes the growth of healthier plants and increases overall crop yields. Furthermore, implementing home composting practices results in substantial financial savings as it prevents the necessity to purchase costly synthetic fertilizers and soil amendments. By doing so, one not only mitigates horticultural expenditures but also guarantees that plants acquire vital nutrients necessary for their maximum development. In addition to these advantages, residential composting significantly reduces greenhouse gas emissions linked to decomposing organic waste in landfills. Additionally, it aids in water conservation by enhancing soil moisture retention, consequently diminishing the need for irrigation. Fundamentally, the adoption of home composting arises as an environmentally conscious action that yields substantial benefits for soil health, waste management, and the overall state of the environment.

Many decomposition techniques are employed to meet the requirements of residential gardens. Vermicomposting, an approach that utilizes worms such as Eisenia fetida or red wigglers, is particularly advantageous for individuals with spatial constraints, as it can be carried out indoors and outdoors in compact containers. Establishing a vermicomposting system encompasses several key components: worm selection, receptacle construction, and vermicompost feeding and maintenance management. Regarding more expansive areas, trench composting presents a direct and well-established resolution. Implementing this method involves excavating a trench within the garden to decompose organic waste efficiently. Pit composting is particularly well-suited for expansive parks or zones designated for decomposition. The process entails setting up a trench, stratifying organic matter, and supervising the decomposition activity to obtain a soil amendment that is abundant in nutrients.

In contrast, sheet composting, also referred to as lasagna composting, offers a pragmatic methodology through the direct deposition of organic matter onto the surface of the soil. This approach streamlines the process of transforming garden waste into compost, including leaves, grass clippings, and plant residue. A comprehensive comprehension of the layering technique and careful consideration of critical elements are imperative for establishing and maintaining an effective sheet composting system. By examining these various decomposition techniques, gardeners can select the strategy most suitable for their environment and requirements. This contributes to the efficient reduction of waste and the improvement of nutrient-rich soil in residential settings.

Several critical steps are required to establish a successful home composting system, guaranteeing nutrient-rich soil and effective waste management. Selecting an appropriate decomposition method considering the space and resources is paramount. In addition to ensuring that the composting structure or container is proper, its placement in an optimal location further improves the process's efficiency. Maintaining a healthy compost pile is critical to ensuring that composting materials are in harmony, including a proportionate amount of kitchen waste, yard refuse, and carbon-rich materials such as leaves or shredded paper. Ensuring accurate carbon-to-nitrogen conversion is a critical factor to be mindful of when choosing feedstock. For effective management of residential composting, routine monitoring and maintenance are required. Selecting feedstock entails comprehending viable alternatives and ensuring an appropriate equilibrium. Consistently turning or aerating the compost pile facilitates oxygen circulation and reinforces decomposition. We shall offer guidance on the optimal methods and frequency to optimize this procedure. By monitoring temperature, moisture, and decomposition rate, ideal conditions are maintained for composting. We will guide the decomposition process on optimizing the outcomes by evaluating and modifying these variables. Individuals can incorporate composting practices into their residential and garden spaces by acquiring knowledge regarding the advantages of home composting, investigating appropriate methodologies, establishing a composting system, and applying strategies for efficient compost administration. This practice facilitates the development of nutrient-rich soil and supports sustainable waste management, thereby advocating for a comprehensive and ecologically conscious approach to managing household refuse.

## 4. LARGE-SCALE COMPOSTING

Composting on a large scale presents many compelling benefits, establishing it as an essential element of sustainable waste management systems. Among the principal advantages is effective waste management. Large-scale facilities can handle significant quantities of organic waste, thereby preventing its disposal in landfills and reducing the burden on conventional waste management systems. This practice promotes environmentally responsible waste management and contributes to reducing environmental impacts, harmonizing with the increasing focus on sustainability. Furthermore, greater control over the decomposition process is achieved through large-scale composting, which produces high-quality compost. This compost, which is reliable and of exceptional quality, has various applications, including landscaping, agriculture, and soil remediation. Its use contributes to the improvement of soil health and fertility. Significant economic benefits accompany large-scale composting as well. By generating revenue from the production and sale of compost and generating employment opportunities, these facilities may contribute to the growth of the local economy. In addition to its economic and waste reduction advantages, large-scale composting is paramount in promoting resource recovery and sustainability. These facilities implement the circular economy by extracting valuable resources from organic refuse. Compost is generated as a nutrient-dense soil amendment, which promotes sustainable agricultural practices and decreases dependence on synthetic fertilizers. Fundamentally, large-scale composting presents itself as a comprehensive resolution that tackles obstacles related to waste management, promotes sustainable resource utilization, and supports local economies.

Composting operations on a large scale utilize various techniques to manage considerable quantities of organic waste. Windrow composting is a widely used method that entails the formation of elongated, slender piles of organic matter, which are rotated at regular intervals to promote decomposition and aeration. This approach is highly compatible with outdoor facilities and effectively treats substantial volumes of organic waste. An additional sophisticated method is aerated static pile composting, in which organic waste is arranged in sizable, oxygenated mounds via pipelines or compressors. By increasing microbial activity, controlled aeration accelerates the decomposition process. This approach demonstrates notable efficacy and applies to decomposition operations conducted on a large scale.

Conversely, in-vessel decomposition occurs exclusively within confined containers or vessels that maintain controlled environmental conditions, including temperature, humidity, and airflow. This approach offers several benefits, particularly for congested urban areas or facilities with restricted areas, as it mitigates unpleasant aromas and facilitates decomposition throughout the year. Each of these composting techniques accommodates the distinct needs of extensive operations, offering adaptability and effectiveness in managing organic waste at a substantial magnitude.

Designing and constructing a large-scale composting facility requires a holistic approach, which entails meticulous attention to detail and careful planning of multiple components. Site selection, layout, apparatus selection, and infrastructure requirements are all fundamental components of facility design. Critical factors in this process include the management of incoming waste, the implementation of efficient decomposition procedures, and the assurance of appropriate discharge and effluent management. Adherence to best management practices is critical to optimize large-scale decomposition operations' environmental impact, productivity, and efficiency. The practices above consist of the following: feedstock management, temperature and moisture monitoring and control, the application of efficient compost turning or combining techniques, and the implementation of pollution and effluent control measures. By implementing these strategies, a large-scale composting facility can enhance its efficiency and positively contribute to the waste management field.

Furthermore, large-scale composting facilities must navigate the regulatory environment effectively to safeguard public health and the environment. Comprehending the permit prerequisites and regulatory structures that govern air quality, leachate management, refuse management, odour control, and compliance with local, regional, and national regulations is imperative for implementing responsible and sustainable practices. By providing an all-encompassing analysis that encompasses design deliberations, optimal management strategies, and adherence to regulatory requirements, this book furnishes its audience with the information necessary to execute and maintain large-scale composting facilities effectively. Consequently, this phenomenon fosters efficient resource recovery and waste management on a regional or local scale.

## 5. APPLICATIONS OF COMPOST

Compost, a valuable resource derived from organic matter, offers a myriad of benefits for both agriculture and landscaping. In agriculture, compost is a powerful soil enricher, providing essential organic matter, nutrients, and beneficial microorganisms that enhance soil fertility, structure, and moisture retention. This, in turn, promotes healthier plant growth and increased yields. Compost's role in erosion control is noteworthy, stabilizing soil in erosion-prone areas and reducing surface discharge, contributing to enhanced soil stability. Water conservation is another key advantage, as compost's ability to increase water-holding capacity minimizes the need for irrigation. This water use efficiency benefits agriculture and landscaping by retaining moisture in the root zone. Compost further proves valuable in soil remediation efforts, aiding in the degradation of pollutants, enhancing soil structure, and supporting the growth of plants that can extract and mitigate toxins from contaminated or degraded soils.

Compost finds versatile applications depending on specific needs. As a fertilizer, it acts as a natural and nutrient-rich source, steadily supplying plants with nourishment over time. When used as a soil amendment, compost can be incorporated into the soil or top-dressed onto existing beds during land preparation. For erosion control, compost applications on slopes or construction sites stabilize soil, reduce erosion, and encourage plant growth. Additionally, in bioremediation initiatives, compost aids in the decomposition and immobilization of contaminants, contributing to the restoration of soil health. Guidelines for effective compost use vary across applications, including recommendations for application rates, timing, methods, and considerations for specific crops or landscaping practices. The economic and environmental benefits of compost utilization are substantial. Beyond cost savings by reducing the need for artificial fertilizers and water, compost contributes to sustainability by mitigating ecological impacts. It aids in carbon sequestration, reducing greenhouse gas emissions and fostering healthy soil. It also supports closed-loop waste management, transforming organic waste into a valuable resource and contributing to a circular economy. In essence, the multifaceted advantages of compost make it an indispensable tool for fostering sustainable and efficient practices in agriculture and landscaping.

## 6. COMPOSTING AND CLIMATE CHANGE

Composting has gained significant attention as a potent instrument in the worldwide endeavour to alleviate climate change, presenting numerous benefits and prospects for reducing greenhouse gas (GHG) emissions. Carbon sequestration is an essential function of decomposition, as it converts organic waste into stable organic matter that is subsequently incorporated into the soil. This process effectively diminishes atmospheric concentrations of CO2, a significant greenhouse gas. In addition, composting is a crucial component in reducing methane emissions, as it diverts organic waste from landfills, thereby preventing the release of this highly potent greenhouse gas. Moreover, by regulating the nitrogen cycle via decomposition, the emission of nitrous oxide (N2O), an additional potent greenhouse gas, is reduced, thereby contributing to climate change mitigation.

Composting has a carbon sequestration potential that transcends the immediate sequestration within the compost. Compost promotes plant growth and biomass production by enhancing soil health and fertility, increasing carbon sequestration in vegetation and soil organic matter. Compared to alternative waste management methods, decomposition distinguishes itself through significant emission reductions, energy conservation in contrast to incineration and anaerobic digestion, carbon offsetting capabilities, and adherence to the principles of a circular economy. Composting effectively transforms organic waste into valuable compost that finds applications in landscaping, agriculture, and soil restoration. This practice not only diminishes dependence on synthetic fertilizers but also advocates for sustainable resource management.

In pursuit of promoting the uptake of composting, a multitude of policies and incentives have been established. Refuse management regulations may impose penalties for noncompliance, landfill bans, or recycling obligations, all designed to encourage the diversion of organic refuse—grants and subsidies, among other financial incentives, aid in constructing decomposition facilities. The public is educated on the advantages of composting through outreach and education programs, whereas market development initiatives advocate for the utilization of compost products. These policies and incentives collectively promote the extensive implementation of decomposition methods. By comprehending the various advantages of composting, including its capacity for carbon sequestration, comparative benefits concerning alternative waste management approaches, and the enabling policies that facilitate its use, readers acquire a deeper understanding of the critical significance of composting as a sustainable waste management solution and its constructive contribution towards mitigating the effects of climate change.

## 7. COMPOSTING CHALLENGES AND SOLUTIONS

Composting, while a sustainable waste management solution, has challenges. Common issues include odours, pests, weed seeds, and pathogens, each potentially affecting the success and acceptability of composting operations. Odours arise from improper compost pile management, while problems like flies, rodents, and insects may disrupt decomposition and pose sanitary risks. Weed seeds surviving the decomposition process can lead to increased vegetation growth, and pathogens in organic debris may persist if decomposition conditions are inadequate. To address these challenges, effective solutions can be implemented. Proper aeration, maintaining correct moisture levels, covering the compost pile, and employing biofilters prove effective for odour management. Pest control involves securing compost piles with fences or covers, proper waste management to avoid attractants, and using pest control measures if necessary. Weed seed issues can be tackled through temperature management during decomposition, screening and sieving, and allowing compost sufficient time to mature. Pathogen control requires proper temperature management, composting duration, maturation, and careful selection of source materials.

Education and training are crucial components of successful composting. Operators and workers need training on composting techniques, best practices, and safety measures to ensure efficient waste management. Public education and awareness campaigns promote composting best practices, compostable materials, and the overall significance of composting for sustainable waste management. Composting challenges can be effectively managed by implementing these solutions and fostering education and awareness. This ensures the production of high-quality compost, minimizing environmental and health hazards associated with odours, pests, weed seeds, and pathogens. Ultimately, these strategies contribute to composting practices' successful and sustainable integration in waste management systems.

Seattle, Washington, has successfully implemented a community-based composting programme titled "Food Plus Yard Waste." This initiative encourages residents to compost their food remnants and yard refuse rather than dispose of them in landfills. Residents are provided with curbside food waste and yard waste collection, then transported to a large-scale composting facility. The programme has gained significant success in diverting organic waste from landfills. By 2020, nearly 60 per cent of Seattle's residential refuse was reported to be compostable. This program's compost enriches the soil in public parks and community gardens, reducing the need for synthetic fertilizers and increasing soil fertility.

Due to its limited land area, Singapore has instituted innovative decomposition methods to manage its organic refuse. The city-state has constructed the Tuas Nexus Integrated Waste Management Facility, one of the world's most extensive integrated waste management facilities. This facility contains a substantial composting facility. The decomposition facility converts culinary refuse and garden waste into high-quality compost. The compost is used for landscaping and soil enhancement throughout the city's parks and green spaces. The facility has contributed to Singapore's aim of becoming a zero-waste nation by substantially reducing the quantity of organic refuse sent to landfills.

The BioCycle Farm in California is a paradigm for integrating sustainable agriculture and decomposition. Composting is a fundamental component of the farm's regenerative agricultural practices. It incorporates food refuse, agricultural residues, and vegetative waste to create compost that is abundant in nutrients. Applying compost to the farm's fields improves soil health, promotes carbon sequestration, and reduces the need for synthetic fertilizers. The farm has instituted large-scale decomposition techniques, such as windrow composting and in-vessel composting, to manage various organic materials effectively.

The future of composting is poised to witness transformative trends and innovations driven by a commitment to enhance efficiency, sustainability, and overall process efficacy. Advanced composting technologies are on the horizon, with the development of automated systems incorporating real-time monitoring and control of temperature, moisture, and oxygen levels. In-vessel composting, conducted within enclosed containers, is gaining prominence for its ability to mitigate fragrance emissions, control pests and weed seeds, and expedite decomposition, making it scalable for various operation sizes.

Carbon footprint tracking is emerging as a crucial focus, aligning with the growing emphasis on reducing carbon emissions. Innovations in carbon monitoring aim to quantify the environmental benefits of composting, contributing to climate change mitigation efforts. Vermicomposting is evolving with advancements in worm species selection, substrate materials, and management techniques, enhancing its efficiency and adaptability for diverse applications. Integrating biochar, a by-product of organic waste pyrolysis, into the composting process holds promise for improving nutrient retention, soil structure, and carbon sequestration potential. Data analytics and artificial intelligence (AI) are expected to optimize composting operations by providing real-time insights and automating adjustments for optimal conditions, ultimately reducing costs and waste.

Innovations in feedstock management, including pre-processing techniques, enhanced sifting and separation methods, and alternative feedstocks, such as agricultural residues, aim to increase decomposition efficiency. Furthermore, integrating composting with renewable energy production, such as combining anaerobic digestion with decomposition to produce biogas, presents an opportunity for resource recovery and reduced reliance on fossil fuels. Decentralized composting solutions are gaining traction, focusing on small-scale operations at or near the source of organic waste production. This approach not only reduces transportation costs but also minimizes waste disposal-related emissions. Compact, on-site composting systems and community-based initiatives are expected to be pivotal in improving localized waste management and enhancing soil fertility. Overall, these future trends and innovations underscore a dynamic and progressive landscape for composting, contributing to its evolving role in sustainable waste management practices.

## 8. DISCUSSION

Composting is unquestionably vital to sustainable agriculture and waste management. It offers many benefits, including recycling organic refuse, reducing landfill waste, and improving soil health (Sharma et al., 2019). Composting helps close the nutrient cycle, reduces reliance on synthetic fertilizers, and promotes sustainable resource management by converting organic materials to nutrient-rich compost (Ayilara et al., 2020). Composting's ability to divert organic waste from landfills is one of its most essential advantages (Wei et al., 2017). It reduces methane emissions, a potent greenhouse gas, and the pressure on landfill capacities (Composting et al.: Keeping a Good Thing Going, 2020). By decomposing organic waste, we can reduce the environmental impact of waste disposal and contribute to the circular economy by turning refuse into a valuable resource (Zhang et al., 2022).

Composting is essential to sustainable agriculture because it increases soil fertility and promotes nutrient cycling (Sequi, 1996). Compost's organic matter and nutrients improve soil structure, water retention, and microbial activity, resulting in healthier and more productive soils. It promotes plant growth, decreases the need for synthetic fertilizers, and reduces the danger of nutrient discharge and water contamination (Evanylo et al., 2008). Composting substantially impacts climate change mitigation (Elbasiouny et al., 2019). Compost-amended soils have demonstrated the capacity to sequester carbon, mitigating greenhouse gas emissions. Through organic waste management and carbon sequestration, composting provides a practical and scalable solution for addressing climate change.

Despite the apparent benefits of composting, some obstacles and challenges must be overcome. These include more infrastructure and resources for large-scale composting, limited awareness and education regarding composting practices, and regulatory barriers. To overcome these obstacles, governments, communities, businesses, and individuals must collaborate to implement supportive policies, provide the necessary resources and infrastructure, and raise awareness about the significance of decomposition. Composting is essential for sustainable agriculture and waste management (Ayilara et al., 2020). It is necessary for a sustainable future due to its capacity to recycle organic waste, enhance soil health, mitigate climate change, and promote resource conservation. By accepting decomposition and working together, we can make our society more resilient and environmentally friendly, where organic waste is seen as a helpful resource instead of a nuisance.

## 9. CONCLUSION

This chapter has examined the water-soil-plant-animal nexus in the era of climate change, with a particular emphasis on decomposition as a crucial solution for a sustainable future. The following are the major topics covered in this chapter: Composting is the natural decomposition of organic refuse into a nutrient-rich soil amendment. It has a long history and has developed into numerous techniques and practices. Composting is essential for sustainable agriculture and waste management because it recycles organic waste, reduces landfill waste, and enhances soil health. For successful decomposition, it is necessary to understand the physics of composting, including the factors that influence the process and the characteristics of compost materials. Composting can be accomplished at home via vermicomposting, trench composting, and sheet composting, providing numerous benefits to home gardens and reducing domestic waste. Composting on a large scale utilizes techniques such as windrow composting, aerated static pile composting, and in-vessel composting, all requiring meticulous planning, construction, and administration. Compost has a variety of uses in agriculture, landscaping, erosion control, and bioremediation, and it provides numerous benefits for soil fertility, plant growth, and environmental sustainability. Carbon sequestration and composting's comparative advantages over other waste management methods are essential for mitigating climate change.

For a sustainable future, composting is of paramount importance. It addresses various environmental issues, including waste management, soil degradation, and climate change. Composting diverts organic waste from landfills, reducing greenhouse gas emissions and releasing hazardous substances. It enriches the soil with organic matter, nutrients, and beneficial microorganisms, enhancing its health. Composting conserves water, increases biodiversity, and decreases the demand for synthetic fertilizers and pesticides.

In addition, composting is a scalable and accessible practice that can be implemented on various scales, from individual households to large-scale operations, making it an inclusive and effective solution for a sustainable future.

To create a sustainable future, we must act to promote decomposition and reduce waste. There is a role for governments, communities, enterprises, and individuals. Governments should enact favourable policies, regulations, and incentives to promote composting and waste reduction initiatives. Communities can establish composting programmes, educate residents about the advantages of composting, and provide infrastructure and resources to facilitate local decomposition. To responsibly manage their organic waste, businesses should implement composting practices, divert organic waste from landfills, and investigate partnership opportunities with composting facilities. To reduce food waste and recycle organic materials, individuals can compost at home, participate in community composting initiatives, and make conscientious decisions. Campaigns for education and awareness should emphasize the significance of decomposition, its benefits, and its role in fostering a sustainable future. We can preserve the environment, improve soil health, mitigate climate change, and create a more sustainable and resilient future for future generations if we collectively embrace decomposition and reduce waste. Composting is a simple yet effective solution to the problem of global waste management. It provides a sustainable alternative to conventional waste management methods and numerous environmental and economic benefits. Composting reduces greenhouse gas emissions, mitigates climate change, promotes sustainable agriculture, and sustains biodiversity. We hope this chapter will inspire individuals, businesses, and communities to adopt composting as a sustainable waste management strategy. It can reduce our environmental imprint, promote sustainable agriculture, and contribute to a more sustainable future. Everyone is encouraged to act and begin composting immediately.

## REFERENCES

Amorim Orrico, A. C., de Oliveira, J. D., Viana Leite, B. K., da Silva Vilela, R. N., Previdelli Orrico, M. A. Junior, Aspilcueta Borquis, R. R., & Macena, I. A. (2023). Effects of aeration and season of the year on fish waste composting and compost quality. *Environmental Technology*, 1–32. doi:10.1080/09593330.2023.2230350 PMID:37401255

Ayilara, M. S., Olanrewaju, O. S., Babalola, O. O., & Odeyemi, O. (2020). Waste Management through Composting: Challenges and Potentials. *Sustainability (Basel)*, *12*(11), 4456. doi:10.3390u12114456

Azis, F. A., Choo, M., Suhaimi, H., & Abas, P. E. (2023). The effect of initial carbon to nitrogen ratio on kitchen waste composting maturity. *Sustainability (Basel)*, *15*(7), 6191. doi:10.3390u15076191

Bai, S., Chen, J., Guo, M., Ren, N., & Zhao, X. (2023). Vertical-scale spatial influence of radial oxygen loss on rhizosphere microbial community in a constructed wetland. *Environment International*, *171*, 107690. doi:10.1016/j.envint.2022.107690 PMID:36516673

Barakat, N. A., Hameed, M. M. A., Fadali, O. A., Abdelraheem, O. H., Hefny, R. A., & Moustafa, H. M. (2023). Graphitized corncob 3D Biomass-driven anode for high-performance batch and continuous modes Air–Cathode microbial fuel cells working by domestic wastewater. *International Journal of Hydrogen Energy*, *48*(98), 38854–38869. doi:10.1016/j.ijhydene.2023.06.231

Cao, T., Zheng, Y., & Dong, H. (2023). Control of odour emissions from livestock farms: A review. *Environmental Research*, *225*, 115545. doi:10.1016/j.envres.2023.115545 PMID:36822532

Elbasiouny, H., Elbanna, B. A., Al-Najoli, E., Alsherief, A., Negm, S., El-Nour, E. A., Nofal, A., & Sharabash, S. (2020). Agricultural Waste Management for Climate Change Mitigation: Some Implications to Egypt. In A. Negm & N. Shareef (Eds.), *Waste Management in MENA Regions. Springer Water*. Springer. doi:10.1007/978-3-030-18350-9_8

Evanylo, G. K., Sherony, C. A., Spargo, J., Starner, D. E., Brosius, M., & Haering, K. (2008). Soil and water environmental effects of fertilizer-, manure-, and compost-based fertility practices in an organic vegetable cropping system. *Agriculture, Ecosystems & Environment*, *127*(1–2), 50–58. doi:10.1016/j.agee.2008.02.014

Folino, A., Karageorgiou, A., Calabrò, P. S., & Komilis, D. (2020). Biodegradation of wasted bioplastics in natural and industrial environments: A review. *Sustainability (Basel)*, *12*(15), 6030. doi:10.3390u12156030

Gerba, C. P., & Smith, J. E. (2005). Sources of pathogenic microorganisms and their fate during land application of wastes. *Journal of Environmental Quality*, *34*(1), 42–48. doi:10.2134/jeq2005.0042a PMID:15647533

Howard, A. (1947). *The Soil and Health: A Study of Organic Agriculture*. University Press of Kentucky. https://www.jstor.org/stable/j.ctt2jcp63

Ju, X., Zhang, F., Xuemei, B., Römheld, V., & Roelcke, M. (2005). Utilization and management of organic wastes in Chinese agriculture: Past, present and perspectives. *Science in China. Series C, Life Sciences*, *48*(S2, Suppl 2), 965–979. doi:10.1007/BF03187135 PMID:20549451

Karmaoui, A. (Ed.). (2023). *Climate Change and the Economic Importance and Damages of Insects* (pp. 1–279). IGI Global. doi:10.4018/978-1-6684-4824-3

Khan, S., Das, P., Thaher, M. I., AbdulQuadir, M., Mahata, C., & Al Jabri, H. (2023). Utilization of nitrogen-rich agricultural waste streams by microalgae to produce protein and value-added compounds. *Current Opinion in Green and Sustainable Chemistry*, *100797*, 100797. doi:10.1016/j.cogsc.2023.100797

Li, F., Ghanizadeh, H., Cui, G., Liu, J., Miao, S., Liu, C., & Wang, A. (2023). Microbiome-based agents can optimize composting of agricultural wastes by modifying microbial communities. *Bioresource Technology*, *374*, 128765. doi:10.1016/j.biortech.2023.128765 PMID:36822555

Martínez-Gallardo, M. R., Estrella-González, M. J., Suárez-Estrella, F., López-González, J. A., Jurado, M. M., Toribio, A. J., & López, M. J. (2023). Effect of upstream bioactivation of plant residues to accelerate the composting process and improve product quality. *Agronomy (Basel)*, *13*(6), 1638. doi:10.3390/agronomy13061638

Pedrosa, M. M., Cuadrado, C., Burbano, C., Allaf, K., Haddad, J., Gelencsér, E., Takács, K., Guillamón, E., & Muzquiz, M. (2012). Effect of instant controlled pressure drop on the oligosaccharides, inositol phosphates, trypsin inhibitors and lectins contents of different legumes. *Food Chemistry*, *131*(3), 862–868. doi:10.1016/j.foodchem.2011.09.061

Sequi, P. (1996). The role of composting in sustainable agriculture. In M. de Bertoldi, P. Sequi, B. Lemmes, & T. Papi (Eds.), *The Science of Composting*. Springer. doi:10.1007/978-94-009-1569-5_3

Sharma, B., Vaish, B., Monika, S., Singh, U. K., Singh, P., & Singh, R. P. (2019). Recycling of organic wastes in agriculture: An environmental perspective. *International Journal of Environmental Research*, *13*(2), 409–429. doi:10.100741742-019-00175-y

Sohil, A., & Kichloo, M. A. (2023). Sustainable Solutions to Animal Waste: Climate Change Mitigation and Bioproduct Harvest. In *Climate Changes Mitigation and Sustainable Bioenergy Harvest Through Animal Waste: Sustainable Environmental Implications of Animal Waste* (pp. 301-332). Cham: Springer Nature Switzerland. https://link.springer.com/chapter/10.1007/978-3-031-26224-1_12

Temel, F. A. (2023). Evaluation of the influence of rice husk amendment on compost quality in sewage sludge composting. *Bioresource Technology*, *373*, 128748. doi:10.1016/j.biortech.2023.128748 PMID:36791979

Wang, H., Dai, Z., Krauss, K. W., Trettin, C. C., Noe, G. B., Burton, A. J., & Ward, E. J. (2023). Modelling impacts of saltwater intrusion on methane and nitrous oxide emissions in tidal forested wetlands. *Ecological Applications*, *33*(5), e2858. doi:10.1002/eap.2858 PMID:37084186

Wei, Y., Li, J., Shi, D., Liu, G., Zhao, Y., & Shimaoka, T. (2017). Environmental challenges impeding the composting of biodegradable municipal solid waste: A critical review. *Resources, Conservation and Recycling*, *122*, 51–65. doi:10.1016/j.resconrec.2017.01.024

Wu, K. (2013). *Memo 2: The History of Composting*. Shaping the Future of New York City. https://eportfolios.macaulay.cuny.edu/macbride13/research/timelines/memo-2-the-history-of-composting/

Yu, Y., Hou, J., Jahangir, A., Cao, X., Balsalobre-Lorente, D., Radulescu, M., & Jiang, T. (2023). Decomposition analysis of China's chemical sector energy-related CO2 emissions: From an extended SDA approach perspective. *Energy & Environment*, 0958305X2311516. doi:10.1177/0958305X231151682

Zhang, L., Su, J., Ali, A., Huang, T., Chen, C., & Yang, W. (2023). A biofilm reactor based on a slow-release carbon source effectively improved the continuous denitrification capacity of slightly polluted surface water at a low carbon-to-nitrogen ratio. *Journal of Environmental Chemical Engineering*, *11*(2), 109552. doi:10.1016/j.jece.2023.109552

Zhang, Z., Malik, M. I., Khan, A., Ali, N., Malik, S., & Bilal, M. (2022). Environmental impacts of hazardous waste, and management strategies to reconcile circular economy and eco-sustainability. *The Science of the Total Environment*, *807*, 150856. doi:10.1016/j.scitotenv.2021.150856 PMID:34627923

## KEY TERMS AND DEFINITIONS

**Composting:** The natural process of decomposing organic waste into nutrient-rich soil conditioner, enhancing soil fertility and reducing the volume of garbage.

**Plants:** Living organisms capable of photosynthesis, converting sunlight into energy, essential for oxygen production, food webs, and ecosystem balance.

**Soil:** The upper layer of Earth's crust, rich in minerals and organic matter, supporting plant growth and providing a foundation for terrestrial ecosystems.

**Sustainable Future:** A collective effort to preserve resources, reduce environmental impact, and foster social equity to ensure a resilient and thriving future for all.

**Sustainable:** Practices that meet present needs without compromising the ability of future generations to meet their needs, promoting long-term environmental health.

**Turning Waste:** Transforming discarded materials into valuable resources through recycling, composting, or other methods, minimizing environmental impact and promoting resource efficiency.

**Water:** A vital liquid essential for life, agriculture, and various industrial processes, crucial for sustaining ecosystems and human well-being.

Section 3

# Animals in the Context of Climate Change

# Chapter 14
# Protozoan, Animal, Socio-Ecological Factors, and Vectorial Diseases:
## The Leishmaniasis in the Province of Errachidia

**Saadia Achichaou**
*Faculty of Science and Technology, Errachidia, Morocco*

**Hajar Ayach**
*Moulay Ismail University, Morocco*

**Fatima Ezzahra Akhatar**
*Moulay Ismail University, Morocco*

**Yassmine Modrak**
*Moulay Ismail University, Morocco*

**Sawsane Zahir**
*Moulay Ismail University, Morocco*

**Ahmed Karmaoui**
https://orcid.org/0000-0003-3881-4029
*Moulay Ismail University, Meknès, Morocco & Moroccan Center for Culture and Science, Morocco*

## ABSTRACT

*Leishmaniasis is a parasitosis caused by a flagellate protozoan of the genus Leishmania, transmitted by a vector (Phlebotomus) from a reservoir, rodent. It is endemic in several countries in the Mediterranean region including Morocco. This is a retrospective and descriptive study of the epidemiological profile of leishmaniasis in the province of Errachidia over the period 2012-2022. The total number of cases recorded during this period is 6726 cases. This number is very high, and despite control programs, the province of Errachidia continues to record cases, reaching 105 cases in 2022. The seasonal distribution of leishmaniasis cases was more concentrated in the autumn-winter seasons of each year. The majority of cases were recorded in rural areas, and Goulmima was the most affected district in the whole province. The authorities concerned have set up a surveillance and control program aimed at reducing the number of cases and limiting the geographical spread of leishmaniasis.*

DOI: 10.4018/978-1-6684-9838-5.ch014

## INTRODUCTION

Parasitic diseases like Leishmaniasis pose a significant challenge to human health, being widespread and progressively more hazardous. Leishmaniases are illnesses transmitted by certain species of sand flies and are prevalent across tropical regions worldwide (Tonelli et al., 2021). Leishmaniasis is a common pathology affecting both humans and specific mammals. to humans and certain mammals. It is caused by flagellate protozoa of the genus *Leishmania*, which are carried by the bite of a dipteran insect vector, the female sandfly (Rosenthal & Marty, 2009). Phlebotomine are nematoceran insects belonging to the phylum Arthropoda, Order Diptera, Family Psychodidae and Subfamily Phlebotominae (Kertesz, 1904). This Subfamily is divided into five genera, *Phlebotomus* and *Sergentomyia* in the Old World, and *Lutzomyia*, *Warileya*, and *Brumptomyia* in the New World, according to the first classification presented by Lewis et al., (1977), based on the external morphological characteristics of sandflies. The second classification was presented by Galati (1995), who divided the Phlebotominae Subfamily into two genera, *Hertigiini* and *Phlebotomini* (Galati, 1995). Sandflies are tiny insects distributed worldwide (Zé-Zé et al., 2021) mesuring 1 to 3 mm long for adults, yellowish-black with a humped appearance; the body and wings have a hairy appearance (Boussa, 2008). Nearly 1000 species of sandflies are spread across six genera. Among them, only three genera—Phlebotomus, Lutzomyia, and Sergentomyia—include hematophagous species (Zé-Zé et al., 2021). The life span of sandflies varies from 20 to 75 days from egg hatching to adult emergence (Berthet-Beaufils, 2010). While exhibiting limited mobility, certain vectors have extensive geographical distributions, posing challenges for the implementation of control measures (Tonelli et al., 2021). The flight capacity of sand flies can be affected by climatic factors such as temperature, wind, and humidity. If the temperature is sufficiently high (19-20°C), if there is no wind and the humidity is high, they don't start to move until nightfall (Mazelet, 2004). There are several clinical forms in humans: Cutaneous leishmaniasis (LC), Visceral leishmaniasis also known as kala-azar (LV), and Cutaneous-mucosal leishmaniasis (LCM) (Rguioui, 2016). Like all Kinetoplastida, leishmania is characterized by the presence of a single mitochondrion comprising circular kinetoplastic DNA (Simpson, 1987). According to the World Health Organization (WHO), it is considered one of the main causes of public health problems worldwide. This vector-borne disease is endemic in 88 countries (in Latin America, Africa, Asia and Southern Europe) (Boussaa, 2008). cutanée et pour la leishmaniose viscérale (LV) étaient *Donovani* (Espinosa et al., 2018). Seasonal disease known as zoonotic cutaneous leishmaniasis (ZCL) is associated with Phlebotomus papatasi, Scopoli, 1786, the main confirmed vector of *L. major* dynamics in North African countries (Karmaoui et al., 2022a). Morocco is one of the countries affected by leishmaniasis in this Mediterranean Basin. The disease used to be confined mainly to the arid regions of the Sahara, but outbreaks have become more frequent. This increase is probably due to climate change and the geographical spread of the vector. Morocco has four types of leishmaniasis including the rural cutaneous leishmaniasis caused by *Leishmania major*, the urban cutaneous leishmaniasis caused by *Leishmania tropica*, the utaneous leishmaniasis caused by *Leishmania infantum*, and the visceral leishmaniasis caused by *Leishmania infantum* (Boussaa, 2008).

The disease was notifiable under Ministerial Order No. 683-95 of March 31, 1995. Later, a leishmaniasis control program was launched in 1997 and has been continually renewed (Ait Antar, 2021). However, leishmaniasis continues to exist, as many factors increase the risk of contracting the disease: poverty, malnutrition, environmental changes, and climatic conditions (WHO, 2016). According to the Koppen-Geiger classification, the Province of Errachidia, which is characterized by a warm Mediterranean climate with dry summers, is affected by cutaneous disease (*L. major*). A study on the epidemiological profile, actual state of the disease distribution in the the province is needed to show the associations between animal, protozoan, environmental and anthropogenic elements. In this context, the aim of this work is to carry out a retrospective epidemiological study of cutaneous and visceral leishmaniasis in the Errachidia Province, based on cases reported by the "Service des Réseaux des Etablissements de Santé (SRES)" in theriod 2010-2022. This study aims also to identify factors associated with the distribution of LC and LV.

## MATERIALS AND METHODS

### Study Area

The province of Errachidia is part of Morocco's Drâa-Tafilalet region (Figure 1A & B). Geographically, this province is bordered to the north by the Province of Midelt and to the Northeast by the Province of Figuig and to the Southeast area by Algeria, to the West by the two Provinces of Tinghir and Zagora. According to the new administrative division of 2015, the province of Errachidia included 29 communes, 22 of which are rural and seven urbans (Figure 1C).

Regarding the anthropogenic characteristics, the population of Errachidia Province represents 25.6% of the total population of the Draa-Tafilalet region and 1.2% of the national population. The total population of Errachidia province is 418,451, including 194,084 in urban areas and 224,367 in rural areas, with 50.9% of females versus 49.1% of males. This province is a vast administrative unit covering an area of 59,585 $Km^2$ (Figure 1C), with zones corresponding to four distinct morphological units: the High Atlas range to the north, characterized by its imposing mountains, the Hamadas domain and the Tafilalet plain. In addition, a large part of the rural communes of Ferkla Oulia and Melaab are part of the Eastern Anti-Atlas region. The climate of the study region is semi-desert with significant continental influences (Ben Salem et al., 2023). It is characterized by scorching dry summers and cold winters, and experiences infrequent and unpredictable rainfall, occurring sporadically in both space and time (Karmaoui et al., 2023). The climate, primarly, precipitation, is one of the determining factors of land use and land cover, wich makes the province more exposed to drought and anthropogenic factors (Karmaoui et al., 2022b). According to the Provincial Monograph of Errachidia (PME) (2019), the province is characterized by asubstantial temperature variations between extremely high summer temperatures and notably low winter temperatures, a rainfall that is scarce and irregularly distributed across time and space, and a wind speeds that surpass 57.6 km/h (during May, June, July, and August). Concerning the water resources in the Province of Errachidia, they are made up of run-off water from wadis in the High Atlas, and groundwater spread over several basins.. Several aquifers were identified in province of Errachidia. There are Quaternary aquifers in Errachidia, Tinjdad, Goulmima, Boudnib-Bouanan and Tafilalet (Arfoud, Rissani, Fezna-Jorf), as well as a Cretaceous Basin aquifer (Infra-Cenomanian, Turonian and Senonian) (PME, 2019). The aquifer systems play an essential role in the management of the region's water resources and in the supply of drinking water to local communities. However, the biological and ecological characteristics include a variety of drought-adapted plants, such as *Tamarix aphylla* and *Acacia* genera. Pine and Cedar forests can also be found in the mountains area. The fauna is equally rich and diverse including Mammals such as gazelles, jackals, foxes, hares, and wild boar andbirds like eagles, vultures, partridges, and falcons. The Errachidia region is also known for its oases, which are fertile agroecosystems, home to a wide variety of plants and animals, including date palms, frogs and migratory birds.

*Figure 1. (a) Morocco; (b) Deraa Tafilalet region; (c) Errachidia province*

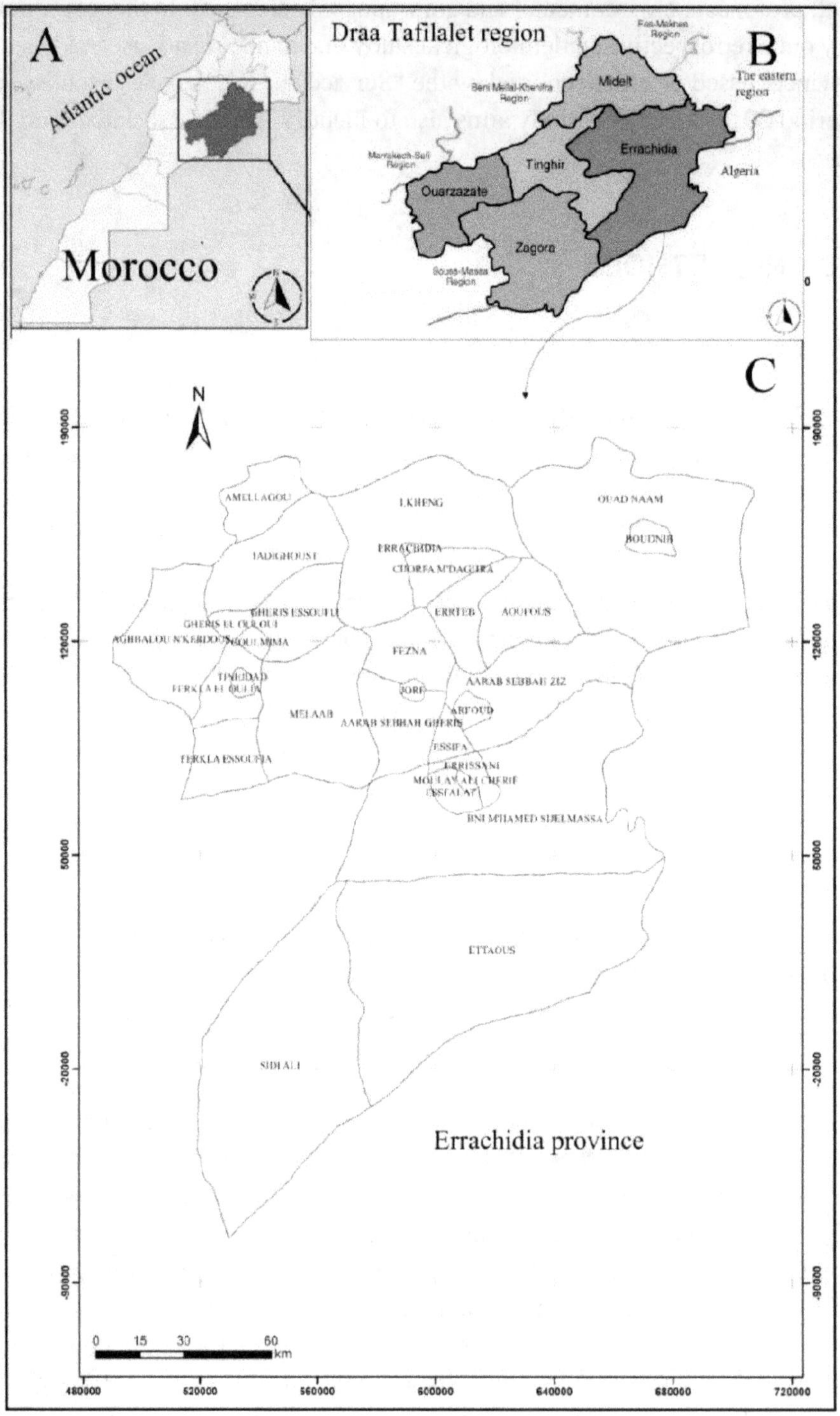

## Methodology

The data was gatherd following two methods, a collection from the city's Health Center and through survey in the University Moulay Ismail, Morocco.

### Data collection from the city's Health Center

Data were collected from registers compiled by the SRES (SRES, 2023) of the provincial delegation of the Ministry of Health in Errachidia, after requesting authorization from the Provincial Health Delegate. The parameters studied were the number of cases, the patient age, gender, rural or urban geographical origin, commune of origin, geographical and temporal distribution of cases.

### Data collection through a survey in the University

The survey is an online form designed and sent via Google Forms, which allows students to choose their answers easily. The survey focuses on students' perceptions of leishmaniasis. The questionnaire comprises 21 questions covering variables such as the age, gender, education, and region of origin.The survey includes also many questions about the disease including knowledge about the disease, types, causes, vectors, reservoirs, most-infected genus, active time of the sandfly, factors in the development of sandflies. It gouped alos some questions like family history, parts of the body affected by the disease, risk factors for the disease, and symptoms.

## RESULTS

### Analysis of Data collected From Official Archives of the Provincial Health Administration

The findings showed that in 2012, the number of leishmaniasis cases recorded is very high and after a decrease until 2015. In 2012 a peak value of leishmaniasis cases was recorded 4109 cases then 2016; and in 2013 and 2014, a limited number of cases were recorded, decreasing each year. Since 2015, this number has increased until 2022; but by 2020, the number has decreased (Figure 2A). However, the monthly distribution of leishmaniasis cases shows low numbers during the spring and summer seasons. The lowest number of cases was recorded in June (33 cases) (Figure 2B). The majority of cases were recorded during the autumn and winter seasons, with a maximum number of cases during the months of November, December, January and February, representing 69.03% (3651 cases) of the total number of cases recorded.

The data collected shows that cases of leishmaniasis are distributed in a polymorphic way in the province of Errachidia (Figure 3). Some communes recorded high numbers of cases, while others have fewer cases. The circle most affected by LC is the circle of Goulmima (2256 cases), followed by the circle of Rissani (753 cases), and the least affected is the circle of Errachidia (220 cases). The most affected communes are Melaab (586 cases), Ferkla El Oulia (672 cases), and Ferkla Es-soufla (380 cases). The least affected zones are Chorfa M'Daghra and Er-rteb with seven cases for each commune while Es-sfalat and Et-taous with 0 cases (Figure 3).

*Figure 2. The annual distribution of leishmaniasis cases in: (a) The province of Errachidia; (b) The monthly distribution of leishmaniasis cases throughout the study period (2012-2022)*
*Source: authors*

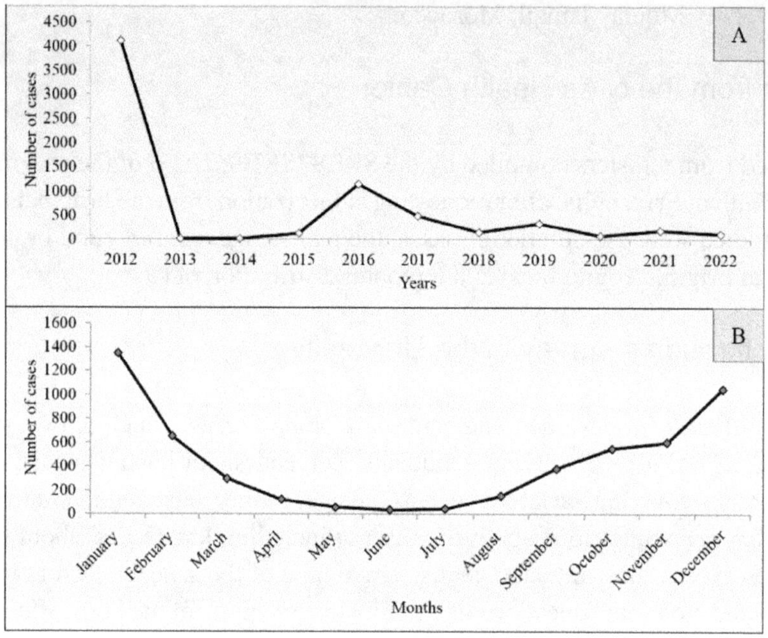


*Figure 3. Geographical distribution of leishmaniasis cases in all communes of Errachidia province*
*Source: Authors using data from the Health Center Archives*

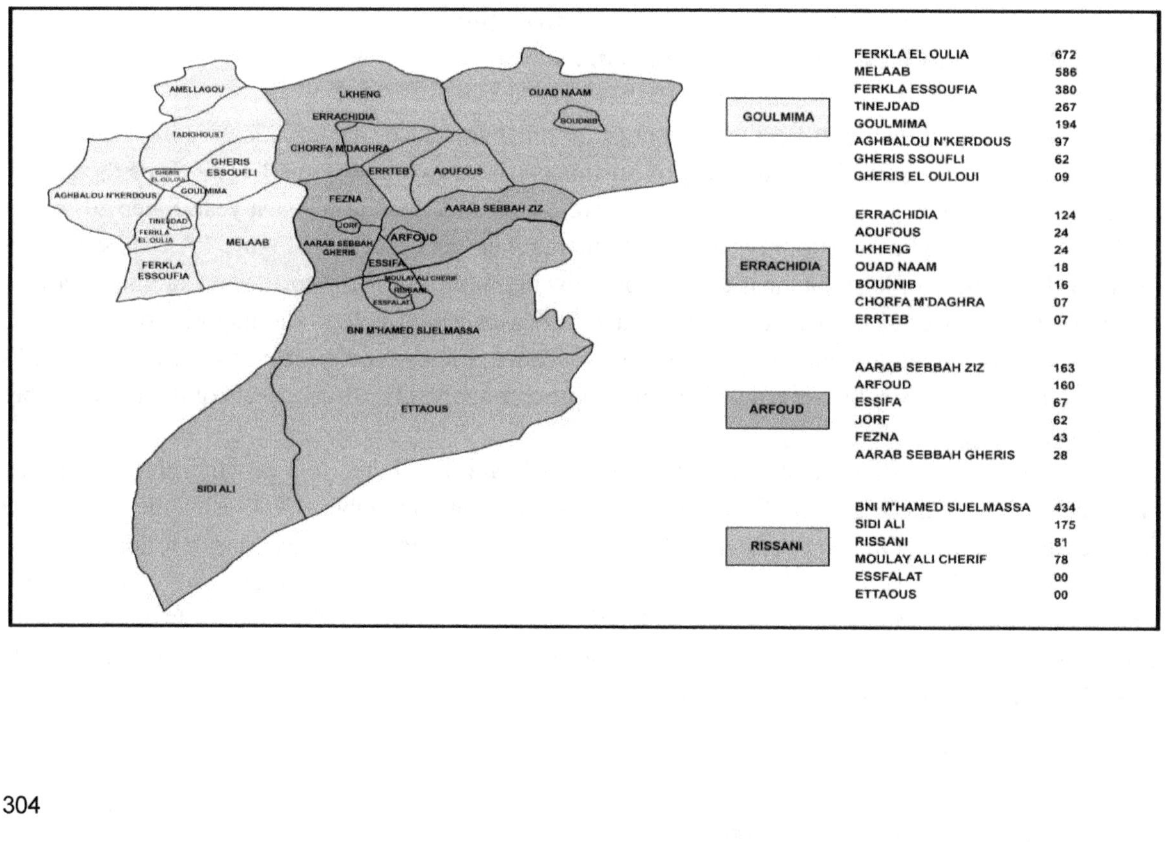

*Figure 4. Annual distribution of LC cases by (a) patient gender and (b) age group*
*Source: The authors, using data from the health center archives*

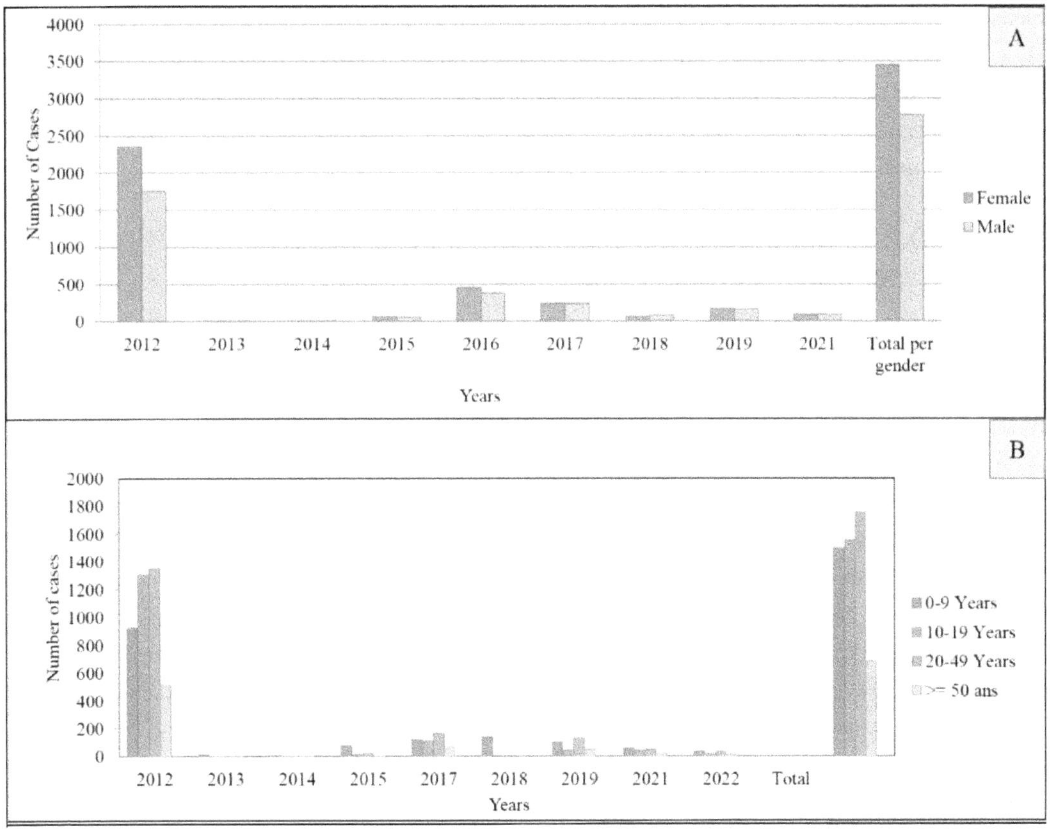


*Figure 5. Classification (Indigenous and imported cases)*
*Source: The authors, using data from the health center archives*

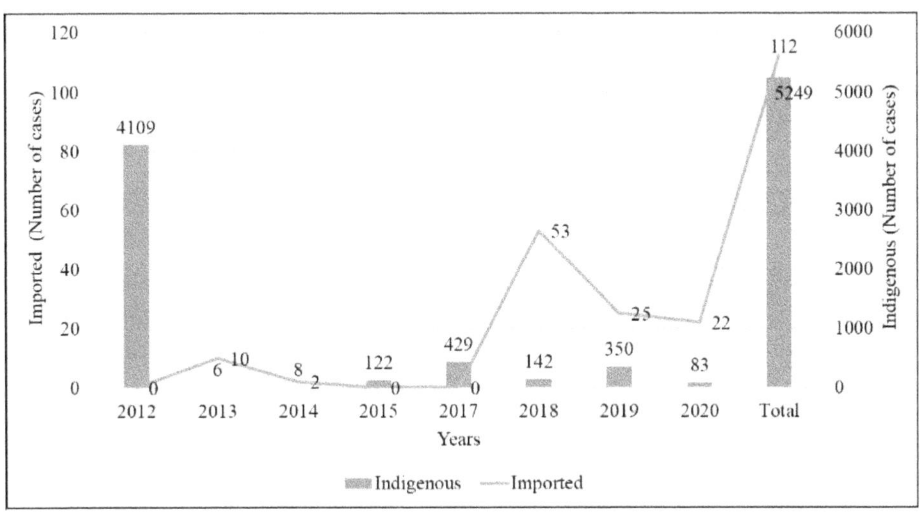

Regarding the gender aspect, the results show that feamles are more affected by leishmaniasis than males in all the years of this study (Figure 4A). Figure 4B shows that the age groups most infected by the disease are those estimated to be between 20 and 49 years of age, followed by those between 10 and 19 years, then those between 0 and 9 years, and people over 50 years of age are less infected by the disease (Figure 4B).

Most registered cases are indigenous with a total number of 5249 compared to the imported cases (112 during the period from 2012 to 2019) (Figure 5).

## Analysis of Data Collected From the Survey in the Faculty of Sciences and Techniques of Errachidia

### Students' Gender and Housing Sector

The majority of respondents were female (71%) and 29% were male (Figure 6B) while 60% of students were from urban areas and 40% from rural areas (Figure 6B).

Regarding the knowledge of respondents about the leishmaniasis disease, 69% were aware of it (Figure 6C). The type most familiar to students is cutaneous leishmaniasis (155 students), followed by visceral leishmaniasis (82 students), and 62 students responded that they know cutaneous mucosal leishmaniasis (Figure 6D). However, 55% of respondents correctly answered the question that the main cause of leishmaniasis is a protozoan, 27% think it is a bacterium, and 18% think it is a virus (Figure 6E). Concerning the vector of the disease, 82% of students think that mosquitoes are the vector of the disease, 10% answered dogs, and 8% rodents (Figure 6F).

The reservoir of cutaneous and visceral leishmaniasis

Among the respondents, 76 students correctly answered the question that the reservoir of cutaneous leishmaniasis is rodents and 60 students find that the reservoir is Humans while 53 believe that the reservoir is a dog and 14 students think it is a cat (Figure 7A). The majority of students questioned responded that dogs are the reservoir of visceral leishmaniasis (62 students), while 41 chose rodents, 39 thought the reservoir was a Human being, while 41 chose rodents and 19 students considered cats as the primary reservoir (Figure 7B).

### The genus most infected by leishmaniasis the active time and seasonality of sandflies

According to students' responses, the sex most infected by the disease is children (57%), followed by women (27%) and men (16%) (Figure 8A). With regard to active time, 79% of students noted that sandflies were active at night, while 21% replied that they were active during the day (Figure 8B). Among the total number of respondents, 71% of found that Summer is the season of vector activity, 15% of students think that Spring is the season of phlebotomine activity, and 14% selected the Winter and Autumn seasons (Figure 8C).

### Factors Favouring the Development of Sandflies and Reservoirs

For the body part the most affected by the infection, among the respondents, 85 respondents answered that the face was the part the most affected by leishmaniasis, followed by the hands (81 students) and

*Figure 6. Responses of the students. (a) Gender percentages; (b) residential area: (c) Students' knowledge of the disease; (d) Types of leishmaniasis known to students; (e) The main causes of leishmaniasis; the disease vector (D)*
*Source: Authors*

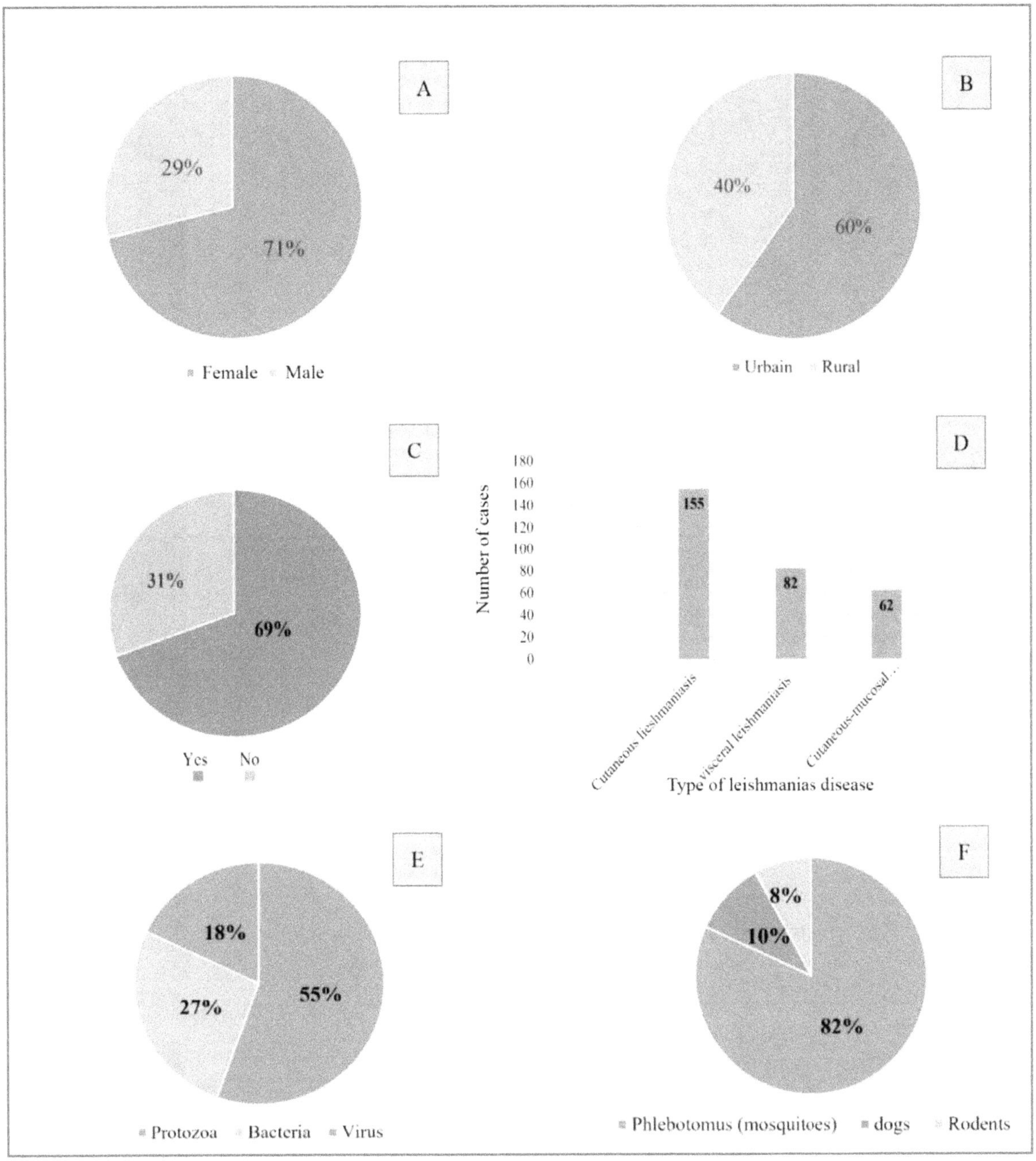

*Figure 7. The reservoir of (a) cutaneous leishmaniasis and (b) visceral leishmaniasis*
*Source: Authors*

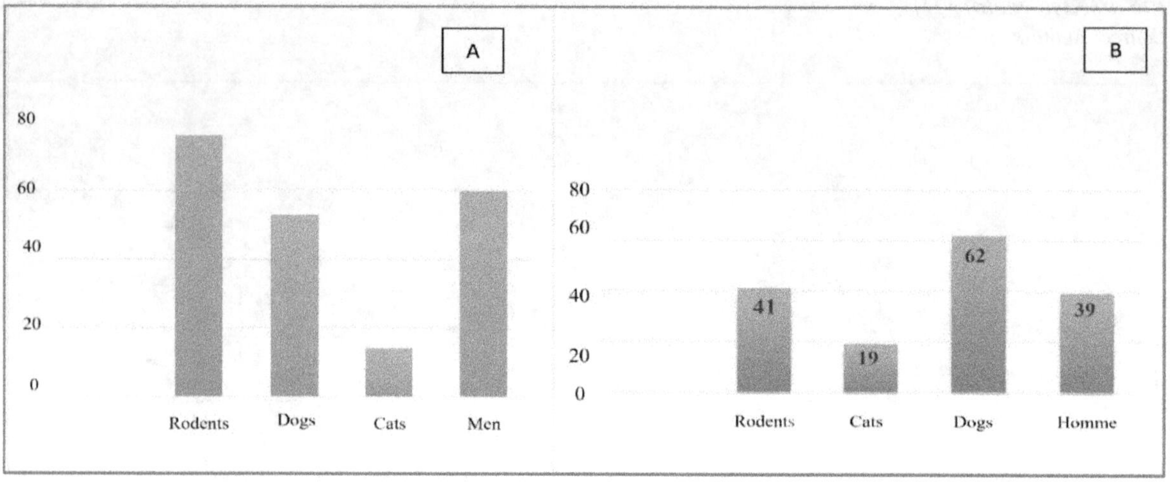


*Figure 8. (a) The active time; (b) the persons the most affected by leishmaniasis; (c) the seasonality of sandflies*
*Source: authors*

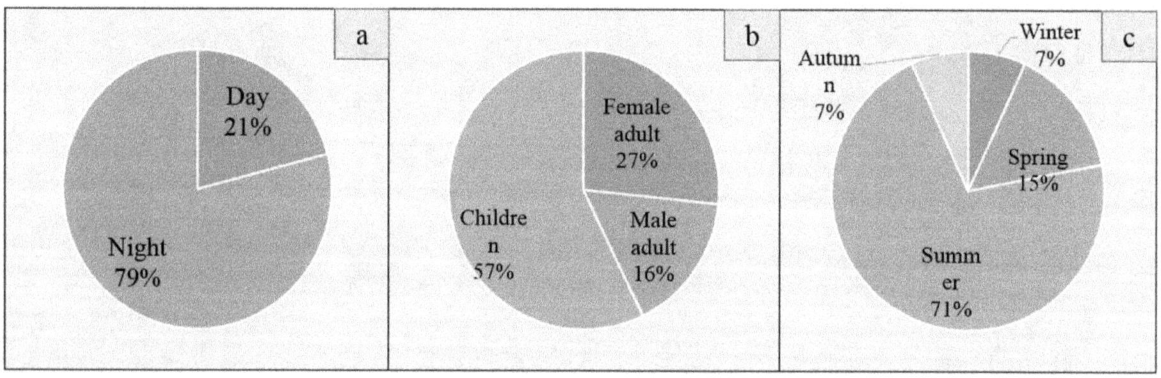


*Figure 9. (a) Body parts affected by leishmaniasis; (b) risk factors for leishmaniasis*
*Source: authors*

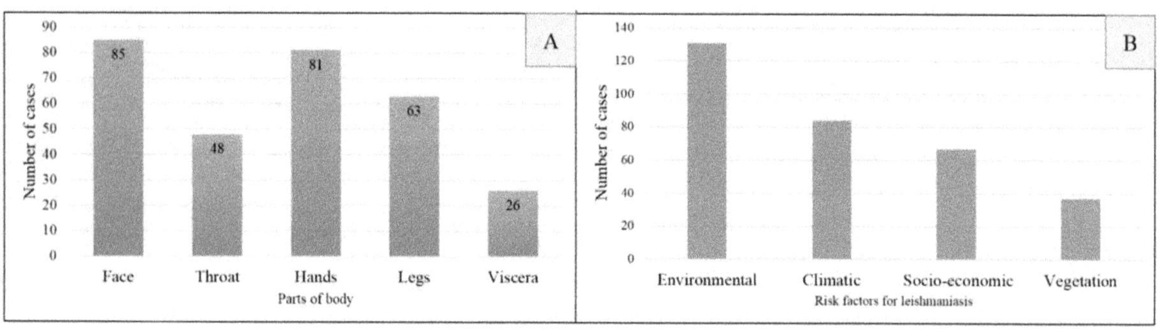

legs (63 students), while 48 students answered the throat, and 26 the viscera (Figure 9A). However, of 200, 131 students answered that the distribution of leishmaniasis depends on environmental factors, 84 students slected climatic factors, while 67 mentioned the socio-economic factor as the primary factors that favour the spread of disease, and 37 indicated vegetation (Figure 9B).

## DISCUSSIONS AND CONCLUSION

This is a descriptive study based on data recorded in public health facilities (SRES) and the results of a survey carried out in the province of Errachidia on students' perception of leishmaniasis disease. The data collected from the health center archives are cases clinically diagnosed by health center from 2012 to 2022. The variables studied are age, gender, annual and monthly distribution, distribution by commune, and type of environment. In the province of Errachidia, leishmaniasis is mainly represented by cutaneous and visceral leishmaniasis. Cases of cutaneous leishmaniasis are more frequent, while visceral leishmaniasis is less widespread but more serious. The sandflies responsible for transmitting the *Leishmania* parasite are found in rural areas, where environmental conditions favor their development. To assess the epidemiological profile of the disease, a retrospective descriptive study in the province of Er-Rachidia on data of the period 2012-2022 was carried out. According to this data, there was a peak in 2012 with 4109 cases recorded, followed by a drop in 2013 and a minimum of cases (13) recorded in 2014, then an increase until 2016. In 2020, there was a decrease in cases compared to 2019 due to the containment caused by the Corona Virus. Between 2021 and 2022, the number of cases remains relatively stable. The total number of cases is 6726, of which 5257 have cutaneous leishmaniasis. Comparing these results with those from the northern region of Morocco (1376 cases) during 2009-2018 (Nabah, 2022), we note a major difference between, which can be attributed to the type of climate, for example. Increased rainfall stimulates vegetation, which favors the proliferation of reservoir hosts and the vector (Yates et al., 2002) and increased humidity decreases vector maturation time (Kasap and Alten, 2006), and aridity: cutaneous leishmaniasis occurs almost exclusively in arid Saharan regions (Rioux et al., 1986). These associations between climatic parameters, vectors, and disease incidence was also found by Karmaoui et al., (2022a). Since seasonal patterns are influenced by temperature change, the adult sand flies' activity period usually follows a seasonal pattern (Karmaoui et al., 2022c). It is worth to mention that the weak eosio-economic conditions including poverty favours the increases of leishmaniasis risk. Unsanitary housing and inadequate domestic sanitation infrastructure (Lack of waste management systems or open sewers) can also foster the development of breeding and resting sites for vectors and host reservoirs. Poor hygiene was also linked rothe incidence of ZCL (Kahime et al., 2014).

Geographically speaking, the most affected circle in Errachidia province is the Goulmima circle with a total nuber of 2256 cases. The age group most infected by the disease is between 20 and 49 in the province. According to the data, the gender most infected in the province is female. This can be explained by the fact that women are in charge of the activities of household chores, so they are in contact with waste and are therefore more likely to be infected by the disease. This gender distribution was also found in a study conducted by Hajji & Amal, (2010) in Ouarzazate, another province of the same region. It was recorded that from a total total number of 5418 cases, 55% are female in Ouarzazate. However, in Taza in the period from 2007 to 2014, Hakkour et al., (2016) showed that the number of cases of cutaneous leishmaniasis has increased since 2010 and the most cases were recorded in both urban and rural areas, with 34% of cases detected in two urban localities, Bab Zitouna and Bab-tété with 297 and 106 cases, respectively. This evolution can be explained by the fact that Taza belongs to sub-arid and sub-humid zones located in a region where water is scarce and an altitude of 600 m and mountains up to 3250 m. These caractersitics are also found in Ouarzazate, which suggests that the same climatic conditions in Ouarzazate, Errachidia, and Taza can explain the evolution of the disease.

According to our survey about the perceptions of the students, of 200 respondents, 31%

of the survey participants were unaware of the disease in the region yet 69% responded affirmatively to the survey. Concerning the body parts the most affected, the responses indicated that facial areas are predominantly affected by lesions, a phenomenon likely attributed to the exposure of the face to bites from Anopheles mosquitoes.

The most popular vector is the sandfly with a percentage of 82%, 10% of respondents chose the dog as a vector that transmits the disease and 8% for rodents. On the one hand, the sex preferred by the vector according to the respondents are children with a percentage of 57%, followed by women with 27% and men with 16%. From the responses of the female students, we can say that the dominance of Anopheles in Summer is due to the high temperatures and low rainfall rates where Anopheles can lay its eggs, particularly in fresh water.

From the results of the questionnaire, we can see that most of the students' knowledge of leishmaniasis is neither correct nor complete. The polymorphic distribution of leishmaniasis can be explained by the vector-climate relationship. Extreme climatic events (droughts or floods); the number of cases is boosted by agricultural development and the exploitation of new water resources (dams, hill lakes, artesian wells), which are at the origin of ecological modifications creating humid microclimates favourable to the development of sandflies and therefore to leishmaniasis cycles. These ecological changes can be also responsible for the geographical spread of the parasite.

As a conclusion, leishmaniasis is a disease that manifests itself in many different ways and affects every continent. The disease is widespread in Morocco, mainly in the form of cutaneous and visceral leishmaniasis. The Darà-Tafilalt region continues to record high levels of leishmaniasis, particularly in the province of Errachidia, where the climate, the geographical and socioeconomical conditions are ideal for the spread, especially of leishmaniasis caused by *L.major*, despite the efforts of the Ministry of Health since the beginning of the epidemic in the area. Numerous measures were taken by the government to remedy the increase of the number of cases in the region including the fight against the vector, the host reservoir, education and awareness of citizens, and students in schools. Raising the population's awareness of hygiene methods and the fight against poverty are also key elements to be taken into account in order to eradicate the disease. In terms of outlook, we suggest that it is vital to continue the scientific research to fully understand this disease in its entirety.

## ACKNOWLEDGMENT

This book chapter was written based on the results of the end-of-study project in Physiology and Health at the Faculty of Science and Technology of Errachidia (University Moulay Ismail (Meknes, Morocco)) of the students S. Achichaou, H. Ayach, and F.E. Akhatar and supervised by Prof. A. Karmaoui. The students Y. Modrak and S. Zahir have transformed the report into a chapter form and provide some scientific background. The first author thanks all the members of the Jury who agreed to judge and comment this work. She would like to thank also Mrs. Hasna Tifa from the "Service de Réseau d'Etablissement Sociaux d'Errachidia" for helping to retrieve the epidemiological data of the Errachidia Province. Sincere thanks also go to Mr. Hassan Ibrahimi for giving the opportunity to observe the parasite under a microscope at the Abd Arrahmed Sahraoui hospital (Errachidia).

## REFERENCES

Abonnenc, E. (1972). Les phlébotomes de la région éthiopienne (Diptera, Psychodidae). *Cahiers de l'ORSTOM, série Entomologie médicale et Parasitologie, 55*, 1-239.

Adl, S. M., Simpson, A. G., Farmer, M. A., Andersen, R. A., Anderson, O. R., Barta, J. R., Bowser, S. S., Brugerolle, G., Fensome, R. A., Fredericq, S., James, T., Karpov, S., Kugrens, P., Krug, J., Lane, C., Lewis, L. A., Lodge, J., Lynn, D. H., Mann, D. G., & Taylor, M. F. (2005). The new higher level classification of eukaryotes with emphasis on the taxonomy of protists. *The Journal of Eukaryotic Microbiology, 52*(5), 399–451. doi:10.1111/j.1550-7408.2005.00053.x PMID:16248873

Ait Antar, D. (2021). *Le profil épidémiologique de leishmaniose cutanée et viscérale dans la province Al Haouz.* [Thèse de doctorat, Faculté de médecine et de pharmacie. Université Cadi Ayad].

Banuls, A. L., Hide, M., & Prugnolle, F. (2007). Leishmania and the leishmaniases: A parasite genetic update and advances in taxonomy, epidemiology and pathogenicity in humans. *Advances in Parasitology, 64*, 1–458. doi:10.1016/S0065-308X(06)64001-3 PMID:17499100

Belazzoug, S., Addadi, K., Mokrani, T., Hafirassou, N., Hamrioui, B., & Belkaid, M. (1985). La leishmaniose viscérale en Algérie. Etude des cas hospitalisés entre 1975 et 1984. In Annales de la Société Belge de Médecine tropicale (Vol. 65, pp. 329-335.

Ben Salem, S., Ben Salem, A., Karmaoui, A., & Khebiza, M. Y. (2023). Vulnerability of Water Resources to Drought Risk in Southeastern of Morocco: Case Study of Ziz Basin. *Water (Basel), 2023*(15), 4085. doi:10.3390/w15234085

Bennai, K. (2018). *Surveillance et contrôle des leishmanioses dans le nord de l'Algérie.* [Thèse de doctorat, université M'hamed Bougara, Boumerdas, Algérie]

Berthet-Beaufils, A. (2010). *Manisfestations dermatologiques associés aux diptères chez le chien et le chat* [Doctoral dissertation].

Biezunski, N. (2013). *Co-infection leishmaniose viscérale-infection VIH: A propos de cinq cas suivis dans le Service des Maladies Infectieuses et Tropicales du CHU de Toulouse, 2007- 2013.* [Thèse de doctorat. Université Toulouse III].

Boulkenafet, F. (2006). Contribution à l'étude de la biodiversité des Phlébotomes (Diptera: Psychodidae) et appréciation de la faune Culicidienne (Diptera: Culicidae) dans la région de Skikda. Mémoire de master-Université Mentouri Constantine-Faculté des Sciences de la Nature et de la Vie.

Boussaa, S. (2008). *Epidémiologie des leishmanioses dans la région de Marrakech, Maroc: effet de l'urbanisation sur la répartition spatio-temporelle des Phlébotomes et caractérisation moléculaire de leurs populations* [Doctoral dissertation, Université Louis Pasteur (Strasbourg)(1971-2008))].

Britto, C., Ravel, C., Bastien, P., Blaineau, C., Pagès, M., Dedet, J. P., & Wincker, P. (1998). Conserved linkage groups associated with large-scale chromosomal rearrangements between Old World and New World Leishmania genomes. *Gene*, *222*(1), 107–117. doi:10.1016/S0378-1119(98)00472-7 PMID:9813266

Dedet, J. P. (1999). *Les leishmanioses*. Edition Ellipses.

Dedet, J. P. (2001). Leishmanies, leishmanioses. Biologie, clinique et thérapeutique. *Encycl. Med. Chir. Maladies Infectieuses*, *8*, 11.

Dedet, J. P. (2009). *Leishmanies, leishmanioses: biologie, clinique et thérapeutique*. Editions Scientifiques et Médicales, Elsevier SAS.

Desjeux, P. (1999). Global control and Leishmania HIV co-infection. *Clinics in Dermatology*, *17*(3), 317–325. doi:10.1016/S0738-081X(99)00050-4 PMID:10384871

Desjeux, P. (2004). Leishmaniasis: Current situation and new perspectives. *Comparative Immunology, Microbiology and Infectious Diseases*, *27*(5), 305–318. doi:10.1016/j.cimid.2004.03.004 PMID:15225981

Diallo, K. (2014). *Leishmaniose cutanée: profil épidémiologique, clinique, thérapeutique et évolutif. A propos de, 87 cas recensés dans le service de dermatologie de l'institut d'hygiène sociale de Dakar*. [Thèse de doctorat en médecine, université des sciences des techniques et des technologies de Bamako, Mali].

Dolmatova, A. V., Demina, N. A., & Kobylansky, A. (1971). *Les phlébotomes (Phlebotominae) et les maladies qu'ils transmettent* (No. 18). Office de la Recherche Scientifique et Technique Outre-. *La Mer (Paris)*, *1971*, 168p.

Elbaroudi, H. (2014). *Eco-épidémiologie de la leishmaniose viscérale au Maroc* [Thèse de doctorat en pharmacie, faculté de médecine et de pharmacie, Université Mohammed V- Souissi. Maroc].

Espinosa, O. A., Serrano, M. G., Camargo, E. P., Teixeira, M. M. G., & Shaw, J. J. (2018). An appraisal of the taxonomy and nomenclature of trypanosomatids presently classified as Leishmania and Endotrypanum. *Parasitology*, *145*(4), 430–442. doi:10.1017/S0031182016002092 PMID:27976601

Foley, H., & Vialatte, C. (1914). Existence dans le Sud marocain du bouton d'orient à l'état endémique. *Bulletin de la Société de Pathologie Exotique*, *7*, 114–115.

Galati, E. A. B. (1995). Phylogenetic systematics of Phlebotominae (Diptera, Psychodidae) with emphasis on American groups. *Boletim de la Direccion de Malariologia Y Saneamiento Ambiental, 35*(1), 133-42.

Hajji, R., & Amal, S. (2010). Le profil épidémiologique de la leishmaniose cutanée dans la province de Ouarzazate entre 2002 et 2009 [Doctoral dissertation, Facuté de Medecine et Pharmacie de Marrakech]. http://wd.fmpm.uca.ma/biblio/theses/annee-htm/art/2010/article30-10.pdf

Hakkour, M., Hmamouch, A., El Alem, M. M., Rhalem, A., Amarir, F., Touzani, M., Sadak, A., Fellah, H., & Sebti, F. (2016). New epidemiological aspects of visceral and cutaneous leishmaniasis in Taza, Morocco. *Parasites & Vectors*, *9*(1), 1–9. doi:10.118613071-016-1910-x PMID:27899126

Hamida, A. B. (2012). Climat, environnement et maladies à transmission vectorielle: cas de la Leishmaniose Cutanée Zoonotique (LCZ) dans la région de Sidi Bouzid, Tunisie. *mémoire d'agrégation en géographie*.

Izri, A., Depaquit, J., & Parola, P. (2006). Phlébotomes et transmission d'agents pathogènes autour du bassin méditerranéen. *Médecine Tropicale*, *66*(5), 429–435. PMID:17201284

Kahime, K., Boussaa, S., Bounoua, L., Fouad, O., Messouli, M., & Boumezzough, A. (2014). Leishmaniasis in Morocco: Diseases and vectors. *Asian Pacific Journal of Tropical Disease*, *4*, S530–S534. doi:10.1016/S2222-1808(14)60671-X

Karmaoui, A., Ben Salem, A., El Jaafari, S., Chaachouay, H., Moumane, A., & Hajji, L. (2022b). Exploring the land use and land cover change in the period 2005–2020 in the province of Errachidia, the pre-sahara of Morocco. *Frontiers in Earth Science (Lausanne)*, *10*, 962097. doi:10.3389/feart.2022.962097

Karmaoui, A., Moumane, A., El Jaafari, S., Menouni, A., Al Karkouri, J., Yacoubi, M., & Hajji, L. (2023). Thirty Years of Change in the Land Use and Land Cover of the Ziz Oases (Pre-Sahara of Morocco) Combining Remote Sensing, GIS, and Field Observations. *Land (Basel)*, *12*(12), 2127. doi:10.3390/land12122127

Karmaoui, A., Sereno, D., El Jaafari, S., & Hajji, L. (2022a). Seasonal patterns of zoonotic cutaneous Leishmaniasis caused by L. major and transmitted by Phlebotomus papatasi in the North Africa Region, a systematic review and a meta-analysis. *Microorganisms*, *10*(12), 2391. doi:10.3390/microorganisms10122391 PMID:36557644

Karmaoui, A., Sereno, D., El Jaafari, S., & Hajji, L. (2022c). A systematic review and global analysis of the seasonal activity of Phlebotomus (Paraphlebotomus) sergenti, the primary vectors of L. tropica. *PLoS Neglected Tropical Diseases*, *16*(12), e0010886. doi:10.1371/journal.pntd.0010886 PMID:36469546

Kasap, O. E., & Alten, B. (2006). Comparative demography of the sand fly Phlebotomus papatasi (Diptera: Psychodidae) at constant temperatures. *Journal of Vector Ecology*, *31*(2), 378–385. doi:10.3376/1081-1710(2006)31[378:CDOTSF]2.0.CO;2 PMID:17249356

Killick-Kendrick, R. (2013). The race to discover the insect vector of Kala-azar: A great saga of tropical medicine 1903-1942. *Bulletin de la Société de Pathologie Exotique*, *106*(2), 131–137. doi:10.100713149-013-0285-x PMID:23516011

Lainson, R., & Shaw, J. J. (1987). *Evolution, classification and geographical distribution* (pp. 1-120). Academic Press.

Lawyer, P. G., Ngumbi, P. M., Anjili, C. O., Odongo, S. O., Mebrahtu, Y. B., Githure, J. I., & Roberts, C. R. (1990). Development of Leishmania major in Phlebotomus duboscqi and Sergentomyia schwetzi (Diptera: Psychodidae). *The American Journal of Tropical Medicine and Hygiene*, *43*(1), 31–43. doi:10.4269/ajtmh.1990.43.31 PMID:2382763

Levine, N. D., Corliss, J. O., Cox, F. E. G., Deroux, G., Grain, J., Honigberg, B. M., Leedale, G. F., Loeblich, A. R., Lom, I. J., Lynn, D., Merinfeld, E. G., Page, F. C., Poljansky, G., Sprague, V., Vavra, J., & Wallace, F. G. (1980). A Newly Revised Classification of the Protozoa* The committee on systematics evolution of the society of protozoologists. *The Journal of Protozoology*, *27*(1), 37–58. doi:10.1111/j.1550-7408.1980.tb04228.x PMID:6989987

Lewis, D. J., Young, D. G., Fairchild, G. B., & Minter, D. M. (1977). Proposals for a stable classification of the phlebotomine sandflies (Diptera: Psychodidae). *Systematic Entomology*, *2*(4), 319–332. doi:10.1111/j.1365-3113.1977.tb00381.x

Lightburn, E., Meynard, J. B., Morand, J., Garnotel, E., Kraemer, P., Hovette, P., & Boutin, J. P. (2002). Surveillance épidémiologique des leishmanioses tégumentaires en Guyane Synthèse de données militaires sur 10 ans. *Médecine Tropicale*, *62*, 545–553. PMID:12616949

Mahjour, J., Akalay, O., & Saddiki, A. (1992). Les leishmanioses au Maroc de l'analyse éco-épidémiologique à la prévention. Direction d'épidémiologie et de lutte contre les maladies. *Bulletin Épidémiologique*, *3*, 4–5.

Marsden, P. D. (1986). Mucosal leishmaniasis. *Transactions of the Royal Society of Tropical Medicine and Hygiene*, *80*(6), 859–876. doi:10.1016/0035-9203(86)90243-9 PMID:3037735

Mazelet, L. (2004). La leishmaniose canine dans le bassin méditerranéen français. Mémoire de Maîtrise de Biologie des Populations et des Ecosystèmes. Université Pierre et Marie Curie, PARIS VI, France..

Moziane, F., Benalioua, M., Trebycka, K., Bouhass, R., Belazzoug, S., & Aguercif, M. (1983). Kalar-Azar chez l'enfant dans l'ouest de l'Algérie. A propos de 14 observations. *Medecine et Hygiene*, *41*(1536), 3649–3652.

Nabah, K., Rharrabe, K., & Mezzoug, N. (2022). Profil épidémiologique de la leishmaniose cutanée au niveau de la région Nord du Maroc. *Revue d'Epidemiologie et de Sante Publique*, *70*, S237. doi:10.1016/j.respe.2022.06.264

Pearson, R. D., Jeronimo, S. M., & De Sousa, A. (2001). Leishmaniasis, pp: 287-331. Principles and Practice of Parasitology. John Wiley & Son, New York.

PME, (2019). *Provincial Monograph of Errachidia*. Monographie de la province d'Errachidia de l'année 2019.

Rguioui, L. (2016). *Epidémiologie de la leishmaniose cutanée à SEFROU* [thèse de doctorat en pharmacie]. Faculté de médecine et de pharmacie. Université Mohammed V-Rabat. Maroc.

Rhajaoui, M. (2011). Les leishmanioses humaines au Maroc: Une diversité nosogéographique. *Pathologie Biologie*, *59*(4), 226–229. doi:10.1016/j.patbio.2009.09.003 PMID:19942371

Rioux, J. A., Lanotte, G., Petter, F., Dereure, J., Akalay, O., Pratlong, F., & Serres, E. (1986). Les leishmanioses cutanées du bassin Méditerranéen occidental. De l'identification enzymatique à l'analyse éco-épidémiologique. L'exemple de trois foyers, tunisien, marocain et français. CollintCNRS/INSERM) IMEE, Montpellier, 1986, 365-395.

Rioux, J. A., Lanotte, G., Serres, E., Pratlong, F., Bastien, P., & Perieres, J. (1990). Taxonomy of Leishmania. Use of isoenzymes. Suggestions for a new classification. *Annales de Parasitologie Humaine et Comparee*, *65*(3), 111–125. doi:10.1051/parasite/1990653111 PMID:2080829

Rosenthal, E., & Marty, P. (2009). Actualités sur la leishmaniose viscérale méditerranéenne. *Revue de Médecine Interne*, *30*, S24–S28. https://www.sciencedirect.com/science/article/pii/S0248866309001398. doi:10.1016/j.revmed.2009.03.016

Rostan, O. (2013). *Place de l'Interleukine-33 dans la réponse immune du foie au cours de la leishmaniose viscérale* [Doctoral dissertation, Université Rennes 1].

Sergent, E., Sergent, E., Parrot, L., Donatien, A., & Beguet, M. (1926). Transmission expérimentale du bouton d'orient (clou de Biskra) à l'homme par Phlebotomus papatasi (Scopoli). *Annales de l'Institut Pasteur*, *40*(5), 411–430.

Simpson, L. (1987). The mitochondrial genome of kinetoplastid protozoa: Genomic organization, transcription, replication, and evolution. *Annual Review of Microbiology*, *41*(1), 363–380. doi:10.1146/annurev.mi.41.100187.002051 PMID:2825587

Tonelli, G. B., Binder, C., Margonari, C., & Andrade Filho, J. D. (2023). Archives if the « Service des réseaux des établissements de santé ». The provincial delegation of the Ministry of Health in Errachidia
Tonelli, G. B., Binder, C., Margonari, C., & Andrade Filho, J. D. (2021). Sand fly behavior: Much more than weak-flying. *Memorias do Instituto Oswaldo Cruz*, *116*, e210230. doi:10.1590/0074-02760210230

Villela, F., Pestana, B. R., & Pessoa, S. B. (1939). Presença de Leishmania braziliensis na mucosa nasal sem lesão aparente em casos recentes de leishmaniose cutânea. *O Hospital*, *16*, 953-960.

World Health Organization. (2010). Control of the leishmaniases: report of a meeting of the WHO Expert Commitee on the Control of Leishmaniases, Geneva, 22-26 March 2010. World Health Organization.

Yates, T. L., Mills, J. N., Parmenter, C. A., Ksiazek, T. G., Parmenter, R. R., Vande Castle, J. R., Calisher, C. H., Nichol, S. T., Abbott, K. D., Young, J. C., Morrison, M. L., Beaty, B. J., Dunnum, J. L., Baker, R. J., Salazar-Bravo, J., & Peters, C. J. (2002). The ecology and evolutionary history of an emergent disease: hantavirus pulmonary syndrome: evidence from two El Niño episodes in the American southwest suggests that El Niño–driven precipitation, the initial catalyst of a trophic cascade that results in a delayed density-dependent rodent response, is sufficient to predict heightened risk for human contraction of hantavirus pulmonary syndrome. *Bioscience*, *52*(11), 989–998. doi:10.1641/0006-3568(2002)052[0989:TEAEHO]2.0.CO;2

Zé-Zé, L., Amaro, F., Osório, H. C., Giovanetti, M., Lourenço, J., & Alves, M. J. (2022). Molecular Identification and Ecology of Portuguese Wild-Caught Phlebotomine Sandfly Specimens. *Zoonotic Diseases*, *2*(1), 19–31. doi:10.3390/zoonoticdis2010003

Section 4

# Sustainable Alternatives in the Environment and Agriculture

# Chapter 15
# Impact of Agrochemicals on the Environment and Ecological Alternatives

**Atin Kumar**
https://orcid.org/0000-0002-1653-2146
*Uttaranchal University, India*

**Nilotpal Das**
*Uttaranchal University, India*

**Himani Sharma**
*Sher-e-Kashmir University of Agricultural Sciences and Technology of Jammu, India*

**Rajat Singh**
*Uttaranchal University, India*

## ABSTRACT

*To promote plant development and enhance agricultural yield, farmers utilise a variety of agrochemicals. Insecticides, herbicides, fungicides, weedicides, and other pesticides are among the agrochemicals. The agrochemicals and their classification are mentioned in this chapter. The pros and cons of agrochemicals are also discussed in the report. Focus is also placed on how agrochemicals affect the ecology and how they work. Included are the risks and protective measures related to the usage of agrochemicals. These crucial agrochemical issues will be covered in full in the study. Future research into the molecular basis of agrochemical mechanisms must be done.*

## 1. INTRODUCTION

Agrochemicals are compounds used to boost crop or plant growth, such as fertilisers, fungicides, insecticides, and nematicides (Biswas et al., 2014). The extensive use of pesticides has had an impact on soil microorganisms as well as the entire ecosystem (Onder et al., 2011). The primary decreasing biotic fac-

DOI: 10.4018/978-1-6684-9838-5.ch015

tors in agriculture are weeds and insects, which reduce crop production, productivity, and resource use effectiveness (Oliveira et al., 2014). These agricultural chemicals are expensive and polluting (Bahadur et al., 2015). These agrochemicals pollute water resources by getting into bodies of water. Pesticide residues continue to build up in the soil and biomagnify in plant and animal tissues. Health issues may result from this, which is hazardous to people (Hans and Farooq, 2000). This is a severe problem since the bacteria develop a resistance to the insecticides. Fish have also died as a result of pesticide use in numerous water bodies. In general, the dosage of the chemical, the characteristics of the soil, and many environmental conditions will affect the effects of pesticides on microorganisms differently. As a result of the decline in organic matter, the soil's microbiota and microbial enzymes are also impacted (Kalia and Gosal, 2011). Rhizobium-legume symbiosis has been impacted by the usage of herbicides. According to Kalia and Gosal (2011), the application of fungicide has caused a decrease in the population of soil fungi.

## 2. AGROCHEMICAL TYPES

Fungicides:

i) **Inorganic fungicides:**

Inorganic fungicides play a crucial role in plant disease management. Mercuric chloride ($HgCl_2$) disrupts fungal cell membranes and enzymes through sulfhydryl binding, impairing vital processes. However, its high toxicity limits its use. Sulfur, a widely employed fungicide, works by inhibiting spore germination and fungal growth. It's effective against a range of diseases and has low toxicity to humans and animals. The Bordeaux mixture, combining copper sulfate ($CuSO_4$) and lime, offers a dual defense mechanism. Copper ions disrupt fungal enzymes and generate toxic compounds, while lime enhances adhesion and prolongs efficacy. These inorganic fungicides are cost-effective and diverse in action, reducing the risk of resistance development. However, their environmental impact and potential harm to non-target organisms necessitate judicious use and exploration of safer alternatives in modern agricultural practices.

ii) **Organic fungicides:**

Organic fungicides offer effective alternatives for plant disease control. Dithanes like S-21 and Z-78 belong to the ethylenebisdithiocarbamate group, inhibiting fungal enzyme activity by disrupting cell membrane integrity. Oxathiins, represented by Vitavax, function as protective fungicides, forming a shield on plant surfaces to prevent fungal invasion. Mercury compounds like Agrosan and Tillex exert antifungal effects through multiple mechanisms, inhibiting enzymes crucial for fungal growth. Benzimidazole derivatives like Benlate interfere with fungal cell division and DNA synthesis, impeding disease progression. These organic fungicides provide diverse modes of action, reducing the likelihood of resistance development. However, their persistence in the environment and potential ecological impacts warrant careful application. Integrating these fungicides into integrated pest management strategies can enhance disease control while minimizing risks to non-target organisms and the ecosystem.

**Herbicides:** following are the different examples of herbicides

**Methyl Bromide ($CH_3Br$):** Methyl bromide was commonly used as a broad-spectrum soil fumigant to control a wide range of pests and diseases. However, due to its negative environmental impact and role in ozone depletion, its use has been largely phased out under international agreements like the Montreal Protocol.

**Ethylene Bromide ($C_2H_4Br_2$):** Ethylene bromide was used as a fumigant as well, particularly for soil disinfection and pest control. Like methyl bromide, its use has also been restricted due to environmental concerns.

**Chloropicrin ($CCl_3NO_2$):** Chloropicrin is a soil fumigant and a powerful irritant. It's often used in combination with other fumigants to enhance their efficacy and to mitigate the development of resistance in pests and pathogens. It acts by damaging cellular proteins and enzymes.

**Aldirab (Propylene Oxide and Dazomet):** Aldirab is a combination of propylene oxide and dazomet. Propylene oxide acts as a fumigant, while dazomet, upon contact with moisture, releases methyl isothiocyanate gas which provides pesticidal activity. It's used for soil-borne pest and disease management.

**Rodenticides:** following are the different examples of rodenticides

**Organophosphates:** These pesticides, including malathion and chlorpyrifos, target the nervous system of pests by inhibiting acetylcholinesterase, an enzyme critical for nerve signal transmission. They have been effective but also pose risks to humans and non-target organisms due to their toxicity and potential for environmental persistence.

**Pyrethroids:** Derived from natural pyrethrins, pyrethroids like permethrin and cypermethrin affect pests' nervous systems, leading to paralysis and death. They are less toxic to mammals but can harm aquatic life and beneficial insects.

**Carbamates:** Pesticides like carbaryl and methomyl act similarly to organophosphates but with a shorter duration of action. They inhibit acetylcholinesterase, causing nerve dysfunction in pests. They are less persistent but still need careful use.

**Organochlorines:** This class includes DDT, which was once widely used. Organochlorines disrupt pests' nervous systems and can persist in the environment for years, leading to bioaccumulation and potential harm to ecosystems and wildlife.

**Ethylene Dibromide:** Used as a fumigant, ethylene dibromide combats soil-borne pests. However, its use has declined due to its high toxicity and concerns about groundwater contamination.

The use of these pesticides has raised environmental and health concerns, leading to stricter regulations and the development of alternative, more environmentally friendly pest management strategies. Integrated pest management (IPM) approaches aim to minimize pesticide use while maximizing sustainable pest control.

Algicides:

Dichlorophen, Bluestone, and cupric sulphate

Pesticide:

**Organochlorines:** BHC (Benzene Hexachloride): BHC, also known as hexachlorocyclohexane (HCH), was used as an insecticide to control pests. However, its persistence in the environment, potential for bioaccumulation, and adverse health impacts led to its ban in many countries.

**Endosulfan:** Endosulfan was an insecticide and acaricide used on various crops. Its environmental persistence, bioaccumulation, and toxic effects on non-target organisms led to its global phase-out due to concerns about its impact on human health and wildlife.

**Aldrin:** Aldrin is an organochlorine insecticide used to control soil-dwelling pests. Its conversion to dieldrin, a more toxic compound, and its persistence in the environment contributed to its ban in many countries.

**DDT (Dichlorodiphenyltrichloroethane):** DDT gained fame for its effectiveness against malaria-carrying mosquitoes. However, its widespread use led to environmental contamination, harm to non-target species, and concerns about human health. International agreements, like the Stockholm Convention, have restricted DDT's use to specific public health situations.

These compounds exemplify the negative impacts of unregulated pesticide use, prompting shifts toward safer alternatives and integrated pest management strategies that minimize harm to ecosystems and human well-being.

**Organophosphates:** following are the different types of organophosphates

**Parathion:** Parathion is an extremely toxic insecticide that affects pests by inhibiting acetylcholinesterase, leading to nervous system disruption. Despite its effectiveness, its high toxicity to humans, animals, and the environment has led to its restricted use and, in some cases, complete bans.

**Fenitrothion:** Fenitrothion is less toxic compared to parathion but still poses risks to non-target organisms and the ecosystem. It functions by inhibiting acetylcholinesterase, causing nerve signal interference in pests. It has been used in agriculture, forestry, and pest control programs, but its persistence in the environment has raised concerns.

**Malathion:** Malathion is commonly used in agriculture and mosquito control due to its relatively lower toxicity to humans and animals. It disrupts pests' nervous systems by inhibiting acetylcholinesterase. Its shorter residual activity makes it suitable for situations requiring frequent applications, but care is needed to avoid potential harm to non-target insects and aquatic organisms.

The use of these organophosphates has led to stricter regulations and the exploration of alternative, safer pest management strategies to mitigate their negative impacts on human health and the environment.

**Carbamates:** following are the types of carbamates

**Aldicarb:** Aldicarb acts by inhibiting acetylcholinesterase, causing nerve signals to accumulate and disrupt the nervous system of pests. It's highly effective but poses significant risks to humans and wildlife. Its use has been limited, and in some cases banned, due to its acute toxicity and potential for groundwater contamination.

**Propoxur:** Propoxur is used for indoor and outdoor pest control in homes, gardens, and agriculture. It also inhibits acetylcholinesterase, affecting pests' nerve functions. While less toxic to mammals than some other pesticides, its potential harm to children and non-target organisms has led to restrictions.

**Carbofuran:** Carbofuran is an insecticide and nematicide used on various crops. Its mode of action is similar to other carbamates, causing nerve signal interference in pests. Due to its high toxicity to birds and aquatic organisms, as well as potential risks to human health, its use has been curtailed in many areas.

Efforts to reduce the impact of these carbamate insecticides on the environment and human health have prompted regulatory actions, encouraging the adoption of more sustainable and less harmful pest management strategies.

**Pyrethroids**: Pyrethroids are synthetic insecticides modeled after natural pyrethrins from chrysanthemum flowers. They target pests' nervous systems, causing paralysis and death. Pyrethroids are widely used due to their effectiveness against a variety of insects, including mosquitoes, agricultural pests, and household pests. They are often considered less toxic to humans and animals compared to older insecticides. However, their widespread use has raised concerns about their impact on non-target organisms, including beneficial insects and aquatic life. Additionally, some pests have developed resistance to pyrethroids. To mitigate these issues, integrated pest management strategies recommend judicious pyrethroid application and rotation with other insecticides to minimize resistance development and environmental impact.

**Triazines:** following are the types of triazines

**Atrazine:** Atrazine is known for its pre- and post-emergence activity against broadleaf and grassy weeds. It disrupts photosynthesis by inhibiting electron transport in plants. Despite its effectiveness, atrazine has raised environmental concerns due to its persistence in soil and potential to leach into groundwater, leading to contamination. It is subject to regulatory scrutiny, with restrictions in some regions to protect water quality and aquatic ecosystems.

**Simazine:** Simazine, like atrazine, inhibits photosynthesis in weeds. It has been used in orchards, vineyards, and various crops. Similar to atrazine, simazine's environmental impact and potential health risks have led to regulatory limitations on its use.

Both atrazine and simazine exemplify the need for responsible pesticide application to balance effective weed control with minimizing their impact on the environment and human health.

**Urea derivatives:** Diuron, linuron, monuron, metoxuron, and fenuron are herbicides belonging to the urea class, widely used for weed control in various agricultural and non-agricultural settings.

**Diuron:** Diuron inhibits photosynthesis by disrupting electron transport in plants, making it effective against broadleaf and grassy weeds. Its residual activity provides prolonged weed control, but it can also pose risks to non-target plants and aquatic organisms. Its use is regulated to prevent water contamination.

**Linuron:** Linuron acts similarly to diuron, affecting photosynthesis. It's used primarily in vegetable and fruit crops but can impact sensitive crops and aquatic ecosystems due to its persistence and potential to leach into water sources.

**Monuron, Metoxuron, and Fenuron:** These urea herbicides also interfere with photosynthesis. Monuron is effective against many weed species, while metoxuron and fenuron have similar modes of action. Their use has diminished due to environmental concerns and regulatory restrictions.

Balancing the efficacy of these urea herbicides with their environmental impact is essential. Integrated pest management strategies emphasize their judicious use and consideration of alternative, more sustainable weed management approaches.

## 3. THE DIVISION OF PESTICIDES (LUSHCHAK ET AL., 2018)

**3.1 Assignment -** e.g., herbicides, fungicides, etc.
**3.2 Pesticide effect method** - Contact
**3.3 Chemical nature of the pesticides -** Another crucial factor is the chemical makeup of insecticides. Bolognesi and Merlo (2011) investigate the chemical makeup of insecticides.

## 4. ADVANTAGES OF AGROCHEMICALS

- From 50 million tonnes to 198 million tonnes were produced as food grains in 1997. This is a result of better irrigation practises, the use of agrochemicals, and enhanced seed quality.
- The use of fertilisers has enhanced crop productivity in several nations.
- Crops are now protected from nematodes, pests, and weeds thanks to the usage of pesticides. When pesticides go through photochemical change and produce non-toxic metabolites, they become less dangerous (Kole et al., 1999).
- Herbicides also boost the economy and the workforce.

- Insecticides assist in controlling insects.

## 5. AGROCHEMICALS' IMPACT ON THE ECOSYSTEM

a. Effects on non-target wildlife:

Neurotoxic compounds include carbamates and organophosphorus cholinesterase inhibitors, pyrethroids, and neonicotinoid insecticides. Arthropods and different animals are affected by these on the nervous system.

b. Effects on flora that is not the goal:

The variety of the flora is impacted by herbicides (Gove et al., 2007). Herbicide spraying causes more damage to young plants. When plants are subjected to the highest concentrations of herbicides, their reproductive systems are impacted. The application of agrochemicals drift begins to affect the species balance after a second year of exposure. This also has an impact on the variety of plant species. Insecticides and fungicides have a negative impact on non-target flora. A decrease in floral visits and a change in the community of floral visitors both have an impact on pollination effectiveness. According to Jennings et al. (2012), pesticides also have an impact on plant growth and germination.

c. Impacts on vertebrates:

Organophosphates and carbamates impair vertebrates' ability to regulate body temperature, consume nutrients, forage for food, reproduce, and other metabolic processes (Story and Cox, 2001). In addition to having an impact on the thyroid function in rodents, birds, amphibians, and fish (Brucker-Davis, 1998), insecticides, herbicides, and fungicides also have an impact on the immune system. According to Sanchez-Bayo (2012), terrestrial vertebrates are less vulnerable to pesticides made from plant or fungal toxins than are birds and mammals. The population of frugivorous and granivorous animals is affected by herbicides. According to Wagner and Bruhl (2017), there aren't enough reports on the effects of herbicides on amphibians and reptiles. Agrochemicals have a greater effect on amphibians and reptiles than other species (Lange et al., 2009).

d. Impacts on invertebrates:

Acetylcholine and insecticides have comparable modes of action. They cause the insects to become paralysed and die. According to Sanchez-Bayo et al. (2016), neonicotinoids are the most hazardous compounds that have an impact on immunological function and reproduction, making people more susceptible to infections. Bees, butterflies, and moths are their primary targets (Gilburn et al., 2015). The neonicotinoid residues in the soil diminish the number of fly larvae (Zhang et al., 2016).

## 6. HOW AGROCHEMICALS WORK

Ecotoxicological research on the environment is required. Each pesticide has a unique mechanism through which it affects both living things and people. These chemicals linger in the environment for a very long period and keep steadily piling up. Pesticides infiltrate surface waters by runoff and erosion, according to Giddings et al. (2005). Pesticides enter aquatic animals through the skin, gills, or food contamination. Pesticides are ingested by terrestrial wildlife through contaminated food or water (Solomon et al., 2008). Pesticides continue to build up, enter the food chain, and have an impact on the aquatic creatures' food chains. The birds also ingest the chemicals.

### 6.1 The Way That Fungicides Work

#### 6.1.1 Blockers of Ergosterol Biosynthesis

Ergosterol, a vital component of the fungal cell membrane, is cycled through by fungicides known as conazoles.

#### 6.1.2 Inhibitors of Protein Biosynthesis

These include dithion, which inhibits proteins. The sulfhydryl group, which is found in proteins is altered by the dithianon, which also prevents germination and expansion of the germ tube, as reported by Lushchal et al. (2018).

#### 6.1.3. Preventing Agents for Mitochondrial Respiration

Azoxystrobin prevents the transfer of electrons at the cytochrome bc1 complex's ubiquinol oxidising, sometimes referred to as the Qo site or quinone "outside" site. This prevents the generation of ATP, the energy source (Casida & Durkin, 2017).

### 6.2 Herbicides' Modes of Action

Herbicides have distinct modes of action that each affect different places in plants (Peterson et al., 2013). The following are these modes of action:

a. Growth control agents:

The growth regulator 2, 4-dichlorophenoxiacetic acid is an example. The auxin-like capacity is the foundation of the 2, 4-D mechanism. Auxin analogues like 2, 4-D will be recognised by the auxin receptor (Jablonkai, 2011).

b. Growth inhibitors for seeds:

Thiocarbamates and acid amides are two examples of substances that hinder seed development. The plant's growth and cell division are impacted by these seed growth inhibitors.

c. Photosynthesis inhibitors:

Examples of herbicides that limit photosynthesis include triazines and insecticides that include copper. These disrupt biomembranes and prevent photosynthesis from occurring. Pesticides with atrazine and copper disrupt the photosystem II electron transport chain. Carbon dioxide ($CO_2$) builds up inside the plant cell as a result of this inhibition (Giddings et al., 2005). Additionally, additional ions take the place of the magnesium ion ($Mg^{2+}$) in the chlorophyll.

d. Anti-amino acid biosynthesis agents:

These herbicides prevent amino acids from being made in the body. For instance, glyphosate [N-(phosphonomethyl) glycine] prevents the production of the amino acid's phenylalanine, tryptophan, and tyrosine. Numerous herbicides inhibit acetohydroxyacid synthase, the first enzyme to restrict the biosynthesis of valine, among others leucine, and isoleucine, as well as 4-hydroxyphenylpyruvate dioxygenase, an essential enzyme in tyrosine catabolism as well as carotenoid synthesis (Garcia et al., 2017). Additionally, several substances prevent glutamine synthase from attaching ammonia ($NH_3$) to glutamate (Tarazona et al., 2017).

## 6.3 Insecticides' Mode of Operation (Lushchak et al., 2018)

6.3.1 Target Muscles and Nerves

a. Inhibition of cholinesterase

Insects are eliminated with insecticides like carbamates and organophosphates. These render acetyl choline esterase inactive, which kills the insects.

### 6.3.2 Targets for Development and Growth

b. Inhibitors of chitin synthesis

These prevent insects from producing chitin, which causes early insect death.

### 6.3.3 Growth Regulators for Insects

The insect endocrine system is impacted by the insect growth regulators.

### 6.3.4 Nonspecific Growth Regulators

Unspecific growth regulators' manner of action is unknown.

### 6.3.5 Objectives for Energy Production

a. Inhibition of electron transport

Organochlorine-based insecticides obstruct electron transport. As a consequence, the synthesis of ATP, a source of energy, is also hampered.

## 7. RISK AND MITIGATION STRATEGIES FOR AGROCHEMICAL USAGE

According to Arts et al. (2015), using pesticides may have a negative impact on the ecology. The tests are carried out in accordance with the United States Environmental Protection Agency's (USEPA) standard procedures in order to evaluate the risk of pesticide exposure (USEPA, 2003). These might not fully account for the intricately intertwined impacts of repeatedly exposing communities to complex pesticide mixes. Agrochemical exposure doses should be chosen so that hazardous pests, nematodes, weeds, etc. are eliminated. The European (EU) framework for agri-environmental units creates plans that utilise the fewest agrochemicals possible and so aid in pest population management (EU Council, 2005). To reduce the risk to the environment and to human health, the use of neonicotinoids and fipronil should be discontinued (European Commission, 2013a and 2013b). Regulations, technical considerations, and best practises are all utilised by the Organisation for Economic Co-operation and Development (OECD) (OECD, 2009). Sustainable methods, like Integrated Pest Management, should be used to reduce pesticide consumption. All nations must create National Action Plans (NAPs) to establish their own agrochemical use reduction policies, laws, regulations, and goals. Therefore, progress must be made towards a sustainable strategy.

## 8. CONCLUSION

The adverse ecological repercussions of agrochemical usage underscore the urgency of seeking alternative approaches. Embracing biological solutions presents a viable substitute for agrochemicals. The influence of agrochemicals on plant growth cannot be underestimated. While these chemicals effectively eliminate undesirable elements such as weeds, nematodes, and insects, their non-eco-friendly nature gives rise to environmental contamination. The necessity to delve into the intricate molecular mechanisms underlying their action becomes evident.

Shifting towards biological methods addresses the concerns surrounding ecosystem disturbance caused by agrochemicals. This transition acknowledges the need to harmonize agricultural practices with environmental preservation. The profound impact of agrochemicals on plant development emphasizes the imperative to adopt more sustainable methodologies that support both crop growth and ecological balance. Although agrochemicals confer immediate pest control benefits, their long-term consequences manifest in ecological destabilization and pollution, which further accentuates the importance of exploring alternative, eco-conscious strategies.

Furthermore, comprehending the intricate molecular pathways through which agrochemicals function is imperative. This deeper understanding facilitates the development of precise, targeted interventions that are less invasive and more environmentally benign. In conclusion, the pursuit of eco-friendly alternatives to agrochemicals, coupled with a comprehensive grasp of their molecular mechanisms, promises a more harmonious coexistence between agriculture and the environment.

## REFERENCES

Arts, G. P., Dollinger, M., Kohlschmid, E., Maltby, L., Ochoa-Acuna, H., & Poulsen, V. (2015). An ecosystem services approach to pesticide risk assessment and risk management of non-target terrestrial plants: Recommendations from a SETAC Europe workshop. *Environmental Science and Pollution Research International, 22*(3), 2350–2355. doi:10.100711356-014-3637-6 PMID:25300179

Bahadur, S., Verma, S. K., Prasad, S. K., Madane, A. J., Maurya, S. P., Verma, V. K., & Sihag, S. K. (2015). Eco-friendly weed management for sustainable crop production - A review. *Journal of Crop and Weed, 11*, 181–189.

Biswas, S. K., Rahman, S., Kobir, S. A., Ferdous, T., & Banu, N. A. (2014). A review on impact of agrochemicals on human health and environment: Bangaldesh perspective. *Plant Environ Develop, 3*, 31–35.

Bolognesi, C., & Merlo, F. D. (2011). Pesticides: human health effects. In J. O. Nriagu (Ed.), *Encyclopedia of environmental health* (pp. 438–453). Elsevier. doi:10.1016/B978-0-444-52272-6.00592-4

Brucker-Davis, F. (1998). Effects of environmental synthetic chemicals on thyroid function. *Thyroid, 8*(9), 827–856. doi:10.1089/thy.1998.8.827 PMID:9777756

Casida, J. E., & Durkin, K. A. (2017). Pesticide chemical research in toxicology: Lessons from nature. *Chemical Research in Toxicology, 30*(1), 94–104. doi:10.1021/acs.chemrestox.6b00303 PMID:27715053

Council, E. U. (2005). *Council Regulation (EC) No 1698/2005 on support for rural development by the European Agricultural Fund for Rural Development (EAFRD)*. Official J Eur Union.

European Commission. (2013a). COMMISSION IMPLEMENTING REGULATION (EU) No 485/2013 amending Implementing Regulation (EU) No 540/2011, as regards the conditions of approval of the active substances clothianidin, thiamethoxam and imidacloprid, and prohibiting the use. *Official J Euro Union, 139*, 12–26.

European Commission. (2013b). COMMISSION IMPLEMENTING REGULATION (EU) No 781/2013 of 14 August 2013 amending Implementing Regulation (EU) No 540/2011, as regards the conditions of approval of the active substance fipronil, and prohibiting the use and sale of seeds treated with plant p. *Official J Eur Union, 219*, 22–25.

Giddings, J. M., Anderson, T. A., Hall, L. W., Kendall, R. J., Richards, R. P., Solomon, K. R., & ... (2005). *A probabilistic aquatic ecological risk assessment of atrazine in North American surface waters*. SETAC Press.

Gilburn, A. S., Bunnefeld, N., Wilson, J. M., Botham, M. S., Brereton, T. M., Fox, R., & Goulson, D. (2015). Are neonicotinoid insecticides driving declines of widespread butterflies? *PeerJ, 3*, e1402. doi:10.7717/peerj.1402 PMID:26623186

Gove, B., Power, S., Buckley, G., & Ghazoul, J. (2007). Effects of herbicide spray drift and fertilizer overspread on selected species of woodland ground flora: Comparison between short-term and long-term impact assessments and field surveys. *Journal of Applied Ecology, 44*(2), 374–384. doi:10.1111/j.1365-2664.2007.01261.x

Hans, R. K., & Farooq, M. (2000). Dissipation and accumulation kinetics of lindane in soil and earthworms. *Pollution Research*, *19*, 394–396.

Jablonkai, I. (2011). Molecular mechanism of function of herbicides. In M. N. A. E.-G. Hasaneen (Ed.), *Herbicides - Mechanisms and mode of action* (pp. 3–24). InTech. doi:10.5772/31251

Jennings, D. E., Congelosi, A. M., & Rohr, J. R. (2012). Insecticides reduce survival and the expression of traits associated with carnivory of carnivorous plants. *Ecotoxicology (London, England)*, *21*(2), 569–575. doi:10.100710646-011-0817-8 PMID:22076028

Kalia, A., & Gosal, S. K. (2011). Effect of pesticide application on soil microorganisms. *Archives of Agronomy and Soil Science*, *57*(6), 569–596. doi:10.1080/03650341003787582

Kole, R. K., Banerjee, H., Bhattacharyya, A., Chowdhury, A., & Chaudhury, N. (1999). Photo transformation of some pesticides. *Journal of the Indian Chemical Society*, *76*, 595–600.

Lange, H. D., Lahr, J., Van der Pol, J. C., Wessels, Y., & Faber, J. H. (2009). Ecological vulnerability in wildlife: An expert judgment and multicriteria analysis tool using ecological traits to assess relative impact of pollutants. *Environmental Toxicology and Chemistry*, *28*(10), 2233–2240. doi:10.1897/08-626.1 PMID:19432506

Lushchak, V. I., Matviishyn, T. M., Husak, V. V., Storey, J. M., & Storey, K. B. (2018). Pesticide toxicity: A mechanistic approach. *EXCLI Journal*, *17*, 1101–1136. PMID:30564086

OECD. (2009). Report of the seminar on pesticide risk reduction through spray drift reduction strategies as part of national risk management. ENV/JM/MONO 36, Paris, France.

Oliveira, C. M., Auad, A. M., Mendes, S. M., & Frizzas, M. R. (2014). Crop losses and the economic impact of insect pests on Brazilian agriculture. *Crop Protection (Guildford, Surrey)*, *56*, 50–54. doi:10.1016/j.cropro.2013.10.022

Onder, M., Ceyhan, E., & Kahraman, A. (2011). Effects of agricultural practices on environment. *Biol Environ Chem.*, *24*, 28–32.

Peterson, D. E., Shoup, D. E., Thompson, C. R., & Olson, B. L. (2013). *Herbicide mode of action, cooperative extension service*. Kansas State University.

Petit, A. N., Fontaine, F., Clément, C., & Vaillant-Gaveau, N. (2008). Photosynthesis limitations of grapevine after treatment with the fungicide fludioxonil. *Journal of Agricultural and Food Chemistry*, *56*(15), 6761–6767. doi:10.1021/jf800919u PMID:18598040

Saladin, G., Magne, C., & Clement, C. (2003). Effects of fludioxonil and pyrimethanil, two fungicides used against Botrytis cinerea, on carbohydrate physiology in Vitis vinifera L. *Pest Management Science*, *59*(10), 1083–1092. doi:10.1002/ps.733 PMID:14561065

Sanchez-Bayo F. (2012). Insecticides mode of action in relation to their toxicity to non-target organisms. *J Environ Anal Toxicol, 2*.

Sanchez-Bayo, F., Goulson, D., Pennacchio, F., Nazzi, F., Goka, K., & Desneux, N. (2016). Are bee diseases linked to pesticides? A brief review. *Environment International*, *89-90*, 7–11. doi:10.1016/j.envint.2016.01.009 PMID:26826357

Sathiakumar, N., MacLennan, P. A., Mandel, J., & Delzell, E. (2011). A review of epidemiologic studies of triazine herbicides and cancer. *Critical Reviews in Toxicology*, *41*(sup1), 1–34. doi:10.3109/10408444.2011.554793 PMID:21425949

Solomon, K. R., Carr, J. A., Du Preez, L. H., Giesy, J. P., Kendall, R. J., Smith, E. E., & Van Der Kraak, G. (2008). Effects of atrazine on fish, amphibians, and aquatic reptiles: A critical review. *Critical Reviews in Toxicology*, *38*(9), 721–772. doi:10.1080/10408440802116496 PMID:18941967

Story, P., & Cox, M. (2001). Review of the effects of organophosphorus and carbamate insecticide vertebrates. Are there implications for locust management in Australia? *Wildlife Research*, *28*(2), 179–193. doi:10.1071/WR99060

USEPA. (2003). *Guidelines for ecological risk assessment. EPA/630/R-95/002F, 1998*. United States Environmental Protection Agency.

Wagner, N., & Bruhl, C. (2017). Ecotoxicology and Genotoxicology - Non-traditional terrestrial models. In: Marcelo L Larramendy (ed.). The use of terrestrial life-stages of European amphibians in toxicological studies. Publisher The Royal Society of Chemistry.

Zhang, Z., Zhang, X., Wang, Y., Zhao, Y., Lin, J., Liu, F., & Mu, W. (2016). Nitenpyram, dinotefuran and thiamethoxam used as seed treatments act as efficient controls against Aphis gossypii via high residues in cotton leaves. *Journal of Agricultural and Food Chemistry*, *64*(49), 9276–9285. doi:10.1021/acs.jafc.6b03430 PMID:27960287

# Chapter 16
# Sustainable Integrated Farming in Agriculture

**Sandeep Poddar**
https://orcid.org/0000-0001-9771-877X
*Lincoln University College, Malaysia*

## ABSTRACT

*Advancement of human civilization has resulted in depletion of resources and the deterioration of living conditions in certain regions of the world. Human activity on Earth has taken a hostile turn that threatens the existence of species and mankind. Circular agriculture focuses primarily on using minimal external input, closing nutrient cycles, regenerating soils, and minimizing environmental impact. Agriculture chains must be efficient in order to maximize the use of resources and return nutrients to the soil for a sustainable future. AI technology will lead to substantial changes in agricultural practices. The current discussion was explicitly related to United Nations Sustainable Development Goal 2: Zero Hunger, as well as other UNSDGs. To generate synergies and effectively manage trade-offs between agriculture, water, energy, land use, and climate change, the international community is urged to increase the number of integrated national and regional decision-making processes.*

## INTRODUCTION

The United Nations General Assembly adopted 17 Sustainable Development Goals (UNSDGs) to be achieved by 2030 in the year 2015. Leveraging the "leaving no one behind" principle, the new agenda advocates for a comprehensive strategy that encompasses all individuals in order to attain sustainable growth.. The SDGs are a road map to a better, more sustainable future for all people and an answer to the global problems we face today. The 17 Goals are interconnected and must all be accomplished by 2030 in order to ensure that no one is left behind. UN SDG 2- Zero Hunger focuses on end hunger, achieve food security and improved nutrition and promote sustainable agriculture. Sustainable integrated farming specially application of circular agriculture is directly supporting the UNSDG 2.

Depletion of resources and living circumstances begin to deteriorate in certain areas of the globe, human activity on Earth has taken a hostile turn that threatens the life of species and man himself. Rapid

DOI: 10.4018/978-1-6684-9838-5.ch016

environmental degradation over the last century has made it clear that we must act now (Pete et al., 2019). The response of countries to achieve carbon neutrality is indicative of a growing awareness of the gravity of the problem, particularly in regions where ecological foundations are gaining power. The Paris Agreement is the global community's plan to cut carbon emissions and other greenhouse gases by half by the year 2050. Agriculture and its role in the transition to sustainable communities throughout the globe has received a lot of attention (Yang et al. 2021; Bender et al. 2016; Dumortier et al. 2020).

The environment has suffered greatly as a result of the recent enormous increase in global food production. Approximately half of the land that is habitable is now used for agriculture. One-third of the world's forest cover has disappeared during the past century due to the rapid increase in agricultural productivity, and between 1990 and 2015, 20% of the standing forest was degraded.

## Circular Agriculture

The circular agriculture principally emphasize on using as little external input as possible, closing nutrient loops, regenerating soils, and reducing environmental effect. Circular agriculture can lessen the amount of resources needed and the environmental impact of farming if it is widely used. Additionally, it can contribute to a decrease in the usage of land, chemical fertilizers, and garbage, all of which assist lower world $CO_2$ emissions (Figure 1). According to estimates from the Ellen MacArthur Foundation (2016), a circular approach to food systems in Europe may cut the usage of chemical fertilizers by 80%.

As more and more people learn about the financial and ecological advantages of recycling, the practice is gaining popularity. Integrating useful end-products into the circular economy through the creation of a closed cycle to reduce the negative effects of production inputs and outputs has been addressed by scientists and farmers for decades (Boulding, 2013). Some researchers apply similar concepts to cycles and agriculture, such as circular bioeconomy (Zabaniotou, et al., 2015, Egea, et al., 2018) and rural ecological economy (Kristensen, et al.; Thorse, H. 2016), agroecology (Dumont, et al., 2013; Mier, et al. 2018) and climate-intelligent regenerative agriculture (Gosnell, et al., 2019). These terms intend to transform agriculture's linear economy into a circular economy (Grumbine, et al., 2022). The goals of this approach include the *development of sustainable landscapes, the promotion of healthy diets, the facilitation of agricultural digitalization, and the development of multifunctional landscapes.*

Animal manure as compost livestocks, pig, pullets, etc. may play a major role in circular agriculture. In recent years, new methods for generating electricity from agricultural refuse have emerged due to the development of a circular economy. Specifically, biogas is the central component of the cycle model (Barros et al., 2020). Biogas is being used in certain areas of the globe to reap environmental benefits and turn garbage into something useful (Vaneeckhaute et al., 2018; Molina-Moreno et al., 2017; Chen et al., 2017) (Figure 2).

Numerous nations have experimented with and adopted CA, or Circular Agriculture, an implementation of the *Circular Economy* in the agricultural sector. To lessen reliance on crude oil, CA is often biorefined from agricultural waste sources (including straw, husk, and husks of maize and rice wastes) (Khoo et al., 2019). Many countries, including those in Europe, Africa (Adenle et al., 2019), Asia (Priyadarshini et al., 2020) and North America (Kremen, 2020), have taken into account the development of a circular bioeconomy (Atinkut et al., 2020). Among them, China and the EU are the main supporters (D'Amato et al., 2017). China has made significant strides in enhancing the effectiveness with which its resources are used since the country's 2013 formulation of a national circular economy policy. China's national sustainable agricultural plan has been in effect since 2015 (Mathews & Tan, 2016). As part of its 2018

*Figure 1. Circular agriculture*

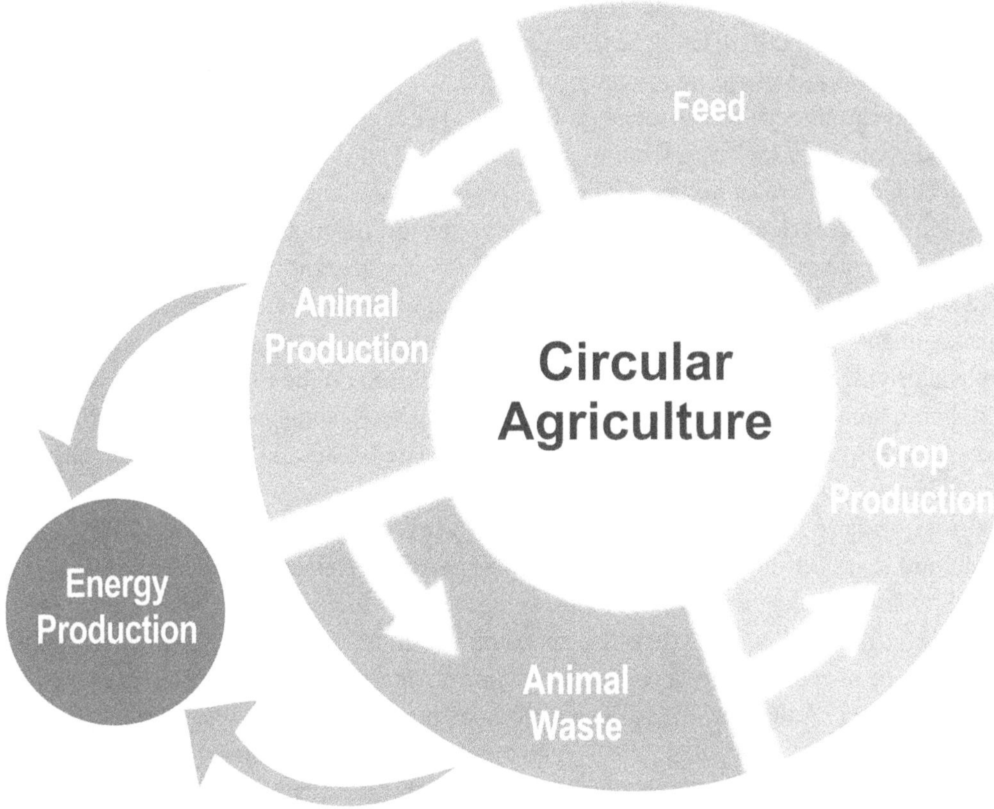


*Figure 2. Agricultural waste management by Biogas production*
*Source: the authors*

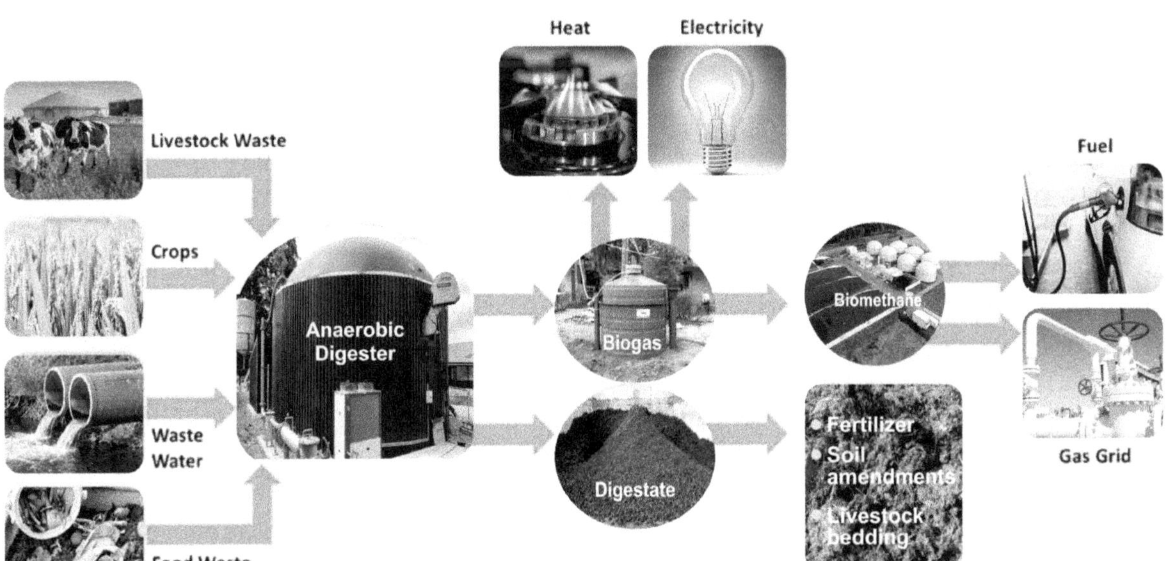

agricultural strategy, the European Union (EU) released a complete list of CA practices (v 2020). The World Economic Forum (WEF) has highlighted one such program, Shaping the Future of Food, which aims to create environmentally friendly, economically viable, and nutritionally sound food production and distribution networks (World Economic Forum 2021).

Different countries have different social, political, and economic conditions, thus their development models for, and goals for, circular agriculture will be unique. Top-down management structures, such as those seen in China and Peru's circular agricultural programs, are a good instance. To promote a synergistic transition among agriculture and bioenergy, Peruvian law mandates the use of circular agriculture in border regions (Quezada et al., 2016). Specifically, biogas generation is included into the circular agriculture system to facilitate a variety of goals, including but not limited to agricultural development, hunger management, and reduced carbon emissions. In both Italy and Germany, the market has played a larger role in driving the growth of the circular economy. Based on the EU's circular economy plastics strategy, Italy implements programs like the Extended Producer Responsibility scheme (EPR) and other conventional market tools to cut down on agricultural plastic waste and carbon emissions (Pazienza, De Lucia, 2020).

Several reviews have included and studied various facets of circular agriculture. Key terms in circular agriculture research include "sustainable development," "sustainability," "recycling," "agricultural robots," "waste management," and "biomass," according to a literature study covering the years 2005 to 2012. (Batlles-dela et al., 2022). The academic literature on the subject of agricultural residue management has been evaluated by few experts Altamira-Algarra, et al. 2022; Rosemarin, et al., 2020; Vasa, & Chacko, 2021. Anaerobic digestion and composting were identified to be the most researched processes in the literature (Song, et al., 2021), and composting chicken dung is an emerging method of reuse (Kamar Zaman, & Yaacob, 2022; Thomson, et al., 2021). Researchers analysed data from 2003 to 2019 to determine what influences farmers' involvement in biological recycling, attitudes toward the environment, economic perspectives, age, educational level, and informational resources in agriculture (Serebrennikov et al., 2020).

Muhie evaluated two Brazilian estates in 2023 and found the confined animals excreted more than 85% of the nitrogen consumed by them as organic decomposition. In general, circular crop management practices facilitated a high rate of N recovery, a reduction in environmental impacts, and an increase in food production at lower costs.

Based on the identification of factors that influence the implementation of local circular agriculture concentrates at the western portion of Shandong Province, AMOS was used to design and evaluate a structural equation model and investigate the interaction mechanism between each variable based on the identification of factors that influence the implementation of local circular agriculture concentrates at the western portion of Shandong Province (Xu, Y. et al. 2003). Xu Y. et al. proposes pertinent countermeasures and suggestions to promote the integrated planting and breeding mode in accordance with local conditions, which can serve as a useful reference for government departments to formulate suitable policies.

When purchasing organic foods, the perspectives of consumers on circular agriculture can be applied by examining a variety of aspects like climate change, trust, and eco-labels indicate a stronger preference among surveyed consumers for purchasing organic food, according to the study's findings (Akter et al 2023).

Bioeconomy in organic agriculture under excessive nutrient levels (Goat manure and arbuscular mycorrhizal fungi), a case study conducted in Indonesia in 2021 demonstrates that the implementation of circular bioeconomy via integrated farming of *P. angulata* was not an immediate solution for economic and environmental optimization, but it can be considered as a means of addressing the environmental issue caused by excessive livestock residues (Kurniawati et al., 2021). Specifically addresses the development of such a *Circular Precision Farming System*, including the essential components and prerequisites for its application to the two mainstays of agriculture (cereal production and dairy farming). Due to the intricate relationship between the animals and the crops used to feed them, such systems can help farmers and agricultural businesses make informed decisions at every stage of the supply chain, thereby increasing profits, decreasing risks, and improving the environment (Lampridi et al., 2019).

Including several forage species may improve the ecological and agricultural sustainability of a landscape (Franzluebbers, A.J., & Martin, G. 2022). When crop and animal activities are combined, ecological synergies may be enhanced via the use of diversified agricultural systems that make use of both annual and perennial forages. Internalizing nutrient cycles, providing cultural control of pests, parasites, and illnesses, and sharing resources within a closed-loop agroecosystem are all possible with integrated crop-livestock systems. *Rice fish farming* is used over a thousand years, humans have practiced rice-fish (RF) co-culture, which combines the cultivation of rice with the rearing of fish. Ge et al. 2023 also reviewed a theoretical framework for designing sustainable agricultural systems that combine ancient wisdom with cutting-edge science and technology.

In case of Tilapia farming with low water use seeks to reduce inputs and reuse waste and effluents, thus closing flows or links between economic and ecological resources and decentralizing production systems approaches is described by Zimmermann, S., Kiessling, A., Zhang, (2023) is similar to circular bioeconomy.

Dominici et al. 2022 argue that the diversification of the marketing strategy and the reduction of operating costs are crucial to the strategy's success. Incorporating the role of constructed wetlands in the provision of ecosystem services in rice agroecosystems into an ecologically mindful plan to implement a self-regulating and circular farming system.

*Indoor Vertical Farms (IVF)* may help contribute to urban circular food systems by decreasing food waste and increasing resource usage efficiency, according to a Life Cycle Assessment of a unit built on a university campus in Portugal to grow broccoli microgreens for salads. Building integration offers the ability to reduce its well-known high energy usage (Parkes et al.'s 2022). Martin, M., Weidner, T., and Gullstrom, C 2022 in their study found that educating residential property owners on the advantages of utilizing unused space for urban food provisioning and expanding the use of vertical farming and circular economy principles in urban settings.

The circular bio-economy (waste reuse) is a viable model for safeguarding human and animal health as well as the environment. Diverse applications for the utilization of aquaculture byproducts and wastes have been considered. Multi-trophic integrated aquaculture is one of the most efficient techniques for reducing metabolic waste. Proteins (collagen, gelatine), polysaccharides (chitosan), lipids (omega 3), and pigments (astaxanthin, beta-carotene) are among the most useful biomolecules that may be extracted from aquaculture waste using the most efficient method. The pharmaceutical business isn't the only one that may benefit from biomolecules; they can also be used in the food industry, cosmetics, and even on animals (Fraga-Corral et al. 2022).

Derived from discarded agricultural materials, biochar is a key "player" on the path to sustainable and transparent food production. Biochar production is particularly intriguing since it may aid in the generation of local energy and the overall valorization of biomass. As a result of the interconnected nature of agroecosystems, climate change, and sustainable resource management, collaborative efforts at the national and local levels are increasingly essential (Nair et al 2017, Sessions et al. 2019).

In order to maximize the use of resources and return the nutrients to the soil for sustainable future agriculture chains must be efficient. Ruohomaa, H., Poykko, T. and Salminen, V (2021) presents a case study for the development of an innovation ecosystem concept to facilitate the creation of a more efficient bio-based value chain and circular economy.

The direction and viability of circular agriculture are determined by economics and markets. Generally, it is believed that market imperfections cause environmental contamination. Nonetheless, environmental deterioration offers substantial business opportunities (Cohen, & Winn 2007). As global concern for sustainability grows, "eco-business" has been devised to advocate for policy and institutional changes. The challenge for organizations from the perspective of "eco-entrepreneurs" is to more effectively incorporate environmental performance into their economic business logic (Hockerts, & Wüstenhagen, 2010). Institutional players such as governments, officials, regional governments, and research organizations are often overlooked even while studying bottom-up processes (Jolly, 2017). Sustainable entrepreneurs in the Netherlands were shown to engage in institutional change over the course of a two-year comparative case study by creating new symbols and ideas, constructing new metrics, building agreement, and establishing new connections (Thompson et al., 2015). In addition, sustainable entrepreneurs work to disseminate inventions that minimize negative impacts on the environment and the community (Dean & McMullen 2007). To be clear, studying these groups is not meant to inspire the launch of environmentally and economically responsible businesses; rather, it is intended to argue for and support changes in existing policies and institutions.

Water quality is an important factor in integrated firming. The work on *Agricultural Water Scarcity Risk* (AWSR) by Chen et al 2017 elucidates the importance of water quality in the assessments of AWSR.

## Insect Pollinator and Integrated Farming

According to research done in apple orchards by Hulsmans et al in 2023, the supply of ecosystem services in agroecosystems is in jeopardy owing to insect pollinator losses, which are mostly caused by the increase and intensification of agricultural land use. The most promising approaches for pollinator conservation involve maintaining and expanding orchard edges through smaller fields, conserving and restoring semi-natural habitats, preserving and expanding organic farming as an alternative to conventional farming, and fostering landscape diversity.

## AI Application in Agriculture

The most recent developments in agricultural sensing and communication technologies have opened the door to creating and implementing more effective management plans that take into account the temporal and spatial variability of field and crop attributes for prompt and precise field applications (Anagnostis, et al. 2021). The creation of interconnected systems with user-friendly interfaces is a requirement in order to benefit from these cutting-edge sensing and management tools. In order to provide the integrated system with real-time data, precision farming technologies can be used in conjunction with Internet of Things (IoT) sensing systems in both crop and animal production operations (Lampridi, et al., 2019; Rodias, et al. 2020).

Sing & Kaur (2022) emphasized the significance of artificial intelligence (AI) in Precision agriculture (PA). PA generates large data from numerous sensors. AI technologies help to accumulate and compile the data. The knowledge derived from big data enables farmers to make more informed decisions, such as the optimal time to sow a crop, the exact quantity of fertilizer and water required, the type of crop best suited to the soil quality, and the optimal time for harvesting. Thus, AI-based precision agriculture results in increased crop quality and yield at a lower cost. Artificial intelligence for sustainable agriculture Intelligent devices perform numerous repetitive tasks without human intervention. Intelligent irrigation systems help to conserve water. AI aids in the reduction of pesticide use, thereby safeguarding our food and the environment. This disruptive technology will lead to significant changes in agricultural practices, including the reduction or elimination of agrochemicals and the use of less land with increased yield. With a vast application potential in agriculture, we must surmount a number of obstacles, some of which are listed below. In developed nations, precision agriculture is ubiquitous, whereas in developing nations, conventional farming practices are still prevalent. For these nations to develop intelligent agricultural systems, experts in artificial intelligence, data analytics, and machine learning are in high demand. Compiling large-scale datasets is required. In addition, numerous machine learning prototype models are created for agricultural applications; however, these models frequently fail in real-world environments due to uncontrollable factors such as weather and soil conditions (Wu et al., 2019).

5G technology's maturation increases the likelihood of field process automation and the acquisition of real-time data. 5G is a significant performer during the fifth industrial revolution, with the potential to facilitate not only high-speed connectivity but also faster data acquisition and analysis. Google Earth, a free geolocation search engine, integrates NDVI-generating processing code (Singh, & Frazier 2018).

According to Muhie S H. (2023), biodynamics is the construction of a farm that functions as an integrated organism. Biodynamic agriculture is scientifically proven to contribute to the sustainability of agriculture by improving soil quality, increasing crop yield and nutritional value, and controlling pests. Therefore, biodynamics is regarded as a promising route towards the future of sustainable and integrated agriculture.

By combining a geographical information system (GIS) cluster analysis with a suitability model and the notion of a circular economy, this research hopes to evaluate the energy production potential and site locations of agricultural residue treatment plants in Myanmar. This study's multi-criteria and GIS-integrated evaluation methodology may help policymakers in Myanmar promote bioenergy in a way that is compatible with sustainable agriculture practices (Min Htoo, T., Yabar, H., Mizunoya, T. 2022).

The suggested model is built on a framework of stochastic multi-objective programming, and it includes features such triangular fuzzy numbers, fuzzy credibility restricted programming, the Stewart model, the analytic hierarchy process, and the opportunity carbon footprint to make it more realistic and applicable. A case study in Zhanghe Irrigation District, Hubei Province, China, confirms the accuracy of the model. Through the reduction of chemical fertilizers, the recycling of agricultural residues, and the substitution of nonrenewable energy, the recycling system reduces 4.26 108 kg of carbon dioxide. Yue et al. 2022 demonstrate that circular agriculture creates a new path for environmentally friendly and high-efficiency production.

## Quality and Quantity of the Production by Using Computer Vision, IOT

Efficient, productive, and resilient agriculture is predicated on a number of factors, two of which are monitoring crop health and weeding. Crops get infected as a consequence of persistent insect and pathogen assaults, lowering both yield and quality. It is difficult to keep an eye on the health of crops without using some type of information and communication technology (ICT). This has led to the development of precision agriculture, which relies heavily on IoT and AI methods (Devi et al 2023).

Researchers have tried many different methods throughout the years in an effort to accurately identify and diagnose plant diseases. Plant disease identification using ML techniques is only one example of their widespread use in modern agriculture (Pantazi et al., 2019). Some of the most popular ML methods used in farming today include the Support Vector Machine (SVM), Random Forest (RF), K-Nearest Neighbor (KNN), Artificial Neural Network (ANN), and Naive Bayes (Harakannanavar et al., 2022).

AI application in weed identification and process control is the suatinable improved way to save money without sacrificing quality. To further aid in water saving, the sensor pack specifies in detail the conditions under which sprinklers should be activated.

Anagnostis et al. 2021 demonstrates that an integrated system employing smart farming instruments and smart processing methodologies based on the concept of a circular economy has been proposed to mitigate the impact of agricultural activity on climate change. Circular agriculture, precision farming, and the use of appropriate crop management instruments may contribute to a more resource-efficient and sustainable agricultural system.

## Policies to Support Circular Agriculture

The adoption of circular agriculture intends to solve the growing concerns about the sustainability of global food production, the depletion, deterioration, and pollution of water and land resources, as well as the destroying of soils, loss of biodiversity, and degradation of soils. Instead of farming against nature, the goal is to work with it. Adopting circular practices in rural areas is only feasible in conjunction with other changes that affect the nation's and cities' overall demand for the food system, such as reducing food waste, altering diets, and being willing to pay more for organic goods. To encourage the adoption of circular agricultural techniques in rural regions, a comprehensive set of policies, technology, and institutions is required,

## International Networking and Collaboration to Support Sustainable Circular Agriculture

All nations have an interest in the adoption of the circularity concept in this sector because of the considerable impact that agriculture and food production have on society and the environment, both nationally and internationally. Through international cooperation for technology transfer and capacity development, smallholder farmers can have better access to new technologies and skills. By establishing uniform guidelines and standards for waste management and reduction, environmentally friendly purchasing methods, agricultural value chains, and reporting on food loss and waste, international collaboration can help promote circular agriculture. International assistance is also necessary for many emerging nations to spend heavily in their infrastructure. Global collaboration is required in order to prevent a race to the bottom in terms of environmental norms and safeguards, given the growing significance of fostering sustainable value chains in agricultural systems. Particularly when it comes to the production of agricultural crops, the COVID-19 pandemic has also prompted countries to consider shorter and more localized value chains of necessary goods and services, which has increased the need for international cooperation to improve technological, financial, and marketing skills as well as other extension services.

## CONCLUSION

The world's agricultural systems must become more efficient and less wasteful. From a comprehensive and integrated standpoint, sustainable agricultural practices and food systems, including both production and consumption, must be pursued. Rural areas are the primary location of extreme poverty and hunger, and smallholder farmers and their families make up a sizable share of those who live in these conditions. As a result, increasing food production, agricultural productivity, and rural incomes is essential to ending poverty and hunger. It is crucial to use and manage land, healthy soils, water, and plant genetic resources sustainably because they are essential components of food production and are becoming increasingly scarce in many regions of the world. Another way to reduce pressure to remove land is to increase yields on currently used agricultural fields, including restoring damaged sites. Future food needs must also be met by halting and reversing land degradation. Given the current level of global land degradation, there are significant potential benefits from land restoration for ensuring food security and reducing climate change. It is acknowledged, nonetheless, that our knowledge of the causes of desertification, land degradation, and drought is still developing. By utilizing sound and sustainable soil, land, water, nutrient, and pest management techniques as well as a greater use of organic fertilizers, traditional farmer knowledge may support productive food systems. This is made possible by combining this knowledge with the most recent advances in science.

The idea of circular agriculture is not new, and pre-industrial societies used it extensively. The large-scale, monoculture, and highly intensive practices of contemporary farming, which frequently prioritize increasing profit over environmental protection, have pushed it aside. Circular agriculture is typically not a good fit for large-scale, specialized agricultural enterprises' economic models. In order to transition to circular agriculture, it is important to put more of an emphasis on promoting smallholder farming that is rooted in organic, mixed, and agroforestry methods. Contrary to export-oriented monocrop farming, which frequently results in greater food insecurity, circular agriculture with higher diversity of production is linked to better health and nutrition.

The international community is urged to generate synergies and effectively handle trade-offs among agriculture, water, energy, land use, and climate change, there has to be an increase in integrated decision-making procedures at the national and regional level. And enhance investment in research, development, and demonstration of technology to improve the sustainability of food systems worldwide in light of anticipated changes in temperatures, precipitation, and pests related to climate change. Building local food system resilience will be essential to preventing future widespread shortages, guaranteeing food security, and promoting healthy eating for all.

The present chapter describes integrated farming and circular economy which will play a pivotal role in near future to combat "Zero Hunger" strengthen "Economy" and support the United Nations Sustainability move.

## REFERENCES

Adenle, A. A., Wedig, K., & Azadi, H. (2019). Sustainable agriculture and food security in Africa: The role of innovative technologies and international organizations. *Technology in Society*, *58*, 101143. doi:10.1016/j.techsoc.2019.05.007

Akter, S., Ali, S., Fekete-Farkas, M., Fogarassy, C., & Lakner, Z. (2023). Why organic food? Factors influence the organic food purchase intension in an emerging country (study from northern part of Bangladesh). *Resources*, *12*(1), 5. doi:10.3390/resources12010005

Altamira-Algarra, B., Puigagut, J., Day, J. W., Mitsch, W. J., Vymazal, J., Hunter, R. G., & García, J. (2022). A review of technologies for closing the P loop in agriculture runoff: Contributing to the transition towards a circular economy. *Ecological Engineering*, *177*, 106571. doi:10.1016/j.ecoleng.2022.106571

Anagnostis, A., Tagarakis, A. C., Kateris, D., Moysiadis, V., Sørensen, C. G., Pearson, S., & Bochtis, D. (2021). Orchard mapping with deep learning semantic segmentation. *Sensors (Basel)*, *21*(11), 3813. doi:10.339021113813 PMID:34072975

Atinkut, H. B., Yan, T., Zhang, F., Qin, S., Gai, H., & Liu, Q. (2020). Cognition of agriculture waste and payments for a circular agriculture model in Central China. *Scientific Reports*, *10*(1), 10826. doi:10.103841598-020-67358-y PMID:32616725

Barros, M. V., Salvador, R., de Francisco, A. C., & Piekarski, C. M. (2020). Mapping of research lines on circular economy practices in agriculture: From waste to energy. *Renewable & Sustainable Energy Reviews*, *131*, 109958. doi:10.1016/j.rser.2020.109958

Batlles-delaFuente, A., Abad-Segura, E., González-Zamar, M. D., & Cortés-García, F. J. (2022). An evolutionary approach on the framework of circular economy applied to agriculture. *Agronomy (Basel)*, *12*(3), 620. doi:10.3390/agronomy12030620

Bender, S. F., Wagg, C., & van der Heijden, M. G. (2016). An underground revolution: Biodiversity and soil ecological engineering for agricultural sustainability. *Trends in Ecology & Evolution*, *31*(6), 440–452. doi:10.1016/j.tree.2016.02.016 PMID:26993667

Boulding, K. E. (2013). The economics of the coming spaceship earth. In *Environmental quality in a growing economy* (pp. 3–14). RFF Press.

Chen, L., Cong, R. G., Shu, B., & Mi, Z. F. (2017). A sustainable biogas model in China: The case study of Beijing Deqingyuan biogas project. *Renewable & Sustainable Energy Reviews*, *78*, 773–779. doi:10.1016/j.rser.2017.05.027

Cohen, B., & Winn, M. I. (2007). Market imperfections, opportunity and sustainable entrepreneurship. *Journal of Business Venturing*, *22*(1), 29–49. doi:10.1016/j.jbusvent.2004.12.001

D'Amato, D., Droste, N., Allen, B., Kettunen, M., Lähtinen, K., Korhonen, J., Leskinen, P., Matthies, B. D., & Toppinen, A. (2017). Green, circular, bio economy: A comparative analysis of sustainability avenues. *Journal of Cleaner Production*, *168*, 716–734. doi:10.1016/j.jclepro.2017.09.053

Dean, T. J., & McMullen, J. S. (2007). Toward a theory of sustainable entrepreneurship: Reducing environmental degradation through entrepreneurial action. *Journal of Business Venturing*, *22*(1), 50–76. doi:10.1016/j.jbusvent.2005.09.003

Devi, N., Sarma, K. K., & Laskar, S. (2023). Design of an intelligent bean cultivation approach using computer vision, IoT and spatio-temporal deep learning structures. *Ecological Informatics*, *75*, 102044. doi:10.1016/j.ecoinf.2023.102044

Dominici, L., Magi, E., Leidi, B., Pastore, M. E., & Comino, E. (2022). Ecologically-oriented business strategy for a small-size rice farm: Integrated wetland management for the improvement of environmental benefits and economic feasibility. *The Science of the Total Environment*, *838*, 156604. doi:10.1016/j.scitotenv.2022.156604 PMID:35691347

Dumont, B., Fortun-Lamothe, L., Jouven, M., Thomas, M., & Tichit, M. (2013). Prospects from agroecology and industrial ecology for animal production in the 21st century. *animal, 7*(6), 1028-1043. doi:10.1017/S1751731112002418

Dumortier, J., Dokoohaki, H., Elobeid, A., Hayes, D. J., Laird, D., & Miguez, F. E. (2020). Global land-use and carbon emission implications from biochar application to cropland in the United States. *Journal of Cleaner Production*, *258*, 120684. doi:10.1016/j.jclepro.2020.120684

Egea, F. J., Torrente, R. G., & Aguilar, A. (2018). An efficient agro-industrial complex in Almería (Spain): Towards an integrated and sustainable bioeconomy model. *New Biotechnology*, *40*, 103–112. doi:10.1016/j.nbt.2017.06.009 PMID:28698129

Ellen MacArthur Foundation. (2016). *Cities and circular economy for food*. Ellen MacArthur Foundation. https://www. ellenmacarthurfoundation.org/publications/cities-andcircular-economy-for-food

Fraga-Corral, M., Ronza, P., Garcia-Oliveira, P., Pereira, A. G., Losada, A. P., Prieto, M. A., Quiroga, M. I., & Simal-Gandara, J. (2022). Aquaculture as a circular bio-economy model with Galicia as a study case: How to transform waste into revalorized by-products. *Trends in Food Science & Technology*, *119*, 23–35. doi:10.1016/j.tifs.2021.11.026

Franzluebbers, A. J., & Martin, G. (2022). Farming with forages can reconnect crop and livestock operations to enhance circularity and foster ecosystem services. *Grass and Forage Science*, *77*(4), 270–281. doi:10.1111/gfs.12592

Ge, L., Sun, Y., Li, Y., Wang, L., Guo, G., Song, L., Wang, C., Wu, G., Zang, X., Cai, X., Li, S., & Li, P. (2023). Ecosystem sustainability of rice and aquatic animal co-culture systems and a synthesis of its underlying mechanisms. *The Science of the Total Environment*, *880*, 163314. doi:10.1016/j.scitotenv.2023.163314 PMID:37030380

Gosnell, H., Gill, N., & Voyer, M. (2019). Transformational adaptation on the farm: Processes of change and persistence in transitions to 'climate-smart'regenerative agriculture. *Global Environmental Change*, *59*, 101965. doi:10.1016/j.gloenvcha.2019.101965

Grumbine, R. E., Xu, J., & Ma, L. (2021). An overview of the problems and prospects for circular agriculture in sustainable food systems in the Anthropocene. *Circular Agricultural Systems*, *1*(1), 1–11. doi:10.48130/CAS-2021-0003

Harakannanavar, S. S., Rudagi, J. M., Puranikmath, V. I., Siddiqua, A., & Pramodhini, R. (2022). Plant leaf disease detection using computer vision and machine learning algorithms. *Global Transitions Proceedings*, *3*(1), 305–310. doi:10.1016/j.gltp.2022.03.016

Hockerts, K., & Wüstenhagen, R. (2010). Greening Goliaths versus emerging Davids—Theorizing about the role of incumbents and new entrants in sustainable entrepreneurship. *Journal of Business Venturing*, *25*(5), 481–492. doi:10.1016/j.jbusvent.2009.07.005

Hulsmans, E., Daelemans, R., Cuypers, V., Van Der Straeten, E., Vanderlinden, M., De Blanck, T., Vertommen, W., Boeraeve, M., Proesmans, W., & Honnay, O. (2023). Cascading effects of management and landscape on insect pollinators, pollination services and yield in apple orchards. *Agriculture, Ecosystems & Environment, 352*, 108509. doi:10.1016/j.agee.2023.108509

Jolly, S. (2017). Role of institutional entrepreneurship in the creation of regional solar PV energy markets: Contrasting developments in Gujarat and West Bengal. *Energy for Sustainable Development, 38*, 77–92. doi:10.1016/j.esd.2016.10.004

Kamar Zaman, A. M., & Yaacob, J. S. (2022). Exploring the potential of vermicompost as a sustainable strategy in circular economy: Improving plants' bioactive properties and boosting agricultural yield and quality. *Environmental Science and Pollution Research International, 29*(9), 1–17. doi:10.100711356-021-18006-z PMID:35034296

Khoo, H. H., Eufrasio-Espinosa, R. M., Koh, L. S., Sharratt, P. N., & Isoni, V. (2019). Sustainability assessment of biorefinery production chains: A combined LCA-supply chain approach. *Journal of Cleaner Production, 235*, 1116–1137. doi:10.1016/j.jclepro.2019.07.007

Kremen, C. (2020). Ecological intensification and diversification approaches to maintain biodiversity, ecosystem services and food production in a changing world. *Emerging Topics in Life Sciences, 4*(2), 229–240. doi:10.1042/ETLS20190205 PMID:32886114

Kristensen, D. K., Kjeldsen, C., & Thorsøe, M. H. (2016). Enabling sustainable agro-food futures: Exploring fault lines and synergies between the integrated territorial paradigm, rural eco-economy and circular economy. *Journal of Agricultural & Environmental Ethics, 29*(5), 749–765. doi:10.100710806-016-9632-9

Kurniawati, A., Windriyati, R. D. H., Wulansari, N. K., Toth, G., & Toth, Z. (2021). Alternatives for circular bioeconomy in organic farming under excessive nutrients (goat manure and arbuscular mycorrhizal fungi): A case study in Indonesia. *Sustainability (Basel), 13*(22), 12333. doi:10.3390u132212333

Lampridi, M. G., Sørensen, C. G., & Bochtis, D. (2019). Agricultural sustainability: A review of concepts and methods. *Sustainability (Basel), 11*(18), 5120. doi:10.3390u11185120

Martin, M., Weidner, T., & Gullström, C. (2022). Estimating the potential of building integration and regional synergies to improve the environmental performance of urban vertical farming. *Frontiers in Sustainable Food Systems, 6*, 849304. doi:10.3389/fsufs.2022.849304

Mathews, J. A., & Tan, H. (2016). Circular economy: Lessons from China. *Nature, 531*(7595), 440–442. doi:10.1038/531440a PMID:27008954

Mier y Terán Giménez Cacho, M., Giraldo, O. F., Aldasoro, M., Morales, H., Ferguson, B. G., Rosset, P., Khadse, A., & Campos, C. (2018). Bringing agroecology to scale: Key drivers and emblematic cases. *Agroecology and Sustainable Food Systems, 42*(6), 637–665. doi:10.1080/21683565.2018.1443313

Min Htoo, T., Yabar, H., & Mizunoya, T. (2022). GIS-Based Cluster and Suitability Analysis of Crop Residues: A Case Study in Yangon Region, Myanmar. *Applied Sciences (Basel, Switzerland), 12*(22), 11822. doi:10.3390/app122211822

Molina-Moreno, V., Leyva-Díaz, J. C., Llorens-Montes, F. J., & Cortés-García, F. J. (2017). Design of indicators of circular economy as instruments for the evaluation of sustainability and efficiency in wastewater from pig farming industry. *Water (Basel)*, *9*(9), 653. doi:10.3390/w9090653

Muhie, S. H. (2023). Concepts, principles, and application of biodynamic farming: A review. *Circular Economy and Sustainability*, *3*(1), 291–304. doi:10.100743615-022-00184-8

Nair, V. D., Nair, P. R., Dari, B., Freitas, A. M., Chatterjee, N., & Pinheiro, F. M. (2017). Biochar in the agroecosystem–climate-change–sustainability nexus. *Frontiers in Plant Science*, *8*, 2051. doi:10.3389/fpls.2017.02051 PMID:29312364

Pantazi, X. E., Moshou, D., & Tamouridou, A. A. (2019). Automated leaf disease detection in different crop species through image features analysis and One Class Classifiers. *Computers and Electronics in Agriculture*, *156*, 96–104. doi:10.1016/j.compag.2018.11.005

Parkes, M. G., Cubillos Tovar, J. P., Dourado, F., Domingos, T., & Teixeira, R. F. (2022). Life Cycle Assessment of a Prospective Technology for Building-Integrated Production of Broccoli Microgreens. *Atmosphere (Basel)*, *13*(8), 1317. doi:10.3390/atmos13081317

Pazienza, P., & De Lucia, C. (2020). For a new plastics economy in agriculture: Policy reflections on the EU strategy from a local perspective. *Journal of Cleaner Production*, *253*, 119844. doi:10.1016/j.jclepro.2019.119844

Pete, S., Johnson, N., Katherine, C., Donovan, C., Francesco, C., Giacomo, G., . . . Vizzarri, M. (2019). Interlinkages between desertification, land degradation, food security and greenhouse gas fluxes: synergies, trade-offs and integrated response options. In Climate Change and Land: IPCC Special Report on Climate Change, Desertification, Land Degradation, Sustainable Land Management, Food Security, and Greenhouse Gas Fluxes in Terrestrial Ecosystems (pp. 551-672). Cambridge University Press. doi:10.1017/9781009157988.008

Priyadarshini, P., & Abhilash, P. C. (2020). Policy recommendations for enabling transition towards sustainable agriculture in India. *Land Use Policy*, *96*, 104718. doi:10.1016/j.landusepol.2020.104718

Quezada, C. A. V., Fonseca, M. B., & Romero, H. (2016). The circular agriculture applied in neighboring countries: The case of biogas on the border between Ecuador and Perú. *New Biotechnology*, *33*, S66–S67. doi:10.1016/j.nbt.2016.06.953

Rodias, E., Aivazidou, E., Achillas, C., Aidonis, D., & Bochtis, D. (2020). Water-energy-nutrients synergies in the agrifood sector: A circular economy framework. *Energies*, *14*(1), 159. doi:10.3390/en14010159

Rosemarin, A., Macura, B., Carolus, J., Barquet, K., Ek, F., Järnberg, L., Lorick, D., Johannesdottir, S., Pedersen, S. M., Koskiaho, J., Haddaway, N. R., & Okruszko, T. (2020). Circular nutrient solutions for agriculture and wastewater–a review of technologies and practices. *Current Opinion in Environmental Sustainability*, *45*, 78–91. doi:10.1016/j.cosust.2020.09.007

Ruohomaa, H., Pöykkö, T., & Salminen, V. (2021). Digitalization of Bio-Based Value Chains. In *Advances in Human Factors, Business Management and Leadership: Proceedings of the AHFE 2021 Virtual Conferences on Human Factors, Business Management and Society, and Human Factors in Management and Leadership,* (pp. 111-118). Springer International Publishing. 10.1007/978-3-030-80876-1_16

Serebrennikov, D., Thorne, F., Kallas, Z., & McCarthy, S. N. (2020). Factors influencing adoption of sustainable farming practices in Europe: A systemic review of empirical literature. *Sustainability (Basel), 12*(22), 9719. doi:10.3390u12229719

Sessions, J., Smith, D., Trippe, K. M., Fried, J. S., Bailey, J. D., Petitmermet, J. H., & Campbell, J. D. (2019). *Can biochar link forest restoration with commercial agriculture?.* doi:10.1016/j.biombioe.2019.02.015

Singh, K. K., & Frazier, A. E. (2018). A meta-analysis and review of unmanned aircraft system (UAS) imagery for terrestrial applications. *International Journal of Remote Sensing, 39*(15-16), 5078–5098. doi:10.1080/01431161.2017.1420941

Singh, P., & Kaur, A. (2022). A systematic review of artificial intelligence in agriculture. *Deep Learning for Sustainable Agriculture*, 57-80. doi:10.1016/B978-0-323-85214-2.00011-2

Song, J., Wang, Y., Zhang, S., Song, Y., Xue, S., Liu, L., Lvy, X., Wang, X., & Yang, G. (2021). Coupling biochar with anaerobic digestion in a circular economy perspective: A promising way to promote sustainable energy, environment and agriculture development in China. *Renewable & Sustainable Energy Reviews, 144*, 110973. doi:10.1016/j.rser.2021.110973

Thompson, N. A., Herrmann, A. M., & Hekkert, M. P. (2015). How sustainable entrepreneurs engage in institutional change: Insights from biomass torrefaction in the Netherlands. *Journal of Cleaner Production, 106*, 608–618. doi:10.1016/j.jclepro.2014.08.011

Thomson, A., Price, G. W., Arnold, P., Dixon, M., & Graham, T. (2022). Review of the potential for recycling CO2 from organic waste composting into plant production under controlled environment agriculture. *Journal of Cleaner Production, 333*, 130051. doi:10.1016/j.jclepro.2021.130051

Vaneeckhaute, C., Styles, D., Prade, T., Adams, P., Thelin, G., Rodhe, L., Gunnarsson, I., & D'Hertefeldt, T. (2018). Closing nutrient loops through decentralized anaerobic digestion of organic residues in agricultural regions: A multi-dimensional sustainability assessment. *Resources, Conservation and Recycling, 136*, 110–117. doi:10.1016/j.resconrec.2018.03.027

Vasa, T. N., & Chacko, S. P. (2021). Recovery of struvite from wastewaters as an eco-friendly fertilizer: Review of the art and perspective for a sustainable agriculture practice in India. *Sustainable Energy Technologies and Assessments, 48*, 101573. doi:10.1016/j.seta.2021.101573

World Economic Forum. (2021). *Shaping the Future of Food.* WEF. https://www.weforum.org/system-initiatives/shaping-the-future-of-food-security-and-agriculture (accessed on 19 July 2021).

Wu, L., Peng, Y., Fan, J., & Wang, Y. (2019). Machine learning models for the estimation of monthly mean daily reference evapotranspiration based on cross-station and synthetic data. *Nordic Hydrology, 50*(6), 1730–1750. doi:10.2166/nh.2019.060

Xu, Y., Wang, G., Yang, S., Liu, M., Wang, Q., Tian, S., Zhang, Y., Xu, Y., Yuan, X., Ma, Q., Ma, H., & Zhang, H. (2023). How to further develop the integrated planting and breeding mode: A case study of the western region of Shandong Province. *Sustainable Development*, 1– 16. doi:10.1002/sd.2573

Yang, Q., Zhou, H., Bartocci, P., Fantozzi, F., Mašek, O., Agblevor, F. A., Wei, Z., Yang, H., Chen, H., Lu, X., Chen, G., Zheng, C., Nielsen, C. P., & McElroy, M. B. (2021). Prospective contributions of biomass pyrolysis to China's 2050 carbon reduction and renewable energy goals. *Nature Communications*, *12*(1), 1698. doi:10.103841467-021-21868-z PMID:33727563

Yue, Q., Guo, P., Wu, H., Wang, Y., & Zhang, C. (2022). Towards sustainable circular agriculture: An integrated optimization framework for crop-livestock-biogas-crop recycling system management under uncertainty. *Agricultural Systems*, *196*, 103347. doi:10.1016/j.agsy.2021.103347

Zabaniotou, A., Rovas, D., Libutti, A., & Monteleone, M. (2015). Boosting circular economy and closing the loop in agriculture: Case study of a small-scale pyrolysis–biochar based system integrated in an olive farm in symbiosis with an olive mill. *Environmental Development*, *14*, 22–36. doi:10.1016/j.envdev.2014.12.002

Zimmermann, S., Kiessling, A., & Zhang, J. (2023). The future of intensive tilapia production and the circular bioeconomy without effluents: Biofloc technology, recirculation aquaculture systems, bio-RAS, partitioned aquaculture systems and integrated multitrophic aquaculture. *Reviews in Aquaculture*, *15*(S1), 22–31. doi:10.1111/raq.12744

## KEY TERMS AND DEFINITIONS

**Circular Agriculture:** A system where agricultural biomass, waste, and leftovers from food preparation are kept as reusable resources. Recycling is done for agricultural food waste.

**Indoor vertical farming:** (IVF): The process of growing food in a controlled, enclosed environment while it is piled one on top of the other.

**Insect Pollination:** This refers to a form of pollination wherein pollen is transferred from plants to insects, predominantly but not exclusively from flowering plants.

**Integrated Farming:** Cropping techniques and other agricultural production systems that meet ecological and commercial requirements.

**Sustainable Agriculture:** A practice that aims to meet the needs of the present generation without compromising the ability of future generations to meet their own needs.

# Chapter 17
# Ecological Thought in Economics:
## Exploring the Intersection of Environmentalism and Economic Theory

**Mehmet Gökhan Özdemir**
https://orcid.org/0000-0002-6756-7285
*Kirikkale University, Turkey*

## ABSTRACT

*This chapter explores the intersection of ecological thought and economics, examining the evolution of ecological economics and its potential implications for economic theory and policy. Ecological economics offers a more comprehensive framework for understanding and addressing the relationship between the natural environment and economic systems. Key concepts, such as externalities, natural capital, and ecosystem services, are discussed, as well as challenges to incorporating ecological considerations into economic decision-making. By promoting more sustainable and just economic systems that prioritize environmental sustainability and social equity, ecological economics can inform and transform economic policy and governance.*

## INTRODUCTION

There are no clear distinctions between us (humans) and other animals, which is one of the truths that biologists have taught us over the past 150 years. In addition, the idea of the individual itself has become hazy as a result of research in genetics, evolutionary biology, and studies of the social behavior of primates and other animals (Diamond, 2006).

These two observations are frequently used as uplifting, positive, or explicative remarks about how things are in the world. Nonetheless, they have significant ramifications for our value theories and recommendations for public policy.

DOI: 10.4018/978-1-6684-9838-5.ch017

One strategy is to apply to other species the standard moral notions of rights, obligations, value, etc. that apply to humans. Yet, this strategy has a flaw because these conventional moral ideas are individualistic. It is needed that we should go the extension the opposite way and consider what happens to conventional moral ideas when the boundaries between individuals and species are blurred.

The intersection of ecological thought and economics has become an increasingly important topic in recent years, as concerns about environmental sustainability and climate change have intensified. Ecological thought challenges the traditional economic model of unlimited growth and resource consumption, and advocates for a more holistic approach that considers the interconnectedness of economic systems and the natural environment. This proposal outlines a research that aims to explore the evolution of ecological thought in economics and its potential implications for economic theory and policy.

Limits of the environment and limits of the human psyche are two boundary constraints that in the long term govern all economic systems, although neoclassical economic theory only partially acknowledges them. Because of this short-term thinking, we now must deal with long-term effects on the environment and society. We need a fresh perspective on the goals and effects of economic activity. To achieve this, it will be necessary to do more than simply adapt our economic thinking to consider environmental and sociopsychological constraints. By creating a unified set of values that considers cost-benefit trade-offs in the environment and our social structures, we cannot resolve our looming issues. These trade-offs rely on comparisons of quantified aggregate results, which presuppose some sort of easy substitutability among all the identifiable components in the ledger. The complexity of ecological and social systems, however, makes such comparisons difficult and sometimes impossible. Therefore, a more holistic approach is needed, which recognizes the interconnectedness and interdependence of economic, social, and ecological systems. This approach requires us to re-evaluate our values and goals as a society, and to prioritize sustainability, equity, and resilience in our economic systems. Only then can we hope to create a future that is both prosperous and sustainable for all humans.

Ecological economic system sustainability depends on our capacity to align local, short-term incentives (such economic growth and private interests) with global, long-term objectives (like sustainability and global welfare). Traditional sustainable cultures have incorporated long-term objectives and limits into local, short-term decision-making processes using taboos, religious mores, etc. Because of the severity of the global environmental catastrophe, we lack the time and flexibility to experiment or to fully establish the necessary taboos and mores. To replace the current Gross National Product[1] craze, we should establish a hierarchy of national and international economic planning goals (with sustainability as the major long-term goal). In this hierarchy, economic growth is only a legitimate objective to the extent as it is compatible with sustainability and other goals higher on the list. Ecological economists can assist in creating and popularizing these objectives. These can be put into practice by being included into political discourse and the decision-making processes of institutions that have an impact on the world economy and ecosystem. Improve our capacity for global ecological economic modeling so that we can see the full range of potential effects of our current efforts. Ecological economists need to be heavily involved in this. Current prices should be adjusted to reflect long-term, global costs, including uncertainty. We should, to paraphrase the catchphrase, "model globally, adapt local incentives accordingly." A flexible assurance bonding system is suggested in addition to the conventional measures of education, regulation, and user fees to explicitly address uncertainty. It is known as an approach that in economics it is not allowed the stock of natural capital to continue to diminish. This approach will support the technological advancement that optimists anticipate while preserving resources if they are mistaken. In other words, we need to balance economic growth with environmental sustainability by implementing policies that promote innovation and conservation. By doing so, we can ensure a better future for ourselves and future generations.

## KEY PRINCIPLES AND CONCEPTS OF ECOLOGICAL THOUGHT IN ECONOMICS

According to Daly and Farley (2010), the goal of ecological economics is to create a sustainable and just economy that balances economic, social, and ecological concerns. The authors argue that this requires a shift towards a steady-state economy, in which resource use is kept within ecological limits and economic growth is no longer the primary objective (Daly & Farley, 2010).

A transdisciplinary field is ecological economics. It doesn't pretend to be a branch of ecology or economics, but rather a link between those two fields and others like psychology, anthropology, archaeology, and history. To get a more complete picture of how people have interacted with their environment in the past and how they might interact going forward, that is required. It aims to view people as part of their ecological life support system rather than as a separate entity from the environment. In terms of how we may create a sustainable future, it also contains certain design components. Not only is the past being examined, but also how that analysis might be used to create something better and new. Ecological economics has three interconnected objectives: sustainable scale, equitable distribution, and effective allocation. These three (sustainable scale, equitable distribution, and effective allocation) all improve sustainability and human well-being (Costanza, 2010).

According to Martinez-Alier et al. (2010), ecological economics emphasizes the importance of systems thinking, recognizing the economy as a complex system embedded within the larger ecosystem. The authors also highlight ecological limits, interdisciplinary approaches, valuing nature, distributional equity, the precautionary principle, and building resilience in economic systems as key principles and concepts of ecological thought in economics (Martínez-Alier et al., 2010).

According to Daly and Farley (2011), ecological economics recognizes the concept of ecological limits, which refers to the idea that the Earth's natural resources and ecosystems are finite and have limits to their capacity to provide goods and services. They argue that economic activities must be designed to operate within these limits to ensure sustainability (Daly & Farley, 2010). According to Costanza (2010), ecological economics adopts an interdisciplinary approach that draws on insights and knowledge from various fields, including ecology, biology, anthropology, and psychology, to understand the complex relationships between the economy, society, and the environment (Costanza, 2010).

In addition to this, according to Constanza et al. (1997), valuing nature is a core principle of ecological economics, which recognizes the intrinsic value of nature and the services it provides, including clean water, air, and soil, biodiversity, and ecosystem stability. The article argues that these values must be considered in economic decision-making to ensure sustainability (Costanza et al., 1997).

According to Martinez-Alier (2002), ecological economics emphasizes the importance of distributional equity, which involves ensuring equitable distribution of resources and benefits across society, both within and between generations. The article argues that this requires addressing social and environmental justice issues, including the disproportionate impacts of environmental degradation on marginalized communities (Martinez-Alier, 2002).

According to Costanza et al. (2007), ecological economics supports the application of the precautionary principle, which means taking action to prevent harm to the environment and society, even in the face of scientific uncertainty. The book argues that this principle is necessary to address the complex and uncertain risks associated with human impacts on the environment (Costanza et al., 2007).

According to Walker and Salt (2006), resilience is a key concept in ecological economics, and refers to the ability of economic systems to adapt and persist in the face of social, economic, and environmental change. The authors argue that building resilience requires a shift away from conventional economic models and towards more holistic approaches that account for the complex and interconnected nature of ecological and social systems (Walker & Salt, 2006).

## ECOLOGICAL THOUGHT'S INFLUENCE ON ECONOMIC THEORY AND ECONOMIC POLICY OVER TIME

Economic theory has historically progressed at a painfully sluggish rate, which has been supported by a status quo that is inextricably linked to the elite of the industrialized world. The neoclassical era, which began more than 100 years ago with the exquisite mathematical explanation of free market capitalism by individuals like Marshall, Walras, and Pareto, is barely discernible from the economic theory taught in undergraduate and graduate programs of the twenty-first century. Most challenges to the status quo in the 20th century were seen as anomalies to assumptions about consumer and corporate behavior, or as problems that could be solved with little government involvement in line with the general equilibrium framework (Gowdy & Erickson, 2005).

The economy is viewed as a closed system in traditional neo-classical economic analysis, and enterprises sell goods and services before compensating the production factors (land, labour and capital). Real economy began to be viewed as an open system that required resource extraction from the environment and extensive waste disposal back into the environment to function. Environmental economics can be thought of as a subfield of neo-classical economics that focuses on two key issues (Munda, 1997):

- environmental externalities, which is a problem,
- proper natural resource management (in particular, the optimal intergenerational allocation of non-renewable resources).

One of the key contributions of ecological thought has been the recognition of environmental externalities, which are costs or benefits that are not reflected in market prices. For example, air pollution from factories imposes costs on people's health and the environment, but these costs are not factored into the prices of the goods produced by those factories. Ecological economists have argued that market failures like this require government intervention to internalize these externalities through policies such as pollution taxes, cap-and-trade systems, and other regulatory measure (Daly, 1999).

Ecological thought has also influenced the concept of sustainability in economic theory and policy. Ecological economists have argued that economic growth cannot continue indefinitely, given the finite nature of natural resources and the carrying capacity of the planet. This has led to the development of the concept of sustainable development, which seeks to balance economic growth with environmental protection and social equity (Elliott, 2010).

According to Spash (2013), ecological thought has led to the recognition of ecosystem services as benefits that people derive from nature, such as clean air, water, and soil. Ecological economists have

*Table 1. Main differences between ecological economics and environmental economics*

| Ecological Economics | Environmental Economics |
|---|---|
| 1. Ideal scale | 1. Externalities and optimal allocation |
| 2. Sustainability as a top priority | 2. Efficiency as a top priority |
| 3. Needs equitable and fulfilled distribution | 3. Optimal welfare or Pareto efficiency |
| 4. Sustainable development, globally and North/South | 4. Sustainable growth in abstract models |
| 5. Growth pessimism and difficult choices | 5. Growth optimism and "win-win" options |
| 6. Unpredictable co-evolution | 6. Deterministic optimisation of intertemporal welfare |
| 7. Long-term focus | 7. Short to medium term focus |
| 8. Complete, integrative and descriptive | 8. Partial, monodisciplinary and analytical |
| 9. Concrete and specific | 9. Abstract and general |
| 10. Physical and biological indicators | 10. Monetary indicators |
| 11. Systems analysis | 11. External costs and economic valuation |
| 12. Multidimensional evaluation | 12. Cost-benefit analysis |
| 13. Integrated models with cause-effect relationships | 13. Applied general equilibrium models with external costs |
| 14. Bounded individual rationality and uncertainty | 14. Maximisation of utility or profit |
| 15. Local communities | 15. Global market and isolated individuals |
| 16. Environmental ethics | 16. Utilitarianism and functionalism |

(van den Bergh, 2001)

argued that these services have economic value and need to be considered in economic decision-making (Spash, 2013).

Ecological thought has also influenced the concept of degrowth, which challenges the idea that economic growth is always desirable and necessary. Degrowth advocates argue that economic growth is a major contributor to environmental degradation and social inequality, and that we need to transition to a steady-state or shrinking economy that operates within ecological limits and prioritizes social and ecological well-being over economic growth (Kaivo-oja et al., 2014).

It is fact that the 1950s saw the founding of Resources for the Future in Washington, DC, which is where environmental economics first emerged. It is a non-profit organization that does independent study on a wide range of environmental challenges. Even a basic history of environmental economics would be incomplete without mentioning the controversy that surrounds the field. The current biggest point of contention is probably about how much ecological economics, as opposed to environmental economics, constitutes a new paradigm (Pearce, 2002).

At table 1, it can be clearly seen the differences of two school of economic thought which are Ecological Economics and Environmental Economics.

Ecological economics is a more recent subfield that challenges traditional economic assumptions and incorporates principles from ecology, while environmental economics focuses on the economic analysis of environmental policies and issues. The debate over which approach is more effective in addressing environmental problems continues to be a topic of discussion among scholars and policymakers (Barbier, 2007).

There has always been a small subset of recognized economists working to present a mainstream environmental critique. Resource and environmental economists' preference for neo-classical economics and ecological economists' seeming acceptance of more radical critiques plainly differ from one another. As a result, the progressive aspects of ecological economics are becoming increasingly incompatible with neo-classical environmental economists who want to simplify all notions to fit them into their models. Here, it is shown that ecological economics is synthesizing several schools of economic thought (such as socialist, institutional, and environmental economics) and returning to the explicit inclusion of moral questions in the method of classical political economy. (Spash, 1999).

Since Pigou, who dealt extensively with the examination of "negative externality" within the neoclassical framework to repair the "market failure," environmental economics has progressively developed into a full-fledged sub-discipline of economics. The analytical framework of all areas of environmental economics, including the Coasian solution, the "second-best solution" for pollution control, non-market valuation within micro cost-benefit analysis, sustainable development, and environmental accounting within macroeconomics, still adheres to the Pigouvian neoclassical tradition (Venkatachalam, 2007). Because it is a branch of economics, environmental economics involves applying conventional economic principles to the environment. It can be said that mainstream economics is primarily concerned with markets, and while it acknowledges that there are externalities, these externalities are real and exist in the world (Costanza, 2010).

Ecological Economics, which has only recently come into existence, views the economic system as a component of the planet Earth's wider ecological system. It acknowledges the independence of the economic and environmental systems before studying their interaction with the aid of advancements in the natural sciences (thermodynamics) and ecology over the past 200 years (Halkos, 2011). The goal of ecological economics is to combine the study of everything that is both inside and outside the market. The fact that we live on a finite planet or that the economy, as a subsystem, cannot develop indefinitely into this bigger, contained system are both concepts that conventional economics doesn't fully acknowledge as being important. There are biophysical restrictions in place. The prevalent viewpoint either ignores these constraints or holds that technology is capable of resolving all resource-constrained issues (Costanza, 2010).

*Figure 1. Theoretical schools interested in the environment-economy dynamic*
*Source: (Gendron, 2014)*

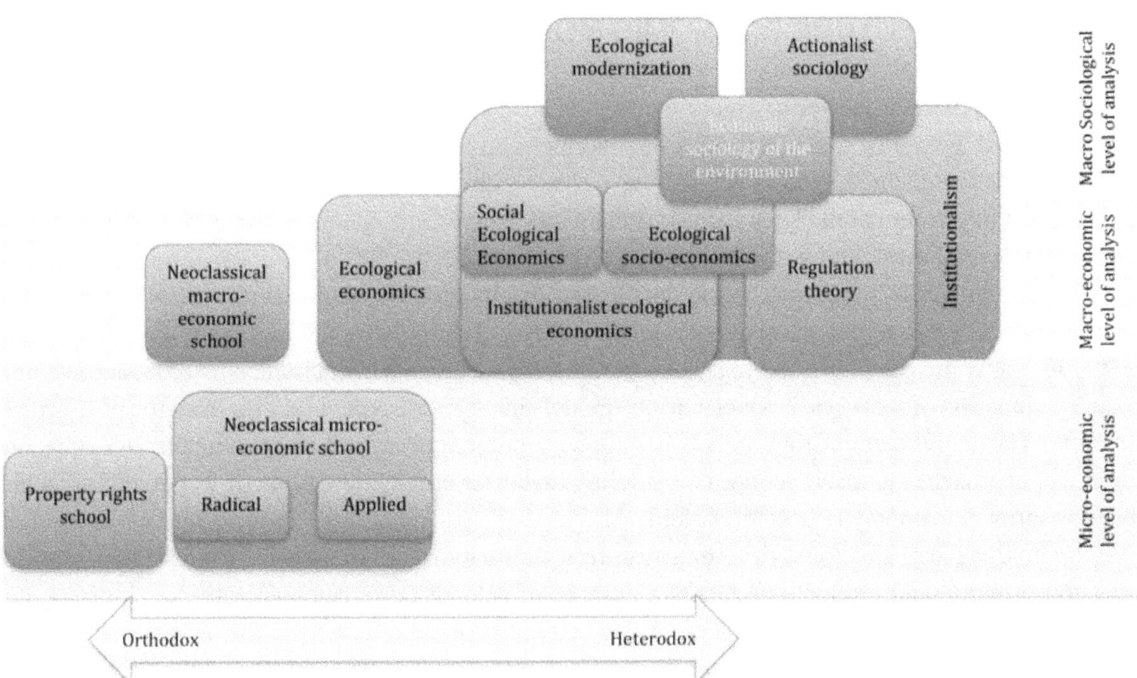

At figure 1, several schools of thought that are closely related to the diagnosis of the ecological misinterpretation of the economic system have emerged from this research agenda. Spash (1999) gave an intriguing historical overview of these schools, demonstrating how ecological economics followed the first environmental economics school and how the two later developed into separate institutions. Gendron (2014) proposes a mapping of the various schools of thought that makes explicit their ideological roots, building on his analysis as well as Jacobs' work and recent European typologies. These schools are presented from the more orthodox (left) to the more heterodox (right), as well as in terms of their level of analysis (from the bottom to the top: micro-economic, macro-economic, and societal). On the more traditional side, the property rights school contends that the environmental catastrophe is due to an improper structure of environmental property rights. This school of thinking, which draws its inspiration from a Coasian viewpoint suggests privatizing the environment so that its owners may take care of it and safeguard it. It assumes that an interindividual dynamic will result in environmental conservation due to a good framework of property rights and does not rely on any official action after an initial privatization. This privatization strategy has received criticism from ethical, theoretical, and methodological angles, but it is still just a minor factor. The neoclassical school, which recognises and shares some concepts with this latter method (Gendron, 2014).

According to Hardin's the Tragedy of the Commons theory, the environment is overused because it is underappreciated. The environment is an excellent illustration of a common good; as such, and in accordance with Pigou's reasoning, government involvement is necessary to change prices to reflect its true value or cost. This method suggests that a market dynamic may be recreated to better manage natural resources rather than that a market dynamic exists that must be exposed. Internalizing environmental costs—that is, altering prices to reflect the true cost of goods—accomplishes this, helping to shape consumer preferences in markets (Gendron, 2014).

## WHAT ARE THE POTENTIAL IMPLICATIONS OF ECOLOGICAL THOUGHT FOR FUTURE ECONOMIC MODELS AND POLICIES?

Ecological Economics is viewed as a pluralistic approach to environmental research that can be contrasted with traditional environmental and resource economics and has in fact evolved in response to them. To make clear the key distinctions between the two fields, a comparison is offered. Also, particular themes are looked at in greater detail. They include dynamic processes, international trade, the growth debate, sustainable development, and behavior and policy (van den Bergh, 2001).

This requirement is being met by the developing discipline of ecological macroeconomics, which includes a few fresh macroeconomic modeling techniques. Based on the literature and researcher interviews, this paper provides (1) a summary of modeling advancements in ecological macroeconomics and (2) an examination of the ways in which various models incorporate policy themes from the post-growth literature. The analysis of existing models has shown a number of unique and promising analytical and numerical approaches to macroeconomic modeling that offer crucial new tools for achieving a sustainable post-growth economy (Hardt & O'Neill, 2017).

Economic value chains must first transition from linearity to circularity. Retro distribution, which links consumption and production, entails a number of smaller tasks, such as gathering, sorting, and reprocessing different resources. Reducing waste that is returned to nature and the usage of new natural resources are both made possible through circular value chains. Governments, manufacturers, distributors, and consumers must work together to create effective material cycles in real life. The basis for the relationship between economy and nature is circularity. Our capacity to identify the relationships between the input and output of natural resources in the economic value chains is essential to sustainability. Cooperation with nature is required on both sides in order to create an economy that improves quality of life. According to this viewpoint, $CO_2$ is a symptom of a carbon-based economy rather than a cause. To put it another way, we need to be aware of the patterns of interconnectivity between the input and output sides of the economy if we are to effectively address climate change. The two are inextricably linked in both the economy and culture. To create an economy that improves quality of life, knowledge and values are crucial. Knowledge influences the economy's inventive processes on the one hand, while changes in the economy also have an impact on cultural development. Contextual thinking—considering nature and culture as well as the economy—is essential (Capra & Jakobsen, 2017).

## HOW CAN ECOLOGICAL THOUGHT BE INTEGRATED INTO ECONOMIC DECISION-MAKING PROCESSES AT THE INDIVIDUAL, ORGANIZATIONAL, AND GOVERNMENTAL LEVELS?

Strategic Environmental Assessment (SEA) in integrating the environment into strategic decision-making is crucial in this regard. In the long run, decisions made based on local, short-term criteria might have severe effects. However, there is growing recognition that traditional economic and ecological models and conceptions are inadequate for addressing these issues (Costanza et al., 1990)

A comprehensive and multifaceted approach is necessary to incorporate ecological awareness into economic decision-making processes at various levels. The following are some strategies for integrating ecological factors at the individual, organizational, and governmental levels:

1. Personal Level:
    a. Awareness and education: People can learn more about ecological ideas and the connections between the economy and the environment by educating themselves. Their decision-making can be influenced by this knowledge.
    b. Sustainable Consumption: People can choose wisely by favoring ecologically friendly goods, cutting back on trash, and choosing energy-efficient substitutes.
    c. Support Ecological Sustainability: People can help companies and organizations that put an emphasis on ecological sustainability, which can have an impact on market dynamics and inspire others to do the same.
    d. Participate as a Citizen: People can participate in grassroots movements, advocacy work, and neighborhood projects to advance ecological principles and shape public policy.
2. Administrative Level:
    a. Organizations can create sustainability strategies that take ecological considerations into account in their day-to-day operations, supply networks, and decision-making procedures. This could entail cutting back on resource usage, switching to renewable energy sources, or putting circular economy principles into effect.

b. Organizations can carry out life cycle analyses to determine how their goods or services will affect the environment. Making decisions based on this information can reduce ecological footprints.
c. Green innovation: Promoting the study and creation of eco-friendly processes, goods, and technology can boost sustainable economic growth and competitiveness.
d. Organizations can actively involve stakeholders, such as staff members, clients, and communities, in decision-making processes to incorporate different viewpoints and make sure environmental concerns are taken into account.

3. Govermental Level:
   a. Governments can create policies that encourage ecological sustainability and enforce them, such as environmental standards, pollution restrictions, and carbon pricing schemes.
   b. Economic Tools: To encourage companies and people to embrace environmentally friendly practices and technologies, governments can use economic tools like grants, tax breaks, and subsidies.
   c. Governments can promote sustainable practices and reduce environmental consequences by integrating ecological considerations into urban planning, infrastructure construction, and land-use legislation.
   d. Collaboration and Partnerships: To share information, assets, and best practices for ecological sustainability, governments can work with other nations, NGOs, and the commercial sector.

It's vital to keep in mind that these recommendations are not all-inclusive and that the precise tactics may change based on regional conditions and priorities. However, we may move toward a more sustainable and resilient future by incorporating ecological ideas into economic decision-making processes at all levels.

## CONCLUSION

The intersection of ecological thought and economics represents an important area of research and inquiry, with significant implications for economic theory and policy. This brief summary on ecological economics aims to contribute to the existing scholarship in this area by exploring the evolution of ecological thought in economics and its potential implications for future economic models and policies. Ultimately, this research will contribute to a more comprehensive understanding of the relationship between the natural environment and economic systems, and the potential for more sustainable and just economic systems in the future. This understanding can lead to the development of policies that prioritize environmental sustainability and social equity, which are essential for creating a better future for ourselves and future generations. It is important to recognize that our economic decisions have real-world consequences, and by incorporating ecological thought into our economic models, we can create a more sustainable and just world.

## REFERENCES

Barbier, E. B. (2007). Valuing ecosystem services as productive inputs. *Economic Policy*, *22*(49), 178–229. doi:10.1111/j.1468-0327.2007.00174.x

Capra, F., & Jakobsen, O. D. (2017). A conceptual framework for ecological economics based on systemic principles of life. *International Journal of Social Economics*, *44*(6), 831–844. doi:10.1108/IJSE-05-2016-0136

Costanza, R. (2010, May 11). *What is ecological economics?* Yale Insights. https://insights.som.yale.edu/insights/what-is-ecological-economics

Costanza, R., d'Arge, R., de Groot, R., Farber, S., Grasso, M., Hannon, B., Limburg, K., Naeem, S., O'Neill, R. V., Paruelo, J., Raskin, R. G., Sutton, P., & van den Belt, M. (1997). The value of the world's ecosystem services and natural capital. *Nature*, *387*(6630), 253–260. doi:10.1038/387253a0

Costanza, R., Haskell, B., Cornwell, L., Daly, H., & Johnson, T. (1990). *The Ecological Economics of Sustainability: Making Local and Short-Term Goals Consistent with Global and Long-Term Goals* (Environment Working Paper No. 32).

Costanza, R., Kubiszewski, I., & Franco, C. (Eds.). (2007). *Introduction to Ecological Economics*. CRC Press.

Daly, H. E. (1999). *Ecological economics and the ecology of economics: Essays in criticism*. Edward Elgar Publishing.

Daly, H. E., & Farley, J. (2010). *Ecological economics: Principles and applications* (2nd ed.). Island Press.

Diamond, J. M. (2006). *The third chimpanzee: The evolution and future of the human animal* (Reissued, [repr.]). Harper Perennial.

Elliott, J. A. (2010). *An Introduction to Sustainable Development* (3rd ed.).

Gendron, C. (2014). Beyond environmental and ecological economics: Proposal for an economic sociology of the environment. *Ecological Economics*, *105*, 240–253. doi:10.1016/j.ecolecon.2014.06.012

Gowdy, J., & Erickson, J. (2005). Ecological economics at a crossroads. *Ecological Economics*, *53*(1), 17–20. doi:10.1016/j.ecolecon.2005.01.010

Halkos, G. (2011). *The evolution of environmental thinking in economics* (MPRA Paper No. 35580; Munich Personal Repec Archive). https://mpra.ub.uni-muenchen.de/35580/

Hardt, L., & O'Neill, D. W. (2017). Ecological Macroeconomic Models: Assessing Current Developments. *Ecological Economics*, *134*, 198–211. doi:10.1016/j.ecolecon.2016.12.027

Kaivo-oja, J., Vehmas, J., & Luukkanen, J. (2014). A Note: De-Growth Debate and New Scientific Analysis of Economic Growth. *Journal of Environmental Protection*, *05*(15), 1477–1481. doi:10.4236/jep.2014.515140

Martinez-Alier, J. (2002). Environmental justice and economic degrowth: An alliance between two movements. *Capitalism, Nature, Socialism*, *13*(1), 51–73. doi:10.1080/10455752.2011.648839

Martínez-Alier, J., Pascual, U., Vivien, F.-D., & Zaccai, E. (2010). Sustainable de-growth: Mapping the context, criticisms and future prospects of an emergent paradigm. *Ecological Economics*, *69*(9), 1741–1747. doi:10.1016/j.ecolecon.2010.04.017

Munda, G. (1997). Environmental Economics, Ecological Economics, and the Concept of Sustainable Development. *Environmental Values*, *6*(2), 213–233. doi:10.1177/096327199700600206

Pearce, D. (2002). An Intellectual History of Environmental Economics. *Annual Review of Energy and the Environment*, *27*(1), 57–81. doi:10.1146/annurev.energy.27.122001.083429

Spash, C. (1999). The Development of Environmental Thinking in Economics. *Environmental Values*, *8*(4), 413–435. doi:10.3197/096327199129341897

Spash, C. L. (2013). The shallow or the deep ecological economics movement? *Ecological Economics*, *93*, 351–362. doi:10.1016/j.ecolecon.2013.05.016

van den Bergh, J. C. J. M. (2001). Ecological economics: Themes, approaches, and differences with environmental economics. *Regional Environmental Change*, *2*(1), 13–23. doi:10.1007101130000020

Venkatachalam, L. (2007). Environmental economics and ecological economics: Where they can converge? *Ecological Economics*, *61*(2–3), 550–558. doi:10.1016/j.ecolecon.2006.05.012

Walker, B., & Salt, D. (2006). *Resilience thinking: Sustaining ecosystems and people in a changing world*. Island Press.

## ENDNOTE

1 The gross national product (GNP) is an estimation of the total worth of all finished goods and services produced over a specific time period using resources owned by the citizens of a nation. Commonly, the gross national product (GNP) is determined by adding up all domestic private investment, domestic government spending, net exports, and any resident income from overseas investments, then deducting foreign resident income. The difference between a country's exports and any imports of products and services is what is known as its net exports.

# Chapter 18
# A Good Assurance of Food Security Requires a Good Understanding of the Plant–Soil–Water–Living Being and Climate Change Interaction

**Nassima Amiri**
*Cadi Ayyad University, Morocco*

**Mohammed Yacoubi Khebiza**
*Cadi Ayyad University, Morocco*

**Mohammed Messouli**
*Cadi Ayyad University, Morocco*

## ABSTRACT

*The impacts of climate change on agriculture can have repercussions on global food security. Losses in agricultural yield can lead to food shortages and increases in food prices, particularly affecting the most vulnerable populations. Climate change can result in food insecurity due to water resource scarcity, while soil degradation may worsen production problems. A quality harvest depends on optimal climatic conditions. However, climate change, characterized by infrequent or irregular precipitation and high temperatures, leads to a decrease in yield on one hand. On the other hand, these climate changes have an additional negative impact by deteriorating soils and reducing water resources through evaporation, reaching yield limits. Within this chapter, the authors explore the different interactions between plants, soil, water, and climate change. The objective is to design strategies specifically tailored to the changing climatic context, thus enabling the development of an agricultural sector that is both resilient and sustainable.*

DOI: 10.4018/978-1-6684-9838-5.ch018

## INTRODUCTION

Plants are closely connected to the soil and water, forming an interconnected ecosystem (Huntley, 2023). Climate variations have a direct impact on plant growth (Raza et al., 2019), soil health (Mondal, 2021), and water availability (Konapala et al., 2020). For example, higher temperatures can affect plant metabolism and alter their growth cycle (Nievola et al., 2017), while precipitation scarcity can lead to a decrease in soil moisture (Soares & Lima, 2022), thereby affecting the ability of plants to absorb the necessary nutrients and water (Seleiman et al., 2021).

Furthermore, climate change can also have indirect consequences on water resources (Hughes et al., 2021). For instance, temperature increase can result in higher evaporation of water from soils and reservoirs (Si et al., 2022), reducing their availability for agriculture. This water scarcity, coupled with soil degradation caused by phenomena like erosion and salinization (Wang et al., 2023), can compromise agricultural productivity and worsen food insecurity.

In the face of these challenges, the primary goal is to clarify the interactions among plants, soil, water, living beings, and climate change, by identifying the relationship of each component with climate change, to build a resilient and sustainable agricultural sector capable of adapting to climate change. This requires developing appropriate strategies that consider the interactions among plants, soil, water, and climate change. For example, it is essential to promote sustainable agricultural practices aimed at improving ecosystem health and enhancing crop resilience against climate fluctuations (El Chami et al., 2020). These practices include agroforestry, conservation agriculture, mixed cropping systems, efficient irrigation, composting, agricultural waste recycling, crop rotation, and organic and integrated farming systems. These approaches promote biodiversity, preserve soil fertility, and mitigate risks associated with climate change, offering long-term benefits for more resilient and sustainable agricultural systems.

The deployment of sustainable agricultural practices faces several challenges, including resistance to change among farmers, lack of knowledge and training, financial constraints, limited access to resources and technologies, as well as unfavorable policies and regulations.

Farmers may hesitate to adopt new practices requiring initial investments or significant changes. Moreover, the lack of adequate knowledge and training can be a major obstacle. High initial costs and restricted access to essential resources and technologies also complicate the transition to sustainable practices. Unfavorable policies and regulations can hinder the adoption of these practices in the absence of financial support or incentives.

To promote the adoption of these practices, various stakeholders play a crucial role. Governments and international organizations can design supportive policies and programs offering financial incentives. Research and training institutions can develop training programs and conduct research to improve these practices. NGOs can provide direct support to farmers, propose local initiatives, and raise awareness of the benefits of sustainable practices. Agricultural businesses and agro-food industries can encourage the adoption of sustainable practices by promoting responsible supply chains and rewarding environmentally friendly practices.

By collaborating and coordinating their efforts, these stakeholders can overcome implementation obstacles and promote widespread adoption of sustainable agricultural practices.

This chapter underscores the complex interdependence among plants, soil, water, living beings, and climate change. Plants play a key role in absorbing $CO_2$ and producing oxygen through photosynthesis, while soil provides vital habitat and nutrients for plants and soil organisms. Living beings, including humans, depend on plants for food and oxygen, while plants require water and soil nutrients to survive.

However, climate change disrupts this delicate interaction with rising temperatures, more frequent droughts, and extreme weather events that alter natural cycles and water availability.

To address these challenges, it is crucial to promote sustainable soil and water management practices, preserve natural ecosystems, and collaborate to mitigate the effects of climate change. By adopting sustainable agricultural practices, preserving wetlands and forests, and reducing greenhouse gas emissions, we can maintain the balance among plants, soil, living beings, water, and climate. In doing so, we protect our environment and the well-being of future generations.

## BACKGROUND

Plants, soils, water, and living beings maintain close and interdependent relationships. Plants rely on soils for nutrients and water, while soils benefit from plant organic waste and the activities of living beings to maintain their fertility (Urra et al., 2019). However, climate change significantly impacts these interactions.

Climate change manifests through variations in temperatures, precipitation, weather patterns (Amiri et al., 2021), and overall climate conditions. These changes directly affect plant growth, water availability, soil composition, and the health of living beings. For instance, prolonged droughts can reduce agricultural yields (Eze et al., 2022), while excessive rainfall can lead to floods (Breinl et al., 2021) and alter soil quality (Meng et al., 2021).

Understanding these complex interactions among different elements is essential to identify necessary actions. For instance, by understanding how plants can adapt to climate change, scientists can develop varieties resistant to heat or drought, contributing to food security under changing climatic conditions.

Plants, like all forms of life, possess adaptation mechanisms that enable them to cope with environmental variations, including those induced by climate change. The adaptation of plants to these changing conditions is a crucial issue for ensuring global food security. Scientists strive to understand these mechanisms and leverage this knowledge to develop plant varieties resistant to heat, drought, and other environmental stresses.

One approach involves identifying genes responsible for resistance to stressful conditions. Scientists study plants thriving in extreme environments, such as deserts or arid zones, to comprehend the molecular mechanisms enabling them to survive and thrive under such conditions. These genetic studies help identify specific traits, such as the ability to retain water, tolerate high temperatures, or maintain stable metabolism despite challenging environmental conditions.

Another major approach is the selection and crossbreeding of plant varieties to reinforce these resistance traits. Scientists use selective breeding techniques to introduce these stress-tolerance characteristics into existing agricultural varieties. For instance, they can crossbreed drought-resistant plants with food crops to create varieties that better retain water or can thrive with less water input.

Furthermore, advancements in biotechnology enable scientists to genetically modify plants to impart better resistance to extreme climatic conditions. This targeted genetic modification approach aims to introduce or enhance genes responsible for stress resistance, providing plants with a better ability to survive and yield satisfactory harvests even under unfavorable weather conditions.

These efforts aim to develop crops capable of adapting and thriving in changing environmental conditions. By enhancing plants' resilience to climate constraints, scientists contribute to ensuring sustainable food security, even in the face of challenges posed by climate change. These resilient varieties can play a crucial role in ensuring sufficient and stable harvests in a world where climatic conditions are becoming increasingly unpredictable and extreme.

Furthermore, comprehending how soils can be preserved and improved allows the implementation of sustainable agricultural practices that promote soil health and minimize adverse effects on the environment. This includes techniques such as crop rotation, agroforestry, and efficient water management.

This visual clarification work on the complex interactions among plants, soils, water, living beings, and climate change is of crucial importance. It will enhance our understanding of these relationships and help determine appropriate interventions to combat climate change, ensure food security, and promote sustainable development in the agricultural sector.

## MAIN FOCUS OF THE CHAPTER

This work focuses on the complex interactions between plants, soils, water, and climate change to deepen our understanding of the connections among these essential elements. By studying these interactions, we can better grasp the effects of climate change on the agricultural sector and identify suitable strategies to address these challenges.

The aim of this study is to provide a clear and visually accessible explanation of the intricate relationships between plants, soils, water, living beings, and climate change. The primary objective is to facilitate a better identification of these relationships and to understand how to intervene effectively in combating climate change, ensuring food security, and promoting the agricultural sector.

### Issues, Controversies, Problems

In Morocco, agriculture is a very important economic sector; it ranks first (Amiri et al., 2021),. Good yields guarantee food security, but the negative impacts of climate change on this sector, both directly and indirectly, could reduce the profitability of agricultural products (Tanure et al., 2020). This, in turn, worsens food security, especially in developing countries that heavily rely on agriculture as a significant economic force. Agriculture serves as a major economic pillar by providing jobs, income, and agricultural exports, thereby contributing to economic growth.

Understanding the interactions between plants, soil, water, living beings, and climate change will play a crucial role in addressing the challenges of climate change and promoting the agricultural sector.

This work will provide a clear visual representation to comprehend the complex interactions between plants, soils, water, living beings, and climate change, aiming to identify these relationships and determine the necessary interventions to combat climate change, ensure food security, and most importantly, promote the agricultural sector.

## COMPONENTS OF AGRICULTURE

1. Plants

Planets provide the necessary environment for food production and the development of sustainable agricultural systems (Hofman-Bergholm, 2022), Plants play a crucial role in the nutrient cycle. They absorb nutrients from the soil for their growth (Hewitt & Smith, 1974), which are then transferred to animals and humans when they consume agricultural products. In return, organic waste from agriculture serves as natural fertilizers for crops (Cen et al., 2020).

Plants, especially cover crops and perennial crops, help prevent soil erosion (Scavo et al., 2022).. Plant roots maintain soil structure and prevent degradation from wind and heavy rains (Lynch et al., 2022).

Plants contribute to capturing atmospheric carbon dioxide ($CO_2$) through the process of photosynthesis (Griffin & Seemann, 1996). This helps mitigate climate change by reducing the amount of $CO_2$ in the atmosphere (Naorem, 2022).

Plants provide a wide variety of foods, such as cereals, vegetables, fruits, and legumes, which form the basis of human and animal diets.

## 2. Soil

The soil plays an essential role in agriculture and is one of the most valuable natural resources for humanity. Its importance lies in several key aspects that contribute to agricultural productivity, food security, and environmental sustainability. Its significance is highlighted by acting as a reservoir of essential nutrients such as nitrogen, phosphorus, potassium, and other elements necessary for plant growth (Jiaying et al., 2022). These elements are absorbed by the roots of crops, supporting their development and productivity (Xu et al., 2020). Moreover, a healthy soil structure promotes the circulation of air and water (Abuarab et al., 2019), which is crucial for root respiration and nutrient absorption (Neira et al., 2015). Additionally, good water retention allows crops to survive during drought periods and ensures an adequate water supply for their growth (Rane et al., 2022).

Soil is a habitat for biodiversity as it harbors a wide variety of microorganisms, insects, and small animals that play a crucial role in biological cycles and decomposition processes (Wurst et al., 2012; Bennett, 2010), thus promoting the overall health of the agricultural ecosystem. Additionally, the soil acts as a natural filter, removing contaminants and pollutants from runoff water before they reach groundwater or waterways (Cheng et al., 2021), thereby contributing to the protection of drinking water and aquatic ecosystems (Gavrilescu, 2021).

Soils can store vast amounts of organic carbon and play a role in preventing erosion (Berryman et al., 2020). If the soil is healthy and well-managed, it is less likely to be eroded by water or wind (Vaca et al., 2023).

## 3. Water

Water plays an essential role in agriculture, and its importance cannot be overstated for several reasons. Water is crucial in this field, particularly in irrigation through various methods, where it is used to irrigate crops when natural precipitation is insufficient to meet the plants water needs (Amiri et al., 2021). Irrigation helps maintain crop growth (Amiri et al., 2023a), increases yields, and promotes food production. Additionally, livestock require potable water for their survival, growth, and production (Wagner & Engle, 2021). Water is used to provide drinking water to cattle, poultry, and other livestock. In numerous food processing procedures (Bekele et al., 2017), water is used for washing, cleaning, and cooking (Bhagwat, 2019). Agricultural areas often depend on local water sources such as rivers, lakes, or groundwater (Pueppke et al., 2020). The availability of water in these ecosystems is vital to support plant and animal life, as well as the long-term sustainability of agricultural practices (Çakmakçı et al.,

2023). Water is essential for plants to absorb nutrients through their roots (Gul et al., 2023). An adequate amount of water enables plants to maintain their health and vitality (Brendel, 2021).

4. Living being

The importance of living beings in agriculture is fundamental and multifaceted, as they play essential roles in the productivity (Ortiz et al., 2021), ecological balance, and sustainability of this key sector. Here are some points that highlight this importance: pollination by pollinating insects, such as bees, butterflies, and other animals, plays a crucial role in crop pollination (Katumo et al., 2022). Without them, many plants could not reproduce and produce fruits or seeds (Picarella & Mazzucato, 2019). Also, natural predators, like certain insect and bird species, help control populations of harmful pests for crops (Mayne et al., 2023). They act as biological control agents, reducing infestations without resorting to harmful pesticides (Bale et al., 2008). Another crucial role is the decomposition of organic matter; soil microorganisms, such as bacteria and fungi, are responsible for decomposing organic materials, such as crop residues and plant debris (Rashid et al., 2016). This process contributes to the formation of humus, thereby improving soil fertility (Figueiredo et al., 2010). Additionally, some animals, such as birds and bats, contribute to natural fertilization by depositing their excrement on the soil, which aids in the natural fertilization of agricultural lands (Dimande et al., 2023). Furthermore, concerning the nutrient cycle, living beings, including earthworms and soil microorganisms, play a key role in the nutrient cycle (Edwards & Arancon, 2022). They decompose organic matter and release essential nutrients, such as nitrogen and phosphorus, making these elements available for plants (Budiastuti et al., 2023).

## DISCUSSION: INTERACTIONS BETWEEN PLANTS, SOIL, WATER, LIVING BEING AND CLIMATE CHANGE

1. Plants and climate change

Climate change affects plants in various ways, and the consequences ripple throughout the entire ecosystem (Shivanna, 2022). Climate change can trigger positive feedback loops that amplify the initial effect (Bajželj & Richards, 2014). For example, permafrost melting in Arctic regions can release significant amounts of carbon dioxide and methane, thereby accelerating global warming, which also impacts the plant ecosystems in the region (Schuur et al., 2022; Miner et al., 2021). Additionally, climate change can lead to a decrease in plant biodiversity (Bellard et al., 2012), as certain species may not be able to adapt quickly enough or migrate to new suitable habitats (Aitken et al., 2008). This could result in significant losses of plant species. Plants are sensitive to climate variations, including temperatures, precipitation, and seasons (Parmesan & Hanley, 2015). Climate change can affect their life cycles (Raza et al., 2019), reproductive behavior (Hacket-Pain & Bogdziewicz, 2021), and geographic distribution (Kosanic et al., 2018).

Certain crops may be more sensitive to temperature or precipitation variations (Porter & Semenov, 2005; Craufurd & Wheeler, 2009). Changes in precipitation patterns can lead to droughts or floods, thereby affecting agricultural yields (Blanc & Noy, 2023). Each crop has optimal climatic conditions for its growth and development (Minoli et al., 2022). Some crops thrive in hot and humid climates (Horie, 2019; Craufurd & Wheeler, 2009), while others prefer cooler and drier climates (Cotrina Cabello et al.,

2023). When the climate changes, these optimal conditions can be disrupted, affecting the health and productivity of crops (Singh et al., 2023).

Global warming leads to increased average temperatures worldwide (Abbass et al., 2022). For some crops, higher temperatures can be damaging (Hatfield & Prueger, 2015). For instance, cereal crops like wheat and rice may be sensitive to high temperatures during their flowering phase, leading to reduced production (Shrestha et al., 2022). Climate change can also bring about changes in precipitation patterns (Tabari, 2020). Some regions may experience more frequent and severe droughts (Ebi et al., 2021), as a decrease in precipitation or prolonged drought can lead to a reduced water supply for irrigation (Lankford et al., 2023), directly impacting crop growth. Water-intensive crops, such as vegetables, are particularly vulnerable during drought periods (Giordano et al., 2021), while others may face more frequent and intense floods, damaging crops by submerging fields, destroying crops, and causing significant losses (Wang et al., 2022). Floods can also promote the spread of plant diseases and the proliferation of pests (Skendžić et al., 2021). These extreme weather events can have devastating consequences for crops (Elahi et al., 2022).

Climate change can also result in shifts in seasons, such as milder winters or earlier springs (Pfeifer & Otto, 2023). These variations can disrupt crop growth cycles and lead to issues with synchronization with pollinators and biological cycles (Freimuth et al., 2022).

To address these challenges, it is essential to take action to mitigate climate change by reducing greenhouse gas emissions. Additionally, agricultural and conservation practices need to be adapted to help plants cope with the new environmental conditions and adopt more resilient and less resource-dependent agricultural practices, such as efficient irrigation techniques, promoting heat and drought-resistant crops, and diversifying crops to reduce risks associated with climate change.

## 2. Soil and climate change

Climate change has a significant impact on the soil, and these effects manifest in various ways (Hamidov et al., 2018). Extreme weather events such as heavy rains and storms can increase soil erosion (Jentsch & Beierkuhnlein, 2008). The intensity of precipitation can exceed the soil's infiltration capacity, leading to floods and carrying away fertile soil (Saco et al., 2021). Additionally, higher temperatures and drought conditions can accelerate the decomposition of organic matter in the soil, reducing soil fertility and its ability to retain water (Kotroczó et al., 2023).

Prolonged droughts can result in the accumulation of salts in the soil as water evaporates (Khamidov et al., 2022). This leads to increased salinity, rendering the soil unsuitable for cultivation and disrupting natural ecosystems (Machado & Serralheiro, 2017). Speaking about biodiversity loss; Climate change affects the composition of ecosystems, leading to shifts in the distribution of plant and animal species (Weiskopf et al., 2020). This can impact the complex interactions between soil organisms, in turn disrupting nutrient cycles and overall soil health. Furthermore, global warming can accelerate the release of stored carbon into the atmosphere in the form of carbon dioxide and methane, thereby contributing to climate change (Shivanna, 2022).

To address these impacts, it is crucial to implement climate change adaptation and mitigation strategies. This includes promoting sustainable agricultural practices, conserving forested areas, managing water resources, and reducing greenhouse gas emissions to limit global warming. Protecting and restoring soils play a critical role in combating climate change and preserving biodiversity.

Soils play a crucial role in carbon sequestration and biodiversity preservation (Rumpel et al., 2022). Protecting soils from erosion, degradation, and excessive urbanization is essential to maintain their capacity to store carbon and support plant growth. Additionally, restoring degraded soils can improve their fertility and resilience to climate change.

3. Water and climate change

Climate change has a significant impact on water resources worldwide (Du et al., 2021). In many regions, it leads to more frequent and intense droughts, thereby reducing the availability of freshwater (Kubiak-Wójcicka & Machula, 2020). The melting of glaciers and ice caps also results in a long-term decrease in water reserves in certain areas (Laurent et al., 2020). Climate change has heightened the intensity and frequency of extreme weather events such as floods and storms (Ebi et al., 2021). These events can cause river overflows, damage infrastructure, and pollute water, compromising its quality (Loudyi et al., 2022; Manzoor et al., 2022). Furthermore, climate change disrupts precipitation patterns in many regions (López et al., 2023). Some areas may experience prolonged periods of drought (Fathi-Taperasht et al., 2022), while others face excessive rainfall and floods (Tabari, 2020). This can make water resource management challenging and lead to issues with potable water supply (Goonetilleke & Vithanage, 2017).

Moreover, the global temperature rise contributes to the melting of glaciers and ice caps (Wunderling et al., 2020), as well as thermal expansion of ocean water (Widlansky et al., 2020). This leads to a rise in sea levels, which can result in the salinization of coastal groundwater, affecting freshwater sources and aquatic ecosystems (Vineis et al., 2011). Water is essential for agriculture, and climate change can reduce water availability for crop irrigation (Tian et al., 2023). This may lead to agricultural yield losses, jeopardizing the food security of local and international populations (Martínez-Valderrama et al., 2023).

In the face of these challenges, it is crucial to implement sustainable water management policies, adopt more water-efficient agricultural practices, and take measures to reduce greenhouse gas emissions to limit future impacts of climate change on water. International collaboration is also essential to address these global-scale problems. The first challenge involves sustainable water management. With a growing global population and increased pressures on water resources, it is essential to manage this resource effectively to meet the needs of all sectors while preserving the health of aquatic ecosystems.

4. Living beings and climate change

Climate change has significant consequences on living beings, whether they are animals, plants, or humans (Shivanna, 2022). We can observe a loss of biodiversity as climate change leads to degradation or destruction of natural habitats (Weiskopf et al., 2020), resulting in the loss of animal and plant species. Species unable to adapt quickly to the new climatic conditions may disappear (Visser, 2008). Additionally, global warming accelerates the pace of extinctions (Panetta et al., 2018). Animals and plants dependent on specific ecosystems are particularly vulnerable (Schleuning et al., 2016). Numerous species, including polar and endemic species, are threatened (Liao et al., 2023).

Climate change also causes the migration and displacement of animals and plants as they move to regions more favorable for their survival (McLeman & Hunter, 2010). This can disrupt existing ecosystems and lead to conflicts when invasive species compete with local species for resources (Burgiel & Muir, 2010). Climate change affects the life cycles of many species (O. Levy et al., 2015). For instance, some plant species bloom earlier in the spring due to warmer temperatures (Kehrberger & Holzschuh, 2019), while the migrations and reproductive habits of animals may be disrupted (Seebacher & Post, 2015). Some species can evolve and adapt to the new conditions, but this process typically takes time (Aitken et al., 2008). The current rates of climate change may exceed the adaptive capacity of many species (Bellard et al., 2012).

Climate change has a significant impact on pests, which are harmful organisms to crops, forests, domestic animals, and the environment in general (Skendžić et al., 2021). Pests encompass a wide variety of insects, mites, fungi, bacteria, viruses, weeds, etc., that can cause substantial damage to ecosystems and agriculture (Tudi et al., 2021). These pests can damage agricultural crops, reducing yields and increasing economic losses for farmers (Amiri, 2023b).

With global warming, temperatures in certain regions become more favorable for the survival and reproduction of pests that were previously limited to warmer areas (Schneider et al., 2022). They can, therefore, expand into new regions where winters were previously too cold for their survival (Lawton et al., 2022). Pests may experience an increase in reproduction rates due to the lengthening of warm seasons and the absence of winter frosts, allowing continuous reproduction throughout the year (Bale & Hayward, 2010; Skendžić et al., 2021). Additionally, climate changes can lead to mismatches between the life cycles of pests and those of their natural predators (Tougeron et al., 2020; Schneider et al., 2022). If pests emerge earlier in the season than their predators, this can lead to an increase in pest populations (Gagic et al., 2021).

Some pests can evolve and develop resistance to pesticides and traditional control methods in response to environmental changes induced by climate change (Ma et al., 2021). Pests can disrupt ecosystems by destroying certain plant or animal species (Guo et al., 2023; Balla et al., 2021), which can have repercussions on the entire food chain and biodiversity.

To address these challenges, it is essential to develop adaptation and pest management strategies that take climate change into account. This may include sustainable agricultural practices, the use of pest-resistant crop varieties, biological control methods, and monitoring and prevention programs to detect and control emerging pests early on. A comprehensive and concerted approach is necessary to minimize the adverse effects of climate change on pests and to protect our ecosystems and livelihoods. Also It is essential to take measures to mitigate climate change and to help living beings adapt to new conditions. This involves reducing greenhouse gas emissions, protecting ecosystems, implementing conservation policies, and promoting sustainable use of natural resources.

## 5. Social and economic impacts of climate change

In addition to the environmental impacts of climate change, there are also significant social and economic impacts which is already significantly felt worldwide, especially in developing countries where its effects are deeply felt (Lawrence et al., 2020). These nations are often more vulnerable to the consequences of climate change due to their reliance on natural resources, less resilient infrastructure, and limited capacity to cope with environmental shocks (Castaño-Rosa et al., 2022).

Food security is one of the most affected areas. Changes in climate patterns lead to prolonged droughts, devastating floods, heatwaves, and other extreme weather events. These disruptions hinder agricultural production, reduce crop yields and harvests, thus compromising the availability of staple foods. Populations relying on agriculture for sustenance are particularly vulnerable, as decreased yields can lead to food insecurity, malnutrition, and even famine in the most affected regions.

Livelihoods of populations are also heavily impacted. Communities dependent on natural resources such as fishermen, farmers, and forest-dependent populations experience significant economic losses. Fragile ecosystems are disrupted, affecting the income and livelihoods of those directly reliant on them (Race et al., 2023).

Moreover, climate change has implications for migration patterns. Extreme weather events and environmental degradation push some populations to migrate, whether internally within their countries or towards other regions. Displaced individuals can become climate refugees, seeking refuge away from areas affected by natural disasters or inhospitable environmental conditions (Marcus et al., 2023).

The consequences of climate change on populations in developing countries highlight the urgent need for global and concerted action. Measures for adaptation and mitigation are necessary to strengthen the resilience of vulnerable communities, improve food security, support livelihoods, and implement policies and initiatives to mitigate environmental impacts (Awuni et al., 2023). International aid and cooperation between nations are also essential to assist the most affected regions in facing the challenges of climate change and adapting to a more sustainable future (Islam & Kieu, 2020).

Indeed, climate change does not affect everyone in the same way, and this is where equity and justice come into play in responding to these changes. The consequences of climate change are disproportionate, often hitting the most vulnerable populations and those least responsible for the climate crisis (B. S. Levy & Patz, 2015).

Countries and communities most affected by the adverse effects of climate change are often those with relatively low carbon footprints and historically less contribution to greenhouse gas emissions. Developing nations, often facing less robust infrastructure and limited resources to cope with climate-induced natural disasters, bear the brunt of climate change effects.(Cappelli et al., 2021)

Within countries, the poorest populations, ethnic minorities, women, and children are often the hardest hit. Socioeconomic inequalities and unequal access to resources exacerbate the impacts of climate change. For instance, people living in disadvantaged neighborhoods may be more exposed to risks from natural disasters due to inadequate infrastructure or precarious housing.

Equity in adapting to climate change, therefore, involves recognizing these disparities and ensuring that responses and solutions are fair and just (Araos et al., 2021). This means prioritizing the protection of the most vulnerable populations, ensuring their access to resources and infrastructure needed to cope with climate shocks, and actively involving them in decision-making processes related to adaptation.

An approach centered on justice in adapting to climate change also requires shared responsibility among nations to reduce greenhouse gas emissions and to financially support the most vulnerable countries in adapting to climate impacts (Dwivedi et al., 2022). This involves technology transfers, increased financial support, and long-term commitment to enhance the resilience of the most affected communities.

Ultimately, equity and justice are essential components of an effective response to climate change. Equitable climate action aims to ensure that no one is left behind in combating climate impacts and to build a more resilient and fair future for all populations, regardless of their geographical or socio-economic situation.

## 6. Mitigation and Adaptation to Climate Change

The use of water in agriculture represents concrete examples of adaptation to climate change. These strategies aim to mitigate the negative effects of climate change on agricultural systems and enhance their resilience in the face of increasingly unpredictable and extreme weather conditions (Santos et al., 2022)

Drought-resistant crop varieties result from genetic research aiming to identify and develop specific traits that enable plants to survive and thrive in water-stressed conditions (Liu et al., 2023). These traits include an increased water retention capacity, better tolerance to prolonged drought periods, and the ability to maintain satisfactory yields despite irregular precipitation. These resilient crops offer significant potential to ensure stable yields even in environments where water is becoming increasingly scarce (Khatun et al., 2021).

Simultaneously, water management practices are crucial for optimizing the use of available water resources. This may involve adopting more efficient and targeted irrigation systems, collecting and storing rainwater, or implementing soil management techniques to retain water and minimize losses due to evaporation (Levidow et al., 2014). These practices allow farmers to maximize water use while reducing pressure on already limited water resources (Mallareddy et al., 2023).

These adaptation strategies extend beyond agriculture to other sectors such as natural resource management, urban planning, and even infrastructure. For instance, cities are adopting adaptation measures like building urban parks to mitigate the effects of heatwaves or implementing drainage systems to cope with increased flood risks (Seyedi Rezvani et al., 2023).

These efforts aim to mitigate vulnerabilities and strengthen system resilience against climate change. However, they require an integrated approach and collaboration among governmental stakeholders, local communities, researchers, and businesses to be fully effective (Kyriakopoulos & Sebos, 2023). By developing adaptive technologies, policies, and practices, we are better equipped to address the imminent challenges of climate change and build a more sustainable and resilient future (Goldstein, 2015).

## SOLUTIONS AND RECOMMENDATIONS

In order to mitigate the effects of climate change on agriculture, it is crucial to implement appropriate adaptation measures. One of these measures is improving the efficiency of irrigation. As water becomes an increasingly limited resource, it is essential to optimize its use by adopting techniques such as drip irrigation or using sensors to determine the actual water needs of crops. This helps reduce water losses through evaporation and provides plants with the right amount of water for their optimal growth.

Drip irrigation and smart irrigation systems are efficient technologies for water conservation and enhancing irrigation effectiveness. However, their high costs, both for initial installation and ongoing maintenance, can pose significant challenges, especially for farmers in developing countries.

The installation of these systems requires substantial investments in specialized equipment, infrastructure, and sometimes even training for farmers. Costs can vary based on farm sizes, technology employed, and the availability of local resources. Furthermore, regular upkeep and maintenance of the equipment can represent a continual financial burden, which may be challenging for farmers with limited means.

In numerous developing countries, where farmers often operate under precarious economic conditions, financial accessibility to these technologies remains a significant challenge. Farmers might struggle to secure financing or subsidies to invest in these costly systems. Additionally, the required infrastructure and support services necessary to maintain these technologies might be limited, further complicating their adoption.

These financial obstacles can discourage farmers, prompting them to continue using traditional irrigation methods, which, though less effective, are more affordable in the short term. Finding solutions to make these technologies more financially accessible, perhaps through public-private partnerships, subsidy programs, or low-interest loans, is crucial to encourage their adoption by farmers, particularly in regions where they could significantly impact food security and water resource sustainability.

Farmers may sometimes encounter obstacles when adopting new technologies or altering their agricultural practices. These resistances can stem from various factors. Initially, concerns often revolve around the initial costs associated with integrating new technologies, along with uncertainties about their long-term profitability. Moreover, traditional methods are deeply rooted in local culture and history, making their replacement with modern approaches challenging. Additionally, inadequate training on new technology usage or lack of support for adoption may dissuade farmers from embracing change. There is also a natural resistance to change and adjusting habitual work methods, which might be exacerbated by technical challenges. Finally, in some regions, limited access to equipment and technical assistance further complicates the adoption of these technologies.

However, to overcome these reluctances, it is crucial to implement targeted educational programs. These programs should emphasize the benefits of new technologies, highlighting tangible outcomes such as increased yields, water savings, or better resilience to climate change. Moreover, providing initial financial support or incentives can facilitate the initial adoption of these innovations. Encouraging experience sharing and establishing support networks among farmers to exchange successes and challenges with these new practices can also play a crucial role in their acceptance.

Another important adaptation strategy is the use of drought-resistant crops. By developing and cultivating plant varieties that can tolerate increased drought conditions, farmers can minimize yield losses caused by water scarcity. These drought-resistant varieties can be selected through traditional breeding methods or targeted genetic modification, depending on local needs and constraints.

Furthermore, the adoption of innovative water management practices can play a key role in climate change adaptation. This may include collecting and storing rainwater, reusing treated wastewater, managing watersheds to recharge groundwater, and implementing smart drainage and irrigation systems. These practices optimize water use, prevent unnecessary losses, and improve water availability for agriculture, even under changing climatic conditions.

The private sector often possesses the necessary resources for research, development, and commercialization of new agricultural technologies. By investing in research and development, private enterprises can design innovative tools, enhanced seeds, efficient irrigation systems, and other technological solutions aimed at boosting agricultural productivity while reducing the ecological footprint.

Moreover, these companies often have the capacity to provide financial and technical support to farmers. They can offer financing programs, loans at preferential rates, or grants to enable farmers to more easily access these new technologies. This financial support can be crucial, especially for farmers in developing countries, where the initial investment in new technologies can pose a significant barrier.

Furthermore, the private sector is often capable of providing technical training to farmers to help them understand and effectively use these new technologies. They can also offer ongoing support to ensure that farmers make the most of the innovations available to them.

Through public-private partnerships, the private sector can collaborate with governments and non-governmental organizations to promote the adoption of these technologies. Together, they can create incentive programs, training initiatives, and support structures that encourage farmers to adopt these new practices.

In summary, the private sector plays a crucial role in the research, development, commercialization, and technical and financial support of agricultural innovations, thereby facilitating their adoption by farmers and contributing to the overall improvement of productivity and sustainability in the agricultural sector.

However, to effectively implement these adaptation measures, an integrated approach is necessary. This means close collaboration between researchers, farmers, and policymakers is essential. Researchers can provide scientific knowledge and technical innovations to support the development of adaptation practices. Farmers, on the other hand, play a crucial role in the practical implementation of these strategies in the field by sharing their knowledge and experience. Policymakers must facilitate the implementation of policies and incentives that promote the adoption of these adaptation practices.

## FUTURE RESEARCH DIRECTIONS

Investing in scientific research on the complex interaction between plants, soil, water, living beings, and climate change is a crucial step to preserve our environment, enhance food security, and address the challenges posed by climate change. This requires international cooperation, adequate funding, and recognition of the importance of scientific research in formulating sustainable policies. This can be achieved through:

1. Food Security: Understanding the interactions between plants, soil, and water is fundamental to improving agriculture and increasing global food security. By identifying sustainable agricultural practices and developing crop varieties resistant to climate change, agricultural productivity can be enhanced, and vulnerability to climate fluctuations can be reduced.
2. Ecosystem Understanding: Scientific research will lead to a better understanding of how plants, soil, water, and living beings interact within ecosystems. This understanding is crucial for maintaining ecological balance and protecting biodiversity.
3. Technological Innovation: Research in this field can stimulate technological innovation by encouraging the development of new technologies and techniques to monitor and manage ecosystems more effectively and sustainably.
4. Climate Change Mitigation and Adaptation: Research can contribute to a better grasp of how interactions between different elements of the ecosystem influence greenhouse gas emissions and carbon storage in the soil. This can lead to more efficient land management strategies for mitigating climate change. Studying plant species that are more resilient to changing conditions can also help in better planning of natural resource management and protecting fragile ecosystems.
5. Informed Decision-Making: Policymakers need solid scientific information to develop effective environmental policies. By investing in research, we can ensure that decisions are based on reliable and up-to-date data.

Overall, investing in scientific research on the complex interactions in the natural world can have wide-ranging benefits for the environment, food security, and climate change mitigation and adaptation. It provides the knowledge needed to make informed decisions and develop sustainable policies for a better future.

## CONCLUSION

The interaction between plants, soil, water, living beings, and climate change constitutes a complex and interdependent ecosystem. Plants play a crucial role by absorbing atmospheric $CO_2$ and converting it into oxygen through photosynthesis. The soil provides a vital habitat for plants, soil organisms, and microorganisms involved in decomposing organic matter and releasing essential nutrients for plant growth. Living beings, including humans, depend on plants for food and oxygen, while plants rely on water and nutrients from the soil to survive. Water is a crucial element in this interaction, as it facilitates the transport of nutrients in the soil and plants, regulates temperature, and promotes plant growth.

However, climate change poses an increasing threat to this delicate interaction. Rising temperatures, more frequent droughts, and extreme weather events can disrupt natural cycles, affect water availability, and alter the natural habitats of plants and living beings.

To address these challenges, it is crucial to promote sustainable soil and water management practices, conserve natural ecosystems, and engage in collective efforts to mitigate the effects of climate change. By adopting sustainable agricultural practices, preserving wetlands and forests, and reducing greenhouse gas emissions, we can contribute to maintaining the delicate balance between plants, soil, living beings, water, and climate. In doing so, we not only protect our environment but also our own well-being and that of future generations.

## FUNDING

This research received no specific grant from any funding agency in the public, commercial, or not-for-profit sectors.

## ACKNOWLEDGMENT

This research was supported by the Laboratory of Water, Biodiversity and Climate Changes, Faculty of Sciences Semlalia, Cadi Ayyad University.

## REFERENCES

Abbass, K., Qasim, M. Z., Song, H., Murshed, M., Mahmood, H., & Younis, I. (2022). A review of the global climate change impacts, adaptation, and sustainable mitigation measures. *Environmental Science and Pollution Research International*, *29*(28), 42539–42559. doi:10.100711356-022-19718-6 PMID:35378646

Abuarab, M. E., El-Mogy, M. M., Hassan, A. M., Abdeldaym, E. A., Abdelkader, N. H., & El-Sawy, B. I., M. (. (2019). The Effects of Root Aeration and Different Soil Conditioners on the Nutritional Values, Yield, and Water Productivity of Potato in Clay Loam Soil. *Agronomy (Basel)*, *9*(8), 8. Advance online publication. doi:10.3390/agronomy9080418

Aitken, S. N., Yeaman, S., Holliday, J. A., Wang, T., & Curtis-McLane, S. (2008). Adaptation, migration or extirpation: Climate change outcomes for tree populations. *Evolutionary Applications*, *1*(1), 95–111. doi:10.1111/j.1752-4571.2007.00013.x PMID:25567494

Amiri, N., Lahlali, R., Amiri, S. E. L., Jarroudi, M., Khebiza, M. Y., & Messouli, M. (2021). Development of an Integrated Model to Assess the Impact of Agricultural Practices and Land Use on Agricultural Production in Morocco under Climate Stress over the Next Twenty Years. *Sustainability (Basel)*, *13*(21), 21. doi:10.3390u132111943

Araos, M., Jagannathan, K., Shukla, R., Ajibade, I., Coughlan de Perez, E., Davis, K., Ford, J. D., Galappaththi, E. K., Grady, C., Hudson, A. J., Joe, E. T., Kirchhoff, C. J., Lesnikowski, A., Alverio, G. N., Nielsen, M., Orlove, B., Pentz, B., Reckien, D., Siders, A. R., & Turek-Hankins, L. L. (2021). Equity in human adaptation-related responses: A systematic global review. *One Earth*, *4*(10), 1454–1467. doi:10.1016/j.oneear.2021.09.001

Awuni, S., Adarkwah, F., Ofori, B. D., Purwestri, R. C., Huertas Bernal, D. C., & Hajek, M. (2023). Managing the challenges of climate change mitigation and adaptation strategies in Ghana. *Heliyon*, *9*(5), e15491. doi:10.1016/j.heliyon.2023.e15491 PMID:37131451

Bajželj, B., & Richards, K. S. (2014). The Positive Feedback Loop between the Impacts of Climate Change and Agricultural Expansion and Relocation. *Land (Basel)*, *3*(3), 3. doi:10.3390/land3030898

Bale, J. S., & Hayward, S. A. L. (2010). Insect overwintering in a changing climate. *The Journal of Experimental Biology*, *213*(6), 980–994. doi:10.1242/jeb.037911 PMID:20190123

Bale, J. S., van Lenteren, J. C., & Bigler, F. (2008). Biological control and sustainable food production. *Philosophical Transactions of the Royal Society of London. Series B, Biological Sciences*, *363*(1492), 761–776. doi:10.1098/rstb.2007.2182 PMID:17827110

Balla, A., Silini, A., Cherif-Silini, H., Chenari Bouket, A., Moser, W. K., Nowakowska, J. A., Oszako, T., Benia, F., & Belbahri, L. (2021). The Threat of Pests and Pathogens and the Potential for Biological Control in Forest Ecosystems. *Forests*, *12*(11), 11. doi:10.3390/f12111579

Bekele, M., Mengistu, A., & Tamir, B. (2017). Livestock and feed water productivity in the mixed crop-livestock system. *Animal*, *11*(10), 1852–1860. doi:10.1017/S1751731117000416 PMID:28222817

Bellard, C., Bertelsmeier, C., Leadley, P., Thuiller, W., & Courchamp, F. (2012). Impacts of climate change on the future of biodiversity. *Ecology Letters*, *15*(4), 365–377. doi:10.1111/j.1461-0248.2011.01736.x PMID:22257223

Bennett, A. (2010). The role of soil community biodiversity in insect biodiversity. *Insect Conservation and Diversity*, *3*(3), 157–171. doi:10.1111/j.1752-4598.2010.00086.x

Berryman, E., Hatten, J., Page-Dumroese, D. S., Heckman, K. A., D'Amore, D. V., Puttere, J., SanClements, M., Connolly, S. J., & Perry, C. H. (Hobie), & Domke, G. M. (2020). Soil Carbon. In R. V. Pouyat, D. S. Page-Dumroese, T. Patel-Weynand, & L. H. Geiser (Eds.), Forest and Rangeland Soils of the United States Under Changing Conditions: A Comprehensive Science Synthesis (pp. 9–31). Springer International Publishing. doi:10.1007/978-3-030-45216-2_2

Bhagwat, V. R. (2019). Safety of Water Used in Food Production. *Food Safety and Human Health,* 219–247. doi:10.1016/B978-0-12-816333-7.00009-6

Blanc, E., & Noy, I. (2023). Impacts of droughts and floods on agricultural productivity in New Zealand as measured from space. *Environmental Research: Climate*, *2*(3), 035001. doi:10.1088/2752-5295/acd4da

Breinl, K., Lun, D., Müller-Thomy, H., & Blöschl, G. (2021). Understanding the relationship between rainfall and flood probabilities through combined intensity-duration-frequency analysis. *Journal of Hydrology (Amsterdam)*, *602*, 126759. doi:10.1016/j.jhydrol.2021.126759

Brendel, O. (2021). The relationship between plant growth and water consumption: A history from the classical four elements to modern stable isotopes. *Annals of Forest Science*, *78*(2), 2. doi:10.100713595-021-01063-2

Budiastuti, M. T. S., Purnomo, D., Pujiasmanto, B., & Setyaningrum, D. (2023). Response of Maize Yield and Nutrient Uptake to Indigenous Organic Fertilizer from Corn Cobs. *Agriculture*, *13*(2), 2. doi:10.3390/agriculture13020309

BurgielS.MuirA. (2010). Invasive Species, Climate Change and Ecosystem-Based Adaptation: Addressing Multiple Drivers of Global Change Global Invasive Species Programme. doi:10.13140/2.1.1460.8161

Çakmakçı, R., Salık, M. A., & Çakmakçı, S. (2023). Assessment and Principles of Environmentally Sustainable Food and Agriculture Systems. *Agriculture*, *13*(5), 5. doi:10.3390/agriculture13051073

Cappelli, F., Costantini, V., & Consoli, D. (2021). The trap of climate change-induced "natural" disasters and inequality. *Global Environmental Change*, *70*, 102329. doi:10.1016/j.gloenvcha.2021.102329

Castaño-Rosa, R., Pelsmakers, S., Järventausta, H., Poutanen, J., Tähtinen, L., Rashidfarokhi, A., & Toivonen, S. (2022). Resilience in the built environment: Key characteristics for solutions to multiple crises. *Sustainable Cities and Society*, *87*, 104259. doi:10.1016/j.scs.2022.104259

Cen, Y., Guo, L., Liu, M., Gu, X., Li, C., & Jiang, G. (2020). Using organic fertilizers to increase crop yield, economic growth, and soil quality in a temperate farmland. *PeerJ*, *8*, e9668. doi:10.7717/peerj.9668 PMID:32879791

Cheng, K., Xu, X., Cui, L., Li, Y., Zheng, J., Wu, W., Sun, J., & Pan, G. (2021). The role of soils in regulation of freshwater and coastal water quality. Philosophical Transactions of the *Royal Society B: Biological Sciences, 376*(1834). doi:10.1098/rstb.2020.0176

Cotrina Cabello, G. G., Ruiz Rodriguez, A., Husnain Gondal, A., Areche, F. O., Flores, D. D. C., Astete, J. A. Q., Camayo-Lapa, B. F., Yapias, R. J. M., Jabbar, A., Yovera Saldarriaga, J., Salas-Contreras, W. H., & Cruz Nieto, D. D. (2023). Plant adaptability to climate change and drought stress for crop growth and production. *Perspectives in Agriculture, Veterinary Science, Nutrition and Natural Resources*, *2023*, cabireviews.2023.0004. doi:10.1079/cabireviews.2023.0004

Craufurd, P. Q., & Wheeler, T. R. (2009). Climate change and the flowering time of annual crops. *Journal of Experimental Botany*, *60*(9), 2529–2539. doi:10.1093/jxb/erp196 PMID:19505929

Dimande, P., Arrobas, M., & Rodrigues, M. Â. (2023). Under a Tropical Climate and in Sandy Soils, Bat Guano Mineralises Very Quickly, Behaving More like a Mineral Fertiliser than a Conventional Farmyard Manure. *Agronomy (Basel)*, *13*(5), 5. doi:10.3390/agronomy13051367

Du, P., Xu, M., & Li, R. (2021). Impacts of climate change on water resources in the major countries along the Belt and Road. *PeerJ*, *9*, e12201. doi:10.7717/peerj.12201 PMID:34721962

Dwivedi, Y. K., Hughes, L., Kar, A. K., Baabdullah, A. M., Grover, P., Abbas, R., Andreini, D., Abumoghli, I., Barlette, Y., Bunker, D., Chandra Kruse, L., Constantiou, I., Davison, R. M., De', R., Dubey, R., Fenby-Taylor, H., Gupta, B., He, W., Kodama, M., & Wade, M. (2022). Climate change and COP26: Are digital technologies and information management part of the problem or the solution? An editorial reflection and call to action. *International Journal of Information Management*, *63*, 102456. doi:10.1016/j.ijinfomgt.2021.102456

Ebi, K. L., Vanos, J., Baldwin, J. W., Bell, J. E., Hondula, D. M., Errett, N. A., Hayes, K., Reid, C. E., Saha, S., Spector, J., & Berry, P. (2021). Extreme Weather and Climate Change: Population Health and Health System Implications. *Annual Review of Public Health*, *42*(1), 293–315. doi:10.1146/annurev-publhealth-012420-105026 PMID:33406378

Edwards, C. A., & Arancon, N. Q. (2022). The Role of Earthworms in Organic Matter and Nutrient Cycles. In C. A. Edwards & N. Q. Arancon (Eds.), *Biology and Ecology of Earthworms* (pp. 233–274). Springer US. doi:10.1007/978-0-387-74943-3_8

El Chami, D., Daccache, A., & El Moujabber, M. (2020). How Can Sustainable Agriculture Increase Climate Resilience? A Systematic Review. *Sustainability (Basel)*, *12*(8), 8. doi:10.3390u12083119

Elahi, E., Khalid, Z., Tauni, M. Z., Zhang, H., & Lirong, X. (2022). Extreme weather events risk to crop-production and the adaptation of innovative management strategies to mitigate the risk: A retrospective survey of rural Punjab, Pakistan. *Technovation*, *117*, 102255. doi:10.1016/j.technovation.2021.102255

Eze, E., Girma, A., Zenebe, A., Okolo, C. C., Kourouma, J. M., & Negash, E. (2022). Predictors of drought-induced crop yield/losses in two agroecologies of southern Tigray, Northern Ethiopia. *Scientific Reports*, *12*(1), 6284. doi:10.103841598-022-09862-x PMID:35428373

Fathi-Taperasht, A., Shafizadeh-Moghadam, H., Minaei, M., & Xu, T. (2022). Influence of drought duration and severity on drought recovery period for different land cover types: Evaluation using MODIS-based indices. *Ecological Indicators*, *141*, 109146. doi:10.1016/j.ecolind.2022.109146

Figueiredo, C. C. D., Resck, D. V. S., & Carneiro, M. A. C. (2010). Labile and stable fractions of soil organic matter under management systems and native cerrado. *Revista Brasileira de Ciência do Solo*, *34*(3), 907–916. doi:10.1590/S0100-06832010000300032

Freimuth, J., Bossdorf, O., Scheepens, J. F., & Willems, F. M. (2022). Climate warming changes synchrony of plants and pollinators. *Proceedings of the Royal Society B: Biological Sciences, 289*(1971), 20212142. 10.1098/rspb.2021.2142

Gagic, V., Holding, M., Venables, W. N., Hulthen, A. D., & Schellhorn, N. A. (2021). Better outcomes for pest pressure, insecticide use, and yield in less intensive agricultural landscapes. *Proceedings of the National Academy of Sciences of the United States of America*, *118*(12), e2018100118. doi:10.1073/pnas.2018100118 PMID:33731476

Gavrilescu, M. (2021). Water, Soil, and Plants Interactions in a Threatened Environment. *Water (Basel)*, *13*(19), 19. doi:10.3390/w13192746

Giordano, M., Petropoulos, S. A., & Rouphael, Y. (2021). Response and Defence Mechanisms of Vegetable Crops against Drought, Heat and Salinity Stress. *Agriculture*, *11*(5), 5. doi:10.3390/agriculture11050463

Goldstein, D. (2015). Climate-adaptive technological change in a small region: A resource-based scenario approach. *Technological Forecasting and Social Change*, *99*, 168–180. doi:10.1016/j.techfore.2015.04.014

Goonetilleke, A., & Vithanage, M. (2017). Water Resources Management: Innovation and Challenges in a Changing World. *Water (Basel)*, *9*(4), 4. Advance online publication. doi:10.3390/w9040281

Griffin, K. L., & Seemann, J. R. (1996). Plants, $CO_2$ and photosynthesis in the 21st century. *Chemistry & Biology*, *3*(4), 245–254. doi:10.1016/S1074-5521(96)90104-0 PMID:8807852

Gul, M. U., Paul, A. S. M., & Chehri, A. (2023). Hydrotropism: Understanding the Impact of Water on Plant Movement and Adaptation. *Water (Basel)*, *15*(3), 3. doi:10.3390/w15030567

Guo, Q., Potter, K. M., Ren, H., & Zhang, P. (2023). Impacts of Exotic Pests on Forest Ecosystems: An Update. *Forests*, *14*(3), 3. doi:10.3390/f14030605

Hacket-Pain, A., & Bogdziewicz, M. (1839). Climate change and plant reproduction: Trends and drivers of mast seeding change. *Philosophical Transactions of the Royal Society of London. Series B, Biological Sciences*, *376*(1839), 20200379. doi:10.1098/rstb.2020.0379 PMID:34657461

Hamidov, A., Helming, K., Bellocchi, G., Bojar, W., Dalgaard, T., Ghaley, B. B., Hoffmann, C., Holman, I., Holzkämper, A., Krzeminska, D., Kværnø, S. H., Lehtonen, H., Niedrist, G., Øygarden, L., Reidsma, P., Roggero, P. P., Rusu, T., Santos, C., Seddaiu, G., & Schönhart, M. (2018). Impacts of climate change adaptation options on soil functions: A review of European case-studies. *Land Degradation & Development*, *29*(8), 2378–2389. doi:10.1002/ldr.3006 PMID:30393451

Hatfield, J. L., & Prueger, J. H. (2015). Temperature extremes: Effect on plant growth and development. *Weather and Climate Extremes*, *10*, 4–10. doi:10.1016/j.wace.2015.08.001

Hewitt, E. J., & Smith, T. A. (1974). Plant mineral nutrition. *Plant Mineral Nutrition.* https://www.cabdirect.org/cabdirect/abstract/19760340035

Hofman-Bergholm, M. (2022). A Transition towards a Food and Agricultural System That Includes Both Food Security and Planetary Health. *Foods*, *12*(1), 12. doi:10.3390/foods12010012 PMID:36613228

Horie, T. (2019). Global warming and rice production in Asia: Modeling, impact prediction and adaptation. *Proceedings of the Japan Academy. Series B, Physical and Biological Sciences*, *95*(6), 211–245. doi:10.2183/pjab.95.016 PMID:31189777

Hughes, J., Cowper-Heays, K., Olesson, E., Bell, R., & Stroombergen, A. (2021). Impacts and implications of climate change on wastewater systems: A New Zealand perspective. *Climate Risk Management*, *31*, 100262. doi:10.1016/j.crm.2020.100262

Huntley, B. J. (2023). Soil, Water and Nutrients. In B. J. Huntley (Ed.), *Ecology of Angola: Terrestrial Biomes and Ecoregions* (pp. 127–147). Springer International Publishing. doi:10.1007/978-3-031-18923-4_6

Islam, M. S., & Kieu, E. (2020). Tackling Regional Climate Change Impacts and Food Security Issues: A Critical Analysis across ASEAN, PIF, and SAARC. *Sustainability (Basel)*, *12*(3), 3. doi:10.3390u12030883

Jentsch, A., & Beierkuhnlein, C. (2008). Research frontiers in climate change: Effects of extreme meteorological events on ecosystems. *Comptes Rendus Geoscience*, *340*(9), 621–628. doi:10.1016/j.crte.2008.07.002

Jiaying, M., Tingting, C., Jie, L., Weimeng, F., Baohua, F., Guangyan, L., Hubo, L., Juncai, L., Zhihai, W., Longxing, T., & Guanfu, F. (2022). Functions of Nitrogen, Phosphorus and Potassium in Energy Status and Their Influences on Rice Growth and Development. *Rice Science*, *29*(2), 166–178. doi:10.1016/j.rsci.2022.01.005

Katumo, D. M., Liang, H., Ochola, A. C., Lv, M., Wang, Q.-F., & Yang, C.-F. (2022). Pollinator diversity benefits natural and agricultural ecosystems, environmental health, and human welfare. *Plant Diversity*, *44*(5), 429–435. doi:10.1016/j.pld.2022.01.005 PMID:36187551

Kehrberger, S., & Holzschuh, A. (2019). How does timing of flowering affect competition for pollinators, flower visitation and seed set in an early spring grassland plant? *Scientific Reports*, *9*(1), 1. doi:10.103841598-019-51916-0 PMID:31666567

Khamidov, M., Ishchanov, J., Hamidov, A., Donmez, C., & Djumaboev, K. (2022). Assessment of Soil Salinity Changes under the Climate Change in the Khorezm Region, Uzbekistan. *International Journal of Environmental Research and Public Health*, *19*(14), 8794. doi:10.3390/ijerph19148794 PMID:35886646

Khatun, M., Sarkar, S., Era, F. M., Islam, A. K. M. M., Anwar, M. P., Fahad, S., Datta, R., & Islam, A. K. M. A. (2021). Drought Stress in Grain Legumes: Effects, Tolerance *Mechanisms and Management. Agronomy (Basel)*, *11*(12), 12. doi:10.3390/agronomy11122374

Konapala, G., Mishra, A. K., Wada, Y., & Mann, M. E. (2020). Climate change will affect global water availability through compounding changes in seasonal precipitation and evaporation. *Nature Communications*, *11*(1), 3044. doi:10.103841467-020-16757-w PMID:32576822

Kosanic, A., Anderson, K., Harrison, S., Turkington, T., & Bennie, J. (2018). Changes in the geographical distribution of plant species and climatic variables on the West Cornwall peninsula (South West UK). *PLoS One*, *13*(2), e0191021. doi:10.1371/journal.pone.0191021 PMID:29401494

Kotroczó, Z., Makádi, M., Kocsis, T., Béni, Á., Várbíró, G., & Fekete, I. (2023). Long-Term Changes in Organic Matter Content and Soil Moisture Determine the Degree of Root and Soil Respiration. *Plants*, *12*(2), 251. doi:10.3390/plants12020251 PMID:36678968

Kubiak-Wójcicka, K., & Machula, S. (2020). Influence of Climate Changes on the State of Water Resources in Poland and Their Usage. *Geosciences*, *10*(8), 8. doi:10.3390/geosciences10080312

Kyriakopoulos, G. L., & Sebos, I. (2023). Enhancing Climate Neutrality and Resilience through Coordinated Climate Action: Review of the Synergies between Mitigation and Adaptation Actions. *Climate (Basel)*, *11*(5), 5. doi:10.3390/cli11050105

Lankford, B., Pringle, C., McCosh, J., Shabalala, M., Hess, T., & Knox, J. W. (2023). Irrigation area, efficiency and water storage mediate the drought resilience of irrigated agriculture in a semi-arid catchment. *The Science of the Total Environment*, *859*, 160263. doi:10.1016/j.scitotenv.2022.160263 PMID:36402330

Laurent, L., Buoncristiani, J.-F., Pohl, B., Zekollari, H., Farinotti, D., Huss, M., Mugnier, J.-L., & Pergaud, J. (2020). The impact of climate change and glacier mass loss on the hydrology in the Mont-Blanc massif. *Scientific Reports*, *10*(1), 10420. doi:10.103841598-020-67379-7 PMID:32591640

Lawrence, J., Blackett, P., & Cradock-Henry, N. A. (2020). Cascading climate change impacts and implications. *Climate Risk Management*, *29*, 100234. doi:10.1016/j.crm.2020.100234

Lawton, D., Huseth, A. S., Kennedy, G. G., Morey, A. C., Hutchison, W. D., Reisig, D. D., Dorman, S. J., Dillard, D., Venette, R. C., Groves, R. L., Adamczyk, J. J., Barbosa Dos Santos, I., Baute, T., Brown, S., Burkness, E., Dean, A., Dively, G. P., Doughty, H. B., Fleischer, S. J., & Zuefle, M. (2022). Pest population dynamics are related to a continental overwintering gradient. *Proceedings of the National Academy of Sciences of the United States of America*, *119*(37), e2203230119. doi:10.1073/pnas.2203230119 PMID:36067290

Levidow, L., Zaccaria, D., Maia, R., Vivas, E., Todorovic, M., & Scardigno, A. (2014). Improving water-efficient irrigation: Prospects and difficulties of innovative practices. *Agricultural Water Management*, *146*, 84–94. doi:10.1016/j.agwat.2014.07.012

Levy, B. S., & Patz, J. A. (2015). Climate Change, Human Rights, and Social Justice. *Annals of Global Health*, *81*(3), 310–322. doi:10.1016/j.aogh.2015.08.008 PMID:26615065

Levy, O., Buckley, L. B., Keitt, T. H., Smith, C. D., Boateng, K. O., Kumar, D. S., & Angilletta, M. J. (2015). Resolving the life cycle alters expected impacts of climate change. *Proceedings of the Royal Society B: Biological Sciences, 282*(1813), 20150837. 10.1098/rspb.2015.0837

Liao, Y., Song, X., Ye, Y., Gu, J., Wang, R., Zhuogabayong, Z., Zhao, D., & Shao, X. (2023). Climate Change May Pose Additional Threats to the Endangered Endemic Species Encalypta buxbaumioidea in China. *Diversity*, *15*(2), 2. doi:10.3390/d15020269

Liu, S., Wang, H., & Qin, F. (2023). Genetic dissection of drought resistance for trait improvement in crops. *The Crop Journal*, *11*(4), 975–985. doi:10.1016/j.cj.2023.05.002

López, J. J., Ayuso-Muñoz, J. L., Goñi, M., & Gimena, F. N. (2023). Analysis of the Effect of Climate Change on the Characteristics of Rainfall in Igeldo-Gipuzkoa (Spain). *Water (Basel)*, *15*(8), 8. Advance online publication. doi:10.3390/w15081529

Loudyi, D., Hasnaoui, M. D., & Fekri, A. (2022). Flood Risk Management Practices in Morocco: Facts and Challenges. In T. Sumi, S. A. Kantoush, & M. Saber (Eds.), Wadi Flash Floods: Challenges and Advanced Approaches for Disaster Risk Reduction (pp. 35–94). Springer. doi:10.1007/978-981-16-2904-4_2

Lynch, J. P., Mooney, S. J., Strock, C. F., & Schneider, H. M. (2022). Future roots for future soils. *Plant, Cell & Environment*, *45*(3), 620–636. doi:10.1111/pce.14213 PMID:34725839

Ma, C.-S., Zhang, W., Peng, Y., Zhao, F., Chang, X.-Q., Xing, K., Zhu, L., Ma, G., Yang, H.-P., & Rudolf, V. H. W. (2021). Climate warming promotes pesticide resistance through expanding overwintering range of a global pest. *Nature Communications*, *12*(1), 5351. doi:10.103841467-021-25505-7 PMID:34504063

Machado, R. M. A., & Serralheiro, R. P. (2017). Soil Salinity: Effect on Vegetable Crop Growth. Management Practices to Prevent and Mitigate Soil Salinization. *Horticulturae*, *3*(2), 2. doi:10.3390/horticulturae3020030

Mallareddy, M., Thirumalaikumar, R., Balasubramanian, P., Naseeruddin, R., Nithya, N., Mariadoss, A., Eazhilkrishna, N., Choudhary, A. K., Deiveegan, M., Subramanian, E., Padmaja, B., & Vijayakumar, S. (2023). Maximizing Water Use Efficiency in Rice Farming: A Comprehensive Review of Innovative Irrigation Management Technologies. *Water (Basel)*, *15*(10), 10. doi:10.3390/w15101802

Manzoor, Z., Ehsan, M., Khan, M. B., Manzoor, A., Akhter, M. M., Sohail, M. T., Hussain, A., Shafi, A., Abu-Alam, T., & Abioui, M. (2022). Floods and flood management and its socio-economic impact on Pakistan: A review of the empirical literature. *Frontiers in Environmental Science*, *10*, 1021862. https://www.frontiersin.org/articles/10.3389/fenvs.2022.1021862. doi:10.3389/fenvs.2022.1021862

Marcus, H., Hanna, L., Tait, P., Stone, S., & Wannous, C. (2023). Climate change and the public health imperative for supporting migration as adaptation. *Journal of Migration and Health*, *7*, 100174. doi:10.1016/j.jmh.2023.100174 PMID:36968560

Martínez-Valderrama, J., Olcina, J., Delacámara, G., Guirado, E., & Maestre, F. T. (2023). Complex Policy Mixes are Needed to Cope with Agricultural Water Demands Under Climate Change. *Water Resources Management*, *37*(6), 2805–2834. doi:10.100711269-023-03481-5

Mayne, S. J., King, D. I., Andersen, J. C., & Elkinton, J. S. (2023). Pest control services on farms vary among bird species on diversified, low-intensity farms. *Global Ecology and Conservation*, *43*, e02447. doi:10.1016/j.gecco.2023.e02447

McLeman, R. A., & Hunter, L. M. (2010). Migration in the context of vulnerability and adaptation to climate change: Insights from analogues. *Wiley Interdisciplinary Reviews: Climate Change*, *1*(3), 450–461. doi:10.1002/wcc.51 PMID:22022342

Meng, X., Zhu, Y., Yin, M., & Liu, D. (2021). The impact of land use and rainfall patterns on the soil loss of the hillslope. *Scientific Reports*, *11*(1), 1. doi:10.103841598-021-95819-5 PMID:34381135

Miner, K. R., D'Andrilli, J., Mackelprang, R., Edwards, A., Malaska, M. J., Waldrop, M. P., & Miller, C. E. (2021). Emergent biogeochemical risks from Arctic permafrost degradation. *Nature Climate Change*, *11*(10), 10. doi:10.103841558-021-01162-y

Minoli, S., Jägermeyr, J., Asseng, S., Urfels, A., & Müller, C. (2022). Global crop yields can be lifted by timely adaptation of growing periods to climate change. *Nature Communications*, *13*(1), 7079. doi:10.103841467-022-34411-5 PMID:36400762

Mondal, S. (2021). *Impact of Climate Change on Soil Fertility*., doi:10.1007/978-3-030-76863-8_28

Naorem, A. (2022). Carbon Sequestration. *Resonance*, *27*(7), 1237–1245. doi:10.100712045-022-1416-0

Neira, J., Ortiz, M., Morales, L., & Acevedo, E. (2015). Oxygen diffusion in soils: Understanding the factors and processes needed for modeling. *Chilean Journal of Agricultural Research*, *75*, 35–44. doi:10.4067/S0718-58392015000300005

Nievola, C. C., Carvalho, C. P., Carvalho, V., & Rodrigues, E. (2017). Rapid responses of plants to temperature changes. Temperature: *Multidisciplinary. Biomedical Journal*, *4*(4), 371–405. doi:10.1080/23328940.2017.1377812 PMID:29435478

Ortiz, A. M. D., Outhwaite, C. L., Dalin, C., & Newbold, T. (2021). A review of the interactions between biodiversity, agriculture, climate change, and international trade: Research and policy priorities. *One Earth*, *4*(1), 88–101. doi:10.1016/j.oneear.2020.12.008

Panetta, A. M., Stanton, M. L., & Harte, J. (2018). Climate warming drives local extinction: Evidence from observation and experimentation. *Science Advances, 4*(2), eaaq1819. doi:10.1126/sciadv.aaq1819

Parmesan, C., & Hanley, M. E. (2015). Plants and climate change: Complexities and surprises. *Annals of Botany*, *116*(6), 849–864. doi:10.1093/aob/mcv169 PMID:26555281

Pfeifer, L., & Otto, I. M. (2023). Changing seasonal temperature offers a window of opportunity for stricter climate policy. *Environmental Science & Policy*, *140*, 35–45. doi:10.1016/j.envsci.2022.11.010

Picarella, M. E., & Mazzucato, A. (2019). The Occurrence of Seedlessness in Higher Plants; Insights on Roles and Mechanisms of Parthenocarpy. *Frontiers in Plant Science*, *9*, 1997. doi:10.3389/fpls.2018.01997 PMID:30713546

Porter, J. R., & Semenov, M. A. (2005). Crop responses to climatic variation. *Philosophical Transactions of the Royal Society of London. Series B, Biological Sciences*, *360*(1463), 2021–2035. doi:10.1098/rstb.2005.1752 PMID:16433091

Pueppke, S. G., Nurtazin, S., & Ou, W. (2020). Water and Land as Shared Resources for Agriculture and Aquaculture: Insights from Asia. *Water (Basel)*, *12*(10), 10. doi:10.3390/w12102787

Race, D., Gentle, P., & Mathew, S. (2023). Living on the margins: Climate change impacts and adaptation by remote communities living in the Pacific Islands, the Himalaya and desert Australia. *Climate Risk Management*, *40*, 100503. doi:10.1016/j.crm.2023.100503

Rane, J., Singh, A. K., Tiwari, M., Prasad, P. V. V., & Jagadish, S. V. K. (2022). Effective Use of Water in Crop Plants in Dryland Agriculture: Implications of Reactive Oxygen Species and Antioxidative System. *Frontiers in Plant Science*, *12*, 778270. doi:10.3389/fpls.2021.778270 PMID:35082809

Rashid, M. I., Mujawar, L. H., Shahzad, T., Almeelbi, T., Ismail, I. M. I., & Oves, M. (2016). Bacteria and fungi can contribute to nutrients bioavailability and aggregate formation in degraded soils. *Microbiological Research, 183*, 26–41. doi:10.1016/j.micres.2015.11.007 PMID:26805616

Raza, A., Razzaq, A., Mehmood, S. S., Zou, X., Zhang, X., Lv, Y., & Xu, J. (2019). Impact of Climate Change on Crops Adaptation and Strategies to Tackle Its Outcome: A Review. *Plants*, *8*(2), 34. doi:10.3390/plants8020034 PMID:30704089

Rumpel, C., Amiraslani, F., Bossio, D., Chenu, C., Henry, B., Espinoza, A. F., Koutika, L.-S., Ladha, J., Madari, B., Minasny, B., Olaleye, A. O., Shirato, Y., Sall, S. N., Soussana, J.-F., & Varela-Ortega, C. (2022). The role of soil carbon sequestration in enhancing human resilience in tackling global crises including pandemics. *Soil Security*, *8*, 100069. doi:10.1016/j.soisec.2022.100069

Saco, P. M., McDonough, K. R., Rodriguez, J. F., Rivera-Zayas, J., & Sandi, S. G. (2021). The role of soils in the regulation of hazards and extreme events. *Philosophical Transactions of the Royal Society B: Biological Sciences*, *376*(1834), 20200178. doi:10.1098/rstb.2020.0178

Santos, F. D., Ferreira, P. L., & Pedersen, J. S. T. (2022). The Climate Change Challenge: A Review of the Barriers and Solutions to Deliver a Paris Solution. *Climate (Basel)*, *10*(5), 5. doi:10.3390/cli10050075

Scavo, A., Fontanazza, S., Restuccia, A., Pesce, G. R., Abbate, C., & Mauromicale, G. (2022). The role of cover crops in improving soil fertility and plant nutritional status in temperate climates. A review. *Agronomy for Sustainable Development*, *42*(5), 93. doi:10.100713593-022-00825-0

Schleuning, M., Fründ, J., Schweiger, O., Welk, E., Albrecht, J., Albrecht, M., Beil, M., Benadi, G., Blüthgen, N., Bruelheide, H., Böhning-Gaese, K., Dehling, D. M., Dormann, C. F., Exeler, N., Farwig, N., Harpke, A., Hickler, T., Kratochwil, A., Kuhlmann, M., ... Hof, C. (2016). Ecological networks are more sensitive to plant than to animal extinction under climate change. *Nature Communications*, *7*(1), 1. doi:10.1038/ncomms13965 PMID:28008919

Schneider, L., Rebetez, M., & Rasmann, S. (2022). The effect of climate change on invasive crop pests across biomes. *Current Opinion in Insect Science*, *50*, 100895. doi:10.1016/j.cois.2022.100895 PMID:35240333

Schuur, E. A. G., Abbott, B. W., Commane, R., Ernakovich, J., Euskirchen, E., Hugelius, G., Grosse, G., Jones, M., Koven, C., Leshyk, V., Lawrence, D., Loranty, M. M., Mauritz, M., Olefeldt, D., Natali, S., Rodenhizer, H., Salmon, V., Schädel, C., Strauss, J., & Turetsky, M. (2022). Permafrost and Climate Change: Carbon Cycle Feedbacks From the Warming Arctic. *Annual Review of Environment and Resources*, *47*(1), 343–371. doi:10.1146/annurev-environ-012220-011847

Seebacher, F., & Post, E. (2015). Climate change impacts on animal migration. *Climate Change Responses*, *2*(1), 5. doi:10.118640665-015-0013-9

Seleiman, M. F., Al-Suhaibani, N., Ali, N., Akmal, M., Alotaibi, M., Refay, Y., Dindaroglu, T., Abdul-Wajid, H. H., & Battaglia, M. L. (2021). Drought Stress Impacts on Plants and Different Approaches to Alleviate Its Adverse Effects. *Plants*, *10*(2), 259. doi:10.3390/plants10020259 PMID:33525688

Seyedi Rezvani, S. M. H., Almeida, N., & Falcão, M. (2023). Climate Adaptation Measures for Enhancing Urban Resilience. *Buildings*, *13*(9), 2163. doi:10.3390/buildings13092163

Shivanna, K. R. (2022). Climate change and its impact on biodiversity and human welfare. *Proceedings of the Indian National Science Academy. Part A, Physical Sciences*, *88*(2), 160–171. doi:10.100743538-022-00073-6

Shrestha, S., Mahat, J., & Shrestha, J. K. M., & Paudel, K. (2022). Influence of high-temperature stress on rice growth and development. A review. *Heliyon*, *8*(12), e12651. doi:10.1016/j.heliyon.2022.e12651 PMID:36643304

Si, M., Guo, X., Lan, Y., Fan, B., & Cao, G. (2022). Effects of Climatic Variability on Soil Water Content in an Alpine Kobresia Meadow, Northern Qinghai–Tibetan Plateau, China. *Water (Basel)*, *14*(17), 17. doi:10.3390/w14172754

Singh, B. K., Delgado-Baquerizo, M., Egidi, E., Guirado, E., Leach, J. E., Liu, H., & Trivedi, P. (2023). Climate change impacts on plant pathogens, food security and paths forward. *Nature Reviews. Microbiology*, *21*(10), 1–17. doi:10.103841579-023-00900-7 PMID:37131070

Skendžić, S., Zovko, M., Živković, I. P., Lešić, V., & Lemić, D. (2021). The Impact of Climate Change on Agricultural Insect Pests. *Insects*, *12*(5), 440. doi:10.3390/insects12050440 PMID:34066138

Soares, P. M. M., & Lima, D. C. A. (2022). Water scarcity down to earth surface in a Mediterranean climate: The extreme future of soil moisture in Portugal. *Journal of Hydrology (Amsterdam)*, *615*, 128731. doi:10.1016/j.jhydrol.2022.128731

Tabari, H. (2020). Climate change impact on flood and extreme precipitation increases with water availability. *Scientific Reports*, *10*(1), 1. doi:10.103841598-020-70816-2 PMID:31913322

Tanure, T. M., Miyajima, D. N., Magalhães, A. S., Domingues, E. P., & Carvalho, T. S. (2020). The Impacts of Climate Change on Agricultural Production, Land Use and Economy of the Legal Amazon Region Between 2030 and 2049. *Economía*, *21*(1), 73–90. doi:10.1016/j.econ.2020.04.001

Tian, X., Dong, J., Jin, S., He, H., Yin, H., & Chen, X. (2023). Climate change impacts on regional agricultural irrigation water use in semi-arid environments. *Agricultural Water Management*, *281*, 108239. doi:10.1016/j.agwat.2023.108239

Tougeron, K., Brodeur, J., Le Lann, C., & van Baaren, J. (2020). How climate change affects the seasonal ecology of insect parasitoids. *Ecological Entomology*, *45*(2), 167–181. doi:10.1111/een.12792

Tudi, M., Daniel Ruan, H., Wang, L., Lyu, J., Sadler, R., Connell, D., Chu, C., & Phung, D. T. (2021). Agriculture Development, Pesticide Application and Its Impact on the Environment. *International Journal of Environmental Research and Public Health*, *18*(3), 1112. doi:10.3390/ijerph18031112 PMID:33513796

Urra, J., Alkorta, I., & Garbisu, C. (2019). Potential Benefits and Risks for Soil Health Derived From the Use of Organic Amendments in Agriculture. *Agronomy (Basel)*, *9*(9), 9.. doi:10.3390/agronomy9090542

Vaca, R., Del Águila, P., Yañez-Ocampo, G., Lugo, J. A., & De la Portilla-López, N. (2023). Soil Quality Assessment in Response to Water Erosion and Mining Activity. *Agriculture*, *13*(7), 7. doi:10.3390/agriculture13071380

Vineis, P., Chan, Q., & Khan, A. (2011). Climate change impacts on water salinity and health. *Journal of Epidemiology and Global Health*, *1*(1), 5–10. doi:10.1016/j.jegh.2011.09.001 PMID:23856370

Visser, M. E. (2008). Keeping up with a warming world; assessing the rate of adaptation to climate change. *Proceedings. Biological Sciences*, *275*(1635), 649–659. doi:10.1098/rspb.2007.0997 PMID:18211875

Wagner, J. J., & Engle, T. E. (2021). Invited Review: Water consumption, and drinking behavior of beef cattle, and effects of water quality. *Applied Animal Science*, *37*(4), 418–435. doi:10.15232/aas.2021-02136

Wang, J., Zhen, J., Hu, W., Chen, S., Lizaga, I., Zeraatpisheh, M., & Yang, X. (2023). Remote sensing of soil degradation: Progress and perspective. *International Soil and Water Conservation Research*, *11*(3), 429–454. doi:10.1016/j.iswcr.2023.03.002

Wang, X., Liu, Z., & Chen, H. (2022). Investigating Flood Impact on Crop Production under a Comprehensive and Spatially Explicit Risk Evaluation Framework. *Agriculture*, *12*(4), 4. Advance online publication. doi:10.3390/agriculture12040484

Weiskopf, S. R., Rubenstein, M. A., Crozier, L. G., Gaichas, S., Griffis, R., Halofsky, J. E., Hyde, K. J. W., Morelli, T. L., Morisette, J. T., Muñoz, R. C., Pershing, A. J., Peterson, D. L., Poudel, R., Staudinger, M. D., Sutton-Grier, A. E., Thompson, L., Vose, J., Weltzin, J. F., & Whyte, K. P. (2020). Climate change effects on biodiversity, ecosystems, ecosystem services, and natural resource management in the United States. *The Science of the Total Environment*, *733*, 137782. doi:10.1016/j.scitotenv.2020.137782 PMID:32209235

Widlansky, M. J., Long, X., & Schloesser, F. (2020). Increase in sea level variability with ocean warming associated with the nonlinear thermal expansion of seawater. *Communications Earth & Environment*, *1*(1), 1. Advance online publication. doi:10.103843247-020-0008-8

Wunderling, N., Willeit, M., Donges, J. F., & Winkelmann, R. (2020). Global warming due to loss of large ice masses and Arctic summer sea ice. *Nature Communications*, *11*(1), 1. doi:10.103841467-020-18934-3 PMID:33110092

Wurst, S., De Deyn, G. B., & Orwin, K. (2012). Soil Biodiversity and Functions. In D. H. Wall, R. D. Bardgett, V. Behan-Pelletier, J. E. Herrick, T. H. Jones, K. Ritz, J. Six, D. R. Strong, & W. H. van der Putten (Eds.), *Soil Ecology and Ecosystem Services (p. 0)*. Oxford University Press. doi:10.1093/acprof:oso/9780199575923.003.0004

Xu, X., Du, X., Wang, F., Sha, J., Chen, Q., Tian, G., Zhu, Z., Ge, S., & Jiang, Y. (2020). Effects of Potassium Levels on Plant Growth, Accumulation and Distribution of Carbon, and Nitrate Metabolism in Apple Dwarf Rootstock Seedlings. *Frontiers in Plant Science*, *11*, 904. https://www.frontiersin.org/articles/10.3389/fpls.2020.00904. doi:10.3389/fpls.2020.00904 PMID:32655607

## ADDITIONAL READING

Furtak, K., & Wolińska, A. (2023). The impact of extreme weather events as a consequence of climate change on the soil moisture and on the quality of the soil environment and agriculture – A review. *Catena*, *231*, 107378. doi:10.1016/j.catena.2023.107378

Omotoso, A. B., Letsoalo, S., Olagunju, K. O., Tshwene, C. S., & Omotayo, A. O. (2023). Climate change and variability in sub-Saharan Africa: A systematic review of trends and impacts on agriculture. *Journal of Cleaner Production*, *414*, 137487. doi:10.1016/j.jclepro.2023.137487

Sun, W., Li, S., Zhang, G., Fu, G., Qi, H., & Li, T. (2023). Effects of climate change and anthropogenic activities on soil pH in grassland regions on the Tibetan Plateau. *Global Ecology and Conservation*, *45*, e02532. doi:10.1016/j.gecco.2023.e02532

## KEY TERMS AND DEFINITIONS

**Agricultural Pests:** Also known as parasites or agricultural nuisances, refer to harmful organisms that attack crops and cultivated plants, leading to damages and significant losses for agriculture. These organisms can include insects, mites, fungi, bacteria, viruses, nematodes, or rodents, among others.

**Climate Change:** This encompasses long-term changes in climatic data within a specific region or on a global scale, including variations in temperatures, precipitation, winds, and other features. These fluctuations are commonly linked to influences such as human activities, natural cycles, or other external factors.

**Interaction:** This refers to a dynamic process where two entities, elements, or individuals interact with each other by mutually acting upon and influencing one another. It is a reciprocal exchange that can take various forms depending on the context, often involving communication or a relationship between the involved parties.

**Living Being:** A living being, also referred to as an organism, is a sophisticated and structured entity that demonstrates the fundamental attributes of life. These attributes encompass the capacity for growth, reproduction, response to stimuli, maintenance of internal stability (homeostasis), and engagement in metabolism (the acquisition and utilization of energy).

**Plants:** A plant is a living organism belonging to the plant kingdom, characterized by its ability to produce its own food through the process of photosynthesis. Plants are multicellular beings, generally immobile, and typically grow rooted in the soil.

**Soil:** Soil is a naturally occurring, intricate, and ever-changing blend of minerals, organic substances, water, air, and living organisms that blankets the Earth's land surface. It develops through the interplay of geological, biological, climatic, and topographical forces over extended periods.

**Water:** A colorless, odorless, and flavorless liquid that is indispensable for the existence of all known living organisms. It is the most abundant substance found on the Earth's surface and serves a crucial function in a wide range of biological, chemical, and physical activities. Depending on temperature and pressure, water can exist in three states: solid (ice), liquid (water), and gas (water vapor).

# Compilation of References


Abbasi, H., M. Jamil, A. Haq, S. Ali, R. Ahmad, and Z. Malik. (2016). Salt stress manifestation on plants, mechanism of salt tolerance and potassium role in alleviating it: a review. *Zemdirbyste- Agricul. 103*, 229– 238. doi:10.13080/z-a.2016.103.030

Abbass, K., Qasim, M. Z., Song, H., Murshed, M., Mahmood, H., & Younis, I. (2022). A review of the global climate change impacts, adaptation, and sustainable mitigation measures. *Environmental Science and Pollution Research International, 29*(28), 42539–42559. doi:10.100711356-022-19718-6 PMID:35378646

Abdallah, M. B., Methenni, K., Nouairi, I., Zarrouk, M., & Youssef, N. B. (2017). Drought priming improves subsequent more severe drought in a drought- sensitive cultivar of olive cv. Chétoui. *Scientia Horticulturae, 221*, 43–52. doi:10.1016/j.scienta.2017.04.021 PMID:28713194

Abdel Latef, A. A. H., & Miransari, M. (2014). The role of arbuscular mycorrhizal fungi in alleviation of salt stress. *Use of Microbes for the Alleviation of Soil Stresses: Volume 2: Alleviation of Soil Stress by PGPR and Mycorrhizal Fungi*, 23-38.

Abihudi, S. A., Venkataramana, P. B., de Boer, H. J., & Treydte, A. C. (2020). Species-specific responses of "Critically Endangered" and "Least Concern" Aloe seed germination to environmental conditions in Tanzania. *Global Ecology and Conservation, 24*, e01241. doi:10.1016/j.gecco.2020.e01241

Abobatta, W. F. (2023). The influence of climate change on interactions between environmental stresses and plants. In *Plant Stress Mitigators* (pp. 425–434). Elsevier. doi:10.1016/B978-0-323-89871-3.00021-5

Abonnenc, E. (1972). Les phlébotomes de la région éthiopienne (Diptera, Psychodidae). *Cahiers de l'ORSTOM, série Entomologie médicale et Parasitologie, 55*, 1-239.

Abou-Tammame, D., Zouhri, A., Boutarfa, A., Fathi, J., & Aboutayeb, R. (2022). The effect of purified wastewater on the physicochemical properties of agricultural soils in Chaouia in Morocco. *Journal of Ecological Engineering, 23*(1), 34–42. doi:10.12911/22998993/143868

Aboutayeb, R. (2015). *Valorisation des fumiers de volailles de chair: Compostage, épandage et étude des effets des fumiers de volailles et leurs composts sur les propriétés physicochimiques du sol sous cultures de maïs fourrager et de menthe verte.* [Thèse de doctorat, Université Hassan I].

Aboutayeb, R., El Yousfi, B., & El Gharras, O. (2020). Impact of No-Till on physicochemical properties of Vertisols in Chaouia region of Morocco. *Eurasian Journal of Soil Science, 9*(2), 119–125. doi:10.18393/ejss.663502

Aboutayeb, R., El-Mrini, S., Zouhri, A., & Azim, K. (2023). Effect of carbon to nitrogen ratio and aeration rate on phosphorus and exchangeable cation contents and their leaching in the soil during olive pomace and turkey manure co-composting. *Journal of Engineering and Applied Sciences (Asian Research Publishing Network), 70*(1), 12.

Aboutayeb, R., El-Mrini, S., Zouhri, A., Idrissi, O., & Azim, K. (2021). Hygienization assessment during heap co-composting of Turkey manure and olive mill pomace. *Eurasian Journal of Soil Science*, *10*(4), 332–342. doi:10.18393/ejss.972062

Abuarab, M. E., El-Mogy, M. M., Hassan, A. M., Abdeldaym, E. A., Abdelkader, N. H., & El-Sawy, B. I., M. (. (2019). The Effects of Root Aeration and Different Soil Conditioners on the Nutritional Values, Yield, and Water Productivity of Potato in Clay Loam Soil. *Agronomy (Basel)*, *9*(8), 8. Advance online publication. doi:10.3390/agronomy9080418

Abushal, L. T., Salama, M., Essa, M. M., & Qoronfleh, M. W. (2021). Agricultural biotechnology: Revealing insights about ethical concerns. *Journal of Biosciences*, *46*(3), 81. doi:10.100712038-021-00203-0 PMID:34423783

Acquaah, G. (2015). Conventional plant breeding principles and techniques. In Springer eBooks (pp. 115–158). doi:10.1007/978-3-319-22521-0_5

Adegbeye, M. J., Reddy, P. R. K., Obaisi, A. I., Elghandour, M. M. M. Y., Oyebamiji, K. J., Salem, A. Z. M., & Camacho-Díaz, L. M. (2020). Sustainable agriculture options for production, greenhouse gasses and pollution alleviation, and nutrient recycling in emerging and transitional nations-An overview. *Journal of Cleaner Production*, *242*, 118319. doi:10.1016/j.jclepro.2019.118319

Adeleke, B. S., & Babalola, O. O. (2021). The endosphere microbial communities, a great promise in agriculture. *International Microbiology*, *24*(1), 1–17. doi:10.100710123-020-00140-2 PMID:32737846

Adenle, A. A., Azadi, H., & Arbiol, J. (2015). Global assessment of technological innovation for climate change adaptation and mitigation in the developing world. *Journal of Environmental Management*, *161*, 261–275. doi:10.1016/j.jenvman.2015.05.040 PMID:26189184

Adenle, A. A., Wedig, K., & Azadi, H. (2019). Sustainable agriculture and food security in Africa: The role of innovative technologies and international organizations. *Technology in Society*, *58*, 101143. doi:10.1016/j.techsoc.2019.05.007

Adhikari, U., Nejadhashemi, A. P., & Woznicki, S. A. (2015). Climate change and eastern Africa: A review of impact on major crops. *Food and Energy Security*, *4*(2), 110–132. doi:10.1002/fes3.61

Adli, M. (2018). The CRISPR tool kit for genome editing and beyond. *Nature Communications*, *9*(1), 1911. doi:10.103841467-018-04252-2 PMID:29765029

Adl, S. M., Simpson, A. G., Farmer, M. A., Andersen, R. A., Anderson, O. R., Barta, J. R., Bowser, S. S., Brugerolle, G., Fensome, R. A., Fredericq, S., James, T., Karpov, S., Kugrens, P., Krug, J., Lane, C., Lewis, L. A., Lodge, J., Lynn, D. H., Mann, D. G., & Taylor, M. F. (2005). The new higher level classification of eukaryotes with emphasis on the taxonomy of protists. *The Journal of Eukaryotic Microbiology*, *52*(5), 399–451. doi:10.1111/j.1550-7408.2005.00053.x PMID:16248873

Agnihotri, P., & Mitra, A. K. (2023). *Understanding the Impact of Global Climate Change on Abiotic Stress in Plants and the Supportive Role of PGPR.*

Aguiar-Menezes, E. L. (2005). *Inseticidas botânicos: seus princípios ativos, modo de ação e uso agrícola*. Embrapa Agrobiologia.

Ahmed, M., Hayat, R., Ahmad, M., Ul-Hassan, M., Kheir, A. M., Ul-Hassan, F., Ur-Rehman, M. H., Shaheen, F. A., Raza, M. A., & Ahmad, S. (2022). Impact of climate change on dryland agricultural systems: A review of current status, potentials, and further work need. *International Journal of Plant Production*, *16*(3), 341–363. doi:10.100742106-022-00197-1 PMID:35614974

Ahuchaogu, I. I., Ehiomogue, P. O., & Udoumoh, U. I. (2022). Effects of soil and water conservation measures on the environment: A review. *Poljoprivredna tehnika, 47*(3), 42-55.

Ait Antar, D. (2021). *Le profil épidémiologique de leishmaniose cutanée et viscérale dans la province Al Haouz.* [Thèse de doctorat, Faculté de médecine et de pharmacie. Université Cadi Ayad].

Aitken, S. N., Yeaman, S., Holliday, J. A., Wang, T., & Curtis-McLane, S. (2008). Adaptation, migration or extirpation: Climate change outcomes for tree populations. *Evolutionary Applications, 1*(1), 95–111. doi:10.1111/j.1752-4571.2007.00013.x PMID:25567494

Akhatar, J., Kaur, H., & Kumar, H. (2022). Conventional Plant Breeding to Modern Biotechnological Approaches in Crop Improvement. In Technologies in Plant Biotechnology and Breeding of Field Crops (pp. 1–21). doi:10.1007/978-981-16-5767-2_1

Akhtar, M. S., Oki, Y., Nakashima, Y., Nishigaki, M., Adachi, T., & Kamigaki, T. (2016). Microcosm investigation on differential potential of free-floating Azolla macrophytes for phytoremediation of P-controlled water eutrophication. *International Journal of Agriculture and Biology, 18*(1), 204–212. doi:10.17957/IJAB/15.0119

Akter, M., Fan, L., Rahman, M. M., Geissen, V., & Ritsema, C. J. (2018). Vegetable farmers' behaviour and knowledge related to pesticide use and related health problems: A case study from Bangladesh. *Journal of Cleaner Production, 200*, 122–133. doi:10.1016/j.jclepro.2018.07.130

Akter, S., Ali, S., Fekete-Farkas, M., Fogarassy, C., & Lakner, Z. (2023). Why organic food? Factors influence the organic food purchase intension in an emerging country (study from northern part of Bangladesh). *Resources, 12*(1), 5. doi:10.3390/resources12010005

Albert, C. H., Grassein, F., Schurr, F. M., Vieilledent, G., & Violle, C. (2011). When and how should intraspecific variability be considered in trait-based plant ecology? *Perspectives in Plant Ecology, Evolution and Systematics, 13*(3), 217–225. doi:10.1016/j.ppees.2011.04.003

Albert, C. H., Thuiller, W., Yoccoz, N. G., Douzet, R., Aubert, S., & Lavorel, S. (2010). A multi-trait approach reveals the structure and the relative importance of intra-vs. interspecific variability in plant traits. *Functional Ecology, 24*(6), 1192–1201. doi:10.1111/j.1365-2435.2010.01727.x

Alengebawy, A., Abdelkhalek, S. T., Qureshi, S. R., & Wang, M.-Q. (2021). Heavy metals and pesticides toxicity in agricultural soil and plants: Ecological risks and human health implications. *Toxics, 9*(3), 42. doi:10.3390/toxics9030042 PMID:33668829

Aliste, M., Pérez-Lucas, G., Garrido, I., Fenoll, J., & Navarro, S. (2021). Mobility of insecticide residues and main intermediates in a clay-loam soil, and impact of leachate components on their photocatalytic degradation. *Chemosphere, 274*, 129965. doi:10.1016/j.chemosphere.2021.129965 PMID:33979936

Allen, C. D., Macalady, A. K., Chenchouni, H., Bachelet, D., McDowell, N., Vennetier, M., Kitzberger, T., Rigling, A., Breshears, D. D., Hogg, E. H. T., Gonzalez, P., Fensham, R., Zhang, Z., Castro, J., Demidova, N., Lim, J.-H., Allard, G., Running, S. W., Semerci, A., & Cobb, N. (2010). A global overview of drought and heat-induced tree mortality reveals emerging climate change risks for forests. *Forest Ecology and Management, 259*(4), 660–684. doi:10.1016/j.foreco.2009.09.001

Allen, R. G., Pereira, L. S., Raes, D., & Smith, M. (1998). FAO Penman–Monteith equation. *Crop Evapotranspiration: Guidelines for Computing Crop Water Requirements. FAO Irrigation and Drainage Paper, 56*, 17–28.

Allen, R. G., Pruitt, W. O., Wright, J. L., Howell, T. A., Ventura, F., Snyder, R., Itenfisu, D., Steduto, P., Berengena, J., Yrisarry, J. B., Smith, M., Pereira, L. S., Raes, D., Perrier, A., Alves, I., Walter, I., & Elliott, R. (2006). A recommendation on standardized surface resistance for hourly calculation of reference ETo by the FAO56 Penman-Monteith method. *Agricultural Water Management*, *81*(1–2), 1–22. doi:10.1016/j.agwat.2005.03.007

Almeida, F., Rodrigues, M. L., & Coelho, C. (2019). The still underestimated problem of fungal diseases worldwide. *Frontiers in Microbiology*, *10*, 214. doi:10.3389/fmicb.2019.00214 PMID:30809213

Altamira-Algarra, B., Puigagut, J., Day, J. W., Mitsch, W. J., Vymazal, J., Hunter, R. G., & García, J. (2022). A review of technologies for closing the P loop in agriculture runoff: Contributing to the transition towards a circular economy. *Ecological Engineering*, *177*, 106571. doi:10.1016/j.ecoleng.2022.106571

Altieri, M. A. (2009). Agroecology, small farms, and food sovereignty. *Monthly Review (New York, N.Y.)*, *61*(3), 102–113. doi:10.14452/MR-061-03-2009-07_8

Altieri, M. A., & Nicholls, C. I. (2017). The adaptation and mitigation potential of traditional agriculture in a changing climate. *Climatic Change*, *140*(1), 33–45. doi:10.100710584-013-0909-y

Altieri, M. A., Nicholls, C. I., Henao, A., & Lana, M. A. (2015). Agroecology and the design of climate change1resilient farming systems. *Agronomy for Sustainable Development*, *35*(3), 869–890. doi:10.100713593-015-0285-2

Alyokhin, A., Nault, B., & Brown, B. (2020). Soil conservation practices for insect pest management in highly disturbed agroecosystems–a review. *Entomologia Experimentalis et Applicata*, *168*(1), 7–27. doi:10.1111/eea.12863

Amanullah, Khalid, S., Imran, Khan, H. A., Arif, M., Altawaha, A. R., Parmar, B. (2020). Effects of climate change on irrigation water quality. *Environment, climate, plant and vegetation growth*, 123-132.

Amar, F. B. (2017). Etude comparative de la biologie florale des variétés d ' olive de table (Olea europaea L. ) Meski et Picholine Languedoc dans la région de Sfax (Tunisie) Floral biology comparative study of Meski and Picholine languedoc table olive varieties. Bulletin de l'Institut Scientifique, Rabat. Section Sciences de La Vie, 55–61. doi:10.2166/wcc.2018.224

Amiri, N., Lahlali, R., Amiri, S. E. L., Jarroudi, M., Khebiza, M. Y., & Messouli, M. (2021). Development of an Integrated Model to Assess the Impact of Agricultural Practices and Land Use on Agricultural Production in Morocco under Climate Stress over the Next Twenty Years. *Sustainability (Basel)*, *13*(21), 21. doi:10.3390u132111943

Amorati, R., Foti, M. C., & Valgimigli, L. (2013). Antioxidant activity of essential oils. *Journal of Agricultural and Food Chemistry*, *61*(46), 10835–10847. doi:10.1021/jf403496k PMID:24156356

Amorim Orrico, A. C., de Oliveira, J. D., Viana Leite, B. K., da Silva Vilela, R. N., Previdelli Orrico, M. A. Junior, Aspilcueta Borquis, R. R., & Macena, I. A. (2023). Effects of aeration and season of the year on fish waste composting and compost quality. *Environmental Technology*, 1–32. doi:10.1080/09593330.2023.2230350 PMID:37401255

Anagnostis, A., Tagarakis, A. C., Kateris, D., Moysiadis, V., Sørensen, C. G., Pearson, S., & Bochtis, D. (2021). Orchard mapping with deep learning semantic segmentation. *Sensors (Basel)*, *21*(11), 3813. doi:10.339021113813 PMID:34072975

Anandakumar, S., Bakhoum, N., Chinnadurai, C., Malarkodi, M., Arulmozhiselvan, K., Karthikeyan, S., & Balachandar, D. (2022). Impact of long-term nutrient management on sequestration and dynamics of soil organic carbon in a semi-arid tropical Alfisol of India. *Applied Soil Ecology*, *177*, 104549. doi:10.1016/j.apsoil.2022.104549

Anjum, S. A., Ashraf, U., Zohaib, A., Tanveer, M., Naeem, M., Ali, I., Tabassum, T., & Nazir, U. (2017). Growth and development responses of crop plants under drought stress: A review. *Zemdirbyste*, *104*(3), 267–276. doi:10.13080/z-a.2017.104.034

Annabi, M., Le Bissonnais, Y., Le Villio–Poitrenaud, M., & Houot, S. (2011). Improvement of soil aggregate stability by repeated applications of organic amendments to a cultivated silty loam soil. *Agriculture, Ecosystems & Environment, 144*(1), 382–389. doi:10.1016/j.agee.2011.07.005

Aragüés, R., Puy, J., & Isidoro, D. (2004). Vegetative growth response of young olive trees (*Olea europaea* L., cv. Arbequina) to soil salinity and waterlogging. *Plant and Soil, 258*(1), 69–80. doi:10.1023/B:PLSO.0000016537.61832.6e

Arambourg, Y. (1986). Traité d'entomologie oléicole (pp. 360-pp). COI.

Arampatzis, G., Hatzigiannakis, E., Pisinaras, V., Kourgialas, N., Psarras, G., Kinigopoulou, V., Panagopoulos, A., & Koubouris, G. (2018). Soil water content and olive tree yield responses to soil management, irrigation, and precipitation in a hilly Mediterranean area. Journal of Water and Climate Change, 9(4), 672–678. https://doi.org/10.2166/wcc.2018.224

Araos, M., Jagannathan, K., Shukla, R., Ajibade, I., Coughlan de Perez, E., Davis, K., Ford, J. D., Galappaththi, E. K., Grady, C., Hudson, A. J., Joe, E. T., Kirchhoff, C. J., Lesnikowski, A., Alverio, G. N., Nielsen, M., Orlove, B., Pentz, B., Reckien, D., Siders, A. R., & Turek-Hankins, L. L. (2021). Equity in human adaptation-related responses: A systematic global review. *One Earth, 4*(10), 1454–1467. doi:10.1016/j.oneear.2021.09.001

Arfanuzzaman, M., & Atiq Rahman, A. (2017). Sustainable water demand management in face of rapid urbanization and ground water depletion for social-ecological resilience building. *Global Ecology and Conservation, 10*(1), 9–22. doi:10.1016/j.gecco.2017.01.005

Armendáriz, A., De La Iglesia, L., Santiago, Y., Campillo, G., Alberte, C., Barroso, L. M., & Sanz, A. P. (2007). Ciclo del prays del olivo (" Prays oleae" Bern.) en Arribes del Duero. *Boletin de Sanidad Vegetal, Plagas, 33*(4), 443–456.

Armstrong, W., & Drew, M. C. (2002). Root growth and metabolism under oxygen defciency. In Y. Waisel, A. Eshel, & U. Kafkaf (Eds.), *Plant roots: the hidden half* (3rd ed., pp. 729–761). Marcel Dekker.

Arts, G. P., Dollinger, M., Kohlschmid, E., Maltby, L., Ochoa-Acuna, H., & Poulsen, V. (2015). An ecosystem services approach to pesticide risk assessment and risk management of non-target terrestrial plants: Recommendations from a SETAC Europe workshop. *Environmental Science and Pollution Research International, 22*(3), 2350–2355. doi:10.100711356-014-3637-6 PMID:25300179

Asch, R. G., Stock, C. A., & Sarmiento, J. L. (2019). Climate change impacts on mismatches between phytoplankton blooms and fish spawning phenology. *Global Change Biology, 25*(8), 2544–2559. doi:10.1111/gcb.14650 PMID:31152499

Asif, Z., Chen, Z., Sadiq, R., & Zhu, Y. (2023). Climate Change Impacts on Water Resources and Sustainable Water Management Strategies in North America. *Water Resources Management, 37*(6-7), 1–16. doi:10.100711269-023-03474-4

Atinkut, H. B., Yan, T., Zhang, F., Qin, S., Gai, H., & Liu, Q. (2020). Cognition of agriculture waste and payments for a circular agriculture model in Central China. *Scientific Reports, 10*(1), 10826. doi:10.103841598-020-67358-y PMID:32616725

Avellan, A., Yun, J., Morais, B. P., Clement, E. T., Rodrigues, S. M., & Lowry, G. V. (2021). Critical review: Role of inorganic nanoparticle properties on their foliar uptake and in planta translocation. *Environmental Science & Technology, 55*(20), 13417–13431. doi:10.1021/acs.est.1c00178 PMID:33988374

Awiti, H. A., Gido, E. O., & Obare, G. A. (2022). Smallholder Farmers Climate-Smart Crop Diversification Cost Structure: Empirical Evidence From Western Kenya. *Frontiers in Sustainable Food Systems, 6*, 86. doi:10.3389/fsufs.2022.842987

Awuni, S., Adarkwah, F., Ofori, B. D., Purwestri, R. C., Huertas Bernal, D. C., & Hajek, M. (2023). Managing the challenges of climate change mitigation and adaptation strategies in Ghana. *Heliyon, 9*(5), e15491. doi:10.1016/j.heliyon.2023.e15491 PMID:37131451

Axelsson, E., Ljungvall, I., Bhoumik, P., Conn, L. B., Murén, E., Ohlsson, Å., Olsen, L. H., Engdahl, K., Hagman, R., Hanson, J., Kryvokhyzha, D., Pettersson, M. E., Grenet, O., Moggs, J. G., Del Rio-Espinola, A., Epe, C., Taillon, B. E., Tawari, N. R., Mane, S., & Lindblad-Toh, K. (2021). The genetic consequences of dog breed formation—Accumulation of deleterious genetic variation and fixation of mutations associated with myxomatous mitral valve disease in cavalier King Charles spaniels. *PLOS Genetics*, *17*(9), e1009726. doi:10.1371/journal.pgen.1009726 PMID:34473707

Ayilara, M. S., Olanrewaju, O. S., Babalola, O. O., & Odeyemi, O. (2020). Waste Management through Composting: Challenges and Potentials. *Sustainability (Basel)*, *12*(11), 4456. doi:10.3390u12114456

Azeem, A., Abbas, N., Azeem, S., Iqbal, Z., & Ul-Allah, S. (2023). Physiological and Molecular Mechanism of Nanoparticles Induced Tolerance in Plants. In Emerging Contaminants and Plants: Interactions, Adaptations and Remediation Technologies (pp. 233-248). Springer. doi:10.1007/978-3-031-22269-6_9

Azeem, A., Ul-Allah, S., Sher, A., Ijaz, M., Sattar, A., Ahmad, W., & Qayyum, A. (2023). Biosafety and Biosecurity in Genetically Modified Crops. In Sustainable Agriculture in the Era of the OMICs Revolution (pp. 501-510). doi:10.1007/978-3-031-15568-0_23

Azenzem, R., Al Feddy, M.-N., & Kassout, J. (2023). Climate Change and Insect Pests: Economic Damages and Adaptive Strategies in Mediterranean Olive Groves. In Climate Change and the Economic Importance and Damages of Insects (pp. 14-47). IGI Global.

Azenzem, R., Alfeddy, M. N., & Kassout, J. (2023). Climate Change and Insect Pests: Economic Damages and Adaptive Strategies in Mediterranean Olive Groves. In Climate Change and the Economic Importance and Damages of Insects (pp. 14-47). IGI Global.

Azis, F. A., Choo, M., Suhaimi, H., & Abas, P. E. (2023). The effect of initial carbon to nitrogen ratio on kitchen waste composting maturity. *Sustainability (Basel)*, *15*(7), 6191. doi:10.3390u15076191

Azra, M. N., Okomoda, V. T., & Ikhwanuddin, M. (2022). Breeding technology as a tool for sustainable aquaculture production and ecosystem services. *Frontiers in Marine Science*, *9*, 679529. doi:10.3389/fmars.2022.679529

Bacelar, E. A., Moutinho-Pereira, J. M., Gonçalves, B. C., Ferreira, H. F., & Correia, C. M. (2007). Changes in growth, gas exchange, xylem hydraulic properties and water use effciency of three olive cultivars under contrasting water availability regimes. *Environmental and Experimental Botany*, *60*(2), 183–192. doi:10.1016/j.envexpbot.2006.10.003

Bacelar, E. A., Santos, D. L., Moutinho-Pereira, J. M., Gonçalves, B. C., Ferreira, H. F., & Correia, C. M. (2006). Immediate responses and adaptative strategies of three olive cultivars under contrasting water availability regimes: Changes on structure and chemical composition of foliage and oxidative damage. *Plant Science*, *170*(3), 596–605. doi:10.1016/j.plantsci.2005.10.014

Bahadur, S., Verma, S. K., Prasad, S. K., Madane, A. J., Maurya, S. P., Verma, V. K., & Sihag, S. K. (2015). Eco-friendly weed management for sustainable crop production - A review. *Journal of Crop and Weed*, *11*, 181–189.

Baibakova, E. V., Nefedjeva, E. E., Suska-Malawska, M., Wilk, M., Sevriukova, G. A., & Zheltobriukhov, V. F. (2019). Modern fungicides: Mechanisms of action, fungal resistance and phytotoxic effects. *Annual Research & Review in Biology*, 1–16. doi:10.9734/arrb/2019/v32i330083

Bailey-Serres, J., Parker, J. E., Ainsworth, E. A., Oldroyd, G. E., & Schroeder, J. I. (2019). Genetic strategies for improving crop yields. *Nature*, *575*(7781), 109–118. doi:10.103841586-019-1679-0 PMID:31695205

Bai, S., Chen, J., Guo, M., Ren, N., & Zhao, X. (2023). Vertical-scale spatial influence of radial oxygen loss on rhizosphere microbial community in a constructed wetland. *Environment International*, *171*, 107690. doi:10.1016/j.envint.2022.107690 PMID:36516673

Bajželj, B., & Richards, K. S. (2014). The Positive Feedback Loop between the Impacts of Climate Change and Agricultural Expansion and Relocation. *Land (Basel)*, *3*(3), 3. doi:10.3390/land3030898

Balafoutis, A., Beck, B., Fountas, S., Vangeyte, J., Van der Wal, T., Soto, I., & Eory, V. (2017). Precision agriculture technologies positively contributing to GHG emissions mitigation, farm productivity and economics. *Sustainability (Basel)*, *9*(8), 1339. doi:10.3390u9081339

Bale, J. S., & Hayward, S. A. L. (2010). Insect overwintering in a changing climate. *The Journal of Experimental Biology*, *213*(6), 980–994. doi:10.1242/jeb.037911 PMID:20190123

Bale, J. S., Masters, G. J., Hodkinson, I. D., Awmack, C., Bezemer, T. M., Brown, V. K., Butterfield, J., Buse, A., Coulson, J. C., Farrar, J., Good, J. E. G., Harrington, R., Hartley, S., Jones, T. H., Lindroth, R. L., Press, M. C., Symrnioudis, I., Watt, A. D., & Whittaker, J. B. (2002). Herbivory in global climate change research: Direct effects of rising temperature on insect herbivores. *Global Change Biology*, *8*(1), 1–16. doi:10.1046/j.1365-2486.2002.00451.x

Bale, J. S., van Lenteren, J. C., & Bigler, F. (2008). Biological control and sustainable food production. *Philosophical Transactions of the Royal Society of London. Series B, Biological Sciences*, *363*(1492), 761–776. doi:10.1098/rstb.2007.2182 PMID:17827110

Balestra, G. M., Lamichhane, J. R., Kshetri, M. B., Mazzaglia, A., & Varvaro, L. (2009). First report of olive knot caused byPseudomonas savastanoipv.savastanoiin Nepal. *Plant Pathology*, *58*(2), 393–393. doi:10.1111/j.1365-3059.2008.02007.x

Balla, A., Silini, A., Cherif-Silini, H., Chenari Bouket, A., Moser, W. K., Nowakowska, J. A., Oszako, T., Benia, F., & Belbahri, L. (2021). The Threat of Pests and Pathogens and the Potential for Biological Control in Forest Ecosystems. *Forests*, *12*(11), 11. doi:10.3390/f12111579

Banuls, A. L., Hide, M., & Prugnolle, F. (2007). Leishmania and the leishmaniases: A parasite genetic update and advances in taxonomy, epidemiology and pathogenicity in humans. *Advances in Parasitology*, *64*, 1–458. doi:10.1016/S0065-308X(06)64001-3 PMID:17499100

Barakat, N. A., Hameed, M. M. A., Fadali, O. A., Abdelraheem, O. H., Hefny, R. A., & Moustafa, H. M. (2023). Graphitized corncob 3D Biomass-driven anode for high-performance batch and continuous modes Air–Cathode microbial fuel cells working by domestic wastewater. *International Journal of Hydrogen Energy*, *48*(98), 38854–38869. doi:10.1016/j.ijhydene.2023.06.231

Barbier, E. B. (2007). Valuing ecosystem services as productive inputs. *Economic Policy*, *22*(49), 178–229. doi:10.1111/j.1468-0327.2007.00174.x

Barçante, B., Nascimento, N. O., Silva, T. F. G., Reis, L. A., & Giani, A. (2020). Cyanobacteria dynamics and phytoplankton species richness as a measure of waterbody recovery: Response to phosphorus removal treatment in a tropical eutrophic reservoir. *Ecological Indicators*, *117*, 106702. doi:10.1016/j.ecolind.2020.106702

Barriuso, E., Benoit, P., & Dubus, I. G. (2008). Formation of pesticide nonextractable (bound) residues in soil: Magnitude, controlling factors and reversibility. *Environmental Science & Technology*, *42*(6), 1845–1854. doi:10.1021/es7021736 PMID:18409603

Barros, M. V., Salvador, R., de Francisco, A. C., & Piekarski, C. M. (2020). Mapping of research lines on circular economy practices in agriculture: From waste to energy. *Renewable & Sustainable Energy Reviews*, *131*, 109958. doi:10.1016/j.rser.2020.109958

Barton, D. N., & Taron, A. (2010). Valuing irrigation water using survey-based methods in the Tungabhadra River Basin, India. *Irrigation and Drainage Systems*, *24*(1), 265–277. doi:10.100710795-010-9098-2

Bashyal, S., Poudel, D., & Gautam, B. (2022). A Review On Cultural Practice As An Effective Pest Management Approach Under Integrated Pest Management. [TAEC]. *Tropical Agroecosystems*, *3*(1), 34–40. doi:10.26480/taec.01.2022.34.40

Basri, W. N. F., Daud, H., Lam, M. K., Cheng, C. K., Oh, W. D., Tan, W. N., Shaharun, M. S., Yeong, Y. F., Paman, U., Kusakabe, K., Kadir, E., Show, & Lim. (2019). A sugarcane-bagasse-based adsorbent employed for mitigating eutrophication threats and producing biodiesel simultaneously. *Processes (Basel, Switzerland)*, *7*(9), 572. doi:10.3390/pr7090572

Batlles-delaFuente, A., Abad-Segura, E., González-Zamar, M. D., & Cortés-García, F. J. (2022). An evolutionary approach on the framework of circular economy applied to agriculture. *Agronomy (Basel)*, *12*(3), 620. doi:10.3390/agronomy12030620

Baumel, A., Mirleau, P., Viruel, J., Bou Dagher Kharrat, M., La Malfa, S., Ouahmane, L., Diadema, K., Moakhar, M., Sanguin, H., & Médail, F. (2018). Assessment of plant species diversity associated with the carob tree (Ceratonia siliqua, Fabaceae) at the Mediterranean scale. *Plant Ecology and Evolution*, *151*(2), 185–193. doi:10.5091/plecevo.2018.1423

Baumel, A., Nieto Feliner, G., Médail, F., La Malfa, S., Di Guardo, M., Bou Dagher Kharrat, M., Lakhal-Mirleau, F., Frelon, V., Ouahmane, L., Diadema, K., Sanguin, H., & Viruel, J. (2022). Genome-wide footprints in the carob tree (Ceratonia siliqua) unveil a new domestication pattern of a fruit tree in the Mediterranean. *Molecular Ecology*, *31*(15), 4095–4111. doi:10.1111/mec.16563 PMID:35691023

Baweja, P., Kumar, S., & Kumar, G. (2020). Fertilizers and pesticides: Their impact on soil health and environment. In Soil Biology (pp. 265–285). doi:10.1007/978-3-030-44364-1_15

Bazzana, D., Foltz, J., & Zhang, Y. (2022). Impact of climate smart agriculture on food security: An agent-based analysis. *Food Policy*, *111*, 102304. doi:10.1016/j.foodpol.2022.102304

Bebber, D. P., Holmes, T., & Gurr, S. J. (2014). The global spread of crop pests and pathogens. *Global Ecology and Biogeography*, *23*(12), 1398–1407. doi:10.1111/geb.12214

Beck, L. L., Young, K. R., Sutherland, C. A., & Smallidge, S. T. (2021). *Pest Management Overview*. College of Agricultural, Consumer and Environmental Sciences.

Beckman, N. G., Aslan, C. E., Rogers, H. S., Kogan, O., Bronstein, J. L., Bullock, J. M., Hartig, F., HilleRisLambers, J., Zhou, Y., Zurell, D., Brodie, J. F., Bruna, E. M., Cantrell, R. S., Decker, R. R., Efiom, E., Fricke, E. C., Gurski, K., Hastings, A., Johnson, J. S., & Zambrano, J. (2020). Advancing an interdisciplinary framework to study seed dispersal ecology. *AoB Plants*, *12*(2), plz048. doi:10.1093/aobpla/plz048 PMID:32346468

Behnassi, M., Baig, M. B., Sraïri, M. T., Alsheikh, A. A., & Risheh, A. W. A. A. (2022). Food Security and Climate-Climate Smart Food Systems—An Introduction. In Food Security and Climate-Smart Food Systems: Building Resilience for the Global South (pp. 1-13). Springer.

Bejarano-Alcázar, J., Blanco-López, M. A., Melero-Vara, J. M., & Jiménez-Díaz, R. M. (1996). Etiology, importance, and distribution of Verticillium wilt of cotton in southern Spain. *Plant Disease*, *80*(11), 1233–1238. doi:10.1094/PD-80-1233

Bekele, M., Mengistu, A., & Tamir, B. (2017). Livestock and feed water productivity in the mixed crop-livestock system. *Animal*, *11*(10), 1852–1860. doi:10.1017/S1751731117000416 PMID:28222817

Belazzoug, S., Addadi, K., Mokrani, T., Hafirassou, N., Hamrioui, B., & Belkaid, M. (1985). La leishmaniose viscérale en Algérie. Etude des cas hospitalisés entre 1975 et 1984. In Annales de la Société Belge de Médecine tropicale (Vol. 65, pp. 329-335.

Belete, T., Wato, T., & Bonga, E. (2020). Challenges and Prospects in the Adoption of Modern Agricultural Inputs in Gimbo District, Kaffa Region, Southern Ethiopia. *Public Policy and Administration Research*, *10*, 1–13.

Bellard, C., Bertelsmeier, C., Leadley, P., Thuiller, W., & Courchamp, F. (2012). Impacts of climate change on the future of biodiversity. *Ecology Letters*, *15*(4), 365–377. doi:10.1111/j.1461-0248.2011.01736.x PMID:22257223

Bell, R., & Bennett, A. (2022). The pros of integrated pest management. *Farmer's Weekly*, *2022*(22019), 46–48.

Ben Salem, S., Ben Salem, A., Karmaoui, A., & Khebiza, M. Y. (2023). Vulnerability of Water Resources to Drought Risk in Southeastern of Morocco: Case Study of Ziz Basin. *Water (Basel)*, *2023*(15), 4085. doi:10.3390/w15234085

Bender, S. F., Wagg, C., & van der Heijden, M. G. (2016). An underground revolution: Biodiversity and soil ecological engineering for agricultural sustainability. *Trends in Ecology & Evolution*, *31*(6), 440–452. doi:10.1016/j.tree.2016.02.016 PMID:26993667

Benitez, C., Tejada, M., & Gonzalez, J. (2003). Kinetics of the mineralization of nitrogen in a pig slurry compost applied to soils. *Composites Science and Technology*, *11*, 72–80.

Bennai, K. (2018). *Surveillance et contrôle des leishmanioses dans le nord de l'Algérie.* [Thèse de doctorat, université M'hamed Bougara, Boumerdas, Algérie]

Bennett, A. (2010). The role of soil community biodiversity in insect biodiversity. *Insect Conservation and Diversity*, *3*(3), 157–171. doi:10.1111/j.1752-4598.2010.00086.x

Bentzon-Tarp, A., Helgadóttir, D., Van den Meersche, K., Gay, F., Priemé, A., Roupsard, O., Mages, C., & Elberling, B. (2023). Spatial-temporal variations of nitrous oxide emissions in coffee agroforestry systems in Costa Rica. *Agriculture, Ecosystems & Environment*, *343*, 108257. doi:10.1016/j.agee.2022.108257

Benzidia, B., Barbouchi, M., Hammouch, H., Belahbib, N., Zouarhi, M., Erramli, H., Ait Daoud, N., Badrane, N., & Hajjaji, N. (2019). Chemical composition and antioxidant activity of tannins extract from green rind of Aloe vera (L.) Burm. F. *Journal of King Saud University. Science*, *31*(4), 1175–1181. doi:10.1016/j.jksus.2018.05.022

Bernal, M. P., Alburquerque, J. A., & Moral, R. (2009). Composting of animal manures and chemical criteria for compost maturity assessment. A review. *Bioresource Technology*, *100*(22), 5444–5453. doi:10.1016/j.biortech.2008.11.027

Bernstein, L. (1965). Salt tolerance of fruit crops (No. 292). Agriculture Research Service, US Department of Agriculture.

Berryman, E., Hatten, J., Page-Dumroese, D. S., Heckman, K. A., D'Amore, D. V., Puttere, J., SanClements, M., Connolly, S. J., & Perry, C. H. (Hobie), & Domke, G. M. (2020). Soil Carbon. In R. V. Pouyat, D. S. Page-Dumroese, T. Patel-Weynand, & L. H. Geiser (Eds.), Forest and Rangeland Soils of the United States Under Changing Conditions: A Comprehensive Science Synthesis (pp. 9–31). Springer International Publishing. doi:10.1007/978-3-030-45216-2_2

Berthet-Beaufils, A. (2010). *Manisfestations dermatologiques associés aux diptères chez le chien et le chat* [Doctoral dissertation].

Besnard, G., Terral, J. F., & Cornille, A. (2018). On the origins and domestication of the olive: A review and perspectives. *Annals of Botany*, *121*(3), 385–403. doi:10.1093/aob/mcx145 PMID:29293871

Bezemer, T. M., Jones, T. H., & Knight, K. J. (1998). Long-term effects of elevated CO 2 and temperature on populations of the peach potato aphid Myzus persicae and its parasitoid Aphidius matricariae. *Oecologia*, *116*(1-2), 128–135. doi:10.1007004420050571 PMID:28308517

Bhagwat, V. R. (2019). Safety of Water Used in Food Production. *Food Safety and Human Health*, 219–247. doi:10.1016/B978-0-12-816333-7.00009-6

Bhat, I. A., Fayaz, M., Rafiq, S., Guleria, K., Qadir, J., Wani, T. A., & Kaloo, Z. A. (2023). Predicting potential distribution and range dynamics of Aquilegia fragrans under climate change: Insights from ensemble species distribution modelling. *Environmental Monitoring and Assessment*, *195*(5), 623. doi:10.100710661-023-11245-2 PMID:37115430

Bhat, J. A., Ali, S., Salgotra, R. K., Mir, Z. A., Dutta, S., Jadon, V., Tyagi, A., Mushtaq, M., Jain, N., Singh, P. K., Singh, G. P., & Prabhu, K. V. (2016). Genomic selection in the era of next-generation sequencing for complex traits in plant breeding. *Frontiers in Genetics*, *7*. doi:10.3389/fgene.2016.00221 PMID:28083016

Bhattacharya, A., Parkhi, V., & Char, B. (2021). Genome editing for crop improvement: A perspective from India. *In Vitro Cellular & Developmental Biology. Plant*, *57*(4), 565–573. doi:10.100711627-021-10184-2 PMID:34075289

Biezunski, N. (2013). *Co-infection leishmaniose viscérale-infection VIH: A propos de cinq cas suivis dans le Service des Maladies Infectieuses et Tropicales du CHU de Toulouse, 2007- 2013*. [Thèse de doctorat. Université Toulouse III].

Biswas, D., & Venkatachalam, L. (2015). Farmers' willingness to pay for improved irrigation water – A case study of Malaprabha irrigation project in Karnataka, India. *Water Economics and Policy*, *1*(1), 1–24. doi:10.1142/S2382624X14500040

Biswas, S. K. (2016). Does the interdependence between oxidative stress and inflammation explain the antioxidant paradox? *Oxidative Medicine and Cellular Longevity*, *2016*, 1–9. doi:10.1155/2016/5698931 PMID:26881031

Biswas, S. K., Rahman, S., Kobir, S. A., Ferdous, T., & Banu, N. A. (2014). A review on impact of agrochemicals on human health and environment: Bangaldesh perspective. *Plant Environ Develop*, *3*, 31–35.

Bjornlund, H., & McKay, J. (1998). Factors affecting water price in a rural water market: A South Australian experience. *Water Resources Research*, *34*(6), 1563–1570. doi:10.1029/97WR03470

Bladon, K. D., Emelko, M. B., Silins, U., & Stone, M. (2014). Wildfire and the future of water supply. In: ACS Publications. doi:10.1021/es500130g

Blanc, E., & Noy, I. (2023). Impacts of droughts and floods on agricultural productivity in New Zealand as measured from space. *Environmental Research: Climate*, *2*(3), 035001. doi:10.1088/2752-5295/acd4da

Blenkinsop, S., Fowler, H., Dubus, I. G., Nolan, B. T., & Hollis, J. (2008). Developing climatic scenarios for pesticide fate modelling in Europe. *Environmental Pollution*, *154*(2), 219–231. doi:10.1016/j.envpol.2007.10.021 PMID:18045756

Blondel, J., Aronson, J., Bodiou, J.-Y., & Boeu, G. (2010). *The Mediterranean region: biological diversity through time and space*. Oxford University Press.

Blum, A. G., Null, C., & Hoffmann, V. (2014). Marketing households water treatment: Willingness to pay results from an experiment in Rural Kenya. *Water (Basel)*, *6*(7), 1873–1886. doi:10.3390/w6071873

Bolognesi, C., & Merlo, F. D. (2011). Pesticides: human health effects. In J. O. Nriagu (Ed.), *Encyclopedia of environmental health* (pp. 438–453). Elsevier. doi:10.1016/B978-0-444-52272-6.00592-4

Bongiorno, G., Bünemann, E. K., Oguejiofor, C. U., Meier, J., Gort, G., Comans, R., Mäder, P., Brussaard, L., & de Goede, R. (2019). Sensitivity of labile carbon fractions to tillage and organic matter management and their potential as comprehensive soil quality indicators across pedoclimatic conditions in Europe. *Ecological Indicators*, *99*, 38–50. doi:10.1016/j.ecolind.2018.12.008

Boopathi, N. M. (2020). Marker-Assisted Selection (MAS). In Springer eBooks (pp. 343–388). doi:10.1007/978-981-15-2949-8_9

Borgoyary, M. (2013). Valuation of water: Options for sustainable development in developing countries. *Water and Development*, *2*(1), 1-18. https://www.eolss.net/sample-chapters/c07/E2-24M-04-01.pdf

Bose, D. (2017). Big data analytics in Agriculture. *preparation of papers for IEEE Transactions and Journals*. IEEE.

Boulding, K. E. (2013). The economics of the coming spaceship earth. In *Environmental quality in a growing economy* (pp. 3–14). RFF Press.

Boulkenafet, F. (2006). Contribution à l'étude de la biodiversité des Phlébotomes (Diptera: Psychodidae) et appréciation de la faune Culicidienne (Diptera: Culicidae) dans la région de Skikda. Mémoire de master-Université Mentouri Constantine-Faculté des Sciences de la Nature et de la Vie.

Boullis, A., Francis, F., & Verheggen, F. J. (2015). Climate change and tritrophic interactions: Will modifications to greenhouse gas emissions increase the vulnerability of herbivorous insects to natural enemies? *Environmental Entomology*, *44*(2), 277–286. doi:10.1093/ee/nvu019 PMID:26313181

Boussaa, S. (2008). *Epidémiologie des leishmanioses dans la région de Marrakech, Maroc: effet de l'urbanisation sur la répartition spatio-temporelle des Phlébotomes et caractérisation moléculaire de leurs populations* [Doctoral dissertation, Université Louis Pasteur (Strasbourg)(1971-2008))].

Bréda, N., Huc, R., Granier, A., & Dreyer, E. (2006). Temperate forest trees and stands under severe drought: A review of ecophysiological responses, adaptation processes and long-term consequences. *Annals of Forest Science*, *63*(6), 625–644. doi:10.1051/forest:2006042

Breinl, K., Lun, D., Müller-Thomy, H., & Blöschl, G. (2021). Understanding the relationship between rainfall and flood probabilities through combined intensity-duration-frequency analysis. *Journal of Hydrology (Amsterdam)*, *602*, 126759. doi:10.1016/j.jhydrol.2021.126759

Brendel, O. (2021). The relationship between plant growth and water consumption: A history from the classical four elements to modern stable isotopes. *Annals of Forest Science*, *78*(2), 2. doi:10.100713595-021-01063-2

Brito, C., Dinis, L. T., Moutinho-Pereira, J., & Correia, C. M. (2019). Drought stress effects and olive tree acclimation under a changing climate. *Plants*, *8*(7), 232. doi:10.3390/plants8070232 PMID:31319621

Britto, C., Ravel, C., Bastien, P., Blaineau, C., Pagès, M., Dedet, J. P., & Wincker, P. (1998). Conserved linkage groups associated with large-scale chromosomal rearrangements between Old World and New World Leishmania genomes. *Gene*, *222*(1), 107–117. doi:10.1016/S0378-1119(98)00472-7 PMID:9813266

Brodribb, T. J., Powers, J., Cochard, H., & Choat, B. (2020). Hanging by a thread? Forests and drought. *Science*, *368*(6488), 261–266. doi:10.1126cience.aat7631 PMID:32299945

Brodt, S., Six, J., Feenstra, G., Ingels, C., & Campbell, D. (2011). Sustainable agriculture. *Nat. Educ. Knowl*, *3*(1). https://charlesscd.com/wp-content/uploads/2019/03/agriculture-and-the-environment-2019-Pennsylvania-Envirothon.pdf

Brown, M., De Beurs, K., & Marshall, M. (2012). Global phenological response to climate change in crop areas using satellite remote sensing of vegetation, humidity and temperature over 26 years. *Remote Sensing of Environment*, *126*, 174–183. doi:10.1016/j.rse.2012.08.009

Brucker-Davis, F. (1998). Effects of environmental synthetic chemicals on thyroid function. *Thyroid*, *8*(9), 827–856. doi:10.1089/thy.1998.8.827 PMID:9777756

Buck, H. J., & Palumbo-Compton, A. (2022). Soil carbon sequestration as a climate strategy: What do farmers think? *Biogeochemistry*, *161*(1), 59–70. doi:10.100710533-022-00948-2

Budhlakoti, N., Kushwaha, A. K., Rai, A., Chaturvedi, K. K., Kumar, A., Pradhan, A. K., Kumar, U., Kumar, R. R., Juliana, P., Mishra, D. C., & Kumar, S. (2022). Genomic selection: a tool for accelerating molecular breeding efficiency for developing Climate-Resilient crops. Frontiers in Genetics. doi:10.3389/fgene.2022.832153

Budiastuti, M. T. S., Purnomo, D., Pujiasmanto, B., & Setyaningrum, D. (2023). Response of Maize Yield and Nutrient Uptake to Indigenous Organic Fertilizer from Corn Cobs. *Agriculture*, *13*(2), 2. doi:10.3390/agriculture13020309

Bueno, A. M., & Jones, O. (2002). Alternative methods for controlling the olive fly, Bactrocera oleae, involving semiochemicals. *IOBC/WPRS Bulletin*, *25*(9), 147–156.

Burbi, S., & Olave, R. J. (2018). Supporting farmers in the transition to agroecology to promote carbon sequestration from silvopastoral systems. *International Journal of Agricultural Extension*, *6*(3), 17–27.

Burgarella, C., Barnaud, A., Kane, N. A., Jankowski, F., Scarcelli, N., Billot, C., Vigouroux, Y., & Berthouly-Salazar, C. (2019). Adaptive introgression: An untapped evolutionary mechanism for crop adaptation. *Frontiers in Plant Science*, *10*, 4. doi:10.3389/fpls.2019.00004 PMID:30774638

BurgielS.MuirA. (2010). Invasive Species, Climate Change and Ecosystem-Based Adaptation: Addressing Multiple Drivers of Global Change Global Invasive Species Programme. doi:10.13140/2.1.1460.8161

Burke, J. J., & Chen, J. (2015). Enhancement of reproductive heat tolerance in plants. *PLoS One*, *10*(4), e0122933. doi:10.1371/journal.pone.0122933 PMID:25849955

Burri, N. M., Weatherl, R., Moeck, C., & Schirmer, M. (2019). A review of threats to groundwater quality in the anthropocene. *The Science of the Total Environment*, *684*, 136–154. doi:10.1016/j.scitotenv.2019.05.236 PMID:31153063

Cai, D., Zhang, Z., Zhao, L., Liu, J., & Chen, H. (2022). A novel hybrid seed production technology based on a unilateral cross-incompatibility gene in maize. *Science China. Life Sciences*, *66*(3), 595–601. doi:10.100711427-022-2191-7 PMID:36190647

Çakmakçı, R., Salık, M. A., & Çakmakçı, S. (2023). Assessment and Principles of Environmentally Sustainable Food and Agriculture Systems. *Agriculture*, *13*(5), 5. doi:10.3390/agriculture13051073

Calicioglu, Ö., Flammini, A., Bracco, S., Bellú, L. G., & Sims, R. (2019). The Future Challenges of Food and Agriculture: An Integrated Analysis of Trends and Solutions. *Sustainability (Basel)*, *11*(1), 222. doi:10.3390u11010222

Cao, T., Zheng, Y., & Dong, H. (2023). Control of odour emissions from livestock farms: A review. *Environmental Research*, *225*, 115545. doi:10.1016/j.envres.2023.115545 PMID:36822532

Cappelli, F., Costantini, V., & Consoli, D. (2021). The trap of climate change-induced "natural" disasters and inequality. *Global Environmental Change*, *70*, 102329. doi:10.1016/j.gloenvcha.2021.102329

Capra, F., & Jakobsen, O. D. (2017). A conceptual framework for ecological economics based on systemic principles of life. *International Journal of Social Economics*, *44*(6), 831–844. doi:10.1108/IJSE-05-2016-0136

Cardinale, M., Luvisi, A., Meyer, J. B., Sabella, E., De Bellis, L., Cruz, A. C., Ampatzidis, Y., & Cherubini, P. (2018). Specific fluorescence in situ hybridization (FISH) test to highlight colonization of xylem vessels by Xylella fastidiosa in naturally infected olive trees (Olea europaea L.). *Frontiers in Plant Science*, *9*, 431. doi:10.3389/fpls.2018.00431 PMID:29681910

Caruso, G., Gucci, R., Urbani, S., Esposto, S., Taticchi, A., Di Maio, I., Selvaggini, R., & Servili, M. (2014). Effect of different irrigation volumes during fruit development on quality of virgin olive oil of cv. Frantoio. *Agricultural Water Management*, *134*, 94–103. doi:10.1016/j.agwat.2013.12.003

Carvajal-Yepes, M., Cardwell, K., Nelson, A., Garrett, K. A., Giovani, B., Saunders, D. G., Kamoun, S., Legg, J. P., Verdier, V., Lessel, J., Neher, R. A., Day, R., Pardey, P., Gullino, M. L., Records, A. R., Bextine, B., Leach, J. E., Staiger, S., & Tohme, J. (2019). A global surveillance system for crop diseases. *Science*, *364*(6447), 1237–1239. doi:10.1126cience.aaw1572 PMID:31249049

Casida, J. E., & Durkin, K. A. (2017). Pesticide chemical research in toxicology: Lessons from nature. *Chemical Research in Toxicology*, *30*(1), 94–104. doi:10.1021/acs.chemrestox.6b00303 PMID:27715053

Castaño-Rosa, R., Pelsmakers, S., Järventausta, H., Poutanen, J., Tähtinen, L., Rashidfarokhi, A., & Toivonen, S. (2022). Resilience in the built environment: Key characteristics for solutions to multiple crises. *Sustainable Cities and Society*, *87*, 104259. doi:10.1016/j.scs.2022.104259

Castroverde, C. D. M., & Dina, D. (2021). Temperature regulation of plant hormone signaling during stress and development. *Journal of Experimental Botany*, *72*(21), 7436–7458. doi:10.1093/jxb/erab257 PMID:34081133

Cavicchioli, R., Ripple, W. J., Timmis, K. N., Azam, F., Bakken, L. R., Baylis, M., Behrenfeld, M. J., Boetius, A., Boyd, P. W., Classen, A. T., Crowther, T. W., Danovaro, R., Foreman, C. M., Huisman, J., Hutchins, D. A., Jansson, J. K., Karl, D. M., Koskella, B., Mark Welch, D. B., & Webster, N. S. (2019). Scientists' warning to humanity: Microorganisms and climate change. *Nature Reviews. Microbiology*, *17*(9), 569–586. doi:10.103841579-019-0222-5 PMID:31213707

Cen, Y., Guo, L., Liu, M., Gu, X., Li, C., & Jiang, G. (2020). Using organic fertilizers to increase crop yield, economic growth, and soil quality in a temperate farmland. *PeerJ*, *8*, e9668. doi:10.7717/peerj.9668 PMID:32879791

Chakraborty, S., & Datta, S. (2003). How will plant pathogens adapt to host plant resistance at elevated CO2 under a changing climate? *The New Phytologist*, *159*(3), 733–742. doi:10.1046/j.1469-8137.2003.00842.x PMID:33873600

Chakroun, M., Banyuls, N., Bel, Y., Escriche, B., & Ferré, J. (2016). Bacterial vegetative insecticidal proteins (Vip) from entomopathogenic bacteria. *Microbiology and Molecular Biology Reviews*, *80*(2), 329–350. doi:10.1128/MMBR.00060-15 PMID:26935135

Chambers, A., Lal, R., & Paustian, K. (2016). Soil carbon sequestration potential of US croplands and grasslands: Implementing the 4 per Thousand Initiative. *Journal of Soil and Water Conservation*, *71*(3), 68A–74A. doi:10.2489/jswc.71.3.68A

Chandler, A. J., Drummond, F. A., Collins, J. A., Lund, J., & Alnajjar, G. (2020). Exposure of the common Eastern bumble bee, Bombus impatiens (Cresson), to sub-lethal doses of acetamiprid and propiconazole in wild blueberry. *Journal of Agricultural and Urban Entomology*, *36*(1), 1–23. doi:10.3954/1523-5475-36.1.1

Chandrasekara, A., & Josheph Kumar, T. (2016). Roots and tuber crops as functional foods: A review on phytochemical constituents and their potential health benefits. *International Journal of Food Sciences*, *2016*, 2016. doi:10.1155/2016/3631647 PMID:27127779

Chandrasekaran, K., Devarajulu, S., & Kuppannan, P. (2009). Farmers' willingness to pay for irrigation water: A case of tank irrigation systems in south India. *Water (Basel)*, *1*(1), 5–18. doi:10.3390/w1010005

Chartzoulakis, K. S. (2005). Salinity and olive: growth, salt tolerance, photosynthesis and yield. *Agricultural Water Management, 78*(1– 2), 108– 121.

Chartzoulakis, K., Loupassaki, M., Bertaki, M., & Androulakis, I. (2002). Effects of NaCl salinity on growth, ion content and C02 assimilation rate of six olive cultivars. *Scientia Horticulturae*, *96*(1-4), 235–247. doi:10.1016/S0304-4238(02)00067-5

Chartzoulakis, K., Psarras, G., Vemmos, S., Loupassaki, M., & Bertaki, M. (2006). Response of two olive cultivars to salt stress and potassium supplement. *Journal of Plant Nutrition*, *29*(11), 2063–2078. doi:10.1080/01904160600932682

Cheng, K., Xu, X., Cui, L., Li, Y., Zheng, J., Wu, W., Sun, J., & Pan, G. (2021). The role of soils in regulation of freshwater and coastal water quality. Philosophical Transactions of the *Royal Society B: Biological Sciences, 376*(1834). doi:10.1098/rstb.2020.0176

Cheng, Y. T., Zhang, L., & He, S. Y. (2019). Plant-microbe interactions facing environmental challenge. *Cell Host & Microbe*, *26*(2), 183–192. doi:10.1016/j.chom.2019.07.009 PMID:31415751

Chen, L., Cong, R. G., Shu, B., & Mi, Z. F. (2017). A sustainable biogas model in China: The case study of Beijing Deqingyuan biogas project. *Renewable & Sustainable Energy Reviews*, *78*, 773–779. doi:10.1016/j.rser.2017.05.027

Chen, M., Wang, Q. Y., Cheng, X. G., Xu, Z. S., Li, L. C., Ye, X. G., Xu, Z. S., Li, L. C., Ye, X. G., Xia, L. Q., & Ma, Y. Z. (2007). GmDREB2, a soybean DRE-binding transcription factor, conferred drought and high-salt tolerance in transgenic plants. *Biochemical and Biophysical Research Communications*, *353*(2), 299–305. doi:10.1016/j.bbrc.2006.12.027 PMID:17178106

Chen, M., Zhang, H., Liu, W., & Zhang, W. (2013). The global pattern of urbanization and economic growth: Evidence from the last three decades. *PLoS One*, *9*(8), 1–15. doi:10.1371/journal.pone.0103799 PMID:25099392

Chen, W. B., Lu, G. Q., Cheng, H. M., Liu, C. X., Xiao, Y. T., Xu, C., Shen, Z. C., & Wu, K. M. (2017). Transgenic cotton coexpressing Vip3A and Cry1Ac has a broad insecticidal spectrum against lepidopteran pests. *Journal of Invertebrate Pathology*, *149*, 59–65. doi:10.1016/j.jip.2017.08.001 PMID:28782511

Chermiti, B. (1989). Dynamique des populations du psylle de l'olivier E. olivina, en conditions méditerranéennes [Thèse Doctorat Es-Sciences, Université d'AixMarseille, France].

Chiabai, A., Travisi, C. M., Markandya, A., Ding, H., & Nunes, P. A. L. D. (2011). Economic assessment of forest ecosystem services losses: Cost of policy inaction. *Environmental and Resource Economics*, *50*(3), 405–445. doi:10.100710640-011-9478-6

Chidawanyika, F., Mudavanhu, P., & Nyamukondiwa, C. (2012). Biologically based methods for pest management in agriculture under changing climates: Challenges and future directions. *Insects*, *3*(4), 1171–1189. doi:10.3390/insects3041171 PMID:26466733

Chinchilla, N., Carrera, C., Durán, A. G., Macías, M., Torres, A., & Macías, F. A. (2013). Aloe barbadensis: How a miraculous plant becomes reality. *Phytochemistry Reviews*, *12*(4), 581–602. doi:10.100711101-013-9323-3

Chloupek, O., & Hrstková, P. (2005). Adaptation of crops to the environment. *Theoretical and Applied Genetics*, *111*(7), 1316–1321. doi:10.100700122-005-0060-x PMID:16187121

Cho, R. (2018). Agriculture and Climate: How Climate Change will Alter Our Food, State of the Planet. *Earth Institute, Colombia University.* https://blogs. ei. columbia. edu/2018/07/25/climate-change-food-agriculture.

Choat, B., Brodribb, T. J., Brodersen, C. R., Duursma, R. A., López, R., & Medlyn, B. E. (2018). Triggers of tree mortality under drought. *Nature*, *558*(7711), 531–539. doi:10.103841586-018-0240-x PMID:29950621

Choat, B., Jansen, S., Brodribb, T. J., Cochard, H., Delzon, S., Bhaskar, R., Bucci, S. J., Feild, T. S., Gleason, S. M., Hacke, U. G., Jacobsen, A. L., Lens, F., Maherali, H., Martínez-Vilalta, J., Mayr, S., Mencuccini, M., Mitchell, P. J., Nardini, A., Pittermann, J., & Zanne, A. E. (2012). Global convergence in the vulnerability of forests to drought. *Nature*, *491*(7426), 752–755. doi:10.1038/nature11688 PMID:23172141

Cho, G., Lee, J. H., & Choi, D. S. (2021). First interception of olive psyllid Euphyllura olivina (Costa, 1839)(Hemiptera: Psylloidea) in the East Palaearctic region. *Phytoparasitica*, *49*(5), 865–868. doi:10.100712600-021-00924-w

Choi, H. K., Iandolino, A., da Silva, F. G., & Cook, D. R. (2013). Water deficit modulates the response of Vitis vinifera to the Pierce's disease pathogen Xylella fastidiosa. *Molecular Plant-Microbe Interactions*, *26*(6), 643–657. doi:10.1094/MPMI-09-12-0217-R PMID:23425100

Choi, Y. M., Shin, D. H., & Kim, C. H. (2016). An optimized methodology to observe internal microstructures of Aloe vera by cryo-scanning electron microscope. *Applied Microscopy*, *46*(2), 76–78. doi:10.9729/AM.2016.46.2.76

Chopra, R., Burow, G., Burke, J. J., Gladman, N., & Xin, Z. (2017). Genome-wide association analysis of seedling traits in diverse Sorghum germplasm under thermal stress. *BMC Plant Biology*, *17*(1), 1–15. doi:10.118612870-016-0966-2 PMID:28086798

Choudhary, J. R., Tripathi, A., Kaldate, R., Rana, M., Mehta, S., Ahlawat, J., & Wani, S. H. (2022). Breeding efforts for crop productivity in abiotic stress environment. In *Augmenting crop productivity in stress environment* (pp. 63–103). Springer Nature Singapore. doi:10.1007/978-981-16-6361-1_5

Choukri, H., El Haddad, N., Aloui, K., Hejjaoui, K., El-Baouchi, A., Smouni, A., Thavarajah, D., Maalouf, F., & Kumar, S. (2022). Effect of High Temperature Stress During the Reproductive Stage on Grain Yield and Nutritional Quality of Lentil (Lens culinaris Medikus). *Frontiers in Nutrition*, *9*, 725. doi:10.3389/fnut.2022.857469 PMID:35495922

Chowdhury, K., & Behera, B. (2021). Economic significance of provisioning ecosystem services of traditional water bodies: Empirical evidence from West Bengal, India. *Resources. Environmental Sustainability*, *5*(1), 1–10. doi:10.1016/j.resenv.2021.100033

Cipollini, D. (2004). Stretching the limits of plasticity: Can a plant defend against both competitors and herbivores? *Ecology*, *85*(1), 28–37. doi:10.1890/02-0615

Côa, F., Bortolozzo, L. S., Petry, R., Da Silva, G. H., Martins, C. H., de Medeiros, A. M., Delite, F. S. (2020). Environmental toxicity of nanopesticides against non-target organisms: the state of the art. *Nanopesticides: From research and development to mechanisms of action and sustainable use in agriculture*, 227-279.

Coakley, S. M., Scherm, H., & Chakraborty, S. (1999). Climate change and plant disease management. *Annual Review of Phytopathology*, *37*(1), 399–426. doi:10.1146/annurev.phyto.37.1.399 PMID:11701829

Cockerton, H. M., Karlström, A., Johnson, A. W., Li, B., Stavridou, E., Hopson, K., Whitehouse, A. B., & Harrison, R. J. (2021). Genomic informed breeding strategies for strawberry yield and fruit quality traits. *Frontiers in Plant Science*, *12*, 724847. doi:10.3389/fpls.2021.724847 PMID:34675948

Cohen, B., & Winn, M. I. (2007). Market imperfections, opportunity and sustainable entrepreneurship. *Journal of Business Venturing*, *22*(1), 29–49. doi:10.1016/j.jbusvent.2004.12.001

Cohen, I., Zandalinas, S. I., Huck, C., Fritschi, F. B., & Mittler, R. (2021). Meta- Analysis of Drought and Heat Stress Combination Impact on Crop Yield and Yield Components. *Physiologia Plantarum*, *171*(1), 66–76. doi:10.1111/ppl.13203 PMID:32880977

Cohen, S. P., & Leach, J. E. (2020). High temperature-induced plant disease susceptibility: More than the sum of its parts. *Current Opinion in Plant Biology*, *56*, 235–241. doi:10.1016/j.pbi.2020.02.008 PMID:32321671

Connor, D. J., & Fereres, E. (2005). The physiology of adaptation and yield expression in olive. *Horticultural Reviews*, *31*, 155–229.

Conway, G. (2019). *The doubly green revolution: food for all in the twenty-first century*. Cornell University Press.

Coolen, S., Van Pelt, J. A., Van Wees, S. C., & Pieterse, C. M. (2019). Mining the natural genetic variation in Arabidopsis thaliana for adaptation to sequential abiotic and biotic stresses. *Planta*, *249*(4), 1087–1105. doi:10.100700425-018-3065-9 PMID:30547240

Cooper, J., & Dobson, H. (2007). The benefits of pesticides to mankind and the environment. *Crop Protection (Guildford, Surrey)*, *26*(9), 1337–1348. doi:10.1016/j.cropro.2007.03.022

Corbeels, M., Cardinael, R., Naudin, K., Guibert, H., & Torquebiau, E. (2019). The 4 per 1000 goal and soil carbon storage under agroforestry and conservation agriculture systems in sub-Saharan Africa. *Soil & Tillage Research, 188*, 16–26. doi:10.1016/j.still.2018.02.015

Cornara, D., Saponari, M., Zeilinger, A. R., de Stradis, A., Boscia, D., Loconsole, G., Bosco, D., Martelli, G. P., Almeida, R. P. P., & Porcelli, F. (2017). Spittlebugs as vectors of Xylella fastidiosa in olive orchards in Italy. *Journal of Pest Science, 90*(2), 521–530. doi:10.100710340-016-0793-0 PMID:28275326

Cosgrove, W. J., & Loucks, D. P. (2015). Water management: Current and future challenges and research directions. *Water Resources Research, 51*(1), 4823–4839. doi:10.1002/2014WR016869

Costa, L. G., Giordano, G., Guizzetti, M., & Vitalone, A. (2008). Neurotoxicity of pesticides: A brief review. *Frontiers in Bioscience, 13*(4), 1240–1249. doi:10.2741/2758 PMID:17981626

Costanza, R. (2010, May 11). *What is ecological economics?* Yale Insights. https://insights.som.yale.edu/insights/what-is-ecological-economics

Costanza, R., Haskell, B., Cornwell, L., Daly, H., & Johnson, T. (1990). *The Ecological Economics of Sustainability: Making Local and Short-Term Goals Consistent with Global and Long-Term Goals* (Environment Working Paper No. 32).

Costanza, R., d'Arge, R., de Groot, R., Farber, S., Grasso, M., Hannon, B., Limburg, K., Naeem, S., O'Neill, R. V., Paruelo, J., Raskin, R. G., Sutton, P., & van den Belt, M. (1997). The value of the world's ecosystem services and natural capital. *Nature, 387*(6630), 253–260. doi:10.1038/387253a0

Costanza, R., Kubiszewski, I., & Franco, C. (Eds.). (2007). *Introduction to Ecological Economics*. CRC Press.

Cotrina Cabello, G. G., Ruiz Rodriguez, A., Husnain Gondal, A., Areche, F. O., Flores, D. D. C., Astete, J. A. Q., Camayo-Lapa, B. F., Yapias, R. J. M., Jabbar, A., Yovera Saldarriaga, J., Salas-Contreras, W. H., & Cruz Nieto, D. D. (2023). Plant adaptability to climate change and drought stress for crop growth and production. *Perspectives in Agriculture, Veterinary Science, Nutrition and Natural Resources, 2023*, cabireviews.2023.0004. doi:10.1079/cabireviews.2023.0004

Council, E. U. (2005). *Council Regulation (EC) No 1698/2005 on support for rural development by the European Agricultural Fund for Rural Development (EAFRD)*. Official J Eur Union.

Cousins, S. R., & Witkowski, E. T. F. (2012). African aloe ecology: A review. *Journal of Arid Environments, 85*, 1–17. doi:10.1016/j.jaridenv.2012.03.022

Cramer, L. K. (2019). Access to early generation seed: Obstacles for delivery of climate-smart varieties. *The climate-smart agriculture papers: Investigating the business of a productive, resilient and low emission future*, 87-98. doi:10.1007/978-3-319-92798-5_8

Cramer, G. R., Urano, K., Delrot, S., Pezzotti, M., & Shinozaki, K. (2011). Effects of abiotic stress on plants: A systems biology perspective. *BMC Plant Biology, 11*(1), 1–14. doi:10.1186/1471-2229-11-163 PMID:22094046

Cramer, W., Guiot, J., Fader, M., Garrabou, J., Gattuso, J. P., Iglesias, A., Lange, M. A., Lionello, P., Llasat, M. C., Paz, S., Peñuelas, J., Snoussi, M., Toreti, A., Tsimplis, M. N., & Xoplaki, E. (2018). Climate change and interconnected risks to sustainable development in the Mediterranean. *Nature Climate Change, 8*(11), 972–980. doi:10.103841558-018-0299-2

Craufurd, P. Q., & Wheeler, T. R. (2009). Climate change and the flowering time of annual crops. *Journal of Experimental Botany, 60*(9), 2529–2539. doi:10.1093/jxb/erp196 PMID:19505929

Crickmore, N., Zeigler, D. R., Schnepf, E., Van Rie, J., Lereclus, D., Baum, J., Bravo, A., & Dean, D. H. (2016). *Bacillus thuringiensis toxin nomenclature*.

Cristiano, G., Murillo-Amador, B., & De Lucia, B. (2016). Propagation techniques and agronomic requirements for the cultivation of Barbados Aloe (Aloe vera (L.) Burm. F.)—A Review. *Frontiers in Plant Science*, *7*, 1410. doi:10.3389/fpls.2016.01410 PMID:27721816

Cristofano, F., El-Nakhel, C., & Rouphael, Y. (2021). Biostimulant substances for sustainable agriculture: Origin, operating mechanisms and effects on Cucurbits, leafy greens, and nightshade vegetables species. *Biomolecules*, *11*(8), 1103. doi:10.3390/biom11081103 PMID:34439770

Crossa, J., Fritsche-Neto, R., Montesinos-Lopez, O. A., Costa-Neto, G., Dreisigacker, S., Montesinos-Lopez, A., & Bentley, A. R. (2021). The modern plant breeding Triangle: Optimizing genomics, phenomics, and environmental data. *Frontiers in Plant Science*, *12*, 651480. doi:10.3389/fpls.2021.651480 PMID:33936136

Cuevas, N., Martins, M., & Costa, P. M. (2018). Risk assessment of pesticides in estuaries: A review addressing the persistence of an old problem in complex environments. *Ecotoxicology (London, England)*, *27*(7), 1008–1018. doi:10.100710646-018-1910-z PMID:29450674

D'Amato, D., Droste, N., Allen, B., Kettunen, M., Lähtinen, K., Korhonen, J., Leskinen, P., Matthies, B. D., & Toppinen, A. (2017). Green, circular, bio economy: A comparative analysis of sustainability avenues. *Journal of Cleaner Production*, *168*, 716–734. doi:10.1016/j.jclepro.2017.09.053

Daba, M. H., & Dejene, S. W. (2018). The role of biodiversity and ecosystem services in carbon sequestration and its implication for climate change mitigation. *Environmental Sciences and Natural Resources*, *11*(2), 1–10.

Dabbou, S., Dabbou, S., Chehab, H., Taticchi, A., Servili, M., & Hammami, M. (2015). Content of Fatty Acids and Phenolics in *Coratina* Olive Oil from Tunisia: Influence of Irrigation and Ripening. *Chemistry & Biodiversity*, *12*(3), 397–406. doi:10.1002/cbdv.201400142 PMID:25766913

Dag, A., Bustan, A., Avni, A., Tzipori, I., Lavee, S., & Riov, J. (2010). Timing of fruit removal affects concurrent vegetative growth and subsequent return bloom and yield in olive (Olea europaea L.). *Scientia Horticulturae*, *123*(4), 469–472. doi:10.1016/j.scienta.2009.11.014

Dagnew, Y., Urge, M., Tadesse, Y., & Gizaw, S. (2018). Conservation-based breeding program designed for genetic improvement in Gumz sheep in the western lowlands of Ethiopia. *Agriculture & Food Security*, *7*(1), 34. doi:10.118640066-018-0173-5

Daly, H. E. (1999). *Ecological economics and the ecology of economics: Essays in criticism*. Edward Elgar Publishing.

Daly, H. E., & Farley, J. (2010). *Ecological economics: Principles and applications* (2nd ed.). Island Press.

Dar, M. H., Zaidi, N. W., Waza, S. A., Verulkar, S. B., Ahmed, T., Singh, P. K., Roy, S. K. B., Chaudhary, B., Yadav, R., Islam, M. M., Iftekharuddaula, K. M., Roy, J. K., Kathiresan, R. M., Singh, B. N., Singh, U. S., & Ismail, A. M. (2018). No yield penalty under favorable conditions paving the way for successful adoption of flood tolerant rice. *Scientific Reports*, *8*(1), 1–7. doi:10.103841598-018-27648-y PMID:29915310

Daryanto, S., Wang, L., & Jacinthe, P. A. (2016). Global synthesis of drought effects on maize and wheat production. *PLoS One*, *11*(5), e0156362. doi:10.1371/journal.pone.0156362 PMID:27223810

Das Dangol, S., Hashmi, M. H., Saeed, F., Yel, I., Öztürk, A., & Bakhsh, A. (2021). Utilizing RNA-Based Approaches to Understand Plant-Insect Interactions. *RNA-Based Technologies for Functional Genomics in Plants*, 393-428. doi:10.1007/978-3-030-64994-4_17

Dasgupta, P., & Dasgupta, R. (2004). Economic value of safe water for infrastructurally disadvamtaged urban household: A case study in Delhi, India. *Water Resources Research*, *40*(11401), 1–10. doi:10.1029/2003WR002461

Daszkowska-Golec, A., & Szarejko, I. (2013). Open or close the gate–stomata action under the control of phytohormones in drought stress conditions. *Frontiers in Plant Science*, *4*, 138. doi:10.3389/fpls.2013.00138 PMID:23717320

Datta, A. (2013). Genetic engineering for improving the quality and productivity of crops. *Agriculture & Food Security*, *2*(1), 15. doi:10.1186/2048-7010-2-15

Davis, G. W., Richardson, D. M., Keeley, J. E., & Hobbs, R. J. (1996). Mediterranean-type ecosystems: the influence of biodiversity on their functioning. *Functional roles of biodiversity: a global perspective.*, 151-183.

Davis, K. F., Downs, S., & Gephart, J. A. (2021). Towards food supply chain resilience to environmental shocks. *Nature Food*, *2*(1), 54–65. doi:10.103843016-020-00196-3 PMID:37117650

De Buyck, P.-J., Van Hulle, S. W., Dumoulin, A., & Rousseau, D. P. (2021). Roof runoff contamination: A review on pollutant nature, material leaching and deposition. *Reviews in Environmental Science and Biotechnology*, *20*(2), 549–606. doi:10.100711157-021-09567-z

De Jong, P., Tanajura, C. A. S., Sánchez, A. S., Dargaville, R., Kiperstok, A., & Torres, E. A. (2018). Hydroelectric production from Brazil's São Francisco River could cease due to climate change and inter-annual variability. *The Science of the Total Environment*, *634*, 1540–1553. doi:10.1016/j.scitotenv.2018.03.256 PMID:29710652

De Pascali, M., Vergine, M., Negro, C., Greco, D., Vita, F., Sabella, E., De Bellis, L., & Luvisi, A. (2022). Xylella fastidiosa and drought stress in olive trees: A complex relationship mediated by soluble sugars. *Biology (Basel)*, *11*(1), 112. doi:10.3390/biology11010112 PMID:35053110

De Rodrıguez, D. J., Hernández-Castillo, D., Rodrıguez-Garcıa, R., & Angulo-Sánchez, J. L. (2005). Antifungal activity in vitro of Aloe vera pulp and liquid fraction against plant pathogenic fungi. *Industrial Crops and Products*, *21*(1), 81–87. doi:10.1016/j.indcrop.2004.01.002

De Schampheleire, M., Spanoghe, P., Brusselman, E., & Sonck, S. (2007). Risk assessment of pesticide spray drift damage in Belgium. *Crop Protection (Guildford, Surrey)*, *26*(4), 602–611. doi:10.1016/j.cropro.2006.05.013

Dean, T. J., & McMullen, J. S. (2007). Toward a theory of sustainable entrepreneurship: Reducing environmental degradation through entrepreneurial action. *Journal of Business Venturing*, *22*(1), 50–76. doi:10.1016/j.jbusvent.2005.09.003

Debnath, T., Ghosh, M., Lee, Y. M., Nath, N. C. D., Lee, K. G., & Lim, B. O. (2018). Identification of phenolic constituents and antioxidant activity of Aloe barbadensis flower extracts. *Food and Agricultural Immunology*, *29*(1), 27–38. doi:10.1080/09540105.2017.1358254

DeBoe, G. (2020). *Economic and environmental sustainability performance of environmental policies in agriculture.*

Dedet, J. P. (1999). *Les leishmanioses*. Edition Ellipses.

Dedet, J. P. (2001). Leishmanies, leishmanioses. Biologie, clinique et thérapeutique. *Encycl. Med. Chir. Maladies Infectieuses*, *8*, 11.

Dedet, J. P. (2009). *Leishmanies, leishmanioses: biologie, clinique et thérapeutique*. Editions Scientifiques et Médicales, Elsevier SAS.

Deguine, J.-P., Aubertot, J.-N., Flor, R. J., Lescourret, F., Wyckhuys, K. A., & Ratnadass, A. (2021). Integrated pest management: Good intentions, hard realities. A review. *Agronomy for Sustainable Development*, *41*(3), 38. doi:10.100713593-021-00689-w

Delcour, I., Spanoghe, P., & Uyttendaele, M. (2015). Literature review: Impact of climate change on pesticide use. *Food Research International*, *68*, 7–15. doi:10.1016/j.foodres.2014.09.030

Delgado-Baquerizo, M., Guerra, C. A., Cano-Díaz, C., Egidi, E., Wang, J.-T., Eisenhauer, N., Singh, B. K., & Maestre, F. T. (2020). The proportion of soil-borne pathogens increases with warming at the global scale. *Nature Climate Change*, *10*(6), 550–554. doi:10.103841558-020-0759-3

DeLonge, M. S., Ryals, R., & Silver, W. L. (2013). A lifecycle model to evaluate carbon sequestration potential and greenhouse gas dynamics of managed grasslands. *Ecosystems (New York, N.Y.)*, *16*(6), 962–979. doi:10.100710021-013-9660-5

DeLucia, E. H., Casteel, C. L., Nabity, P. D., & O'Neill, B. F. (2008). Insects take a bigger bite out of plants in a warmer, higher carbon dioxide world. *Proceedings of the National Academy of Sciences of the United States of America*, *105*(6), 1781–1782. doi:10.1073/pnas.0712056105 PMID:18256185

Den Bieman, C. F. M., Malenovský, I., Burckhardt, D., & Heijerman, T. (2019). First checklist of the Dutch jumping plant lice since 93 years (Hemiptera: Psylloidea). *Nederlandse Faunistische Mededelingen*, *53*, 55–118.

Denancé, N., Legendre, B., Briand, M., Olivier, V., De Boisseson, C., Poliakoff, F., & Jacques, M. A. (2017). Several subspecies and sequence types are associated with the emergence of Xylella fastidiosa in natural settings in France. *Plant Pathology*, *66*(7), 1054–1064. doi:10.1111/ppa.12695

Des Roches, S., Post, D. M., Turley, N. E., Bailey, J. K., Hendry, A. P., Kinnison, M. T., Schweitzer, J. A., & Palkovacs, E. P. (2018). The ecological importance of intraspecific variation. *Nature Ecology & Evolution*, *2*(1), 57–64. doi:10.103841559-017-0402-5 PMID:29203921

Desjeux, P. (1999). Global control and Leishmania HIV co-infection. *Clinics in Dermatology*, *17*(3), 317–325. doi:10.1016/S0738-081X(99)00050-4 PMID:10384871

Desjeux, P. (2004). Leishmaniasis: Current situation and new perspectives. *Comparative Immunology, Microbiology and Infectious Diseases*, *27*(5), 305–318. doi:10.1016/j.cimid.2004.03.004 PMID:15225981

Deutsch, C. A., Tewksbury, J. J., Tigchelaar, M., Battisti, D. S., Merrill, S. C., Huey, R. B., & Naylor, R. L. (2018). Increase in crop losses to insect pests in a warming climate. *Science*, *361*(6405), 916–919. doi:10.1126cience.aat3466 PMID:30166490

DeVay, J. E., Forrester, L. L., Garber, R. H., & Butterfield, E. J. (1974). Characteristics and Concentration of Propagules of. *Phytopathology*, *64*(1), 22–29. doi:10.1094/Phyto-64-22

Devi, J. S., Joseph, B., Karunakaran, K., Anurdha, B., & Devi, K. R. (2009). People's attitudes towards paying for water. *Current Science*, *97*(9), 1296–1302.

Devi, N., Sarma, K. K., & Laskar, S. (2023). Design of an intelligent bean cultivation approach using computer vision, IoT and spatio-temporal deep learning structures. *Ecological Informatics*, *75*, 102044. doi:10.1016/j.ecoinf.2023.102044

Devi, P. I., Manjula, M., & Bhavani, R. (2022). Agrochemicals, environment, and human health. *Annual Review of Environment and Resources*, *47*(1), 399–421. doi:10.1146/annurev-environ-120920-111015

Dhungana, A. R., & Baral, B. (2016). Factors affecting willingness to pay for improved water supply system in rural Tanahu, Nepal. *Janapriya Journal of Interdsciplinary Studies*, *5*(1), 1–13. doi:10.3126/jjis.v5i0.17836

Dhyani, S. K., Ram, A., Newaj, R., Handa, A. K., & Dev, I. (2020). Agroforestry for carbon sequestration in tropical India. *Carbon management in tropical and sub-tropical terrestrial systems*, 313-331.

Diacono, M., & Montemurro, F. (2011). Long-term effects of organic amendments on soil fertility. *Sustainable agriculture volume 2*, 761-786.

Diallo, K. (2014). *Leishmaniose cutanée: profil épidémiologique, clinique, thérapeutique et évolutif. A propos de, 87 cas recensés dans le service de dermatologie de l'institut d'hygiène sociale de Dakar.* [Thèse de doctorat en médecine, université des sciences des techniques et des technologies de Bamako, Mali].

Diamond, J. M. (2006). *The third chimpanzee: The evolution and future of the human animal* (Reissued, [repr.]). Harper Perennial.

Diaz, S., Hodgson, J. G., Thompson, K., Cabido, M., Cornelissen, J. H., Jalili, A., Montserrat-Martí, G., Grime, J. P., Zarrinkamar, F., Asri, Y., Band, S. R., Basconcelo, S., Castro-Díez, P., Funes, G., Hamzehee, B., Khoshnevi, M., Pérez-Harguindeguy, N., Pérez-Rontomé, M. C., Shirvany, F. A., & Zak, M. R. (2004). The plant traits that drive ecosystems: Evidence from three continents. *Journal of Vegetation Science*, *15*(3), 295–304. doi:10.1111/j.1654-1103.2004.tb02266.x

Díaz, S., Kattge, J., Cornelissen, J. H., Wright, I. J., Lavorel, S., Dray, S., Reu, B., Kleyer, M., Wirth, C., Colin Prentice, I., Garnier, E., Bönisch, G., Westoby, M., Poorter, H., Reich, P. B., Moles, A. T., Dickie, J., Gillison, A. N., Zanne, A. E., & Gorné, L. D. (2016). The global spectrum of plant form and function. *Nature*, *529*(7585), 167–171. doi:10.1038/nature16489 PMID:26700811

Dichio, B., Xiloyannis, C., Sofo, A., & Montanaro, G. (2006). Osmotic regulation in leaves and roots of olive trees during a water defcit and rewatering. *Tree Physiology*, *26*(2), 179–185. doi:10.1093/treephys/26.2.179 PMID:16356914

Dignac, M. F., Derrien, D., Barré, P., Barot, S., Cécillon, L., Chenu, C., Chevallier, T., Freschet, G. T., Garnier, P., Guenet, B., Hedde, M., Klumpp, K., Lashermes, G., Maron, P.-A., Nunan, N., Roumet, C., & Basile-Doelsch, I. (2017). Increasing soil carbon storage: Mechanisms, effects of agricultural practices and proxies. A review. *Agronomy for Sustainable Development*, *37*(2), 1–27. doi:10.100713593-017-0421-2

Dimande, P., Arrobas, M., & Rodrigues, M. Â. (2023). Under a Tropical Climate and in Sandy Soils, Bat Guano Mineralises Very Quickly, Behaving More like a Mineral Fertiliser than a Conventional Farmyard Manure. *Agronomy (Basel)*, *13*(5), 5. doi:10.3390/agronomy13051367

Dimkpa, C. O., Fugice, J., Singh, U., & Lewis, T. D. (2020). Development of fertilizers for enhanced nitrogen use efficiency–Trends and perspectives. *The Science of the Total Environment*, *731*, 139113. doi:10.1016/j.scitotenv.2020.139113 PMID:32438083

Dimnwobi, S. K., Okere, K. I., Onuoha, F. C., & Ekesiobi, C. (2023). Energy poverty, environmental degradation and agricultural productivity in Sub-Saharan Africa. *International Journal of Sustainable Development and World Ecology*, *30*(4), 428–444. doi:10.1080/13504509.2022.2158957

Dolmatova, A. V., Demina, N. A., & Kobylansky, A. (1971). *Les phlébotomes (Phlebotominae) et les maladies qu'ils transmettent* (No. 18). Office de la Recherche Scientifique et Technique Outre-. *La Mer (Paris)*, *1971*, 168p.

Dominici, L., Magi, E., Leidi, B., Pastore, M. E., & Comino, E. (2022). Ecologically-oriented business strategy for a small-size rice farm: Integrated wetland management for the improvement of environmental benefits and economic feasibility. *The Science of the Total Environment*, *838*, 156604. doi:10.1016/j.scitotenv.2022.156604 PMID:35691347

Drakvik, E., Altenburger, R., Aoki, Y., Backhaus, T., Bahadori, T., Barouki, R., & Bennekou, S. H. (2020). Statement on advancing the assessment of chemical mixtures and their risks for human health and the environment. *Environment International*, *134*, 105267. doi:10.1016/j.envint.2019.105267 PMID:31704565

Driouech, F., M., Déqué, & E. Sánchez- Gómez. (2010). Weather regimes— Moroccan precipitation link in a regional climate change simulation. *Global and Planetary Change, 72*(1– 2), 1– 10.

Driouech, F., Déqué, M., & Sánchez-Gómez, E. (2010). Weather regimes—Moroccan precipitation link in a regional climate change simulation. *Global and Planetary Change*, *72*(1-2), 1–10. doi:10.1016/j.gloplacha.2010.03.004

Du Jardin, P. (2015). Plant biostimulants: Definition, concept, main categories and regulation. *Scientia Horticulturae, 196*, 3–14. doi:10.1016/j.scienta.2015.09.021

Dumont, B., Fortun-Lamothe, L., Jouven, M., Thomas, M., & Tichit, M. (2013). Prospects from agroecology and industrial ecology for animal production in the 21st century. *animal, 7*(6), 1028-1043. doi:10.1017/S1751731112002418

Dumortier, J., Dokoohaki, H., Elobeid, A., Hayes, D. J., Laird, D., & Miguez, F. E. (2020). Global land-use and carbon emission implications from biochar application to cropland in the United States. *Journal of Cleaner Production, 258*, 120684. doi:10.1016/j.jclepro.2020.120684

Du, P., Xu, M., & Li, R. (2021). Impacts of climate change on water resources in the major countries along the Belt and Road. *PeerJ, 9*, e12201. doi:10.7717/peerj.12201 PMID:34721962

Dutta, A., Trivedi, A., Nath, C. P., Gupta, D. S., & Hazra, K. K. (2022). A comprehensive review on grain legumes as climate-smart crops: Challenges and prospects. *Environmental Challenges, 7*, 100479. doi:10.1016/j.envc.2022.100479

Dutta, V., Chander, S., & Srivastava, L. (2005). Public support for water supply improvements: Empirical evidence from unplanned settlements of Delhi, India. *Journal of Environment & Development, 14*(4), 439–462. doi:10.1177/1070496505281841

Dutta, V., & Tiwari, A. P. (2005). Cost of service and willingness to pay for reliable urban water supply: A study from Delhi, India. *Water Science and Technology: Water Supply, 5*(6), 135–144. doi:10.2166/ws.2005.0058

Du, Z. L., Zhao, J. K., Wang, Y. D., & Zhang, Q. Z. (2017). Biochar addition drives soil aggregation and carbon sequestration in aggregate fractions from an intensive agricultural system. *Journal of Soils and Sediments, 17*(3), 581–589. doi:10.100711368-015-1349-2

Du, Z., Angers, D. A., Ren, T., Zhang, Q., & Li, G. (2017). The effect of no-till on organic C storage in Chinese soils should not be overemphasized: A meta-analysis. *Agriculture, Ecosystems & Environment, 236*, 1–11. doi:10.1016/j.agee.2016.11.007

Dwivedi, Y. K., Hughes, L., Kar, A. K., Baabdullah, A. M., Grover, P., Abbas, R., Andreini, D., Abumoghli, I., Barlette, Y., Bunker, D., Chandra Kruse, L., Constantiou, I., Davison, R. M., De', R., Dubey, R., Fenby-Taylor, H., Gupta, B., He, W., Kodama, M., & Wade, M. (2022). Climate change and COP26: Are digital technologies and information management part of the problem or the solution? An editorial reflection and call to action. *International Journal of Information Management, 63*, 102456. doi:10.1016/j.ijinfomgt.2021.102456

Ebi, K. L., Vanos, J., Baldwin, J. W., Bell, J. E., Hondula, D. M., Errett, N. A., Hayes, K., Reid, C. E., Saha, S., Spector, J., & Berry, P. (2021). Extreme Weather and Climate Change: Population Health and Health System Implications. *Annual Review of Public Health, 42*(1), 293–315. doi:10.1146/annurev-publhealth-012420-105026 PMID:33406378

Edwards, C. A., & Arancon, N. Q. (2022). The Role of Earthworms in Organic Matter and Nutrient Cycles. In C. A. Edwards & N. Q. Arancon (Eds.), *Biology and Ecology of Earthworms* (pp. 233–274). Springer US. doi:10.1007/978-0-387-74943-3_8

Egea, F. J., Torrente, R. G., & Aguilar, A. (2018). An efficient agro-industrial complex in Almería (Spain): Towards an integrated and sustainable bioeconomy model. *New Biotechnology, 40*, 103–112. doi:10.1016/j.nbt.2017.06.009 PMID:28698129

Eggli, U., & Nyffeler, R. (2009). Living under temporarily arid conditions-succulence as an adaptive strategy. *Bradleya, 2009*(27), 13–36. doi:10.25223/brad.n27.2009.a10

Eid, H. M., El-Marsafawy, S. M., & Ouda, S. A. (2007). Assessing the economic impacts of climate change on agriculture in Egypt: a Ricardian approach. *World Bank Policy Research Working Paper*(4293).

El Chami, D., Daccache, A., & El Moujabber, M. (2020). How Can Sustainable Agriculture Increase Climate Resilience? A Systematic Review. *Sustainability (Basel)*, *12*(8), 8. doi:10.3390u12083119

El Handi, K., Hafidi, M., Sabri, M., Frem, M., El Moujabber, M., Habbadi, K., Haddad, N., Benbouazza, A., Abou Kubaa, R., & Achbani, E. H. (2022). Continuous pest surveillance and monitoring constitute a tool for sustainable agriculture: Case of Xylella fastidiosa in Morocco. *Sustainability (Basel)*, *14*(3), 1485. doi:10.3390u14031485

Elad, Y., & Pertot, I. (2014). Climate Change Impacts on Plant Pathogens and Plant Diseases. *Journal of Crop Improvement*, *28*(1), 99–139. doi:10.1080/15427528.2014.865412

Elahi, E., Khalid, Z., Tauni, M. Z., Zhang, H., & Lirong, X. (2022). Extreme weather events risk to crop-production and the adaptation of innovative management strategies to mitigate the risk: A retrospective survey of rural Punjab, Pakistan. *Technovation*, *117*, 102255. doi:10.1016/j.technovation.2021.102255

Elbaroudi, H. (2014). *Eco-épidémiologie de la leishmaniose viscérale au Maroc* [Thèse de doctorat en pharmacie, faculté de médecine et de pharmacie, Université Mohammed V- Souissi. Maroc].

Elbasiouny, H., Elbanna, B. A., Al-Najoli, E., Alsherief, A., Negm, S., El-Nour, E. A., Nofal, A., & Sharabash, S. (2020). Agricultural Waste Management for Climate Change Mitigation: Some Implications to Egypt. In A. Negm & N. Shareef (Eds.), *Waste Management in MENA Regions. Springer Water*. Springer. doi:10.1007/978-3-030-18350-9_8

Elbasiouny, H., El-Ramady, H., Elbehiry, F., Rajput, V. D., Minkina, T., & Mandzhieva, S. (2022). Plant nutrition under climate change and soil carbon sequestration. *Sustainability (Basel)*, *14*(2), 914. doi:10.3390u14020914

Elisha, I. L., & Viljoen, A. (2021). Trends in Rooibos Tea (Aspalathus linearis) research (1994–2018): A scientometric assessment. *South African Journal of Botany*, *137*, 159–170. doi:10.1016/j.sajb.2020.10.004

Ellen MacArthur Foundation. (2016). *Cities and circular economy for food*. Ellen MacArthur Foundation. https://www.ellenmacarthurfoundation.org/publications/cities-andcircular-economy-for-food

Elliott, J. A. (2010). *An Introduction to Sustainable Development* (3rd ed.).

El-mrini S, Aboutayeb R (2022) Co-composting of olive mill pomace and turkey manure: theoretical risk analysis towards a HACCP plan model. *Environ Eng Manag J 21*(12),1987–2001 doi:.2022.176. doi:10.30638/eemj

El-mrini, S., Aboutayeb, R., & Zouhri, A. (2022). Effect of initial C/N ratio and turning frequency on quality of final compost of turkey manure and olive pomace. *Journal of Engineering and Applied Sciences (Asian Research Publishing Network)*, *69*(1), 1–20.

El-Salam, A., Ezzat, A. M., Salem, S. A. W., Abdel-Rahman, R. S., El-Behery, H. H., & Magd Elden, M. A. (2019). Effects of climatic changes on olive fly, Bactrocera oleae (Rossi) population dynamic with respect to the efficacy of its larval parasitoid in Egyptian olive trees. *Bulletin of the National Research Center*, *43*(1), 1–9. doi:10.118642269-019-0220-9

El-ssawy, W., Darwish, A., & Hassanien, A. E. (2023). The Role of Artificial Intelligence in Water Management in Agriculture for Climate Change Impacts. In The Power of Data: Driving Climate Change with Data Science and Artificial Intelligence Innovations (pp. 221-238). Springer. doi:10.1007/978-3-031-22456-0_13

Embuscado, M. E. (2015). Spices and herbs: Natural sources of antioxidants–a mini review. *Journal of Functional Foods*, *18*, 811–819. doi:10.1016/j.jff.2015.03.005

Epple, C., Rangel, García, Jenkins, S., & Guth, M, M. (2016). Managing ecosystems in the context of climate change mitigation: A review of current knowledge and recommendations to support ecosystem-based mitigation actions that look beyond terrestrial forests. *CBD Technical Series*, *86*, 5.

Erenso, F., & Andemo, A. (2022). Vulnerability of Agrobiodiversity and Agroforestry Settings to Climate Change in Gedeo Zone, Ethiopia. *International Journal of Agronomy*, *2022*, 2022. doi:10.1155/2022/8738482

Ergon, Å., Skøt, L., Sæther, V. E., & Rognli, O. A. (2019). Allele Frequency Changes Provide Evidence for Selection and Identification of Candidate Loci for Survival in Red Clover (Trifolium pratense L.). *Frontiers in Plant Science*, *10*, 718. doi:10.3389/fpls.2019.00718 PMID:31244867

Er-Raki, S., Chehbouni, A., Khabba, S., Simonneaux, V., Jarlan, L., Ouldbba, A., Rodriguez, J. C., & Allen, R. (2010). Assessment of reference evapotranspiration methods in semi-arid regions: Can weather forecast data be used as alternate of ground meteorological parameters? *Journal of Arid Environments*, *74*(12), 1587–1596. doi:10.1016/j.jaridenv.2010.07.002

Espinosa, O. A., Serrano, M. G., Camargo, E. P., Teixeira, M. M. G., & Shaw, J. J. (2018). An appraisal of the taxonomy and nomenclature of trypanosomatids presently classified as Leishmania and Endotrypanum. *Parasitology*, *145*(4), 430–442. doi:10.1017/S0031182016002092 PMID:27976601

Esua, M. F., & Rauwald, J. W. (2006). Novel bioactive maloyl glucans from Aloe vera gel: Isolation, structure elucidation and in vitro bioassays. *Carbohydrate Research*, *341*(3), 355–364. doi:10.1016/j.carres.2005.11.022 PMID:16343466

European Commission. (2013a). COMMISSION IMPLEMENTING REGULATION (EU) No 485/2013 amending Implementing Regulation (EU) No 540/2011, as regards the conditions of approval of the active substances clothianidin, thiamethoxam and imidacloprid, and prohibiting the use. *Official J Euro Union*, *139*, 12–26.

European Commission. (2013b). COMMISSION IMPLEMENTING REGULATION (EU) No 781/2013 of 14 August 2013 amending Implementing Regulation (EU) No 540/2011, as regards the conditions of approval of the active substance fipronil, and prohibiting the use and sale of seeds treated with plant p. *Official J Eur Union*, *219*, 22–25.

European Commission. (2020). *Genetic resources and breeding*. Research and Innovation. https://research-and-innovation.ec.europa.eu/research-area/agriculture-forestry-and-rural-areas/genetic-resources-and-breeding_en

European Food Safety Authority (EFSA), Delbianco, A., Gibin, D., Pasinato, L., & Morelli, M. (2021). Update of the Xylella spp. host plant database–systematic literature search up to 31 December 2020. *EFSA Journal*, *19*(6), e06674. PMID:34188716

Evangelista, C., Olden, J. D., Lecerf, A., & Cucherousset, J. (2019). Scale-dependent patterns of intraspecific trait variations in two globally invasive species. *Oecologia*, *189*(4), 1083–1094. doi:10.100700442-019-04374-4 PMID:30824999

Evanylo, G. K., Sherony, C. A., Spargo, J., Starner, D. E., Brosius, M., & Haering, K. (2008). Soil and water environmental effects of fertilizer-, manure-, and compost-based fertility practices in an organic vegetable cropping system. *Agriculture, Ecosystems & Environment*, *127*(1–2), 50–58. doi:10.1016/j.agee.2008.02.014

Eze, E., Girma, A., Zenebe, A., Okolo, C. C., Kourouma, J. M., & Negash, E. (2022). Predictors of drought-induced crop yield/losses in two agroecologies of southern Tigray, Northern Ethiopia. *Scientific Reports*, *12*(1), 6284. doi:10.103841598-022-09862-x PMID:35428373

Fageria, N. K. (2012). Role of soil organic matter in maintaining sustainability of cropping systems. *Communications in Soil Science and Plant Analysis*, *43*(16), 2063–2113. doi:10.1080/00103624.2012.697234

Fahad, S., & Wang, J. (2018). Farmers' risk perception, vulnerability, and adaptation to climate change in rural Pakistan. *Land Use Policy*, *79*, 301–309. doi:10.1016/j.landusepol.2018.08.018

Fajardo, A., & Siefert, A. (2018). *Intraspecific trait variation and the leaf economics spectrum across resource gradients and levels of organization.*

Fajardo, A., & Piper, F. I. (2011). Intraspecific trait variation and covariation in a widespread tree species (Nothofagus pumilio) in southern Chile. *The New Phytologist, 189*(1), 259–271. doi:10.1111/j.1469-8137.2010.03468.x PMID:21039558

Fajardo, A., & Siefert, A. (2016). Phenological variation of leaf functional traits within species. *Oecologia, 180*(4), 951–959. doi:10.100700442-016-3545-1 PMID:26796408

Falkenmark, M., & Rockström, J. (2006). *The new blue and green water paradigm: Breaking new ground for water resources planning and management* (Vol. 132). American Society of Civil Engineers.

Fan, W., Chong, C., Xiaoling, G., Hua, Y., & Juyun, W. (2015, December). Prediction of crop yield using big data. In *2015 8th International Symposium on Computational Intelligence and Design (ISCID)* (Vol. 1, pp. 255-260). IEEE. 10.1109/ISCID.2015.191

Fand, B. B., Kamble, A. L., & Kumar, M. (2012). Will climate change pose serious threat to crop pest management: A critical review. *International journal of scientific and Research publications, 2*(11), 1-14. https://www.academia.edu/download/40340132/ijsrp-nov-2012-print.pdf#page=52

Fang, X., Turner, N. C., Yan, G., Li, F., & Siddique, K. H. (2010). Flower numbers, pod production, pollen viability, and pistil function are reduced and flower and pod abortion increased in chickpea (*Cicer arietinum* L.) under terminal drought. *Journal of Experimental Botany, 61*(2), 335–345. doi:10.1093/jxb/erp307 PMID:19854801

Fan, L., Wang, F., Liu, G., Yang, X., & Qin, W. (2014). Public perception of water consumption and its effects on water conservation behavior. *Water (Basel), 6*(1), 1771–1784. doi:10.3390/w6061771

FAO. (2010). *The second report on the state of the World's Plant Genetic Resources for Food and Agriculture*. Commission on Genetic Resources for Food.

FAO. (2020). *NSP – Integrated Pest Management, FAO definition*. FAO. http:// www.fao.org/agriculture/crops/thematic-sitemap/theme/pests/ipm/ en/

FAO. (2023). *How to Practice Integrated Pest Management*. FAO. https://www.fao.org/agriculture/crops/thematic-sitemap/theme/compendium/tools-guidelines/how-to-ipm/en/

FAOSTAT. (2021). Food and Agriculture Organization of the United Nations Statistical Dataset. FAO: Rome, Italy.

Farigoule, P., Chartois, M., Mesmin, X., Lambert, M., Rossi, J. P., Rasplus, J. Y., & Cruaud, A. (2022). Vectors as Sentinels: Rising Temperatures Increase the Risk of Xylella fastidiosa Outbreaks. *Biology (Basel), 11*(9), 1299. doi:10.3390/biology11091299 PMID:36138778

Farooq, M., M. Hussain, A. Wahid, & K. H. M. Siddique. (2012). Drought stress in plants: an overview. *Plant responses to drought stress* 1– 33

Farraro, P. J., Lawlor, K., Mullan, K. L., & Pattanayak, S. K. (2011). Forest figures: Ecosystem services valuation and policy evaluation in developing countries. *Review of Environmental Economics and Policy, 1*(1), 1–26. doi:10.1093/reep/rer019

Fathi-Taperasht, A., Shafizadeh-Moghadam, H., Minaei, M., & Xu, T. (2022). Influence of drought duration and severity on drought recovery period for different land cover types: Evaluation using MODIS-based indices. *Ecological Indicators, 141*, 109146. doi:10.1016/j.ecolind.2022.109146

Favoino, E., & Hogg, D. (2008). The potential role of compost in reducing greenhouse gases. *Waste Management & Research*, *26*(1), 61–69. doi:10.1177/0734242X08088584

Fedoroff, N. V., Battisti, D. S., Beachy, R. N., Cooper, P. J., Fischhoff, D. A., Hodges, C. N. et al. (2010). Radically rethinking agriculture for the 21st century. *science, 327*(5967), 833-834.

Fereres, E., Goldhamer, D., Sadras, V., & ... (2012). Yield response to water of fruit trees and vines: Guidelines. *FAO Irrigation and Drainage Paper*, *66*, 246–497.

Fernandez-Cornejo, J., Wechsler, S., Livingston, M., & Mitchell, L. (2014). Genetically engineered crops in the United States. *SSRN*, (162). Advance online publication. doi:10.2139srn.2503388

Fernández, J. E. (2014). Understanding olive adaptation to abiotic stresses as a tool to increase crop performance. *Environmental and Experimental Botany*, *103*, 158–179. doi:10.1016/j.envexpbot.2013.12.003

Fernández, J. E., Moreno, F., Cabrera, F., Arrue, J. L., & Martín-Aranda, J. (1991). Drip irrigation, soil characteristics and the root distribution and root activity of olive trees. *Plant and Soil*, *133*(2), 239–251. doi:10.1007/BF00009196

Fernández, J. E., Perez-Martin, A., Torres-Ruiz, J. M., Cuevas, M. V., Rodriguez-Dominguez, C. M., Elsayed-Farag, S., Morales-Sillero, A., García, J. M., Hernandez-Santana, V., & Diaz-Espejo, A. (2013). A regulated deficit irrigation strategy for hedgerow olive orchards with high plant density. *Plant and Soil*, *372*(1–2), 279–295. doi:10.100711104-013-1704-2

Figueiredo, C. C. D., Resck, D. V. S., & Carneiro, M. A. C. (2010). Labile and stable fractions of soil organic matter under management systems and native cerrado. *Revista Brasileira de Ciência do Solo*, *34*(3), 907–916. doi:10.1590/S0100-06832010000300032

Fletcher, B. S. (1989). Temperature-development rate relationships of immature stages and adult of tephritid fruit flies. In A. S. Robinson and G. Hooper [eds.], Fruit flies: their biology, natural enemies, and control. Elsevier, Amsterdam.

Foley, H., & Vialatte, C. (1914). Existence dans le Sud marocain du bouton d'orient à l'état endémique. *Bulletin de la Société de Pathologie Exotique*, *7*, 114–115.

Folino, A., Karageorgiou, A., Calabrò, P. S., & Komilis, D. (2020). Biodegradation of wasted bioplastics in natural and industrial environments: A review. *Sustainability (Basel)*, *12*(15), 6030. doi:10.3390u12156030

Fones, H. N., Bebber, D. P., Chaloner, T. M., Kay, W. T., Steinberg, G., & Gurr, S. J. (2020). Threats to global food security from emerging fungal and oomycete crop pathogens. *Nature Food*, *1*(6), 332–342. doi:10.103843016-020-0075-0 PMID:37128085

Forrer, H.-R., Musa, T., Schwab, F., Jenny, E., Bucheli, T. D., Wettstein, F. E., & Vogelgsang, S. (2014). Fusarium head blight control and prevention of mycotoxin contamination in wheat with botanicals and tannic acid. *Toxins*, *6*(3), 830–849. doi:10.3390/toxins6030830 PMID:24577585

Forzieri, G., Miralles, D. G., Ciais, P., Alkama, R., Ryu, Y., Duveiller, G., Zhang, K., Robertson, E., Kautz, M., Martens, B., Jiang, C., Arneth, A., Georgievski, G., Li, W., Ceccherini, G., Anthoni, P., Lawrence, P., Wiltshire, A., Pongratz, J., & Cescatti, A. (2020). Increased control of vegetation on global terrestrial energy fluxes. *Nature Climate Change*, *10*(4), 356–362. doi:10.103841558-020-0717-0

Fouque, F., & Reeder, J. C. (2019). Impact of past and on-going changes on climate and weather on vector-borne diseases transmission: A look at the evidence. *Infectious Diseases of Poverty*, *8*(03), 1–9. doi:10.118640249-019-0565-1 PMID:31196187

Fraga, H., Pinto, J. G., & Santos, J. A. (2020b). Olive tree irrigation as a climate change adaptation measure in Alentejo, Portugal. *Agricultural Water Management, 237*, 106193. https://doi.org/ doi:10.1016/j.agwat.2020.106193Girona

Fraga-Corral, M., Ronza, P., Garcia-Oliveira, P., Pereira, A. G., Losada, A. P., Prieto, M. A., Quiroga, M. I., & Simal-Gandara, J. (2022). Aquaculture as a circular bio-economy model with Galicia as a study case: How to transform waste into revalorized by-products. *Trends in Food Science & Technology, 119*, 23–35. doi:10.1016/j.tifs.2021.11.026

Fraga, H., Moriondo, M., Leolini, L., & Santos, J. A. (2020a). Mediterranean Olive Orchards under Climate Change: A Review of Future Impacts and Adaptation Strategies. *Agronomy (Basel), 11*(1), 56. doi:10.3390/agronomy11010056

Fraga, H., Pinto, J. G., Viola, F., & Santos, J. A. (2020). Climate change projections for olive yields in the Mediterranean Basin. *International Journal of Climatology, 40*(2), 769–781. doi:10.1002/joc.6237

Frampton, G. K., & Dorne, J. L. C. (2007). The effects on terrestrial invertebrates of reducing pesticide inputs in arable crop edges: A meta-analysis. *Journal of Applied Ecology, 44*(2), 362–373. doi:10.1111/j.1365-2664.2007.01277.x

Francis, M. R., Nagarajan, G., Sarkar, R., Mohan, V. R., Kang, G., & Balraj, V. (2015). Perception of drinking water safety and factors influencing acceptance and sustainability of a water quality intervention in rural southern India. *BMC Public Health, 15*(731), 1–9. doi:10.118612889-015-1974-0 PMID:26223687

Franzluebbers, A. J., & Martin, G. (2022). Farming with forages can reconnect crop and livestock operations to enhance circularity and foster ecosystem services. *Grass and Forage Science, 77*(4), 270–281. doi:10.1111/gfs.12592

Franzoni, G., Cocetta, G., Prinsi, B., Ferrante, A., & Espen, L. (2022). Biostimulants on crops: Their impact under abiotic stress conditions. *Horticulturae, 8*(3), 189. doi:10.3390/horticulturae8030189

Freedman, D. H. (2013). Are engineered foods evil? *Scientific American, 309*(3), 80–89. https://www.jstor.org/stable/26017991. doi:10.1038cientificamerican0913-80 PMID:24003560

Freibauer, A., Rounsevell, M. D., Smith, P., & Verhagen, J. (2004). Carbon sequestration in the agricultural soils of Europe. *Geoderma, 122*(1), 1–23. doi:10.1016/j.geoderma.2004.01.021

Freimuth, J., Bossdorf, O., Scheepens, J. F., & Willems, F. M. (2022). Climate warming changes synchrony of plants and pollinators. *Proceedings of the Royal Society B: Biological Sciences, 289*(1971), 20212142. 10.1098/rspb.2021.2142

Fridahl, M., & Lehtveer, M. (2018). Bioenergy with carbon capture and storage (BECCS): Global potential, investment preferences, and deployment barriers. *Energy Research & Social Science, 42*, 155–165. doi:10.1016/j.erss.2018.03.019

Fukao, T., Barrera-Figueroa, B. E., Juntawong, P., & Peña-Castro, J. M. (2019). Submergence and waterlogging stress in plants: A review highlighting research opportunities and understudied aspects. *Frontiers in Plant Science, 10*, 340. doi:10.3389/fpls.2019.00340 PMID:30967888

Funk, C., Peterson, P., Landsfeld, M., Pedreros, D., Verdin, J., Shukla, S., Husak, G., Rowland, J., Harrison, L., Hoell, A., & Michaelsen, J. (2015). The climate hazards infrared precipitation with stations—A new environmental record for monitoring extremes. *Scientific Data, 2*(1), 1–21. doi:10.1038data.2015.66 PMID:26646728

Gagic, V., Holding, M., Venables, W. N., Hulthen, A. D., & Schellhorn, N. A. (2021). Better outcomes for pest pressure, insecticide use, and yield in less intensive agricultural landscapes. *Proceedings of the National Academy of Sciences of the United States of America, 118*(12), e2018100118. doi:10.1073/pnas.2018100118 PMID:33731476

Gagic, V., Kleijn, D., Báldi, A., Boros, G., Jørgensen, H. B., Elek, Z., & Kovács-Hostyánszki, A. (2017). Combined effects of agrochemicals and ecosystem services on crop yield across Europe. *Ecology Letters, 20*(11), 1427–1436. doi:10.1111/ele.12850 PMID:28901046

Galati, E. A. B. (1995). Phylogenetic systematics of Phlebotominae (Diptera, Psychodidae) with emphasis on American groups. *Boletim de la Direccion de Malariologia Y Saneamiento Ambiental, 35*(1), 133-42.

Galluzzi, G., Seyoum, A., Halewood, M., Noriega, I. L., & Welch, E. W. (2020). Genetic resources' role in climate change breeding: The case of public breeding programmes in eighteen developing countries. *Plants*, *9*(9), 1129. doi:10.3390/plants9091129 PMID:32878309

Gao, X., & Zhao, D. (2022). Impacts of climate change on vegetation phenology over the Great Lakes Region of Central Asia from 1982 to 2014. *The Science of the Total Environment*, *845*, 157227. doi:10.1016/j.scitotenv.2022.157227 PMID:35809736

Garantonakis, N., Varikou, K., & Birouraki, A. (2017). Parasitism of Psytallia concolor (Hymenoptera: Braconidae) on Bactrocera oleae (Diptera: Tephritidae) infesting different olive varieties. *Phytoparasitica*, *45*(4), 461–469. doi:10.100712600-017-0597-7

Garibaldi, L. A., Gemmill-Herren, B., D'Annolfo, R., Graeub, B. E., Cunningham, S. A., & Breeze, T. D. (2017). Farming approaches for greater biodiversity, livelihoods, and food security. *Trends in Ecology & Evolution*, *32*(1), 68–80. doi:10.1016/j.tree.2016.10.001 PMID:27793463

Garland, T. Jr, Downs, C. J., & Ives, A. R. (2022). Trade-offs (and constraints) in organismal biology. *Physiological and Biochemical Zoology*, *95*(1), 82–112. doi:10.1086/717897 PMID:34905443

Garnier, E., Navas, M. L., & Grigulis, K. (2016). *Plant functional diversity: organism traits, community structure, and ecosystem properties*. Oxford University Press.

Garrigou, A., Laurent, C., Berthet, A., Colosio, C., Jas, N., Daubas-Letourneux, V., & Baldi, I. (2020). Critical review of the role of PPE in the prevention of risks related to agricultural pesticide use. *Safety Science*, *123*, 104527. doi:10.1016/j.ssci.2019.104527

Gavahi, K., Abbaszadeh, P., & Moradkhani, H. (2021). DeepYield: A combined convolutional neural network with long short-term memory for crop yield forecasting. *Expert Systems with Applications*, *184*, 115511. doi:10.1016/j.eswa.2021.115511

Gavrilescu, M. (2021). Water, soil, and plants interactions in a threatened environment. *Water (Basel)*, *13*(19), 2746. doi:10.3390/w13192746

Geber, M. A., & Griffen, L. R. (2003). Inheritance and natural selection on functional traits. *International Journal of Plant Sciences*, *164*(S3), S21–S42. doi:10.1086/368233

Ge, L., Sun, Y., Li, Y., Wang, L., Guo, G., Song, L., Wang, C., Wu, G., Zang, X., Cai, X., Li, S., & Li, P. (2023). Ecosystem sustainability of rice and aquatic animal co-culture systems and a synthesis of its underlying mechanisms. *The Science of the Total Environment*, *880*, 163314. doi:10.1016/j.scitotenv.2023.163314 PMID:37030380

Gelybó, G., Tóth, E., Farkas, C., Horel, Á., Kása, I., & Bakacsi, Z. (2018). Potential impacts of climate change on soil properties. *Agrokémia és Talajtan*, *67*(1), 121–141. doi:10.1556/0088.2018.67.1.9

Gendron, C. (2014). Beyond environmental and ecological economics: Proposal for an economic sociology of the environment. *Ecological Economics*, *105*, 240–253. doi:10.1016/j.ecolecon.2014.06.012

Gerba, C. P., & Smith, J. E. (2005). Sources of pathogenic microorganisms and their fate during land application of wastes. *Journal of Environmental Quality*, *34*(1), 42–48. doi:10.2134/jeq2005.0042a PMID:15647533

Gerber, P., Key, N., Portet, F., & Steinfeld, H. (2010). Policy options in addressing livestock's contribution to climate change. *Animal*, *4*(3), 393–406. doi:10.1017/S1751731110000133

Getnet, A., Tesfaye, E., Ahmed, Y., & Ahmed, M. (2022). Economic valuation and its determinates of improved irrigation water use; evidence based on South Gondar Zone, Ethiopia. *Cogent Economics & Finance*, *10*(1), 1–22. doi:10.1080/23322039.2022.2090663

Gheysen, G., & Custers, R. (2017). Why Organic Farming Should Embrace Co-Existence with Cisgenic Late Blight–Resistant Potato. *Sustainability (Basel)*, *9*(2), 172. doi:10.3390u9020172

Ghosh, N., & Bandyopadhyay, J. (2017). Methods of valuation of water resources: A review. *Sawas Journal*, *1*(1), 19–50.

Ghosh, P. (2021). Water stress and water crisis in large cities of India. In R. K. Mishra, R. B. Singh, & A. Dubey (Eds.), *Sustainable Climate Action and Water Management. Advances in Geographical and Environmental Sciences*. Springer. doi:10.1007/978-981-15-8237-0_11

Giddings, J. M., Anderson, T. A., Hall, L. W., Kendall, R. J., Richards, R. P., Solomon, K. R., & ... (2005). *A probabilistic aquatic ecological risk assessment of atrazine in North American surface waters*. SETAC Press.

Gilburn, A. S., Bunnefeld, N., Wilson, J. M., Botham, M. S., Brereton, T. M., Fox, R., & Goulson, D. (2015). Are neonicotinoid insecticides driving declines of widespread butterflies? *PeerJ*, *3*, e1402. doi:10.7717/peerj.1402 PMID:26623186

Ginwala, R., Bhavsar, R., Chigbu, D. G. I., Jain, P., & Khan, Z. K. (2019). Potential role of flavonoids in treating chronic inflammatory diseases with a special focus on the anti-inflammatory activity of apigenin. *Antioxidants*, *8*(2), 35. doi:10.3390/antiox8020035 PMID:30764536

Giordano, M., Petropoulos, S. A., & Rouphael, Y. (2021). Response and Defence Mechanisms of Vegetable Crops against Drought, Heat and Salinity Stress. *Agriculture*, *11*(5), 5. doi:10.3390/agriculture11050463

Girish, L., Gangadhar, S., Bharath, T. R., Balaji, K. S., & Abhishek, K. T. (2018). Crop yield and rainfall prediction in Tumakuru district using machine learning. *International Journal for Research in Engineering Application and Management (IJREAM)*, (pp. 61-65). IEEE.

Godde, C. M., Mason-D'Croz, D., Mayberry, D., Thornton, P. K., & Herrero, M. (2021). Impacts of climate change on the livestock food supply chain; a review of the evidence. *Global Food Security*, *28*, 100488. doi:10.1016/j.gfs.2020.100488 PMID:33738188

Godefroid, M., Morente, M., Schartel, T., Cornara, D., Purcell, A., Gallego, D., & Fereres, A. (2021). Climate tolerances of Philaenus spumarius should be considered in risk assessment of disease outbreaks related to Xylella fastidiosa. *Journal of Pest Science*, 1–14.

Godwin, I. D., Rutkoski, J., Varshney, R. K., & Hickey, L. T. (2019). Technological perspectives for plant breeding. *Theoretical and Applied Genetics*, *132*(3), 555–557. doi:10.100700122-019-03321-4 PMID:30888430

Gojon, A., Nussaume, L., Luu, D. T., Murchie, E. H., Baekelandt, A., Rodrigues Saltenis, V. L., Cohan, J., Desnos, T., Inzé, D., Ferguson, J. N., Guiderdonni, E., Krapp, A., Klein Lankhorst, R., Maurel, C., Rouached, H., Parry, M. A. J., Pribil, M., Scharff, L. B., & Nacry, P. (2023). Approaches and determinants to sustainably improve crop production. *Food and Energy Security*, *12*(1), e369. doi:10.1002/fes3.369

Goldhamer, D. A. (1999). Regulated deficit irrigation for canning olives. *Acta Horticulturae*, (474), 369–372. doi:10.17660/ActaHortic.1999.474.76

Gold, K. M., Townsend, P. A., Herrmann, I., & Gevens, A. J. (2020). Investigating potato late blight physiological differences across potato cultivars with spectroscopy and machine learning. *Plant Science*, *295*, 110316. doi:10.1016/j.plantsci.2019.110316 PMID:32534618

Goldstein, D. (2015). Climate-adaptive technological change in a small region: A resource-based scenario approach. *Technological Forecasting and Social Change*, *99*, 168–180. doi:10.1016/j.techfore.2015.04.014

Golubev, G. N. (1983). Economic activity, water resources and the environment: A challenge for hydrology. *Hydrological Sciences Journal*, *28*(1), 57–75. doi:10.1080/02626668309491143

Gomes, H. O., Menezes, J. M. C., da Costa, J. G. M., Coutinho, H. D. M., Teixeira, R. N. P., & do Nascimento, R. F. (2020). A socio-environmental perspective on pesticide use and food production. *Ecotoxicology and Environmental Safety*, *197*, 110627. doi:10.1016/j.ecoenv.2020.110627 PMID:32302856

Gong, H., Meng, F., Wang, G., Hartmann, T. E., Feng, G., Wu, J., & Zhang, F. (2022). Toward the sustainable use of mineral phosphorus fertilizers for crop production in China: From primary resource demand to final agricultural use. *The Science of the Total Environment*, *804*, 150183. doi:10.1016/j.scitotenv.2021.150183 PMID:34520915

Gong, X., & Liu, J. (2012). Genetic transformation and genes for resistance to abiotic and biotic stresses in Citrus and its related genera. *Plant Cell, Tissue and Organ Culture*, *113*(2), 137–147. doi:10.100711240-012-0267-x

González, M., Caetano, P., & Sánchez, M. (2017). Testing systemic fungicides for control of Phytophthora oak root disease. *Forest Pathology*, *47*(4), e12343. doi:10.1111/efp.12343

González-Sánchez, E. J., Ordóñez-Fernández, R., Carbonell-Bojollo, R., Veroz-González, O., & Gil-Ribes, J. A. (2012). Meta-analysis on atmospheric carbon capture in Spain through the use of conservation agriculture. *Soil & Tillage Research*, *122*, 52–60. doi:10.1016/j.still.2012.03.001

Goonetilleke, A., & Vithanage, M. (2017). Water Resources Management: Innovation and Challenges in a Changing World. *Water (Basel)*, *9*(4), 4. Advance online publication. doi:10.3390/w9040281

Gora, M. M., Mir, M., Mehraj, S., Sharma, R., Hussain, A., Nazim, N., & Khan, I. (2023). Biodiversity for Sustainable Development. In biodiversity (pp. 369-379). CRC Press.

Gornall, J., Betts, R., Burke, E., Clark, R., Camp, J., Willett, K., & Wiltshire, A. (2010). Implications of climate change for agricultural productivity in the early twenty-first century. *Philosophical Transactions of the Royal Society of London. Series B, Biological Sciences*, *365*(1554), 2973–2989. doi:10.1098/rstb.2010.0158 PMID:20713397

Gosnell, H., Gill, N., & Voyer, M. (2019). Transformational adaptation on the farm: Processes of change and persistence in transitions to 'climate-smart'regenerative agriculture. *Global Environmental Change*, *59*, 101965. doi:10.1016/j.gloenvcha.2019.101965

Goswami, K., & Bisht, P. S. (2017). The role of water resources in socio-economic development. *International Journal for Research in Applied Science and Engineering Technology*, *5*(12), 669–1674.

Gove, B., Power, S., Buckley, G., & Ghazoul, J. (2007). Effects of herbicide spray drift and fertilizer overspread on selected species of woodland ground flora: Comparison between short-term and long-term impact assessments and field surveys. *Journal of Applied Ecology*, *44*(2), 374–384. doi:10.1111/j.1365-2664.2007.01261.x

Gowdy, J., & Erickson, J. (2005). Ecological economics at a crossroads. *Ecological Economics*, *53*(1), 17–20. doi:10.1016/j.ecolecon.2005.01.010

Grace, O. M. (2011). Current perspectives on the economic botany of the genus Aloe L.(Xanthorrhoeaceae). *South African Journal of Botany*, *77*(4), 980-987. doi:10.1016/j.sajb.2011.07.002

Grace, M. A., Achick, T.-F. E., Bonghan, B. E., Bih, M. E., Ngo, N. V., Ajeck, M. J., & Ntungwen, F. C. (2019). An overview of the impact of climate change on pathogens, pest of crops on sustainable food biosecurity. *International Journal of Ecotoxicology and Ecobiology*, *4*(4), 114–119. doi:10.11648/j.ijee.20190404.15

Graniti, A., Faedda, R., Cacciola, S. O., & di San Lio, G. M. (2011). *Olive diseases in a changing ecosystem.*

Grant, C. A., Clarke, J. M., Duguid, S., & Chaney, R. L. (2008). Selection and breeding of plant cultivars to minimize cadmium accumulation. *The Science of the Total Environment*, *390*(2-3), 301–310. doi:10.1016/j.scitotenv.2007.10.038 PMID:18036635

Grattapaglia, D., Silva-Junior, O. B., Resende, R. T., Cappa, E. P., Müller, B. S. F., Tan, B., Isik, F., Wang, T., & El-Kassaby, Y. A. (2018). Quantitative Genetics and Genomics Converge to Accelerate Forest Tree Breeding. *Frontiers in Plant Science*, *9*, 1693. doi:10.3389/fpls.2018.01693 PMID:30524463

Gray, S. B., & Brady, S. M. (2016). Plant developmental responses to climate change. *Developmental Biology*, *419*(1), 64–77. doi:10.1016/j.ydbio.2016.07.023 PMID:27521050

Gregory, P. J., Johnson, S. N., Newton, A. C., & Ingram, J. S. (2009). Integrating pests and pathogens into the climate change/food security debate. *Journal of Experimental Botany*, *60*(10), 2827–2838. doi:10.1093/jxb/erp080 PMID:19380424

Griffin, K. L., & Seemann, J. R. (1996). Plants, $CO_2$ and photosynthesis in the 21st century. *Chemistry & Biology*, *3*(4), 245–254. doi:10.1016/S1074-5521(96)90104-0 PMID:8807852

Grillo, R., Fraceto, L. F., Amorim, M. J., Scott-Fordsmand, J. J., Schoonjans, R., & Chaudhry, Q. (2021). Ecotoxicological and regulatory aspects of environmental sustainability of nanopesticides. *Journal of Hazardous Materials*, *404*, 124148. doi:10.1016/j.jhazmat.2020.124148 PMID:33059255

Grime, J. P. (2001). *Plant strategies, vegetation processes, and ecosystem properties* (2nd ed.). Wiley.

Grumbine, R. E., Xu, J., & Ma, L. (2021). An overview of the problems and prospects for circular agriculture in sustainable food systems in the Anthropocene. *Circular Agricultural Systems*, *1*(1), 1–11. doi:10.48130/CAS-2021-0003

Gucci, R., Mancuso, S., & Sebastiani, L. (2003). Resistenza agli stress ambientali. In: Fiorino P (ed) Olea— Trattato di Olivicoltura. Edagricole. Edizioni Agricole de Il Sole 24 ORE Edagricole Srl, pp 91– 111.

Gucci, R., Caruso, G., Gennai, C., Esposto, S., Urbani, S., & Servili, M. (2019). Fruit growth, yield and oil quality changes induced by deficit irrigation at different stages of olive fruit development. *Agricultural Water Management*, *212*, 88–98. doi:10.1016/j.agwat.2018.08.022

Gucci, R., & Tattini, M. (2010). Salinity tolerance in olive. In J. Janick (Ed.), *Horticultural Reviews* (Vol. 21, pp. 177–214). John Wiley & Sons, Inc.

Guerfel, M., Baccouri, O., Boujnah, D., Chaïbi, W., & Zarrouk, M. (2009). Impacts of water stress on gas exchange, water relations, chlorophyll content and leaf structure in the two main Tunisian olive (*Olea europaea* L.) cultivars. *Scientia Horticulturae*, *119*(3), 257–263. doi:10.1016/j.scienta.2008.08.006

Guimaraes, P. R. Jr, Pires, M. M., Jordano, P., Bascompte, J., & Thompson, J. N. (2017). Indirect effects drive coevolution in mutualistic networks. *Nature*, *550*(7677), 511–514. doi:10.1038/nature24273 PMID:29045396

Gul, M. U., Paul, A. S. M., & Chehri, A. (2023). Hydrotropism: Understanding the Impact of Water on Plant Movement and Adaptation. *Water (Basel)*, *15*(3), 3. doi:10.3390/w15030567

Guo, Q., Potter, K. M., Ren, H., & Zhang, P. (2023). Impacts of Exotic Pests on Forest Ecosystems: An Update. *Forests*, *14*(3), 3. doi:10.3390/f14030605

Guo, S.-H., Yang, B.-H., Wang, X.-W., Li, J.-N., Li, S., Yang, X., Ren, R.-H., Fang, Y.-L., Xu, T.-F., Zhang, Z.-W., & Meng, J.-F. (2021). ABA signaling plays a key role in regulated deficit irrigation-driven anthocyanins accumulation in 'Cabernet Sauvignon' grape berries. *Environmental and Experimental Botany*, *181*, 104290. doi:10.1016/j.envexpbot.2020.104290

Gupta, A., Sarkar, A. K., & Senthil-Kumar, M. (2016). Global transcriptional analysis reveals unique and shared responses in Arabidopsis thaliana exposed to combined drought and pathogen stress. *Frontiers in Plant Science*, *7*, 686. doi:10.3389/fpls.2016.00686 PMID:27252712

Gutierrez, A. P., Ponti, L., & Cossu, Q. A. (2009). Effects of climate warming on olive and olive fly (Bactrocera oleae (Gmelin)) in California and Italy. *Climatic Change*, *95*(1), 195–217. doi:10.100710584-008-9528-4

Hacket-Pain, A., & Bogdziewicz, M. (1839). Climate change and plant reproduction: Trends and drivers of mast seeding change. *Philosophical Transactions of the Royal Society of London. Series B, Biological Sciences*, *376*(1839), 20200379. doi:10.1098/rstb.2020.0379 PMID:34657461

Hadiarto, T., & Tran, L. S. P. (2011). Progress studies of drought-responsive genes in rice. *Plant Cell Reports*, *30*(3), 297–310. doi:10.100700299-010-0956-z PMID:21132431

Hahn, D. A., & Denlinger, D. L. (2007). Meeting the energetic demands of insect diapause: Nutrient storage and utilization. *Journal of Insect Physiology*, *53*(8), 760–773. doi:10.1016/j.jinsphys.2007.03.018 PMID:17532002

Hahn, M. (2014). The rising threat of fungicide resistance in plant pathogenic fungi: Botrytis as a case study. *Journal of Chemical Biology*, *7*(4), 133–141. doi:10.100712154-014-0113-1 PMID:25320647

Haigh, R., Ianson, D., Holt, C. A., Neate, H. E., & Edwards, A. M. (2015). Effects of ocean acidification on temperate coastal marine ecosystems and fisheries in the Northeast Pacific. *PLoS One*, *10*(2), e0117533. doi:10.1371/journal.pone.0117533 PMID:25671596

Hajek, A. E., & Eilenberg, J. (2018). *Natural enemies: an introduction to biological control*. Cambridge University Press. doi:10.1017/9781107280267

Hajimirzajan, A., Vahdat, M., Sadegheih, A., Shadkam, E., & El Bilali, H. (2021). An integrated strategic framework for large-scale crop planning: Sustainable climate-smart crop planning and agri-food supply chain management. *Sustainable Production and Consumption*, *26*, 709–732. doi:10.1016/j.spc.2020.12.016

Hajji, R., & Amal, S. (2010). Le profil épidémiologique de la leishmaniose cutanée dans la province de Ouarzazate entre 2002 et 2009 [Doctoral dissertation, Facuté de Medecine et Pharmacie de Marrakech]. http://wd.fmpm.uca.ma/biblio/theses/annee-htm/art/2010/article30-10.pdf

Hakkour, M., Hmamouch, A., El Alem, M. M., Rhalem, A., Amarir, F., Touzani, M., Sadak, A., Fellah, H., & Sebti, F. (2016). New epidemiological aspects of visceral and cutaneous leishmaniasis in Taza, Morocco. *Parasites & Vectors*, *9*(1), 1–9. doi:10.118613071-016-1910-x PMID:27899126

Halkos, G. (2011). *The evolution of environmental thinking in economics* (MPRA Paper No. 35580; Munich Personal Repec Archive). https://mpra.ub.uni-muenchen.de/35580/

Hall, B. H., Cother, E. J., Whattam, M., Noble, D., Luck, J., & Cartwright, D. (2004). First report of olive knot caused by Pseudomonas savastanoi pv. savastanoi on olives (Olea europaea) in Australia. *Australasian Plant Pathology*, *33*(3), 433–436. doi:10.1071/AP04031

Hamida, A. B. (2012). Climat, environnement et maladies à transmission vectorielle: cas de la Leishmaniose Cutanée Zoonotique (LCZ) dans la région de Sidi Bouzid, Tunisie. *mémoire d'agrégation en géographie*.

Hamidov, A., Helming, K., Bellocchi, G., Bojar, W., Dalgaard, T., Ghaley, B. B., Hoffmann, C., Holman, I., Holzkämper, A., Krzeminska, D., Kværnø, S. H., Lehtonen, H., Niedrist, G., Øygarden, L., Reidsma, P., Roggero, P. P., Rusu, T., Santos, C., Seddaiu, G., & Schönhart, M. (2018). Impacts of climate change adaptation options on soil functions: A review of European case-studies. *Land Degradation & Development*, *29*(8), 2378–2389. doi:10.1002/ldr.3006 PMID:30393451

Hamilton, J. G., Dermody, O., Aldea, M., Zangerl, A. R., Rogers, A., Berenbaum, M. R., & Delucia, E. H. (2005). Anthropogenic changes in tropospheric composition increase susceptibility of soybean to insect herbivory. *Environmental Entomology*, *34*(2), 479–485. doi:10.1603/0046-225X-34.2.479

Hamman, J. H. (2008). Composition and applications of Aloe vera leaf gel. *Molecules (Basel, Switzerland)*, *13*(8), 1599–1616. doi:10.3390/molecules13081599 PMID:18794775

Han, F., Guo, R., Hussain, S., Guo, S., Cai, T., Zhang, P., Jia, Z., Naseer, M. A., Saqib, M., Chen, X., & Ren, X. (2023). Rotation of planting strips and reduction in nitrogen fertilizer application can reduce nitrogen loss and optimize its balance in maize–peanut intercropping. *European Journal of Agronomy*, *143*, 126707. doi:10.1016/j.eja.2022.126707

Hans, R. K., & Farooq, M. (2000). Dissipation and accumulation kinetics of lindane in soil and earthworms. *Pollution Research*, *19*, 394–396.

Harakannanavar, S. S., Rudagi, J. M., Puranikmath, V. I., Siddiqua, A., & Pramodhini, R. (2022). Plant leaf disease detection using computer vision and machine learning algorithms. *Global Transitions Proceedings*, *3*(1), 305–310. doi:10.1016/j.gltp.2022.03.016

Hardt, L., & O'Neill, D. W. (2017). Ecological Macroeconomic Models: Assessing Current Developments. *Ecological Economics*, *134*, 198–211. doi:10.1016/j.ecolecon.2016.12.027

Hargreaves, J. C., Adl, M. S., & Warman, P. R. (2008). A review of the use of composted municipal solid waste in agriculture. *Agriculture, Ecosystems & Environment*, *123*(1-3), 1–14. doi:10.1016/j.agee.2007.07.004

Harvey, J. A., Heinen, R., Gols, R., & Thakur, M. P. (2020). Climate change-mediated temperature extremes and insects: From outbreaks to breakdowns. *Global Change Biology*, *26*(12), 6685–6701. doi:10.1111/gcb.15377 PMID:33006246

Hasan, N., Choudhary, S., Naaz, N., Sharma, N., & Laskar, R. A. (2021). Recent advancements in molecular marker-assisted selection and applications in plant breeding programmes. *Journal of Genetic Engineering and Biotechnology*, *19*(1), 128. doi:10.118643141-021-00231-1 PMID:34448979

Hasanuzzaman, M., Nahar, K., & Fujita, M. (2013). Extreme temperature responses, oxidative stress and antioxidant defense in plants. *Abiotic stress-plant responses and applications in agriculture, 13*, 169-205.

Hasanuzzaman, M., Mohsin, S. M., Bhuyan, M. B., Bhuiyan, T. F., Anee, T. I., Masud, A. A. C., & Nahar, K. (2020). *Phytotoxicity, environmental and health hazards of herbicides: Challenges and ways forward Agrochemicals detection, treatment and remediation*. Elsevier.

Hassan, M. M., & Seif, S. A. (1990). Response of seven olive cultivars to water logging. *Gartenbauwissenschaft*, *55*, 223–225.

Hatfield, J. L., & Prueger, J. H. (2015). Temperature extremes: Effect on plant growth and development. *Weather and Climate Extremes*, *10*, 4–10. doi:10.1016/j.wace.2015.08.001

Hawkes, J. G. (1991). 1. The importance of genetic resources in plant breeding. *Biological Journal of the Linnean Society. Linnean Society of London*, *43*(1), 3–10. doi:10.1111/j.1095-8312.1991.tb00578.x

Hazrati, S., Tahmasebi-Sarvestani, Z., Nicola, S., Kashkooli, A. B., Habibzadeh, F., Mohammadi, H., & Mokhtassi-Bidgoli, A. (2020). Effect of light and water deficiency on growth and concentration of various primary and secondary metabolites of aloe vera l. *Journal of Agricultural Science and Technology, 22*(5), 1343-1358.

Hazrati, S., Tahmasebi-Sarvestani, Z., Mokhtassi-Bidgoli, A., Modarres-Sanavy, S. A. M., Mohammadi, H., & Nicola, S. (2017). Effects of zeolite and water stress on growth, yield and chemical compositions of Aloe vera L. *Agricultural Water Management, 181*, 66–72. doi:10.1016/j.agwat.2016.11.026

Hazrati, S., Tahmasebi, Z., Modares Sanevi, S. A. M., & Mokhtasi, A. (2016). Effect of Water Deficit Stress on Growth, Yield and Aloin Concentration of Aloe vera L. at different harvesting dates. *Journal of Plant Protection Research, 23*(3), 63–80. doi:10.22069/jopp.2017.9304.1886

Heagle, A. S., Spencer, S., & Letchworth, M. B. (1979). Yield response of winter wheat to chronic doses of ozone. *Canadian Journal of Botany, 57*(19), 1999–2005. doi:10.1139/b79-250

Heap, I. (2023). *The international survey of herbicide resistant weeds*. Online. Internet.

Hecht, S. B. (2018). The evolution of agroecological thought. In *Agroecology* (pp. 1–19). CRC Press. doi:10.1201/9780429495465-1

Heeb, L., Jenner, E., & Cock, M. J. (2019). Climate-smart pest management: Building resilience of farms and landscapes to changing pest threats. *Journal of Pest Science, 92*(3), 951–969. doi:10.100710340-019-01083-y

Henricot, B., Gorton, C., Denton, J., & Denton, G. (2009). Pseudocercospora cladosporioides, the cause of leaf spot on olive, a pathogen new to the United Kingdom. *Plant Pathology, 58*(4), 803–803. doi:10.1111/j.1365-3059.2009.02079.x

Hernandez-Santana, V., Rodriguez-Dominguez, C. M., Fernández, J. E., & Diaz-Espejo, A. (2016). Role of leaf hydraulic conductance in the regulation of stomatal conductance in almond and olive in response to water stress. *Tree Physiology, 36*(6), 725–735. doi:10.1093/treephys/tpv146 PMID:26846979

Hęś, M., Dziedzic, K., Górecka, D., Jędrusek-Golińska, A., & Gujska, E. (2019). Aloe vera (L.) Webb.: Natural sources of antioxidants–a review. *Plant Foods for Human Nutrition (Dordrecht, Netherlands), 74*(3), 255–265. doi:10.100711130-019-00747-5 PMID:31209704

Hewitt, E. J., & Smith, T. A. (1974). Plant mineral nutrition. *Plant Mineral Nutrition.* https://www.cabdirect.org/cabdirect/abstract/19760340035

Hewlings, S. J., & Kalman, D. S. (2017). Curcumin: A review of its effects on human health. *Foods, 6*(10), 92. doi:10.3390/foods6100092 PMID:29065496

Hirel, B., Le Gouis, J., Ney, B., & Gallais, A. (2007). The challenge of improving nitrogen use efficiency in crop plants: Towards a more central role for genetic variability and quantitative genetics within integrated approaches. *Journal of Experimental Botany, 58*(9), 2369–2387. doi:10.1093/jxb/erm097 PMID:17556767

Hockerts, K., & Wüstenhagen, R. (2010). Greening Goliaths versus emerging Davids—Theorizing about the role of incumbents and new entrants in sustainable entrepreneurship. *Journal of Business Venturing, 25*(5), 481–492. doi:10.1016/j.jbusvent.2009.07.005

Hofman-Bergholm, M. (2022). A Transition towards a Food and Agricultural System That Includes Both Food Security and Planetary Health. *Foods, 12*(1), 12. doi:10.3390/foods12010012 PMID:36613228

Hok, L., de Moraes Sá, J. C., Boulakia, S., Reyes, M., Leng, V., Kong, R., Tivet, F. E., Briedis, C., Hartman, D., Ferreira, L. A., Magno, T., & Pheav, S. (2015). Short-term conservation agriculture and biomass-C input impacts on soil C dynamics in a savanna ecosystem in Cambodia. *Agriculture, Ecosystems & Environment, 214*, 54–67. doi:10.1016/j.agee.2015.08.013

Hoosbeek, M. R., Remme, R. P., & Rusch, G. M. (2018). Trees enhance soil carbon sequestration and nutrient cycling in a silvopastoral system in south-western Nicaragua. *Agroforestry Systems, 92*, 263–273.

Hoppin, J. A., Umbach, D. M., London, S. J., Alavanja, M. C., & Sandler, D. P. (2002). Chemical predictors of wheeze among farmer pesticide applicators in the Agricultural Health Study. *American Journal of Respiratory and Critical Care Medicine, 165*(5), 683–689. doi:10.1164/ajrccm.165.5.2106074 PMID:11874814

Horie, T. (2019). Global warming and rice production in Asia: Modeling, impact prediction and adaptation. *Proceedings of the Japan Academy. Series B, Physical and Biological Sciences, 95*(6), 211–245. doi:10.2183/pjab.95.016 PMID:31189777

Hossain, M. (2015). Recent perspective of herbicide: Review of demand and adoption in world agriculture. *Journal of the Bangladesh Agricultural University, 13*(452-2016-35850), 13-24.

Hou, B., & Wu, L. (2010). Safety impact and farmer awareness of pesticide residues. *Food and Agricultural Immunology, 21*(3), 191–200. doi:10.1080/09540105.2010.484858

Hougardy, E., Wang, X., Hogg, B. N., Johnson, M. W., Daane, K. M., & Pickett, C. H. (2020). Current distribution of the olive psyllid, Euphyllura olivina, in California and initial evaluation of the Mediterranean parasitoid Psyllaephagus euphyllurae as a biological control candidate. *Insects, 11*(3), 146. doi:10.3390/insects11030146 PMID:32110919

Hou, H., Atlihan, N., & Lu, Z. (2014). New biotechnology enhances the application of cisgenesis in plant breeding. *Frontiers in Plant Science, 5*. doi:10.3389/fpls.2014.00389 PMID:25157261

Houtven, G. L. V., Pattanayak, S. K., Usmani, F., & Yang, J. C. (2017). What are households willing to pay for improved water access? Results from a meta-analysis. *Ecological Economics, 136*(1), 126–135. doi:10.1016/j.ecolecon.2017.01.023

Howard, A. (1947). *The Soil and Health: A Study of Organic Agriculture*. University Press of Kentucky. https://www.jstor.org/stable/j.ctt2jcp63

Hsu, P. K., Dubeaux, G., Takahashi, Y., & Schroeder, J. I. (2021). Signaling mechanisms in abscisic acid-mediated stomatal closure. *The Plant Journal, 105*(2), 307–321. doi:10.1111/tpj.15067 PMID:33145840

Hsu, Y.-T., Shen, T.-C., & Hwang, S.-Y. (2009). Soil fertility management and pest responses: A comparison of organic and synthetic fertilization. *Journal of Economic Entomology, 102*(1), 160–169. doi:10.1603/029.102.0123 PMID:19253632

Hueso, A., Camacho, G., & Gómez-del-Campo, M. (2021). Spring deficit irrigation promotes significant reduction on vegetative growth, flowering, fruit growth and production in hedgerow olive orchards (cv. Arbequina). *Agricultural Water Management, 248*, 106695. doi:10.1016/j.agwat.2020.106695

Hufnagel, J., Reckling, M., & Ewert, F. (2020). Diverse approaches to crop diversification in agricultural research. A review. *Agronomy for Sustainable Development, 40*(2), 14. doi:10.100713593-020-00617-4

Hughes, J., Cowper-Heays, K., Olesson, E., Bell, R., & Stroombergen, A. (2021). Impacts and implications of climate change on wastewater systems: A New Zealand perspective. *Climate Risk Management, 31*, 100262. doi:10.1016/j.crm.2020.100262

Hulsmans, E., Daelemans, R., Cuypers, V., Van Der Straeten, E., Vanderlinden, M., De Blanck, T., Vertommen, W., Boeraeve, M., Proesmans, W., & Honnay, O. (2023). Cascading effects of management and landscape on insect pollinators, pollination services and yield in apple orchards. *Agriculture, Ecosystems & Environment*, *352*, 108509. doi:10.1016/j.agee.2023.108509

Huntley, B. J. (2023). Soil, Water and Nutrients. In B. J. Huntley (Ed.), *Ecology of Angola: Terrestrial Biomes and Ecoregions* (pp. 127–147). Springer International Publishing. doi:10.1007/978-3-031-18923-4_6

Hussein, A., & Workeneh, S. (2021). Modeling the impacts of climate changes on the distribution of Aloe vera species in Ethiopia. *Journal of Earth Science & Climatic Change*, *12*(7). Advance online publication. doi:10.4172/2157-7617.1000567

Hu, Y., Zhang, Q., Hu, S., Xiao, G., Chen, X., Wang, J., Qi, Y., Zhang, L., & Han, L. (2022). Research progress and prospects of ecosystem carbon sequestration under climate change (1992–2022). *Ecological Indicators*, *145*, 109656. doi:10.1016/j.ecolind.2022.109656

Iavorivska, L., Veith, T. L., Cibin, R., Preisendanz, H. E., & Steinman, A. D. (2021). Mitigating Lake Eutrophication through stakeholder-driven hydrologic modeling of agricultural conservation practices: A case study of Lake Macatawa, Michigan. *Journal of Great Lakes Research*, *47*(6), 1710–1725. doi:10.1016/j.jglr.2021.10.001

Ibba, K., Kassout, J., Boselli, V., Er-Raki, S., Oulbi, S., Mansouri, L. E., Bouizgaren, A., Sikaoui, L., & Hadria, R. (2023). Assessing the impact of deficit irrigation strategies on agronomic and productive parameters of Menara olive cultivar: Implications for operational water management. *Frontiers in Environmental Science*, *11*, 1100552. doi:10.3389/fenvs.2023.1100552

IFOAM. (2010). *Definition of organic agriculture*. IFOAM. http://www.ifoam.org/growingorganic/definitions/doa/index.html

IFOAM. (international movement of organic agriculture movements). (2008). The world of organic agriculture - statistics and emerging trends 2008. International federation of organic agriculture movements -ifoam, Bonn, Germany.

Iglesias, A., & Garrote, L. (2015). Adaptation strategies for agricultural water management under climate change in Europe. *Agricultural Water Management*, *155*, 113–124. doi:10.1016/j.agwat.2015.03.014

Inouye, D. W. (2022). Climate change and phenology. *Wiley Interdisciplinary Reviews: Climate Change*, *13*(3), e764. doi:10.1002/wcc.764

Insam, H., & De Bertoldi, M. (2007). Microbiology of the composting process. *Waste management series*, *8*, 25-48.

Intergovernmental Panel on Climate Change (IPCC). (2021). Summary for policymakers. In V. Masson-Delmotte, P. Zhai, A. Pirani, S. L. Connors, C. Péan, & S. Berger (Eds.), *Climate change 2021: The physical science basis. Contribution of working group I to the sixth assessment report of the intergovernmental panel on climate change*. Cambridge University Press.

Intergovernmental Panel on Climate Change (IPCC). (2022). *Sixth assessment report*. IPCC. https://www.ipcc.ch/assessment-report/ar6/

Isayenkov, S. V. (2012). Physiological and molecular aspects of salt stress in plants. *Cytology and Genetics*, *46*(5), 302–318. doi:10.3103/S0095452712050040

Islam, M. S., & Kieu, E. (2020). Tackling Regional Climate Change Impacts and Food Security Issues: A Critical Analysis across ASEAN, PIF, and SAARC. *Sustainability (Basel)*, *12*(3), 3. doi:10.3390u12030883

Ivits, E., Horion, S., Erhard, M., & Fensholt, R. (2016). Assessing European ecosystem stability to drought in the vegetation growing season: Ecosystem stability to drought. *Global Ecology and Biogeography*, *25*(9), 1131–1143. doi:10.1111/geb.12472

Izri, A., Depaquit, J., & Parola, P. (2006). Phlébotomes et transmission d'agents pathogènes autour du bassin méditerranéen. *Médecine Tropicale*, *66*(5), 429–435. PMID:17201284

Jablonkai, I. (2011). Molecular mechanism of function of herbicides. In M. N. A. E.-G. Hasaneen (Ed.), *Herbicides - Mechanisms and mode of action* (pp. 3–24). InTech. doi:10.5772/31251

Jacobsen, E., & Schouten, H. J. (2008). Cisgenesis, a new tool for traditional plant breeding, should be exempted from regulating genetically modified organisms step-by-step. *Potato Research*, *51*(1), 75–88. https://link.springer.com/article/10.1007/s11540-008-9097-y. doi:10.100711540-008-9097-y

Jactel, H., Koricheva, J., & Castagneyrol, B. (2019). Responses of forest insect pests to climate change: Not so simple. *Current Opinion in Insect Science*, *35*, 103–108. doi:10.1016/j.cois.2019.07.010 PMID:31454625

Jain, S., Rathod, N., Nagi, R., Sur, J., Laheji, A., & Gupta, N. (2016). Antibacterial Effect of Aloe Vera Gel against Oral Pathogens: An In-vitro Study. *Journal of Clinical and Diagnostic Research : JCDR*, *10*(11), ZC41. doi:10.7860/JCDR/2016/21450.8890 PMID:28050502

Jala, & Nandagiri, L. (2015). Evaluation of economic value of Pilikula Lake using travel cost and contingent valuation methods. *Aquatic Procedia*, *4*(1), 1315-1321. doi:10.1016/j.aqpro.2015.02.171

Jama, A. A., & Mourad, K. A. (2019). Water services sustainability: Institutional arrangements and shared responsibilities. *Sustainability (Basel)*, *11*(3), 1–15. doi:10.3390u11030916

Jamel, L., & Derbali, A. (2016). Do energy consumption and economic growth lead to ED? Evidence from Asian economies. *Cogent Economics & Finance*, *4*(11), 1–19. doi:10.1080/23322039.2016.1170653

Janda, M., Lamparová, L., Zubíková, A., Burketová, L., Martinec, J., & Krčková, Z. (2019). Temporary heat stress suppresses PAMP-triggered immunity and resistance to bacteria in Arabidopsis thaliana. *Molecular Plant Pathology*, *20*(7), 1005–1012. https://bsppjournals.onlinelibrary.wiley.com/doi/pdfdirect/10.1111/mpp.12799. doi:10.1111/mpp.12799 PMID:30924595

Janni, M., Gullì, M., Maestri, E., Marmiroli, M., Valliyodan, B., Nguyen, H. T., & Marmiroli, N. (2020). Molecular and genetic bases of heat stress responses in crop plants and breeding for increased resilience and productivity. *Journal of Experimental Botany*, *71*(13), 3780–3802. doi:10.1093/jxb/eraa034 PMID:31970395

Jarecki, M. K., & Lal, R. (2003). Crop management for soil carbon sequestration. *Critical Reviews in Plant Sciences*, *22*(6), 471–502. doi:10.1080/713608318

Javaid, M., Haleem, A., Singh, R. P., & Suman, R. (2022). Enhancing smart farming through the applications of Agriculture 4.0 technologies. *International Journal of Intelligent Networks*, *3*, 150–164. doi:10.1016/j.ijin.2022.09.004

Jennings, D. E., Congelosi, A. M., & Rohr, J. R. (2012). Insecticides reduce survival and the expression of traits associated with carnivory of carnivorous plants. *Ecotoxicology (London, England)*, *21*(2), 569–575. doi:10.100710646-011-0817-8 PMID:22076028

Jensen, H. G., Jacobsen, L.-B., Pedersen, S. M., & Tavella, E. (2012). Socioeconomic impact of widespread adoption of precision farming and controlled traffic systems in Denmark. *Precision Agriculture*, *13*(6), 661–677. doi:10.100711119-012-9276-3

Jentsch, A., & Beierkuhnlein, C. (2008). Research frontiers in climate change: Effects of extreme meteorological events on ecosystems. *Comptes Rendus Geoscience*, *340*(9), 621–628. doi:10.1016/j.crte.2008.07.002

Jeong, S. T., Cho, S. R., Lee, J. G., Kim, P. J., & Kim, G. W. (2019). Composting and compost application: Trade-off between greenhouse gas emission and soil carbon sequestration in whole rice cropping system. *Journal of Cleaner Production*, *212*, 1132–1142. doi:10.1016/j.jclepro.2018.12.011

Jepson, P. C., Murray, K., Bach, O., Bonilla, M. A., & Neumeister, L. (2020). Selection of pesticides to reduce human and environmental health risks: A global guideline and minimum pesticides list. *The Lancet. Planetary Health*, *4*(2), 56–63. doi:10.1016/S2542-5196(19)30266-9 PMID:32112748

Jha, A. K., Chakraborty, S., Kumari, K., & Bauddh, K. (2020). Ecological Consequences of Genetically Modified Crops on Soil Biodiversity. In Ecological and Practical Applications for Sustainable Agriculture. doi:10.1007/978-981-15-3372-3_5

Jianjun, J., Wenyu, W., Ying, F., & Xiaomin, W. (2016). Measuring the willingness to pay for drinking water quality improvements: Results of a contingent valuation survey in Songzi, China. *Journal of Water and Health*, *14*(3), 504–512. doi:10.2166/wh.2016.247 PMID:27280614

Jiao, L., Zhang, P., Zhang, H., Shao, D., Yang, W., & Sun, T. (2023). Density-dependent dispersal strategy of pollinator moderates the adverse effect of habitat loss on plant reproduction: An integrated model based on pollinators' behavioural response. *Functional Ecology*, *37*(3), 732–747. doi:10.1111/1365-2435.14252

Jiaying, M., Tingting, C., Jie, L., Weimeng, F., Baohua, F., Guangyan, L., Hubo, L., Juncai, L., Zhihai, W., Longxing, T., & Guanfu, F. (2022). Functions of Nitrogen, Phosphorus and Potassium in Energy Status and Their Influences on Rice Growth and Development. *Rice Science*, *29*(2), 166–178. doi:10.1016/j.rsci.2022.01.005

Jiménez-Díaz, R. M., Tjamos, E. C., & Cirulli, M. (1998). Verticillium wilt of major tree hosts: olive. *A compendium of Verticillium wilts in tree species*, 13-16.

Jimenez-Diaz, R. M., Cirulli, M., Bubici, G., del Mar Jimenez-Gasco, M., Antoniou, P. P., & Tjamos, E. C. (2012). Verticillium wilt, a major threat to olive production: Current status and future prospects for its management. *Plant Disease*, *96*(3), 304–329. doi:10.1094/PDIS-06-11-0496 PMID:30727142

Jiménez-Díaz, R. M., Olivares-García, C., Landa, B. B., del Mar Jiménez-Gasco, M., & Navas-Cortés, J. A. (2011). Region-wide analysis of genetic diversity in Verticillium dahliae populations infecting olive in southern Spain and agricultural factors influencing the distribution and prevalence of vegetative compatibility groups and pathotypes. *Phytopathology*, *101*(3), 304–315. doi:10.1094/PHYTO-07-10-0176 PMID:20942654

Jindra, M., Tumova, S., Milacek, M., & Bittova, L. (2021). A decade with the juvenile hormone receptor [): Elsevier.]. *Advances in Insect Physiology*, *60*, 37–85. doi:10.1016/bs.aiip.2021.03.001

Johnson, J. A., Runge, C. F., Senauer, B., Foley, J., & Polasky, S. (2014). Global agriculture and carbon trade-offs. *Proceedings of the National Academy of Sciences of the United States of America*, *111*(34), 12342–12347. doi:10.1073/pnas.1412835111

Johnson, M. S., Reddy, G., & Desai, M. M. (2023). Epistasis and evolution: Recent advances and an outlook for prediction. *BMC Biology*, *21*(1), 120. doi:10.118612915-023-01585-3 PMID:37226182

Jolly, S. (2017). Role of institutional entrepreneurship in the creation of regional solar PV energy markets: Contrasting developments in Gujarat and West Bengal. *Energy for Sustainable Development*, *38*, 77–92. doi:10.1016/j.esd.2016.10.004

Jose, S. (2009). Agroforestry for ecosystem services and environmental benefits: An overview. *Agroforestry Systems*, *76*(1), 1–10. doi:10.100710457-009-9229-7

Jung, V., Albert, C. H., Violle, C., Kunstler, G., Loucougaray, G., & Spiegelberger, T. (2014). Intraspecific trait variability mediates the response of subalpine grassland communities to extreme drought events. *Journal of Ecology*, *102*(1), 45–53. doi:10.1111/1365-2745.12177

Juroszek, P., & Von Tiedemann, A. (2011). Potential strategies and future requirements for plant disease management under a changing climate. *Plant Pathology*, *60*(1), 100–112. doi:10.1111/j.1365-3059.2010.02410.x

Ju, X., Zhang, F., Xuemei, B., Römheld, V., & Roelcke, M. (2005). Utilization and management of organic wastes in Chinese agriculture: Past, present and perspectives. *Science in China. Series C, Life Sciences*, *48*(S2, Suppl 2), 965–979. doi:10.1007/BF03187135 PMID:20549451

Jyoti, B., & Singh, A. K. (2020). Projected sugarcane yield in different climate change scenarios in Indian states: A state-wise panel data exploration. *International Journal of Food and Agricultural Economics*, *8*(1), 343–365.

Jyoti, B., Singh, A. K., & Ashraf, S. N. (2023). Appropriate technologies and their implications in the agricultural sector. In P. H. Alex Khang (Ed.), *Advanced Technologies and AI-Equipped IoT Applications in High-Tech Agriculture*. IGI Global. doi:10.4018/978-1-6684-9231-4.ch004

Kabtni, S., Sdouga, D., Bettaib Rebey, I., Save, M., Trifi-Farah, N., Fauconnier, M. L., & Marghali, S. (2020). Influence of climate variation on phenolic composition and antioxidant capacity of Medicago minima populations. *Scientific Reports*, *10*(1), 1–15. doi:10.103841598-020-65160-4 PMID:32427946

Kafy, A. A., Bakshi, A., Saha, M., Al Faisal, A., Almulhim, A. I., Rahaman, Z. A., & Mohammad, P. (2023). Assessment and prediction of index based agricultural drought vulnerability using machine learning algorithms. *The Science of the Total Environment*, *867*, 161394. doi:10.1016/j.scitotenv.2023.161394 PMID:36634773

Kahaduwa, A., & Rajapakse, L. (2022). Review of climate change impacts on reservoir hydrology and long-term basin-wide water resources management. *Building Research and Information*, *50*(5), 515–526. doi:10.1080/09613218.2021.1977908

Kahime, K., Boussaa, S., Bounoua, L., Fouad, O., Messouli, M., & Boumezzough, A. (2014). Leishmaniasis in Morocco: Diseases and vectors. *Asian Pacific Journal of Tropical Disease*, *4*, S530–S534. doi:10.1016/S2222-1808(14)60671-X

Kahlon, J. B., Kemp, M. C., Carpenter, R. H., McAnalley, B. H., McDaniel, H. R., & Shannon, W. M. (1991). Inhibition of AIDS virus replication by acemannan in vitro. *Molecular Biotherapy*, *3*(3), 127–135. https://europepmc.org/article/med/1768365 PMID:1768365

Kahramanoğlu, İ., Chen, C., Chen, J., & Wan, C. (2019). Chemical constituents, antimicrobial activity, and food preservative characteristics of Aloe vera Gel. *Agronomy (Basel)*, *9*(12), 831. doi:10.3390/agronomy9120831

Kaivo-oja, J., Vehmas, J., & Luukkanen, J. (2014). A Note: De-Growth Debate and New Scientific Analysis of Economic Growth. *Journal of Environmental Protection*, *05*(15), 1477–1481. doi:10.4236/jep.2014.515140

Kalia, A., & Gosal, S. K. (2011). Effect of pesticide application on soil microorganisms. *Archives of Agronomy and Soil Science*, *57*(6), 569–596. doi:10.1080/03650341003787582

Kamar Zaman, A. M., & Yaacob, J. S. (2022). Exploring the potential of vermicompost as a sustainable strategy in circular economy: Improving plants' bioactive properties and boosting agricultural yield and quality. *Environmental Science and Pollution Research International*, *29*(9), 1–17. doi:10.100711356-021-18006-z PMID:35034296

Kang, S., Zhao, X., Yue, L., & Liu, L. (2017). Main anthraquinone components in Aloe vera and their inhibitory effects on the formation of advanced glycation end-products. *Journal of Food Processing and Preservation*, *41*(5), e13160. doi:10.1111/jfpp.13160

Kaniewski, D., Van Campo, E., Boiy, T., Terral, J. F., Khadari, B., & Besnard, G. (2012). Primary domestication and early uses of the emblematic olive tree: Palaeobotanical, historical and molecular evidence from the Middle East. *Biological Reviews of the Cambridge Philosophical Society*, *87*(4), 885–899. doi:10.1111/j.1469-185X.2012.00229.x PMID:22512893

Kansiime, M. K., & Mastenbroek, A. (2016). Enhancing resilience of farmer seed system to climate-induced stresses: Insights from a case study in West Nile region, Uganda. *Journal of Rural Studies*, *47*, 220–230. doi:10.1016/j.jrurstud.2016.08.004

Kapur, R. (2016). Natural resources and environmental issues. *Journal of Ecosystem & Ecology*, *6*(2), 1–4.

Karmaoui, A. (2023). The Future of Artificial Intelligence in Agricultural Field: A Bibliometric Analysis. In A. Karmaoui (Ed.), *Intelligent Solutions for Optimizing Agriculture and Tackling Climate Change: Current and Future Dimensions*. doi:10.4018/978-1-6684-4649-2.ch003

Karmaoui, A. (Ed.). (2023). *Climate Change and the Economic Importance and Damages of Insects* (pp. 1–279). IGI Global. doi:10.4018/978-1-6684-4824-3

Karmaoui, A., Ben Salem, A., El Jaafari, S., Chaachouay, H., Moumane, A., & Hajji, L. (2022b). Exploring the land use and land cover change in the period 2005–2020 in the province of Errachidia, the pre-sahara of Morocco. *Frontiers in Earth Science (Lausanne)*, *10*, 962097. doi:10.3389/feart.2022.962097

Karmaoui, A., El Jaafari, S., Chaachouay, H., & Hajji, L. (2023a). A bibliometric review of geospatial analyses and artificial intelligence literature in agriculture. *GeoJournal*. doi:10.100710708-023-10859-w

Karmaoui, A., Minucci, G., Messouli, M., Khebiza, M. Y., Ifaadassan, I., & Babqiqi, A. (2019). Climate change impacts the water supply system of the Middle Draa Valley in South Morocco. In *Climate change, food security and natural resource management* (pp. 163–178). Springer. doi:10.1007/978-3-319-97091-2_8

Karmaoui, A., Moumane, A., El Jaafari, S., Menouni, A., Al Karkouri, J., Yacoubi, M., & Hajji, L. (2023). Thirty Years of Change in the Land Use and Land Cover of the Ziz Oases (Pre-Sahara of Morocco) Combining Remote Sensing, GIS, and Field Observations. *Land (Basel)*, *12*(12), 2127. doi:10.3390/land12122127

Karmaoui, A., Moumane, A., Minucci, G., El Jaafari, S., Chaachouay, H., & Hajji, L. (2023b). Affordable Methods and Techniques to Optimize Farming Productivity in a Context of Climate Change in Arid Areas. In A. Karmaoui (Ed.), *Intelligent Solutions for Optimizing Agriculture and Tackling Climate Change: Current and Future Dimensions*. doi:10.4018/978-1-6684-4649-2.ch006

Karmaoui, A., Sereno, D., El Jaafari, S., & Hajji, L. (2022a). Seasonal patterns of zoonotic cutaneous Leishmaniasis caused by L. major and transmitted by Phlebotomus papatasi in the North Africa Region, a systematic review and a meta-analysis. *Microorganisms*, *10*(12), 2391. doi:10.3390/microorganisms10122391 PMID:36557644

Karmaoui, A., Sereno, D., El Jaafari, S., & Hajji, L. (2022c). A systematic review and global analysis of the seasonal activity of Phlebotomus (Paraphlebotomus) sergenti, the primary vectors of L. tropica. *PLoS Neglected Tropical Diseases*, *16*(12), e0010886. doi:10.1371/journal.pntd.0010886 PMID:36469546

Karthikeyan, C. (2010). *Economic and social value of irrigation water: Implications for sustainability*. Fourteenth International Water Technology Conference, IWTC 14 2010, Cairo, Energy.

Kasap, O. E., & Alten, B. (2006). Comparative demography of the sand fly Phlebotomus papatasi (Diptera: Psychodidae) at constant temperatures. *Journal of Vector Ecology*, *31*(2), 378–385. doi:10.3376/1081-1710(2006)31[378:CDOTSF]2.0.CO;2 PMID:17249356

Kashyap, A., Garg, P., Tanwar, K., Sharma, J., Gupta, N. C., Ha, P. T. T., & Rao, M. (2022). Strategies for utilization of crop wild relatives in plant breeding programs. *Theoretical and Applied Genetics*, *135*(12), 4151–4167. https://link.springer.com/article/10.1007/s00122-022-04220-x. doi:10.100700122-022-04220-x PMID:36136128

Kassam, A., Friedrich, T., & Derpsch, R. (2022). Successful experiences and lessons from conservation agriculture worldwide. *Agronomy (Basel)*, *12*(4), 769. doi:10.3390/agronomy12040769

Kassout, J., Terral, J. F., El Ouahrani, A., Houssni, M., Ivorra, S., Kadaoui, K., et al. (2022b). Species distribution based-modelling under climate change: the case of two native wild *Olea europaea* subspecies in Morocco, *O. e.* subsp. *europaea* var. *sylvestris* and *O. e.* subsp. *maroccana. Climate change in the mediterranean and middle eastern region*, 21-43.

Kassout, J., Ater, M., Ivorra, S., Barbara, H., Limier, B., Ros, J., Girard, V., Paradis, L., & Terral, J. F. (2021). Resisting aridifcation: Adaptation of sap conduction performance in Moroccan wild olive subspecies distributed over an aridity gradient. *Frontiers in Plant Science*, *12*, 1288. doi:10.3389/fpls.2021.663721 PMID:34276726

Kassout, J., Hmimsa, Y., El Fatehi, S., El Ouahrani, A., Kadaoui, K., Chakkour, S., Ariza-Mateos, D., Palacios-Rodríguez, G., Navarro-Cerrillo, R., & Ater, M. (2022a). Image analysis of Moroccan carob seeds (*Ceratonia siliqua* L.) revealed substantial intraspecific variations depending on climate and geographic origin. *Ecological Processes*, *11*(1), 1–14. doi:10.118613717-022-00378-w

Kassout, J., Hmimsa, Y., Fatehi, S. E., Kadaoui, K., Houssni, M., Chakkour, S., Sahli, A., El Chami, M. A., Ariza-Mateos, D., Palacios-Rodríguez, G., Navarro-Cerrillo, R. M., & Ater, M. (2023b). Aridity Gradients Shape Intraspecific Variability of Morphological Traits in Native *Ceratonia siliqua* L. of Morocco. *Plants*, *12*(19), 3447. doi:10.3390/plants12193447 PMID:37836187

Kassout, J., Terral, J. F., Azenzem, R., El Ouahrani, A., Sahli, A., Houssni, M., & Ater, M. (2023a). *Olea europaea* and Stressful Conditions. In *Medicinal Plant Responses to Stressful Conditions* (pp. 243–272). CRC Press. doi:10.1201/9781003242963-13

Kassout, J., Terral, J. F., El Ouahrani, A., Houssni, M., & Ivorra, S. (2022). Species Distribution Based- Modelling Under Climate Change: The Case of Two Native Wild *Olea europaea* Subspecies in Morocco, *O. e.* subsp. *europaea* var. *sylvestris* and *O. e.* subsp. *maroccana*. In *Climate Change in the Mediterranean and Middle Eastern Region* (pp. 21–43). Springer. doi:10.1007/978-3-030-78566-6_2

Kassout, J., Terral, J.-F., Hodgson, J., & Ater, M. (2019). Trait- based plant ecology a flawed tool in climate studies? The leaf traits of wild olive that pattern with climate are not those routinely measured. *PLoS One*, *14*(7), e0219908. doi:10.1371/journal.pone.0219908 PMID:31314789

Katko, T. S., & Hukka, J. J. (2015). Social and economic importance of water services in the build environment: Need for more structural thinking. *Procedia Economics and Finance*, *21*(1), 217–223. doi:10.1016/S2212-5671(15)00170-7

Kattge, J., Diaz, S., Lavorel, S., Prentice, I. C., Leadley, P., Bönisch, G., Garnier, E., Westoby, M., Reich, P. B., Wright, I. J., Cornelissen, J. H. C., Violle, C., Harrison, S. P., Van BODEGOM, P. M., Reichstein, M., Enquist, B. J., Soudzilovskaia, N. A., Ackerly, D. D., Anand, M., & Wirth, C. (2011). TRY–a global database of plant traits. *Global Change Biology*, *17*(9), 2905–2935. doi:10.1111/j.1365-2486.2011.02451.x

Katumo, D. M., Liang, H., Ochola, A. C., Lv, M., Wang, Q.-F., & Yang, C.-F. (2022). Pollinator diversity benefits natural and agricultural ecosystems, environmental health, and human welfare. *Plant Diversity*, *44*(5), 429–435. doi:10.1016/j.pld.2022.01.005 PMID:36187551

Kaur, G., Singh, G., Motavalli, P. P., Nelson, K. A., Orlowski, J. M., & Golden, B. R. (2020). Impacts and management strategies for crop production in waterlogged or flooded soils: A review. *Agronomy Journal*, *112*(3), 1475–1501. doi:10.1002/agj2.20093

Kaur, T., & Sinha, A. K. (2019). Pesticides in agricultural run offs affecting water resources: A study of Punjab (India). *Agricultural Sciences*, *10*(10), 1381–1395. doi:10.4236/as.2019.1010101

Kaushal, N., Bhandari, K., Siddique, K. H., & Nayyar, H. (2016). Food crops face rising temperatures: An overview of responses, adaptive mechanisms, and approaches to improve heat tolerance. *Cogent Food & Agriculture*, *2*(1), 1134380. doi:10.1080/23311932.2015.1134380

Kaya, A., & Doganlar, Z. B. (2016). Exogenous jasmonic acid induces stress tolerance in tobacco (Nicotiana tabacum) exposed to imazapic. *Ecotoxicology and Environmental Safety*, *124*, 470–479. doi:10.1016/j.ecoenv.2015.11.026 PMID:26629659

Kaye, J. P., & Quemada, M. (2017). Using cover crops to mitigate and adapt to climate change. A review. *Agronomy for Sustainable Development*, *37*(1), 1–17. doi:10.100713593-016-0410-x

Keesstra, S. D., Bouma, J., Wallinga, J., Tittonell, P., Smith, P., Cerdà, A., & Fresco, L. O. (2016). Forum paper: The significance of soils and soil science towards realization of the UN sustainable development goals (SDGS). *Soil Discussions*, *2016*, 1–28.

Keesstra, S., Mol, G., De Leeuw, J., Okx, J., Molenaar, C., De Cleen, M., & Visser, S. (2018). Soil-related sustainable development goals: Four concepts to make land degradation neutrality and restoration work. *Land (Basel)*, *7*(4), 133. doi:10.3390/land7040133

Kehrberger, S., & Holzschuh, A. (2019). How does timing of flowering affect competition for pollinators, flower visitation and seed set in an early spring grassland plant? *Scientific Reports*, *9*(1), 1. doi:10.103841598-019-51916-0 PMID:31666567

Keikotlhaile, B. M. (2011). *Influence of the processing factors on pesticide residues in fruits and vegetables and its application in consumer risk assessment*. Ghent University.

Kenis, M., Hurley, B. P., Hajek, A. E., & Cock, M. J. (2017). Classical biological control of insect pests of trees: Facts and figures. *Biological Invasions*, *19*(11), 3401–3417. doi:10.100710530-017-1414-4

Kerr, R. B., Postigo, J. C., Smith, P., Cowie, A., Singh, P. K., Rivera-Ferre, M., & Neufeldt, H. (2023). Agroecology as a transformative approach to tackle climatic, food, and ecosystemic crises. *Current Opinion in Environmental Sustainability*, *62*, 101275. doi:10.1016/j.cosust.2023.101275

Keyhanian, S., & Stahl-Biskup, E. (2007). Phenolic constituents in dried flowers of Aloe vera (Aloe barbadensis) and their in vitro antioxidative capacity. *Planta Medica*, *73*(06), 599–602. doi:10.1055-2007-967202 PMID:17520524

Khamidov, M., Ishchanov, J., Hamidov, A., Donmez, C., & Djumaboev, K. (2022). Assessment of Soil Salinity Changes under the Climate Change in the Khorezm Region, Uzbekistan. *International Journal of Environmental Research and Public Health*, *19*(14), 8794. doi:10.3390/ijerph19148794 PMID:35886646

Khan, N. I., Brouwer, R., & Yang, H. (2014). Household's willingness to pay for arsenic safe drinking water in Bangladesh. *Journal of Environmental Management*, *143*(1), 151–161. doi:10.1016/j.jenvman.2014.04.018 PMID:24905645

Khan, S., Das, P., Thaher, M. I., AbdulQuadir, M., Mahata, C., & Al Jabri, H. (2023). Utilization of nitrogen-rich agricultural waste streams by microalgae to produce protein and value-added compounds. *Current Opinion in Green and Sustainable Chemistry*, *100797*, 100797. doi:10.1016/j.cogsc.2023.100797

Khatun, M., Sarkar, S., Era, F. M., Islam, A. K. M. M., Anwar, M. P., Fahad, S., Datta, R., & Islam, A. K. M. A. (2021). Drought Stress in Grain Legumes: Effects, Tolerance *Mechanisms and Management. Agronomy (Basel)*, *11*(12), 12. doi:10.3390/agronomy11122374

Khoo, H. H., Eufrasio-Espinosa, R. M., Koh, L. S., Sharratt, P. N., & Isoni, V. (2019). Sustainability assessment of biorefinery production chains: A combined LCA-supply chain approach. *Journal of Cleaner Production*, *235*, 1116–1137. doi:10.1016/j.jclepro.2019.07.007

Kiem, A. S., & Austin, E. K. (2013). Drought and the future of rural communities: Opportunities and challenges for climate change adaptation in regional Victoria, Australia. *Global Environmental Change*, *23*(5), 1307–1316. doi:10.1016/j.gloenvcha.2013.06.003

Killick-Kendrick, R. (2013). The race to discover the insect vector of Kala-azar: A great saga of tropical medicine 1903-1942. *Bulletin de la Société de Pathologie Exotique*, *106*(2), 131–137. doi:10.100713149-013-0285-x PMID:23516011

Kim, D. G. (2012). Estimation of net gain of soil carbon in a nitrogen-fixing tree and crop intercropping system in sub-Saharan Africa: Results from re-examining a study. *Agroforestry Systems*, *86*(2), 175–184. doi:10.100710457-011-9477-1

Kitsaki, C. K., & Drossopoulos, J. B. (2005). Environmental effect on ABA concentration and water potential in olive leaves (*Olea europaea* L. cv "Koroneiki") under non- irrigated feld conditions. *Environmental and Experimental Botany*, *54*(1), 77–89. doi:10.1016/j.envexpbot.2004.06.002

Kitture, R., Ghosh, S., More, P. A., Gaware, S., Datar, S., Chopade, B. A., & Kale, S. N. (2015). Curcumin-loaded, self-assembled aloe vera template for superior antioxidant activity and trans-membrane drug release. *Journal of Nanoscience and Nanotechnology*, *15*(6), 4039–4045. doi:10.1166/jnn.2015.10322 PMID:26369010

Klümper, W., & Qaim, M. (2014). A Meta-Analysis of the impacts of Genetically Modified crops. *PLoS One*, *9*(11), e111629. doi:10.1371/journal.pone.0111629 PMID:25365303

Kniss, A. R. (2017). Long-term trends in the intensity and relative toxicity of herbicide use. *Nature Communications*, *8*(1), 1–7. doi:10.1038/ncomms14865 PMID:28393866

Kolanowska, M. (2023). Loss of fungal symbionts and changes in pollinator availability caused by climate change will affect the distribution and survival chances of myco-heterotrophic orchid species. *Scientific Reports*, *13*(1), 6848. doi:10.103841598-023-33856-y PMID:37100884

Kole, R. K., Banerjee, H., Bhattacharyya, A., Chowdhury, A., & Chaudhury, N. (1999). Photo transformation of some pesticides. *Journal of the Indian Chemical Society*, *76*, 595–600.

Konapala, G., Mishra, A. K., Wada, Y., & Mann, M. E. (2020). Climate change will affect global water availability through compounding changes in seasonal precipitation and evaporation. *Nature Communications*, *11*(1), 3044. doi:10.103841467-020-16757-w PMID:32576822

Kosanic, A., Anderson, K., Harrison, S., Turkington, T., & Bennie, J. (2018). Changes in the geographical distribution of plant species and climatic variables on the West Cornwall peninsula (South West UK). *PLoS One*, *13*(2), e0191021. doi:10.1371/journal.pone.0191021 PMID:29401494

Kotir, J. H. (2011). Climate change and variability in Sub-Saharan Africa: A review of current and future trends and impacts on agriculture and food security. *Environment, Development and Sustainability*, *13*(3), 587–605. doi:10.100710668-010-9278-0

Kotroczó, Z., Makádi, M., Kocsis, T., Béni, Á., Várbíró, G., & Fekete, I. (2023). Long-Term Changes in Organic Matter Content and Soil Moisture Determine the Degree of Root and Soil Respiration. *Plants*, *12*(2), 251. doi:10.3390/plants12020251 PMID:36678968

Kouadio, L., Deo, R. C., Byrareddy, V., Adamowski, J. F., Mushtaq, S., & Phuong Nguyen, V. (2018). Artificial intelligence approach for the prediction of Robusta coffee yield using soil fertility properties. *Computers and Electronics in Agriculture*, *155*, 324–338. doi:10.1016/j.compag.2018.10.014

Koul, B. (2022). Cisgenics and Crop Improvement. In Cisgenics and Transgenics (pp. 107–129). doi:10.1007/978-981-19-2119-3_3

Kowalska, A., Grobelak, A., Almås, Å. R., & Singh, B. R. (2020). Effect of biowastes on soil remediation, plant productivity and soil organic carbon sequestration: A review. *Energies*, *13*(21), 5813. doi:10.3390/en13215813

Kremen, C. (2020). Ecological intensification and diversification approaches to maintain biodiversity, ecosystem services and food production in a changing world. *Emerging Topics in Life Sciences*, *4*(2), 229–240. doi:10.1042/ETLS20190205 PMID:32886114

Kremen, C., & Miles, A. (2012). Ecosystem services in biologically diversified versus conventional farming systems: Benefits, externalities, and trade-offs. *Ecology and Society*, *17*(4), art40. doi:10.5751/ES-05035-170440

Kremer, R. J., & Means, N. E. (2009). Glyphosate and glyphosate-resistant crop interactions with rhizosphere microorganisms. *European Journal of Agronomy*, *31*(3), 153–161. doi:10.1016/j.eja.2009.06.004

Kristensen, D. K., Kjeldsen, C., & Thorsøe, M. H. (2016). Enabling sustainable agro-food futures: Exploring fault lines and synergies between the integrated territorial paradigm, rural eco-economy and circular economy. *Journal of Agricultural & Environmental Ethics*, *29*(5), 749–765. doi:10.100710806-016-9632-9

Ksantini, M. (2003). *Contribution à l'étude de la dynamique des populations du psylle de l'olivier Euphylluraolivina Costa (Homoptera, Aphalaridae) et de sa nuisibilité dans la région de Sfax* [Thèse de Doctorat en Sciences biologiques]. Fac. Sc. Sfax.

Ksantini, M., Jardak, T., & BouAïn, A. (2002). Temperature Effect on the Biology of Euphylluraolivina Costa. *Acta Horticulturae*, (586), 827–829. doi:10.17660/ActaHortic.2002.586.179

Kubiak-Wójcicka, K., & Machula, S. (2020). Influence of Climate Changes on the State of Water Resources in Poland and Their Usage. *Geosciences*, *10*(8), 8. doi:10.3390/geosciences10080312

Kumar, A. (2015). *Effect of climatic factors on agricultural productivity and food security in India: An econometric analysis*. [PhD Thesis, IIT Indore, India].

Kumar, A., & Sharma, P. (2013). Impact of climate variation on agricultural productivity and food security in rural India. Economics Discussion Papers, No. 2013-43. Publisher: Kiel Institute for the World Economy Germany.

Kumar, S., David Raj, A., Kalambukattu, J. G., & Chatterjee, U. (2023). Climate Change Impact on Land Degradation and Soil Erosion in Hilly and Mountainous Landscape: Sustainability Issues and Adaptation Strategies. In Ecological Footprints of Climate Change: Adaptive Approaches and Sustainability (pp. 119-155). Springer.

Kumar, A., Ahmad, M. M., & Sharma, P. (2017). Influence of climatic and non-climatic factors on sustainable food security in India: A statistical investigation. *International Journal of Sustainable Agricultural Management and Informatics*, *3*(1), 1–30. doi:10.1504/IJSAMI.2017.082917

Kumar, A., Sengar, R., Pathak, R. K., & Singh, A. K. (2023). Integrated approaches to develop drought-tolerant rice: Demand of era for global food security. *Journal of Plant Growth Regulation*, *42*(1), 96–120. doi:10.100700344-021-10561-6

Kumar, A., & Sharma, P. (2014). Climate change and sugarcane productivity in India: An econometric analysis. *Journal of Social and Development Sciences*, *5*(2), 111–122. doi:10.22610/jsds.v5i2.811

Kumar, A., Sharma, P., & Ambrammal, S. K. (2015a). Climatic effects on sugarcane productivity in India: A stochastic production function application. *International Journal of Economics and Business Research*, *10*(2), 179–203. doi:10.1504/IJEBR.2015.070984

Kumar, A., Sharma, P., & Joshi, S. (2015b). Effects of climatic factors on agricultural productivity in India: A state-wise panel data analysis. *International Journal of Basic and Life Sciences*, *3*(1), 48–67.

Kumar, A., Sharma, P., & Joshi, S. (2016). Assessing the impacts of climate change on land productivity in Indian crop agriculture: An evidence from panel data analysis. *Journal of Agricultural Science and Technology*, *18*(1), 1–13. https://jast.modares.ac.ir/article-23-4491-en.pdf

Kumar, A., Singh, B. J., & Sharma, P. (2020). Assessing the climate change impact on rice and wheat production in Uttar Pradesh and Haryana states of India. *Climatic Change*, *6*(21), 74–93.

Kumararaja, P., Suvana, S., Saraswathy, R., Lalitha, N., & Muralidhar, M. (2019). Mitigation of eutrophication through phosphate removal by aluminum pillared bentonite from aquaculture discharge water. *Ocean and Coastal Management*, *182*, 104951. doi:10.1016/j.ocecoaman.2019.104951

Kumar, D., Punetha, A., Verma, P. P., & Padalia, R. (2022). Micronutrient based approach to increase yield and quality of essential oil in aromatic crops. *Journal of Applied Research on Medicinal and Aromatic Plants*, *26*, 100361. doi:10.1016/j.jarmap.2021.100361

Kumari, S. (2021). Chapter-5 Global Warming: Causes, Effects and Solutions. *Emerging Trends in Agriculture Sciences*, 67.

Kumari, M., & Rai, S. (2020). Hydrogeochemical evaluation of groundwater quality for drinking and irrigation purposes using water quality index in semi arid region of India. *Journal of the Geological Society of India*, *95*(2), 159–168. doi:10.100712594-020-1405-4

Kumar, K., Gambhir, G., Dass, A., Tripathi, A. K., Singh, A., Jha, A. K., & Rakshit, S. (2020). Genetically modified crops: Current status and future prospects. *Planta*, *251*(4), 1–27. doi:10.100700425-020-03372-8 PMID:32236850

Kumar, N. H., Murali, M., Girish, H., Chandrashekar, S., Amruthesh, K., Sreenivasa, M., & Jagannath, S. (2021). Impact of climate change on biodiversity and shift in major biomes. In *Global Climate Change* (pp. 33–44). Elsevier. doi:10.1016/B978-0-12-822928-6.00007-1

Kumar, R., Kumar, N., Rajput, V. D., Mandzhieva, S., Minkina, T., Saharan, B. S., & Duhan, J. S. (2022). Advances in biopolymeric nanopesticides: A new eco-friendly/eco-protective perspective in precision agriculture. *Nanomaterials (Basel, Switzerland)*, *12*(22), 3964. doi:10.3390/nano12223964 PMID:36432250

Kumar, S., & Singh, A. K. (2023). Modeling the effects of climate change on agricultural productivity: Evidence from Himachal Pradesh. *Asia-Pacific Journal of Regional Science*, *7*(2), 1–28. doi:10.100741685-023-00291-w

Kumar, S., Yadav, A., Yadav, M., & Yadav, J. P. (2017a). Effect of climate change on phytochemical diversity, total phenolic content and in vitro antioxidant activity of Aloe vera (L.) Burm. f. *BMC Research Notes*, *10*(1), 1–12. doi:10.118613104-017-2385-3 PMID:28118858

Kumar, S., Yadav, M., Yadav, A., & Yadav, J. P. (2017b). Impact of spatial and climatic conditions on phytochemical diversity and in vitro antioxidant activity of Indian Aloe vera (L.) Burm. f. *South African Journal of Botany*, *111*, 50–59. doi:10.1016/j.sajb.2017.03.012

Kumar, V., Paillard, S., Fopa-Fomeju, B., Falentin, C., Deniot, G., Baron, C., Vallée, P., & Delourme, R. (2018). Multi-year linkage and association mapping confirm the high number of genomic regions involved in oilseed rape quantitative resistance to blackleg. *Theoretical and Applied Genetics*, *131*(8), 1627–1643. doi:10.100700122-018-3103-9 PMID:29728747

Kuppler, J., Albert, C. H., Ames, G. M., Armbruster, W. S., Boenisch, G., Boucher, F. C., Campbell, D. R., Carneiro, L. T., Chacón-Madrigal, E., Enquist, B. J., Fonseca, C. R., Gómez, J. M., Guisan, A., Higuchi, P., Karger, D. N., Kattge, J., Kleyer, M., Kraft, N. J. B., Larue-Kontić, A.-A. C., & Junker, R. R. (2020). Global gradients in intraspecific variation in vegetative and floral traits are partially associated with climate and species richness. *Global Ecology and Biogeography*, *29*(6), 992–1007. doi:10.1111/geb.13077

Kurniawati, A., Windriyati, R. D. H., Wulansari, N. K., Toth, G., & Toth, Z. (2021). Alternatives for circular bioeconomy in organic farming under excessive nutrients (goat manure and arbuscular mycorrhizal fungi): A case study in Indonesia. *Sustainability (Basel)*, *13*(22), 12333. doi:10.3390u132212333

Kushwaha, A. K., & Bhattacharya, S. (2015). Crop yield prediction using Argo Algorithm in Hadoop. [IJCSITS]. *International Journal of Computer Science and Information Technology & Security*, *5*(2), 271–274.

Kutos, S., Stricker, E., Cooper, A., Ryals, R., Creque, J., Machmuller, M., Kroegar, M., & Silver, W. L. (2023). Compost amendment to enhance carbon sequestration in rangelands. *Journal of Soil and Water Conservation*, *78*(2), 163–177. doi:10.2489/jswc.2023.00072

Kyriakopoulos, G. L., & Sebos, I. (2023). Enhancing Climate Neutrality and Resilience through Coordinated Climate Action: Review of the Synergies between Mitigation and Adaptation Actions. *Climate (Basel)*, *11*(5), 5. doi:10.3390/cli11050105

Laforest-Lapointe, I., Martínez-Vilalta, J., & Retana, J. (2014). Intraspecific variability in functional traits matters: Case study of Scots pine. *Oecologia*, *175*(4), 1337–1348. doi:10.100700442-014-2967-x PMID:24850418

Lainson, R., & Shaw, J. J. (1987). *Evolution, classification and geographical distribution* (pp. 1-120). Academic Press.

Lal, R. (2004). Soil carbon sequestration to mitigate climate change. *Geoderma*, *123*(1-2), 1–22. doi:10.1016/j.geoderma.2004.01.032

Lal, R. (2015). Sequestering carbon and increasing productivity by conservation agriculture. *Journal of Soil and Water Conservation*, *70*(3), 55A–62A. doi:10.2489/jswc.70.3.55A

Lal, R., Delgado, J., Groffman, P., Millar, N., Dell, C., & Rotz, A. (2011). Management to mitigate and adapt to climate change. *Journal of Soil and Water Conservation*, *66*(4), 276–285. doi:10.2489/jswc.66.4.276

Lal, R., & Kimble, J. M. (1997). Conservation tillage for carbon sequestration. *Nutrient Cycling in Agroecosystems*, *49*(1/3), 243–253. doi:10.1023/A:1009794514742

Lal, R., Mohtar, R. H., Assi, A. T., Ray, R., Baybil, H., & Jahn, M. (2017). Soil as a basic nexus tool: Soils at the center of the food–energy–water nexus. *Current Sustainable/Renewable. Energy Reports*, *4*, 117–129.

Lal, R., Negassa, W., & Lorenz, K. (2015). Carbon sequestration in soil. *Current Opinion in Environmental Sustainability*, *15*, 79–86. doi:10.1016/j.cosust.2015.09.002

Lamers, J., Van Der Meer, T., & Testerink, C. (2020). How plants sense and respond to stressful environments. *Plant Physiology*, *182*(4), 1624–1635. doi:10.1104/pp.19.01464 PMID:32132112

Lampridi, M. G., Sørensen, C. G., & Bochtis, D. (2019). Agricultural sustainability: A review of concepts and methods. *Sustainability (Basel)*, *11*(18), 5120. doi:10.3390u11185120

Landa, B. B., Castillo, A. I., Giampetruzzi, A., Kahn, A., Román-Écija, M., Velasco-Amo, M. P., Navas-Cortés, J. A., Marco-Noales, E., Barbé, S., Moralejo, E., Coletta-Filho, H. D., Saldarelli, P., Saponari, M., & Almeida, R. P. (2020). Emergence of a plant pathogen in Europe associated with multiple intercontinental introductions. *Applied and Environmental Microbiology*, *86*(3), e01521–e19. doi:10.1128/AEM.01521-19 PMID:31704683

Lange, H. D., Lahr, J., Van der Pol, J. C., Wessels, Y., & Faber, J. H. (2009). Ecological vulnerability in wildlife: An expert judgment and multicriteria analysis tool using ecological traits to assess relative impact of pollutants. *Environmental Toxicology and Chemistry*, *28*(10), 2233–2240. doi:10.1897/08-626.1 PMID:19432506

Lanka, S. (2018). A review on Aloe vera-The wonder medicinal plant. *Journal of Drug Delivery and Therapeutics*, *8*(5-s), 94–99. doi:10.22270/jddt.v8i5-s.1962

Lankford, B., Pringle, C., McCosh, J., Shabalala, M., Hess, T., & Knox, J. W. (2023). Irrigation area, efficiency and water storage mediate the drought resilience of irrigated agriculture in a semi-arid catchment. *The Science of the Total Environment*, *859*, 160263. doi:10.1016/j.scitotenv.2022.160263 PMID:36402330

Larkan, N. J., Ma, L., & Borhan, M. H. (2015). The *Brassica napus* receptor-like protein RLM 2 is encoded by a second allele of the L ep R 3/R lm2 blackleg resistance locus. *Plant Biotechnology Journal*, *13*(7), 983–992. https://onlinelibrary.wiley.com/doi/pdf/10.1111/pbi.12341. doi:10.1111/pbi.12341 PMID:25644479

Lasco, R. D., Delfino, R. J. P., & Espaldon, M. L. O. (2014). Agroforestry systems: Helping smallholders adapt to climate risks while mitigating climate change. *Wiley Interdisciplinary Reviews: Climate Change*, *5*(6), 825–833. doi:10.1002/wcc.301

Laurent, L., Buoncristiani, J.-F., Pohl, B., Zekollari, H., Farinotti, D., Huss, M., Mugnier, J.-L., & Pergaud, J. (2020). The impact of climate change and glacier mass loss on the hydrology in the Mont-Blanc massif. *Scientific Reports*, *10*(1), 10420. doi:10.103841598-020-67379-7 PMID:32591640

Lau, S. E., Teo, W. F. A., Teoh, E. Y., & Tan, B. C. (2022). Microbiome engineering and plant biostimulants for sustainable crop improvement and mitigation of biotic and abiotic stresses. *Discover Food*, *2*(1), 9. doi:10.100744187-022-00009-5

Lavorel, S., & Garnier, E. (2002). Predicting changes in community composition and ecosystem functioning from plant traits: Revisiting the Holy Grail. *Functional Ecology*, *16*(5), 545–556. doi:10.1046/j.1365-2435.2002.00664.x

Lawrence, J., Blackett, P., & Cradock-Henry, N. A. (2020). Cascading climate change impacts and implications. *Climate Risk Management*, *29*, 100234. doi:10.1016/j.crm.2020.100234

Lawton, D., Huseth, A. S., Kennedy, G. G., Morey, A. C., Hutchison, W. D., Reisig, D. D., Dorman, S. J., Dillard, D., Venette, R. C., Groves, R. L., Adamczyk, J. J., Barbosa Dos Santos, I., Baute, T., Brown, S., Burkness, E., Dean, A., Dively, G. P., Doughty, H. B., Fleischer, S. J., & Zuefle, M. (2022). Pest population dynamics are related to a continental overwintering gradient. *Proceedings of the National Academy of Sciences of the United States of America*, *119*(37), e2203230119. doi:10.1073/pnas.2203230119 PMID:36067290

Lawyer, P. G., Ngumbi, P. M., Anjili, C. O., Odongo, S. O., Mebrahtu, Y. B., Githure, J. I., & Roberts, C. R. (1990). Development of Leishmania major in Phlebotomus duboscqi and Sergentomyia schwetzi (Diptera: Psychodidae). *The American Journal of Tropical Medicine and Hygiene*, *43*(1), 31–43. doi:10.4269/ajtmh.1990.43.31 PMID:2382763

Lebrazi, S., & Fikri-Benbrahim, K. (2022). *Potential of tree legumes in agroforestry systems and soil conservation Advances in Legumes for Sustainable Intensification*. Elsevier.

Leifeld, J., & Fuhrer, J. (2010). Organic farming and soil carbon sequestration: What do we really know about the benefits? *Ambio*, *39*(8), 585–599. doi:10.100713280-010-0082-8

Lelieveld, J., Hadjinicolaou, P., Kostopoulou, E., Chenoweth, J., El Maayar, M., Giannakopoulos, C., Hannides, C., Lange, M. A., Tanarhte, M., Tyrlis, E., & Xoplaki, E. (2012). Climate change and impacts in the Eastern Mediterranean and the Middle East. *Climatic Change*, *114*(3–4), 667–687. doi:10.100710584-012-0418-4 PMID:25834296

Lenoir, J., Gégout, J. C., Marquet, P. A., de Ruffray, P., & Brisse, H. (2008). A significant upward shift in plant species optimum elevation during the 20th century. *science, 320*(5884), 1768-1771.

Levidow, L., Zaccaria, D., Maia, R., Vivas, E., Todorovic, M., & Scardigno, A. (2014). Improving water-efficient irrigation: Prospects and difficulties of innovative practices. *Agricultural Water Management*, *146*, 84–94. doi:10.1016/j.agwat.2014.07.012

Levin, A. G., Lavee, S., & Tsror, L. (2003). Epidemiology of Verticillium dahliae on olive (cv. Picual) and its effect on yield under saline conditions. *Plant Pathology*, *52*(2), 212–218. doi:10.1046/j.1365-3059.2003.00809.x

Levine, N. D., Corliss, J. O., Cox, F. E. G., Deroux, G., Grain, J., Honigberg, B. M., Leedale, G. F., Loeblich, A. R., Lom, I. J., Lynn, D., Merinfeld, E. G., Page, F. C., Poljansky, G., Sprague, V., Vavra, J., & Wallace, F. G. (1980). A Newly Revised Classification of the Protozoa* The committee on systematics evolution of the society of protozoologists. *The Journal of Protozoology*, *27*(1), 37–58. doi:10.1111/j.1550-7408.1980.tb04228.x PMID:6989987

Levy, O., Buckley, L. B., Keitt, T. H., Smith, C. D., Boateng, K. O., Kumar, D. S., & Angilletta, M. J. (2015). Resolving the life cycle alters expected impacts of climate change. *Proceedings of the Royal Society B: Biological Sciences, 282*(1813), 20150837. 10.1098/rspb.2015.0837

Levy, B. S., & Patz, J. A. (2015). Climate Change, Human Rights, and Social Justice. *Annals of Global Health*, *81*(3), 310–322. doi:10.1016/j.aogh.2015.08.008 PMID:26615065

Lewis, D. J., Young, D. G., Fairchild, G. B., & Minter, D. M. (1977). Proposals for a stable classification of the phlebotomine sandflies (Diptera: Psychodidae). *Systematic Entomology*, *2*(4), 319–332. doi:10.1111/j.1365-3113.1977.tb00381.x

Leydesdorff, L., & Rafols, I. (2012). Interactive overlays: A new method for generating global journal maps from Web-of-Science data. Journal of Informetric, 6(2), 318-332. https://doi.org/doi:10.1016/j.joi.2011.11.003

Leyk, S., Binder, C. R., & Nuckols, J. R. (2009). Spatial modeling of personalized exposure dynamics: The case of pesticide use in small-scale agricultural production landscapes of the developing world. *International Journal of Health Geographics*, *8*(1), 1–16. doi:10.1186/1476-072X-8-17 PMID:19331690

Liao, Y., Song, X., Ye, Y., Gu, J., Wang, R., Zhuogabayong, Z., Zhao, D., & Shao, X. (2023). Climate Change May Pose Additional Threats to the Endangered Endemic Species Encalypta buxbaumioidea in China. *Diversity*, *15*(2), 2. doi:10.3390/d15020269

Li, F., Ghanizadeh, H., Cui, G., Liu, J., Miao, S., Liu, C., & Wang, A. (2023). Microbiome-based agents can optimize composting of agricultural wastes by modifying microbial communities. *Bioresource Technology*, *374*, 128765. doi:10.1016/j.biortech.2023.128765 PMID:36822555

Lightburn, E., Meynard, J. B., Morand, J., Garnotel, E., Kraemer, P., Hovette, P., & Boutin, J. P. (2002). Surveillance épidémiologique des leishmanioses tégumentaires en Guyane Synthèse de données militaires sur 10 ans. *Médecine Tropicale*, *62*, 545–553. PMID:12616949

Liliane, T. N., & Charles, M. S. (2020). Factors affecting yield of crops. *Agronomy-climate change & food security*, *9*. doi:10.1186/s12870-016-0877-2

Lima, F. A., Córcoles, J. I., Tarjuelo, J. M., & Martínez-Romero, A. (2019). Model for management of an on-demand irrigation network based on irrigation scheduling of crops to minimize energy use (Part II): Financial impact of regulated deficit irrigation. *Agricultural Water Management, 215*, 44–54. doi:10.1016/j.agwat.2019.01.006

Linnakoski, R., Kasanen, R., Dounavi, A., & Forbes, K. M. (2019). Editorial: Forest Health Under Climate Change: Effects on Tree Resilience, and Pest and Pathogen Dynamics. *Frontiers in Plant Science, 10*, 1157. doi:10.3389/fpls.2019.01157 PMID:31681344

Linneck, S., Parker, A. H., Smeets, P., Cruddas, P. H., & Jeffrey, P. (2015). *Social and economic factors influencing the 'choice' of water supply in rural Karnataka, India.* 38th WEDC International Conference on Water, Sanitation and Hygiene Beyond 2015: Improving Access and Sustainability, Loughborough University, UK. https://wedc-knowledge.lboro.ac.uk/resources/conference/38/Linneck-2136.pdf

Lin, S. S., Shen, S. L., Zhou, A., & Lyu, H. M. (2021). Assessment and management of Lake Eutrophication: A case study in Lake Erhai, China. *The Science of the Total Environment, 751*, 141618. doi:10.1016/j.scitotenv.2020.141618 PMID:33167190

Lionello, P., & Scarascia, L. (2018). The relation between climate change in the Mediterranean region and global warming. *Regional Environmental Change, 18*(5), 1481–1493. doi:10.100710113-018-1290-1

Liontakis, A., & Tzouramani, I. (2016). Economic sustainability of organic Aloe vera farming in Greece under risk and uncertainty. Sustainability, 8(4), 338. https://doi.org/ doi:10.3390u8040338

Liu, J. G., Zhou, Q., Guo, Q., Yang, Z. H., Xie, F., & Han, J. T. (2017). Knowledge diffusion of dynamical network in terms of interaction frequency. *Scientific Reports, 7*(1), 1–7. doi:10.103841598-017-11057-8 PMID:28883456

Liu, S., Wang, H., & Qin, F. (2023). Genetic dissection of drought resistance for trait improvement in crops. *The Crop Journal, 11*(4), 975–985. doi:10.1016/j.cj.2023.05.002

Li, Y., Li, Z., Chang, S. X., Cui, S., Jagadamma, S., Zhang, Q., & Cai, Y. (2020). Residue retention promotes soil carbon accumulation in minimum tillage systems: Implications for conservation agriculture. *The Science of the Total Environment, 740*, 140147. doi:10.1016/j.scitotenv.2020.140147

Li, Y., Shang, J., Zhang, C., Zhang, W., Niu, L., Wang, L., & Zhang, H. (2021). The role of freshwater eutrophication in greenhouse gas emissions: A review. *The Science of the Total Environment, 768*, 144582. doi:10.1016/j.scitotenv.2020.144582 PMID:33736331

Li, Z. (2023). Modeling Environmental Fate, Transport, and Transformation of Pesticides: First-Order Kinetic Models for Regional and Global Applications. *Reviews of Environmental Contamination and Toxicology, 261*(1), 14. doi:10.100744169-023-00040-2

Lloret, F., de la Riva, E. G., Pérez-Ramos, I. M., Marañón, T., Saura-Mas, S., Díaz-Delgado, R., & Villar, R. (2016). Climatic events inducing die-off in Mediterranean shrublands: Are species' responses related to their functional traits? *Oecologia, 180*(4), 961–973. doi:10.100700442-016-3550-4 PMID:26801493

Logan, J. A., Régnière, J., & Powell, J. A. (2003). Assessing the impacts of global warming on forest pest dynamics. *Frontiers in Ecology and the Environment, 1*(3), 130–137. doi:10.1890/1540-9295(2003)001[0130:ATIOGW]2.0.CO;2

López-Escudero, F. J., & Blanco-López, M. A. (2005). Effects of drip irrigation on population of Verticillium dahliae in olive orchards. *Journal of Phytopathology, 153*(4), 238–239. doi:10.1111/j.1439-0434.2005.00961.x

López-Escudero, F. J., & Blanco-López, M. A. (2007). Relationship between the inoculum density of Verticillium dahliae and the progress of Verticillium wilt of olive. *Plant Disease*, *91*(11), 1372–1378. doi:10.1094/PDIS-91-11-1372 PMID:30780740

López, J. J., Ayuso-Muñoz, J. L., Goñi, M., & Gimena, F. N. (2023). Analysis of the Effect of Climate Change on the Characteristics of Rainfall in Igeldo-Gipuzkoa (Spain). *Water (Basel)*, *15*(8), 8. Advance online publication. doi:10.3390/w15081529

Loudyi, D., Hasnaoui, M. D., & Fekri, A. (2022). Flood Risk Management Practices in Morocco: Facts and Challenges. In T. Sumi, S. A. Kantoush, & M. Saber (Eds.), Wadi Flash Floods: Challenges and Advanced Approaches for Disaster Risk Reduction (pp. 35–94). Springer. doi:10.1007/978-981-16-2904-4_2

Łozowicka, B., Kaczyński, P., Mojsak, P., Rusiłowska, J., Beknazarova, Z., Ilyasova, G., & Absatarova, D. (2020). Systemic and non-systemic pesticides in apples from Kazakhstan and their impact on human health. *Journal of Food Composition and Analysis*, *90*, 103494. doi:10.1016/j.jfca.2020.103494

Lu, H. P., Luo, T., Fu, H. W., Wang, L., Tan, Y. Y., Huang, J. Z., Wang, Q., Ye, G., Gatehouse, A. M. R., Lou, Y., & Shu, Q. Y. (2018). Resistance of rice to insect pests mediated by suppression of serotonin biosynthesis. *Nature Plants*, *4*(6), 338–344. doi:10.103841477-018-0152-7 PMID:29735983

Lushchak, V. I., Matviishyn, T. M., Husak, V. V., Storey, J. M., & Storey, K. B. (2018). Pesticide toxicity: A mechanistic approach. *EXCLI Journal*, *17*, 1101–1136. PMID:30564086

Lynch, J. P., Mooney, S. J., Strock, C. F., & Schneider, H. M. (2022). Future roots for future soils. *Plant, Cell & Environment*, *45*(3), 620–636. doi:10.1111/pce.14213 PMID:34725839

Maas, E. V., & Hoffman, G. J. (1977). Crop salt tolerance— Current assessment. *Journal of the Irrigation and Drainage Division*, *103*(2), 115–134. doi:10.1061/JRCEA4.0001137

Ma, C. S., Ma, G., & Pincebourde, S. (2021). Survive a warming climate: Insect responses to extreme high temperatures. *Annual Review of Entomology*, *66*(1), 163–184. doi:10.1146/annurev-ento-041520-074454 PMID:32870704

Ma, C.-S., Zhang, W., Peng, Y., Zhao, F., Chang, X.-Q., Xing, K., Zhu, L., Ma, G., Yang, H.-P., & Rudolf, V. H. W. (2021). Climate warming promotes pesticide resistance through expanding overwintering range of a global pest. *Nature Communications*, *12*(1), 5351. doi:10.103841467-021-25505-7 PMID:34504063

Machado, R. M. A., & Serralheiro, R. P. (2017). Soil Salinity: Effect on Vegetable Crop Growth. Management Practices to Prevent and Mitigate Soil Salinization. *Horticulturae*, *3*(2), 2. doi:10.3390/horticulturae3020030

Maharjan, S. K., & Maharjan, K. L. (2018). Roles and contributions of community seed banks in climate adaptation in Nepal. *Development in Practice*, *28*(2), 292–302. doi:10.1080/09614524.2018.1418838

Maheswari, M., Sarkar, B., Vanaja, M., Rao, M. S., Rao, C. S., Venkateswarlu, B., & Sikka, A. K. (2015). *Climate resilient crop varieties for sustainable food production under aberrant weather conditions*. Krishi. https://krishi.icar.gov.in/jspui/bitstream/123456789/2586/1/CCRV_11th%20August%202015.pdf

Mahjour, J., Akalay, O., & Saddiki, A. (1992). Les leishmanioses au Maroc de l'analyse éco-épidémiologique à la prévention. Direction d'épidémiologie et de lutte contre les maladies. *Bulletin Épidémiologique*, *3*, 4–5.

Maingi, F. M., & Mbuvi, H. M. (2020). The Effect of potassium dosage on selected growth parameters and yield response modeling on potatoes grown in Molo, Kenya. *Journal of Scientific Agriculture, 4*, 0101-0107.

Mairech, H., López-Bernal, Á., Moriondo, M., Dibari, C., Regni, L., Proietti, P., Villalobos, F. J., & Testi, L. (2020). Is new olive farming sustainable? A spatial comparison of productive and environmental performances between traditional and new olive orchards with the model OliveCan. *Agricultural Systems*, *181*, 102816. doi:10.1016/j.agsy.2020.102816

Majumdar, C., & Gupta, G. (2009). Willingness to pay and municipal water pricing in transition: A case study. *Journal of Integrative Environmental Sciences*, *6*(4), 247–260. doi:10.1080/19438150903068224

Mallareddy, M., Thirumalaikumar, R., Balasubramanian, P., Naseeruddin, R., Nithya, N., Mariadoss, A., Eazhilkrishna, N., Choudhary, A. K., Deiveegan, M., Subramanian, E., Padmaja, B., & Vijayakumar, S. (2023). Maximizing Water Use Efficiency in Rice Farming: A Comprehensive Review of Innovative Irrigation Management Technologies. *Water (Basel)*, *15*(10), 10. doi:10.3390/w15101802

Mallikarjuna, G., Mallikarjuna, K., Reddy, M. K., & Kaul, T. (2011). Expression of OsDREB2A transcription factor confers enhanced dehydration and salt stress tolerance in rice (Oryza sativa L.). *Biotechnology Letters*, *33*(8), 1689–1697. doi:10.100710529-011-0620-x PMID:21528404

Maltas, A., Kebli, H., Oberholzer, H. R., Weisskopf, P., & Sinaj, S. (2018). The effects of organic and mineral fertilizers on carbon sequestration, soil properties, and crop yields from a long-term field experiment under a Swiss conventional farming system. *Land Degradation & Development*, *29*(4), 926–938. doi:10.1002/ldr.2913

Mammadov, J., Buyyarapu, R., Guttikonda, S. K., Parliament, K., Abdurakhmonov, I. Y., & Kumpatla, S. P. (2018). Wild relatives of maize, rice, cotton, and soybean: Treasure troves for tolerance to biotic and abiotic stresses. *Frontiers in Plant Science*, *9*, 886. doi:10.3389/fpls.2018.00886 PMID:30002665

Mandal, G., & Das, A. (1980). Structure of the D-galactan isolated from Aloe barbadensis Miller. *Carbohydrate Research*, *86*(2), 247–257. doi:10.1016/S0008-6215(00)85902-9

Mansour, A. A., Ouanaimi, F., Chemseddine, M., & Boumezzough, A. (2017). Study of the flight dynamics of Prays oleae (Lepidoptera: Yponomeutidae) using sexual trapping in olive orchards of Essaouira region, Morocco. *Journal of Entomology and Zoology Studies*, *5*(2), 943–952.

Manvitha, K., & Bidya, B. (2014). Aloe vera: a wonder plant its history, cultivation and medicinal uses. *Journal of Pharmacognosy and Phytochemistry, 2*(5), 85-88. Available on https://www.phytojournal.com/archives?year=2014&vol=2&issue=5&part=B&ArticleId=249 accessed on Aprtil 04, 2023.

Manzoor, Z., Ehsan, M., Khan, M. B., Manzoor, A., Akhter, M. M., Sohail, M. T., Hussain, A., Shafi, A., Abu-Alam, T., & Abioui, M. (2022). Floods and flood management and its socio-economic impact on Pakistan: A review of the empirical literature. *Frontiers in Environmental Science*, *10*, 1021862. https://www.frontiersin.org/articles/10.3389/fenvs.2022.1021862. doi:10.3389/fenvs.2022.1021862

Mao, Y. B., Cai, W. J., Wang, J. W., Hong, G. J., Tao, X. Y., Wang, L. J., & Chen, X. Y. (2007). Silencing a cotton bollworm P450 monooxygenase gene by plant-mediated RNAi impairs larval tolerance of gossypol. *Nature Biotechnology*, *25*(11), 1307–1313. doi:10.1038/nbt1352 PMID:17982444

Marchi, S., Guidotti, D., Ricciolini, M., & Petacchi, R. (2016). Towards understanding temporal and spatial dynamics of Bactrocera oleae (Rossi) infestations using decade-long agrometeorological time series. *International Journal of Biometeorology*, *60*(11), 1681–1694. doi:10.100700484-016-1159-2 PMID:27044274

Marchi, S., Tognetti, R., Minnocci, A., Borghi, M., & Sebastiani, L. (2008). Variation in mesophyll anatomy and photosynthetic capacity during leaf development in a deciduous mesophyte fruit tree (*Prunus persica*) and an evergreen sclerophyllous Mediterranean shrub (*Olea europaea*). *Trees (Berlin)*, *22*(4), 559–571. doi:10.100700468-008-0216-9

Marcus, H., Hanna, L., Tait, P., Stone, S., & Wannous, C. (2023). Climate change and the public health imperative for supporting migration as adaptation. *Journal of Migration and Health*, *7*, 100174. doi:10.1016/j.jmh.2023.100174 PMID:36968560

Marengo, J. A., Galdos, M. V., Challinor, A., Cunha, A. P., Marin, F. R., Vianna, M. D. S., Alvala, R. C. S., Alves, L. M., Moraes, O. L., & Bender, F. (2022). Drought in Northeast Brazil: A review of agricultural and policy adaptation options for food security. *Climate Resilience and Sustainability*, *1*(1), e17. doi:10.1002/cli2.17

Marsden, P. D. (1986). Mucosal leishmaniasis. *Transactions of the Royal Society of Tropical Medicine and Hygiene*, *80*(6), 859–876. doi:10.1016/0035-9203(86)90243-9 PMID:3037735

Martinez-Alier, J. (2002). Environmental justice and economic degrowth: An alliance between two movements. *Capitalism, Nature, Socialism*, *13*(1), 51–73. doi:10.1080/10455752.2011.648839

Martínez-Alier, J., Pascual, U., Vivien, F.-D., & Zaccai, E. (2010). Sustainable de-growth: Mapping the context, criticisms and future prospects of an emergent paradigm. *Ecological Economics*, *69*(9), 1741–1747. doi:10.1016/j.ecolecon.2010.04.017

Martínez-Gallardo, M. R., Estrella-González, M. J., Suárez-Estrella, F., López-González, J. A., Jurado, M. M., Toribio, A. J., & López, M. J. (2023). Effect of upstream bioactivation of plant residues to accelerate the composting process and improve product quality. *Agronomy (Basel)*, *13*(6), 1638. doi:10.3390/agronomy13061638

Martínez-Valderrama, J., Olcina, J., Delacámara, G., Guirado, E., & Maestre, F. T. (2023). Complex Policy Mixes are Needed to Cope with Agricultural Water Demands Under Climate Change. *Water Resources Management*, *37*(6), 2805–2834. doi:10.100711269-023-03481-5

Martin, M., Weidner, T., & Gullström, C. (2022). Estimating the potential of building integration and regional synergies to improve the environmental performance of urban vertical farming. *Frontiers in Sustainable Food Systems*, *6*, 849304. doi:10.3389/fsufs.2022.849304

Mart, M. (2018). *Pesticides, a Love Story: America's Enduring Embrace of Dangerous Chemicals*. University Press of Kansas.

Maschler, J., Bialic-Murphy, L., Wan, J., Andresen, L. C., Zohner, C. M., Reich, P. B., Lüscher, A., Schneider, M. K., Müller, C., Moser, G., Dukes, J. S., Schmidt, I. K., Bilton, M. C., Zhu, K., & Crowther, T. W. (2022). Links across ecological scales: Plant biomass responses to elevated CO2. *Global Change Biology*, *28*(21), 6115–6134. doi:10.1111/gcb.16351 PMID:36069191

Masclaux-Daubresse, C., Daniel-Vedele, F., Dechorgnat, J., Chardon, F., Gaufichon, L., & Suzuki, A. (2010). Nitrogen uptake, assimilation and remobilization in plants: Challenges for sustainable and productive agriculture. *Annals of Botany*, *105*(7), 1141–1157. doi:10.1093/aob/mcq028 PMID:20299346

Mateos Fernández, R., Petek, M., Gerasymenko, I., Juteršek, M., Baebler, Š., Kallam, K., & Patron, N. J. (2022). Insect pest management in the age of synthetic biology. *Plant Biotechnology Journal*, *20*(1), 25–36. https://onlinelibrary.wiley.com/doi/pdf/10.1111/pbi.13685. doi:10.1111/pbi.13685 PMID:34416790

Matesanz, S., & Valladares, F. (2014). Ecological and evolutionary responses of Mediterranean plants to global change. *Environmental and Experimental Botany*, *103*, 53–67. doi:10.1016/j.envexpbot.2013.09.004

Mathews, J. A., & Tan, H. (2016). Circular economy: Lessons from China. *Nature*, *531*(7595), 440–442. doi:10.1038/531440a PMID:27008954

Mathivanan, S. (2021). Abiotic Stress-Induced Molecular and Physiological Changes and Adaptive Mechanisms in Plants. *Abiotic Stress in Plants*, 315.

Mathur, O. P., & Thakur, S. (2003). *Urban water pricing: Setting the stage for reforms*. National Institute of Public Finance and Policy. https://www.nipfp.org.in/publications/reports/1068/

Mattos, L., & Moretti, C. (2015). Oxidative stress in plants under drought conditions and the role of different enzymes. *Enzyme Engineering*, *5*, 1–6.

Ma, Y., Dias, M. C., & Freitas, H. (2020). Drought and salinity stress responses and microbe-induced tolerance in plants. *Frontiers in Plant Science*, *11*, 591911. doi:10.3389/fpls.2020.591911 PMID:33281852

Mayne, S. J., King, D. I., Andersen, J. C., & Elkinton, J. S. (2023). Pest control services on farms vary among bird species on diversified, low-intensity farms. *Global Ecology and Conservation*, *43*, e02447. doi:10.1016/j.gecco.2023.e02447

Mazelet, L. (2004). La leishmaniose canine dans le bassin méditerranéen français. Mémoire de Maîtrise de Biologie des Populations et des Ecosystèmes. Université Pierre et Marie Curie, PARIS VI, France..

McCouch, S. R., & Rieseberg, L. H. (2023). Harnessing crop diversity. In *National Acad Sciences, 120*. https://www.pnas.org/doi/full/10.1073/pnas.2221410120

Mcelrone, A. J., Reid, C. D., Hoye, K. A., Hart, E., & Jackson, R. B. (2005). Elevated CO2 reduces disease incidence and severity of a red maple fungal pathogen via changes in host physiology and leaf chemistry. *Global Change Biology*, *11*(10), 1828–1836. doi:10.1111/j.1365-2486.2005.001015.x

McKeown, E. (1987). Aloe vera. *Cosmet Toiletries*, *102*, 64–65.

McLeman, R. A., & Hunter, L. M. (2010). Migration in the context of vulnerability and adaptation to climate change: Insights from analogues. *Wiley Interdisciplinary Reviews: Climate Change*, *1*(3), 450–461. doi:10.1002/wcc.51 PMID:22022342

Médail, F., & Diadema, K. (2009). Glacial refugia influence plant diversity patterns in the Mediterranean Basin. *Journal of Biogeography*, *36*(7), 1333–1345. doi:10.1111/j.1365-2699.2008.02051.x

Mediavilla, S., Gallardo-López, V., González-Zurdo, P., & Escudero, A. (2012). Patterns of leaf morphology and leaf N content in relation to winter temperatures in three evergreen tree species. *International Journal of Biometeorology*, *56*(5), 915–926. doi:10.100700484-011-0498-2 PMID:21969112

Mehdizadeh, M., Mushtaq, W., Siddiqui, S. A., Ayadi, S., Kaur, P., Yeboah, S., & Tampubolon, K. (2021). Herbicide residues in agroecosystems: Fate, detection, and effect on non-target plants. *Reviews in Agricultural Science*, *9*, 157–167. doi:10.7831/ras.9.0_157

Mekonnen, T. W., Gerrano, A. S., Mbuma, N. W., & Labuschagne, M. T. (2022). Breeding of vegetable cowpea for nutrition and climate resilience in Sub-Saharan Africa: Progress, opportunities, and challenges. *Plants*, *11*(12), 1583. doi:10.3390/plants11121583 PMID:35736733

Mendel, Z., Podoler, H., & Rosen, D. (1984). Population dynamics of the mediterranean black scale, Saissetia oleae (Olivier), on citrus in Israel. 4. The natural enemies. *Journal of the Entomological Society of Southern Africa*, *47*(1), 1–21.

Meng, X., Zhu, Y., Yin, M., & Liu, D. (2021). The impact of land use and rainfall patterns on the soil loss of the hillslope. *Scientific Reports*, *11*(1), 1. doi:10.103841598-021-95819-5 PMID:34381135

Messier, J., McGill, B. J., Enquist, B. J., & Lechowicz, M. J. (2017). Trait variation and integration across scales: Is the leaf economic spectrum present at local scales? *Ecography*, *40*(6), 685–697. doi:10.1111/ecog.02006

Messier, J., McGill, B. J., & Lechowicz, M. J. (2010). How do traits vary across ecological scales? A case for trait-based ecology. *Ecology Letters*, *13*(7), 838–848. doi:10.1111/j.1461-0248.2010.01476.x PMID:20482582

Michaletz, S. T., Weiser, M. D., Zhou, J., Kaspari, M., Helliker, B. R., & Enquist, B. J. (2015). Plant Thermoregulation: Energetics, Trait– Environment Interactions, and Carbon Economics. *Trends in Ecology & Evolution*, *30*(12), 714–724. doi:10.1016/j.tree.2015.09.006 PMID:26476814

Mier y Terán Giménez Cacho, M., Giraldo, O. F., Aldasoro, M., Morales, H., Ferguson, B. G., Rosset, P., Khadse, A., & Campos, C. (2018). Bringing agroecology to scale: Key drivers and emblematic cases. *Agroecology and Sustainable Food Systems*, *42*(6), 637–665. doi:10.1080/21683565.2018.1443313

Mila, F. A., & Parvin, M. (2019). Forecasting area, production and yield of onion in Bangladesh by using ARIMA model. *Asian Journal of Agricultural Extension. Economia e Sociologia*, *37*(4), 1–12.

Millennium Ecosystem Assessment (MEA). (2005). *Ecosystems and Human Well-being: Synthesis*. Island Press, Washington DC. https://www.millenniumassessment.org/documents/document.356.aspx.pdf

Min Htoo, T., Yabar, H., & Mizunoya, T. (2022). GIS-Based Cluster and Suitability Analysis of Crop Residues: A Case Study in Yangon Region, Myanmar. *Applied Sciences (Basel, Switzerland)*, *12*(22), 11822. doi:10.3390/app122211822

Miner, K. R., D'Andrilli, J., Mackelprang, R., Edwards, A., Malaska, M. J., Waldrop, M. P., & Miller, C. E. (2021). Emergent biogeochemical risks from Arctic permafrost degradation. *Nature Climate Change*, *11*(10), 10. doi:10.103841558-021-01162-y

Minoli, S., Jägermeyr, J., Asseng, S., Urfels, A., & Müller, C. (2022). Global crop yields can be lifted by timely adaptation of growing periods to climate change. *Nature Communications*, *13*(1), 7079. doi:10.103841467-022-34411-5 PMID:36400762

Miranda, M. A., Miquel, M., Terrassa, J., Melis, N., & Monerris, M. (2008). Parasitism of Bactrocera oleae (Diptera; Tephritidae) by Psyttalia concolor (Hymenoptera; Braconidae) in the Balearic Islands (Spain). *Journal of Applied Entomology*, *132*(9-10), 798–805. doi:10.1111/j.1439-0418.2008.01358.x

Mizoi, J., Shinozaki, K., & Yamaguchi-Shinozaki, K. (2012). AP2/ERF family transcription factors in plant abiotic stress responses. *Biochimica et Biophysica Acta (BBA)-. Biochimica et Biophysica Acta. Gene Regulatory Mechanisms*, *1819*(2), 86–96. doi:10.1016/j.bbagrm.2011.08.004 PMID:21867785

Mkala, E. M., Mutinda, E. S., Wanga, V. O., Oulo, M. A., Oluoch, W. A., Waswa, E. N., & Wang, Q. F. (2022). Modeling impacts of climate change on the potential distribution of three endemic Aloe species critically endangered in East Africa. *Ecological Informatics*, *71*, 101765. doi:10.1016/j.ecoinf.2022.101765

Mohammadi, K., Heidari, G., Khalesro, S., & Sohrabi, Y. (2011). Soil management, microorganisms and organic matter interactions: A review. *African Journal of Biotechnology*, *10*(86), 19840.

Mohee, R., Jumnoodoo, V., Sobratee, N., Mudhoo, A., & Unmar, G. (2009). Assessing the suitability of the composting process in treating contaminating pesticides and pathogenic wastes. *Compost II. Dynamic Soil, Dynamic Plant*, *3*, 103–114.

Molina-Moreno, V., Leyva-Díaz, J. C., Llorens-Montes, F. J., & Cortés-García, F. J. (2017). Design of indicators of circular economy as instruments for the evaluation of sustainability and efficiency in wastewater from pig farming industry. *Water (Basel)*, *9*(9), 653. doi:10.3390/w9090653

Mollier, L., Seyler, F., Chotte, J. L., & Ringler, C. (2017). *End hunger, achieve food security and improved nutrition and promote sustainable agriculture: SDG 2. A Guide to SDG Interactions: From Science to Implementation*. ICSU.

Mondal, S. (2021). *Impact of Climate Change on Soil Fertility*., doi:10.1007/978-3-030-76863-8_28

Montagnini, F., & del Fierro, S. (2022). Functions of agroforestry systems as biodiversity islands in productive landscapes. In Biodiversity Islands: Strategies for Conservation in Human-Dominated Environments (pp. 89-116). Springer. doi:10.1007/978-3-030-92234-4_4

Montzka, S. (2022). *The NOAA Annual Greenhouse Gas Index (AGGI).* NOAA Global Monitoring Laboratory Website. https://gml.noaa.gov/aggi/aggi.html

Moore, T. E., Jones, C. S., Chong, C., & Schlichting, C. D. (2020). Impact of rainfall seasonality on intraspecific trait variation in a shrub from a Mediterranean climate. *Functional Ecology*, *34*(4), 865–876. doi:10.1111/1365-2435.13533

Moraes, W. B., Madden, L. V., Baik, B.-K., Gillespie, J., & Paul, P. A. (2023). Environmental Conditions After Fusarium Head Blight Visual Symptom Development Affect Contamination of Wheat Grain with Deoxynivalenol and Deoxynivalenol-3-Glucoside. *Phytopathology*, *113*(2), 206–224. doi:10.1094/PHYTO-06-22-0199-R PMID:36131392

Moran, E. V., Hartig, F., & Bell, D. M. (2016). Intraspecific trait variation across scales: Implications for understanding global change responses. *Global Change Biology*, *22*(1), 137–150. doi:10.1111/gcb.13000 PMID:26061811

Moreira, M. F., Mansur, J. F., & Mansur, J. (2012). Resistência e inseticidas: estratégias, desafios e perspectivas no controle de insetos. Resistência e Inseticidas: Estratégias, Desafios e Perspectivas no Controle de Insetos. 1ed. Rio de Janeiro: INCT-EM, 1-23.

Moriana, A., Orgaz, F., Pastor, M., & Fereres, E. (2003). Yield responses of a mature olive orchard to water deficits. *Journal of the American Society for Horticultural Science*, *128*(3), 425–431. doi:10.21273/JASHS.128.3.0425

Mota-Ituarte, M., Pedroza-Sandoval, A., Minjares-Fuentes, R., Trejo-Calzada, R., Zegbe, J. A., & Quezada-Rivera, J. J. (2023). Water deficit and salinity modify some morphometric, physiological, and productive attributes of Aloe vera (L.). *Botanical Sciences*, *101*(2), 463–475. doi:10.17129/botsci.3175

Moumane, A., Al Karkouri, J., Benmansour, A., El Ghazali, F. E., Fico, J., Karmaoui, A., & Batchi, M. (2022). Monitoring long-term land use, land cover change, and desertification in the Ternata Oasis, Middle Draa Valley, Morocco. *Remote Sensing Applications: Society and Environment*, *26*, 100745. doi:10.1016/j.rsase.2022.100745

Moziane, F., Benalioua, M., Trebycka, K., Bouhass, R., Belazzoug, S., & Aguercif, M. (1983). Kalar-Azar chez l'enfant dans l'ouest de l'Algérie. A propos de 14 observations. *Medecine et Hygiene*, *41*(1536), 3649–3652.

Mrabet, R., & Moussadek, R. (2022). Development of climate smart agriculture in Africa. In Conservation Agriculture in Africa: Climate Smart Agricultural Development (pp. 17-65). GB: CABI. doi:10.1079/9781789245745.0002

Mrabet, R. (2023). Sustainable agriculture for food and nutritional security. In *Sustainable Agriculture and the Environment* (pp. 25–90). Academic Press. doi:10.1016/B978-0-323-90500-8.00013-0

Mrabet, R., Moussadek, R., Fadlaoui, A., & Van Ranst, E. (2012). Conservation agriculture in dry areas of Morocco. *Field Crops Research*, *132*, 84–94. doi:10.1016/j.fcr.2011.11.017

Mrabet, R., Saber, N., El-Brahli, A., Lahlou, S., & Bessam, F. (2001). Total, particulate organic matter and structural stability of a Calcixeroll soil under different wheat rotations and tillage systems in a semiarid area of Morocco. *Soil & Tillage Research*, *57*(4), 225–235. doi:10.1016/S0167-1987(00)00180-X

Muhie, S. H. (2023). Concepts, principles, and application of biodynamic farming: A review. *Circular Economy and Sustainability*, *3*(1), 291–304. doi:10.100743615-022-00184-8

Mukherjee, S., Keswani, K., Nath, P., & Paul, S. (2023). A study on the correlation of anthropogenic activities with climate change. *Environmental Quality Management*, tqem.21984. doi:10.1002/tqem.21984

Mumby, P. J., Chollett, I., Bozec, Y.-M., & Wolff, N. H. (2014). Ecological resilience, robustness and vulnerability: How do these concepts benefit ecosystem management? *Current Opinion in Environmental Sustainability*, *7*, 22–27. doi:10.1016/j.cosust.2013.11.021

Munaweera, T., Jayawardana, N., Rajaratnam, R., & Dissanayake, N. (2022). Modern plant biotechnology as a strategy in addressing climate change and attaining food security. *Agriculture & Food Security*, *11*(1), 1–28. doi:10.118640066-022-00369-2

Munda, G. (1997). Environmental Economics, Ecological Economics, and the Concept of Sustainable Development. *Environmental Values*, *6*(2), 213–233. doi:10.1177/096327199700600206

Murillo-Amador, B., Nieto-Garibay, A., Troyo-Diéguez, E., García-Hernández, J. L., Hernández-Montiel, L., & Valdez-Cepeda, R. D. (2015). Estrés salino moderado en los rasgos fisiológicos y morfológicos del Aloe vera L. *Botanical Sciences*, *93*(3), 639–648. doi:10.17129/botsci.73

Nabah, K., Rharrabe, K., & Mezzoug, N. (2022). Profil épidémiologique de la leishmaniose cutanée au niveau de la région Nord du Maroc. *Revue d'Epidemiologie et de Sante Publique*, *70*, S237. doi:10.1016/j.respe.2022.06.264

Nair, P. R., Kumar, B. M., Nair, V. D., Nair, P. R., Kumar, B. M., & Nair, V. D. (2021). Definition and concepts of agroforestry. *An Introduction to Agroforestry: Four Decades of Scientific Developments*, 21-28.

Nair, V. D., Nair, P. R., Dari, B., Freitas, A. M., Chatterjee, N., & Pinheiro, F. M. (2017). Biochar in the agroecosystem–climate-change–sustainability nexus. *Frontiers in Plant Science*, *8*, 2051. doi:10.3389/fpls.2017.02051 PMID:29312364

Najafi, E., Devineni, N., Khanbilvardi, R., & Kogan, F. (2018). Understanding the changes in global crop yields through changes in climate and technology. *Earth's Future*, *6*(3), 410–427. doi:10.1002/2017EF000690

Naorem, A. (2022). Carbon Sequestration. *Resonance*, *27*(7), 1237–1245. doi:10.100712045-022-1416-0

Napier, J. A., Haslam, R. P., Tsalavouta, M., & Sayanova, O. (2019). The challenges of delivering genetically modified crops with nutritional enhancement traits. *Nature Plants*, *5*(6), 563–567. doi:10.103841477-019-0430-z PMID:31160704

NASA. (2022). Global Climate Change. Carbon dioxide. *Latest Measurement June 2022*. Carbon Dioxide|Vital Signs—Climate Change: Vital Signs of the Planet.

Navas-Cortés, J. A., Landa, B. B., Mercado-Blanco, J., Trapero-Casas, J. L., Rodríguez-Jurado, D., & Jiménez-Díaz, R. M. (2008). Spatiotemporal analysis of spread of infections by Verticillium dahliae pathotypes within a high tree density olive orchard in southern Spain. *Phytopathology*, *98*(2), 167–180. doi:10.1094/PHYTO-98-2-0167 PMID:18943193

Nazir, T., Khan, S., & Qiu, D. (2019). Biological control of insect pest. *Pests Control and Acarology*, *21*.

Neelam, S., & Sood, S. K. (2020). A scientometric review of global research on smart disaster management. *IEEE Transactions on Engineering Management*, *68*(1), 317–329. doi:10.1109/TEM.2020.2972288

Negatu, B., Dugassa, S., & Mekonnen, Y. (2021). Environmental and health risks of pesticide use in Ethiopia. *Journal of Health & Pollution*, *11*(30), 210601. doi:10.5696/2156-9614-11.30.210601 PMID:34267988

Neira, J., Ortiz, M., Morales, L., & Acevedo, E. (2015). Oxygen diffusion in soils: Understanding the factors and processes needed for modeling. *Chilean Journal of Agricultural Research*, *75*, 35–44. doi:10.4067/S0718-58392015000300005

Nelson, A. R. L. E., Ravichandran, K., & Antony, U. (2019). The impact of the Green Revolution on indigenous crops of India. *Journal of Ethnic Foods*, *6*(1), 8. doi:10.118642779-019-0011-9

Nguyen, H., Chen, Z., Fries, A., Berlin, M., Hallingbäck, H. R., & Wu, H. X. (2021). Effect of additive, dominant and epistatic variances on breeding and deployment strategy in Norway spruce. *Forestry*, *95*(3), 416–427. doi:10.1093/forestry/cpab052

Nievola, C. C., Carvalho, C. P., Carvalho, V., & Rodrigues, E. (2017). Rapid responses of plants to temperature changes. Temperature: *Multidisciplinary. Biomedical Journal*, *4*(4), 371–405. doi:10.1080/23328940.2017.1377812 PMID:29435478

Niggli, U., Fließbach, A., Hepperly, P., & Scialabba, N. (2009). Low greenhouse gas agriculture: Mitigation and adaptation potential of sustainable farming systems. *Ökologie und Landbau*, *141*, 32–33.

Ninan, K. N., & Inoue, M. (2013). Valuing forest ecosystem services: Case study of a forest reserve in Japan. *Ecosystem Services*, *5*(1), 78–87. doi:10.1016/j.ecoser.2013.02.006

Ning, K., Chen, J., Li, Z., Liu, C., Nie, X., Liu, Y., Wang, L., & Hu, X. (2021). Land use change induced by the implementation of ecological restoration Programs increases future terrestrial ecosystem carbon sequestration in red soil hilly region of China. *Ecological Indicators*, *133*, 108409. doi:10.1016/j.ecolind.2021.108409

Ni, Y., Turner, D., Yates, K. Á., & Tizard, I. (2004). Isolation and characterization of structural components of Aloe vera L. leaf pulp. *International Immunopharmacology*, *4*(14), 1745–1755. doi:10.1016/j.intimp.2004.07.006 PMID:15531291

Noyes, P. D., McElwee, M. K., Miller, H. D., Clark, B. W., Van Tiem, L. A., Walcott, K. C., & Levin, E. D. (2009). The toxicology of climate change: Environmental contaminants in a warming world. *Environment International*, *35*(6), 971–986. doi:10.1016/j.envint.2009.02.006 PMID:19375165

Nuccio, M. L., Wu, J., Mowers, R., Zhou, H. P., Meghji, M., Primavesi, L. F., Paul, M. J., Chen, X., Gao, Y., Haque, E., & Lagrimini, L. M. (2015). Expression of trehalose-6-phosphate phosphatase in maize ears improves yield in well-watered and drought conditions. *Nature Biotechnology*, *33*(8), 862–869. doi:10.1038/nbt.3277 PMID:26473199

Nyenje, P. M., Foppen, J. W., Uhlenbrook, S., Kulabako, R., & Muwanga, A. (2010). Eutrophication and nutrient release in urban areas of sub-Saharan Africa—A review. *The Science of the Total Environment*, *408*(3), 447–455. doi:10.1016/j.scitotenv.2009.10.020 PMID:19889445

Obanor, F. O., Walter, M., Jones, E. E., & Jaspers, M. V. (2008). Effect of temperature, relative humidity, leaf wetness and leaf age on Spilocaea oleagina conidium germination on olive leaves. *European Journal of Plant Pathology*, *120*(3), 211–222. doi:10.100710658-007-9209-6

OECD. (2009). Report of the seminar on pesticide risk reduction through spray drift reduction strategies as part of national risk management. ENV/JM/MONO 36, Paris, France.

Okamura, N., Hine, N., Tateyama, Y., Nakazawa, M., Fujioka, T., Mihashi, K., & Yagi, A. (1998). Five chromones from Aloe vera leaves. *Phytochemistry*, *49*(1), 219–223. doi:10.1016/S0031-9422(97)01071-6

Oladele, O., & Braimoh, A. (2014). Potential of agricultural land management activities for increased soil carbon sequestration in Africa–a review. *Applied Ecology and Environmental Research*, *12*(3), 741–751. doi:10.15666/aeer/1203_741751

Oliva, J., Stenlid, J., & Martinez-Vilalta, J. (2014). The effect of fungal pathogens on the water and carbon economy of trees: Implications for drought-induced mortality. *The New Phytologist*, *203*(4), 1028–1035. doi:10.1111/nph.12857 PMID:24824859

Oliveira, C. M., Auad, A. M., Mendes, S. M., & Frizzas, M. R. (2014). Crop losses and the economic impact of insect pests on Brazilian agriculture. *Crop Protection (Guildford, Surrey)*, *56*, 50–54. doi:10.1016/j.cropro.2013.10.022

Oliveira, T. C., Cabral, J. S. R., Santana, L. R., Tavares, G. G., Santos, L. D. S., Paim, T. P., Müller, C., Silva, F. G., Costa, A. C., Souchie, E. L., & Mendes, G. C. (2022). The arbuscular mycorrhizal fungus Rhizophagus clarus improves physiological tolerance to drought stress in soybean plants. *Scientific Reports*, *12*(1), 9044. doi:10.103841598-022-13059-7 PMID:35641544

Oliverio, A. M., Bradford, M. A., & Fierer, N. (2017). Identifying the microbial taxa that consistently respond to soil warming across time and space. *Global Change Biology*, *23*(5), 2117–2129. doi:10.1111/gcb.13557 PMID:27891711

Olmo, D., Nieto, A., Borràs, D., Montesinos, M., Adrover, F., Pascual, A., Gost, P. A., Quetglas, B., Urbano, A., García, J. D., Velasco-Amo, M. P., Olivares-García, C., Beidas, O., Juan, A., Marco-Noales, E., Gomila, M., Rita, J., Moralejo, E., & Landa, B. B. (2021). Landscape epidemiology of Xylella fastidiosa in the Balearic Islands. *Agronomy (Basel)*, *11*(3), 473. doi:10.3390/agronomy11030473

Omidi, M., Fatehinya, A., Farahani, M., Akbari, Z., Shahmoradi, S., & Yazdian, F. (2017). Vashaee, D. Characterization of biomaterials. In *Biomaterials for oral and dental tissue engineering* (pp. 97–115). Woodhead Publishing. doi:10.1016/B978-0-08-100961-1.00007-4

Onaga, G., Wydra, K. D., Koopmann, B., Séré, Y., & von Tiedemann, A. (2016). Elevated temperature increases in planta expression levels of virulence related genes in *Magnaporthe oryzae* and compromises resistance in *Oryza sativa* cv. Nipponbare. *Functional Plant Biology*, *44*(3), 358–371. doi:10.1071/FP16151 PMID:32480570

Onder, M., Ceyhan, E., & Kahraman, A. (2011). Effects of agricultural practices on environment. *Biol Environ Chem.*, *24*, 28–32.

Orgaz, F., Villalobos, F. J., Testi, L., & Fereres, E. (2007). A model of daily mean canopy conductance for calculating transpiration of olive canopies. *Functional Plant Biology*, *34*(3), 178. doi:10.1071/FP06306 PMID:32689344

Ortiz, A. M. D., Outhwaite, C. L., Dalin, C., & Newbold, T. (2021). A review of the interactions between biodiversity, agriculture, climate change, and international trade: Research and policy priorities. *One Earth*, *4*(1), 88–101. doi:10.1016/j.oneear.2020.12.008

Oulbi, S. Kassout, J., Bouizgaren, A., El Khatri, N., Sikaoui, L., & Hadria, R. (2023). Assessment of Carbon Balance in Olive Cultivation Systems in Morocco: Toward an Intelligent Solution in Climate Change Mitigation. In Intelligent Solutions for Optimizing Agriculture and Tackling Climate Change: Current and Future Dimensions (pp. 96-114). IGI Global.

Ovaskainen, O., Skorokhodova, S., Yakovleva, M., Sukhov, A., Kutenkov, A., Kutenkova, N., Shcherbakov, A., Meyke, E., & Delgado, M. M. (2013). Community-level phenological response to climate change. *Proceedings of the National Academy of Sciences of the United States of America*, *110*(33), 13434–13439. doi:10.1073/pnas.1305533110 PMID:23901098

Padedda, B. M., Lugliè, A., Lai, G. G., Giadrossich, F., Satta, C. T., & Pulina, S. (2022). Land-based impact of nutrient loads and eutrophication on an ancient Mediterranean natural lake. *Hydrology*, *9*(1), 7. doi:10.3390/hydrology9010007

Pahwa, R., Goyal, A., Bansal, P., & Jialal, I. (2020). Chronic inflammation. *StatPearls*.

Pallioti, A., Bongi, G., & Rocchi, P. (1994). Peltate trichomes effects on photosynthetic gas exchange of *Olea europaea* L. leaves. *Plant Physiology*, *13*, 35–44.

Palm, C. A., Smukler, S. M., Sullivan, C. C., Mutuo, P. K., Nyadzi, G. I., & Walsh, M. G. (2010). Identifying potential synergies and trade-offs for meeting food security and climate change objectives in sub-Saharan Africa. *Proceedings of the National Academy of Sciences of the United States of America*, *107*(46), 19661–19666. doi:10.1073/pnas.0912248107

Palm, C., Blanco-Canqui, H., DeClerck, F., Gatere, L., & Grace, P. (2014). Conservation agriculture and ecosystem services: An overview. *Agriculture, Ecosystems & Environment*, *187*, 87–105. doi:10.1016/j.agee.2013.10.010

Panetta, A. M., Stanton, M. L., & Harte, J. (2018). Climate warming drives local extinction: Evidence from observation and experimentation. *Science Advances, 4*(2), eaaq1819. doi:10.1126/sciadv.aaq1819

Pan, J., Sharif, R., Xu, X., & Chen, X. (2021). Mechanisms of waterlogging tolerance in plants: Research progress and prospects. *Frontiers in Plant Science, 11*, 627331. doi:10.3389/fpls.2020.627331 PMID:33643336

Pan, T., Sun, X., Liu, Y., Li, H., Deng, G., Lin, H., & Wang, S. (2018). Heat stress alters genome-wide profiles of circular RNAs in Arabidopsis. *Plant Molecular Biology, 96*(3), 217–229. doi:10.100711103-017-0684-7 PMID:29177640

Pantazi, X. E., Moshou, D., & Tamouridou, A. A. (2019). Automated leaf disease detection in different crop species through image features analysis and One Class Classifiers. *Computers and Electronics in Agriculture, 156*, 96–104. doi:10.1016/j.compag.2018.11.005

Pantera, A., Mosquera-Losada, M., Herzog, F., & Den Herder, M. (2021). Agroforestry and the environment. *Agroforestry Systems, 95*(5), 767–774. doi:10.100710457-021-00640-8

Pantin, F., Monnet, F., Jannaud, D., Costa, J. M., Renaud, J., Muller, B., Simonneau, T., & Genty, B. (2013). The dual effect of abscisic acid on stomata. *The New Phytologist, 197*(1), 65–72. doi:10.1111/nph.12013 PMID:23106390

Paradis, E., Claude, J., & Strimmer, K. (2004). APE: Analyses of phylogenetics and evolution in R language. *Bioinformatics (Oxford, England), 20*(2), 289–290. doi:10.1093/bioinformatics/btg412 PMID:14734327

Parihar, A. (2022). *Shiwani.* Molecular Breeding and Marker-Assisted Selection for Crop Improvement. In Plant Genomics for Sustainable Agriculture. doi:10.1007/978-981-16-6974-3_6

Parkes, M. G., Cubillos Tovar, J. P., Dourado, F., Domingos, T., & Teixeira, R. F. (2022). Life Cycle Assessment of a Prospective Technology for Building-Integrated Production of Broccoli Microgreens. *Atmosphere (Basel), 13*(8), 1317. doi:10.3390/atmos13081317

Parmar, N., Singh, K. H., Sharma, D., Singh, L., Kumar, P., Nanjundan, J., & Thakur, A. K. (2017). Genetic engineering strategies for biotic and abiotic stress tolerance and quality enhancement in horticultural crops: a comprehensive review. *3 Biotech, 7*, 1-35. doi:10.1007/s13205-017-0870-y

Parmesan, C., & Hanley, M. E. (2015). Plants and climate change: Complexities and surprises. *Annals of Botany, 116*(6), 849–864. doi:10.1093/aob/mcv169 PMID:26555281

Parra, A. S., de Figueiredo, E. B., de Bordonal, R. O., Moitinho, M. R., Teixeira, D. D. B., & La Scala, N. Jr. (2019). Greenhouse gas emissions in conversion from extensive pasture to other agricultural systems in the Andean region of Colombia. *Environment, Development and Sustainability, 21*(1), 249–262. doi:10.100710668-017-0034-6

Pathak, H., Aggarwal, P. K., & Singh, S. D. (2012). *Climate change impact, adaptation and mitigation in agriculture: methodology for assessment and applications*. Indian Agricultural Research Institute.

Pathirana, R., & Carimi, F. (2022). Management and utilization of plant genetic resources for a sustainable agriculture. *Plants, 11*(15), 2038. doi:10.3390/plants11152038 PMID:35956515

Patil, P., & Kumar, A. K. (2017). Biological carbon sequestration through fruit crops (perennial crops-natural "sponges" for absorbing carbon dioxide from atmosphere). *Plant Archives, 17*(2), 1041–1046.

Patil, R., & Kumar, S. (2020). Bibliometric Survey on Diagnosis of Plant Leaf Diseases Using Artificial Intelligence. *International Journal of Modern Agriculture, 9*(3), 1111–1131. http://modern-journals.com/index.php/ijma/article/view/316

Patni, B., Bhattacharyya, M., kumari, A., & Purohit, V. K. (2021). Alarming influence of climate change and compromising quality of medicinal plants. *Plant Physiology Reports*, 1-10.

Patra, S., Julich, S., Feger, K. H., Jat, M. L., Sharma, P. C., & Schwärzel, K. (2019). Effect of conservation agriculture on stratification of soil organic matter under cereal-based cropping systems. *Archives of Agronomy and Soil Science*, *65*(14), 2013–2028. doi:10.1080/03650340.2019.1588462

Patrick, R. (2015). When the well runs dry: The slow train wreck of global water scarcity. *Journal - American Water Works Association*, *107*(3), 65–76. doi:10.5942/jawwa.2015.107.0042

Paul, M. J., Oszvald, M., Jesus, C., Rajulu, C., & Griffiths, C. A. (2017). Increasing crop yield and resilience with trehalose 6-phosphate: Targeting a feast–famine mechanism in cereals for better source–sink optimization. *Journal of Experimental Botany*, *68*(16), 4455–4462. doi:10.1093/jxb/erx083 PMID:28981769

Pavlidis, G., & Tsihrintzis, V. A. (2018). Environmental benefits and control of pollution to surface water and groundwater by agroforestry systems: A review. *Water Resources Management*, *32*(1), 1–29. doi:10.100711269-017-1805-4

Pazienza, P., & De Lucia, C. (2020). For a new plastics economy in agriculture: Policy reflections on the EU strategy from a local perspective. *Journal of Cleaner Production*, *253*, 119844. doi:10.1016/j.jclepro.2019.119844

Pearce, D. (2002). An Intellectual History of Environmental Economics. *Annual Review of Energy and the Environment*, *27*(1), 57–81. doi:10.1146/annurev.energy.27.122001.083429

Pearson, R. D., Jeronimo, S. M., & De Sousa, A. (2001). Leishmaniasis, pp: 287-331. Principles and Practice of Parasitology. John Wiley & Son, New York.

Pedrosa, M. M., Cuadrado, C., Burbano, C., Allaf, K., Haddad, J., Gelencsér, E., Takács, K., Guillamón, E., & Muzquiz, M. (2012). Effect of instant controlled pressure drop on the oligosaccharides, inositol phosphates, trypsin inhibitors and lectins contents of different legumes. *Food Chemistry*, *131*(3), 862–868. doi:10.1016/j.foodchem.2011.09.061

Peerzada, A. M., Bukhari, S. A. H., Dawood, M., Nawaz, A., Ahmad, S., & Adkins, S. (2019). *Weed management for healthy crop production. Agronomic Crops* (Vol. 2). Management Practices.

Perez-Harguindeguy, N., Diaz, S., Garnier, E., Lavorel, S., Poorter, H., Jaureguiberry, P., Bret-Harte, M. S., Cornwell, W. K., Craine, J. M., Gurvich, D. E., Urcelay, C., Veneklaas, E. J., Reich, P. B., Poorter, L., Wright, I. J., Ray, P., Enrico, L., Pausas, J. G., de Vos, A. C., & Cornelissen, J. H. C. (2013). New handbook for standardised measurement of plant functional traits worldwide. *Australian Journal of Botany*, *61*(3), 167–234. doi:10.1071/BT12225

Pérez-Lucas, G., Gambín, M., & Navarro, S. (2020). Leaching behaviour appraisal of eight persistent herbicides on a loam soil amended with different composted organic wastes using screening indices. *Journal of Environmental Management*, *273*, 111179. doi:10.1016/j.jenvman.2020.111179 PMID:32771853

Pérez-Rodríguez, M., Serrano, N., Arquero, O., Orgaz, F., Moral, J., & López-Escudero, F. J. (2016). The effect of short irrigation frequencies on the development of Verticillium wilt in the susceptible olive cultivar 'Picual' under field conditions. *Plant Disease*, *100*(9), 1880–1888. doi:10.1094/PDIS-09-15-1018-RE PMID:30682984

Pete, S., Johnson, N., Katherine, C., Donovan, C., Francesco, C., Giacomo, G., . . . Vizzarri, M. (2019). Interlinkages between desertification, land degradation, food security and greenhouse gas fluxes: synergies, trade-offs and integrated response options. In Climate Change and Land: IPCC Special Report on Climate Change, Desertification, Land Degradation, Sustainable Land Management, Food Security, and Greenhouse Gas Fluxes in Terrestrial Ecosystems (pp. 551-672). Cambridge University Press. doi:10.1017/9781009157988.008

Petersen, J. K., Saurel, C., Nielsen, P., & Timmermann, K. (2016). The use of shellfish for eutrophication control. *Aquaculture International*, *24*(3), 857–878. doi:10.100710499-015-9953-0

Peterson, C. A., Eviner, V. T., & Gaudin, A. C. (2018). Ways forward for resilience research in agroecosystems. *Agricultural Systems, 162*, 19–27. doi:10.1016/j.agsy.2018.01.011

Peterson, D. E., Shoup, D. E., Thompson, C. R., & Olson, B. L. (2013). *Herbicide mode of action, cooperative extension service*. Kansas State University.

Peterson, M. A., Collavo, A., Ovejero, R., Shivrain, V., & Walsh, M. J. (2018). The challenge of herbicide resistance around the world: A current summary. *Pest Management Science, 74*(10), 2246–2259. doi:10.1002/ps.4821 PMID:29222931

Petit, A. N., Fontaine, F., Clément, C., & Vaillant-Gaveau, N. (2008). Photosynthesis limitations of grapevine after treatment with the fungicide fludioxonil. *Journal of Agricultural and Food Chemistry, 56*(15), 6761–6767. doi:10.1021/jf800919u PMID:18598040

Petridis, A., Therios, I., Samouris, G., Koundouras, S., & Giannakoula, A.. (2012). Effect of water defcit on leaf phenolic composition, gas exchange, oxidative damage and antioxidant activity of four Greek olive (Olea europaea L.) cultivars. Plant physiology and biochemistry, 60, 1– 11.

Petzoldt, C., & Seaman, A. (2006). Climate change effects on insects and pathogens. *Climate change and agriculture: Promoting practical and profitable responses, 3*, 6-16.

Pfeifer, L., & Otto, I. M. (2023). Changing seasonal temperature offers a window of opportunity for stricter climate policy. *Environmental Science & Policy, 140*, 35–45. doi:10.1016/j.envsci.2022.11.010

Picarella, M. E., & Mazzucato, A. (2019). The Occurrence of Seedlessness in Higher Plants; Insights on Roles and Mechanisms of Parthenocarpy. *Frontiers in Plant Science, 9*, 1997. doi:10.3389/fpls.2018.01997 PMID:30713546

Picchi, M. S., Tondini, E., Albertarelli, N., Monteforti, G., & Petacchi, R. (2022). Following the pest outbreak: Preliminary findings on the landscape effect on Dasineura oleae and its parasitoids in central Italy. *Phytoparasitica, 50*(2), 1–15. doi:10.100712600-021-00962-4

Pierantozzi, P., Torres, M., Tivani, M., Contreras, C., Gentili, L., Parera, C., & Maestri, D. (2020). Spring deficit irrigation in olive (cv. Genovesa) growing under arid continental climate: Effects on vegetative growth and productive parameters. *Agricultural Water Management, 238*, 106212. doi:10.1016/j.agwat.2020.106212

Pierce, S., Negreiros, D., Cerabolini, B. E., Kattge, J., Díaz, S., Kleyer, M., Shipley, B., Wright, S. J., Soudzilovskaia, N. A., Onipchenko, V. G., van Bodegom, P. M., Frenette-Dussault, C., Weiher, E., Pinho, B. X., Cornelissen, J. H. C., Grime, J. P., Thompson, K., Hunt, R., Wilson, P. J., & Tampucci, D. (2017). A global method for calculating plant CSR ecological strategies applied across biomes world-wide. *Functional Ecology, 31*(2), 444–457. doi:10.1111/1365-2435.12722

Piñeiro, V., Arias, J., Dürr, J., Elverdin, P., Ibáñez, A. M., Kinengyere, A., & Prager, S. D. (2020). A scoping review on incentives for adoption of sustainable agricultural practices and their outcomes. *Nature Sustainability, 3*(10), 809–820. doi:10.103841893-020-00617-y

Pinheiro, J., Bates, D., DebRoy, S., & Sarkar, D. (2021). Nonlinear mixed-effects models. *R package version, 3*, 1-153.

Pinheiro, L. A., Medina, P., Bento, A., Pereira, J. A., & Santos, S. A. P. (2017). Bioecology of Psyttalia concolor (Szépligeti) (Hymenoptera: Braconidae) and management strategies to increase biological control of the olive fruit fly. In S. A. P. Santos (Ed.), *Natural Enemies: Identification, Protection Strategies and Ecological Impacts* (pp. 141–172). Nova Science Publishers.

Pisante, M., Stagnari, F., Acutis, M., Bindi, M., Brilli, L., Di Stefano, V., & Carozzi, M. (2015). Conservation agriculture and climate change. *Conservation agriculture*, 579-620.

Pizzorno, J. (2014). Glutathione! *Integrative Medicine (Encinitas, Calif.), 13*(1), 8. PMID:26770075

PME, (2019). *Provincial Monograph of Errachidia*. Monographie de la province d'Errachidia de l'année 2019.

Polade, S. D., Pierce, D. W., Cayan, D. R., Gershunov, A., & Dettinger, M. D. (2014). The key role of dry days in changing regional climate and precipitation regimes. *Scientific Reports*, *4*(1), 4364. doi:10.1038rep04364 PMID:24621567

Ponti, L., Gutierrez, A. P., Ruti, P. M., & Dell'Aquila, A. (2014). Fine-scale ecological and economic assessment of climate change on olive in the Mediterranean Basin reveals winners and losers. *Proceedings of the National Academy of Sciences of the United States of America*, *111*(15), 5598–5603. doi:10.1073/pnas.1314437111 PMID:24706833

Porter, J. R., & Semenov, M. A. (2005). Crop responses to climatic variation. *Philosophical Transactions of the Royal Society of London. Series B, Biological Sciences*, *360*(1463), 2021–2035. doi:10.1098/rstb.2005.1752 PMID:16433091

Pörtner, H. O., Roberts, D. C., Poloczanska, E. S., Mintenbeck, K., & Tignor, M. (2022). IPCC, 2022: Summary for policymakers.

Potter, C., Klooster, S., Myneni, R., Genovese, V., Tan, P. N., & Kumar, V. (2003). Continental-scale comparisons of terrestrial carbon sinks estimated from satellite data and ecosystem modeling 1982–1998. *Global and Planetary Change*, *39*(3-4), 201–213. doi:10.1016/j.gloplacha.2003.07.001

Poudel, S., Poudel, B., Acharya, B., & Poudel, P. (2020). Pesticide use and its impacts on human health and environment. *Environ Ecosyst Sci*, *4*(1), 47–51. doi:10.26480/ees.01.2020.47.51

Poulton, P., Johnston, J., Macdonald, A., White, R., & Powlson, D. (2018). Major limitations to achieving "4 per 1000" increases in soil organic carbon stock in temperate regions: Evidence from long-term experiments at Rothamsted Research, United Kingdom. *Global Change Biology*, *24*(6), 2563–2584. doi:10.1111/gcb.14066

Pound, B., Lamboll, R., Croxton, S., Gupta, N., & Bahadur, A. V. (2018). *Climate-Resilient Agriculture in South Asia: An analytical framework and insights from practice*. Action on Climate Today.

Pour, M. T., & Kalashami, M. K. (2012). Applying CVM for economic valuation of drinking water in Iran. *International Journal of Agricultural Management and Development*, *2*(3), 209–214.

Prasad, P. V., Bheemanahalli, R., & Jagadish, S. K. (2017). Field crops and the fear of heat stress—Opportunities, challenges and future directions. *Field Crops Research*, *200*, 114–121. doi:10.1016/j.fcr.2016.09.024

Pretty, J., & Pervez Bharucha, Z. (2015). Integrated pest management for sustainable intensification of agriculture in Asia and Africa. *Insects*, *6*(1), 152–182. doi:10.3390/insects6010152 PMID:26463073

Priori, S. (2018). Effect of organic treatments on soil carbon and nitrogen dynamics in vineyard. *EQA*, *31*, 1–10.

Priya, R., Ramesh, D., & Khosla, E. (2018, September). Crop prediction on the region belts of India: A Naïve Bayes MapReduce precision agricultural model. In 2018 international conference on advances in computing, communications, and informatics (ICACCI) (pp. 99-104). IEEE. doi:10.1109/ICACCI.2018.8554948

Priyadarshini, P., & Abhilash, P. C. (2020). Policy recommendations for enabling transition towards sustainable agriculture in India. *Land Use Policy*, *96*, 104718. doi:10.1016/j.landusepol.2020.104718

Pueppke, S. G., Nurtazin, S., & Ou, W. (2020). Water and Land as Shared Resources for Agriculture and Aquaculture: Insights from Asia. *Water (Basel)*, *12*(10), 10. doi:10.3390/w12102787

Qaim, M. (2020). Role of new plant breeding technologies for food security and sustainable agricultural development. *Applied Economic Perspectives and Policy*, *42*(2), 129–150. doi:10.1002/aepp.13044

Quezada, C. A. V., Fonseca, M. B., & Romero, H. (2016). The circular agriculture applied in neighboring countries: The case of biogas on the border between Ecuador and Perú. *New Biotechnology*, *33*, S66–S67. doi:10.1016/j.nbt.2016.06.953

Quezada-Martinez, D., Nyarko, C. P. A., Schiessl, S., & Mason, A. S. (2021). Using wild relatives and related species to build climate resilience in Brassica crops. *Theoretical and Applied Genetics*, *134*(6), 1711–1728. doi:10.100700122-021-03793-3 PMID:33730183

Quézel, P., & Médail, F. (2003). *Ecologie et biogéographie des forêts du bassin méditerranéen*. Elsevier.

Qu, R. Y., He, B., Yang, J. F., Lin, H. Y., Yang, W. C., Wu, Q. Y., & Yang, G. F. (2021). Where are the new herbicides? *Pest Management Science*, *77*(6), 2620–2625. doi:10.1002/ps.6285 PMID:33460493

R Core Team. (2020). *R: A language and environment for statistical computing*. R.

Race, D., Gentle, P., & Mathew, S. (2023). Living on the margins: Climate change impacts and adaptation by remote communities living in the Pacific Islands, the Himalaya and desert Australia. *Climate Risk Management*, *40*, 100503. doi:10.1016/j.crm.2023.100503

Rai, A., Irulappan, V., & Senthil-Kumar, M. (2022). Dry root rot of chickpea: A disease favored by drought. *Plant Disease*, *106*(2), 346–356. doi:10.1094/PDIS-07-21-1410-FE PMID:34649462

Rajasekaran, S., Sivagnanam, K., & Subramanian, S. (2005). Mineral contents of Aloe vera leaf gel and their role on streptozotocin-induced diabetic rats. *Biological Trace Element Research*, *108*(1), 185–195. doi:10.1385/BTER:108:1-3:185 PMID:16327071

Rakotovao, H. N. (2017). *Impacts environnementaux de l'adoption des pratiques agroécologiques: empreinte car1bone d'exploitation agricole et stocks de carbone, Région Itasy, Madagascar*. [Thèse de Doctorat en Sciences Agro1nomiques, Université d'Antananarivo].

Rakotovao, N. H., Ramifehiarivo, N., Grinand, C., Razakamanarivo, H. R., Albrecht, A., & Razafimbelo, T. M. (2020). *Effects of agroecological practices on soil organic carbon sequestration using synchronic and diachronic approaches in Madagascar*.

Raman, H., Uppal, R. K., & Raman, R. (2019). Genetic solutions to improve resilience of canola to climate change. *Genomic designing of climate-smart oilseed crops*, pp. 75-131. Springer. https://link.springer.com/chapter/10.1007/978-3-319-93536-2_2

Ramegowda, V., Senthil-Kumar, M., Ishiga, Y., Kaundal, A., Udayakumar, M., & Mysore, K. S. (2013). Drought stress acclimation imparts tolerance to Sclerotinia sclerotiorum and Pseudomonas syringae in Nicotiana benthamiana. *International Journal of Molecular Sciences*, *14*(5), 9497–9513. doi:10.3390/ijms14059497 PMID:23644883

Ramesh, T., Bolan, N. S., Kirkham, M. B., Wijesekara, H., Kanchikerimath, M., Rao, C. S., & Freeman, O. W. II. (2019). Soil organic carbon dynamics: Impact of land use changes and management practices: A review. *Advances in Agronomy*, *156*, 1–107. doi:10.1016/bs.agron.2019.02.001

Ramsey, E., Berglund, E. Z., & Goyal, R. (2017). The impact of demographic factors, beliefs, and social influences on residential water consumption and implications for non-price policies in urban India. *Water (Basel)*, *9*(11), 1–21. doi:10.3390/w9110844

Ramya, M. G. P. G., Balaji, C., & Girish, L. (2015). Environment change prediction to adapt climate-smart agriculture using big data analytics. [IJARCET]. *International Journal of Advanced Research in Computer Engineering and Technology*, *4*(5).

Randell, H., Gray, C., & Shayo, E. H. (2022). Climatic conditions and household food security: Evidence from Tanzania. *Food Policy*, *112*, 102362. doi:10.1016/j.foodpol.2022.102362 PMID:37064798

Rane, J., Singh, A. K., Tiwari, M., Prasad, P. V. V., & Jagadish, S. V. K. (2022). Effective Use of Water in Crop Plants in Dryland Agriculture: Implications of Reactive Oxygen Species and Antioxidative System. *Frontiers in Plant Science*, *12*, 778270. doi:10.3389/fpls.2021.778270 PMID:35082809

Rapoport, H. F., Hammami, S. B. M., Martins, P., Pérez-Priego, O., & Orgaz, F. (2012). Influence of water deficits at different times during olive tree inflorescence and flower development. *Environmental and Experimental Botany*, *77*, 227–233. doi:10.1016/j.envexpbot.2011.11.021

Rasheed, A., Ilyas, M., Khan, T. N., Mahmood, A., Riaz, U., Chattha, M. B., Kashgry, N. A. T. A., Binothman, N., Hassan, M. U., Wu, Z., & Qari, S. H. (2023). Study of genetic variability, heritability, and genetic advance for yield-related traits in tomato (Solanum lycopersicon MILL.). *Frontiers in Genetics*, *13*, 1030309. doi:10.3389/fgene.2022.1030309 PMID:36685955

Rashid, M. I., Mujawar, L. H., Shahzad, T., Almeelbi, T., Ismail, I. M. I., & Oves, M. (2016). Bacteria and fungi can contribute to nutrients bioavailability and aggregate formation in degraded soils. *Microbiological Research*, *183*, 26–41. doi:10.1016/j.micres.2015.11.007 PMID:26805616

Rathore, S. S., Chandravanshi, P., Chandravanshi, A., & Jaiswal, K. (2016). Eutrophication: Impacts of excess nutrient inputs on aquatic ecosystem. *IOSR Journal of Agriculture and Veterinary Science*, *09*(10), 89–96. doi:10.9790/2380-0910018996

Ray, R. L., Ampim, P. A., & Gao, M. (2020). Crop protection under drought stress. *Crop protection under changing climate*. 145-170. doi:10.1007/978-3-030-46111-9_6

Raza, A., Razzaq, A., Mehmood, S. S., Zou, X., Zhang, X., Lv, Y., & Xu, J. (2019). Impact of climate change on crops adaptation and strategies to tackle its outcome: A review. *Plants*, *8*(2), 34. doi:10.3390/plants8020034 PMID:30704089

Reddy, P. C., & Babu, A. S. (2017, February). Survey on weather prediction using big data analytics. In *2017 Second international conference on electrical, computer and communication technologies (ICECCT)* (pp. 1-6). IEEE.

Reddy, V. R. (1999). Quenching the thirst: The cost of water in fragile environments. *Development and Change*, *30*(1), 79–113. doi:10.1111/1467-7660.00108

Reich, P. B., Walters, M. B., & Ellsworth, D. S. (1992). Leaf life-span in relation to leaf, plant, and stand characteristics among diverse ecosystems. *Ecological Monographs*, *62*(3), 365–392. doi:10.2307/2937116

Reicosky, D. C. (2003). Conservation agriculture: Global environmental benefits of soil carbon management. *Conservation Agriculture: Environment, Farmers Experiences, Innovations, Socio-economy, Policy*, 3-12.

Reilly, J., Tubiello, F., McCarl, B., Abler, D., Darwin, R., Fuglie, K., & Jones, J. (2003). US agriculture and climate change: New results. *Climatic Change*, *57*(1/2), 43–67. doi:10.1023/A:1022103315424

Repullo-Ruiberriz de Torres, M. A., Moreno-García, M., Ordóñez-Fernández, R., Rodríguez-Lizana, A., Carceles Rodriguez, B., García-Tejero, I. F., Durán Zuazo, V. H., & Carbonell-Bojollo, R. M. (2021). Cover crop contributions to improve the soil nitrogen and carbon sequestration in almond orchards (SW Spain). *Agronomy (Basel)*, *11*(2), 387. doi:10.3390/agronomy11020387

Requena-Mullor, J. M., García-Garrido, J. M., García, P. A., & Rodríguez, E. (2020). Climatic drivers of Verticillium dahliae occurrence in Mediterranean olive-growing areas of southern Spain. *PLoS One*, *15*(12), e0232648. doi:10.1371/journal.pone.0232648 PMID:33378350

Reynolds, M., Tattaris, M., Cossani, C. M., Ellis, M., Yamaguchi-Shinozaki, K., & Pierre, C. S. (2015). Exploring genetic resources to increase adaptation of wheat to climate change. In *Advances in Wheat Genetics: From Genome to Field: Proceedings of the 12th International Wheat Genetics Symposium* (pp. 355-368). Springer Japan. https://library.oapen.org/bitstream/handle/20.500.12657/27833/1002172.pdf?sequence=1#page=336

Rezende-Teixeira, P., Dusi, R. G., Jimenez, P. C., Espindola, L. S., & Costa-Lotufo, L. V. (2022). What can we learn from commercial insecticides? Efficacy, toxicity, environmental impacts, and future developments. *Environmental Pollution, 300*, 118983. doi:10.1016/j.envpol.2022.118983 PMID:35151812

Rguioui, L. (2016). *Epidémiologie de la leishmaniose cutanée à SEFROU* [thèse de doctorat en pharmacie]. Faculté de médecine et de pharmacie. Université Mohammed V-Rabat. Maroc.

Rhajaoui, M. (2011). Les leishmanioses humaines au Maroc: Une diversité nosogéographique. *Pathologie Biologie, 59*(4), 226–229. doi:10.1016/j.patbio.2009.09.003 PMID:19942371

Ribeiro, B. M., Martins, É. S., de Souza Aguiar, R. W., & Corrêa, R. F. T. (2017). Expression of Bacillus thuringiensis toxins in insect cells. *Bacillus thuringiensis and Lysinibacillus sphaericus: Characterization and use in the field of biocontrol*, 99-110. Springer. doi:10.1007/978-3-319-56678-8_8

Ribeiro, H., Cunha, M., & Abreu, I. (2008). Quantitative forecasting of olive yield in Northern Portugal using a bioclimatic model. *Aerobiologia, 24*(3), 141–150. doi:10.100710453-008-9094-2

Richardson, D., Black, A. S., Irving, D., Matear, R. J., Monselesan, D. P., Risbey, J. S., Squire, D. T., & Tozer, C. R. (2022). Global increase in wildfire potential from compound fire weather and drought. *NPJ Climate and Atmospheric Science, 5*(1), 23. doi:10.103841612-022-00248-4

Rios, P. C. S., Deen, T. A., Nagabhatla, N., & Ayala, G. (2018). Explaining water pricing through a water security lens. *Water (Basel), 10*(9), 1–20. doi:10.3390/w10091173

Rioux, J. A., Lanotte, G., Petter, F., Dereure, J., Akalay, O., Pratlong, F., & Serres, E. (1986). Les leishmanioses cutanées du bassin Méditerranéen occidental. De l'identification enzymatique à l'analyse éco-épidémiologique. L'exemple de trois foyers, tunisien, marocain et français. CollintCNRS/INSERM) IMEE, Montpellier, 1986, 365-395.

Rioux, J. A., Lanotte, G., Serres, E., Pratlong, F., Bastien, P., & Perieres, J. (1990). Taxonomy of Leishmania. Use of isoenzymes. Suggestions for a new classification. *Annales de Parasitologie Humaine et Comparee, 65*(3), 111–125. doi:10.1051/parasite/1990653111 PMID:2080829

Rivero, R. M., Mittler, R., Blumwald, E., & Zandalinas, S. I. (2022). Developing climate-resilient crops: Improving plant tolerance to stress combination. *The Plant Journal, 109*(2), 373–389. doi:10.1111/tpj.15483 PMID:34482588

Roberts, T. R., Hutson, D. H., Lee, P. W., Nicholls, P. H., Plimmer, J. R., Roberts, M. C., & Croucher, L. (2007). *Metabolic pathways of agrochemicals: part 1: herbicides and plant growth regulators*. Royal Society of Chemistry.

Robette, N., Genin, E., & Clerget-Darpoux, F. (2022). Heritability: What is the point? What is it not for? A human genetics perspective. *Genetica, 150*(3–4), 199–208. doi:10.100710709-022-00149-7 PMID:35092541

Robinson, E. A., Ryan, G. D., & Newman, J. A. (2012). A meta-analytical review of the effects of elevated CO2 on plant–arthropod interactions highlights the importance of interacting environmental and biological variables. *The New Phytologist, 194*(2), 321–336. doi:10.1111/j.1469-8137.2012.04074.x PMID:22380757

Rodgers, C., & Hellegers, P. J. G. J. (2005). Water pricing and valuation in Indonesia: Case study of the Brantas River Basin. *EPT Discussion Paper 141.* International Food Policy Research Institute, USA. doi:10.22004/ag.econ.58586

Rodias, E., Aivazidou, E., Achillas, C., Aidonis, D., & Bochtis, D. (2020). Water-energy-nutrients synergies in the agrifood sector: A circular economy framework. *Energies*, *14*(1), 159. doi:10.3390/en14010159

Rodrigues, C. R. F., Silva, E. N., Ferreira-Silva, S. L., Voigt, E. L., Viégas, R. A., & Silveira, J. A. G. (2013). High $K^+$ supply avoids $Na^+$ toxicity and improves photosynthesis by allowing favorable $K^+$: $Na^+$ ratios through the inhibition of $Na^+$ uptake and transport to the shoots of Jatropha curcas plants. *Journal of Plant Nutrition and Soil Science*, *176*(2), 157–164. doi:10.1002/jpln.201200230

Romero, F., Cazzato, S., Walder, F., Vogelgsang, S., Bender, S. F., & van der Heijden, M. G. (2022). Humidity and high temperature are important for predicting fungal disease outbreaks worldwide. *The New Phytologist*, *234*(5), 1553–1556. doi:10.1111/nph.17340 PMID:33713447

Romero-Trigueros, C., Cabañero, J. J., Tortosa, P. A., Gambín, J. M., Maestre-Valero, J. F., & Nicolás, E. N. (2020). Medium-long term effects of saline reclaimed water and regulated deficit irrigation on fruit quality of citrus. *Journal of the Science of Food and Agriculture*, *100*(3), 1350–1357. doi:10.1002/jsfa.10091 PMID:31617215

Rosca-Casian, O., Parvu, M., Vlase, L., & Tamas, M. (2007). Antifungal activity of Aloe vera leaves. *Fitoterapia*, *78*(3), 219–222. doi:10.1016/j.fitote.2006.11.008 PMID:17336466

Rosemarin, A., Macura, B., Carolus, J., Barquet, K., Ek, F., Järnberg, L., Lorick, D., Johannesdottir, S., Pedersen, S. M., Koskiaho, J., Haddaway, N. R., & Okruszko, T. (2020). Circular nutrient solutions for agriculture and wastewater– a review of technologies and practices. *Current Opinion in Environmental Sustainability*, *45*, 78–91. doi:10.1016/j.cosust.2020.09.007

Rosenthal, E., & Marty, P. (2009). Actualités sur la leishmaniose viscérale méditerranéenne. *Revue de Médecine Interne*, *30*, S24–S28. https://www.sciencedirect.com/science/article/pii/S0248866309001398. doi:10.1016/j.revmed.2009.03.016

Rosero, A., Berdugo-Cely, J. A., Šamajová, O., Šamaj, J., & Cerkal, R. (2020). A dual strategy of breeding for drought tolerance and introducing drought-tolerant, underutilized crops into production systems to enhance their resilience to water deficiency. *Plants*, *9*(10), 1263. doi:10.3390/plants9101263 PMID:32987964

Rossi, L., Sebastiani, L., Tognetti, R., d'Andria, R., Morelli, G., & Cherubini, P. (2013). Tree- ring wood anatomy and stable isotopes show structural and functional adjustments in olive trees under different water availability. *Plant and Soil*, *372*(1), 567–579. doi:10.100711104-013-1759-0

Rostan, O. (2013). *Place de l'Interleukine-33 dans la réponse immune du foie au cours de la leishmaniose viscérale* [Doctoral dissertation, Université Rennes 1].

Roy, P., Pal, S. C., Chakrabortty, R., Chowdhuri, I., Saha, A., & Shit, M. (2023). Effects of climate change and sea-level rise on coastal habitat: Vulnerability assessment, adaptation strategies and policy recommendations. *Journal of Environmental Management*, *330*, 117187. doi:10.1016/j.jenvman.2022.117187 PMID:36610196

Rugini, E., & Fedeli, E. (1990). Olive (*Olea europaea* L.) as an oilseed crop. In *Legumes and oilseed crops I* (pp. 593–641). Springer. doi:10.1007/978-3-642-74448-8_29

Rumpel, C., Amiraslani, F., Bossio, D., Chenu, C., Henry, B., Espinoza, A. F., Koutika, L.-S., Ladha, J., Madari, B., Minasny, B., Olaleye, A. O., Shirato, Y., Sall, S. N., Soussana, J.-F., & Varela-Ortega, C. (2022). The role of soil carbon sequestration in enhancing human resilience in tackling global crises including pandemics. *Soil Security*, *8*, 100069. doi:10.1016/j.soisec.2022.100069

Ruohomaa, H., Pöykkö, T., & Salminen, V. (2021). Digitalization of Bio-Based Value Chains. In *Advances in Human Factors, Business Management and Leadership: Proceedings of the AHFE 2021 Virtual Conferences on Human Factors, Business Management and Society, and Human Factors in Management and Leadership,* (pp. 111-118). Springer International Publishing. 10.1007/978-3-030-80876-1_16

Saatkamp, A., Poschlod, P., & Venable, D. L. (2014). The functional role of soil seed banks in natural communities. In *Seeds: the ecology of regeneration in plant communities* (pp. 263–295). CABI. doi:10.1079/9781780641836.0263

Sabadin, F., DoVale, J. C., Platten, J. D., & Fritsche-Neto, R. (2022). Optimizing self-pollinated crop breeding employing genomic selection: From schemes to updating training sets. *Frontiers in Plant Science*, *13*, 935885. doi:10.3389/fpls.2022.935885 PMID:36275547

Sabbahi, R. (2022). Effects of Climate Change on Insect Pollinators and Implications for Food Security—Evidence and Recommended Actions. In The Food Security, Biodiversity, and Climate Nexus (pp. 143-163). Springer.

Sabeh, F., Wright, T., & Norton, S. J. (1993). Purification and characterization of a glutathione peroxidase from the Aloe vera plant. *Enzyme & Protein*, *47*(2), 92–98. doi:10.1159/000468662 PMID:8193675

Saco, P. M., McDonough, K. R., Rodriguez, J. F., Rivera-Zayas, J., & Sandi, S. G. (2021). The role of soils in the regulation of hazards and extreme events. *Philosophical Transactions of the Royal Society B: Biological Sciences*, *376*(1834), 20200178. doi:10.1098/rstb.2020.0178

Saeed, F., Chaudhry, U. K., Raza, A., Charagh, S., Bakhsh, A., Bohra, A., Ali, S., Chitikineni, A., Saeed, Y., Visser, R. G. F., Siddique, K. H. M., & Varshney, R. K. (2023). Developing future heat-resilient vegetable crops. *Functional & Integrative Genomics*, *23*(1), 47. doi:10.100710142-023-00967-8 PMID:36692535

Saladin, G., Magne, C., & Clement, C. (2003). Effects of fludioxonil and pyrimethanil, two fungicides used against Botrytis cinerea, on carbohydrate physiology in Vitis vinifera L. *Pest Management Science*, *59*(10), 1083–1092. doi:10.1002/ps.733 PMID:14561065

Salehi, B., Albayrak, S., Antolak, H., Kręgiel, D., Pawlikowska, E., Sharifi-Rad, M., Uprety, Y., Tsouh Fokou, P., Yousef, Z., Amiruddin Zakaria, Z., Varoni, E., Sharopov, F., Martins, N., Iriti, M., & Sharifi-Rad, J. (2018). Aloe genus plants: From farm to food applications and phytopharmacotherapy. *International Journal of Molecular Sciences*, *19*(9), 2843. doi:10.3390/ijms19092843 PMID:30235891

Salgotra, R. K., & Chauhan, B. S. (2023). Genetic diversity, conservation, and utilization of plant genetic resources. *Genes*, *14*(1), 174. doi:10.3390/genes14010174 PMID:36672915

Salinas, C., Handford, M., Pauly, M., Dupree, P., & Cardemil, L. (2016). Structural modifications of fructans in Aloe barbadensis Miller (Aloe vera) grown under water stress. *PLoS One*, *11*(7), e0159819. doi:10.1371/journal.pone.0159819 PMID:27454873

Samui, R. P., Kulkarni, P. S., Kamble, M. V., & Vaidya, N. G. (2014). A critical evaluation of sugarcane yield variation as influenced by climatic parameters in Uttar Pradesh and Maharashtra states of India. *Time Journals of Agriculture and Veterinary Sciences*, *2*(1), 63–69.

Sanchez-Bayo F. (2012). Insecticides mode of action in relation to their toxicity to non-target organisms. *J Environ Anal Toxicol, 2*.

Sánchez-Bayo, F. (2021). Indirect effect of pesticides on insects and other arthropods. *Toxics*, *9*(8), 177. doi:10.3390/toxics9080177 PMID:34437495

Sanchez-Bayo, F., Goulson, D., Pennacchio, F., Nazzi, F., Goka, K., & Desneux, N. (2016). Are bee diseases linked to pesticides? A brief review. *Environment International*, *89-90*, 7–11. doi:10.1016/j.envint.2016.01.009 PMID:26826357

Sanchéz, J. I., & Akdemir, D. (2021). Training set optimization for sparse phenotyping in Genomic Selection: A Conceptual Overview. *Frontiers in Plant Science*, *12*, 715910. doi:10.3389/fpls.2021.715910 PMID:34589099

Sánchez-León, S., Gil-Humanes, J., Ozuna, C. V., Giménez, M. J., Sousa, C., Voytas, D. F., & Barro, F. (2018). Low-gluten, nontransgenic wheat engineered with CRISPR/Cas9. *Plant Biotechnology Journal*, *16*(4), 902–910. doi:10.1111/pbi.12837 PMID:28921815

Saniasiaya, J., Salim, R., Mohamad, I., & Harun, A. (2017). Antifungal effect of Malaysian Aloe vera leaf extract on selected fungal species of pathogenic otomycosis species in in vitro culture medium. *Oman Medical Journal*, *32*(1), 41–46. doi:10.5001/omj.2017.08 PMID:28042402

Santos, F. D., Ferreira, P. L., & Pedersen, J. S. T. (2022). The Climate Change Challenge: A Review of the Barriers and Solutions to Deliver a Paris Solution. *Climate (Basel)*, *10*(5), 5. doi:10.3390/cli10050075

Sanz-Cortés, F., Martinez-Calvo, J., Badenes, M. L., Bleiholder, H., Hack, H., Llácer, G., & Meier, U. (2002). Phenological growth stages of olive trees (Olea europaea). *Annals of Applied Biology*, *140*(2), 151–157. doi:10.1111/j.1744-7348.2002.tb00167.x

Sapkota, T. B., Jat, R. K., Singh, R. G., Jat, M. L., Stirling, C. M., Jat, M. K., Bijarniya, D., Kumar, M., Yadvinder-Singh, Saharawat, Y. S., & Gupta, R. K. (2017). Soil organic carbon changes after seven years of conservation agriculture in a rice–wheat system of the eastern Indo-Gangetic Plains. *Soil Use and Management*, *33*(1), 81–89. doi:10.1111um.12331

Saponari, M., Boscia, D., Nigro, F., & Martelli, G. P. (2013). Identification of DNA sequences related to Xylella fastidiosa in oleander, almond and olive trees exhibiting leaf scorch symptoms in Apulia (Southern Italy). *Journal of Plant Pathology*, *95*(3).

Sasidharan, R., Bailey-Serres, J., Ashikari, M., Atwell, B. J., Colmer, T. D., Fagerstedt, K., Fukao, T., Geigenberger, P., Hebelstrup, K. H., Hill, R. D., Holdsworth, M. J., Ismail, A. M., Licausi, F., Mustroph, A., Nakazono, M., Pedersen, O., Perata, P., Sauter, M., Shih, M.-C., & Voesenek, L. A. C. J. (2017). Community recommendations on terminology and procedures used in flooding and low oxygen stress research. *The New Phytologist*, *214*(4), 1403–1407. doi:10.1111/nph.14519 PMID:28277605

Sathiakumar, N., MacLennan, P. A., Mandel, J., & Delzell, E. (2011). A review of epidemiologic studies of triazine herbicides and cancer. *Critical Reviews in Toxicology*, *41*(sup1), 1–34. doi:10.3109/10408444.2011.554793 PMID:21425949

Sayão, V. M., dos Santos, N. V., de Sousa Mendes, W., Marques, K. P., Safanelli, J. L., Poppiel, R. R., & Demattê, J. A. (2020). Land use/land cover changes and bare soil surface temperature monitoring in southeast Brazil. *Geoderma Regional*, *22*, e00313. doi:10.1016/j.geodrs.2020.e00313

Scavo, A., Fontanazza, S., Restuccia, A., Pesce, G. R., Abbate, C., & Mauromicale, G. (2022). The role of cover crops in improving soil fertility and plant nutritional status in temperate climates. A review. *Agronomy for Sustainable Development*, *42*(5), 93. doi:10.100713593-022-00825-0

Schäffer, A., Fenner, K., Wang, Z., & Scheringer, M. (2022). To be or not to be degraded: In defense of persistence assessment of chemicals. *Environmental Science. Processes & Impacts*, *24*(8), 1104–1109. doi:10.1039/D2EM00213B PMID:35822690

Scherr, S. J., Shames, S., & Friedman, R. (2012). From climate-smart agriculture to climate-smart landscapes. *Agriculture & Food Security*, *1*(1), 1–15. doi:10.1186/2048-7010-1-12

Schindler, D. W., Hecky, R. E., Findlay, D. L., Stainton, M. P., Parker, B. R., Paterson, M. J., Beaty, K. G., Lyng, M., & Kasian, S. E. M. (2008). Eutrophication of lakes cannot be controlled by reducing nitrogen input: Results of a 37-year whole- ecosystem experiment. *Proceedings of the National Academy of Sciences of the United States of America, 105*(32), 11254–11258. doi:10.1073/pnas.0805108105 PMID:18667696

Schirmer, J., Dovers, S., & Clayton, H. (2012). Informing conservation policy design through an examination of landholder preferences: A case study of scattered tree conservation in Australia. *Biological Conservation, 153*, 51–63. doi:10.1016/j.biocon.2012.04.014

Schleuning, M., Fründ, J., Schweiger, O., Welk, E., Albrecht, J., Albrecht, M., Beil, M., Benadi, G., Blüthgen, N., Bruelheide, H., Böhning-Gaese, K., Dehling, D. M., Dormann, C. F., Exeler, N., Farwig, N., Harpke, A., Hickler, T., Kratochwil, A., Kuhlmann, M., ... Hof, C. (2016). Ecological networks are more sensitive to plant than to animal extinction under climate change. *Nature Communications, 7*(1), 1. doi:10.1038/ncomms13965 PMID:28008919

Schleyer, C., Görg, C., Hauck, J., & Winkler, K. J. (2015). Opportunities and challenges for mainstreaming the ecosystem services concept in the multi-level policy-making within the EU. *Ecosystem Services, 16*, 174–181. doi:10.1016/j.ecoser.2015.10.014

Schneider, K., Van der Werf, W., Cendoya, M., Mourits, M., Navas-Cortés, J. A., Vicent, A., & Oude Lansink, A. (2020). Impact of Xylella fastidiosa subspecies pauca in European olives. *Proceedings of the National Academy of Sciences of the United States of America, 117*(17), 9250–9259. doi:10.1073/pnas.1912206117 PMID:32284411

Schneider, L., Rebetez, M., & Rasmann, S. (2022). The effect of climate change on invasive crop pests across biomes. *Current Opinion in Insect Science, 50*, 100895. doi:10.1016/j.cois.2022.100895 PMID:35240333

Schnoor, J. L. (2010). *Water sustainability in a changing world.* National Water Research Institute. https://research.uiowa.edu/sites/research.uiowa.edu/files/files/Clarke Prize Lecture.pdf

Schuur, E. A. G., Abbott, B. W., Commane, R., Ernakovich, J., Euskirchen, E., Hugelius, G., Grosse, G., Jones, M., Koven, C., Leshyk, V., Lawrence, D., Loranty, M. M., Mauritz, M., Olefeldt, D., Natali, S., Rodenhizer, H., Salmon, V., Schädel, C., Strauss, J., & Turetsky, M. (2022). Permafrost and Climate Change: Carbon Cycle Feedbacks From the Warming Arctic. *Annual Review of Environment and Resources, 47*(1), 343–371. doi:10.1146/annurev-environ-012220-011847

Seebacher, F., & Post, E. (2015). Climate change impacts on animal migration. *Climate Change Responses, 2*(1), 5. doi:10.118640665-015-0013-9

Sehgal, A., Sita, K., Kumar, J., Kumar, S., Singh, S., Siddique, K. H., & Nayyar, H. (2017). Effects of drought, heat and their interaction on the growth, yield and photosynthetic function of lentil (Lens culinaris Medikus) genotypes varying in heat and drought sensitivity. *Frontiers in Plant Science, 8*, 1776. doi:10.3389/fpls.2017.01776 PMID:29089954

Seleiman, M. F., Al-Suhaibani, N., Ali, N., Akmal, M., Alotaibi, M., Refay, Y., Dindaroglu, T., Abdul-Wajid, H. H., & Battaglia, M. L. (2021). Drought Stress Impacts on Plants and Different Approaches to Alleviate Its Adverse Effects. *Plants, 10*(2), 259. doi:10.3390/plants10020259 PMID:33525688

Seppelt, R., Klotz, S., Peiter, E., & Volk, M. (2022). Agriculture and food security under a changing climate: An underestimated challenge. *iScience, 105551*(12), 105551. https://www.cell.com/iscience/pdf/S2589-0042(22)01823-5.pdf. doi:10.1016/j.isci.2022.105551 PMID:36458255

Sequi, P. (1996). The role of composting in sustainable agriculture. In M. de Bertoldi, P. Sequi, B. Lemmes, & T. Papi (Eds.), *The Science of Composting*. Springer. doi:10.1007/978-94-009-1569-5_3

Serebrennikov, D., Thorne, F., Kallas, Z., & McCarthy, S. N. (2020). Factors influencing adoption of sustainable farming practices in Europe: A systemic review of empirical literature. *Sustainability (Basel), 12*(22), 9719. doi:10.3390u12229719

Sergent, E., Sergent, E., Parrot, L., Donatien, A., & Beguet, M. (1926). Transmission expérimentale du bouton d'orient (clou de Biskra) à l'homme par Phlebotomus papatasi (Scopoli). *Annales de l'Institut Pasteur*, *40*(5), 411–430.

Serpico, D. (2020). Beyond quantitative and qualitative traits: Three telling cases in the life sciences. *Biology & Philosophy*, *35*(3), 34. doi:10.100710539-020-09750-6

Serrão, J. E., Plata-Rueda, A., Martínez, L. C., & Zanuncio, J. C. (2022). Side-effects of pesticides on non-target insects in agriculture: A mini-review. *Naturwissenschaften*, *109*(2), 17. doi:10.100700114-022-01788-8 PMID:35138481

Sessions, J., Smith, D., Trippe, K. M., Fried, J. S., Bailey, J. D., Petitmermet, J. H., & Campbell, J. D. (2019). *Can biochar link forest restoration with commercial agriculture?*. doi:10.1016/j.biombioe.2019.02.015

Seyedi Rezvani, S. M. H., Almeida, N., & Falcão, M. (2023). Climate Adaptation Measures for Enhancing Urban Resilience. *Buildings*, *13*(9), 2163. doi:10.3390/buildings13092163

Shad, M., Yasmeen, A., Azam, S., Bakhsh, A., Latif, A., Shahid, N., & Shahid, A. A. (2021). Enhancing the resilience of transgenic cotton for insect resistance. *Molecular Biology Reports*, 1–9. doi:10.1007/978-3-319-90318-7_8 PMID:34839448

Shaji, H., Chandran, V., & Mathew, L. (2021). *Organic fertilizers as a route to controlled release of nutrients Controlled release fertilizers for sustainable agriculture*. Elsevier.

Shao, H. B., Chu, L. Y., Shao, M. A., Jaleel, C. A., & Hong-mei, M. (2008). Higher plant antioxidants and redox signaling under environmental stresses. *Comptes Rendus Biologies*, *331*(6), 433–441. doi:10.1016/j.crvi.2008.03.011 PMID:18510996

Sharma, A., Kumar, V., Shahzad, B., Ramakrishnan, M., Singh Sidhu, G. P., Bali, A. S., Handa, N., Kapoor, D., Yadav, P., Khanna, K., Bakshi, P., Rehman, A., Kohli, S. K., Khan, E. A., Parihar, R. D., Yuan, H., Thukral, A. K., Bhardwaj, R., & Zheng, B. (2020). Photosynthetic response of plants under different abiotic stresses: A review. *Journal of Plant Growth Regulation*, *39*(2), 509–531. doi:10.100700344-019-10018-x

Sharma, B., Vaish, B., Monika, S., Singh, U. K., Singh, P., & Singh, R. P. (2019). Recycling of organic wastes in agriculture: An environmental perspective. *International Journal of Environmental Research*, *13*(2), 409–429. doi:10.100741742-019-00175-y

Sharma, M., Kumar, P., Verma, V., Sharma, R., Bhargava, B., & Irfan, M. (2022). Understanding plant stress memory response for abiotic stress resilience: Molecular insights and prospects. *Plant Physiology and Biochemistry*, *179*, 10–24. doi:10.1016/j.plaphy.2022.03.004 PMID:35305363

Sharma, P., & Singh, A. K. (2017). Association of state-wise food security index with climatic factors in India: Evidence from state-wise panel data. *Journal of Global Agriculture and Ecology*, *6*(3), 196–205.

Shehzad, K. (2023). Extreme flood in Pakistan: Is Pakistan paying the cost of climate change? A short communication. *The Science of the Total Environment*, *880*, 162973. doi:10.1016/j.scitotenv.2023.162973 PMID:36958563

Shen, R., Huang, A., Li, B., & Guo, J. (2019). Construction of a drought monitoring model using deep learning based on multi-source remote sensing data. *International Journal of Applied Earth Observation and Geoinformation*, *79*, 48–57. doi:10.1016/j.jag.2019.03.006

Shipley, B., De Bello, F., Cornelissen, J. H. C., Laliberté, E., Laughlin, D. C., & Reich, P. B. (2016). Reinforcing loose foundation stones in trait-based plant ecology. *Oecologia*, *180*(4), 923–931. doi:10.100700442-016-3549-x PMID:26796410

Shishegaran, A., Shishegaran, A., Najari, M., Ghotbi, A., & Boushehri, A. N. (2020). Effect of plants on an environment with high carbon dioxide concentration. *Cleaner Engineering and Technology*, *1*, 100002. doi:10.1016/j.clet.2020.100002

Shivanna, K. R. (2022). Climate change and its impact on biodiversity and human welfare. *Proceedings of the Indian National Science Academy. Part A, Physical Sciences*, *88*(2), 160–171. doi:10.100743538-022-00073-6

Shoaib, M., Banerjee, B. P., Hayden, M., & Kant, S. (2022). Roots' drought adaptive traits in crop improvement. *Plants*, *11*(17), 2256. doi:10.3390/plants11172256 PMID:36079644

Shokri, F., Ziarati, P., & Mousavi, Z. (2016). Removal of Selected Heavy Metals from Pharmaceutical Effluent by Aloe vera L. *Biomedical & Pharmacology Journal*, *9*(2), 705–713. doi:10.13005/bpj/993

Shrestha, J., Subedi, S., Timsina, K. P., Chaudhary, A., Kandel, M., & Tripathi, S. (2020). Conservation agriculture as an approach towards sustainable crop production: A review. *Farming and Management, 5*(1), 7-15. doi:10.31830/2456-8724.2020.002

Shrestha, S. (2019). Effects of climate change in agricultural insect pest. *Acta Sci. Agric*, *3*(12), 74–80. doi:10.31080/ASAG.2019.03.0727

Shrestha, S., Mahat, J., & Shrestha, J. K. M., & Paudel, K. (2022). Influence of high-temperature stress on rice growth and development. A review. *Heliyon*, *8*(12), e12651. doi:10.1016/j.heliyon.2022.e12651 PMID:36643304

Shukla, A., & Ratan, V. (2019). Management of Early Blight of Potato by Using Different Bioagents as Tuber Dressing and its Effect on Germination and Growth. *International Journal of Current Microbiology and Applied Sciences*, *8*(06), 2019. http://www. ijcmas. com. doi:10.20546/ijcmas.2019.806.233

Siedt, M., Schäffer, A., Smith, K. E., Nabel, M., Roß-Nickoll, M., & van Dongen, J. T. (2021). Comparing straw, compost, and biochar regarding their suitability as agricultural soil amendments to affect soil structure, nutrient leaching, microbial communities, and the fate of pesticides. *The Science of the Total Environment*, *751*, 141607. doi:10.1016/j.scitotenv.2020.141607

Siefert, A., Violle, C., Chalmandrier, L., Albert, C. H., Taudiere, A., & Fajardo, A. (2015). A global meta-analysis of the relative extent of intraspecific trait variation in plant communities. *Ecology Letters*, *18*(12), 1406–1419. doi:10.1111/ele.12508 PMID:26415616

Siegner, A. B., Acey, C., & Sowerwine, J. (2020). Producing urban agroecology in the East Bay: From soil health to community empowerment. *Agroecology and Sustainable Food Systems*, *44*(5), 566–593. doi:10.1080/21683565.2019.1690615

Siew, M. K., Yacob, M. R., Radam, A., Adamu, A., & Alias, E. F. (2015). Estimating willingness to pay for wetland conservation: A contingent valuation study of Paya Indah Wetland, Selangor Malaysia. *Procedia Environmental Sciences*, *30*(1), 268–272. doi:10.1016/j.proenv.2015.10.048

Silva-Olaya, A. M., Cerri, C. E. P., La Scala, N. Jr, Dias, C. T. S., & Cerri, C. C. (2013). Carbon dioxide emissions under different soil tillage systems in mechanically harvested sugarcane. *Environmental Research Letters*, *8*(1), 015014. doi:10.1088/1748-9326/8/1/015014

Silva-Parra, A., Trujillo-González, J. M., & Brevik, E. C. (2021). Greenhouse gas balance and mitigation potential of agricultural systems in Colombia: A systematic analysis. *Greenhouse Gases: Science and Technology*, *11*(3), 554–572. doi:10.1002/ghg.2066

Silva, S., Dias, M. C., & Silva, A. M. S. (2022). Titanium and zinc based nanomaterials in agriculture: A promising approach to deal with (a)biotic stresses? *Toxics*, *10*(4), 172. doi:10.3390/toxics10040172 PMID:35448432

Si, M., Guo, X., Lan, Y., Fan, B., & Cao, G. (2022). Effects of Climatic Variability on Soil Water Content in an Alpine Kobresia Meadow, Northern Qinghai–Tibetan Plateau, China. *Water (Basel)*, *14*(17), 17. doi:10.3390/w14172754

Simoglou, K. B., Karataraki, A., Roditakis, N. E., & Roditakis, E. (2012). Euzophera bigella (Zeller)(Lepidoptera: Pyralidae) and Dasineura oleae (F. Low)(Diptera: Cecidomyiidae): emerging olive crop pests in the Mediterranean? *Journal of Pest Science*, *85*(2), 169–177. doi:10.100710340-012-0418-1

Simpson, L. (1987). The mitochondrial genome of kinetoplastid protozoa: Genomic organization, transcription, replication, and evolution. *Annual Review of Microbiology*, *41*(1), 363–380. doi:10.1146/annurev.mi.41.100187.002051 PMID:2825587

Singh, A. K., & Jyoti, B. (2021a). Projected food-grain production and yield in India: An evidence from state-wise panel data investigation during 1977-2014. *Journal of Agricultural Sciences – Sri Lanka*, *16*(1), 108-125. doi:10.4038/jas.v16i1.9188

Singh, A. K., Kumar, S., & Jyoti, B. (2022b). Association of sustainable food security with climatic and non-climatic factors in Gujarat state of India: A district-wise panel data investigation. In: Leal Filho, W., Kovaleva, M., Popkova, E. (eds.), Sustainable Agriculture and Food Security. World Sustainability Series (pp. 259-282). Springer Cham, Springer Nature Switzerland. doi:10.1007/978-3-030-98617-9_16

Singh, P., & Kaur, A. (2022). A systematic review of artificial intelligence in agriculture. *Deep Learning for Sustainable Agriculture*, 57-80. doi:10.1016/B978-0-323-85214-2.00011-2

Singh, A. (2009). A study on farmers' knowledge, perception and intensity of pesticide use in vegetable cultivation in Western Uttar Pradesh. *Pusa AgriScience*, *32*, 63–69.

Singh, A. K. (2017). *Climate Change and Productivity of Food-grain and Cash Crops in India*. LAP LAMBERT Academic Publishing.

Singh, A. K. (2020). Economic valuation of ecosystem services of river and human well-being in developing countries: A research direction. *JOJ Wildlife & Biodiversity*, *3*(1), 1–2.

Singh, A. K., Ahmad, M. M., & Sharma, P. (2017a). Implications of socioeconomic factors on food security in selected economies: An empirical assessment. *Journal of Global Economics. Management and Business Research*, *8*(2), 103–115.

Singh, A. K., Ashraf, S. N., Sharma, S. K., & Kumar, S. (2023b). Factors affecting farm income and climate adaptation strategies in the farming sector of Gujarat state of India. In B. J. Sadhik, J. Adewole, A. Yeneneh, & L. Abidoye (Eds.), *Opportunities and Challenges in Climate-Friendly Clean Water and Energy Technologies* (pp. 122–147). IGI Global. doi:10.4018/978-1-6684-7303-0.ch006

Singh, A. K., & Issac, J. (2018). Impact of climatic and non-climatic factors on sustainable livelihood security in Gujarat state of India: A statistical exploration. *Agriculture and Food Sciences Research*, *5*(1), 30–46. doi:10.20448/journal.512.2018.51.30.46

Singh, A. K., Issac, J., & Narayanan, K. G. S. (2019a). Measurement of environmental sustainability index and its association with socio-economic indicators in selected Asian economies: An empirical investigation. *International Journal of Environment and Sustainable Development*, *18*(1), 58–100. doi:10.1504/IJESD.2019.098641

Singh, A. K., & Jyoti, B. (2019). Measuring the climate variability impact on cash crops farming in India: An empirical investigation. *Agriculture and Food Sciences Research*, *6*(2), 155–165. doi:10.20448/journal.512.2019.62.155.165

Singh, A. K., & Jyoti, B. (2021b). Projected productivity of cash crops in different climate change scenarios in India: Use of marginal impact analysis technique. *Finance & Economics Review*, *3*(1), 63–87.

Singh, A. K., Jyoti, B., Kumar, S., & Lenka, S. K. (2021). Assessment of global sustainable development, environmental sustainability, economic development and social development index in selected economies. *International Journal of Sustainable Development and Planning*, *16*(1), 123–138. doi:10.18280/ijsdp.160113

Singh, A. K., & Kumar, S. (2021). Assessing the performance and factors affecting industrial development in Indian states: An empirical analysis. *Journal of Social Economics Research*, *8*(2), 135–154. https://archive.conscientiabeam.com/index.php/35/article/view/1374/1915. doi:10.18488/journal.35.2021.82.135.154

Singh, A. K., Kumar, S., Ashraf, S. N., & Jyoti, B. (2023c). Implications of farmer's adaptation strategies to climate change in agricultural sector of Gujarat: Evidence from farm level data. *Research on World Agricultural Economy*, *3*(1), 42–57. doi:10.36956/rwae.v3i1.498

Singh, A. K., Kumar, S., & Jyoti, B. (2022a). Influence of climate change on agricultural sustainability in India: A state-wise panel data analysis. *Asian Journal of Agriculture*, *6*(1), 15–27. doi:10.13057/asianjagric/g060103

Singh, A. K., Narayanan, K. G. S., & Sharma, P. (2017b). Effect of climatic factors on cash crop farming in India: An application of Cobb-Douglas production function model. *International Journal of Agricultural Resources, Governance and Ecology*, *13*(2), 175–210. doi:10.1504/IJARGE.2017.086452

Singh, A. K., Narayanan, K. G. S., & Sharma, P. (2019b). Measurement of technical efficiency of climatic and non-climatic factors in sugarcane farming in Indian states: Use of stochastic frontier production function approach. *Climatic Change*, *5*(19), 150–166.

Singh, A. K., Sharma, A. K., & Jyoti, B. (2023a). Does economic development have a causal relationship with environmental degradation? Experience from different income group countries. In G. Sart (Ed.), *Considerations on Education for Economic, Social, and Environmental Sustainability* (pp. 300–333). IGI Global. doi:10.4018/978-1-6684-8356-5.ch015

Singh, A. K., & Sharma, P. (2018a). Implications of climatic and non-climatic variables on food security in developing economies: A conceptual review. *MOJ Food Processing & Technology*, *6*(1), 1–12. doi:10.15406/mojfpt.2018.06.00138

Singh, A. K., & Sharma, P. (2018b). Measuring the productivity of food-grain crops in different climate change scenarios in India: An evidence from time series investigation. *Climatic Change*, *4*(16), 661–673.

Singh, A. K., Sharma, P., & Singh, D. K. (2016). Measuring the influence of weather variables on productivity of food-grain crops in India: An application of Just & Pope's production technique. *ABBS Management Business and Entrepreneurship Review*, *7*(2), 29–46. doi:10.23874/amber/2016/v7/i2/121780

Singh, A. K., Singh, B. J., & Negi, V. (2020). Does sustainable development have a causal relationship with environmental development? Evidence from a country-wise panel data analysis. *International Journal of Technology Management & Sustainable Development*, *19*(2), 147–171. doi:10.1386/tmsd_00020_1

Singh, B. K., Delgado-Baquerizo, M., Egidi, E., Guirado, E., Leach, J. E., Liu, H., & Trivedi, P. (2023). Climate change impacts on plant pathogens, food security and paths forward. *Nature Reviews. Microbiology*, *21*(10), 1–17. doi:10.103841579-023-00900-7 PMID:37131070

Singh, H., Raigar, O. P., & Chahota, R. K. (2021). Estimation of genetic diversity and its exploitation in plant breeding. *Botanical Review*, *88*(3), 413–435. doi:10.100712229-021-09274-y

Singh, J. S., Pandey, V. C., & Singh, D. P. (2011). Efficient soil microorganisms: A new dimension for sustainable agriculture and environmental development. *Agriculture, Ecosystems & Environment*, *140*(3-4), 339–353. doi:10.1016/j.agee.2011.01.017

Singh, K. K., & Frazier, A. E. (2018). A meta-analysis and review of unmanned aircraft system (UAS) imagery for terrestrial applications. *International Journal of Remote Sensing*, *39*(15-16), 5078–5098. doi:10.1080/01431161.2017.1420941

Sinha, R., Fritschi, F. B., Zandalinas, S. I., & Mittler, R. (2021). The impact of stress combination on reproductive processes in crops. *Plant Science*, *311*, 111007. doi:10.1016/j.plantsci.2021.111007 PMID:34482910

Sishodia, R. P., Ray, R. L., & Singh, S. K. (2020). Applications of remote sensing in precision agriculture: A review. *Remote Sensing (Basel)*, *12*(19), 3136. doi:10.3390/rs12193136

Skendžić, S., Zovko, M., Živković, I. P., Lešić, V., & Lemić, D. (2021). The Impact of Climate Change on Agricultural Insect Pests. *Insects*, *12*(5), 440. doi:10.3390/insects12050440 PMID:34066138

Slamini, M., Sbaa, M., Arabi, M., & Darmous, A. (2022). Review on Partial Root-zone Drying irrigation: Impact on crop yield, soil and water pollution. *Agricultural Water Management*, *271*, 107807. doi:10.1016/j.agwat.2022.107807

Smith, P., & Olesen, J. E. (2010). Synergies between the mitigation of, and adaptation to, climate change in agriculture. *Journal of Agricultural Science*, *148*(5), 543–552. doi:10.1017/S0021859610000341

Smyth, S. J. (2020). Regulatory barriers to improving global food security. *Global Food Security*. doi:10.1016/j.gfs.2020.100440

Soares, P. M. M., & Lima, D. C. A. (2022). Water scarcity down to earth surface in a Mediterranean climate: The extreme future of soil moisture in Portugal. *Journal of Hydrology (Amsterdam)*, *615*, 128731. doi:10.1016/j.jhydrol.2022.128731

Sofo, A., Manfreda, S., Fiorentino, M., Dichio, B., & Xiloyannis, C. (2008). The olive tree: A paradigm for drought tolerance in Mediterranean climates. *Hydrology and Earth System Sciences*, *12*(1), 293–301. doi:10.5194/hess-12-293-2008

Sohil, A., & Kichloo, M. A. (2023). Sustainable Solutions to Animal Waste: Climate Change Mitigation and Bioproduct Harvest. In *Climate Changes Mitigation and Sustainable Bioenergy Harvest Through Animal Waste: Sustainable Environmental Implications of Animal Waste* (pp. 301-332). Cham: Springer Nature Switzerland. https://link.springer.com/chapter/10.1007/978-3-031-26224-1_12

Solomon, K. R., Carr, J. A., Du Preez, L. H., Giesy, J. P., Kendall, R. J., Smith, E. E., & Van Der Kraak, G. (2008). Effects of atrazine on fish, amphibians, and aquatic reptiles: A critical review. *Critical Reviews in Toxicology*, *38*(9), 721–772. doi:10.1080/10408440802116496 PMID:18941967

Solomou, A. D., Germani, R., & Georgakopoulou, M. (2020). Ecological value, cultivation, utilization and commercialization of Aloe vera in Greece. *JAPS. The Journal of Animal and Plant Sciences*, *30*(4), 1013–1023. doi:10.36899/JAPS.2020.4.0116

Somanathan, E., & Ravindranath, R. (2006). Measuring the marginal value of water and elasticity of demand for water in agriculture. *Economic and Political Weekly*, *6*(1), 2712–2715. doi:10.2307/4418398

Song, J., Wang, Y., Zhang, S., Song, Y., Xue, S., Liu, L., Lvy, X., Wang, X., & Yang, G. (2021). Coupling biochar with anaerobic digestion in a circular economy perspective: A promising way to promote sustainable energy, environment and agriculture development in China. *Renewable & Sustainable Energy Reviews*, *144*, 110973. doi:10.1016/j.rser.2021.110973

Soria, R., Rodríguez-Berbel, N., Sánchez-Cañete, E. P., Villafuerte, A. B., Ortega, R., & Miralles, I. (2023). Organic amendments from recycled waste promote short-term carbon sequestration of restored soils in drylands. *Journal of Environmental Management*, *327*, 116873. doi:10.1016/j.jenvman.2022.116873

Souguir, D., Abd-Alla, H. I., El Ferjani, E., Larbi Khouja, M., & Hachicha, M. (2015). Aloe vera long-term saline irrigation increases contents of hydrogen peroxide, lipid peroxidation and phenolic compounds. *Acta Agriculturæ Scandinavica. Section B, Soil and Plant Science*, *65*(8), 688–696. doi:10.1080/09064710.2015.1049653

Soussana, J. F., Lutfalla, S., Ehrhardt, F., Rosenstock, T., Lamanna, C., Havlík, P., Richards, M., Wollenberg, E. L., Chotte, J.-L., Torquebiau, E., Ciais, P., Smith, P., & Lal, R. (2019). Matching policy and science: Rationale for the '4 per 1000-soils for food security and climate'initiative. *Soil & Tillage Research, 188*, 3–15. doi:10.1016/j.still.2017.12.002

Spadaro, D., & Gullino, M. L. (2019). Sustainable management of plant diseases. *Innovations in Sustainable Agriculture*, 337-359.

Spaldon, S., Samnotra, R., & Chopra, S. (2015). Climate resilient technologies to meet the challenges in vegetable production. *International Research on Current and Academic Review, 3*(2), 28-47. https://www.researchgate.net/publication/277664533

Spash, C. (1999). The Development of Environmental Thinking in Economics. *Environmental Values, 8*(4), 413–435. doi:10.3197/096327199129341897

Spash, C. L. (2013). The shallow or the deep ecological economics movement? *Ecological Economics, 93*, 351–362. doi:10.1016/j.ecolecon.2013.05.016

Spiertz, J. H. J. (2009). Nitrogen, sustainable agriculture and food security: a review. *Sustainable agriculture*, 635-651.

Stavi, I., & Lal, R. (2013). Agroforestry and biochar to offset climate change: A review. *Agronomy for Sustainable Development, 33*(1), 81–96. doi:10.100713593-012-0081-1

Story, P., & Cox, M. (2001). Review of the effects of organophosphorus and carbamate insecticide vertebrates. Are there implications for locust management in Australia? *Wildlife Research, 28*(2), 179–193. doi:10.1071/WR99060

Streimikiene, D. (2015). Environmental indicators for the assessment of quality of life. *Intellectual Economics, 9*(1), 67–79. doi:10.1016/j.intele.2015.10.001

Sultan, S. E. (2000). Phenotypic plasticity for plant development, function and life history. *Trends in Plant Science, 5*(12), 537–542. doi:10.1016/S1360-1385(00)01797-0 PMID:11120476

Sultan, S., & El–Qassem, A. (2021). Future prospects for sustainable agricultural development. *International Journal of Modern Agriculture and Environment, 1*(2), 54–82. doi:10.21608/ijmae.2023.215952.1012

Sun, G. Q., Li, L., Li, J., Liu, C., Wu, Y. P., Gao, S., Wang, Z., & Feng, G. L. (2022). Impacts of climate change on vegetation pattern: Mathematical modelling and data analysis. *Physics of Life Reviews, 43*, 239–270. doi:10.1016/j.plrev.2022.09.005 PMID:36343569

Sun, J., & Yuan, B. Z. (2020). Mapping of the world rice research: A bibliometric analysis of top papers during 2008–2018. [ALIS]. *Annals of Library and Information Studies, 67*(1), 55–66. http://op.niscair.res.in/index.php/ALIS/article/view/30297

Suprasanna, P. (2020). Plant abiotic stress tolerance: Insights into resilience build-up. *Journal of Biosciences, 45*(1), 1–8. doi:10.100712038-020-00088-5 PMID:33097677

Surjushe, A., Vasani, R., & Saple, D. G. (2008). Aloe vera: A short review. *Indian Journal of Dermatology, 53*(4), 163. doi:10.4103/0019-5154.44785 PMID:19882025

Sutherst, R. W., Yonow, T., Chakraborty, S., O'Donnell, C., & White, N. (1996). A generic approach to defining impacts of climate change on pests, weeds and diseases in Australasia.

Sutherst, R. W., Constable, F., Finlay, K. J., Harrington, R., Luck, J., & Zalucki, M. P. (2011). Adapting to crop pest and pathogen risks under a changing climate. *Wiley Interdisciplinary Reviews: Climate Change, 2*(2), 220–237. doi:10.1002/wcc.102

Swanepoel, C. M., Rötter, R. P., Van der Laan, M., Annandale, J. G., Beukes, D. J., du Preez, C. C., Swanepoel, L. H., van der Merwe, A., & Hoffmann, M. P. (2018). The benefits of conservation agriculture on soil organic carbon and yield in southern Africa are site-specific. *Soil & Tillage Research, 183*, 72–82. doi:10.1016/j.still.2018.05.016

Syromyatnikov, M. Y., Isuwa, M. M., Savinkova, O. V., Derevshchikova, M. I., & Popov, V. N. (2020). The effect of pesticides on the microbiome of animals. *Agriculture, 10*(3), 79. doi:10.3390/agriculture10030079

Szynkowska, M. I., Pawlaczyk, A., & Maćkiewicz, E. (2018). Bioaccumulation and biomagnification of trace elements in the environment. *Recent Advances in Trace Elements*, 251-276.

Tabari, H. (2020). Climate change impact on flood and extreme precipitation increases with water availability. *Scientific Reports, 10*(1), 1. doi:10.103841598-020-70816-2 PMID:31913322

Tabatabaei-Malazy, O., Atlasi, R., Larijani, B., & Abdollahi, M. (2015). Trends in publication on evidence-based antioxidative herbal medicines in management of diabetic nephropathy. *Journal of Diabetes and Metabolic Disorders, 15*(1), 1–8. doi:10.118640200-016-0221-2 PMID:26877997

Taherei Ghazvinei, P., Hassanpour Darvishi, H., Mosavi, A., Yusof, K. B. W., Alizamir, M., Shamshirband, S., & Chau, K. W. (2018). Sugarcane growth prediction based on meteorological parameters using extreme learning machine and artificial neural network. *Engineering Applications of Computational Fluid Mechanics, 12*(1), 738–749. doi:10.1080/19942060.2018.1526119

Taheripour, F., Tyner, W. E., Haqiqi, I., & Sajedinia, E. (2020). *Water Scarcity in Morocco*. World Bank., doi:10.1596/33306

Taiz, L., and E. Zeiger. (2006). *Fisiologia vegetal* (Vol. 10). Universitat Jaume I

Tajnari, H. (2001). *Étude de la nuisibilité du psylle de l'olivier, Euphyllura olivina Costa*. Symposium sur la Protection Intégrée des Cultures dans la région Méditerranéenne, DPVCTRF, Rabat, Maroc

Takimoto, A., Nair, P. R., & Nair, V. D. (2008). Carbon stock and sequestration potential of traditional and improved agroforestry systems in the West African Sahel. *Agriculture, Ecosystems & Environment, 125*(1-4), 159–166. doi:10.1016/j.agee.2007.12.010

Talmadge, J., Chavez, J., Jacobs, L., Munger, C., Chinnah, T., Chow, J. T., & Yates, K. (2004). Fractionation of Aloe vera L. inner gel, purification and molecular profiling of activity. *International Immunopharmacology, 4*(14), 1757-1773. https://doi.org/ doi:10.1016/j.intimp.2004.07.013

Tanaka, M., Yamamoto, Y., Misawa, E., Nabeshima, K., Saito, M., Yamauchi, K., Abe, F., & Furukawa, F. (2016). Aloe sterol supplementation improves skin elasticity in Japanese men with sunlight-exposed skin: A 12-week double-blind, randomized controlled trial. *Clinical, Cosmetic and Investigational Dermatology, 9*, 435–442. doi:10.2147/CCID.S118947 PMID:27877061

Tangu, N. A. (2014). Effects on plant morphology of drought in olive. *Türk Tarım ve Doğa Bilimleri Dergisi, 1*(Özel Sayı-1), 900-904. https://dergipark.org.tr/en/pub/turkjans/issue/13310/160838

Tanure, T. M., Miyajima, D. N., Magalhães, A. S., Domingues, E. P., & Carvalho, T. S. (2020). The Impacts of Climate Change on Agricultural Production, Land Use and Economy of the Legal Amazon Region Between 2030 and 2049. *Economía, 21*(1), 73–90. doi:10.1016/j.econ.2020.04.001

Tátrai, Z. A., Sanoubar, R., Pluhár, Z., Mancarella, S., Orsini, F., & Gianquinto, G. (2016). Morphological and physiological plant responses to drought stress in Thymus citriodorus. *International Journal of Agronomy, 2016*, 1–8. doi:10.1155/2016/4165750

Tattini, M., Bertoni, P., & Caselli, S. (1992). Genotipic responses of olive plants to sodium chloride. *Journal of Plant Nutrition, 15*(9), 1467–1485. doi:10.1080/01904169209364412

Tautges, N. E., Chiartas, J. L., Gaudin, A. C., O'Geen, A. T., Herrera, I., & Scow, K. M. (2019). Deep soil inventories reveal that impacts of cover crops and compost on soil carbon sequestration iffer in surface and subsurface soils. *Global Change Biology, 25*(11), 3753–3766. doi:10.1111/gcb.14762

Taylor, M., Lyons, S. M., Davie-Martin, C. L., Geoghegan, T. S., & Hageman, K. J. (2019). Understanding trends in pesticide volatilization from agricultural fields using the pesticide loss via volatilization model. *Environmental Science & Technology, 54*(4), 2202–2209. doi:10.1021/acs.est.9b04762 PMID:31858785

Tedersoo, L., Bahram, M., Põlme, S., Kõljalg, U., Yorou, N. S., Wijesundera, R., & Abarenkov, K. (2014). Global diversity and geography of soil fungi. *science, 346*(6213), 1256688.

Temel, F. A. (2023). Evaluation of the influence of rice husk amendment on compost quality in sewage sludge composting. *Bioresource Technology, 373*, 128748. doi:10.1016/j.biortech.2023.128748 PMID:36791979

Terdoo, F., & Adekola, O. (2014). Assessing the role of climate-smart agriculture in combating climate change, desertification and improving rural livelihood in Northern Nigeria. *African Journal of Agricultural Research, 9*(15), 1180–1191. doi:10.5897/AJAR2013.7665

Terral, J.-F., Badal, E., Heinz, C., Roiron, P., Thiébault, S., & Figueiral, I. (2004). A hydraulic conductivity model points to post- Neogene survival of the Mediterranean Olive in riparian habitat. *Ecology, 85*(11), 3158–3165. doi:10.1890/03-3081

Teshome, D. T., Zharare, G. E., & Naidoo, S. (2020). The Threat of the Combined Effect of Biotic and Abiotic Stress Factors in Forestry Under a Changing Climate. *Frontiers in Plant Science, 11*, 601009. doi:10.3389/fpls.2020.601009 PMID:33329666

Thierfelder, C., & Wall, P. C. (2010). Rotation in conservation agriculture systems of Zambia: Effects on soil quality and water relations. *Experimental Agriculture, 46*(3), 309–325. doi:10.1017/S001447971000030X

Thomas, H. J., Bjorkman, A. D., Myers-Smith, I. H., Elmendorf, S. C., Kattge, J., Diaz, S., Vellend, M., Blok, D., Cornelissen, J. H. C., Forbes, B. C., Henry, G. H. R., Hollister, R. D., Normand, S., Prevéy, J. S., Rixen, C., Schaepman-Strub, G., Wilmking, M., Wipf, S., Cornwell, W. K., & de Vries, F. T. (2020). Global plant trait relationships extend to the climatic extremes of the tundra biome. *Nature Communications, 11*(1), 1351. doi:10.103841467-020-15014-4 PMID:32165619

Thompson, N. A., Herrmann, A. M., & Hekkert, M. P. (2015). How sustainable entrepreneurs engage in institutional change: Insights from biomass torrefaction in the Netherlands. *Journal of Cleaner Production, 106*, 608–618. doi:10.1016/j.jclepro.2014.08.011

Thomson, A., Price, G. W., Arnold, P., Dixon, M., & Graham, T. (2022). Review of the potential for recycling CO2 from organic waste composting into plant production under controlled environment agriculture. *Journal of Cleaner Production, 333*, 130051. doi:10.1016/j.jclepro.2021.130051

Thomson, L. J., Macfadyen, S., & Hoffmann, A. A. (2010). Predicting the effects of climate change on natural enemies of agricultural pests. *Biological Control, 52*(3), 296–306. doi:10.1016/j.biocontrol.2009.01.022

Thrupp, L. A. (2000). Linking agricultural biodiversity and food security: The valuable role of agrobiodiversity for sustainable agriculture. *International Affairs, 76*(2), 265–281. doi:10.1111/1468-2346.00133

Tian, L., Li, J., Bi, W., Zuo, S., Li, L., Li, W., & Sun, L. (2019). Effects of waterlogging stress at different growth stages on the photosynthetic characteristics and grain yield of spring maize (*Zea mays* L.) Under feld conditions. *Agricultural Water Management, 218*, 250–258. doi:10.1016/j.agwat.2019.03.054

Tian, X., Dong, J., Jin, S., He, H., Yin, H., & Chen, X. (2023). Climate change impacts on regional agricultural irrigation water use in semi-arid environments. *Agricultural Water Management*, *281*, 108239. doi:10.1016/j.agwat.2023.108239

Tiemann, L., Grandy, A., Atkinson, E., Marin-Spiotta, E., & McDaniel, M. (2015). Crop rotational diversity enhances belowground communities and functions in an agroecosystem. *Ecology Letters*, *18*(8), 761–771. doi:10.1111/ele.12453 PMID:26011743

Timofeeva, A., Galyamova, M., & Sedykh, S. (2022). Prospects for using phosphate-solubilizing microorganisms as natural fertilizers in agriculture. *Plants*, *11*(16), 2119. doi:10.3390/plants11162119 PMID:36015422

Tonelli, G. B., Binder, C., Margonari, C., & Andrade Filho, J. D. (2023). Archives if the « Service des réseaux des établissements de santé ». The provincial delegation of the Ministry of Health in Errachidia Tonelli, G. B., Binder, C., Margonari, C., & Andrade Filho, J. D. (2021). Sand fly behavior: Much more than weak-flying. *Memorias do Instituto Oswaldo Cruz*, *116*, e210230. doi:10.1590/0074-02760210230

Torres-Ruiz, J. M., Cochard, H., Mayr, S., Beikircher, B., Diaz-Espejo, A., Rodriguez-Dominguez, C. M., Badel, E., & Fernández, J. E. (2014). Vulnerability to cavitation in *Olea europaea* current—year shoots: Further evidence of an open-vessel artifact associated with centrifuge and air-injection techniques. *Physiologia Plantarum*, *152*(3), 465–474. doi:10.1111/ppl.12185 PMID:24611594

Torres-Ruiz, J. M., Díaz-Espejo, A., Perez-Martin, A., & Hernandez-Santana, V. (2013). Loss of hydraulic functioning at leaf, stem and root level and its role in the stomatal behaviour during drought in olive trees. *Acta Horticulturae*, (991), 333–339. doi:10.17660/ActaHortic.2013.991.41

Tougeron, K., Brodeur, J., Le Lann, C., & van Baaren, J. (2020). How climate change affects the seasonal ecology of insect parasitoids. *Ecological Entomology*, *45*(2), 167–181. doi:10.1111/een.12792

Trębicki, P., Dáder, B., Vassiliadis, S., & Fereres, A. (2017). Insect–plant–pathogen interactions as shaped by future climate: Effects on biology, distribution, and implications for agriculture. *Insect Science*, *24*(6), 975–989. doi:10.1111/1744-7917.12531 PMID:28843026

Trumbore, S. (2000). Age of soil organic matter and soil respiration: Radiocarbon constraints on belowground C dynamics. *Ecological Applications*, *10*(2), 399–411. doi:10.1890/1051-0761(2000)010[0399:AOSOMA]2.0.CO;2

Tschora, H., & Cherubini, F. (2020). Co-benefits and trade-offs of agroforestry for climate change mitigation and other sustainability goals in West Africa. *Global Ecology and Conservation*, *22*, e00919. doi:10.1016/j.gecco.2020.e00919

Tsitsipis, J. A. (1980). Effect of constant temperatures on larval and pupal development of olive fruit flies reared on artificial diet. *Environmental Entomology*, *9*(6), 764–768. doi:10.1093/ee/9.6.764

Tsror, L. (2011). Epidemiology and control of Verticillium wilt on olive. *Israel Journal of Plant Sciences*, *59*(1), 59–69. doi:10.1560/IJPS.59.1.59

Tuc, V. P., Wangsuphachart, V., Tasanapradit, P., Fungladda, W., Van Trong, P., & Nhung, N. T. (2007). Impacts of pesticide use on semen characteristics among rice famers in Kienxuong District, Thaibinh Province, Vietnam. *The Southeast Asian Journal of Tropical Medicine and Public Health*, *38*(3), 569. PMID:17877235

Tudi, M., Daniel Ruan, H., Wang, L., Lyu, J., Sadler, R., Connell, D., & Phung, D. T. (2021). Agriculture development, pesticide application and its impact on the environment. *International Journal of Environmental Research and Public Health*, *18*(3), 1112. doi:10.3390/ijerph18031112 PMID:33513796

Tundisi, J. G. (2008). Water resources in the future: Problems and solutions. *Estudos Avançados*, *22*(63), 1–8.

Twesige, D., & Mbabazize, P. M. (2013). Relationship between environmental accounting indicators, macroeconomic indicators, and sustainable development indicators in developing countries: Case study Rwanda. *Researchjournali's. Journal of Economics*, *1*(1), 1–11. https://www.jkuat.ac.ke/campuses/kigali/wp-content/uploads/2014/06/my-publication-1.pdf

Ul-Allah, S., Azeem, A., Sher, A., Ijaz, M., Sattar, A., Saleem, M. A., Bibi, M., Abbas, N., & Hussain, M. (2021). Assessment of genetic variability and direct-indirect contribution of post-anthesis traits to the grain yield in bread wheat (Triticum aestivum) at different sowing dates. *International Journal of Agriculture and Biology*, *26*(1), 193–200. doi:10.17957/IJAB/15.1824

Upadhyaya, A., Jeet, P., Singh, A. K., & Sundaram, P. K. (2023). Estimation of the economic value of irrigation water in canal and tube well command areas. *$H_2$ Open Journal*, *6*(2), 1 – 9. doi:10.2166/h2oj.2023.011

Urra, J., Alkorta, I., & Garbisu, C. (2019). Potential Benefits and Risks for Soil Health Derived From the Use of Organic Amendments in Agriculture. *Agronomy (Basel)*, *9*(9), 9.. doi:10.3390/agronomy9090542

USEPA. (2003). *Guidelines for ecological risk assessment. EPA/630/R-95/002F, 1998*. United States Environmental Protection Agency.

USGS. (2016). *USGS Pesticide National Synthesis Project*. USGS. https://water.usgs.gov/nawqa/pnsp/usage/maps/

Usman, S., Umar, N., & Abdulhamid, A. (2023). Climate Change and Drought in the Dryland Areas of Nigeria. In Climate Change Impacts on Nigeria: Environment and Sustainable Development (pp. 361-378). Springer. doi:10.1007/978-3-031-21007-5_19

Va'squez, W. F., Mozumder, P., Herna'ndez-Arce, J., & Berrens, R. P. (2009). Willingness to pay for safe drinking water: Evidence from Parral, Mexico. *Journal of Environmental Management*, *90*(1), 3391–3400. doi:10.1016/j.jenvman.2009.05.009 PMID:19525059

Vaca, R., Del Águila, P., Yañez-Ocampo, G., Lugo, J. A., & De la Portilla-López, N. (2023). Soil Quality Assessment in Response to Water Erosion and Mining Activity. *Agriculture*, *13*(7), 7. doi:10.3390/agriculture13071380

Valdés-López, O., Batek, J., Gomez-Hernandez, N., Nguyen, C. T., Isidra-Arellano, M. C., Zhang, N., Joshi, T., Xu, D., Hixson, K. K., Weitz, K. K., Aldrich, J. T., Paša-Tolić, L., & Stacey, G. (2016). Soybean roots grown under heat stress show global changes in their transcriptional and proteomic profiles. *Frontiers in Plant Science*, *7*, 517. https://www.frontiersin.org/articles/10.3389/fpls.2016.00517/full. doi:10.3389/fpls.2016.00517 PMID:27200004

Vallgårda, A. F. (2018). *The value of water – A valuation study of groundwater in Uppsala municipality.* [Degree thesis No 1132, Sveriges lantbruksuniversitet Swedish University of Agricultural Sciences]. http://stud.epsilon.slu.se

Valverde, P., Barranco, D., López-Escudero, F. J., Díez, C. M., & Trapero, C. (2023). Efficiency of breeding olives for resistance to Verticillium wilt. *Frontiers in Plant Science*, *14*, 1149570. doi:10.3389/fpls.2023.1149570 PMID:36909426

Van Bodegom, P. M., Douma, J. C., & Verheijen, L. M. (2014). A fully traits-based approach to modeling global vegetation distribution. *Proceedings of the National Academy of Sciences of the United States of America*, *111*(38), 13733–13738. doi:10.1073/pnas.1304551110 PMID:25225413

van den Bergh, J. C. J. M. (2001). Ecological economics: Themes, approaches, and differences with environmental economics. *Regional Environmental Change*, *2*(1), 13–23. doi:10.1007101130000020

Van Eck, N. J., & Waltman, L. (2010). Software survey: VOSviewer, a computer program for bibliometric mapping. *Scientometrics*, *84*(2), 523–538. doi:10.100711192-009-0146-3 PMID:20585380

Vanclay, F. (2004). Social principles for agricultural extension to assist in the promotion of natural resource management. *Australian Journal of Experimental Agriculture*, *44*(3), 213–222. doi:10.1071/EA02139

Vaneeckhaute, C., Styles, D., Prade, T., Adams, P., Thelin, G., Rodhe, L., Gunnarsson, I., & D'Hertefeldt, T. (2018). Closing nutrient loops through decentralized anaerobic digestion of organic residues in agricultural regions: A multi-dimensional sustainability assessment. *Resources, Conservation and Recycling*, *136*, 110–117. doi:10.1016/j.resconrec.2018.03.027

Vasa, T. N., & Chacko, S. P. (2021). Recovery of struvite from wastewaters as an eco-friendly fertilizer: Review of the art and perspective for a sustainable agriculture practice in India. *Sustainable Energy Technologies and Assessments*, *48*, 101573. doi:10.1016/j.seta.2021.101573

Vasseghian, Y., Arunkumar, P., Joo, S.-W., Gnanasekaran, L., Kamyab, H., Rajendran, S., & Klemeš, J. J. (2022). Metal-organic framework-enabled pesticides are an emerging tool for sustainable cleaner production and environmental hazard reduction. *Journal of Cleaner Production*, *373*, 133966. doi:10.1016/j.jclepro.2022.133966

Vázquez, D. P., Gianoli, E., Morris, W. F., & Bozinovic, F. (2017). Ecological and evolutionary impacts of changing climatic variability. *Biological Reviews of the Cambridge Philosophical Society*, *92*(1), 22–42. doi:10.1111/brv.12216 PMID:26290132

Veettil, P. C., Speelman, S., Frija, A., Buysse, J., Mondelaers, K., & van Huylenbroeck, G. (2011). Price sensitivity of farmer's preferences for irrigation water-pricing method: Evidence from choice model analysis in Krishna River basin, India. *Journal of Water Resources Planning and Management*, *1*(1), 205–214. doi:10.1061/(ASCE)WR.1943-5452.0000103

Vellend, M., Baeten, L., Becker-Scarpitta, A., Boucher-Lalonde, V., McCune, J. L., Messier, J., Myers-Smith, I. H., & Sax, D. F. (2017). Plant biodiversity change across scales during the Anthropocene. *Annual Review of Plant Biology*, *68*(1), 563–586. doi:10.1146/annurev-arplant-042916-040949 PMID:28125286

Venkatachalam, L. (2007). Environmental economics and ecological economics: Where they can converge? *Ecological Economics*, *61*(2–3), 550–558. doi:10.1016/j.ecolecon.2006.05.012

Venkatachalam, L., & Narayanamoorthy, A. (2012). Estimating economic value of irrigation through contingent valuation of method: Results from Bhavani River Basin, Tamil Nadu. *Indian Journal of Agricultural Economics*, *67*(3), 308–315. doi:10.22004/ag.econ.204814

Venkatesh, M. S., Hazra, K. K., Ghosh, P. K., Praharaj, C. S., & Kumar, N. (2013). Long-term effect of pulses and nutrient management on soil carbon sequestration in Indo-Gangetic plains of India. *Canadian Journal of Soil Science*, *93*(1), 127–136. doi:10.4141/cjss2012-072

Venkateswarlu, B., Singh, A.K., Prasad, Y.G., Ravindra C., G, Srinivasa, R. C., Rao, K.V., Ramana, D.B.V. and Rao, V.U.M. (2011). *District level contingency plans for weather aberration in India.* Central Research Institute for Dryland Agriculture, Natural Resource Management Division.

Verrall, B., & Pickering, C. M. (2020). Alpine vegetation in the context of climate change: A global review of past research and future directions. *The Science of the Total Environment*, *141344*, 141344. doi:10.1016/j.scitotenv.2020.141344 PMID:32814293

Vigo, C., Therios, I., & Bosabalidis, A. (2005). Plant growth, nutrient concentration and leaf anatomy in olive plants irrigated with diluted seawater. *Journal of Plant Nutrition*, *28*(6), 101–102. doi:10.1081/PLN-200058897

Vigo, C., Therios, I., Patakas, A., Karatassou, A., & Nastou, A. (2002). Changes in photosynthetic parameters and nutrient distribution of olive plants (*Olea europaea* L.) cultivar 'Chondrolia Chalkidikis' under NaCl, Na2SO4 and KCl salinities. *Agrochimica*, *46*, 33–46.

Vijaya Baskar, P. S., Sreemathi, R., & Keertanaa, E. (2017, March). Crop prediction using predictive analytics. In *2017 International Conference on Computation of Power, Energy Information and Communication (ICCPEIC)* (pp. 370-373). IEEE. 10.1109/ICCPEIC.2017.8290395

Villela, F., Pestana, B. R., & Pessoa, S. B. (1939). Presença de Leishmania braziliensis na mucosa nasal sem lesão aparente em casos recentes de leishmaniose cutânea. *O Hospital, 16*, 953-960.

Vineis, P., Chan, Q., & Khan, A. (2011). Climate change impacts on water salinity and health. *Journal of Epidemiology and Global Health*, *1*(1), 5–10. doi:10.1016/j.jegh.2011.09.001 PMID:23856370

Visser, M. E. (2008). Keeping up with a warming world; assessing the rate of adaptation to climate change. *Proceedings. Biological Sciences*, *275*(1635), 649–659. doi:10.1098/rspb.2007.0997 PMID:18211875

Visser, M. E., & Both, C. (2005). Shifts in phenology due to global climate change: The need for a yardstick. *Proceedings. Biological Sciences*, *272*(1581), 2561–2569. doi:10.1098/rspb.2005.3356 PMID:16321776

Volk, G. M., Carver, D., Irish, B. M., Marek, L., Frances, A., Greene, S., Khoury, C. K., Bamberg, J., del Rio, A., Warburton, M. L., & Bretting, P. K. (2023). Safeguarding plant genetic resources in the United States during global climate change. *Crop Science*, *63*(4), 2274–2296. doi:10.1002/csc2.21003

Volpi, I., Guidotti, D., Mammini, M., Petacchi, R., & Marchi, S. (2020). Managing complex datasets to predict Bactrocera oleae infestation at the regional scale. *Computers and Electronics in Agriculture*, *179*, 105867. doi:10.1016/j.compag.2020.105867

Waajen, G., van Oosterhout, F., Douglas, G., & Lürling, M. (2016). Geo-engineering experiments in two urban ponds to control eutrophication. *Water Research*, *97*, 69–82. doi:10.1016/j.watres.2015.11.070 PMID:26725204

Wagner, N., & Bruhl, C. (2017). Ecotoxicology and Genotoxicology - Non-traditional terrestrial models. In: Marcelo L Larramendy (ed.). The use of terrestrial life-stages of European amphibians in toxicological studies. Publisher The Royal Society of Chemistry.

Wagner, J. J., & Engle, T. E. (2021). Invited Review: Water consumption, and drinking behavior of beef cattle, and effects of water quality. *Applied Animal Science*, *37*(4), 418–435. doi:10.15232/aas.2021-02136

Waha, K., Krummenauer, L., Adams, S., Aich, V., Baarsch, F., Coumou, D., Fader, M., Hoff, H., Jobbins, G., Marcus, R., Mengel, M., Otto, I. M., Perrette, M., Rocha, M., Robinson, A., & Schleussner, C.-F. (2017). Climate change impacts in the Middle East and Northern Africa (MENA) region and their implications for vulnerable population groups. *Regional Environmental Change*, *17*(6), 1623–1638. doi:10.100710113-017-1144-2

Wakelin, S. A., Gomez-Gallego, M., Jones, E., Smaill, S., Lear, G., & Lambie, S. (2018). Climate change induced drought impacts on plant diseases in New Zealand. *Australasian Plant Pathology*, *47*(1), 101–114. doi:10.100713313-018-0541-4

Walker, B., & Salt, D. (2006). *Resilience thinking: Sustaining ecosystems and people in a changing world.* Island Press.

Wang, B. X., Hof, A. R., & Ma, C. S. (2022). Impacts of climate change on crop production, pests and pathogens of wheat and rice. *Frontiers of Agricultural Science and Engineering*, *9*(1), 4–18. doi:10.15302/J-FASE-2021432

Wang, D., Saleh, N. B., Byro, A., Zepp, R., Sahle-Demessie, E., Luxton, T. P., Ho, K. T., Burgess, R. M., Flury, M., White, J. C., & Su, C. (2022). Nano-enabled pesticides for sustainable agriculture and global food security. *Nature Nanotechnology*, *17*(4), 347–360. https://www.ncbi.nlm.nih.gov/pmc/articles/PMC9774002/. doi:10.103841565-022-01082-8 PMID:35332293

Wang, H., Dai, Z., Krauss, K. W., Trettin, C. C., Noe, G. B., Burton, A. J., & Ward, E. J. (2023). Modelling impacts of saltwater intrusion on methane and nitrous oxide emissions in tidal forested wetlands. *Ecological Applications*, *33*(5), e2858. doi:10.1002/eap.2858 PMID:37084186

Wang, H., Xie, J., & Li, H. (2010). Water pricing with household surveys: A study of acceptability and willingness to pay in Chongqing, China. *China Economic Review*, *21*(1), 136–149. doi:10.1016/j.chieco.2009.12.001

Wang, J., Zhen, J., Hu, W., Chen, S., Lizaga, I., Zeraatpisheh, M., & Yang, X. (2023). Remote sensing of soil degradation: Progress and perspective. *International Soil and Water Conservation Research*, *11*(3), 429–454. doi:10.1016/j.iswcr.2023.03.002

Wang, R., Huang, W., Chen, L., Ma, L., Guo, C., & Liu, X. (2011). Anatomical and physiological plasticity in Leymus chinensis (Poaceae) along large-scale longitudinal gradient in northeast China. *PLoS One*, *6*(11), e26209. doi:10.1371/journal.pone.0026209 PMID:22087223

Wang, R., Yu, G., He, N., Wang, Q., Zhao, N., Xu, Z., & Ge, J. (2015). Latitudinal variation of leaf stomatal traits from species to community level in forests: Linkage with ecosystem productivity. *Scientific Reports*, *5*(1), 14454. doi:10.1038rep14454 PMID:26403303

Wang, T., Li, H., & Huang, Y. (2021). The complex ecological network's resilience of the Wuhan metropolitan area. *Ecological Indicators*, *130*, 108101. doi:10.1016/j.ecolind.2021.108101

Wang, W., Pan, Q., He, F., Akhunova, A., Chao, S., Trick, H., & Akhunov, E. (2018). Transgenerational CRISPR-Cas9 activity facilitates multiplex gene editing in allopolyploid wheat. *The CRISPR Journal*, *1*(1), 65–74. https://www.liebertpub.com/doi/full/10.1089/crispr.2017.0010?_ga=2.19975736.1449706263.1533772800-181300196.1533772800. doi:10.1089/crispr.2017.0010 PMID:30627700

Wang, X. G., Johnson, M. W., Daane, K. M., & Nadel, H. (2009). High summer temperatures affect the survival and reproduction of olive fruit fly (Diptera: Tephritidae). *Environmental Entomology*, *38*(5), 1496–1504. doi:10.1603/022.038.0518 PMID:19825305

Wang, X., Liu, Z., & Chen, H. (2022). Investigating Flood Impact on Crop Production under a Comprehensive and Spatially Explicit Risk Evaluation Framework. *Agriculture*, *12*(4), 4. Advance online publication. doi:10.3390/agriculture12040484

Waqas, M., Korres, N. E., Khan, M. D., Nizami, A., Deeba, F., Ali, I., & Hussain, H. (2019). Advances in the concept and methods of seed priming. In Springer eBooks (pp. 11–41). doi:10.1007/978-981-13-8625-1_2

Waqas, M. A., Kaya, C., Riaz, A., Farooq, M., Nawaz, I., Wilkes, A., & Li, Y. (2019). Potential mechanisms of abiotic stress tolerance in crop plants induced by thiourea. *Frontiers in Plant Science*, *10*, 1336. doi:10.3389/fpls.2019.01336 PMID:31736993

War, A. R., Taggar, G. K., War, M. Y., & Hussain, B. (2016). Impact of climate change on insect pests, plant chemical ecology, tritrophic interactions and food production. *International Journal of Clinical and Biological Sciences*, *1*(02), 16–29.

Ware, G. W., & Whitacre, D. M. (2004). An introduction to insecticides. *The pesticide book, 6.*

Warra, A. A., & Prasad, M. N. V. (2020). *African perspective of chemical usage in agriculture and horticulture—their impact on human health and environment Agrochemicals detection, treatment and remediation.* Elsevier.

Warwick, S. I., Beckie, H. J., & Hall, L. M. (2009). Gene flow, invasiveness, and ecological impact of genetically modified crops. *Annals of the New York Academy of Sciences*, *1168*(1), 72–99. doi:10.1111/j.1749-6632.2009.04576.x PMID:19566704

Weiskopf, S. R., Rubenstein, M. A., Crozier, L. G., Gaichas, S., Griffis, R., Halofsky, J. E., Hyde, K. J. W., Morelli, T. L., Morisette, J. T., Muñoz, R. C., Pershing, A. J., Peterson, D. L., Poudel, R., Staudinger, M. D., Sutton-Grier, A. E., Thompson, L., Vose, J., Weltzin, J. F., & Whyte, K. P. (2020). Climate change effects on biodiversity, ecosystems, ecosystem services, and natural resource management in the United States. *The Science of the Total Environment*, *733*, 137782. doi:10.1016/j.scitotenv.2020.137782 PMID:32209235

Wei, Y., Li, J., Shi, D., Liu, G., Zhao, Y., & Shimaoka, T. (2017). Environmental challenges impeding the composting of biodegradable municipal solid waste: A critical review. *Resources, Conservation and Recycling*, *122*, 51–65. doi:10.1016/j.resconrec.2017.01.024

West, J. S., Townsend, J. A., Stevens, M., & Fitt, B. D. L. (2012). Comparative biology of different plant pathogens to estimate effects of climate change on crop diseases in Europe. *European Journal of Plant Pathology*, *133*(1), 315–331. doi:10.100710658-011-9932-x

Westoby, M., Falster, D. S., Moles, A. T., Vesk, P. A., & Wright, I. J. (2002). Plant ecological strategies: Some leading dimensions of variation between species. *Annual Review of Ecology and Systematics*, *33*(1), 125–159. doi:10.1146/annurev.ecolsys.33.010802.150452

Wezel, A., Bellon, S., Doré, T., Francis, C., Vallod, D., & David, C. (2009). Agroecology as a science, a movement and a practice. A review. *Agronomy for Sustainable Development*, *29*(4), 503–515. doi:10.1051/agro/2009004

Wheaton, E., & Kulshreshtha, S. (2017). Environmental sustainability of agriculture stressed by changing extremes of drought and excess moisture: A conceptual review. *Sustainability (Basel)*, *9*(6), 970. doi:10.3390u9060970

Wheeler, T., & Von Braun, J. (2013). Climate change impacts on global food security. *Science*, *341*(6145), 508–513. doi:10.1126cience.1239402 PMID:23908229

Wibbelmann, M., Schmutz, U., Wright, J., Udall, D., Rayns, F., Kneafsey, M., & Turner, M. L. (2013). Mainstreaming agroecology: Implications for global food and farming systems.

Widlansky, M. J., Long, X., & Schloesser, F. (2020). Increase in sea level variability with ocean warming associated with the nonlinear thermal expansion of seawater. *Communications Earth & Environment*, *1*(1), 1. Advance online publication. doi:10.103843247-020-0008-8

Wijerathna-Yapa, A., & Pathirana, R. (2022). Sustainable Agro-Food Systems for Addressing Climate Change and Food Security. *Agriculture*, *12*(10), 1554. https://www.mdpi.com/2077-0472/12/10/1554/pdf. doi:10.3390/agriculture12101554

Wishart, D. S., Bartok, B., Oler, E., Liang, K. Y. H., Budinski, Z., Berjanskii, M., Guo, A., Cao, X., & Wilson, M. (2020). MarkerDB: An online database of molecular biomarkers. *Nucleic Acids Research*, *49*(D1), D1259–D1267. doi:10.1093/nar/gkaa1067 PMID:33245771

Withers, P. J. A., Vadas, P. A., Uusitalo, R., Forber, K. J., Hart, M., Foy, R. H., Delgado, A., Dougherty, W., Lilja, H., Burkitt, L. L., Rubæk, G. H., Pote, D., Barlow, K., Rothwell, S., & Owens, P. R. (2019). A global perspective on integrated strategies to manage soil phosphorus status for eutrophication control without limiting land productivity. *Journal of Environmental Quality*, *48*(5), 1234–1246. doi:10.2134/jeq2019.03.0131 PMID:31589721

Wolka, K., Mulder, J., & Biazin, B. (2018). Effects of soil and water conservation techniques on crop yield, runoff and soil loss in Sub-Saharan Africa: A review. *Agricultural Water Management*, *207*, 67–79. doi:10.1016/j.agwat.2018.05.016

Woodrow, J. E., Gibson, K. A., & Seiber, J. N. (2019). Pesticides and related toxicants in the atmosphere. *Reviews of Environmental Contamination and Toxicology*, *247*, 147–196. PMID:30535549

Worku, D., Gowane, G. R., Kumar, R., Joshi, P. K., Gupta, I. D., & Verma, A. (2021). Estimation of genetic parameters for production and reproductive traits in Indian Karan-Fries cattle using multi-trait Bayesian approach. *Tropical Animal Health and Production*, *53*(3), 369. doi:10.100711250-021-02806-z PMID:34169379

World Economic Forum. (2021). *Shaping the Future of Food*. WEF. https://www.weforum.org/system-initiatives/shaping-the-future-of-food-security-and-agriculture (accessed on 19 July 2021).

World Health Organization. (2010). Control of the leishmaniases: report of a meeting of the WHO Expert Commitee on the Control of Leishmaniases, Geneva, 22-26 March 2010. World Health Organization.

Wright, C. S. (1936). Aloe vera in the treatment of roentgen ulcers and telangiectasis. *Journal of the American Medical Association*, *106*(16), 1363–1364. doi:10.1001/jama.1936.02770160021006

Wright, I. J., Reich, P. B., Westoby, M., Ackerly, D. D., Baruch, Z., Bongers, F., Cavender-Bares, J., Chapin, T., Cornelissen, J. H. C., Diemer, M., Flexas, J., Garnier, E., Groom, P. K., Gulias, J., Hikosaka, K., Lamont, B. B., Lee, T., Lee, W., Lusk, C., & Villar, R. (2004). The worldwide leaf economics spectrum. *Nature*, *428*(6985), 821–827. doi:10.1038/nature02403 PMID:15103368

Wu, K. (2013). *Memo 2: The History of Composting*. Shaping the Future of New York City. https://eportfolios.macaulay.cuny.edu/macbride13/research/timelines/memo-2-the-history-of-composting/

Wu, X., Liu, Y., Zhang, Y., & Gu, R. (2021). Advances in research on the mechanism of heterosis in plants. Frontiers in Plant Science, p. 12. doi:10.3389/fpls.2021.745726

Wu, L., Jiang, Y., Zhao, F., He, X., Liu, H., & Yu, K. (2020). Increased organic fertilizer application and reduced chemical fertilizer application affect the soil properties and bacterial communities of grape rhizosphere soil. *Scientific Reports*, *10*(1), 9568. doi:10.103841598-020-66648-9 PMID:32533037

Wu, L., Peng, Y., Fan, J., & Wang, Y. (2019). Machine learning models for the estimation of monthly mean daily reference evapotranspiration based on cross-station and synthetic data. *Nordic Hydrology*, *50*(6), 1730–1750. doi:10.2166/nh.2019.060

Wunderling, N., Willeit, M., Donges, J. F., & Winkelmann, R. (2020). Global warming due to loss of large ice masses and Arctic summer sea ice. *Nature Communications*, *11*(1), 1. doi:10.103841467-020-18934-3 PMID:33110092

Wurst, S., De Deyn, G. B., & Orwin, K. (2012). Soil Biodiversity and Functions. In D. H. Wall, R. D. Bardgett, V. Behan-Pelletier, J. E. Herrick, T. H. Jones, K. Ritz, J. Six, D. R. Strong, & W. H. van der Putten (Eds.), *Soil Ecology and Ecosystem Services (p. 0)*. Oxford University Press. doi:10.1093/acprof:oso/9780199575923.003.0004

Wu, Y., Zhao, Z., Liu, S., Huang, X., & Wang, W. (2020). Does partial root-zone drying have advantages over regulated deficit irrigation in pear orchard under desert climates? *Scientia Horticulturae*, *262*, 109099. doi:10.1016/j.scienta.2019.109099

Xia, X. J., Huang, Y. Y., Wang, L., Huang, L. F., Yu, Y. L., Zhou, Y. H., & Yu, J. Q. (2006). Pesticides-induced depression of photosynthesis was alleviated by 24-epibrassinolide pretreatment in Cucumis sativus L. *Pesticide Biochemistry and Physiology*, *86*(1), 42–48. doi:10.1016/j.pestbp.2006.01.005

Xiong, W., Reynolds, M., & Xu, Y. (2022). Climate change challenges plant breeding. *Current Opinion in Plant Biology*, *70*, 102308. doi:10.1016/j.pbi.2022.102308 PMID:36279790

Xu, Y., Wang, G., Yang, S., Liu, M., Wang, Q., Tian, S., Zhang, Y., Xu, Y., Yuan, X., Ma, Q., Ma, H., & Zhang, H. (2023). How to further develop the integrated planting and breeding mode: A case study of the western region of Shandong Province. *Sustainable Development*, 1– 16. doi:10.1002/sd.2573

Xu, C., McDowell, N. G., Fisher, R. A., Wei, L., Sevanto, S., Christoffersen, B. O., Weng, E., & Middleton, R. S. (2019). Increasing impacts of extreme droughts on vegetation productivity under climate change. *Nature Climate Change*, *9*(12), 948–953. doi:10.103841558-019-0630-6

Xu, F., Yang, L., Zhang, J., Guo, X., Zhang, X., & Li, G. (2012). Effect of temperature on conidial germination, mycelial growth and aggressiveness of the defoliating and nondefoliating pathotypes of Verticillium dahliae from cotton in China. *Phytoparasitica*, *40*(4), 319–327. doi:10.100712600-012-0232-6

Xu, Q., Yang, C., & Pei, Y. (2022). Editorial: Genetic Pleiotropy in Complex Traits and Diseases. *Frontiers in Genetics*, *13*, 897383. doi:10.3389/fgene.2022.897383 PMID:35547251

Xu, X., Du, X., Wang, F., Sha, J., Chen, Q., Tian, G., Zhu, Z., Ge, S., & Jiang, Y. (2020). Effects of Potassium Levels on Plant Growth, Accumulation and Distribution of Carbon, and Nitrate Metabolism in Apple Dwarf Rootstock Seedlings. *Frontiers in Plant Science*, *11*, 904. https://www.frontiersin.org/articles/10.3389/fpls.2020.00904. doi:10.3389/fpls.2020.00904 PMID:32655607

Yadav, S., Singh, M., & Thakur, L. K. (2023). A Comprehensive Review on the Effect of Sugarcane Trash Ash on the Sorption, Degradation and Leaching of Atrazine and Fipronil in Agricultural Soils. *International Journal of Environment and Climate Change*, *13*(9), 305–323. doi:10.9734/ijecc/2023/v13i92233

Yamagata, K. (2019). Effects of Dietary Polyphenols on Chronic Diseases: Epidemiological Data and Molecular Mechanisms of Action. *Frontiers in Natural Product Chemistry*, *5*(5), 55–94. https://www.sciencedirect.com/topics/chemistry/quercetin. doi:10.2174/9789811423772119050005

Yáñez-López, R., Torres-Pacheco, I., Guevara-González, R. G., Hernández-Zul, M. I., Quijano-Carranza, J. A., & Rico-García, E. (2012). The effect of climate change on plant diseases. *African Journal of Biotechnology*, *11*(10), 2417–2428. doi:10.5897/AJB10.2442

Yang, B., Fu, P., Lu, J., Ma, F., Sun, X., & Fang, Y. (2022). Regulated deficit irrigation: An effective way to solve the shortage of agricultural water for horticulture. *Stress Biology*, *2*(1), 28. doi:10.100744154-022-00050-5 PMID:37676363

Yang, Q., Zhou, H., Bartocci, P., Fantozzi, F., Mašek, O., Agblevor, F. A., Wei, Z., Yang, H., Chen, H., Lu, X., Chen, G., Zheng, C., Nielsen, C. P., & McElroy, M. B. (2021). Prospective contributions of biomass pyrolysis to China's 2050 carbon reduction and renewable energy goals. *Nature Communications*, *12*(1), 1698. doi:10.103841467-021-21868-z PMID:33727563

Yang, Y., Kang, L., Zhao, J., Qi, N., Li, R., Wen, Z., Kassout, J., Peng, C., Lin, G., & Zheng, H. (2021). Quantifying Leaf Trait Covariations and Their Relationships with Plant Adaptation Strategies along an Aridity Gradient. *Biology (Basel)*, *10*(10), 1066. doi:10.3390/biology10101066 PMID:34681167

Yang, Y., Saand, M. A., Huang, L., Abdelaal, W. B., Zhang, J., Wu, Y., Li, J., Sirohi, M. H., & Wang, F. (2021). Applications of Multi-Omics Technologies for crop improvement. *Frontiers in Plant Science*, *12*, 563953. doi:10.3389/fpls.2021.563953 PMID:34539683

Yan, Z., Xiong, C., Liu, H., & Singh, B. K. (2022). Sustainable agricultural practices contribute significantly to One Health. *Journal of Sustainable Agriculture and Environment*, *1*(3), 165–176. doi:10.1002ae2.12019

Yates, T. L., Mills, J. N., Parmenter, C. A., Ksiazek, T. G., Parmenter, R. R., Vande Castle, J. R., Calisher, C. H., Nichol, S. T., Abbott, K. D., Young, J. C., Morrison, M. L., Beaty, B. J., Dunnum, J. L., Baker, R. J., Salazar-Bravo, J., & Peters, C. J. (2002). The ecology and evolutionary history of an emergent disease: hantavirus pulmonary syndrome: evidence from two El Niño episodes in the American southwest suggests that El Niño–driven precipitation, the initial catalyst of a trophic cascade that results in a delayed density-dependent rodent response, is sufficient to predict heightened risk for human contraction of hantavirus pulmonary syndrome. *Bioscience*, *52*(11), 989–998. doi:10.1641/0006-3568(2002)052[0989:TEAEHO]2.0.CO;2

Ye, C., Tenorio, F. A., Argayoso, M. A., Laza, M. A., Koh, H. J., Redoña, E. D., Jagadish, K. S. V., & Gregorio, G. B. (2015). Identifying and confirming quantitative trait loci associated with heat tolerance at flowering stage in different rice populations. *BMC Genetics*, *16*(1), 1–10. https://link.springer.com/article/10.1186/s12863-015-0199-7. doi:10.118612863-015-0199-7 PMID:25895682

Ye, J., Tian, W., & Jin, C. (2022). Nitrogen in plants: From nutrition to the modulation of abiotic stress adaptation. *Stress Biology*, *2*(1), 4. doi:10.100744154-021-00030-1 PMID:37676383

Ye, R., Yang, X., & Rao, Y. (2022). Genetic engineering technologies for improving crop yield and quality. *Agronomy (Basel)*, *12*(4), 759. doi:10.3390/agronomy12040759

Yeung, A. W. K., Tzvetkov, N. T., Durazzo, A., Lucarini, M., Souto, E. B., Santini, A., & Atanasov, A. G. (2020). Natural products in diabetes research: Quantitative literature analysis. *Natural Product Research*, 1-15. doi:10.1080/14786419.2020.1821019

Yeung, A. W. K., Heinrich, M., & Atanasov, A. G. (2018). Ethnopharmacology—A bibliometric analysis of a field of research meandering between medicine and food science? *Frontiers in Pharmacology*, *9*, 215. doi:10.3389/fphar.2018.00215 PMID:29599720

Yihdego, Y., Salem, H. S., & Muhammed, H. H. (2019). Agricultural pest management policies during drought: Case studies in Australia and the state of Palestine. *Natural Hazards Review*, *20*(1), 05018010. doi:10.1061/(ASCE)NH.1527-6996.0000312

Yilmaz, B., Terekeci, H., Sandal, S., & Kelestimur, F. (2020). Endocrine disrupting chemicals: Exposure, effects on human health, mechanism of action, models for testing and strategies for prevention. *Reviews in Endocrine & Metabolic Disorders*, *21*(1), 127–147. doi:10.100711154-019-09521-z PMID:31792807

Yin, Y., Qin, K., Song, X., Zhang, Q., Zhou, Y., Xia, X., & Yu, J. (2018). BZR1 transcription factor regulates heat stress tolerance through FERONIA receptor-like kinase-mediated reactive oxygen species signaling in tomato. *Plant & Cell Physiology*, *59*(11), 2239–2254. https://academic.oup.com/pcp/article/59/11/2239/5071972. doi:10.1093/pcp/pcy146 PMID:30107607

Yuan, P., Wang, J., Pan, Y., Shen, B., & Wu, C. (2019). Review of biochar for the management of contaminated soil: Preparation, application and prospect. *The Science of the Total Environment*, *659*, 473–490. doi:10.1016/j.scitotenv.2018.12.400 PMID:31096377

Yue, Q., Guo, P., Wu, H., Wang, Y., & Zhang, C. (2022). Towards sustainable circular agriculture: An integrated optimization framework for crop-livestock-biogas-crop recycling system management under uncertainty. *Agricultural Systems*, *196*, 103347. doi:10.1016/j.agsy.2021.103347

Yu, H. Y., Kim, S. H., & Kim, J. G. (2022). Carbon sequestration potential in montane wetlands of Korea. *Global Ecology and Conservation*, *37*, e02166. doi:10.1016/j.gecco.2022.e02166

Yu, W., Wang, L., Zhao, R., Sheng, J., Zhang, S., Li, R., & Shen, L. (2019). Knockout of SlMAPK3 enhances tolerance to heat stress involving ROS homeostasis in tomato plants. *BMC Plant Biology*, *19*(1), 1–13. https://bmcplantbiol.biomedcentral.com/articles/10.1186/s12870-019-1939-z. doi:10.118612870-019-1939-z PMID:31412779

Yu, Y., Hou, J., Jahangir, A., Cao, X., Balsalobre-Lorente, D., Radulescu, M., & Jiang, T. (2023). Decomposition analysis of China's chemical sector energy-related CO2 emissions: From an extended SDA approach perspective. *Energy & Environment*, 0958305X2311516. doi:10.1177/0958305X231151682

Zabaniotou, A., Rovas, D., Libutti, A., & Monteleone, M. (2015). Boosting circular economy and closing the loop in agriculture: Case study of a small-scale pyrolysis–biochar based system integrated in an olive farm in symbiosis with an olive mill. *Environmental Development*, *14*, 22–36. doi:10.1016/j.envdev.2014.12.002

Zaidi, S. S. E. A., Mahas, A., Vanderschuren, H., & Mahfouz, M. M. (2020). Engineering crops of the future: CRISPR approaches to develop climate-resilient and disease-resistant plants. *Genome Biology*, *21*(1), 1–19. https://genomebiology.biomedcentral.com/articles/10.1186/s13059-020-02204-y. doi:10.118613059-020-02204-y PMID:33256828

Zakari, S., Ibro, G., Moussa, B., & Abdoulaye, T. (2022). Adaptation Strategies to Climate Change and Impacts on Household Income and Food Security: Evidence from Sahelian Region of Niger. *Sustainability (Basel)*, *14*(5), 2847. doi:10.3390u14052847

Zeferino, L. B., de Souza, L. F. T., do Amaral, C. H., Fernandes Filho, E. I., & de Oliveira, T. S. (2020). Does environmental data increase the accuracy of land use and land cover classification? *International Journal of Applied Earth Observation and Geoinformation*, *91*, 102128. doi:10.1016/j.jag.2020.102128

Zé-Zé, L., Amaro, F., Osório, H. C., Giovanetti, M., Lourenço, J., & Alves, M. J. (2022). Molecular Identification and Ecology of Portuguese Wild-Caught Phlebotomine Sandfly Specimens. *Zoonotic Diseases*, *2*(1), 19–31. doi:10.3390/zoonoticdis2010003

Zhang, G., Bi, Z., Jiang, J., Lu, J., Li, K., Dan, B., Wang, X., Zhao, X., Li, M., Zhao, X., Wang, W., Xu, J., Li, Z., Zhang, F., & Shi, Y. (2023). Genome-wide association and epistasis studies reveal the genetic basis of saline-alkali tolerance at the germination stage in rice. *Frontiers in Plant Science*, *14*, 1170641. doi:10.3389/fpls.2023.1170641 PMID:37251777

Zhang, H., Mittal, N., Leamy, L. J., Barazani, O., & Song, B. H. (2017). Back into the wild—Apply untapped genetic diversity of wild relatives for crop improvement. *Evolutionary Applications*, *10*(1), 5–24. https://onlinelibrary.wiley.com/doi/pdfdirect/10.1111/eva.12434. doi:10.1111/eva.12434 PMID:28035232

Zhang, H., Zhu, J., Gong, Z., & Zhu, J. K. (2022). Abiotic stress responses in plants. *Nature Reviews. Genetics*, *23*(2), 104–119. doi:10.103841576-021-00413-0 PMID:34561623

Zhang, L., Liu, L., Fang, Z., Shen, W., Dai, Y., Jia, R., Liang, J., & Liu, B. (2023). Fitness changes in wild soybeans are caused by gene flow from genetically modified soybeans. *BMC Plant Biology*, *23*(1), 424. doi:10.118612870-023-04398-2 PMID:37710180

Zhang, L., Su, J., Ali, A., Huang, T., Chen, C., & Yang, W. (2023). A biofilm reactor based on a slow-release carbon source effectively improved the continuous denitrification capacity of slightly polluted surface water at a low carbon-to-nitrogen ratio. *Journal of Environmental Chemical Engineering*, *11*(2), 109552. doi:10.1016/j.jece.2023.109552

Zhang, P., Cai, Y., He, Y., Xie, Y., Zhang, X., & Li, Z. (2022). Changes of vegetational cover and the induced impacts on hydrological processes under climate change for a high-diversity watershed of south China. *Journal of Environmental Management*, *322*, 115963. doi:10.1016/j.jenvman.2022.115963 PMID:36041299

Zhang, Z., Malik, M. I., Khan, A., Ali, N., Malik, S., & Bilal, M. (2022). Environmental impacts of hazardous waste, and management strategies to reconcile circular economy and eco-sustainability. *The Science of the Total Environment*, *807*, 150856. doi:10.1016/j.scitotenv.2021.150856 PMID:34627923

Zhang, Z., Zhang, X., Wang, Y., Zhao, Y., Lin, J., Liu, F., & Mu, W. (2016). Nitenpyram, dinotefuran and thiamethoxam used as seed treatments act as efficient controls against Aphis gossypii via high residues in cotton leaves. *Journal of Agricultural and Food Chemistry*, *64*(49), 9276–9285. doi:10.1021/acs.jafc.6b03430 PMID:27960287

Zhao, A., Yu, Q., Feng, L., Zhang, A., & Pei, T. (2020). Evaluating the cumulative and time-lag effects of drought on grassland vegetation: A case study in the Chinese Loess Plateau. *Journal of Environmental Management*, *261*, 110214. doi:10.1016/j.jenvman.2020.110214 PMID:32148284

Zhao, F., Xi, S., Yang, X., Yang, W., Li, J., Gu, B., & He, Z. (2012). Purifying eutrophic river waters with integrated floating island systems. *Ecological Engineering*, *40*, 53–60. doi:10.1016/j.ecoleng.2011.12.012

Zhao, F., Zhan, X., Xu, H., Zhu, G., Zou, W., Zhu, M., Kang, L., Guo, Y., Zhao, X., Wang, Z., & Tang, W. (2022). New insights into eutrophication management: Importance of temperature and water residence time. *Journal of Environmental Sciences (China)*, *111*, 229–239. doi:10.1016/j.jes.2021.02.033 PMID:34949352

Zheng, Q., Majsec, K., & Katagiri, F. (2022). Pathogen-driven coevolution across the CBP60 plant immune regulator subfamilies confers resilience on the regulator module. *The New Phytologist*, *233*(1), 479–495. doi:10.1111/nph.17769 PMID:34610150

Zhong, H., Li, X., Lobell, D., Ermon, S., & Brandeau, M. L. (2018). Hierarchical modelling of seed variety yields and decision making for future planting plans. *Environment Systems & Decisions*, *38*(4), 458–470. doi:10.100710669-018-9695-4

Zhou, W., Chen, F., Meng, Y., Chandrasekaran, U., Luo, X., Yang, W., & Shu, K. (2020). Plant waterlogging/ flooding stress responses: From seed germination to maturation. *Plant Physiology and Biochemistry*, *148*, 228–236. doi:10.1016/j.plaphy.2020.01.020 PMID:31981875

Zhou, Y., Van Leeuwen, S. K., Pieterse, C. M., Bakker, P. A., & Van Wees, S. C. (2019). Effect of atmospheric CO 2 on plant defense against leaf and root pathogens of Arabidopsis. *European Journal of Plant Pathology*, *154*(1), 31–42. doi:10.100710658-019-01706-1

Zho, X., Wang, J., Su, J., & Sun, W. (2021). Ecosystem service value evaluation method in a complex ecological environment: A case study of Gansu province, China. *PLoS One*, *16*(2), 1–25. doi:10.1371/journal.pone.0240272 PMID:33544729

Zhu, J. K. (2002). Salt and drought stress signal transduction in plants. *Annual Review of Plant Biology*, *53*(1), 247–273. doi:10.1146/annurev.arplant.53.091401.143329 PMID:12221975

Zimmermann, S., Kiessling, A., & Zhang, J. (2023). The future of intensive tilapia production and the circular bioeconomy without effluents: Biofloc technology, recirculation aquaculture systems, bio-RAS, partitioned aquaculture systems and integrated multitrophic aquaculture. *Reviews in Aquaculture*, *15*(S1), 22–31. doi:10.1111/raq.12744

Zittis, G., Almazroui, M., Alpert, P., Ciais, P., Cramer, W., Dahdal, Y., et al. (2022). Climate change and weather extremes in the Eastern Mediterranean and Middle East. *Reviews of Geophysics, 60*, e2021RG000762. doi:10.1029/2021RG000762

Zittis, G., Hadjinicolaou, P., Klangidou, M., Proestos, Y., & Lelieveld, J. (2019). A multi-model, multi-scenario, and multi-domain analysis of regional climate projections for the Mediterranean. *Regional Environmental Change*, *19*(8), 2621–2635. doi:10.100710113-019-01565-w

Zohary, D. (2002). Domestication of the carob (*Ceratonia siliqua* L.). *Israel Journal of Plant Sciences, 50*(sup1), 141-145.

Zohary, D., Hopf, M., & Weiss, E. (2012). *Domestication of Plants in the Old World: The origin and spread of domesticated plants in Southwest Asia, Europe, and the Mediterranean Basin*. Oxford University Press. doi:10.1093/acprof:osobl/9780199549061.001.0001

Zougmoré, R. B., Partey, S. T., Ouédraogo, M., Torquebiau, E., & Campbell, B. M. (2018). Facing climate variability in sub-Saharan Africa: Analysis of climate-smart agriculture opportunities to manage climate-related risks. *Cahiers Agricultures (TSI), 27*(3), 1-9..

Zulfiqar, F., Casadesús, A., Brockman, H., & Munné-Bosch, S. (2020). An overview of plant-based natural biostimulants for sustainable horticulture with a particular focus on moringa leaf extracts. *Plant Science*, *295*, 110194. doi:10.1016/j.plantsci.2019.110194 PMID:32534612

Zuo, W., Gu, B., Zou, X., Peng, K., Shan, Y., Yi, S., Shan, Y., Gu, C., & Bai, Y. (2023). Soil organic carbon sequestration in croplands can make remarkable contributions to China's carbon neutrality. Journal of Cleaner Production, 382, 135268

# About the Contributors

**Karmaoui Ahmed** started his academic career as Research Fellow at Cadi Ayyad University and awarded Ph.D. from the Faculty of Semlalia (UCA); graduated from Publons Academy, and awarded Top 1% of international Reviewers (2018) by Publons. He has published various peer-reviewed international journal articles, book chapters, and books on a range of environmental topics. He is currently the president of the Moroccan Center for Culture and Science, Morocco, and an assistant professor at FSTE (UMI). Broad areas of his research interests include ecosystem services assessment.

***

**Mohammed Ater** is Professor of Botany at the Biology Department of the Faculty of Sciences of Tetouan (Morocco). He created and directed the Laboratory Diversity and Conservation of Biological Systems (LDICOSYB) from 2005 to 2012. Currently, coordinator of the Bio-Agrodiversity team and lead of the Applied Botanical Laboratory. He conducted extensive research on genetic resources, agro-biodiversity and ecology.

**Rachid Azenzem** is a plant pathology researcher at the National Institute of Agricultural Research in Morocco.

**Abdeljalil Boutarfa** is a soil scientist.

**Loubna Elmansouri** is Professor at the College of Geomatic Sciences and Surveying Engineering at Institute of Agronomy and Veterinary Medicine (IAV) Hassan II, Rabat, Morocco, where she has been since 2004. From 2014 to 2017 she served as Head of Department of Geodetic Sciences at IAV Hassan 2. During 2009-2013 she was a Ph.D. student at "SURAFCE" Laboratory at Science College of LIEGE University (ULG) in Belgium. From 2010 to 2012 she was a visiting staff at "MONTEFIORE" Institute of Computer Science at ULG. She received an engineering degree in Geomatics and Geodetic Sciences from IAV Hassan 2 in 2004 and enrolled as licensed Surveyor. Holds a Doctorate of Sciences in Remote Sensing from the ULG in 2013. Her research interests span both digital image process and remote sensing. Much of her work has been on improving the application of maching learning technique in land use/ land cover data producing for territory management, more over in agriculture and forest domain. Professor Elmansouri published more than 30 papers, coordinator of many R&D projects, and Co-Advisor of Doctorate Studies.

**Salah Er-Raki** is actually a Full Professor of Applied physics at Faculty of Sciences and Techniques, Cadi Ayyad University, Morocco. He received his PhD in "crop water requirements estimates through combining the modeling, in situ measurements and remote sensing" in 2007, and an ACR (Ability to Conduct Research) in 2014 from Cadi Ayyad University. His main research interests include the application of multispectral remote sensing observations (optical, radar and thermal data) in land-surface-atmosphere models for high spatio-temporal resolution monitoring of crops water needs especially in arid and semi-arid regions. He also contributed in developing a software named « SAMIR, Satellite monitoring of irrigation", which such basic concepts are based on his thesis work. He also participates in several international experiments (e.g. Yaqui-2006-2008 in Mexico) and he was a co-leader of the joint international laboratory "Télédétection et Rèssources en Eau en Méditerranée semi-Aride", LMI-TREMA (). Salah Er-Raki has published more than 85 papers in international peer-reviewed journals, which have been cited more than 2500 times and brought him an H-index of 31 (). He serves as a Guest editor, editorial boards and reviewer of many international scientific journals with high impact factor among them are: Agricultural Water Management, Remote Sensing of Environment, remote sensing and Agricultural and Forest Meteorology, Computers and Electronics in Agriculture, Theoretical and Applied Climatology, Soft Computing, irrigation science, water resources management. He has attended and serves as chair sessions in many international congresses and symposiums. He supervised and co-supervised 8 PhD students and he coordinates several research projects (e.g. PHC Maghreb, PRIMA, H2020 RISE action,…). He is Awarded as a Morocco's top researchers (2016- 2020) in Agricultural Sciences ranked by Essential Science Indicators (ESI) and Web of Sciences, in the Second edition of "Research Excellence Awards", organized by Clarivate and CNRST, June 2021 . He also awarded as International Young Scientist Award, by the Foundation of Innovative Research, Noval Research Foundation and International Research and Academic Excellence Awards (IRAEA - 2021), August 202.

**Rachid Hadria** recived his PhD degree in crop modelling and remote sensing from Caddi Ayyad University in Morocco in 2006 and is now working as a reaserch scientist at the National Institute of Agronomic Research. His resaerchs include the combination of remote sensing, GIS and crop modeling to manage natural resources in arid zones. He has published more than twenty papers (pricipal author or coauthor) in refereed journals.

**Khaoula Ibba** is a Ph.D. student at Cadi Ayyad University and the National Institute of Agricultural Research in Marrakech, Morocco. Her current research focuses on exploring deficit irrigation strategies in olive trees. She aims to leverage these strategies as an effective tool for managing irrigation water, particularly in arid and semi-arid regions. Furthermore, her research also extends to crop modeling, where she is dedicated to using models that contribute to the efficient management of natural resources in arid zones.

**Kassout Jalal** is currently a researcher at the National Institute for Agricultural Research (INRA). JDr. alal work mainly on functional responses of fruit trees such as Olive, Carob and Argan tree. Thus, his research deal with issues related to climate change and anthropogenic effects on naturel agroecosystems.

**Atin Kumar** is currently associated with the School of Agriculture, Uttaranchal University Dehradun (NAAC A+) as an Assistant Professor. He holds a B.Sc. in Agriculture and M.Sc. in Agriculture with a specialization in Agricultural Chemistry & Soil Science from CCS University, Meerut. He obtained his

Ph.D. in Soil Science from Sardar Vallabhbhai Patel University of Agriculture & Technology, Meerut (UP) in 2019. He has actively participated in numerous national and international workshops, conferences, and trainings. His expertise lies in soil fertility, nutrient management, and water quality. With almost 4 years of experience in teaching and research, he has published over 37 research papers, articles, abstracts, and book chapters. Dr. Atin honored with the Young Researcher Award, and the Global Scientist Award for his contributions to agriculture and sustainable development.

**N. Sasikaladevi** received Ph. D. Degree in Computer Science in 2013. She has published more than 42 papers in reputed International journals and conferences including publications in SCI-Indexed Journals. She is a reviewer of more than a dozen of reputed journals including IEEE transactions on Services Computing and IEEE Journal of Internet of Things. s. Her research focus includes the design of machine learning strategies to solve Discrete Logarithm Problem. She published several books and chapters in reputed publisher including Prentice Hall of India, Lambert Academic Publisher, IGI Global, Springer and Science Direct. She has received fund from Department of Science and Technology, Government of India to carry out projects in the domain of Services computing and Security System. She has received young scientist award and women scientist award from DST, India. She is also involved in the security enriched payment system design as a part of Digital India Initiative. Her current research interests include information security, digital authentication, security of wireless sensor networks, computer vision, deep learning and security of vehicular ad-hoc networks.

**Mehmet Gökhan Özdemir** is a researcher based on economic theory. He is Ph.D. candidate with two MBA degree in the field of economics. He has almost 34 publications so far.

**Sandeep Poddar**, Deputy Vice Chancellor (Research & Innovation), Member of Board of Studies, Lincoln University College, Malaysia, has been graduated from University of Calcutta in 1993 with Honours in Zoology, he has obtained Post Graduate Diploma in Dietetics from All India Institute of Hygiene and Public Health 1995, Master of Science in Zoology with specialization in Biochemical Genetics from Dayalbagh Educational Institute 1998 with distinction. He has completed PhD in Zoology from Vivekananda Institute of Medical Sciences on Cytotoxicity in 2004. After completing PhD he pursued Post Doctoral Research in different projects on Hemoglobinopathies and Oral Cancer mutation. Dr Sandeep completed his MBA from Lincoln University College, Malaysia. He has published several research papers, filed patent and organized international conferences, and also edited books in Malaysia, Australia and India. Dr Sandeep is Fellow member of World Researchers Associations and Life members of Different associations including Indian Science Congress Association. Dr Sandeep is founder Assistant Secretary of Dr Tarak Nath Podder Memorial Foundation, Kolkata, India

**Sara Oulbi** is a researcher at the National Institut of Agricultural Research (INRA). Intersted on how olive tree reacts to envrionmental conditions.

**Himani Sharma** did her graduation with B.Sc. (Hons.) Agriculture from Sher-E-Kashmir University of Agricultural Sciences and Technology, Jammu, and Master's degree of M.Sc. Ag. (Agronomy) from School of Agriculture, Uttaranchal University Dehradun (U.K.). She was also gold medalist in master's degree. She also published 12 articles, 3 book chapters (Scopus Indexed), 2 trainings, and attended 3 national/international seminars/conferences. She has a life member of the Indian Society of Agriculture

& Horticulture Research Development (ISAHRD), Chandigarh, Punjab, India. She has also received the "Young Achiever Award" for outstanding contribution to the field of Agronomy. She has expertise in soil fertility, nutrient management, and irrigation water quality.

**Muhammad Sikandar Zaman** is a dedicated scholar currently pursuing a Bachelor of Science (Hons.) degree in Agri-Sciences with a major in Plant Breeding and Genetics. Born with a deep passion for agricultural sciences, he has always been fascinated by the intricacies of plant genetics and the potential they hold for improving crop yield and quality. Growing up in a rural community, Sikandar developed a profound understanding of the challenges faced by farmers in his region. This drove him to pursue higher education and contribute towards sustainable agricultural practices. With a curious and analytical mind, he consistently excelled in his academic endeavors, earning him numerous accolades and scholarships throughout his educational journey. Sikandar's commitment to his field extends beyond the classroom. He actively engages in research projects, where he collaborates with esteemed professors and fellow students to explore innovative techniques in plant breeding. His contributions to these projects have been instrumental in developing new crop varieties that are disease-resistant, high-yielding, and adapted to local environmental conditions. Apart from his academic pursuits, Sikandar is an active member of various agricultural organizations and community development initiatives. He understands the importance of knowledge dissemination and regularly participates in workshops and seminars to share his expertise and learn from industry experts. With a combination of theoretical knowledge and practical experience, Mr. Muhammad Sikandar Zaman is poised to make significant contributions to the field of plant breeding and genetics. His unwavering passion, dedication, and commitment to sustainable agriculture make him a promising young professional with the potential to shape the future of agricultural practices for the betterment of farmers and food security.

**Sikaoui Lhassane** head of olive growing, olive technologies and environnement, IOC Madrid.

**Ajay K. Singh** is working as an Associate Professor (Research) in the Department of Humanities & Social Sciences, Graphic Era Deemed to be University Dehradun. He did Post-Doctorate Research with EDI of India, Ahmedabad, Gujarat (India). He received MPhil (Economics) from Devi Ahilya University Indore (India), and PhD (Economics) from IIT Indore (India). He has published research papers in the area of climate change and agricultural productivity, assessment of food security; estimation of GFSI, assessment of environmental sustainability index and its association with socio-economic indicators; implications of entrepreneurship ecosystem on economic development; and implications of digitalization on economic development at national and global level.

**Rajat Singh** is currently associated with the School of Agriculture, Uttaranchal University Dehradun (NAAC A+) as an Assistant professor. He completed his B.Sc. (Agriculture) in 2016, his M.Sc. (Horticulture) in 2018, and completed his Doctorate in Horticulture with a Specialization in Vegetable Science. Dr. Rajat Singh has authored 1 book, & 1 Practical Manual, 23 Articles, 21 Research Paper and 3 review papers in SCI, Scopus, IEEE, and NAAS-rated journals, and 10 book chapters, National and International Conferences, 6 Workshop and Editorial Board Membership awarding Science Letters (International Journal). He has 4 awards for his excellence in research.

**Santosh Singh**, Assistant Professor English Literature and Language, DIT University, Dehradun, Uttarakhand. received his PhD in English Literature from Gurukul Kangri Vishwavidyalaya, Haridwar in 2015, and an MA in English Literature and BA from HNB Garhwal University. He is having more than 22 years' experience in industry and teaching. His research interest lies in Indian Street theatre, Postcolonial Studies, Folk Studies and Indian Writing in English. He has many research papers and book chapters in his credit in nationally and internationally reputed journals. He has experience working as Assistant Professor at Raffles University Neemrana, Rajasthan and Uttaranchal University, Dehradun, Uttarakhand.

**Ranjit Singha** is a Doctorate Research Fellow at Christ (Deemed to be University) and a distinguished American Psychological Association (APA) member. His expertise lies in research and development across various domains, including Mindfulness, Addiction Psychology, Women Empowerment, UN Sustainable Development Goals, and Data Science. He has earned certifications from renowned institutions, including IBM and The University of Oxford Mindfulness Centre, UK, in Mindfulness. Additionally, he holds certifications as a Microsoft Innovative Educator, Licensed Yoga Professional, Certified Mindfulness Teacher, and CBCT Teachers Training from Emory University, USA. Mr Ranjit's educational qualifications include PGDBA (GM), MBA (IB), MSc in Counseling Psychology, and completion of a Senior Diploma in Tabla (Musical Instrumentation). His dedication to continuous learning is evident through his involvement in the SEE Learning® (Social, Emotional, and Ethical) Learning program. As a committed researcher and educator, Mr Ranjit focuses on mindfulness and compassion-based interventions. He has an impressive publication record, having authored twenty-three research papers, ten chapters, four books, and five edited books. His research interests encompass various aspects of mindfulness, such as assessment, benefits of mindfulness-based programs, change mechanisms, professional training, mindfulness ethics, cognitive and neuropsychology, and studies related to high-risk behaviours. Apart from his research endeavours, Mr Ranjit has extensive teaching experience, instructing courses in diverse subjects like Forensic Psychology, Positive Psychology, Organizational Planning, Strategic Management, Psycho Metric Tests, Counseling Skills, Disaster Management, Basic Computer Science, Business Planning, Business Law, and Auditing. He has mentored numerous Postgraduate and undergraduate research projects, demonstrating his commitment to nurturing young minds in psychology. Furthermore, Mr Ranjit actively provides personal counselling services, showcasing his genuine concern for his students' well-being and academic success. His unwavering dedication to research and education has solidified his position as a valuable contributor to psychology.

**Surjit Singha** is an academician with a broad spectrum of interests, including UN Sustainable Development Goals, Organizational Climate, Workforce Diversity, Organizational Culture, HRM, Marketing, Finance, IB, Global Business, Business, AI, Women Studies, and Cultural Studies. Currently a faculty member at Kristu Jayanti College, Dr. Surjit also serves as an Editor, reviewer, and author for prominent global publications and journals, including being on the Editorial review board of Information Resources Management Journal and a contributor to IGI Global. With over 13 years of experience in Administration, Teaching, and Research, Dr. Surjit is dedicated to imparting knowledge and guiding students in their research pursuits. As a research mentor, Dr. Surjit has nurtured young minds and fostered academic growth. Dr. Surjit has an impressive track record of over 75 publications, including articles, book chapters, and textbooks, holds two US Copyrights, and has successfully completed and published two fully funded minor research projects from Kristu Jayanti College.

**Muhammad Zia Ul Haq**, Ph.D., is an accomplished agricultural scientist specializing in herbicide resistance. Dr. Zia Ul Haq developed a profound passion for agriculture from an early age. He pursued his higher education in the field of weed research and has since become a recognized expert in the area of herbicide resistance. Dr. Zia Ul Haq's academic journey began at the prestigious University of Agriculture in Faisalabad, where he completed his undergraduate studies in Agronomy. His insatiable curiosity and dedication to making a meaningful impact on agricultural practices led him to pursue a Ph.D. at the Weed Research Laboratory, College of Life Science, Nanjing Agricultural University in the People's Republic of China. It was during his doctoral research that Dr. Haq focused on studying the mechanisms of herbicide resistance in weeds, aiming to develop sustainable strategies to combat this pressing challenge faced by farmers worldwide. His groundbreaking research and innovative contributions to the field of weed science earned him widespread recognition. Dr. Haq's work not only shed light on the genetic and physiological mechanisms underlying herbicide resistance but also provided practical solutions to help farmers effectively manage resistant weed populations. His research findings have been published in several reputable scientific journals, further establishing his credibility as a leading expert in the field. Following the completion of his Ph.D., Dr. Haq returned to his alma mater, the University of Agriculture in Faisalabad, where he currently holds the position of Assistant Professor in the Department of Agronomy. In this role, he plays a pivotal role in shaping the minds of future agronomists, instilling in them the importance of sustainable farming practices and the significance of addressing herbicide resistance. Beyond his academic pursuits, Dr. Zia Ul Haq actively engages in collaborative research projects with national and international organizations. His expertise is sought after by both government and private entities, who rely on his insights to develop policies and strategies to overcome herbicide resistance challenges. Dr. Zia Ul Haq's unwavering commitment to agricultural research and education has earned him numerous accolades throughout his career. He is a recipient of the Young Scientist Award for his outstanding contributions to the field of agronomy and weed science. His passion for sharing knowledge and fostering scientific inquiry has led him to mentor and guide numerous students, who have gone on to make their own mark in the agricultural industry. In conclusion, Dr. Muhammad Zia Ul Haq's biography stands as a testament to his unwavering dedication to agricultural research and his tireless efforts to combat herbicide resistance. His expertise, passion, and commitment have positioned him as a leading figure in the field, working towards sustainable agricultural practices and ensuring food security for future generations.

# Index

## O



## P



## R



## S



## T

## U



## W

www.ingramcontent.com/pod-product-compliance
Lightning Source LLC
LaVergne TN
LVHW060352300926
849162LV00006B/1251